Advanced Network Administration

Steve Wisniewski
Greenwich Technology Partners

Prentice
Hall

Upper Saddle River, New Jersey
Columbus, Ohio

Library of Congress Cataloging-in-Publication Data

Wisniewski, Steve.
   Advanced network administration / Steve Wisniewski.
     p. cm.
   ISBN 0-13-097048-4
   1. Computer networks—Management. I. Title.

TK5105.5 .W575 2003
004.6'068—dc21

2002066216

Editor in Chief: Stephen Helba
Assistant Vice President and Publisher: Charles E. Stewart, Jr.
Production Editor: Tricia L. Rawnsley
Production Coordination: Carlisle Publishers Services
Design Coordinator: Diane Ernsberger
Cover Designer: Linda Sorrells Smith
Production Manager: Matthew Ottenweller
Electronic Text Management: Karen L. Bretz
Marketing Manager: Ben Leonard

This book was set in Garamond by Carlisle Publishers Services. It was printed and bound by R.R. Donnelley & Sons Company. The cover was printed by The Lehigh Press, Inc.

Pearson Education Ltd.
Pearson Education Australia Pty. Limited
Pearson Education Singapore Pte. Ltd.
Pearson Education North Asia Ltd.
Pearson Education Canada, Ltd.
Pearson Educación de Mexico, S.A. de C.V.
Pearson Education—Japan
Pearson Education Malaysia Pte. Ltd.
Pearson Education, *Upper Saddle River, New Jersey*

10 9 8 7 6 5 4 3 2 1
ISBN: 0-13-097048-4

*To my mother Helen,
my father Theodore,
and my brother Eugene.*

Preface

The world of internetworking has become challenging and exciting as we begin the 21st century. Gone is the monstrosity of running cable and gone is old technology. We are looking at even faster speeds of transmitting voice and data. Technologies such as SONET and wireless LANs are being used to decrease bandwidth usage and, at the same time, allow more data to be transmitted.

Despite all the advances in technology, the old reliable architectures and protocols are still being implemented in the internetworking arena today because they have proven to be reliable and have worked with many different products. *Advanced Network Administration* covers the concepts of routing, bridging, switching, and network management. This book is recommended for novices who wish to further their knowledge of internetwork design. Chapters 1–3, OSI *Internetworking Basics, Bridging and Switching Fundamentals,* and *Routing Basics,* should be taught as a unit. Chapters 4 and 5 cover *Network Management Fundamentals* and *Simple Network Management Protocol.* From Chapters 6 to 10, each of the main network architectures, along with the various protocols that comprise these architectures, are exposed. Chapters 11 and 12 depart from the architectural viewpoint and discuss *Interdomain Routing Basics.* Open Shortest Path First, explained in Chapter 11 (OSPF), and Border Gateway Protocol, discussed in Chapter 12 (BGP), are the main protocols being used in the Internet to provide reliable routing from host to host. These two chapters should be taught as a unit. Chapter 13, *Advanced IP Routing,* provides a brief explanation of IP addressing and more advanced concepts such as VLSM and VLANs, along with route summarization and redistribution. For a more thorough explanation on IP addressing refer to Chapter 9, *Digital Network Architecture (DNA) DECnet Phase IV.*

Chapter 14 discusses Data Link Protocols and their importance in internetworking. Chapter 15, *Internetworking Design Basics,* is the capstone chapter discussing how to design an internetwork using routing and switching methods.

Advanced Network Administration complements any text or certification program such as Microsoft or Novell Networking Technologies, and will provide a thorough understanding of the architectures and their protocols. Reader comments, questions, and suggestions for improvements are welcome.

Keep on internetworking!

Acknowledgments

My thanks to the following people for their encouragement and assistance in helping me complete this manuscript. Charles E. Stewart, my publisher, gave me the opportunity to undertake this project. Getting one book published, let alone two books, is very difficult. *Advanced Network Administration* is the sequel to *Network Administration,* which covers network architectures and the various protocols that comprise those architectures. Thanks also go to Mayda Bosco, assistant editor, who is very timely and caring.

I would like to thank the following reviewers for their invaluable feedback: Bill H. Liu, DeVry Institute of Technology; Jeffrey L. Rankinen, Pennsylvania College of Technology; and Fred Seals, Blinn College.

I give humble thanks to the faculty at the Stevens Institute of Technology for the wonderful job they did preparing me for my master's degree in telecommunications management and increasing my knowledge of the telecommunications industry. The Stevens graduate program takes a novice in the telecom industry and gives him/her a good, solid, fundamental, working knowledge of the big picture that telecom incorporates.

I would also like to thank the DeVry Institute of Technology for giving me a solid foundation and giving me my start in the telecom field. My special thanks go to Professors Frank Relotto, Robert Brunson, and especially Rick Lika, my first telecom instructor. I also want to thank my algebra professor, Lisa Zimmerman, who inspired me to conquer mathematics.

Thanks also goes to Marshall University for giving me a good education and preparing me to become a teacher. I especially want to remember my friends and teammates who died in an airplane crash on November 14, 1970. I will always remember them.

I would also like to give special thanks to the Dale Carnegie training course for teaching me the principles of human relations and getting me through difficult times, while encouraging me to aspire to greater heights by taking action. Thanks to the instructors and to my fellow students.

I also want to thank Tony Robbins for showing me, through the 30-day personal power program, the way to use my personal power. It gave me the ability to change my life. Thanks, Tony!

Finally, I wish to thank my mother, Helen, and my father, Theodore, for instilling the work ethic in me.

Last but not least, I want to thank the good Lord, Jesus Christ and the Holy Spirit, for being there and carrying me through all the dark periods in my life. Thanks, also, to Father Bill Halbing for giving me the guidance to reach the Lord through Jesus Christ.

Steve Wisniewski
stvwsn@aol.com

About the Author

Steve Wisniewski is currently working as a computer telephony engineer for Greenwich Technology Partners and has been in the internetworking and networking industry for more than 12 years. He is a 1972 graduate of Marshall University, a 1995 graduate of DeVry Technical Institute, and has earned a master of science degree in telecommunications management from the Stevens Institute of Technology.

Mr. Wisniewski is Microsoft, Novell, and Cisco certified and has edited several books for Cisco and Sybex Press, and is currently writing a book for Cisco Press on a CCIE level regarding troubleshooting wireless LANs.

Wisniewski's Wireless Networking and Cellular Communications Texts: Building Blocks of Success

Steve Wisniewski is one of the most prolific wireless technology authors today. His field experience combined with his firsthand knowledge of progressive communications equipment and protocols render these three texts necessary additions to your library. These informative and illustrative volumes provide readers with the knowledge they need to pursue careers in network engineering, wireless technology, cellular communications, and many other fields.

Network Administration (ISBN: 0-13-015882-8, © 2001)

Steve Wisniewski's first book for Prentice Hall, *Network Administration,* provides an overall understanding of modern technologies while de-emphasizing math. This book, designed for readers at beginning or intermediate skill levels, reviews the basics and brings readers to a more thorough understanding of the subject matter.

Each chapter contains brief historical information, major concepts with an explanation of how they operate, standards, and key terms to help illustrate the discussion.

Covering the topics of IP, ISDN, ATM, DWDM, and the frequently overlooked topic of network testing, this text helps readers skillfully navigate through the important issues they might face in the job market as network engineers. Given current security breaches in the public and private sectors, this important title will be a welcome and much needed addition to any telecommunications and networking curriculum.

Advanced Network Administration (ISBN: 0-13-097048-4, © 2003)

The sequel to *Network Administration, Advanced Network Administration* goes beyond its predecessor by teaching the reader how to build and design an internetwork. Readers at an intermediate to advanced level will find that this book takes them several more steps into the ever-changing world of wireless networking.

Anyone seeking to understand internetworking, or who may be preparing for a Cisco Certified Network Associate certificate, should read this book. *Advanced Network Administration* covers all major network protocols, discusses tips on troubleshooting networking equipment and cabling, and describes concrete aspects of layer 2 bridging and switching.

While de-emphasizing math, the text incorporates the features of protocol architecture, design, historical perspective, and the many problems that network engineers may encounter while building a large corporate infrastructure. The protocols discussed are Novell RIP, IBM's SNA, DEC's DNA, EIGRP, IGRP, BGP, OSPF, routing optimization, and VLAN architectures.

Advanced Network Administration is a must for anyone seeking to design a properly scaled network in today's competitive environment.

Wireless and Cellular Communications (forthcoming)

Wisniewski's third book for Prentice Hall, *Wireless and Cellular Communications,* will enrich the reader's knowledge about the latest trends in wireless and cellular networks.

Wireless and Cellular Communications discusses the nuts and bolts of implementing a wireless network in today's corporate network infrastructure. The book discusses wireless transmission, encoding, and standards such as Bluetooth and IEEE 802.11a and b. Cellular communications and different roadblocks that may interfere with building and implementing wireless networks are detailed.

The book also discusses a new technology: Mobile IP. Imagine connecting to the Internet in any city with a different IP address!

An advanced book, *Wireless and Cellular Communications* will bring readers to the cutting edge of technology.

Contents

CHAPTER 1 Internetworking Basics 1

CHAPTER 2 Bridging and Switching Fundamentals 37

CHAPTER 5   Simple Network Management Protocol   169

CHAPTER 6   Systems Network Architecture   213

CHAPTER 7   NetWare Protocols   255

CHAPTER 8   AppleTalk Protocols   299

CHAPTER 9   DECnet Phase IV Digital Network Architecture   361

CHAPTER 10   Open Systems Interconnection Protocols   409

CHAPTER 11   Interdomain Routing Basics, Part I: Open Shortest Path First Routing Protocol   441

CHAPTER 12   Interdomain Routing Basics, Part II: Border Gateway Protocol   487

CHAPTER 13   Advanced IP Routing   525

CHAPTER 14   The Data Link Protocols   561

CHAPTER 15   Internetworking Design Basics   617

CHAPTER 16   ATM   679

CHAPTER 17   T1 Digital Communications   717

Internetworking Basics

Objectives

- Define an internetwork and explain the challenges that confront an internetwork.
- Explain the Open Systems Interconnection (OSI) model, the various OSI layers, and how computers communicate with each other.
- Discuss the ways in which information is transmitted across an internetwork.
- Explain the International Standardization for Organization (ISO) hierarchy of networks
- Define the difference between connection oriented and connectionless networks.
- Explain internetwork addressing.
- Discuss flow control basics, error checking basics, and multiplexing basics.
- Explain a Local Area Network (LAN) and discuss various LAN protocols.
- Define a Wide Area Network (WAN) and discuss various WAN protocols.

Key Terms

Introduction

Internetworking is defined as the communication between two or more networks. It encompasses every aspect of connecting computers. Internetworks have grown to support vastly disparate end system communication requirements. In order to permit scalability and manageability without constant manual intervention, many protocols and features are necessary. Large internetworks consist of the following three distinct components:

- Campus networks, which consist of locally connected users in a building or group of buildings.
- Wide Area Networks (WANs), which connect campus networks.
- Remote connections, which link branch offices and single users (mobile users and telecommuters) to a typical enterprise internetwork.

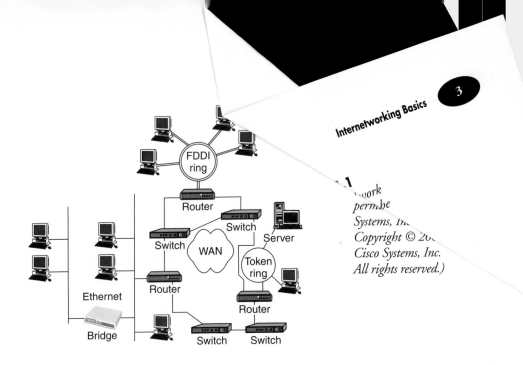

What Is an Internetwork?

An *internetwork* is a collection of individual networks, connected by intermediate networking devices that function as a single large network. Internetworking refers to the industry, products, and procedures that meet the challenge of creating and administering internetworks. Figure 1.1 illustrates different kinds of network technologies that can be interconnected through the use of routers and other networking devices to create an internetwork.

> An **internetwork** is a collection of networks connected by networking devices that function as a single large network.

The History of Internetworking

The first networks were time sharing networks that used mainframes and attached terminals. Such networking environments were implemented by IBM's System Network Architecture (SNA) and Digital's Network Architecture. *Local Area Networks* (*LANs*) evolved around the time of the PC revolution. LANs enable multiple users in a relatively small geographical area, roughly 10 miles or fewer, to exchange files and messages and share resources such as file servers.

Wide Area Networks (*WANs*) connect LANs across telephone lines and other media, thereby connecting geographically dispersed users. Today, high speed LANs and switched internetworks are becoming widely used mainly because they operate at very high speeds and support high-bandwidth applications such as voice and videoconferencing. Internetworking evolved as a solution to three key problems: isolated LANs, duplication of resources, and lack of network management. Isolated LANs made electronic communication between different offices or departments impossible. Duplication of resources meant that the same hardware and software had to be supplied to each office or department. In addition, each department or office had to maintain a separate support

> **Local Area Networks** allow multiple users to share resources and exchange files and messages.

> **Wide Area Networks** interconnect LANs across telephone lines and other media.

resources. The lack of network management meant that no centralized managing and troubleshooting networks existed.

Networking Challenges

Implementing a functional internetwork is no simple task. Some problem areas include connectivity, reliability, network management, and flexibility. Each individual area is a key in establishing an efficient and effective internetwork. The main challenge in connecting various systems together is supporting the communication between disparate technologies. Different locations may use different types of media, or they might operate at varying speeds. Another challenge, reliable service, must be maintained in any internetwork. Individual users and entire organizations depend on consistent, reliable access to network resources. Network management must provide centralized support and troubleshooting capabilities. Configuration, security, performance, fault, and accounting issues must be adequately addressed so that the internetwork will function smoothly. The final challenge, flexibility, is necessary for network expansion and the addition of new applications and services, which are necessary to keep businesses competitive in the 21st century.

The Open System Interconnect OSI Reference Model

Open Systems Interconnect allows information from one computer to move to another computer through a network.

The *Open Systems Interconnection* (*OSI*) reference model describes how information from a software application in one computer moves to a software application in another computer through a network medium. OSI is a conceptual model composed of seven layers, each specifying particular network functions, and is considered the primary architectural model for intercomputer communications. OSI divides tasks, which are involved with moving information between networked computers, into seven smaller and more manageable task groups. A task, or combination of tasks, is then assigned to each of the server layers. Each layer is reasonably self-contained so that the tasks assigned to that particular layer can be independently implemented. The independence of each layer operating separately allows the solutions offered by one layer to be updated without affecting the other layers. Figure 1.2 illustrates the seven-layer OSI reference model.

Characteristics of the OSI Layers

The **upper layers** of the OSI model include layers 5, 6, and 7.

The **lower layers** of the OSI model include layers 1, 2, 3, and 4.

The seven layers that comprise the OSI reference model (OSI model) can be divided into two categories: *upper layers* and *lower layers*. The upper layers of the OSI model deal with application issues and are generally implemented only in software. The application layer, which is the highest layer of the model, is closest to the end user. Users and application layer processes interact with software applications that contain a communications component. The term *upper layer* may be used to refer to any layer above another layer in the OSI model.

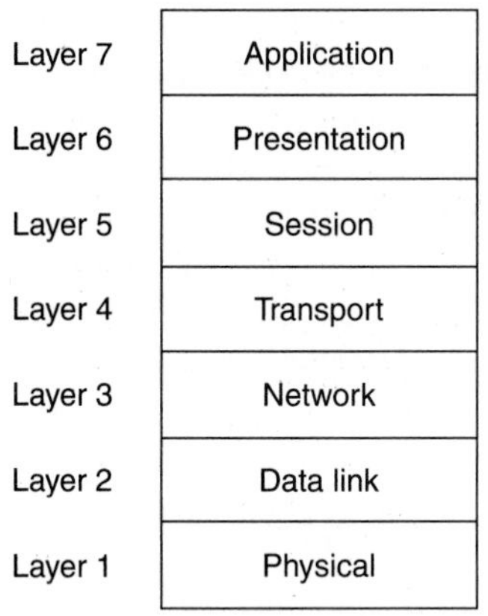

Layer 7	Application
Layer 6	Presentation
Layer 5	Session
Layer 4	Transport
Layer 3	Network
Layer 2	Data link
Layer 1	Physical

FIGURE 1.2

The OSI reference model contains seven independent layers. (Reproduced with permission of Cisco Systems, Inc. Copyright © 2001 Cisco Systems, Inc. All rights reserved.)

Application Layer
{
Layer 7	Application
Layer 6	Presentation
Layer 5	Session
}

Data Transport Layer
{
Layer 4	Transport
Layer 3	Network
Layer 2	Data link
Layer 1	Physical
}

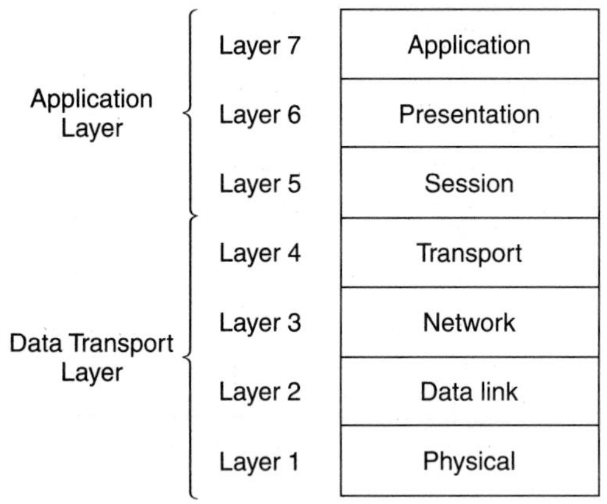

FIGURE 1.3

Two sets of layers make up the OSI model—the upper (application) layer and the lower (data transport) layer. (Reproduced with permission of Cisco Systems, Inc. Copyright © 2001 Cisco Systems, Inc. All rights reserved.)

The lower layers of the OSI model handle data transport issues. The physical and data link layers are implemented in hardware and software. The other lower layers are generally implemented only in software. The physical layer, which is the lowest layer of the model, is closest to the physical network, the network cabling that connects end users to the network. The physical layer is also responsible for actually placing information on the medium. Figure 1.3 illustrates the division between the upper and lower layers of the OSI model.

Protocols

The purpose of the OSI model is to provide a conceptual framework for communications between computers, but the model itself is not a method of communication. Actual communication is made possible by using communication protocols. In the context of data networking, a *protocol* is a formal set of rules and conventions that

Protocols govern the exchange of information over a network.

governs how computers exchange information over a network. A protocol implements the functions of one or more of the OSI layers. A wide variety of communication protocols exist. However, most tend to be categorized into one of four groups: LAN, WAN, network, and routing protocols. LAN protocols operate at the physical and data link layers of the OSI model and define communication over various LAN media. WAN protocols operate at the lower three layers of the OSI model and define communication over the various wide-area media. Routing protocols are network layer protocols that are responsible for path determination and traffic switching. Network protocols are the various upper layer protocols that exist in a given protocol suite.

Communication Between Systems

When an end user wishes to transfer information from a software application in one computer system to a software application in another system, the information must pass through each of the OSI layers. For example, a software application on system *A* has information to transmit to a software application on system *B*. The application program in system *A* will pass its information to the application layer (layer 7) of system *A*. The application layer then passes the information to the presentation layer (layer 6), which relays the data to the session layer (layer 5), and so on all the way down to the physical layer (layer 1). At the physical layer, the information is placed onto the network and is sent to system *B*. The physical layer of system *B* removes the information from the network and passes the information up to the data link layer (layer 2), which passes the information to the network layer (layer 3), and so on until the information reaches the application layer (layer 7). The application layer will then pass the information to the receiver application program to complete the communication process.

Interaction Between OSI Layers

A given layer in the OSI model generally communicates with three other OSI layers: the layer directly above it, the layer directly below it, and its peer layer in other networked computer systems. The data link layer in system *A* communicates with the network layer of system *A*, the physical layer of system *A*, and the data link layer of system *B*. Figure 1.4 illustrates this example.

OSI Services

A **service user** requests services from an adjacent OSI layer.

A **service provider** provides services to service users.

The **service access point** is a conceptual location where one OSI layer requests the service of another.

One OSI layer communicates with another layer to make use of the other layer's services. Services provided by adjacent layers help any given layer communicate with its peer layer in other computer systems. Three basic elements are involved in layer services: the *service user*, the *service provider*, and the *service access point* (SAP). In this context, the service user is the layer that requests services from an adjacent layer. The service provider is the layer that provides services to service users. OSI layers can provide services to multiple service users at once. The SAP is a conceptual location whereby one OSI layer can request the service of another. Figure 1.5 illustrates how these three elements interact at the network and data link layers.

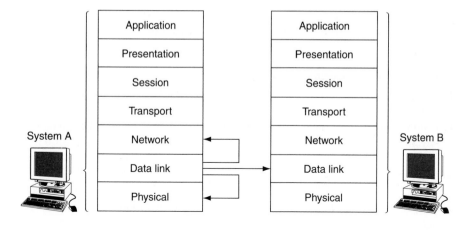

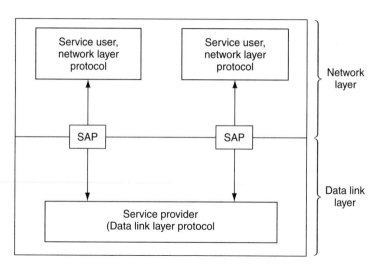

Information Exchange Between OSI Layers

OSI layers use various forms of control information to communicate with their peer layers in other systems. Control information consists of specific requests and instructions exchanged between peer layers. Control information usually takes the form of a header or a trailer. *Headers* are prepended to data that have been passed down from the upper layers. *Trailers* are appended to data that have been passed down from upper layers. An OSI layer, though, is not required to attach a header or a trailer to data from upper layers. Headers, trailers, and data are concepts that are relative to each other depending on the layer that analyzes the information being transmitted. At the network layer, information being sent consists of a layer 3 header and data. At the data link layer, all the information passed down by the network layer (the layer 3 header and data) is treated as data. In other words, the data

Headers are control information prepended to data passed down from upper layers.

Trailers are control information appended to data passed down from upper layers.

Encapsulation happens when data from a higher level, along with its header, are captured into the data at any lower level.

portion of the information being sent at a given OSI layer can contain headers, trailers, and data from all higher layers. This is referred to as *encapsulation*.

The Information Exchange Process

Information exchange occurs between peer OSI layers. Each layer in the source system adds control information to data, and each layer in the destination system analyzes and removes the control information from that data. If system *A* has data from a software application to send to system *B*, the data are passed to the application layer. The application layer in system *A* then communicates any control information required by the application layer in system *B* by prepending a header to the data. The resulting information unit (a header and data) is passed to the presentation layer, which prepends its own header containing control information intended for the presentation layer in system *B*. The information unit grows in size as each layer prepends its own header and trailer that contains control information needed by its peer layer in system *B*. At the physical layer, the entire information unit is placed onto the network medium. The physical layer in system *B* receives the information unit and passes it to the data link layer. The data link layer in system *B* then reads the control information contained in the header prepended by the data link layer in system *A*. The header is then removed, and the remainder of the information unit is passed to the network layer. Each layer performs the same actions: The layer reads the header from its peer layer, strips it off, and passes the remaining information unit to the next highest layer. After the application layer performs these actions, the data are passed to the recipient software application in system *B*, in exactly the form transmitted by the application in system *A*. Figure 1.6 illustrates the ways in which the header and data from one layer are encapsulated into the data of the next lowest layer.

FIGURE 1.6

Headers and data can be encapsulated during information exchange. (Reproduced with permission of Cisco Systems, Inc. Copyright © 2001 Cisco Systems, Inc. All rights reserved.)

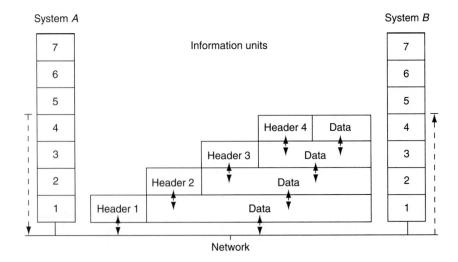

Functions of OSI Layers

The Physical Layer

The physical layer defines the electrical, mechanical, procedural, and functional specifications for activating, maintaining, and deactivating the physical link between communicating network systems. Physical layer specifications define characteristics such as voltage levels, timing of voltage changes, physical data rates, maximum transmission distances, and physical connectors. Physical layer implementations may be categorized as LAN or WAN specifications. Figure 1.7 illustrates common LAN and WAN physical layer implementations.

The Data Link Layer

The data link layer provides a reliable transit of data across a physical network link. Different data link layer specifications define different network and protocol characteristics. Data link layer characteristics include physical addressing, network topology, error notification, frame sequencing, and flow control. Physical addressing defines how devices are addressed at the data link layer. Network topology consists of data link layer specifications, which define how devices are to be physically connected. Network topology can be described as a bus, star, or ring. Error notification alerts upper layer protocols that a transmission error has occurred. The sequencing of data frames reorders frames that are transmitted out of sequence. Flow control moderates the transmission of data so that the receiving device is not overwhelmed with more traffic than it can handle at one time. The Institute of Electrical and Electronics Engineers, Inc. (IEEE) has subdivided the data link layer into two sublayers: the Logical Link Control (LLC) and Media Access Control (MAC) sublayers. Figure 1.8 illustrates the IEEE subdivision of the data link layer.

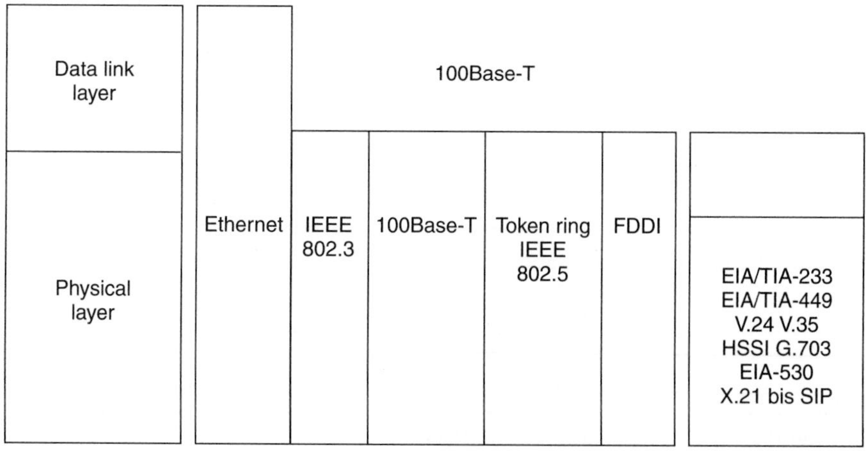

FIGURE 1.7

Common physical layer implementations. (Reproduced with permission of Cisco Systems, Inc. Copyright © 2001 Cisco Systems, Inc. All rights reserved.)

FIGURE 1.8
IEEE data link layer subdivision. (Reproduced with permission of Cisco Systems, Inc. Copyright © 2001 Cisco Systems, Inc. All rights reserved.)

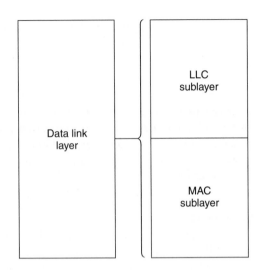

The Network Layer

The network layer provides routing and related functions that enable multiple data links to be combined into an internetwork. This is accomplished by addressing devices logically. The network layer supports connection oriented and connectionless service from higher layer protocols. Network layer protocols are typically routing protocols. However, other types of protocols are implemented at the network layer as well. Common routing protocols include the following: the Border Gateway Protocol (BGP), which is an Internet interdomain routing protocol; the Open Shortest Path First (OSPF) protocol, a link-state interior gateway protocol developed for use in Transmission Control Protocol/Internet Protocol (TCP/IP) networks; and the Routing Information Protocol (RIP), an Internet routing protocol that uses hop count as its metric.

The Transport Layer

The transport layer implements reliable internetwork data transport services that are transparent to upper layers. Transport layer functions typically include flow control, multiplexing, virtual circuit management, and error checking and recovery. Flow control manages data transmission between devices so that the transmitting device does not send more data than the receiving device can process at any given time. Multiplexing enables data from several applications to be transmitted onto a single physical link. Virtual circuits are established, maintained, and terminated by the transport layer. Error checking involves creating various mechanisms for detecting transmission errors, while recovery involves taking an action, such as requesting that the data be retransmitted, to resolve any errors that occurred.

Transport layer protocols include the Transmission Control Protocol (TCP), Name Binding Protocol (NBP), and OSI Transport Protocols. TCP is the protocol, used in the TCP/IP suite of protocols, which provides for reliable transmission of data. NBP is the protocol that associates AppleTalk names with AppleTalk addresses. OSI Transport Protocols are a series of transport protocols in the OSI protocol suite.

The Session Layer

The session layer establishes, manages, and terminates communication sessions between presentation layer entities. Communication sessions consist of service requests and responses that occur between applications located in different network devices. These requests and responses are coordinated by protocols implemented at the session layer. Session layer implementations include the Zone Information Protocol (ZIP); the AppleTalk Protocol, which coordinates the name binding process; and the Session Control Protocol (SCP), the DECnet Phase IV session layer protocol.

The Presentation Layer

The presentation layer provides a variety of coding and conversion functions, which are applied to the application layer data. The coding and conversion functions that the presentation layer provides ensures that information sent from the application layer of one system will be readable by the application layer of another system. Some examples of presentation layer coding and conversion schemes include common data representation formats, conversion of character representation formats, common data compression schemes, and common data encryption schemes.

Common data representation formats, or the use of standard image, sound, and video formats, enable the interchange of application data between different types of computer systems. Conversion schemes are used to exchange information between systems by using different text and data representations, such as Extended Binary Coded Decimal Interchange code (EBCDIC) and American Standard Code for Information Interchange and ASCII. Standard data encryption schemes enable data encrypted at the source device to be properly deciphered at the destination.

Presentation layer implementations are not typically associated with a particular protocol stack. Some well-known standards for video include QuickTime and Moving Picture Experts Group (MPEG). QuickTime is an Apple specification for video and audio, and MPEG is a standard for video compression and coding.

Among well-known graphic image formats are the Graphics Interchange Format (GIF), the Joint Photographic Experts Group (JPEG), and the tag image file format (TIFF). GIF and JPEG are standards for compressing and coding graphic images, while TIFF is solely a standard coding format for graphic images.

The Application Layer

The application layer is the OSI layer closest to the end users. Both the OSI application layer and the user interact directly with the software application, which implements a communication component. Such application programs fall outside the scope of the OSI model. Application layer functions include identifying communication partners, determining resource availability, and synchronizing communication.

When identifying communication partners, the application layer determines the identity and availability of communication partners for an application that has data to transmit. When determining resource availability, the application layer must decide whether sufficient network resources for the requested communications exist. All communication between applications requires cooperation and synchronization that is managed by the application layer.

Two key types of application layer implementations are TCP/IP applications and OSI applications. TCP/IP applications are protocols, such as Telnet, File Transfer Protocol (FTP), and Simple Mail Transfer Protocol (SMTP), that exist in the Internet Protocol Suite (IPS). OSI applications are protocols, such as File Transfer and Access Management (FTAM), Virtual Terminal Protocol (VTP), and Common Management Information Protocol (CMIP), that exist in the OSI suite.

OSI Information Formats

The data and control information that is transmitted through internetworks encompasses a variety of different forms. Common information formats include frames, packets, datagrams, messages, cells, and data units. A *frame* is an information unit whose source and destination are data link entities. A frame is composed of the data link layer header and (possibly) a trailer, as well as upper layer data. The header and trailer contain control information intended for the data link layer entity in the destination system. Data from upper layer entities are encapsulated in the data link header and trailer. Figure 1.9 illustrates the basic components of a data link frame.

A *packet* is an information unit whose source and destination are network layer entities. A packet is composed of the network layer header and (possibly) a trailer, as well as upper layer data. The header and trailer contain control information intended for

A **frame** is an information unit whose source and destination are data link entities.

A **packet** is an information unit whose source and destination are network layer entities.

FIGURE 1.9

Data from upper layer entities make up the data link layer frame. (Reproduced with permission of Cisco Systems, Inc. Copyright © 2001 Cisco Systems, Inc. All rights reserved.)

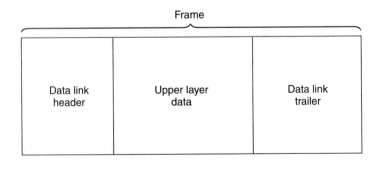

Frame

| Data link header | Upper layer data | Data link trailer |

Packet

| Network layer header | Upper layer data | Network layer trailer |

the network layer entity in the destination system. Data from upper layer entities are encapsulated in the network layer header and trailer. Figure 1.10 shows the three basic components of a network layer packet.

A *datagram* is an information unit whose source and destination are network layer entities that use connectionless network services.

A *message* is an information unit whose source and destination entities exist above the network layer, often in the application layer.

A *cell* is an information unit of a fixed size whose source and destination are data link entities. Cells are used in switched environments, such as in Asynchronous Transfer Mode (ATM) and Switched Multimegabit Data Service (SMDS) networks. A cell is composed of the header and the payload. The header contains control information intended for the destination data link layer entity and is typically five bytes long. The payload contains upper layer data that are encapsulated in the cell header and are typically 48 bytes long. The length of the header and the payload fields are always the same for each cell, which is 53 bytes long. Figure 1.11 depicts the components of a typical cell.

A *data unit* is a compilation of other information units. Some common data units are service data units (SDUs), protocol data units (PDUs), and bridge protocol data units (BPDUs). SDUs are information units from upper layer protocols that define a service request to a lower layer protocol. PDUs are an OSI term for a packet. The spanning tree algorithm uses BPDUs as "hello" messages.

The International Standardization for Organization (ISO) Hierarchy of Networks

Large networks are typically organized as hierarchies. A hierarchical organization provides advantages, which include ease of management, flexibility, and reduced unnecessary traffic. International Standardization Organization (ISO) has adopted a

A **datagram** is an information unit whose source and destination are network layer entities that use connectionless network services.

A **message** is an information unit whose source and destination entities exist above the network layer.

A **cell** is an information unit of fixed size whose source and destination are data link entities.

A **data unit** is a compilation of other information units.

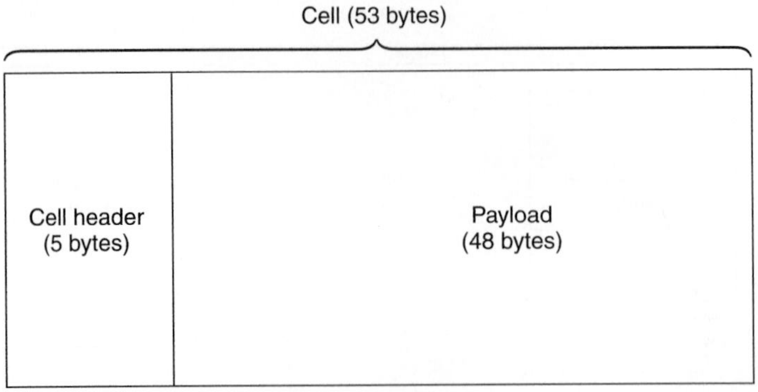

FIGURE 1.11

*The components
that make up a
typical cell.
(Reproduced with
permission of Cisco
Systems, Inc.
Copyright © 2001
Cisco Systems, Inc.
All rights reserved.)*

Cell (53 bytes)

| Cell header (5 bytes) | Payload (48 bytes) |

number of terminology conventions for addressing network entities, and defines those addressing conventions as *end system* (ES), *intermediate system* (IS), *area,* and *autonomous system* (AS).

An ES is a network device that does not route and forward traffic. Typical end systems include devices such as terminals, personal computers, and printers. An IS is a network device that does route and forward traffic. Typical intermediate systems include routers, switches, and bridges. Two types of IS networks exist: the intradomain IS and the interdomain IS. An intradomain IS communicates within a single autonomous system, while an interdomain IS communicates within *and* between autonomous systems. An area is a logical group of network segments and their attached devices. Areas are subdivisions of autonomous systems (AS). An AS is a collection of networks under a common administration that all share a common routing strategy. Autonomous systems are subdivided into areas and are sometimes referred to as domains. Figure 1.12 illustrates a typical hierarchical network and its components.

> An **end system** is a network device that does not route or forward traffic.

> An **intermediate system** is a network device that does route and forward traffic.

> An **area** is a logical group of network segments and its attached devices.

> An **autonomous system** is a collection of networks under a common administration.

Connection Oriented and Connectionless Network Services

In general, networking protocols and the data traffic that they support are classified as either connection-oriented or connectionless. Connection oriented data handling involves using a specific path that is established for the duration of the connection. Connectionless data handling involves passing data through a permanently established connection. Connection oriented service involves three phases: connection establishment, data transfer, and connection termination. During the connection termination phase, an established connection that is no longer needed is terminated. Further communication between source and destination systems requires a new connection.

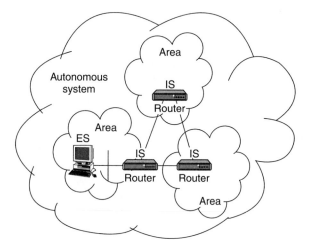

FIGURE 1.12

A hierarchical network contains numerous components. (Reproduced with permission of Cisco Systems, Inc. Copyright © 2001 Cisco Systems, Inc. All rights reserved.)

Connection oriented network service carries two significant disadvantages as compared to connectionless service. The disadvantages are static path selection and the static reservation of network resources. Static path selection can create difficulty because all traffic must travel along the same static path. A failure anywhere along that path causes the connection to fail. Static reservation of resources causes difficulty because it requires a guaranteed rate of throughput and, thus, a commitment of resources that other network users cannot share. Unless the connection uses full, uninterrupted throughput, bandwidth is not used efficiently. Connection oriented services are useful for transmitting data from applications that do not tolerate delays and packet resequencing. Voice and video applications are typically based on connection oriented services. Another disadvantage is that connectionless network service does not predetermine the path from the source to the destination system, and packet sequencing, data throughput, and other network resources are not guaranteed. Each packet must be completely addressed because different paths through the network may be selected for different packets based on a variety of influences. Each packet is transmitted independently by the source system and is handled independently by intermediate network devices.

The important advantages of using connectionless services rather than connection oriented services are dynamic path selection and dynamic bandwidth allocation. Dynamic path selection enables traffic to be routed around network failures because paths are selected on a packet-by-packet basis. Dynamic bandwidth allocation allows bandwidth to be utilized more efficiently because unnecessary network resources are not allocated. Connectionless services are useful for transmitting data from applications, such as data-based applications, that can tolerate delay and resequencing.

Internetwork Addressing

Internetwork addresses identify devices separately or as a member of a group.

Internetwork addresses identify devices separately or as a member of a group. Addressing schemes vary depending on the protocol family and the OSI layer. Three types of internetwork addresses are commonly used: data link layer addresses, MAC addresses, and network layer addresses.

Data Link Layer Addresses

Data link layer addresses are also referred to as **physical** or **hardware addresses.** They are a type of internetwork address.

A *data link layer address* uniquely identifies each physical network connection on a network device. Data link addresses sometimes are referred to as *physical* or *hardware addresses.* They usually exist within a flat address space and have a preestablished and fixed relationship to a specific device. End systems generally have only one physical network connection, thus only one data link address. Routers and other internetworking devices have multiple physical network connections and, therefore, have multiple data link addresses. Figure 1.13 illustrates how each interface on a device is uniquely identified by a data link address.

MAC Addresses

A **MAC address** is one type of internetwork address.

MAC addresses consist of a subset of data link layer addresses. MAC addresses identify LAN entities that implement the IEEE MAC addresses in the data link layer. As with most data link addresses, MAC addresses are unique for each LAN interface. Figure 1.14 illustrates the relationship between the IEEE sublayer, the MAC sublayer, MAC addresses, and data link addresses of the data link layer.

FIGURE 1.13

Each interface on a device is uniquely identified by a data link address. (Reproduced with permission of Cisco Systems, Inc. Copyright © 2001 Cisco Systems, Inc. All rights reserved.)

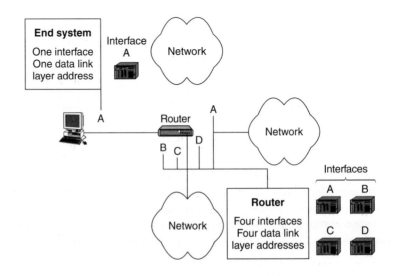

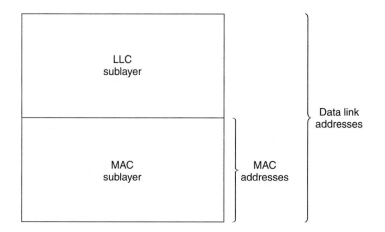

MAC addresses are 48 bits in length and are expressed as 12 hexadecimal digits. The first six hexadecimal digits, which are administered by the IEEE, identify the manufacturer and thus comprise the Organizationally Unique Identifier (OUI). The last six hexadecimal digits comprise the interface serial number, another value administered by the manufacturer. MAC addresses are sometimes called burned-in addresses (BIAs) because they are burned into read only memory (ROM) and are copied into random access memory (RAM) when the interface card initializes. Figure 1.15 illustrates the MAC address format.

Different protocol suites use different methods for determining the MAC address of a device. The most commonly used methods are to map network addresses to MAC addresses using the Address Resolution Protocol (ARP); to use the Hello Protocol, which enables network devices to learn the MAC addresses of other network devices; or to find the MAC addresses, embedded in the network layer address or generated

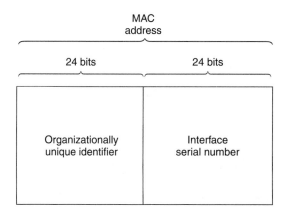

by an algorithm. Address resolution is the process of mapping network addresses to MAC addresses. This process is accomplished by using the ARP, which is implemented in many protocol suites. When a network address is successfully associated with a MAC address, the network device stores the information in the ARP cache. The ARP cache enables devices to send traffic to a destination without creating ARP traffic because the MAC address of the destination is already known.

The process of address resolution differs slightly depending on the network environment. Address resolution on a single LAN begins when end system (ES) *A* broadcasts an ARP request onto the LAN in an attempt to learn the MAC address of ES *B*. The broadcast is received and processed by every device that is attached to the LAN. ES *B* will reply to the ARP request by sending end system *A*, an ARP reply containing its MAC address. ES *A* receives the reply and saves the MAC address of ES *B* in its ARP cache. The ARP cache is where network addresses are associated with MAC addresses. Whenever end system *A* must communicate with ES *B*, it checks the ARP cache, locates the MAC address of ES *B*, and sends the frame directly without first having to use an ARP request.

Address resolution works differently when source and destination devices are attached to different LANs that are connected by a router. ES *X* broadcasts an ARP request onto the LAN in an attempt to learn the MAC address of ES *W*. The broadcast is received and processed by all devices on the LAN, including router *R*, which acts as a proxy for ES *W* by checking its routing table to determine that ES *W* is located on a different LAN. Router *R* then replies to the ARP request from ES *X*, sending an ARP reply containing the MAC address of router *R* in its ARP cache in the entry for ES *W*. When ES *X* must communicate with ES *W*, it checks the ARP cache, locates the MAC address of router *R*, and sends the frame directly without using ARP requests. Router *R* receives the traffic from ES *X* and forwards it to ES *W* on the other LAN.

The Hello Protocol is a network layer protocol that enables network devices to identify one another and indicate that they are still functioning. Devices on the network then return Hello replies, and Hello messages are sent at specific intervals to indicate that the devices are functioning. Network devices can learn the MAC addresses of other devices by examining Hello Protocol packets. Three protocols, use predictable MAC addresses: Xerox Network System (XNS), Novell's Internet Packet Xchange (IPX), and DECnet Phase IV. In these particular protocol suites, MAC addresses are predictable because the network layer either embeds the MAC address in the network layer address or uses an algorithm to determine the MAC address.

Network layer addresses are also referred to as **virtual** or **logical addresses.** They are a type of internetwork address.

Network Layer Addresses

A *network layer address* identifies an entity at the network layer. Network addresses usually exist within a hierarchical address space and sometimes are called *virtual* or *logical addresses*. The relationship between a network address and a device is logical and unfixed, and is based on the physical network characteristics of a device located on a particular network segment or on groupings that have no physical basis. For example, if the device is part of an AppleTalk zone, then ESs will require one network

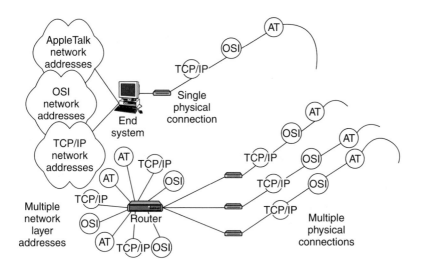

layer address for each network layer protocol that they support. This assumes that the device has only one physical network connection. Routers and other internetworking devices require one network layer address per physical network connection for each network layer protocol supported. A router, for example, with a connection to each of three interfaces that each run a different protocol such as AppleTalk, TCP/IP, and OSI must have three network layer addresses for each connection. The router, therefore, must have nine network layer addresses. Figure 1.16 illustrates how each network interface must be assigned a network address for each protocol supported.

Comparing Hierarchical and Flat Address Space

Internetwork address space takes one of two forms, *hierarchical address space* or *flat address space*. A hierarchical address space is organized into numerous subgroups, each successively narrowing an address until it points to a single device. It is similar to a street address. A flat address space is organized into a single group. It is similar to a Social Security number.

Hierarchical addressing offers the advantage of simplifying address sorting and recall through the use of comparison operations. For example, including "Ireland" in a mailing address eliminates any other country as a possible destination. Figure 1.17 illustrates the difference between a hierarchical and flat address space.

> Internetwork address space is either **hierarchical** or **flat.**

Address Assignments

Addresses assigned to a device can be static, dynamic, or server-assigned addresses. *Static addresses* are assigned by a network administrator according to a preconceived internetwork addressing plan. A static address does not change until the network

> **Static addresses** are assigned by network administrators.

FIGURE 1.17

Hierarchical and flat address spaces differ in comparison operations. (Reproduced with permission of Cisco Systems, Inc. Copyright © 2001 Cisco Systems, Inc. All rights reserved.)

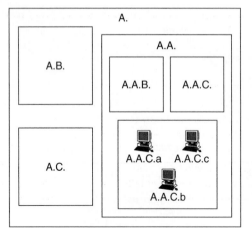

Hierarchical address space

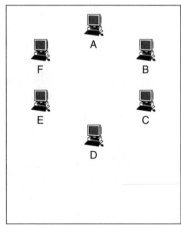

Flat address space

● **Dynamic addresses** are obtained by devices via a protocol-specific process when they attach to a network.

administrator manually changes it. *Dynamic addresses* are obtained by devices when they attach to a network by means of a protocol-specific process. A device using a dynamic address often has a different address each time it connects to the network. *Server-assigned addresses* are given to devices as they connect to the network. These addresses are then recycled for reuse as devices disconnect. A device is likely to have a different address each time it connects to the network.

● **Server-assigned addresses** are given to devices by servers as they connect to a network.

Addresses versus Names

Internetwork devices usually have a name and an address associated with them. Internetwork names typically are independent of location and remain with a device wherever that device moves. Internetwork addresses are usually dependent upon location and change when a device is moved. MAC addresses are the exception to this rule. In MAC addresses, names and addresses represent a logical identifier, which may be a local system administrator or an organization, such as the Internet Assigned Numbers Authority (IANA).

Flow Control, Error Checking, and Multiplexing

Flow Control

● **Flow control** prevents network congestion.

Flow control is a function that prevents network congestion by ensuring that transmitting devices do not overwhelm receiving devices with data. Countless possible causes of network congestion exist. A high speed computer may generate traffic faster than a slow network can transfer it or faster than the destination device can receive and process it. The three commonly used methods for handling network congestion are buffering, transmitting source-quench messages, and windowing.

Buffering is utilized by network devices to store bursts of excess data in memory temporarily until they can be processed. Occasional data bursts are easily handled by buffering, but excess data bursts can exhaust memory, forcing the device to discard any additional datagrams that arrive.

Buffering stores data temporarily until it can be utilized.

Source-quench messages are used by receiving devices to help prevent their buffers from overflowing. The receiving device sends source-quench messages to request that the source reduce its current rate of data transmission. First, the receiving device begins discarding data due to overflowing buffers. Second, the receiving device begins sending source-quench messages to the transmitting device at the rate of one message for each packet dropped. The source device receives the source-quench messages and lowers the data rate until it stops receiving messages. Finally, the source device then gradually increases the data rate as long as no further source-quench requests are received from the destination device.

Source-quench messages help prevent buffers from overflowing.

Windowing is a flow control scheme in which the source device requires an acknowledgment from the destination device after a certain number of packets have been transmitted. With a window size of three, the source requires an acknowledgment after sending three packets. First, the source device sends three packets to the destination device. Second, after receiving the three packets, the destination device sends an acknowledgment to the source. Finally, the source receives the acknowledgment and sends three more packets. If the destination does not receive one or more of the packets for any reason, it will not send an acknowledgment. The source will then retransmit the most recently sent packets at a reduced transmission rate.

Windowing requires acknowledgment from the destination device after a certain number of packets have been transmitted.

Error Checking

Error checking schemes determine whether data have become corrupt or damaged while traveling from the source to the destination. Error checking is implemented at several of the OSI layers. One common error checking scheme is the cyclic redundancy check (CRC). The CRC detects and discards corrupted data. Error checking functions such as data retransmission are left to higher layer protocols. A CRC value is generated by a calculation that is performed at the source device.

Error checking determines whether data became corrupt or damaged while being transmitted.

The source places the calculated value in the packet and transmits the packet to the destination. The destination performs the same predetermined set of calculations over the contents of the packet and compares the calculated value with the value that is contained inside the packet. If the values are equal, the packet is valid, but if the values are unequal, the packet contains errors and is discarded.

Multiple data channels are combined into a single data or physical channel at the source address in **multiplexing.**

Multiplexing

Multiplexing is a process by which several data channels are combined into a single data or physical channel at the source address and can be implemented at any OSI layer. Conversely, *demultiplexing* is the process of separating multiplexed data channels at the

Demultiplexing separates multiplexed data channels at the destination address.

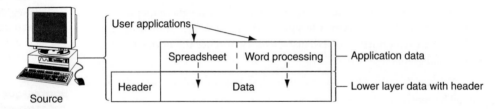

FIGURE 1.18

Several applications can be multiplexed into a single lower layer data packet at the source. (Reproduced with permission of Cisco Systems, Inc. Copyright © 2001 Cisco Systems, Inc. All rights reserved.)

destination address. One example of multiplexing is when data from multiple applications are multiplexed into a single lower layer data packet, as illustrated in Figure 1.18.

Another example of multiplexing is when data from multiple devices are combined into a single physical channel utilizing a device called a multiplexer. A multiplexer is a physical layer device that combines multiple data streams into one or more output channels at the source. Multiplexers can demultiplex a single channel into multiple data streams at the remote end and, thus, maximize the use of bandwidth of the physical medium by enabling it to be shared by multiple traffic sources. This example is illustrated in Figure 1.19. Some methods used for multiplexing data are Time Division Multiplex (TDM), Asynchronous Time Division Multiplexing (ATD), Frequency Division Multiplexing (FDM), and Statistical Multiplexing.

In TDM, information from each data channel is allocated bandwidth based on pre-assigned time slots, even if no data are to be transmitted. In ATD, information from each data channel is allocated bandwidth as needed, by utilizing dynamically assigned time slots. In FDM, information from each data channel is allocated bandwidth based on the signal frequency of the traffic. In statistical multiplexing, bandwidth is dynamically allocated to any data channels that have information to transmit.

FIGURE 1.19

Multiple devices can be multiplexed into a single physical channel. (Reproduced with permission of Cisco Systems, Inc. Copyright © 2001 Cisco Systems, Inc. All rights reserved.)

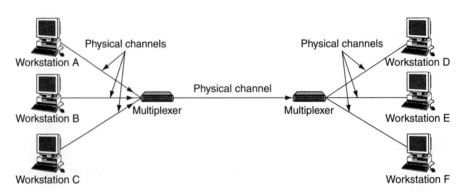

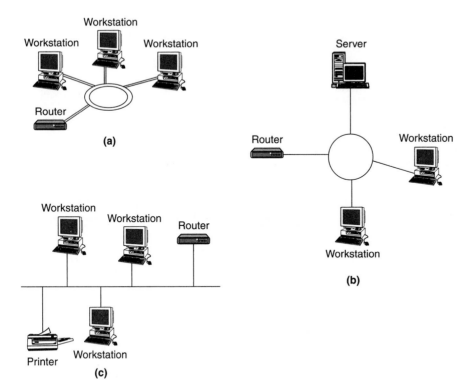

Local Area Networks: An In-Depth Look

This section introduces the various media access methods, transmission methods, topologies, and devices used in a LAN. Topics include methods and devices used in Ethernet/IEEE 802.3, Token Ring/IEEE 802.5, and fiber distributed data interface (FDDI). Figure 1.20 illustrates the basic layout of these three implementations.

What Is a LAN?

A LAN is a high speed, fault-tolerant data network that covers a relatively small geographic area. A LAN typically connects work stations, personal computers, servers, printers, and other devices. LANs offer computer users many advantages, including shared access to devices and applications, file exchange between connected users, and communication between users through electronic mail or other applications. LAN protocols function at the lowest two layers of the OSI model, between the physical layer and the data link layer. Figure 1.21 illustrates how several popular protocols map to the OSI reference model.

FIGURE 1.21

Popular LAN protocols mapped to the OSI reference model. (Reproduced with permission of Cisco Systems, Inc. Copyright © 2001 Cisco Systems, Inc. All rights reserved.)

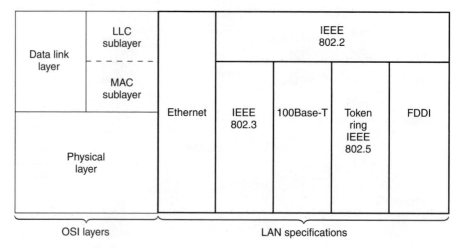

Media Access Methods

CSMA/CD and **token passing** are two methods used by LAN protocols to access a physical network medium. CSMA/CD is also known as **contention access.**

LAN protocols use one of two methods to access the physical network medium, *carrier sense multiple access with collision detection* (*CSMA/CD*) or *token passing.* Using CSMA/CD, network devices contend for the use of the physical network medium or bandwidth. CSMA/CD is, therefore, sometimes called "*contention access.*" Examples of LANs that use CSMA/CD include Ethernet/IEEE 802.3 networks and 100Base-T. In token passing, network devices access the physical medium based on possession of a token. Examples of LANs that use token passing are Token Ring/IEEE 802.5 and FDDI.

Transmission Methods

Three types of data transmission include **unicast, multicast,** and **broadcast.**

There are three different types of LAN data transmission, *unicast, multicast,* and *broadcast.* In each type of transmission, a single, or bundle of data, is sent to one or more *nodes,* or points of connection into a network. In a unicast transmission, a single packet is sent from the source to the destination on a network. The source node addresses the packet by using the address of the destination node. The packet is transmitted onto the network, and the network passes the packet to its destination.

A **node** is a point of connection into a network.

A multicast transmission consists of a single data packet that is copied and sent to a specific subset of nodes on the network. The source node addresses the packet by using a multicast address. The packet is then transmitted onto the network, which makes copies of the packet and transmits a copy to every node attached to the network.

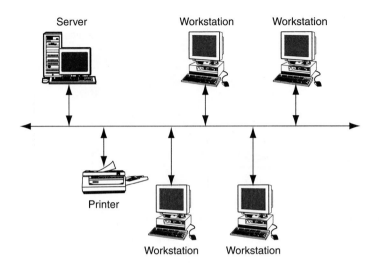

Server   Workstation   Workstation

Printer

Workstation   Workstation

Broadcast transmissions consist of single data packets that are copied and transmitted to all nodes on the network. In broadcast transmissions, the source node addresses the packet by using the broadcast address. The packet is transmitted onto the network, which makes copies of the packet and sends a copy to every node on the network.

Topologies

LAN *topologies* define the manner in which network devices are installed and connected together. The four common LAN topologies are bus, ring, star, and tree.

These topologies describe the logical architecture, but the actual devices need not be physically connected in these configurations. Logical bus and ring topologies are commonly connected as a physical star. A bus topology is a linear LAN architecture whereby transmissions from network stations propagate the length of the medium and are received by all other stations. Ethernet/IEEE 802.3, 100Base-T networks implement a bus topology, which is illustrated in Figure 1.22.

A ring topology is a LAN architecture that consists of a series of devices connected to one another by unidirectional transmission links to form a single closed loop. Token Ring/IEEE 802.5 and FDDI networks implement a ring topology illustrated in Figure 1.23.

A star topology is a LAN architecture in which the endpoints on a network are connected to a common central hub, or switch, by dedicated links. Logical bus and ring topologies are often implemented physically in a star topology, which is depicted in Figure 1.24.

The architecture of a LAN is referred to as its **topology.**

FIGURE 1.23

A logical ring topology. (Reproduced with permission of Cisco Systems, Inc. Copyright © 2001 Cisco Systems, Inc. All rights reserved.)

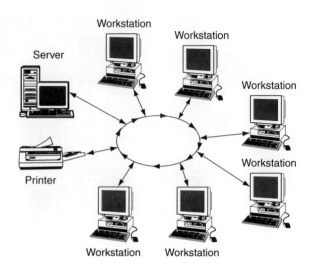

FIGURE 1.24

A physical star topology. (Reproduced with permission of Cisco Systems, Inc. Copyright © 2001 Cisco Systems, Inc. All rights reserved.)

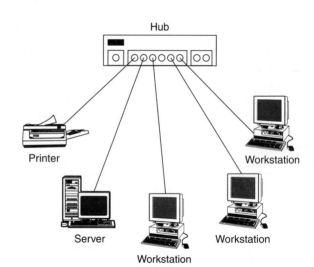

A tree topology is a LAN architecture similar to the bus topology, except that branches with multiple nodes are possible. Figure 1.25 illustrates a logical tree topology with multiple nodes.

A **LAN repeater** is a physical layer device used to connect media segments of an extended network.

LAN Devices

Devices commonly used in LANs include *repeaters, hubs, LAN extenders, bridges, LAN switches,* and *routers.* A *repeater* is a physical layer device used to connect media segments of an extended network. A repeater enables a series of cable segments to be treated as a single cable. It receives signals from one network segment and amplifies,

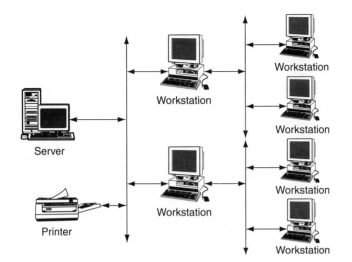

retimes, and retransmits those signals to another segment. These actions prevent signal deterioration caused by long cable lengths and large numbers of connected devices. Repeaters are incapable of performing complex filtering and traffic processing. In addition, all electrical signals, including electrical disturbances and other errors are repeated and amplified. The total number of repeaters and network segments that can be deployed is limited due to timing and other issues. Figure 1.26 illustrates a repeater connecting two network segments together.

A *hub* is a physical layer device that connects multiple user stations, each via a dedicated cable. Electrical connections are established inside the hub. Hubs are used to

Multiple workstations connect via dedicated cables to a **LAN hub.**

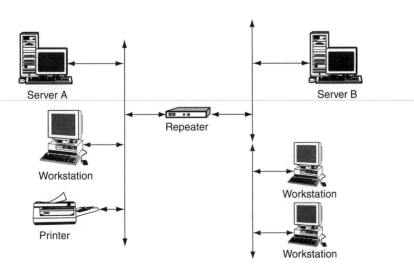

create a physical star network while maintaining the logical bus or ring configuration of the LAN. In some respects, a hub functions as a multiport repeater.

A *LAN extender* is a remote access multiplayer switch that connects to a host router. It forwards traffic from all the standard network layer protocols such as IP, IPX, and AppleTalk, and filters traffic based on the MAC address or network layer protocol type. LAN extenders scale well because the host router filters out unwanted broadcasts and multicasts. LAN extenders are not capable of segmenting traffic or creating security firewalls.

LAN extenders are switches that connect to host routers.

Wide Area Networks: An In-Depth Look

This section introduces the various protocols and technologies used in Wide Area Network (WAN) environments. Topics summarized here include point-to-point links, circuit switching, packet switching, virtual circuits, dial up services, and WAN devices.

What Is a WAN?

A WAN is a data communications network that covers a relatively broad geographic area and often uses transmission facilities provided by common carriers, such as telephone companies. WAN technologies function at the lower three layers of the OSI reference model: the physical layer, the data link layer, and the network layer. Figure 1.27 illustrates the relationship between common WAN technologies and the OSI model.

FIGURE 1.27

WAN technologies operate at the three lowest levels of the OSI model. (Reproduced with permission of Cisco Systems, Inc. Copyright © 2001 Cisco Systems, Inc. All rights reserved.)

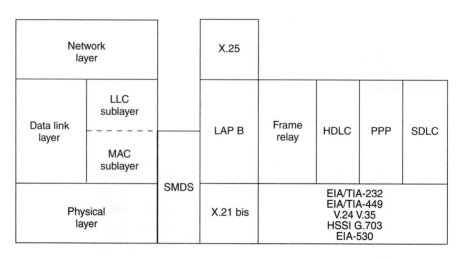

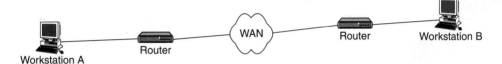

FIGURE 1.28

A typical point-to-point link operates through a WAN to a remote network. (Reproduced with permission of Cisco Systems, Inc. Copyright © 2001 Cisco Systems, Inc. All rights reserved.)

Point-to-Point Links

In a WAN, a *point-to-point link* provides a single, preestablished communication path from the customer premises, through a carrier network such as a telephone company, to a remote network. A point-to-point link is also known as a leased line because its established path is permanent and fixed for each remote network reached through carrier facilities. Carrier companies reserve point-to-point links for the private use of the customer, and the links accommodate two types of transmissions: datagram transmissions, which are composed of individually addressed frames; and data stream transmissions, which are composed of a stream of data for which address checking occurs only once. A typical point-to-point link through a WAN is illustrated in Figure 1.28.

A **point-to-point link** provides a preestablished link from the customer premises to a remote network.

Circuit Switching

Circuit switching is a WAN switching method in which a dedicated physical circuit is established, maintained, and terminated through a carrier network for each communication session. Circuit switching accommodates datagram transmissions and data stream transmissions. Circuit switching is used extensively in telephone company networks and operates much like a normal telephone call. An Integrated Services Digital Network (ISDN) is one example of a circuit-switched WAN technology and is illustrated in Figure 1.29.

In **circuit switching,** a dedicated physical circuit is established, maintained, and terminated for each communication session.

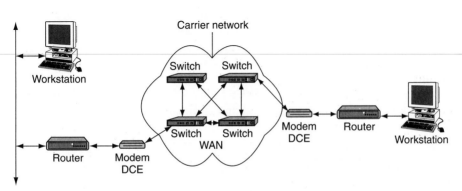

FIGURE 1.29

A circuit-switched WAN undergoes a process similar to that used for a telephone call. (Reproduced with permission of Cisco Systems, Inc. Copyright © 2001 Cisco Systems, Inc. All rights reserved.)

FIGURE 1.30

Packet switching transfers packets across a carrier network. (Reproduced with permission of Cisco Systems, Inc. Copyright © Cisco Systems, Inc. All rights reserved.)

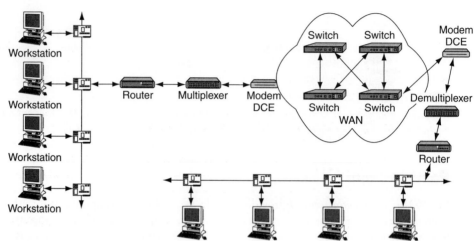

Packet Switching

●
Another WAN switching method is **packet switching.**

Packet switching is a WAN switching method in which network devices share a single point-to-point link to transport packets from a source to a destination across a carrier network. Statistical multiplexing is used to enable devices to share these circuits. Asynchronous Transfer Mode (ATM), frame relay, SMDS, and X.25 are some examples of packet switched WAN technologies. Packet switching is illustrated in Figure 1.30.

Virtual Circuits

●
Virtual circuits such as **switched virtual circuits** and **permanent virtual circuits** enable reliable communication between two network devices.

A *virtual circuit* is a logical circuit created to enable reliable communication between two network devices. The two types of virtual circuits are *switched virtual circuits* (*SVCs*) and *permanent virtual circuits* (*PVCs*). SVCs are virtual circuits that are dynamically established on demand and terminated when transmission is complete. Communication over an SVC consists of three phases: circuit establishment, data transfer, and circuit termination. The establishment phase involves creating the virtual circuit between the source and destination devices. Data transfer is just that, and circuit termination involves tearing down the virtual circuit between the source and destination devices. SVCs are used in situations in which data transmission between devices is sporadic, largely because SVCs increase the amount of bandwidth used. This is due to the circuit establishment and termination phases, but the cost is decreased because of the constant virtual circuit availability.

A PVC is a permanently established virtual circuit that has one mode called a data transfer mode. PVCs are used in situations in which data transfer between devices is constant. PVCs decrease the amount of bandwidth used but increase the cost due to the need for constant virtual circuit availability.

Dial Up Services

Dial up services offer cost effective methods for connectivity across WANs. Two popular dial up implementations are *dial on demand routing (DDR)* and *dial backup.* DDR is a technique whereby a router can dynamically initiate and close a circuit switched session as transmitting end stations demand. A router is configured to consider certain traffic as important (such as traffic from a particular protocol) and other traffic as unimportant. When the router receives important traffic destined for a remote network, a circuit is established and the traffic is transmitted normally. If the router receives unimportant traffic and a circuit is already established, that traffic also is transmitted normally. The router maintains an idle timer that is reset only when important traffic is received. If the router receives no important traffic before the idle timer expires, the circuit is terminated. Likewise, if unimportant traffic is received and no circuit exists, the router drops the traffic. Upon receiving important traffic, the router initiates a new circuit. DDR can be used to replace point-to-point links and switched multi-access WAN services.

Dial backup is a service that activates a backup line under certain conditions. The secondary line can act either as a backup link that is utilized when the primary link fails or as a source of additional bandwidth when the load on the primary link reaches a certain threshold. Dial backup provides protection against WAN performance degradation and downtime.

WAN Devices

WANs use numerous WAN-specific devices. They include *switches, access servers, modems,* CSU/DSUs *ISDN terminal adapters, routers, ATM switches,* and *multiplexers.*

A *switch* is a multiport internetworking device used in carrier networks. Switches typically switch traffic, such as Frame Relay, X.25, and SMDS. Switches operate at the data link layer of the OSI model but can also operate at the network layer. Figure 1.31 illustrates two routers at remote ends of a WAN that are connected by WAN switches.

An *access server* acts as a concentration point for dial in and dial out connections. Figure 1.32 illustrates an access server concentrating dial out connections into a WAN.

A *modem* is a device that interprets digital and analog signals, enabling data to be transmitted over voice grade telephone lines. At the source, digital signals are made suitable for transmission over analog communication facilities. At the destination, these analog signals are returned to their digital form. Figure 1.33 illustrates a simple modem-to-modem connection through a WAN.

A (*CSU/DSU*) is a digital interface device (or sometimes two separate digital devices) that adapts the physical interface on data terminal equipment (DTE), such as a terminal, to the interface of data communications equipment (DCE), such as a switch in a switched

Dial on demand routing and **dial backup** are two types of **dial up services.**

A **WAN switch** is a multiport internetworking device used in carrier networks to switch traffic.

Access servers act as concentration points for dial in and dial out connections.

Modems allow data to be transmitted over voice grade telephone lines.

A **WAN CSU/DSU** adapts the physical interface on DTE devices to the interface of a DCE device in a switched carrier network.

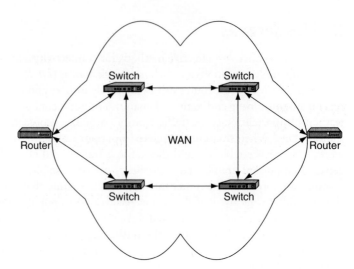

FIGURE 1.31

Two routers at remote ends of a WAN can be connected by WAN switches.
(Reproduced with permission of Cisco Systems, Inc. Copyright © 2001 Cisco Systems, Inc. All rights reserved.)

FIGURE 1.32

An access server concentrates dial in and dial out connections into a WAN.
(Reproduced with permission of Cisco Systems, Inc. Copyright © 2001 Cisco Systems, Inc. All rights reserved.)

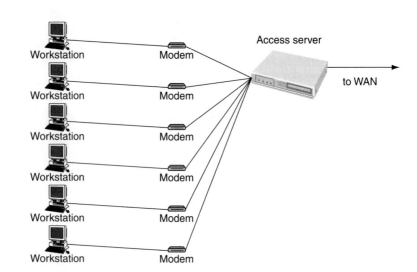

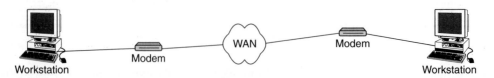

FIGURE 1.33

WAN modems interpret digital signals and convert them to analog signals at the source, send them to the WAN, and convert them back to digital signals at the destination. (Reproduced with permission of Cisco Systems, Inc. Copyright © 2001 Cisco Systems, Inc. All rights reserved.)

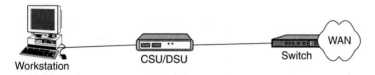

FIGURE 1.34

The CSU/DSU stands between the work station and the switch. (Reproduced with permission of Cisco Systems, Inc. Copyright © 2001 Cisco Systems, Inc. All rights reserved.)

carrier network. The CSU/DSU also provides signal timing for communication between these devices. Figure 1.34 illustrates the placement of the CSU/DSU in a WAN.

An *ISDN terminal adapter* is a device used to connect ISDN Basic Rate Interface (BRI) connections to other interfaces such as EIA/TIA-232. A terminal adapter is essentially an ISDN modem.

A **ISDN terminal adapter** is an ISDN modem.

Summary

Internetworking is defined as the communication between two or more networks and encompasses every aspect of connecting computers together. They have grown to support vastly disparate end system communication requirements. In order to permit scalability and manageability without constant manual intervention, many protocols and features are necessary. Large internetworks consist of the following three distinct components:

- Campus networks, which consist of locally connected users in a building or group of buildings.
- WANs, which connect different campuses.
- Remote connections, which link branch offices and single users (mobile users and/or telecommuters) to a typical enterprise internetwork.

An internetwork is a collection of individual networks, connected by intermediate networking devices that function as a single large network. Internetworking refers to the industry, products, and procedures that meet the challenge of creating and administering internetworks.

The first networks were time sharing networks that used mainframes and attached terminals. Such networking environments were implemented by IBM's SNA and Digital's network architecture. LANs evolved around the PC revolution. They enable multiple users in a relatively small geographical area (roughly 10 miles or fewer) to exchange files and messages and to share resources. WANs connect LANs across normal telephone lines and other media, thereby connecting geographically dispersed users. High speed LANs and switched internetworks are now becoming more widely used, mainly because they operate at very high speeds. High speed LANs support high bandwidth applications such as voice and videoconferencing. Internetworking evolved as a solution to three key problems: isolated LANs, duplication of resources, and lack of network management. Isolated LANs made electronic communication between different offices or departments impossible. Duplication of resources meant that the same hardware and software had to be supplied to each office or department and could not be shared between campuses. In addition, each department or office had to maintain a separate support staff to manage those resources. The lack of network management meant that no centralized method of managing and troubleshooting networks existed.

Implementing a functional internetwork is no simple task. Many challenges must be faced. Some of the problem areas include the areas of connectivity, reliability, network management, and flexibility. Each individual area is a key in establishing an efficient and effective internetwork. The main challenge in connecting various systems together is supporting the communication between disparate technologies. Different locations may use different types of media, or they might operate at varying speeds. Another challenge, reliable service, must be maintained in any internetwork. Individual users and entire organizations depend on consistent, reliable access to network resources. Network management must provide centralized support and troubleshooting capabilities in an internetwork. Configuration, security, performance, fault, and accounting issues must be adequately addressed for the internetwork to function smoothly. The final challenge, flexibility, is necessary for network expansion and the addition of new applications and services, which are necessary to keep businesses competitive in the 21st century.

In Chapter 2, bridging and switching two basic networking devices used to connect networks will be discussed.

Review Questions

1. What is an internetwork?
2. Describe the first networks.
3. What are some challenges you must face when connecting various systems?
4. What is the main objective of network management?
5. What are the seven layers of the OSI model?
6. Which layer of the OSI model is closest to the user?
7. Which layer of the OSI model transmits bits onto the physical network media?
8. What is a protocol?
9. What are the different groups of protocols?
10. How do two computers communicate with each other?
11. What are SAPs?
12. Describe the information exchange process between two computer systems.
13. What are the two sublayers of the data link layer as defined by IEEE?
14. What are some transport layer protocols?
15. What are the three basic components that make up a network layer packet?
16. What are the two components that make up a cell?

17. What is an autonomous system?
18. Define the main difference between a connection oriented network and a connectionless network.
19. What is a MAC address?
20. What are the main differences between hierarchical and flat address space?
21. What is buffering?
22. What is one common form of error checking?

Summary Questions

1. What is multiplexing?
2. What is a LAN?
3. Define *unicast, multicast*, and *broadcast*.
4. What is CSMA/CD?
5. What are some typical LAN devices?
6. What is a WAN?
7. List some WAN devices.
8. What is a point to point link?
9. What are the two types of circuit switching?
10. What is packet switching?

Further Reading

McQuerry, Steve. *Interconnecting Cisco Network Devices*. Indianapolis: Cisco Press, 2000.

Minoli, Daniel. *Telecommunications Technology Handbook*. Norwood: Artech House Inc., 1991.

Teare, Diane. *Designing Cisco Networks*. Indianapolis: Cisco Press, 1999.

Bridging and Switching Fundamentals

Objectives

- Define in which layer of the OSI model bridges and switches operate.
- Discuss the various types of bridges.
- Explain how and why it is necessary to segment LANs.
- Explain the types of LAN segmentation and discuss the differences between segmenting with repeaters, bridges, and switches.
- Explain the 80/20 rule.
- Discuss the various switching modes.
- Define and examine the various bridging technologies.
- Define the Spanning Tree Protocol (STP) and discuss how it helps to alleviate the bridging loops.

Key Terms

Bridges enable packet forwarding between networks.

Switches are data link devices that allow a connection to be established.

Transparent bridging is found primarily in Ethernet environments.

Source route bridging occurs primarily in Token Ring environments.

Translational bridging provides translation between different networks.

Source route transparent bridging enables communication in mixed Ethernet/ Token Ring environments.

Introduction

Bridges and *switches* are data communication devices that operate principally at layer 2 of the OSI reference model. As a result, bridges and switches are widely referred to as data link devices.

Bridges became commercially available in the early 1980s. At the time of their introduction, bridges connected and enabled packet forwarding between homogeneous networks. More recently, bridging between different networks has also been defined and standardized.

Several kinds of bridging schemes have proven important as internetworking devices. *Transparent bridging* is found primarily in Ethernet environments, while *source route bridging (SRB)* occurs primarily in Token Ring environments. *Translational bridging* provides translation between the formats and transit principles of different media types, usually Ethernet and Token Ring. Finally, *source route transparent bridging (SRT)* combines the algorithms of transparent bridging and source route bridging to enable communication in mixed Ethernet/Token Ring environments.

Today, switching technology has emerged as the evolutionary heir to bridging based internetworking solutions. Switching implementations now dominate applications in which bridging technologies were implemented previously. Superior throughput performance, higher port density, lower per port cost, and greater flexibility have contributed to the emergence of switches as replacements for bridges and as complements to routing technology.

Link Layer Devices: An Overview

Bridging and switching occur at the link layer, which controls data flow, handles transmission errors, provides physical as opposed to logical addressing, and manages access to the physical medium. Bridges and switches provide these functions by using various link layer protocols that dictate specific flow control, error handling, addressing, and media access algorithms. Examples of popular link layer protocols include Ethernet, Token Ring, and FDDI.

Bridges and switches are not complicated devices. Bridges and switches analyze incoming frames, make forwarding decisions based on information contained in the frames, and then forward the frames toward the respective destination. In some cases, such as source route bridging, the entire path to the destination is contained in each frame. In other cases, such as transparent bridging, frames are forwarded one hop at a time toward the destination.

Upper layer protocol transparency is a primary advantage of bridging and switching. Because both devices operate at the data link layer they are not required to examine upper layer information. This means that bridges and switches can rapidly forward traffic representing any network layer protocol. A bridge or a switch can move different protocols and other traffic between two or more networks.

Bridges are capable of filtering frames based on layer 2 fields. A bridge can be programmed to reject and not forward all frames sourced from a particular network. Link layer information often includes a reference to an upper layer protocol, which allows a bridge to filter on this parameter. Filters can be helpful in dealing with unnecessary broadcast and multicast packets.

By segmenting large networks into self-contained units, bridges and switches provide several advantages. Because only a certain portion of the traffic is forwarded, a bridge or switch diminishes the traffic experienced by devices on all connected segments. The bridge or switch will act as a firewall for potentially damaging network errors, and together they can accommodate communication between a larger number of devices than would be supported on any single LAN connected only to the bridge. Bridges and switches extend the effective length of a LAN, permitting the attachment of distant workstations that were not permitted previously.

Although bridges and switches share most relevant attributes, several distinctions differentiate these networking technologies. Switches are significantly faster because they switch in hardware, while bridges switch in software. Switches can interconnect LANs of bandwidth whereas a bridge cannot. A 10-Mbps Ethernet LAN and a 100-Mbps Ethernet LAN can be connected using a switch. Switches can also support higher port density than bridges. Some switches support cut-through switching, which reduces latency and delays in the network, while bridges support only store and forward traffic switching. Finally, switches reduce collisions on network segments because switches provide dedicated bandwidth to each network segment.

Types of Bridges

Bridges can be grouped into categories based on various product characteristics. They are classified as either *local* or *remote*. Local bridges provide a direct connection between multiple LAN segments in the same area. Remote bridges connect multiple LAN segments in different areas, usually over telecommunications lines. Figure 2.1 illustrates the local and remote configurations of bridges.

Remote bridges cannot improve WAN speeds, but they can compensate for speed discrepancies through a sufficient buffering capability. If a LAN device, capable of a 10-Mbps transmission rate, wants to communicate with a device on a remote LAN, the local bridge must regulate the 10-Mbps data stream so that it does not overwhelm the 1.544 T1 link. This is done by storing the incoming data in onboard buffers and sending the data over the T1 link at a rate that the T1 link can accommodate. This buffering can be achieved only for short bursts of data that do not overwhelm the bridge's buffering capability.

The IEEE has divided the OSI data link layer into two separate sublayers: the *Media Access Control (MAC)* sublayer and the *Logical Link Control (LLC)* sublayer. The MAC sublayer permits and orchestrates media access, such as contention and token passing, while the LLC sublayer deals with framing, flow control, error control, and MAC sublayer addressing.

Some bridges are MAC layer bridges, which bridge between homogeneous networks such as Ethernet and Ethernet. Other bridges can translate between different data link layer protocols such as Ethernet and Token Ring. The basic mechanics of such a translation are illustrated in Figure 2.2. Figure 2.2 illustrates how an Ethernet host (host *A*) formulates a packet that contains application information and encapsulates the packet in an Ethernet compatible frame for transit over the Ethernet medium to

---

Local bridges connect networks in the same area.

Remote bridges connect networks in different areas.

The **Media Access Control** sublayer handles media access.

The **Logical Link Control** sublayer handles framing, flow control, error control, and so on.

FIGURE 2.1

Local and remote bridges connect LAN segments in specific areas. (Reproduced with permission of Cisco Systems, Inc. Copyright © 2001 Cisco Systems, Inc. All rights reserved.)

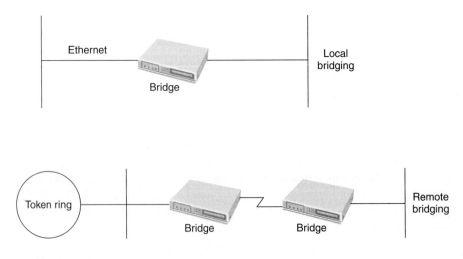

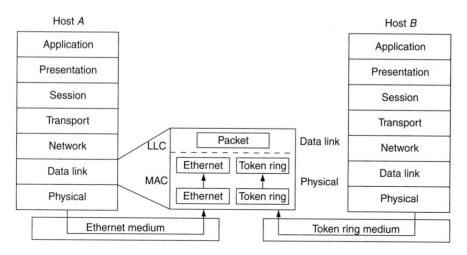

the bridge. At the bridge, the frame is stripped of its Ethernet header at the MAC sublayer of the data link layer and is subsequently passed up to the LLC sublayer for further processing. After this processing, the packet is passed back down to a Token Ring implementation, which encapsulates the packet in a Token Ring header for transmission on the Token Ring network to the Token Ring host (host *B*).

Segmenting LANs

As networks grow larger, network administrators have become more frustrated. They need to increase the number of users and also the amount of bandwidth available to the users. To further confuse the issue, network administrators are confronted with budgetary constraints, which conflict with these two objectives. Network engineers building LAN infrastructures can select from many different internetworking devices to extend networks. These devices include repeaters, bridges, routers, and switches. Each component serves a specific role and is versatile when properly deployed.

Why Segment LANs?

There is a prevalent need to extend the distance of a network, the number of users on the system, and the amount of available bandwidth that is available to users. From a corporate viewpoint, this may result in growth, but to the network administrator, it means a possible nightmare. Segmenting LANs is an approach to provide users additional bandwidth without replacing all user equipment. Segmenting LANs breaks a network into smaller portions and connects them with some type of internetworking equipment. Figure 2.3 illustrates before and after examples for segmenting networks.

Before segmentation, all 500 users share the network's 10-Mbps bandwidth because the segments interconnect with repeaters. The revised network replaces the repeaters

FIGURE 2.3

A network before and after segmentation. (Reproduced with permission of Cisco Systems, Inc. Copyright © 2001 Cisco Systems, Inc. All rights reserved.)

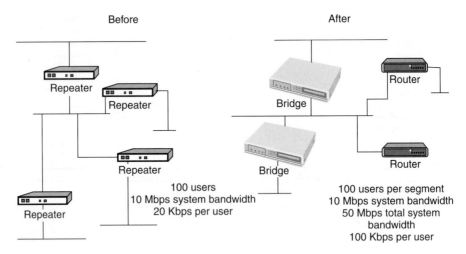

Before

After

Repeater

Repeater

Repeater

Repeater

Bridge

Bridge

Router

Router

100 users
10 Mbps system bandwidth
20 Kbps per user

100 users per segment
10 Mbps system bandwidth
50 Mbps total system
bandwidth
100 Kbps per user

Table 2.1 A Comparison of Collision and Broadcast Domains

Device	*Collision Domains*	*Broadcast Domains*
Repeater	One	One
Bridge	Many	One
Router	Many	Many
Switch	Many	Configurable

with bridges and routers isolating segments and providing more bandwidth for users. Bridges and routers generate bandwidth by creating new collision and broadcast domains, which are summarized in Table 2.1.

Each segment can be divided further with additional bridges and switches providing even more bandwidth for users. In the extreme case, one user can be on each segment and receive full media bandwidth. Switches follow the concept of one user to each segment receiving full media bandwidth. Repeaters do not segment a network and do not create more bandwidth. Repeaters allow the network to be extended to some degree. Bridges and switches are more suitable for LAN segmentation.

Segmenting LANs Using Repeaters

Traditional Ethernet systems such as 10Base-5, 10Base-2, and 10Base-T have distance limitations for segments. Whenever the network has to be extended, a repeater can be used. Repeaters operate at the physical layer of the OSI model and appear as an extension to the cable segment. Workstations have no knowledge of the presence

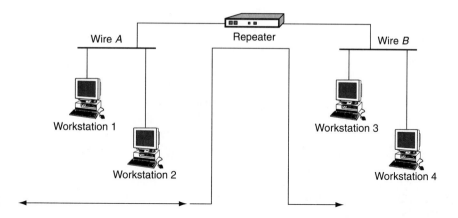

of a repeater, which is completely transparent to the attached devices. Figure 2.4 illustrates a repeater attaching wire segments together.

Repeaters regenerate the signal from one wire to the other. When station 1 transmits to station 2, the frame also appears on wire *B* even though the source and destination device coexist on wire *A*. Repeaters are unintelligent devices and have no insight to the data content. Repeaters blindly perform their responsibility of forwarding signals from one wire to all other wires. If the frame contains errors, the repeater forwards the errors. If the frame violates the minimum or maximum frame sizes specified by the Ethernet protocol, the repeater forwards the abnormal sized frame. If a collision occurs on wire *A,* wire *B* also sees the collision. Repeaters truly act like an extension cable. Although Figure 2.4 shows the interconnection of two segments, repeaters can have many ports to attach several segments as shown in Figure 2.5.

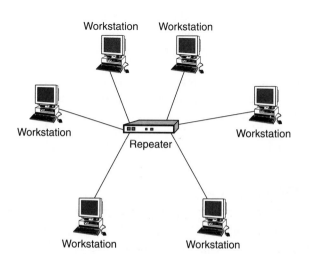

A 10Base-T network is comprised of hubs and twisted-pair cables to interconnect workstations. Hubs are multiport repeaters that forward signals from one interface to all other interfaces. As in Figure 2.4, all stations attached to the hub (in Figure 2.5) see all traffic.

Repeaters perform several duties associated with signal propagation:

1. Repeaters regenerate and retime the signal and create a new preamble. Preamble bits precede the frame destination MAC address and help receivers to synchronize. The last byte of the preamble, which ends in a binary pattern of 10101011, is called the start frame delimiter (SFD). The last two bits indicate to the receiver that data will follow. Repeaters strip all eight preamble bytes from the incoming frame and generate and prepend a new preamble on the frame before transmission through the outbound interface.

2. Repeaters also ensure that collisions are signaled on all ports. If stations 1 and 2 in Figure 2.4 participate in a collision, the collision is enforced through the repeater so that the stations on wire *B* know of the collision. Stations on wire *B* must wait for the collision to clear before transmitting. If stations 3 and 4 do not know of the collision, they might attempt a transmission during station 1 and 2's collision event. Stations 3 and 4 become additional participants in the collision.

Limitations exist in a repeater based network. Limitations arise from different causes and must be considered when extending a network with repeaters. The limitations for repeaters include the following:

- Shared bandwidth between devices
- Specification constraints on the number of stations per segment
- End-to-end distance capability

Shared Bandwidth

A repeater extends the distance of the cable and extends the collision domain. Collisions on one segment affect stations on another repeater connected segment. Collisions extend through a repeater and consume bandwidth on all interconnected segments. If the network uses shared network technology, all stations in the repeater based network share bandwidth. This holds true whether the packet is unicast, multicast, or broadcast. All stations see all frames. Adding more stations to the repeater network could potentially divide the bandwidth more. Traditional Ethernet systems have a shared 10-Mbps bandwidth. The stations take turns using the bandwidth and, as the number of workstations increases, the amount of available bandwidth decreases.

The Number of Stations per Segment

Ethernet imposes a limit on how many workstations can be attached to a cable. These constraints arise because of electrical considerations. As the number of transceivers attached to a cable increases, the cable impedance changes and creates electrical reflections in the system. If the impedance changes too much, the collision detection process fails. Limits for traditional systems include no more than 100 attached devices per segment for a 10Base-5. A 10Base-2 system cannot exceed 30 stations. Repeaters cannot increase the number of stations supported per segment. The limitation is inherent in the bus architectures of 10Base-5 and 10Base-2 networks.

End-to-End Distance Capability

Another limitation on extending networks with repeaters concerns distance. An Ethernet link can extend only so far before the media slotTime specified by Ethernet standards is violated. The slotTime is a function of the network data rate. A 10-Mbps network such as 10Base-T has a slotTime of 51.2 microseconds. A 100-Mbps network slotTime is one-tenth that of 10Base-T. The distance to extend a network using repeaters takes into account the slotTime size, latency through various media such as copper and fiber, and the number of repeaters in the network. In a 10-Mbps Ethernet, the number of repeaters in a network must follow the 5/3/1 rule illustrated in Figure 2.6. The 5/3/1 rule states that up to five segments can be interconnected with repeaters. Only three of the segments can have devices attached. The other two segments interconnect segments and only allow repeaters to attach at the ends of each segment. When using the 5/3/1 rule, one collision domain is created. A collision in the network propagates through all repeaters to all other segments.

When repeaters are correctly used, they extend the collision domain by interconnecting segments at the physical layer of the OSI model. Any transmission in the collision domain propagates to all other stations in the network. If the network needs to extend beyond the 5/3/1 rule, other internetworking devices must be used such as a bridge or a switch.

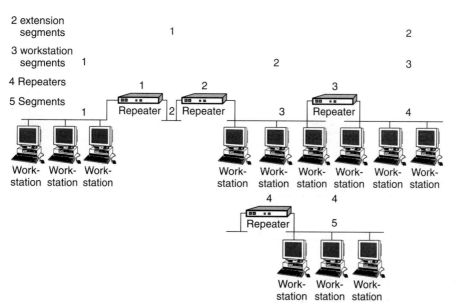

FIGURE 2.6

Interconnecting with the 5/3/1 rule. (Reproduced with permission of Cisco Systems, Inc. Copyright © 2001 Cisco Systems, Inc. All rights reserved.)

Repeaters extend the bounds of broadcast domains and collision domains only to the extent allowed by media repeater rules. The maximum geographical extent constrained by the media slotTime value defines the collision domain bounds. If you extend the collision domain beyond the bounds defined by the media, the network will not function correctly. Ethernet networks will experience late collisions if the network is extended too far. Late collision events occur whenever a station experiences a collision outside of the 51.2 microsecond slotTime.

Segmenting LANs Using Bridges

Ethernet rules limit the overall distance a network segment can be extended and the number of stations that can be attached to a cable segment. Bridges provide a solution if the network needs to be extended further or more devices need to be added. When connecting networks with a bridge, as illustrated in Figure 2.7, significant differences exist compared with repeater connected networks.

Whenever stations on the same segment transmit to each other in a repeater network, the frame appears on all other segments in the repeated network. In a bridged network this does not normally happen. Bridges have a filtering process that determines whether or not to forward a frame to other interfaces.

The filtering process differs for access methods such as Ethernet and Token Ring Ethernet employs a process called transparent bridging that examines the MAC destination address (DA) to determine if a frame should be forwarded, filtered, or flooded. Bridges operate at the data link layer of the OSI model. As a result of operating at layer 2, bridges have the capability to examine the MAC headers of frames. Bridges can make forwarding decisions based on information in the header such as the MAC

FIGURE 2.7

Interconnecting segments with a bridge. (Reproduced with permission of Cisco Systems, Inc. Copyright © 2001 Cisco Systems, Inc. All rights reserved.)

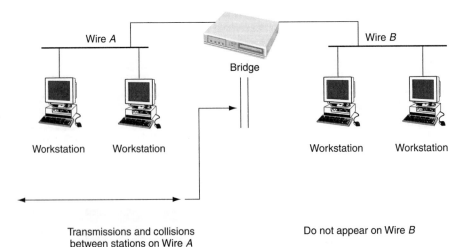

Transmissions and collisions between stations on Wire *A*

Do not appear on Wire *B*

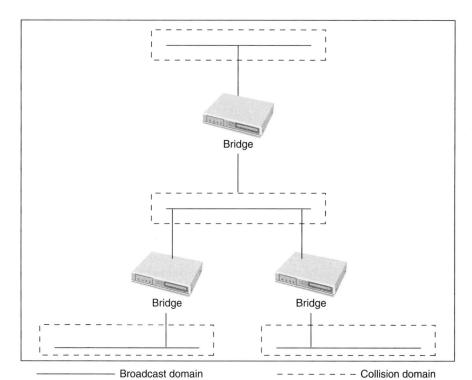

———————— Broadcast domain   – – – – – – – Collision domain

address. Token Ring uses source route bridging, which determines frame flow differently from transparent bridging. Bridges interconnect collision domains allowing independent collision domains to appear as if they were connected without propagating collisions between them. Figure 2.8 illustrates the same network as Figure 2.4 but with bridges interconnecting the segments.

In the repeater based network, all the segments belong to the same collision domain. The network bandwidth was divided between the four segments. In a bridged network, each segment belongs to a different collision domain. If this were a 10-Mbps traditional network, each segment would have its own 10-Mbps bandwidth for a collective bandwidth of 40 Mbps.

The same number of users in a bridged network as opposed to a repeater based network now have more available bandwidth. Another advantage of bridges is a direct result of operating at layer 2. In a repeater based network, an end-to-end distance limitation prevents the network from extending indefinitely. Bridges allow each segment to extend a full distance. Each segment has its own slotTime value. Bridges do not forward collisions between segments, but rather isolate collision domains and reestablish slotTimes. Bridges can extend networks indefinitely, but practical considerations prevent this from occurring.

Bridges filter traffic when the source and destination reside on the same segment. Broadcast and multicast frames are the exception. Whenever a bridge receives a broadcast or multicast, it floods the message out on all ports.

The 80/20 Rule

The **80/20 rule** states that bridges are efficient when 80 percent of segment traffic is local and only 20 percent needs to cross a bridge to another segment.

A rule of thumb when designing networks with bridges is the *80/20 rule.* The 80/20 rule states that bridges are most efficient when 80 percent of the segment traffic is local and only 20 percent needs to cross a bridge to another segment. The 80/20 rule originated from traditional network design where server resources resided on the same segments with the client devices they served, as illustrated in Figure 2.9.

The clients only infrequently needed to access devices on the other side of a bridge. Bridged networks are considered to be well designed when the 80/20 rule is observed. As long as this traffic balance is maintained, each segment in the network appears to have full media bandwidth. If the flow balance shifts such that more traffic will be forwarded through the bridge rather than filtered, the network behaves as if all segments operate on the same shared network. The bridge in this case provides nothing more than the capability to daisychain collision domains to extend distance without any bandwidth improvements.

In the worst case for traffic flow in a bridged network, 0/100, none of the traffic remains local and all sources transmit to destinations on other segments. Using a two-port bridge, the entire system has shared rather than isolated bandwidth. The bridge only extends the geographical bounds of the network and offers no bandwidth gains. Many Intranets see similar traffic patterns with typical ratios of 20/80 rather than 80/20. Violating the 80/20 rule in this instance results from too many users attempting to communicate with and through the Internet. The traffic flows from a local segment to the WAN connection and crosses broadcast domain boundaries.

FIGURE 2.9

The 80/20 rule demonstrated in a traditional network. (Reproduced with permission of Cisco Systems, Inc. Copyright © 2001 Cisco Systems, Inc. All rights reserved.)

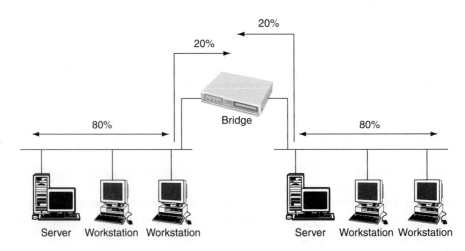

One other advantage of bridges is they prevent error frames from transiting to another segment. If the bridge sees a frame that has errors or that violates the media access method size rules, the bridge drops the frame. In essence, this protects the destination network from bad frames that do nothing more than consume bandwidth. Collisions on shared traditional networks often create frame fragments that are sometimes called runt frames. Runt frames violate the Ethernet minimum frame size rule of 64 bytes. A repeater forwards runt frames, while a bridge blocks them.

Segmenting LANs Using Switches

A LAN switch is a multiport bridge that allows workstations to attach directly to the switch to experience full duplex bandwidth, which enables many workstations to transmit concurrently. Figure 2.10 shows four workstations communicating at the same time, which is impossible in a shared network environment.

Because a switch is nothing more than a complex bridge with several interfaces, all of the ports on a switch belong to one broadcast domain. If station 1 sends a broadcast frame, all devices attached to the switch receive it. The switch floods broadcast transmissions to all other ports. Unfortunately, this makes the switch no more efficient than a shared media interconnected with repeaters or bridges when dealing with broadcast or multicast frames.

It is possible to design the switch so that ports can belong to different broadcast domains as designed by a network administrator, thus providing broadcast isolation. In Figure 2.11, some ports belong to broadcast domain 1 (BD1), some ports to broadcast domain 2 (BD2), and others to broadcast domain 3 (BD3). If a station attached to an interface in BD1 transmits a broadcast frame, the switch forwards the broadcast

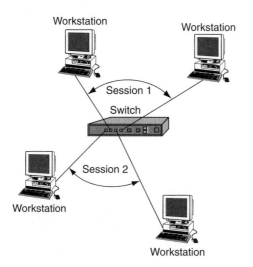

FIGURE 2.10

Several concurrent sessions through a LAN switch. (Reproduced with permission of Cisco Systems, Inc. Copyright © 2001 Cisco Systems, Inc. All rights reserved.)

FIGURE 2.11

A multibroadcast domain capable switch.
(Reproduced with permission of Cisco Systems, Inc. Copyright © 2001 Cisco Systems, Inc. All rights reserved.)

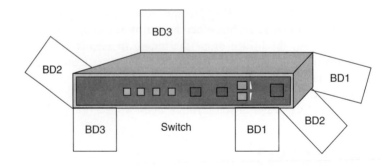

only to the interfaces belonging to the same domain. The other broadcast domains do not experience any bandwidth consumption resulting from BD1's broadcast. In fact, it is impossible for any frame to cross from one broadcast domain to another without the introduction of another external device, such as a router to interconnect the domains.

●

Virtual LANs are devices capable of defining several broadcast domains.

Switches capable of defining several broadcast domains actually define *Virtual LANs* (VLANs). Each broadcast domain equates to one VLAN. A VLAN capable switch is a device that creates multiple isolated bridges, as illustrated in Figure 2.12.

If five VLANs are created, five virtual bridge functions are created within the switch.

Switching Modes

The switch can be configured to behave as multiple bridges by defining internal virtual bridges. Each virtual bridge defines a new broadcast domain because no internal connection exists between them. Broadcasts for one virtual bridge are not seen by any other vir-

FIGURE 2.12

A logical internal representation of a VLAN capable switch.
(Reproduced with permission of Cisco Systems, Inc. Copyright © 2001 Cisco Systems, Inc. All rights reserved.)

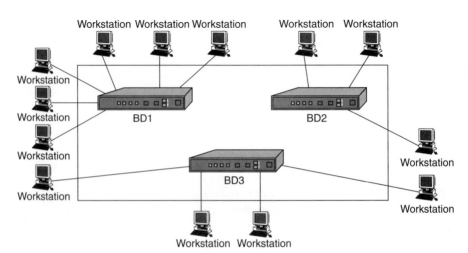

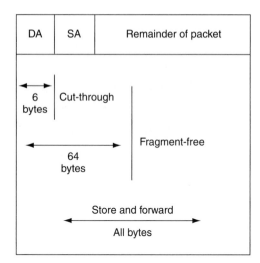

FIGURE 2.13
Switching mode trigger points. (Reproduced with permission of Cisco Systems, Inc. Copyright © 2001 Cisco Systems, Inc. All rights reserved.)

tual bridge. Only routers, either external or internal, should connect broadcast domains together. Using a bridge to interconnect broadcast domains merges the domains and creates one giant domain. This defeats the reason for having individual broadcast domains.

Switches make forwarding decisions the same as a transparent bridge. There are three modes that are available for switches: store and forward, cut-through, and fragment-free. Figure 2.13 illustrates the trigger points for the three methods of switching.

Each has its advantages and disadvantages. As a result of the different trigger points, the effective differences between the modes are in error handling and latency. Table 2.2 compares the various approaches and summarizes how each mode handles errored frames and the associated latency characteristics.

One of the objectives of switching is to provide more bandwidth to the user. Each port on a switch defines a new collision domain that offers full duplex bandwidth. If only one station attaches to an interface, that station has full dedicated bandwidth and does not need to share it with any other device. All the switching modes defined support full dedicated bandwidth.

Table 2.2 Switching Mode Comparison

Switching Mode	Errored Frame Handling	Latency
Store and forward	Drops	Variable
Cut-through	Forwards	Low to fixed
Fragment-free	Drops if error detected in first 64 octets	Moderate to fixed

Store and Forward Switching

The *store and forward switching* mode receives the entire frame before beginning the switching process. After receiving the complete frame, the switch examines it for the source and destination addresses and any errors that it may contain, and then applies any special filters created to modify the default forwarding behavior. If the switch observes any errors in the frame, the frame is discarded. Because the destination device discards errored frames, store and forward switching prevents errored frames from consuming bandwidth on the destination segment. If the network experiences a high rate of frame alignment or *frame check sequence* (FCS) errors, the store and forward switching mode may be the best of the available modes. The best solution is to fix the cause of the errors. Using store and forward switching only acts as a bandage. Adopting the store and forward method should not be the fix.

If the source and destination segments use different transmission media, then you must use the store and forward switching mode. Different media often have issues when transferring data. Store and forward switching is necessary to resolve the problems in a bridged environment.

The switch must receive the entire frame before the switch can start to forward. Transfer latency varies based on the frame size. In a 10Base-T network for example, the minimum frame, 64 octets, takes 51.2 microseconds to receive. At the other extreme, a 1,518-octet frame requires at least 1.2 milliseconds. Latency for 100Base-FX (fast Ethernet) networks is one-tenth the 10Base-T numbers.

Cut-Through Switching

Cut-through switching enables a switch to start the forwarding process as soon as it receives the destination address. Cut-through switching reduces latency to the time necessary to receive the six-octet destination address: 4.8 microseconds. Cut-through switching cannot check for errored frames before it forwards the frame. Errored frames pass through the switch, resulting in wasted bandwidth because the receiving device discards errored frames.

As network and internal processor speeds increase, the latency issues become less relevant. In high speed switching environments, the time to receive and process a frame decreases significantly, thereby minimizing the advantages of cut-through switching. Thus, store and forward switching becomes an attractive choice for most networks.

Some switches support both cut-through and store and forward switching modes. Such switches usually contain a third mode called adaptive cut-through switching. These multimodal switches use cut-through as the default switching mode and selectively activate store and forward. The switch monitors the frame as the frame passes through looking for errors. Although the switch cannot stop an errored frame, the switch counts how many errored frames that it sees. If the switch observes that too many frames contain errors, the switch automatically activates the store and forward switching mode. As previously noted, this type of switching is referred to as adaptive

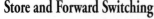

Store and forward switching processes the entire frame before it is forwarded to the appropriate port.

The **frame check sequence** refers to the extra characters added to a frame for error control.

Cut-through switching enables a switch to start the forwarding process as soon as it receives the destination address.

cut-through. Adaptive cut-through has the advantage of providing low latency while the network operates in a healthy state, at the same time, adaptive cut-through switching provides automatic protection for the outbound segment if the inbound segment experiences trouble.

Fragment-Free Switching

Another switching alternative offers some advantages of both cut-through and store and forward switching. *Fragment-free switching* is similar to cut-through switching in that fragment-free switching does not wait for an entire frame before forwarding. Fragment-free switching forwards a frame after it receives the first 64 octets of the frame. Fragment-free switching protects the destination segment from fragments, which are half duplex Ethernet collisions. In a correctly designed Ethernet system, devices detect a collision before the source finishes its transmission of the 64-octet frame. When a collision occurs, a fragment (a frame that is less than 64 octets long) is created. This fragment is a useless Ethernet frame and is discarded by the switch in the store and forward mode. In contrast, in cut-through switching the fragment frame is forwarded if the destination address exists. Because collisions and most frame errors occur during the first 64 octets, the fragment-free switching mode can detect most bad frames and discard rather than forward them. Fragment-free has a higher latency than cut-through because fragment-free switching must wait for an additional 58 octets before forwarding the frame. The advantages of fragment-free switching are minimal given the higher network speeds and faster switch processors.

Fragment-free switching forwards a frame after it receives the first 64 octets of the frame.

Bridging Technologies

Transparent Bridging

Networks are segmented to provide more bandwidth per user. Bridges provide more bandwidth by reducing the number of devices contending for the segmented bandwidth. Bridges also provide additional bandwidth by controlling data flow in a network. Bridges forward traffic only to the interfaces that need to receive the traffic. In the case of unicast traffic, bridges forward the traffic to a single port rather than to all ports.

Transparent bridges were first developed at Digital Equipment Corporation (DEC) in the early 1980s. Digital submitted its work to the IEEE, which incorporated the work into the IEEE 802.1 standard. Transparent bridges are very popular in Ethernet/IEEE 802.3.

Transparent bridges are so named because their presence and operation are transparent to network hosts. When transparent bridges are powered on, they learn the network's topology by analyzing the source address of incoming frames from all attached networks. If the bridge sees a frame arrive on wire *A* from host *A*, the bridge concludes that host *A* can be reached through the network connected to wire *A*. Through this process, transparent bridges build a bridging table similar to the one in Table 2.3.

Table 2.3 Typical Transparent Bridging Table

Host Address	*Network Number*
15	1
17	1
12	2
13	2
18	1
9	1
—	—

The **destination address** is the address of a network device that receives data.

The bridge uses its table as the basis for traffic forwarding. When a frame is received on one of the bridge's interfaces, the bridge looks up the frame's *destination address (DA)* in its internal table. If the table contains an association between the (DA) and any of the bridge's ports aside from the one on which the frame was received, the frame is forwarded out the indicated port. If no association is found, the frame is flooded to all ports except the inbound port. Broadcasts and multicasts also are flooded this way. Transparent bridges successfully isolate intrasegment traffic, thereby reducing the traffic seen on each individual segment. The bridge performing this function generally improves network response times as seen by the user. The extent to which traffic is reduced and improved response times result depends on the volume of intersegment traffic relative to the total traffic including the volume of broadcast and multicast traffic. Transparent bridging describes the following five bridging processes for determining what to do with a frame:

1. Learning
2. Flooding
3. Filtering
4. Forwarding
5. Aging

Figure 2.14 illustrates the five processes involved in transparent bridging.

The **source address** is the address of a network device that sends data.

When a frame enters the transparent bridge, the bridge adds the Ethernet MAC *source address (SA)* and source port to its bridging table. If the SA already exists in the table, the bridge updates the aging timer. The bridge examines the MAC DA. If the DA is a broadcast, multicast, or unknown unicast, the bridge floods the frame out all bridge ports in the spanning tree forwarding state, except for the source port. If the DA and SA are on the same interface, the bridge discards (filters) the frame. Otherwise, the bridge forwards the frame out the interface where the destination is known in its bridging table.

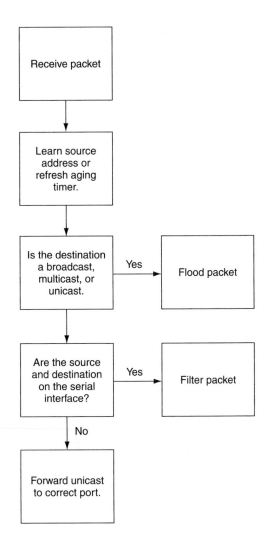

The Learning Stage

Each bridge has a table that records all of the workstations that the bridge knows about on every interface. The bridge records the MAC source address and the source port in the table whenever the bridge sees a frame from a device. This action is referred to as the *bridge learning process.* Bridges learn only unicast source addresses. A station never generates a frame with a broadcast or multicast source address. Bridges learn MAC source addresses in order to intelligently send data to appropriate destination segments. When the bridge receives a frame, the bridge references its table to determine on what port the destination MAC address exists. The bridge uses the information in the table to either filter the traffic or flood the packet out the appropriate interface.

The **bridge learning process** is an action in which the bridge records MAC addresses and the interfaces associated with each address.

FIGURE 2.15
Sample bridged network. (Reproduced with permission of Cisco Systems, Inc. Copyright © 2001 Cisco Systems, Inc. All rights reserved.)

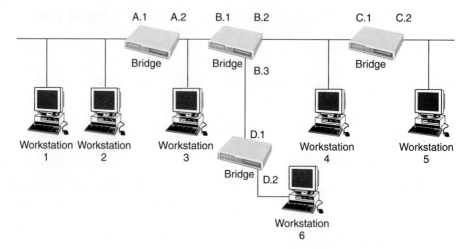

When a bridge is first powered on, the table contains no entries. Assume that the bridges in Figure 2.15 were recently powered on and no station had yet transmitted. Therefore, the tables in all four bridges are empty. If station 1 transmits a unicast frame to station 2, all the stations on that segment, including the bridge, receive the packet frame because of the shared media nature of the segment. Bridge *A* learns that station 1 exists off port A.1. The bridge learns this address by examining the source address in the data link frame header. Bridge *A* enters the MAC source address and bridge port in the table.

The Flooding Stage

Using Figure 2.15 as the sample bridged network, when station 1 transmits, bridge *A* also examines the DA in the data link header to see if there is an entry in the table. At this point bridge *A* only knows about station 1. When a bridge receives a unicast frame no table entry exists for the DA, therefore, the bridge receives an unknown unicast frame. The bridging rules state that a bridge must send an unknown unicast frame out all forwarding ports except the source port. This is commonly referred to as *flooding*. Therefore, bridge *A* floods the unknown unicast frame out all ports even though station 1 and station 2 are on the same port. Bridge *B* receives the frame and undergoes the same process as bridge *A* of learning and flooding. Bridge *B* floods the unknown unicast frame to bridges *C* and *D* and they also learn and flood. The bridges do not know about station 2 because station 2 did not yet transmit. The bridging table for bridge *B* would look like the table shown in Table 2.4.

Flooding is a technique in which a bridge must send an unknown frame out all forwarding ports except the source port.

Table 2.4 Bridging Table After Flooding

Bridge port	A.1	A.2	B.1	B.2	B.3	C.1	C.2	D.1
MAC address	1		1			1		1

Still examining Figure 2.15, all the bridges on the network have an entry for station 1 associated with a port pointing to station 1. Examining bridge C's table, an entry for station 1 is associated with port C.1. This does not mean that station 1 directly attaches to port C.1. It merely reflects that bridge C learned from station 1 on this port. In addition to flooding unknown unicast frames, legacy bridges flood two other frame types: broadcast and multicast. Many multimedia network applications generate broadcast or multicast frames that propagate throughout a bridged network or bridged broadcast domain. As the number of users that interact with multimedia services increases, more broadcast or multicast frames will consume network bandwidth.

The Filtering Stage

What happens in Figure 2.15 when station 2 responds to station 1? All stations on the segment off port A.1, including bridge A, receive the frame. Bridge A learns about the presence of station 2 and adds its MAC address to the bridge table along with the port identifier (A.1). Bridge A also looks at the MAC destination address to determine where to send the frame. Bridge A knows that station 1 and station 2 exist on the same port. Bridge A concludes that it does not need to send the frame anywhere. Therefore, bridge A filters the frame. *Filtering* occurs when the source and destination reside on the same port. Bridge A could send the frame out other ports, but because this wastes bandwidth on other segments, the bridging algorithm specifies to discard the frame. Note that only bridge A knows about the existence of station 2 because no frame from station 2 ever crossed the bridge.

Filtering occurs when the source and destination reside on the same port.

The Forwarding Stage

If in Figure 2.15, station 2 sends a frame to station 6, the bridge floods the frame because no entry exists for station 6. All the bridges learn station 2's MAC address and relative location in the network. When station 6 responds to station 2, bridge D examines its bridging table and sees that in order to reach station 2, it must forward the frame out port D.1. A bridge *forwards* a frame when the DA is a known unicast address because it has an entry in its bridging table, and the SA and DA are on different ports. The frame reaches bridge B, which forwards the frame out port B.1. Bridge A receives the frame and forwards the frame out port A.1. Only bridges A, B, and D learn about station 6. Table 2.5 shows the current updated bridging tables.

Forwarding is the process of sending a frame with a destination address to a given port.

Table 2.5 Bridging Table After Forwarding

Bridge port	A.1	A.2	B.1	B.2	B.3	C.1	C.2	D.1	D.2
MAC address	1		1			1		1	
	2		2				2	2	
		6			6				6

Note that B.1, C.1, and D.1 did not learn about station 2 until station 2 transmitted to station 6.

The Aging Stage

Aging timer is a mechanism whereby a bridge removes an entry from its table after a certain period of time.

When a bridge learns an SA, the bridge time stamps the entry. Every time the bridge sees a frame from that SA, the bridge updates the time stamp. If the bridge does not hear from that SA before an *aging timer* expires, the bridge removes that entry from its bridging table. The aging timer in the bridge table can be modified from the default value of five minutes. The reason for removing SA and DA entries from the bridging tables is because bridges have a finite amount of memory, limiting the number of addresses that can be remembered in the bridging tables.

Higher-end bridges can remember more than 16,000 addresses, while some of the lower-end bridges can remember as few as 4,096 addresses. If all 16,000 spaces are full in the bridging table, and there are 16,001 devices, the bridge will flood all frames from station 16,001 until an opening in the bridging table allows the bridge to learn about station 16,001. Entries become available in the bridging table whenever the aging timer expires for an address. The aging timer helps to limit flooding by remembering the most active stations in the network. If there are fewer devices than the bridge table size, the aging timer can be increased, which will cause the bridge to remember the station longer and reduce flooding.

Bridges also use aging timers to accommodate station moves. In Table 2.5, the bridges know the locations of stations 1, 2, and 6. If station 6 is moved to another location, devices may not be able to reach station 6. If station 6 relocates to port C.2 on bridge *C* and station 1 transmits a frame to station 6, the frame will never reach station 6. Bridge *A* forwards the frame to bridge *B*, but bridge *B* still thinks station 6 is located on port B.3. Aging allows the bridges to forget station 6's entry. After bridge *B* ages the station 6 entry, bridge *B* floods the frames destined to station 6 until bridge *B* learns the new location of station 6. On the other hand, if station 6 initiates the transmission to station 1, the bridges immediately learn the new location of station 6. If the aging timer is set to a high value, this may cause reachability issues within the network before the timer expires.

Token Ring Bridging

Source route translational bridging is a method of bridging that offers some hope of mixed media communication.

When IBM introduced Token Ring an alternative bridging technique called source route bridging (SRB) was initiated as well. The SRB algorithm was developed by IBM and proposed to the IEEE 802.5 committee as a means to bridge all LANs. Although transparent bridging works in a Token Ring environment, IBM networks have unique situations in which transparent bridging creates some obstacles. Many networks have a combination of transparent and SRB devices. Source route transparent bridging (SRT) was developed for hybrid networks allowing them to coexist. SRT bridging eliminates pure SRBs entirely, proposing that the two types of LAN bridges be transparent bridges and SRT bridges. Source route devices cannot inherently communicate with the transparent devices. *Source route translational bridging (SR/TLB),* another bridging technique, offers some hope of mixed media communication. Although SRT bridging has achieved support, SRBs are still widely deployed.

Source Route Bridging (SRB)

SRBs are so named because they assume that the complete source-to-destination route is placed in all inter-LAN frames sent by the source. SRBs store and forward the frames as indicated by the route appearing in the appropriate frame field.

In the Token Ring environment, rings interconnect with bridges. Each ring and bridge has a numeric identifier. The network administrator assigns the values and must follow several rules. Each ring is uniquely identified within a bridged network with a value between 1 and 4,095. Valid bridge identifiers include 1 through 15 and must be unique to the local and remote rings. A ring cannot have two bridges with the same bridge number. Source devices use ring and bridge numbers to specify the path that the frame will travel through the bridged network. Figure 2.16 illustrates an SRB network with several attached workstations.

When station *A* wants to communicate with station *B*, station *A* sends a test frame to determine whether the DA is located on the same ring as the SA. If station *B* responds to the test frame, station *A* knows that they are both on the same ring. The two stations communicate without involving any Token Ring bridges. If station *A* receives no response to the test frame, station *A* attempts to reach the DA on the other rings. The frame now must traverse a bridge. In order to pass through a bridge, the frame includes a *Routing Information Field (RIF)*. One bit in the frame header signals the bridges that an RIF is present and needs to be examined by the bridge. This bit called the *Routing Information Indicator (RII)* is set to zero, when the source address and the destination address are located on the same ring, otherwise the RII is set to one.

> The **Routing Information Field** tells the bridge how to send the frame toward the destination.

> The **Routing Information Indicator** signals the bridges that a RIF is present and needs to be examined by the bridge.

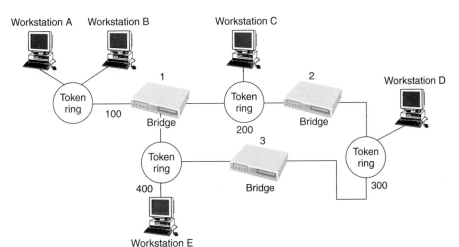

FIGURE 2.16

A source route bridged network. (Reproduced with permission of Cisco Systems, Inc. Copyright © 2001 Cisco Systems, Inc. All rights reserved.)

The RIF tells the bridge how to send the frame toward the destination. When the source host first attempts to contact the destination host, the RIF is empty because the source host does not know any path to the destination host. To complete the RIF, the source host sends an *all routes explorer (ARE)* frame. The ARE passes through all bridges and all rings. As the ARE passes through a bridge, the bridge inserts the local ring and bridge number into the RIF. If in Figure 2.16, station *A* sends an ARE to find the best path to reach station *D*, station *D* will receive two AREs. The RIFs look like the following:

```
Ring 100 - Bridge 1 - Ring 200 - Bridge 2 - Ring 300
Ring 100 - Bridge 1 - Ring 400 - Bridge 3 - Ring 300
```

An **all routes explorer** passes through an entire SRB network following all paths to a specific destination.

Each ring in the network, except for ring 100, sees two AREs. The stations on ring 200 receive two AREs that look like the following:

```
Ring 100 - Bridge 1 - Ring 200
Ring 100 - Bridge 1 - Ring 400 - Bridge 3 - Ring 300
```

The AREs on ring 200 are useless for this session and consume bandwidth unnecessarily. As the Token Ring network becomes more complex with many rings interconnected in a hybrid design, the number of AREs increases dramatically in the network.

Station *D* returns every ARE it receives to the source host. The source host uses the responses to determine the best path to the destination. The source host has the following options to help determine what is the best path to the destination host:

- Use the first response it receives.
- Use the path with the fewest hops.
- Use the path with the largest maximum transmission unit (MTU).
- Use a combination of criteria.
- Most Token Ring implementations choose the first option.

A **specifically routed frame** is used when the RIF specifies the ring and bridge hops to the destination.

Once station *A* knows how to reach station *D*, station *A* transmits each frame as a *specifically routed frame* where the RIF specifies the ring and bridge hops to the destination. When a bridge receives the frame, the bridge examines the RIF to determine if the bridge has any responsibility to forward the frame. If more than one bridge attaches to ring 100, only one of the bridges will forward the specifically routed frame. The other bridge will discard the frame. Station *D* uses the information in the RIF when station *D* transmits back to station *A*. Station *D* creates a frame with the RIF completed in reverse. The source host and the destination host use the same path in both directions.

Frame Format for SRB

The IEEE 802.5 RIF is structured as shown in Figure 2.17. The RIF illustrated in Figure 2.18 consists of two main fields: routing control and routing descriptor.

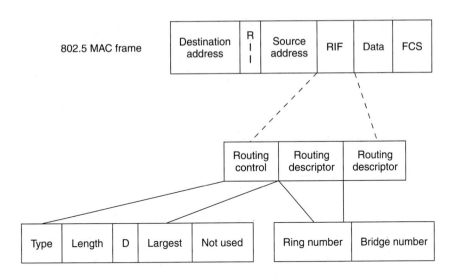

FIGURE 2.17

An IEEE 802.5 RIF is present in frames destined for other LANs. (Reproduced with permission of Cisco Systems, Inc. Copyright © 2001 Cisco Systems, Inc. All rights reserved.)

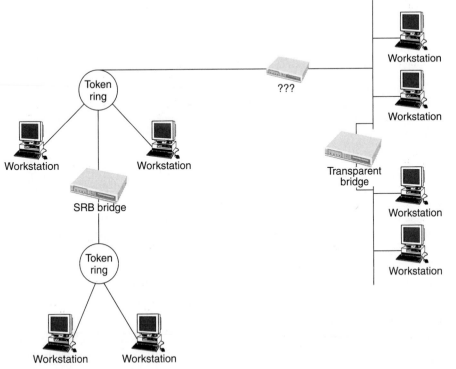

FIGURE 2.18

Bridging mixed media networks. (Reproduced with permission of Cisco Systems, Inc. Copyright © 2001. Cisco Systems, Inc. All rights reserved.)

61

Routing Control Field The routing control field consists of four subfields: type, length, D bit, and largest frame. The fields are summarized below:

1. Type—Consists of three possible routing protocols:

   - Specifically routed: Used when the source node supplies the route in the RIF header. The bridges route the frame by using the route designator field(s).
   - All paths explorer: Used to locate a remote node. The route is collected as the frame traverses the network. Bridges add their bridge number and ring number to the frame, which is forwarded. The first bridge also adds the ring number of the first ring. The remote destination will receive as many frames as routes to that destination.
   - Spanning tree explorer: Used to locate a remote node. Only bridges in the spanning tree forward the frame, adding their bridge number and attached ring number as it is forwarded. The spanning tree explorer reduces the number of frames sent during the discovery process.

2. Length—Indicates the total length in bytes of the RIF. The value can range from 2 to 30 bytes.
3. D bit—Indicates and controls the direction the frame traverses, whether forward or reverse. The D bit affects whether bridges read the ring number and bridge number combinations in the route designators from right to left forward or left to right reverse.
4. Largest frame—Indicates the largest frame size that can be handled along a designator route. The source initially sets the largest frame size, but bridges can lower it if they cannot accommodate the requested size.

Routing Descriptor Field Each routing descriptor field consists of two subfields:

- Ring number (12 bits)—Assigns values that must be unique within the bridged network.
- Bridge number (4 bits)—Assigns values that follow the ring number. This number does not have to be unique unless it is parallel with another bridge connecting two rings.

Bridges add their bridge number and their ring number to the frame, which is forwarded. Routes are alternating sequences of ring and bridge numbers that start and end with ring numbers. A single RIF can contain more than one routing descriptor field and up to a maximum of seven bridges or hops. Many bridge manufacturers followed IBM's implementation. Newer IBM bridge software programs combined with new LAN adapters support 13 hops.

Transparent bridging differs from SRB in significant ways. In SRB, the source host determines what path the frame must follow to reach the destination host. In transparent bridging, the bridge determines the path. In addition, the information used to determine the path differs. SRB uses bridge and ring identifiers, and transparent bridging uses the destination MAC address.

Source Route Transparent Bridging (SRT)

Although many Token Ring networks begin as homogeneous systems, transparently bridged Ethernet works its way into many of these networks. As a result there are hybrid systems, SRB devices, and transparently bridged devices. The SRB devices cannot communicate with the transparently bridged Ethernet devices. Non-source-routed devices do not understand RIFs, SRBs, or any other such frames. To further confuse the issue, some Token Ring protocols operate in transparent mode, typically an Ethernet process. Source route translational bridging (SR/TLB) can overcome some of the limitations of source route transparent bridging (SRT). The best solution is to use a router to connect routed protocols residing on mixed media. SRT supports source route and transparent bridging for Token Ring devices. The SRT bridge uses the RII bit to determine the correct bridging mode. If the bridge sees a frame with the RII set to zero, the SRT bridge treats the frame using transparent bridging methods. The SRT bridge examines the destination MAC address and determines whether to forward, flood, or filter the frame. If the frame contains an RIF with the RII bit set to one, the bridge initiates SRB and uses the RIF to forward the frame. Table 2.6 compares how SRT and SRB react to RII values.

The resulting behavior of the two technologies causes problems in some IBM environments. Whenever an IBM Token Ring attached device wants to connect to another, the device first issues a frame to see whether the destination host resides on the same ring as the source host. If the source host receives no response, it sends an SRB explorer frame.

The SRT deficiency occurs with the test frame. The source host intends for the test frame to remain local to its ring and sets the RII to zero. An RII set to zero signals the SRT bridge to transparently bridge the frame. The bridge floods the frame to all rings. After the test frame reaches the destination host, the source and destination workstations communicate using transparent bridging techniques as if they both reside on the same ring. Transparent bridging does not take advantage of parallel paths like SRB can. Parallel Token Ring backbones can be created to distribute traffic and not overburden any single link. However, transparent bridging selects a single path and does not use another link unless the primary link fails. Therefore, all the traffic passes through the same links, increasing the load on one while another remains unused. This defeats the intent of the parallel Token Rings.

In addition, SRT bridging becomes unsuitable when redundant devices are installed in order to achieve high levels of service. The redundant controllers use the same MAC address to simplify workstation configuration; otherwise, multiple entries need to be entered. If the primary unit fails, the workstation needs to resolve the logical

Table 2.6 SRT and SRB Responses to RII

RII Value	SRB	SRT
Zero	Discard frame	Transparent bridge frame
One	Source-route frame	Source-route frame

address to the new MAC address of the backup unit. By having duplicate MAC addresses, fully automatic recovery is available without needing to resolve a new address.

Duplicate MAC addresses within a transparently bridged network confuse bridging tables. A bridge table can have only one entry for a MAC address; a station cannot appear on two ports. If a device sends a frame to the MAC address for two devices with the same MAC address, how will the transparent bridge know to which device to send the frame? Both devices have the same MAC address, but only one MAC address can exist in the bridging table. Therefore, the resiliency feature will not work in the SRT mode. The solution when using the resiliency feature is to use SRB.

Source Route Translational Bridging (SR/TLB)

Not all networks are homogeneous. Networks may contain a mixture of Ethernet and Token Ring access methods. How can a Token Ring network become attached to an Ethernet network as illustrated in Figure 2.18?

Several obstacles prevent devices on the two networks from communicating with each other. Some of those obstacles are:

- MAC address format
- MAC address representation in the protocol field
- LLC and Ethernet framing
- Routing information field translation
- MTU size mismatch

Translational bridging helps to resolve these issues, allowing the devices to communicate. Presently there is no standard for translational bridging, leaving a number of implementation details to the vendors.

MAC Address Format

When devices transmit frames, the bit sequence varies depending on the access method. An Ethernet frame has a format of DA/SA/Type/Data. When the frame enters the media, which bit is transmitted first? In Ethernet, the least significant bit (LSB) of each octet is transmitted first. Ethernet transmits the LSB first, which is referred to as the canonical address format. Token Ring and *Fiber Distributed Data Interface (FDDI)* use a non-canonical format transmitting the most significant bit (MSB) first and the LSB last. A translational bridge sequences the bits appropriately for the receiving network type so the devices will see the correct MAC address. A canonical source MAC address sent onto a Token Ring segment must have the address transmitted in a non-canonical format. If the bridge does not perform address translation, the Token Ring device will respond to the wrong MAC address.

The **Fiber Distributed Data Interface** uses a non-canonical address format.

Embedded Address Format

MAC address translation may need to occur at layer 3. Some protocols embed the MAC address in the layer 3 protocol header. *Internet Packet EXchange* (IPX) uses a logical IPX address format comprising a network number and a MAC address. In *Transmission Control Protocol/Internet Protocol (TCP/IP),* the ARP includes MAC source and destination addresses in the frame as part of the IP header. Other protocols that embed MAC addresses in the layer 3 header include *DECnet PhaseIV, AppleTalk,* Banyan *Virtual Integrated Network Service (VINES),* and *Xerox Network System (XNS).* Many of these protocols respond to the MAC address contained in the layer 3 header, not the data link layer. A translational bridge must be intelligent and protocol-aware to know when to modify the layer 3 information to represent the MAC address correctly. Normally this activity is relegated to a router, which operates at layer 3. If a new protocol that needs translational bridging is added to the network, the bridge must be updated to know about the translation tasks.

LLC and Ethernet Frame Differences

Ethernet formatted frames use a DA/SA/Protocol_Type format. When bridged onto an IEEE 802 formatted network, the problem arises of what to do with the Protocol_Type value. IEEE 802 headers do not have a provision for the Protocol_Type value. The *Subnetwork Access Protocol (SNAP)* encapsulation approach was devised to carry the Protocol_Type value across IEEE 802 based networks. Figure 2.19 illustrates a translation from Ethernet to Token Ring SNAP.

The frame field illustrated in Figure 2.19 is as follows:

- Access control (AC)
- Frame control (FC)
- Destination address (DA)
- Source address (SA)
- Routing information field (RIF)
- Destination service access point (DSAP)
- Source service access point (SSAP)
- Control (LLC control)
- Organizationally unique identifier (OUI) vendor ID
- Type (Protocol_Type)

RIF Interpretation

When connecting source route devices to transparent devices, another issue involves the RIF. The RIF is absent from transparent devices but is vital to the Token Ring bridging process. How does a source route bridged device specify a path to a transparently bridging device?

A translational bridge assigns a ring number to the transparent segment. To an SRB device, it appears that the destination device resides on a source-routed segment. To the transparent device, the SRB device appears to attach to a transparent segment. The translational bridge keeps a source routing table to reach the Token Ring MAC address. When a transparent device transmits a frame to the Token Ring device, the bridge examines the destination MAC address, locates a source route entry for that address, and creates a frame with a completed RIF.

The **Internet Packet Exchange** uses a logical IPX address format.

The **Transmission Control Protocol/ Internet Protocol** includes MAC source and destination addresses in the frame.

DEC/net Phase IV is a protocol developed and supported by Digital Equipment Corporation.

AppleTalk is a protocol designed by Apple Computer.

Virtual Integrated Network Service is a protocol developed by Banyan Systems.

Xerox Network System is a protocol originally designed by PARC.

The **Subnetwork Access Protocol** carries the Protocol_Type value across networks.

FIGURE 2.19

Frame translation from Ethernet to Token Ring. (Reproduced with permission of Cisco Systems, Inc. Copyright © 2001. Cisco Systems, Inc. All rights reserved.)

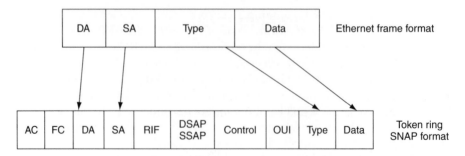

MTU Size

Ethernet, Token Ring, and FDDI support different frame sizes. Table 2.7 lists the minimum and maximum frame sizes for these media access transmission methods.

A frame from one network type cannot violate the frame size constraints of the destination network. If an FDDI device transmits to an Ethernet device, FDDI must not create a frame more than 1,518 octets or less than 64 octets. Otherwise, the bridge will discard the frame. Translational bridges attempt to adjust the frame size to accommodate the *maximum transmission unit (MTU)* mismatch. A translational bridge may fragment an IP frame if the incoming frame exceeds the MTU of the outbound segment. Routers normally perform fragmentation because fragmentation is a layer 3 process. Translational bridges that perform fragmentation actually perform part of the router's responsibility.

●

The **maximum transmission unit** is the maximum frame size in bytes.

Fragmentation is an IP process. Other protocols do not have fragmentation, so the source must create frames appropriately sized for the segment containing the smallest MTU frame size. MTU discovery allows these protocols to work correctly, because the host stations determine the largest allowed frame for the path(s) between the source host and the destination host. This option also exists in IP and is preferred over fragmentation.

Table 2.7 Frame Size Comparison

Media Access Method	*Minimum Frame Size (Octets)*	*Maximum Frame Size (Octets)*
Ethernet	64	1,518
Token Ring (4 Mbps)	21	4,511
Token Ring (16 Mbps)	21	17,839
FDDI	28	4,500

Spanning Tree Algorithm (STA)

The *spanning tree algorithm (STA)* was developed by Digital Equipment Corporation (DEC), a key Ethernet vendor, to preserve the benefits of bridging loops while eliminating the problems caused by bridging loops. Digital's algorithm was subsequently revised by the IEEE 802 committee and published in the IEEE 802.1d specification. The DEC algorithm and the IEEE algorithm are not compatible.

> The **spanning tree algorithm** preserves the benefits of bridging loops while eliminating the problems.

What Is the Spanning Tree Protocol (STP) and Why Use Spanning Tree?

The *Spanning Tree Protocol (STP)* is a loop prevention protocol. STP is a technology that allows bridges to communicate with each other in order to discover bridging loops that may occur in the network. The protocol also specifies an algorithm that bridges can use to create a loop-free logical topology. STP creates a tree structure of loop-free leaves and branches that spans the entire layer 2 network.

> The **Spanning Tree Protocol** is a loop prevention protocol.

Loops occur in networks for a variety of reasons. The most common reason is the result of a deliberate redundancy—in case one link or bridge fails, another bridge or link can take over. Loops can also result by mistake. Figure 2.20 shows a typical bridged network and how loops can be intentionally used to provided redundancy.

Loops are potentially disastrous in a bridged network for two primary reasons: broad bridging loops and bridge table corruption.

> **Bridging loops** can result in inaccurate forwarding and learning in transparent bridging.

Bridging Loops

Without a bridge-to-bridge protocol, the transparent bridge algorithm fails when several bridges and LANs exist between any two LANs in the network. In such a case, a *bridging loop* will occur, which is illustrated in Figure 2.21.

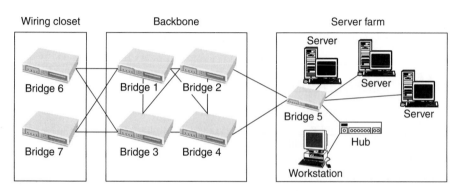

Wiring closet   Backbone   Server farm

Bridge 6, Bridge 7, Bridge 1, Bridge 2, Bridge 3, Bridge 4, Bridge 5, Server, Server, Server, Hub, Workstation

FIGURE 2.20

Networks often include bridging loops to provide redundancy. (Reproduced with permission of Cisco Systems, Inc. Copyright © 2001. Cisco Systems, Inc. All rights reserved.)

FIGURE 2.21
Bridging loops can result in inaccurate forwarding and learning in transparent bridging environments. (Reproduced with permission of Cisco Systems, Inc. Copyright © 2001 Cisco Systems, Inc. All rights reserved.)

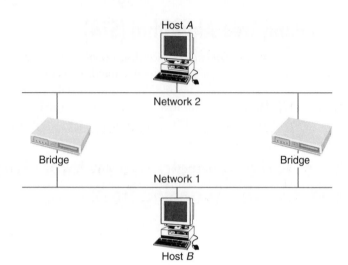

In Figure 2.21, suppose host *A* sends a frame to host *B*. Both bridges receive the frame and correctly conclude that host *A* is located on network 2. Unfortunately, after host *B* receives two copies of host *A*'s frame, both bridges will again receive the frame on their network 1 port because all hosts receive all messages on broadcast LANs. In some cases, the bridges will change their internal bridging tables to indicate that host *A* is located on network 1. If so, when host *B* replies to host *A*'s frame, both bridges will receive and subsequently discard the replies because their bridging tables will indicate that the destination host *A* is located on the same network segment as the SA of the frame.

In addition to basic connectivity problems, the proliferation of broadcast messages in networks with bridging loops represents a potentially serious network problem. Referring to Figure 2.21, assume that host *A*'s initial frame is a broadcast frame. Both bridges will forward the frame endlessly, using all available network bandwidth and blocking transmission of other packets on both segments.

A topology with loops, such as in Figure 2.21, can be useful as well as potentially harmful. A bridging loop implies the existence of multiple paths through the internetwork, and a network with multiple paths from source to destination can increase overall network fault tolerance through improved topological flexibility.

Bridge Table Corruption

A **broadcast storm** causes extreme congestion in a network.

Bridges face the problem of broadcast storms, which can occur in a network. A *broadcast storm* is a phenomenon of extreme congestion, which can be created by bridging loops because bridges waste bandwidth trying to process the amount of traffic that has been generated by the bridging loop. Unicast frames can circulate forever in a network that contains loops. In addition to broadcast storms, unicast ping packets can ruin the network by saturating the bandwidth and creating slower response times between devices.

Two Key STP Concepts

Spanning tree calculations make extensive use of two key concepts when creating a loop-free logical topology: bridge ID (BID) and path cost.

Bridge ID (BID)

A *bridge ID (BID)* is a single 8-byte field that is composed of two subfields as illustrated in Figure 2.22.

The low-order subfield consists of a 6-byte MAC address assigned to the switch. The bridge uses one of the MAC addresses from a pool of 1,024 addresses assigned to the bridged network. This MAC address is a hard-coded number that is not designed to be changed by the user. The MAC address in the BID is expressed in its usual hexadecimal format. The high-order BID subfield is referred to as the *bridge priority.* The bridge priority is a 2-byte (16-bit) value. An unsigned 16-bit integer can have 2^{16} possible values that range from 0–65,535. The default bridge priority is the midpoint value, 32,768. Bridge priorities are typically expressed in decimal (base 10) format.

A **bridge ID** is a single 8-byte field composed of two subfields.

The **bridge priority** is a 2-byte value.

Path Cost

Bridges use the concept of cost to evaluate how close they are to other bridges. 802.1d originally defined cost as 1,000 Mbps divided by the bandwidth of the link in Mbps. A 10Base-T link has a cost of 100 (1,000/10), and fast Ethernet and FDDI use a cost of 10 (1,000/100). With the rise of Gigabit Ethernet and OC-48 ATM (2.4 Gbps), a problem resulted because the cost is stored as an integer value that cannot carry fractional costs. OC-48 ATM results in 1,000 Mbps/2,400 Mbps = .41667, an invalid value. One option is to use a cost of 1 for all links equal to or greater than 1 Gbps. The only problem is that this method prevents STP from accurately choosing the best path

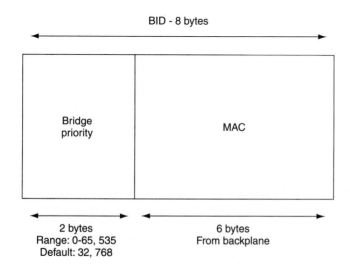

BID - 8 bytes

| Bridge priority | MAC |

2 bytes
Range: 0-65, 535
Default: 32, 768

6 bytes
From backplane

FIGURE 2.22

The bridge ID is composed of bridge priority and a MAC address. (Reproduced with permission of Cisco Systems, Inc. Copyright © 2001 Cisco Systems, Inc. All right reserved.)

Table 2.8 STP Cost Values for Bridged Networks

Bandwidth	STP Cost
4 Mbps	250
10 Mbps	100
16 Mbps	62
45 Mbps	39
100 Mbps	19
155 Mbps	14
622 Mbps	6
1 Gbps	4
10 Gbps	2

in Gigabit networks. As a solution to the above dilemma, the IEEE has decided to modify cost to use a non-linear scale. Table 2.8 lists the newly assigned path cost values.

The values in Table 2.8 were carefully chosen so that the old and new schemes interoperate for the link speeds that are in common use today. The key point concerning STP cost values is that lower costs are better.

Four-Step STP Decision Process

When creating a loop-free logical topology, STP always uses the same four-step decision sequence:

Step 1. Lowest Root BID
Step 2. Lowest Path Cost to Root Bridge
Step 3. Lowest Sender BID
Step 4. Lowest Port ID

Bridge Protocol Data Units are used by bridges to exchange information.

Bridges pass spanning tree information between themselves using special frames known as *Bridge Protocol Data Units (BPDUs)*. A bridge uses this four-step decision sequence to save a copy of the best BPDU seen on every port. When making this evaluation, the bridge considers all of the BPDUs received on the port as well as the BPDU that would be sent on that port. As every BPDU arrives, it is checked against this four-step process to see if it is lower in value than the existing BPDUs saved for that port. If the new BPDU is lower than the existing BPDU value, which was previously selected as the lowest value, then the old value is replaced. Bridges send configuration BPDUs until a lower valued BPDU is received. Saving the best BPDU process also controls the sending of BPDUs. When a bridge first becomes active, all of its ports are sending BPDUs every two seconds when using the default timer values. If a port hears a BPDU from another bridge that has a better lower cost than the BPDU it has been sending, the local port stops sending BPDUs. If the lower valued BPDU stops arriving from a neighbor for a period of time 20 seconds by default, the local port can once again resume the sending of BPDUs.

Three Steps of Initial STP Convergence

STP uses three simple steps in the initial convergence process when a network first comes on line.

Step 1. Elect One Root Bridge
Step 2. Elect Root Ports
Step 3. Elect Designated Ports

When a network first initializes, all of the bridges are announcing a chaotic mix of BPDU information. The bridges immediately begin applying the four-step decision sequence discussed in the previous section. This allows the bridges to hone in on the set of BPDUs that form a single tree spanning the entire network. A single *root bridge* is elected to act as the center point for all bridges in the network (Step 1). All of the remaining bridges calculate a set of root ports (Step 2) and designated ports (Step 3) to build a loop-free topology. The resulting topology can be thought of as a wheel with the root bridge acting as the hub, coupled with loop-free active paths (spokes) radiating outward. In a steady-state network, BPDUs flow from the root bridge outward along these loop-free spokes to every segment in the network. After the network has converged on a loop-free active topology utilizing the three-step process, additional changes are handled using the topology change process. Figure 2.23 illustrates a model network layout for discussion of basic STP operations.

●
A **root bridge** exchanges information with designated bridges in an STP.

The network in Figure 2.23 consists of three bridges connected in a looped configuration. Each bridge has been assigned a fictitious MAC address that corresponds to the device's name. For example, bridge 1 uses MAC address AA-AA-AA-AA-AA-AA-AA.

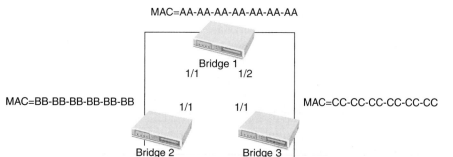

MAC=AA-AA-AA-AA-AA-AA-AA

Bridge 1
1/1   1/2

MAC=BB-BB-BB-BB-BB-BB   1/1   1/1   MAC=CC-CC-CC-CC-CC-CC

Bridge 2   Bridge 3
1/2   1/2

FIGURE 2.23
Model network for basic STP operations. (Reproduced with permission of Cisco Systems, Inc. Copyright © 2001 Cisco Systems. All rights reserved.)

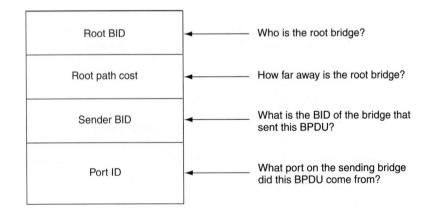

Root BID	← Who is the root bridge?
Root path cost	← How far away is the root bridge?
Sender BID	← What is the BID of the bridge that sent this BPDU?
Port ID	← What port on the sending bridge did this BPDU come from?

Step One: Elect One Root Bridge

The bridges first need to elect a single root bridge by looking for the bridge with the lowest BID. In STP economics, the lowest BID wins. Bridge 1 uses a default ID of 32,768. AA-AA-AA-AA-AA-AA; bridge 2 also assumes a default BID of 32,768.BB-BB-BB-BB-BB-BB; and bridge 3 also assumes a BID of 32,768.CC-CC-CC-CC-CC-CC. Because all three bridges are using the default bridge priority of 32,768, the lowest MAC address AA-AA-AA-AA-AA-AA serves as the tiebreaker, and bridge 1 becomes the root bridge. The other two bridges learned that bridge 1 had the lowest MAC address through the exchange of BPDUs. BPDUs are bridge-to-bridge traffic and do not carry end user traffic. Figure 2.24 illustrates the basic layout of a BPDU.

For the purposes of electing a root bridge, the only items of concern in the BPDU layout are the root BID and the sender BID fields. When a bridge generates a BPDU every two seconds, the bridge places what it thinks is the root bridge at that instant in time in the root BID field. The bridge always places its own BID in the sender BID field. The root BID is the bridge ID of the current root bridge, while the sender BID is the bridge ID of the local bridge. Suppose that bridge 2 boots first and starts sending out BPDUs announcing itself as the Root Bridge every two seconds. A few minutes later, bridge 3 boots and announces itself as the Root Bridge. When bridge 3's BPDUs start arriving at bridge 2, bridge 2 discards the BPDU because bridge 2 has a lower BID saved on its ports (its own BID). As soon as bridge 2 transmits a BPDU, bridge 3 learns that it is not quite as important as it initially assumed. At this point, bridge 3 starts sending BPDUs that list bridge 2 as the Root BID and bridge 3 as the Sender BID. The network is now in agreement that bridge 2 is the Root Bridge.

Five minutes later, bridge 1 boots. As seen with bridge 2 earlier, bridge 1 initially assumes that it is the Root Bridge and starts advertising this fact by sending BPDUs. As soon as these BPDUs arrive at bridges 2 and 3, these bridges decline the root bridge position to bridge 1. All three bridges are now sending out BPDUs that announce bridge 1 as the root bridge and themselves as the sender BID.

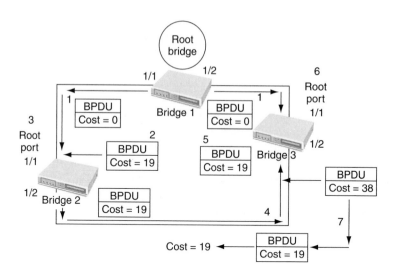

Step Two: Elect Root Ports

After the root bridge is elected, the bridges move on to selecting Root Ports. A bridge's root port is the port that is closest to the root bridge. Every non-root bridge must select one root port. Bridges use the concept of cost to judge closeness. Bridges track something called the root path cost, which is the cumulative cost of all links to the root bridge. Figure 2.25 illustrates how this value is calculated across multiple bridges and the resulting root port election process.

When bridge 1 (the root bridge) sends out BPDUs, they contain a root path cost of 0 (Step 1). When bridge 2 receives these BPDUs, bridge 2 adds the path cost of port 1/1 to the root path cost contained in the received BPDU. If the network is running fast Ethernet as its transmission medium, bridge 2 receives a root path cost of 0 and adds in Port 1/1's cost of 19. Bridge 2 then uses the value of 19 internally and sends BPDUs with a root path cost of 19 out port 1/2 (Step 3). When bridge 3 receives these BPDUs from bridge 2 (Step 4), bridge 3 increases the root path cost to 38 (19 + 19). Bridge 3 is also receiving BPDUs from the root bridge on Port 1/1. These BPDUs enter bridge 3 via port 1/1 with a cost of 0. Bridge 3 increases the cost to 19 internally (Step 5). Bridge 3 has a decision to make Bridge 3 must select a single root port, the port that is closest to the root bridge. Bridge 3 sees a root path cost of 19 on port 1/1 and 38 on Port 1/2. Bridge 3 chooses port 1/1 as its root port (Step 6). Bridge 3 then begins advertising this root path cost of 19 to downstream bridges.

Bridge 2 in Figure 2.25 undergoes a similar set of calculations. Bridge 2 port 1/1 can reach the root bridge at a cost of 19, whereas port 1/2 calculates a cost of 38. Port 1/1 becomes the root port for bridge 2. STP costs are incremented as BPDUs are received on a port not as they are sent out a port. BPDUs arrive on port 1/1 with a cost of 0 and get increased to 19 internally inside bridge 2. The difference between path cost and root path cost is simple. Path cost is a value assigned to each port. The path cost is added to BPDUs received on that particular port to calculate the root path cost. The root path cost is defined as the cumulative cost to the root bridge. In a BPDU, this is the receiving port's path cost to the value contained in the BPDU.

Step Three: Elect Designated Ports

The loop prevention portion of the STP becomes obvious during the third step of initial STP convergence: electing designated ports. Each segment in a bridged network has one designated port. The *designated port* functions as the single bridge port that sends and receives traffic to and from that segment and the root bridge. The idea is that if only one port is designated to handle traffic for each link, all the loops have been broken. The bridge containing the designated port for a given segment is referred to as the *designated bridge* for that segment.

As with the root port selection, the designated ports are chosen based on cumulative root path cost to the root bridge as illustrated in Figure 2.26.

To locate the designated ports, examine each segment in turn. First examine segment 1, the link between bridge 1 and bridge 2. There are two bridges on the segment: bridge 1 port 1/1 and bridge 2 port 1/1. Bridge 1 port 1/1 has a root path cost of 0 because it is the root bridge. Bridge 2 port 1/1 has a root path cost of 19. Bridge 1 port 1/1 has the lower root path cost and becomes the designated port for this link. For segment 2, bridge 1 to bridge 3, a similar election takes place. Bridge 1 port 1/2 has a root path cost of 0, whereas bridge 3 port 1/1 has a root path cost of 19. Bridge 1 port 1/2 has the lower cost and becomes the designated port. Every active port on the Root Bridge becomes a designated port. The only exception to this rule is a layer 1 physical loop to the root bridge, when two ports on the root bridge are connected to the same hub, or two ports are connected with a crossover cable.

The **designated port** sends and receives traffic to and from a segment and the root bridge.

The **designated bridge** is the bridge containing the designated port for a given segment.

FIGURE 2.26

Every segment elects one designated port based on the lowest cost. (Reproduced with permission of Cisco Systems, Inc. Copyright © 2001 Cisco Systems, Inc. All rights reserved.)

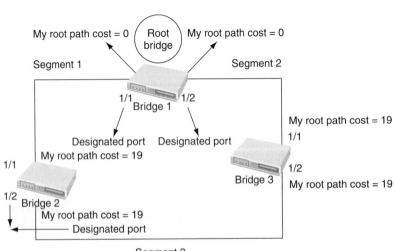

Now let us examine Segment 3. Bridge 2 to bridge 3: Both bridge 2 port 1/2 and bridge 3 port 1/2 have a root path cost of 19. There is a tie between the two ports. When faced with a tie or any other determination, STP always uses the four-step decision sequence of:

Step 1. Lowest Root BID
Step 2. Lowest Path Cost to Root Bridge
Step 3. Lowest Sender BID
Step 4. Lowest Port ID

In Figure 2.26, all three bridges are in agreement that bridge 1 is the root bridge causing root path cost to be evaluated next. Bridge 2 and bridge 3 have a cost of 19. This causes BID, the third decision criteria, to be the deciding factor. Because bridge 2's BID 32, 768:BB-BB-BB-BB-BB-BB is lower than bridge 3's BID 32,768:CC-CC-CC-CC-CC-CC, bridge 2 port 1/2 becomes the designated port for Segment 3. Bridge 3 port 1/2 therefore becomes a non-designated port.

Five STP States

After the bridges have classified their ports as root, designated, or non-designated, creating a loop-free topology is straightforward: root and designated ports forward traffic, whereas non-designated ports block traffic. STP actually has five states even though forwarding and blocking are the only states commonly seen in a stable network. Table 2.9 illustrates the five states of STP.

This list can be viewed as a hierarchy in that bridge ports start at the bottom in the disabled or blocking state and work their way up to forwarding. The disabled state allows network administrators to shut down a port manually. After initialization, ports start in the blocking state where they listen for BPDUs. The bridge can go through a variety of events such as thinking it is a root bridge for a certain period of time, which can cause the bridge to transition into the listening state. During the listening state no user data are being passed as the port is sending and receiving BPDUs in an effort to determine the active topology. During the listening state, the three initial convergence steps take place. Ports that lose the designated port election become non-designated ports and drop back to the blocking state.

Table 2.9 STP States

State	*Purpose*
Forwarding	Sending/receiving user data
Learning	Building bridging table
Listening	Building active topology
Blocking	Receives BPDUs only
Disabled	Administratively down

Ports that remain designated or root ports after 15 seconds progress into the learning state. The learning state is another 15-second period where the bridge is still not passing user data frames. Instead, the bridge is building its bridging table. As the bridge receives frames, the bridge places the MAC SAs and port into the bridging table. The learning state reduces the amount of flooding when data forwarding begins. If a port is still a designated or root port at the end of the learning state period, the port transitions into the forwarding state. At this stage, the bridge starts sending and receiving frames.

Summary

Bridges and switches are data communication devices that operate principally at layer 2 of the OSI reference model. As a result, bridges and switches are widely referred to as data link devices.

Bridges became commercially available in the early 1980s. At the time of their introduction, bridges connected and enabled packet forwarding between homogeneous networks. More recently, bridging between different networks has also been defined and standardized.

Several kinds of bridging have proven important as internetworking devices. Transparent bridging is found primarily in Ethernet environments, while source route bridging (SRB) occurs primarily in Token Ring environments. Translational bridging provides translation between the formats and transit principles of different media types, usually Ethernet and Token Ring. Finally, source route transparent bridging (SRT) combines the algorithms of transparent bridging and SRB to enable communication in mixed Ethernet/Token Ring environments.

Today, switching technology has emerged as the evolutionary heir to bridging based internetworking solutions. Switching implementations now dominate applications in which bridging technologies were implemented previously. Superior throughput performance, higher port density, lower per port cost, and greater flexibility have contributed to the emergence of switches as replacement technology for bridges and as complements to routing technology.

Bridging and switching occur at the link layer, which controls data flow, handles transmission errors, provides physical as opposed to logical addressing, and manages access to the physical medium. Bridges and switches provide these functions by using various link layer protocols that dictate specific flow control, error handling, addressing, and media access algorithms. Examples of popular link layer protocols include Ethernet, Token Ring, and FDDI.

Bridges and switches are not complicated devices. Bridges and switches analyze incoming frames, make forwarding decisions based on information contained in the frames, and then forward the frames toward the respective destination. In some cases, such as SRB, the entire path to the destination is contained in each frame. In other cases, such as transparent bridging, frames are forwarded one hop at a time toward the destination.

Upper layer protocol transparency is a primary advantage of bridging and switching. Because both devices operate at the data link layer they are not required to examine upper layer information. This means that bridges and switches can rapidly forward traffic representing any network layer protocol. A bridge or a switch can move different protocols and other traffic between two or more networks.

Bridges are capable of filtering frames based on layer 2 fields. A bridge can be programmed to reject and not forward all frames sourced from a particular network. Link layer information often includes a reference to an upper layer protocol, which allows a bridge to filter on this parameter. Filters can be helpful in dealing with unnecessary broadcast and multicast packets.

By segmenting large networks into self-contained units, bridges and switches provide several advantages. Because only a certain portion of the traffic is forwarded, a bridge or switch diminishes the traffic experienced by devices on all connected segments. The bridge or switch will act as a firewall for potentially damaging network errors, and together they can accommodate communication between a larger number of devices than would be supported on any single LAN connected only to the bridge. Bridges and switches extend the effective length of a LAN, permitting the attachment of distant workstations that were not permitted previously.

Although bridges and switches share most relevant attributes, several distinctions differentiate these networking technologies. Switches are significantly faster because they switch in hardware, while bridges switch in software. Switches can interconnect LANs of different bandwidth whereas a bridge cannot. A 10-Mbps Ethernet LAN and a 100-Mbps Ethernet LAN can be connected using a switch. Switches can also support higher port density than bridges. Some switches support cut-through switching, which reduces latency and delays in the network, while bridges support only store and forward traffic switching. Finally, switches reduce collisions on network segments because switches provide dedicated bandwidth to each network segment.

Chapter 3 will discuss the basics of routers, another device that is used to interconnect networks.

Review Questions

1. What are the several kinds of bridging that have proven important as internetworking devices?
2. Which layer of the OSI model does bridging and switching operate?
3. What are the two categories that you can group bridges into?
4. What are the two sublayers that comprise the data link layer?
5. What is LAN segmentation?
6. What layer of the OSI model do repeaters operate at?
7. What are the limitations in a repeater based network?
8. What is the number of workstations that can be attached to a 10Base-5 or 10Base-2 network?

9. What is the 5/3/1 rule?
10. What will happen if the Ethernet network is extended too far?
11. What is the 80/20 rule?
12. Explain a LAN switch.
13. What are the three switching modes?
14. If different transmission media are used in the source and destination segments, what type of switching method is best used?
15. Explain a transparent bridge.
16. What are the five processes used by a transparent bridge to determine what to do with a frame?
17. What is the learning process for transparent bridging?
18. What is filtering?
19. What happens if a bridge does not hear from the source address before an aging timer expires?
20. What is SRB?

Summary Questions

1. In SRB, what are the options available to the host to determine the best path?
2. What is SRT?
3. What is SR/TLB?
4. What is STP and why do we use it?
5. What are two key STP concepts when creating a loop-free logical topology?
6. What is the four-step decision sequence used by STP?
7. What are the three steps of initial STP convergence?
8. What are the five states of STP?

Further Reading

Kennedy, Clark. *CISCO LAN Switching*. Indianapolis: Cisco Press, 1999.

Perlman, Radia. *Interconnections Bridges and Routers*. Reading: Addison-Wesley, 1992.

Routing Basics

Objectives

- Explain routing.
- Discuss routing components and what it means to scale large internetworks.
- Discuss the three layer hierarchical model suggested for designing an internetwork.
- Explain path determination.
- Discuss the differences between routing and switching.
- Define a routing algorithm and discuss the common characteristics.
- Explain the different types of routing algorithms.
- Explain how distance vector and link state routing algorithms work.
- Define the various routing metrics and how they are used in routing.
- Explain route calculation and convergence time.
- Explain network congestion and how you can keep it to a minimum without affecting network performance.
- Explain hybrid routing and snapshot routing.

Key Terms

Introduction

Routing is the act of moving information across an internetwork from an SA to a DA. Along the way, at least one intermediate node is encountered. Routing is often contrasted with bridging, which might seem to accomplish the same thing to the casual observer. The primary difference is that bridging occurs at layer 2, the data link layer of the OSI model, whereas routing occurs at layer 3, the network layer. The distinction of where each device transfers data provides routing and bridging with different information to use in the process of transmitting data from the SA to the DA. The two functions accomplish their tasks in different ways.

Routing has been covered in computer science literature for more than two decades, but routing did not achieve commercial popularity until the mid-1980s. Practically every company and every office has at least one router at its location. The primary reason for this time lag is that networks in the 1970s were fairly simple, homogeneous environments. Only recently has large-scale internetworking become popular.

Routing is the process of finding a path to a destination host.

Switching is the process of transporting packets through an internetwork.

A **scalable network** can be adjusted without major modification.

Routing Components

Routing involves two basic activities: determining optimal routing paths and transporting information groups called packets through an internetwork. In the context of the routing process, transporting packets through an internetwork can be referred to as *switching*. Although switching is straightforward, path determination can be complex.

Scaling Large Internetworks

A *scalable network* is one that can be adjusted without major modification as time and resources require. Many of today's internetworks need to be scalable because they are experiencing phenomenal growth. The growth is primarily a result of the increasing demands for connectivity in business and telecommuting.

Scalable internetworks are described as networks that are experiencing constant growth. These networks must be flexible and expandable. The best managed scalable internetworks are designed to follow a hierarchical model of routing. A hierarchical model simplifies the management of the internetwork and allows for controlled growth without overlooking requirements. Figure 3.1 illustrates a three layer hierarchical internetworking model. The layers are defined as follows:

- Core Layer: The *core layer* is the central internetwork for the entire enterprise and may include LAN and WAN backbone. The primary function of the core layer is

The **core layer** is the central internetwork for the entire enterprise.

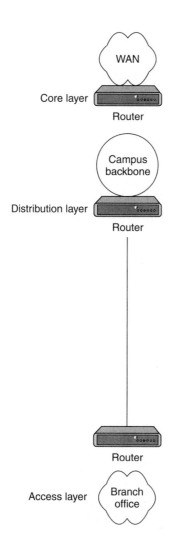

FIGURE 3.1

An internetwork includes the core, distribution, and access layers. (Reproduced with permission of Cisco Systems, Inc. Copyright © 2001 Cisco Systems, Inc. All rights reserved.)

Core routers provide services that optimize communication among routers.

The distribution layer represents the campus backbone.

Quality of service is the measure of performance for a system.

The access layer provides access to corporate resources for a workgroup.

to provide an optimized and reliable transport structure. *Core routers* provide services that optimize communication among routers at different sites or in different logical groupings. In addition, core routers provide maximum availability and reliability. Core routers should be able to maintain connectivity when LAN or WAN circuits fail. A fault-tolerant network design ensures that failures do not have a major impact on network connectivity. Core routers must be reliable because they carry information about all the routes in an internetwork. If one of the core routers fails, that router will affect routing on a larger scale than if an access router fails.

● Distribution Layer: The *distribution layer* represents the campus backbone. The primary function of the distribution layer is to provide access to various parts of the internetwork as well as access to services. Distribution routers control access to resources that are available at the core layer. Distribution routers must make efficient use of bandwidth. In addition, a distribution router must address the *quality of service (QoS)* needs for different protocols by implementing policy based traffic control to isolate backbone and local environments. Policy based traffic control enables you to prioritize traffic to ensure the best performance for the most time-critical and time-dependent applications. Distribution routers need to select the best path to different locations in order to make efficient use of bandwidth.

● Access Layer: The *access layer* provides access to corporate resources for a workgroup on a local segment. Access routers control traffic by localizing broadcasts and service requests to the access media. Access routers must also provide connectivity without compromising network integrity. The routers at the access point must be able to detect if a telecommuter dialing in is legitimate and to require minimal authentication steps for the telecommuter to gain access to the corporate LAN. Access routers are placed where security and filtering must be defined. The primary function of the access layer router is to reduce the amount of overhead by keeping unnecessary traffic out of the core of the network.

A hierarchy simplifies tasks such as addressing and device management. Using an addressing scheme that maps to the hierarchy reduces the need to redo the network addresses as a result of growth. Knowing the placement of devices in the hierarchy enables one to program all routers within one layer in a consistent manner because all routers must perform similar tasks. Figure 3.2 illustrates the concept of making sure that routers are placed in their designated layers so that they can perform their proper functions.

Characteristics of Scalable Internetworks

There are several characteristics that classify internetworks as scalable. The following represent the key characteristics that should be considered when planning an internetwork design strategy:

● Reliable and available: Being reliable and available means being dependable and accessible 24 hours a day, 7 days a week (24×7). Failures need to be isolated and recovery must be invisible to the end user.

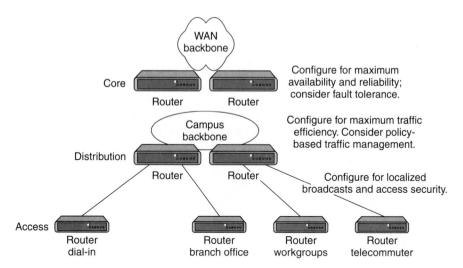

- Responsive: Responsiveness includes managing the QoS needs for the various protocols being utilized without affecting a response from the end user to the help desk. The internetwork must be able to respond to latency issues that might be common for one routing architecture but not for another.
- Efficient: Large internetworks must optimize their use of resources, especially bandwidth. Reducing the amount of overhead traffic such as unnecessary broadcasts, service location, and routing updates results in an increase in data throughput without increasing the cost of hardware or the need for additional WAN services.
- Adaptable: Adaptability includes accommodating disparate networks and interconnecting independent network clusters, as well as integrating legacy technologies.
- Accessible but secure: Accessibility and security must include the capability to enable connections into the internetwork using dedicated, dial up, and switched services while maintaining network integrity.

Path Determination: An Overview

A metric defines a standard of measurement, such as path length, that is utilized by the router algorithms to determine the optimal path to a destination. The process of path determination is aided by the various routing algorithms, which initialize and maintain routing tables. *Routing tables* contain route information that varies depending on the routing algorithms that are programmed into the router.

Routing algorithms fill routing tables with a variety of information. Destination/next hop associations inform a router that a particular destination can be gained optimally by sending the packet to a particular router representing the next hop on the way to

Routing tables contain route information.

FIGURE 3.3

Destination/next hop associations determine the optimal path to a destination. (Reproduced with permission of Cisco Systems, Inc. Copyright © 2001 Cisco Systems, Inc. All rights reserved.)

27	Node A
57	Node B
17	Node C
24	Node A
52	Node A
16	Node B
26	Node A

the final destination. When a router receives an incoming packet, the router checks the destination address and attempts to associate this address with a next hop. Figure 3.3 depicts a sample destination/next hop routing table.

Routing tables also contain other pertinent information, such as data regarding the desirability of a particular path. Routers compare metrics to determine optimal routes, and these metrics differ depending on the design of the routing algorithm being utilized. Routers communicate with one another and maintain their routing tables through the transmission of a variety of messages. The *routing update* message consists of all or a portion of the routing table. By analyzing routing updates from all other routers, a router can build a detailed picture of a network topology. A *link state advertisement (LSA)* is an example of a message sent between routers to inform other routers of the state of the sender's links. Link information can be used to build a complete picture of topology to enable routers to determine optimal routes to network destinations.

The **routing update** is a message sent from a router at regular intervals.

A **link state advertisement** is a message sent between routers to inform other routers of the state of the sender's links.

Switching

Switching algorithms are relatively simple and are basically the same for most routing protocols. In most cases, a host destination determines that it must send a packet to another host. The host acquires a router's address by some means. The source host sends a packet addressed specifically to a router's physical MAC layer address. The address of the source host contains the protocol or network layer address of the destination host. As the router examines the packet's destination protocol address, the router determines that the router either knows or does not know how to forward the packet to the next hop. If the router does not know how to forward the packet, the router will drop the packet. If the router knows how to forward the packet, the router will change the physical MAC destination address to that of the next hop and transmit the packet.

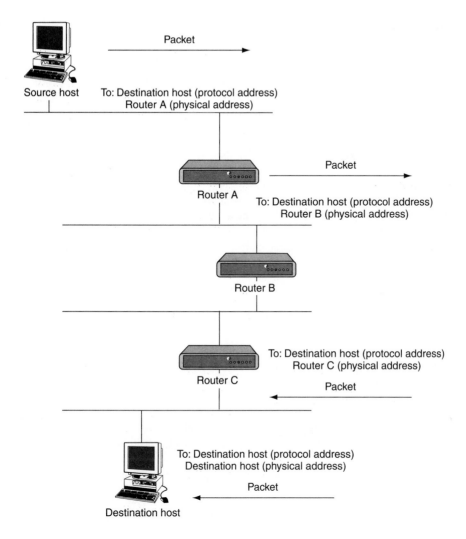

The next hop may be the ultimate destination host. If not, the next hop is usually another router, which executes the same switching decision process. As the packet moves through an internetwork, the packet's physical address changes, but the packet's protocol address remains constant. The addressing of a packet is illustrated in Figure 3.4. The IOS has developed a hierarchical terminology, as shown in Figure 3.4, that describes the process of switching between a source and a destination end system. Network devices without the capability to forward packets between subnetworks are called end systems (ES). Network devices that possess the capability to forward packets are called intermediate systems (IS). Intermediate systems are further divided into those devices that can communicate within routing domains, referred to as intradomain IS, and those devices that communicate within and between routing domains, referred to as interdomain IS. A *routing domain* is considered to be a portion of an internetwork under common administrative authority that is regulated by a particular set of administrative guidelines.

Routing domains are also referred to as autonomous systems (AS). Certain protocols inside a routing domain can be further subdivided into routing areas but intradomain routing protocols are still used for switching both within and between areas.

● A **routing domain** is a portion of an internetwork that is regulated by administrative guidelines.

Routing Algorithms

Routing algorithms can be differentiated based on several key criteria. First, the particular goals of the algorithm designer affect the operation of the resulting routing protocol. Second, various types of routing algorithms exist, and each algorithm has a different impact on network and router resources. Finally, routing algorithms use a variety of metrics that affect the calculation of optimal routes.

● **Routing algorithms** determine the best route from a source to a destination.

Design Goals

Routing algorithms often have one or more of the following design goals:

- Optimality
- Simplicity and low overhead
- Robustness and stability
- Rapid convergence
- Flexibility

Optimality refers to the capability of the routing algorithm to select the best route. The best route depends on the metrics and metric weightings that are utilized in making the calculations. Although a routing algorithm may use a number of hops and delays, it may weigh delay more heavily in the calculation. Routing protocols must strictly define their metric calculation algorithms.

Routing algorithms are also designed to be as simple as possible. The routing algorithm must offer its functionality efficiently, with a minimum of software and utilization overhead. Efficiency is particularly important when the software implementing the routing algorithm must run on a computer with limited physical resources. Routing algorithms must be robust, meaning they should perform correctly in the face of unusual or unforeseen circumstances. Unforeseen circumstances include hardware failures, high traffic load conditions, and incorrect implementations. Because routers are located at network junction points they can cause considerable problems when they fail. The best routing algorithms are often those that have withstood the test of time and have proven stable under a variety of different network stress conditions.

Routing protocols must converge rapidly. Convergence is the process of agreement by all routers on optimal routes. When a network event causes routes either to go down or become available, routers distribute routing update messages that permeate

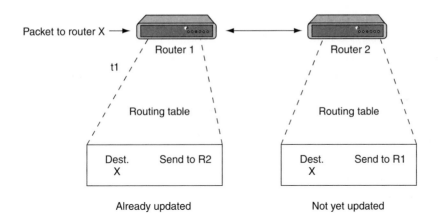

networks. The routing updates stimulate recalculation of optimal routes and eventually cause all routers to agree on the proposed routes. Routing algorithms that converge slowly may cause routing loops or network outages.

The routing loop displayed in Figure 3.5 illustrates a packet arriving at router 1 at time t1. Router 1 has already been updated and knows that the optimal route to the destination calls for router 2 to be the next stop. Router 1 forwards the packet to router 2, but router 2 has not yet updated its routing table; therefore, router 2 believes the optimal next hop is router 1. Router 2 forwards the packet back to router 1. The packet continues to bounce back and forth between the two routers until router 2 receives its routing update or until the packet has been switched the maximum number of allowed times.

Routing algorithms should also be flexible. Flexibility means that the routing algorithm should quickly and accurately adapt to a variety of network circumstances. When a network segment has gone down, many routing algorithms will quickly select the next best path for all routes that are normally using the downed segment. Routing algorithms can be programmed to adapt to changes in network bandwidth, router queue size, network delay, and network topology changes.

Routing Algorithm Types

Routing algorithms can be classified by type of algorithm. Key types include:

- Routed protocol versus routing protocol
- Static versus dynamic
- Single-path versus multipath
- Flat versus hierarchical
- Host-intelligent versus router-intelligent
- Intradomain versus interdomain
- Link state versus distance vector

Routed Protocol versus Routing Protocol

A *routed protocol* is a protocol that contains enough network layer addressing information for user traffic to be directed from one network to another network. Routed protocols define the format and use of the fields within a packet. Packets that use a routed protocol are conveyed from ES to ES through an internetwork. Examples of routed protocols are the Internet protocol (IP) and Novell's IPX.

A *routing protocol* supports a routed protocol by providing mechanisms for sharing routing information. Routing protocol messages move between routers. A routing protocol allows the routers to communicate with other routers to update and maintain routing tables. Routing protocol messages do not carry end user traffic from network to network. A routing protocol uses the routed protocol to pass information between routers. TCP/IP examples of routing protocols are (RIP) Routing Information Protocol and (OSPF) Open Shortest Path First.

Usually, routing protocols function only between routers, but because some routing protocols are unaware of other routers, they rely on data link broadcast messages to provide information to other routers. At times these broadcast messages are used by end systems for their own purposes. An ES receiving a router's update broadcast can record the existence of the router and use of the router at a later time if the ES needs to acquire information about the topology of the internetwork.

Static versus Dynamic

Static routing algorithms are hardly algorithms at all. *Static routing algorithms* are table mappings established by the network administrator prior to the beginning of routing. These static mappings do not change unless the network administrator alters them. Algorithms that use static routes are simple to design and work well in environments where network traffic is relatively predictable and network design is relatively simple.

Because static routing systems cannot react to network changes, static routing is considered unsuitable for today's large, changing networks. Most of the dominant routing algorithms in the 1990s were dynamic routing algorithms. Dynamic routing algorithms adjust to changing network circumstances by analyzing incoming routing update messages. If the message indicates that a network change has occurred, the routing software recalculates routes and sends out new routing update messages. The new routing update messages permeate the network, stimulating routers to rerun their algorithms and change their routing tables accordingly.

Dynamic routing algorithms can be supplemented with static routes where appropriate. A router of last resort, which is a router to which all unroutable packets are sent, can be designated to act as a repository for all unroutable packets, ensuring that all messages are at least processed.

Single-Path versus Multipath

Some sophisticated routing protocols support multiple paths to the same destination. Unlike single-path algorithms, multipath algorithms permit traffic multiplexing over

---

A **routed protocol** defines the format and use of the fields within a packet.

A **routing protocol** provides mechanisms for sharing routing information.

Static routing algorithms take precedence over routes chosen by dynamic routing protocols.

Dynamic routing algorithms adjust automatically to network topology or traffic changes.

multiple lines. The advantages of multipath algorithms are that they provide better throughput and reliability.

Flat versus Hierarchical

Some routing algorithms operate in a flat space. Other routing algorithms operate using hierarchies. In a flat routing system, the routers are peers of all other routers. In a hierarchical routing system, some routers form a routing backbone. Packets from non-backbone routers travel to the backbone routers where the packets are sent through the backbone until they reach the area of their destination. At this point, the packets from non-backbone routers travel from the last backbone router through one or more non-backbone routers to the final destination. Routing systems often designate logical groups of nodes called domains, AS, or areas. In hierarchical systems, some routers in a domain can communicate with routers from other domains. Otherwise, routers can communicate only with routers within their domain. In large networks, additional hierarchical levels may exist. Routers are placed at the highest hierarchical level forming the routing backbone.

The primary advantage of hierarchical routing is that hierarchical routing mimics the organization of most companies and supports their traffic patterns. Most network communication occurs within small company groups referred to earlier as domains. Because intradomain routers need to know only about other routers within their domain, their routing algorithms can be simplified. Depending on the routing algorithm being utilized, routing update traffic can be reduced accordingly.

Host-Intelligent versus Router-Intelligent

Some routing algorithms assume that the source end node will determine the entire route. The concept of the source end node determining the entire route is referred to as source routing. In source routing systems, routers act as store and forward devices, mindlessly sending the packet to the next stop.

Other routing algorithms assume that hosts know nothing about routes. In these types of algorithms, routers determine the path through the internetwork based on their own calculations. In source routing, the hosts have the routing intelligence. In routing algorithms where the hosts know nothing about routes, the routers have the routing intelligence.

The trade-off between host-intelligent and router-intelligent routing is one of path optimality versus traffic overhead. Host-intelligence systems choose the better routes more often because they discover all possible routes to the destination before the packet is actually sent. Host-intelligent systems then choose the best path based on that particular system's definition of optimal. The act of determining all routes often requires substantial discovery traffic and a significant amount of time.

Intradomain versus Interdomain

Some routing algorithms work only within domains, while other routing algorithms work within and between domains. The nature of these two algorithm types is different.

An optimal intradomain routing algorithm would not necessarily be an optimal interdomain routing algorithm.

Link State versus Distance Vector

Most routing protocols can be classified as one of two basic algorithms: distance vector or link state *Link state algorithms,* also referred to as *shortest path first (SPF) algorithms,* flood routing information to all nodes in the internetwork. Each router sends only a portion of the routing table that describes the state of its own links. *Distance vector algorithms,* also referred to as Bellman-Ford algorithms, call for each router to send all or some portion of its routing table, but only to its neighbors. In essence, link state algorithms send small updates everywhere, while distance vector algorithms send larger updates only to neighboring routers.

Link state routing algorithms converge more quickly and are less prone to routing loops than distance vector algorithms. Because link state algorithms require more CPU power and memory than distance vector algorithms they can be more expensive to implement and support.

Distance Vector Routing Algorithms

Distance vector based routing algorithms (Bellman-Ford algorithms) periodically pass copies of a routing table from router to router. Updates between routers also communicate topology changes immediately when they occur. Each router receives a routing table from direct neighbor routers connected to the same network, as illustrated in Figure 3.6.

For example, in Figure 3.6 router B receives information from router A, router B's router neighbor across the WAN link. Router B adds a distance vector number, such

FIGURE 3.6

Distance vector routers periodically pass copies of their routing table to neighboring routers and accumulate distance vectors. (Reproduced with permission of Cisco Systems, Inc. Copyright © 2001 Cisco Systems, Inc. All rights reserved.)

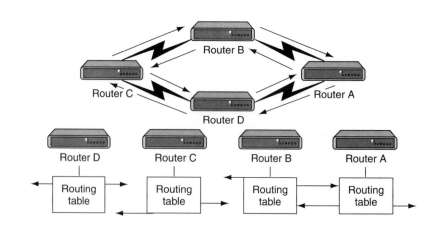

as the number of hops increasing the distance vector, and then passes the routing table to its other neighbor, router C. This step-by-step process occurs in all directions between direct neighbor routers. The algorithm is accumulating network distances so that it can maintain a database of internetwork topology information. Distance vector algorithms do not allow a router to know the exact topology of an internetwork.

Distance Vector Network Discovery

Each router using distance vector routing begins by identifying its own directly connected networks. In Figure 3.7, the interface to each directly connected network is shown in the routing tables as having a distance of 0. As the distance vector discovery process proceeds, routers discover the best path to destination networks based on accumulated metrics from each neighbor.

For example, router A learns about other networks based on information router A receives from router B. Each of the other network entries learned from router B is placed in router A's routing table. An accumulated distance vector shows how far away that learned network is in the given direction.

Distance Vector Topology Changes

Routing table updates communicate topology changes in the network. Topology change updates proceed step by step from router to router as shown in Figure 3.8.

Distance vector algorithms call for each router to send its entire routing table to each of its adjacent neighbors. Distance vector routing tables include information about the total path cost defined by its metric and the logical address of the first router on the path to each network it knows about. In Figure 3.7, the metric of each path is shown in the third column of the routing tables.

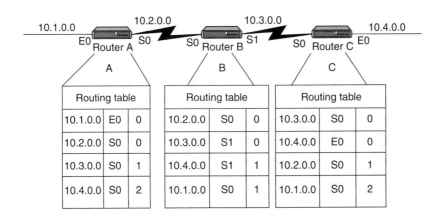

FIGURE 3.7

Distance vectors for routers discover the best path to destinations from each neighbor. (Reproduced with permission of Cisco Systems, Inc. Copyright © 2001 Cisco Systems, Inc. All rights reserved.)

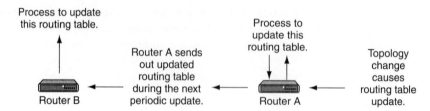

FIGURE 3.8

Updates proceed step by step from router to router. (Reproduced with permission of Cisco Systems, Inc. Copyright © 2001 Cisco Systems, Inc. All rights reserved.)

When a router receives an update from a neighboring router, the router compares the update with its own routing table. If a router learns from its neighbor about a better route to a network the router updates its own routing table. To calculate the new metric, the router adds the cost of reaching the neighbor router to the path cost reported by the neighbor. The new metric is added into the router's routing table.

If router B in Figure 3.8 is one unit of cost from router A, router B would add 1 to all costs reported by router A when router B runs the distance vector processes to update its routing table.

A router will send updates by multicasting or broadcasting its routing table on each configured port. Other methods such as sending the table to only preconfigured neighbors are employed by some routing algorithms. Routing algorithms such as RIP v2 and OSPF use multicast messages. RIP v1 uses broadcast messages.

Problem: Routing Loops

Routing loops can occur if the internetwork's slow convergence on a new configuration causes inconsistent routing tables. Figure 3.9 uses a simplistic network design to show how a routing loop can develop.

FIGURE 3.9

Router A updates its table to reflect a new but erroneous hop count. (Reproduced with permission of Cisco Systems, Inc. Copyright © 2001 Cisco Systems, Inc. All rights reserved.)

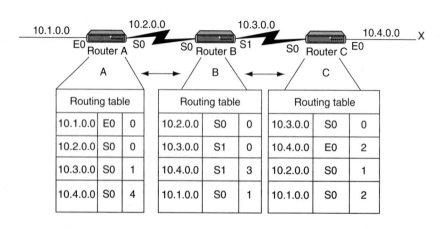

In Figure 3.9, network 10.4.0.0 has failed, initiating a routing loop between routers A, B, and C. The following steps describe the process of the routing loop:

1. Just before the failure of network 10.4.0.0, all routers have consistent knowledge and correct routing tables. The network is said to have converged. The cost function metric is hop count so the cost of each link is 1. Router C is directly connected to network 10.4.0.0 with a distance of 0. Router A's path to network 10.4.0.0 is through router B with a hop count of 2.

2. When network 10.4.0.0 fails, router C detects the failure and stops routing packets out its Ethernet 0 interface. Router A has not yet received notification of the failure and still believes it can access 10.4.0.0 through router B. Router A's routing table reflects a path to network 10.4.0.0 with a distance of 2.

3. Because router B's routing table indicates a path to network 10.4.0.0, router C believes it now has a viable path to network 10.4.0.0 through router B. Router C updates its routing table to reflect a path to network 10.4.0.0 with a hop count of 2.

4. Router A receives the new routing table from router B, detects the modified distance vector to network 10.4.0.0, and recalculates its own distance vector to 10.4.0.0 as 3.

Because routers A, B, and C conclude that the best path to network 10.4.0.0 is through each other, packets destined to network 10.4.0.0 continue to bounce between the three routers.

Counting to Infinity

The invalid updates about network 10.4.0.0 continue to loop, and the hop count increments each time the update packet passes through another router. This process of continually incrementing the hop count is called *counting to infinity*. Without countermeasures to stop the process, the loop and the process of counting to infinity will continue indefinitely.

Counting to infinity is a process in which routers continually increment the hop count.

Solutions

Solution: Defining a Maximum

The countermeasure to counting to infinity is that distance vector protocols define infinity as some maximum number. Such a maximum can be defined for any routing metric, including hop count. For example, RIP has a maximum hop count of 16. Using this approach, the routing protocol permits the routing loop until the metric exceeds its maximum allowed value. Once the metric value exceeds the maximum, network 10.4.0.0 in Figure 3.9 is considered unreachable. The routers will designate the network as unreachable in their routing tables and stop circulating update information.

By defining a maximum, distance vector routing algorithms are self-correcting in response to incorrect routing information. A loop may occur for some finite period of time, until the maximum metric value is exceeded.

Time To Live is a packet parameter that decreases each time a router processes the packet.

Another related concept is the *Time To Live (TTL)* parameter. The TTL is a packet parameter that decreases each time a router processes the packet. When the TTL reaches zero, a router discards or drops the packet without forwarding the packet. A packet caught in a routing loop is removed from the internetwork when its TTL expires. IP uses a TTL counter to stop counting to infinity problems. When the TTL reaches 0, a router discards the packets, thereby preventing the packets from looping forever.

Split Horizon

Split horizon ensures that information about a route is never sent back in the direction from which it originated.

One way to eliminate routing loops and speed up convergence is through the technique referred to as *split horizon*. The concept behind split horizon is that it is never useful to send information about a route back in the direction from which the information originated. In Figure 3.10, router B learns that network 10.4.0.0 has failed through the following steps:

1. Router B has access to network 10.4.0.0 through router C. It makes no sense for router B to announce to router C that router B has access to network 10.4.0.0 through router C because router C will always possess the best information regarding network 10.4.0.0.
2. Given that router B passed the announcement of its route to network 10.4.0.0 to router A, it makes no sense for router A to announce its distance from network 10.4.0.0 to router B.
3. Having no alternate path to network 10.4.0.0, router B concludes that network 10.4.0.0 is inaccessible.

Basically, the split horizon technique does not allow update information to flow out the same interface that it arrived on.

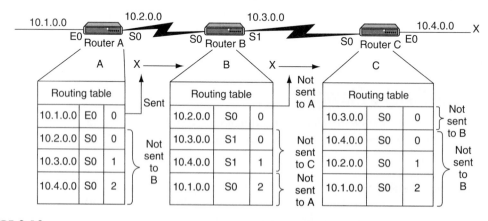

FIGURE 3.10

Split horizon ensures that information about a route is never sent back in the direction from which the packet originated. (Reproduced with permission of Cisco Systems, Inc. Copyright © 2001 Cisco Systems, Inc. All rights reserved.)

Poison Reverse

Poison reverse is a variation of split horizon. Poison reverse attempts to eliminate routing loops caused by inconsistent updates. Using the technique of poison reverse, a router that discovers an inaccessible route sets a table entry that keeps the network state consistent while other routers gradually converge on the topology change. Poison reverse is used with hold-down timers to provide a solution to long loops. When network 10.4.0.0 fails, as shown in Figure 3.11, router C can poison its link to network 10.4.0.0 by recording a table entry for that link as having infinite cost, therefore being unreachable.

By poisoning its route to network 10.4.0.0, router C is not susceptible to other incorrect updates about network 10.4.0.0 coming from neighboring routers that might claim to have a valid alternate path.

When an update shows the metric for an existing route to have increased sufficiently, there is a loop. The route should be removed, poisoned, and put into holddown. Currently, the rule is that a route is removed if the composite metric increases more than a factor of 1.1. It is not safe for just any increase in the composite metric to trigger removal of the route because small metric changes can occur as a result of changes in channel occupancy or reliability. This rule is needed only to break large loops because small ones will be prevented by split horizon, triggered updates, and holddowns.

Hold-Down Timers

Hold-down timers are necessary to prevent regular update messages inappropriately reinstating a route that may have gone bad. Holddowns tell routers to hold any changes that might affect routes for some period of time. The hold-down period is usually calculated

> **Poison reverse** is a technique that attempts to eliminate routing loops caused by inconsistent updates.

> **Hold-down timers** tell routers to hold any changes that may affect routes for some period of time.

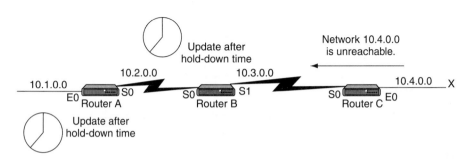

FIGURE 3.11

A router keeps an entry for the network down state, allowing time for other routers to recompute for the topology change. (Reproduced with permission of Cisco Systems, Inc. Copyright © 2001 Cisco Systems, Inc. All rights reserved.)

to be just greater than the period of time necessary to update the entire network with a routing change. Hold-down timers work as follows:

1. When a router receives an update from a neighbor indicating that a previously accessible network is now inaccessible, the router marks the route as inaccessible and starts a hold-down timer, as shown in Figure 3.11. If at any time before the hold-down timer expires an update is received from the same neighbor indicating that the network is again accessible, the router marks the network as accessible and removes the hold-down timer.
2. If an update arrives from a different neighboring router with a better metric than originally recorded for the network, the router marks the network as accessible and removes the hold-down timer.
3. If at any time before the hold-down timer expires an update is received from a different neighboring router with a poorer metric, the update is ignored. Ignoring an update with a poorer metric when a holddown is in effect allows more time for the knowledge of the change to propagate through the entire network.

Triggered Updates

As seen previously in solutions of routing loops, the routing loops were caused by erroneous information calculated as a result of inconsistent updates, slow convergence, and timing of updates. If routers wait for their regularly scheduled updates before notifying neighboring routers of network catastrophes, serious problems may result.

Normally, new routing tables are sent to neighboring routers on a regular basis. IP RIP updates the routing tables every 30 seconds. IPX RIP updates the routing tables every 60 seconds. A *triggered update* is an update that is sent immediately in response to some change in the routing table. The router detects a topology change and immediately sends an update message to adjacent routers that, in turn, generate triggered updates notifying their adjacent neighbors of the change. The wave of exchanging routing updates will propagate throughout that portion of the network where routes connect to the faulty link.

In Figure 3.12, router C immediately announces that network 10.4.0.0 is unreachable. Upon receipt of this information, router B announces through interface S0 that

> A **triggered update** is an update that is sent immediately in response to some change in the routing table.

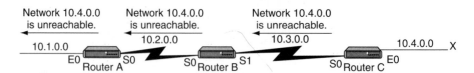

FIGURE 3.12

With the triggered update approach, nodes send messages when they notice a change in the routing table. (Reproduced with permission of Cisco Systems, Inc. Copyright © 2001 Cisco Systems, Inc. All rights reserved.)

network 10.4.0.0 is down. In turn, router A sends an update out interface E0. Triggered updates would be sufficient if you could guarantee that the wave of updates reached every appropriate router in the network immediately. However, there are two problems that can result:

- Packets containing the update message can be dropped or corrupted by some link in the network.
- The triggered updates do not happen instantaneously. It is possible that a router that has not yet received the triggered update will issue a regular update at just the wrong time, causing the bad route to be reinserted in a neighbor that has already received the triggered update.

Utilizing triggered updates in conjunction with hold-down timers is designed to get around these problems. The hold-down rule states that when a route is removed, no new route to the same destination will be accepted for a certain period of time. Thus, the triggered update has time to propagate throughout the network.

Implementing Solutions in Multiple Routes

The individual solutions discussed thus far work together to prevent routing loops in more complex network designs. In complex network designs, these routers have multiple routes interconnecting to each other.

Consider the design shown in Figure 3.13. Routers A, D, and E each have two routes to network 10.4.0.0. When network 10.4.0.0 fails, the following steps must be taken:

1. Poison route: As soon as router B detects the failure of network 10.4.0.0, router B poisons its route to that network by indicating an infinite hop count.

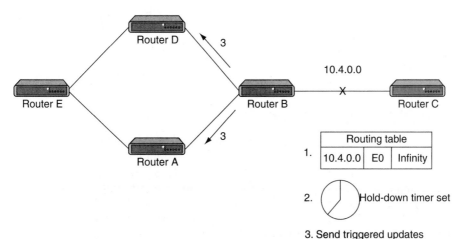

FIGURE 3.13

As soon as router B detects that network 10.4.0.0 is down, router B poisons its route entry in its routing table. (Reproduced with permission of Cisco Systems, Inc. Copyright © 2001 Cisco Systems, Inc. All rights reserved.)

2. Set hold-down timer: Once router B poisons its route to network 10.4.0.0, router B then sets its hold-down timer.

3. Send triggered update: Router B also sends a triggered update to routers D and A, indicating that network 10.4.0.0 is possibly down. New route information propagates through the rest of this network as the series of connected routers set hold-down timers and trigger updates. Routers D and A receive the triggered update and set their own hold-down timers to suppress any route changes for a specific period of time. Routers D and A, in turn, send a triggered update to router E indicating the possible inaccessibility of network 10.4.0.0.

4. Finally, router E receives the triggered update about the status of network 10.4.0.0 from routers D and A. Router E then sets its hold-down timer and waits until one of the following events occurs:

- The hold-down timer expires. In this case, router E knows that network 10.4.0.0 is definitely unavailable.
- Another update is received indicating that the network status has changed. In this case, router E updates its tables with the new information.
- Another update is received indicating a new route with a better metric. In this case, router E updates its tables with the new route information.

During the hold-down period, router E assumes the network status is unchanged from its original state and will attempt to route packets to network 10.4.0.0.

Link State Routing Algorithms

The second basic algorithm used for routing is the link state algorithm. Link state routing algorithms, also known as SPF algorithms, maintain a complex database of topology information. Although the distance vector algorithm contains entries for distant networks and has a metric value to reach those networks but contains no knowledge of distant routers, a link state routing algorithm maintains full knowledge of distant routers and how they interconnect. Examples of link state routing protocols are *NetWare Link Services Protocol* (*NLSP*), OSPF, and Intermediate System to Intermediate System (IS-IS). Link state routing uses link state packets (LSPs), a topological database, the SPF algorithm, the resulting SPF tree, and a routing table of paths and ports to each network.

●
NetWare Link Services Protocol is a link state routing protocol based on IS-IS.

Link State Network Discovery

Link state network discovery mechanisms are used to create a common picture of the entire internetwork. All link state routers share this common view of the internetwork. In Figure 3.14, four networks (W, X, Y, and Z) are connected by three link state routers (A, B, and C). Network discovery for link state routers uses the following process:

1. Routers learn of their neighbors—that is, the other routers that are on directly connected networks with them. This process is often referred to as neighbor no-

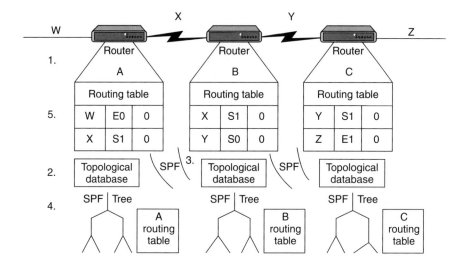

tification. In link state routing, each router connected to a network keeps track of its neighbors.

2. Routers transmit LSPs onto the network. The LSPs contain information about which networks the routers are connected to.

3. Routers construct their topological databases consisting of all the LSPs from the internetwork.

4. The SPF algorithm computes network reachability, determining the shortest path from a router to each network in the link state protocol internetwork. The router uses the Dijkstra algorithm to construct this logical topology of shortest paths as an SPF tree with itself as the root. The SPF tree expresses paths from the router to all destinations.

5. The router lists its best paths and the ports to these destination networks in the routing table.

After the routers dynamically discover the details of their internetwork, they can use the routing table for switching packet traffic.

Link State Topology Changes

Link state algorithms rely on routers having a common view of the network. Whenever a link state topology changes, the routers that first become aware of the change send information to other routers or to a designated router that all routers can use for updates. The above action entails the propagation of common routing information to all routers in the internetwork. In order to achieve network convergence, each router does the following:

- Keeps track of its neighbors such as the neighbor's name, whether the neighbor is up or down, and the cost of the link to the neighbor.
- Constructs an LSP that lists the names and link costs of its neighbor routers. This information includes new neighbors, changes in link costs, and links to neighbors that have gone down.
- Sends out this LSP so that all routers receive it.
- When the router receives an LSP, the router records the LSP in its database so that the router can store the most recently generated LSP from each router.
- Using accumulated LSP data to construct a complete map of the internetwork topology, the router then proceeds from this common starting point to rerun the SPF algorithm and recomputed routes to every network destination.

Each time an LSP causes a change to the link state database, the link state algorithm recalculates the best paths and updates the routing table. Every router then takes the topology change into account as it determines the shortest paths to use for packet switching.

Unlike distance vector algorithms, link state routing algorithms are immediately self-correcting. A link state routing algorithm terminates a loop as soon as the link state database and routing table are updated.

Link State Concerns

No routing protocol is perfect. It is necessary to keep in mind two primary concerns about link state routing: processing and memory requirements and bandwidth requirements.

Running link state routing protocols in most situations requires that routers use more memory and perform more processing. It should be noted that the routers selected must be capable of providing the necessary resources that will be needed to complete routing tasks.

Routers keep track of their neighbors and the networks they reach through other routing nodes. For link state routing, memory must hold information from various link state advertisements, the topology tree, and the routing table.

The processing complexity of computing the SPF is proportional to the number of links in the internetwork times the number of routers in the network. Another cause for consideration is the bandwidth consumed for initial LSP flooding. During the initial discovery process, all routers using link state routing protocols send LSPs to all other routers. This action floods the internetwork as routers make their peak demand and temporarily reduces the bandwidth available for routed traffic that carries user data.

After this initial flooding, link state routing protocols generally require internetwork bandwidth only to send infrequent or event-triggered LSPs that reflect topology changes.

Problems with Link State Updates

The most complex and critical aspect of link state routing is making sure that all routers receive all the LSPs necessary. Routers with different sets of LSPs will calculate routes based on different topological data. Routes then become unreachable as a result of the disagreement among routers about a link. Figure 3.15 provides an example of inconsistent path information.

The following sequence of events occur in Figure 3.15:

- Suppose that network 1 between routers C and D fails. As discussed previously, both routers construct an LSP to reflect this unreachable status.
- Soon afterward, network 1 is restored and another LSP reflecting this next topology change is needed.
- If the original "Network 1, Unreachable" update message from router C uses a slow path, it may arrive at router A after router D's "Network 1, Restored." Now LSP is sent out and received by all routers.
- With unsynchronized LSPs, router A faces a dilemma about which SPF tree to construct: Does router A use paths with or without network 12, which was most recently reported as unreachable?

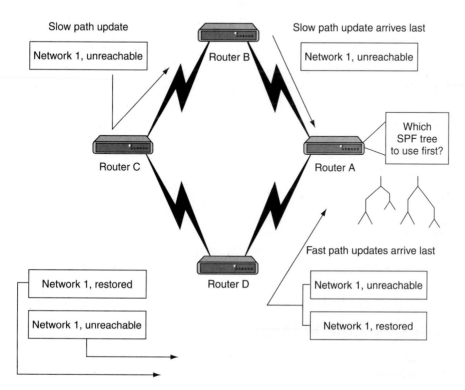

FIGURE 3.15

Unsynchronized updates create inconsistent path decisions. (Reproduced with permission of Cisco Systems, Inc. Copyright © 2001 Cisco Systems, Inc. All rights reserved.)

If LSP distribution to all routers is not synchronized correctly, link state routing can result in invalid routes. Utilizing link state protocols on large internetworks may intensify the problem of faulty LSPs being distributed. For example, if one part of the internetwork restores first with other parts restoring later, as often occurs when a network is in the process of growing, the order for sending and receiving LSPs will vary. This variation may alter and impair convergence of the network. Routers might learn about different versions of the topology before they construct their SPF trees and routing tables. Also, on a large internetwork there is likely to be variation in transmission speed in different parts of the network. Parts of the network that update more quickly can cause problems for parts of the network that update more slowly. Eventually, a partition may split the internetwork into a fast updating part and a slow updating part. The link state complexities must be troubleshooted to restore acceptable connectivity.

A cause-and-effect problem exists for link state routing that is exacerbated on large internetworks. Specifically, correct delivery of LSPs depends on correct routing table entries, but correct routing table entries depend on accurate LSPs. Routers sending out LSPs cannot assume they will be correctly transported because existing routing table entries might not reflect the current topology. With faulty updates, LSPs can multiply as they propagate through the internetwork, unproductively consuming more bandwidth.

Solution: Link State Mechanisms

Link state routing has several techniques for preventing or correcting potential problems arising from resource requirements and LSP distribution:

- The periodic distribution of LSPs can be reduced so that updates occur only after a long, configurable duration. Reducing the rate of periodic updates does not interfere with LSP updates triggered by topology changes.
- LSP updates can go into a multicast group rather than in a flood to all routers. On interconnected LANs, you can use one or more designated routers as the target depository for LSP transmissions. Other routers can use these designated routers as a specialized source of consistent topology data.
- In large networks, you can set up a hierarchy composed of different areas. A router in one area of the hierarchical domain does not need to store and process LSPs from other routers not located in its area.
- For problems of LSP coordination, link state implementations can allow for LSP time stamps, sequence numbers, aging schemes, and other related mechanisms to help avoid inaccurate LSP distribution or uncoordinated updates.

Comparing Distance Vector Routing with Link State Routing

Distance vector routing can be compared with link state routing in several key areas as illustrated.

Distance Vector Routing

- Views network topology from neighbor's perspective.
- Increments metrics as an update; passes from router to router.
- Frequent, periodic updates: slow convergence.
- Passes copies of routing table to neighboring routers.

Link State Routing

- Receives common view of entire network topology.
- Calculates the shortest path to other routers.
- Event-triggered updates: faster convergence.
- Passes link state routing updates to other routers.

The key differences may be summarized as follows:

- Distance vector routing receives all topological data from the perspective it receives from processing the routing table information of its neighbors. Link state routing obtains a wide view of the entire internetwork topology by accumulating all necessary LSPs.
- Distance vector routing determines the best path by adding to the metric value of each route for each router that must be crossed to get to a network as tables are exchanged from router to router. The larger the metric and the farther away a network is, the less suitable the path is. For link state routing, each router works simultaneously to calculate its own shortest path to destinations.
- With most distance vector routing protocols, updates for topology changes come periodically in table updates. Entire tables pass incrementally from router to router, usually resulting in slower convergence than in link state routing. With link state routing protocols, updates are usually triggered by topology changes. Relatively small LSPs passed to all other routers, or a multicast group of routers, usually result in faster convergence.

Designing your network's routing characteristics to meet technical goals—that is, to use the quickest, shortest, cheapest, or most reliable path—is not always the only goal of a network administrator. Business concerns may also influence routing policy. Conformance with the policies, priorities, and partnerships of an organization impacts routing choices. One routing selection might be considered more desirable because it uses the facilities of a partner or avoids the facilities of a competitor. Multivendor support or standards conformity might outweigh technical superiority.

Operational issues such as the concern for network simplicity are also important. For the chosen routing protocol to fit some organizations properly, it must be easy to set up and manage. The chosen routing protocol must handle several routed protocols without requiring several inconsistent and complex configuration templates. Avoiding the risks of unproven technologies may be a factor in designing routing policies and in sustaining one's career with a company.

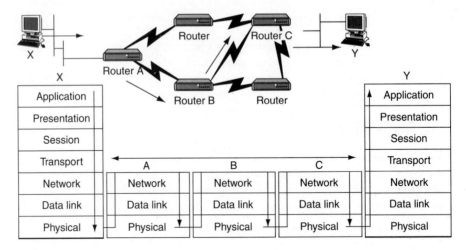

Network Layer Protocol Operations

When a host application sends a packet to a destination on a different network, a data link frame is received on one of a router's interfaces. The router strips off the MAC header and examines the frame's network layer header, such as an IP or IPX header, in order to make a forwarding decision as illustrated in Figure 3.16.

The network layer data are sent to the appropriate network layer process, and the data link layer frame is discarded. The network layer process examines the header to determine the destination network and then references the routing table that associates networks to outgoing interfaces. The packet is again encapsulated in the data link frame for the selected interface and queued for delivery to the next hop in the path. This process occurs each time the packet switches through another router. At the router connected to the network containing the destination host, the packet is again encapsulated in the destination LAN's data link frame type for delivery to the protocol stack on the destination host.

Multiprotocol Routing

**Ships-in-the-
night routing** is
a term that de-
scribes the ability
of protocols to exist
without knowledge
of other protocols.

Routers are capable of supporting multiprotocol routing. This means routers can support multiple independent routing algorithms and maintain associated routing tables for several routed protocols concurrently. This capability allows a router to interleave packets from several routed protocols over the same data links. Each routed and routing protocol has no knowledge of other protocols. This concept is called *ships-in-the-night routing*. For example, in Figure 3.17, router 1 and router 2 handle IP, IPX, AppleTalk, and DECnet traffic. The routing information for each environment is not absorbed by and does not affect the other protocols.

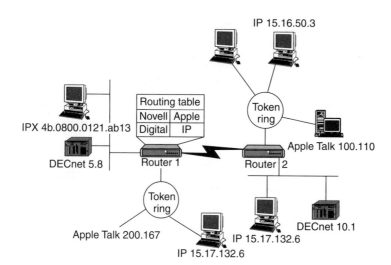

FIGURE 3.17
Routers pass traffic from all routed protocols over the internetwork. (Reproduced with permission of Cisco Systems, Inc. Copyright © 2001 Cisco Systems, Inc. All rights reserved.)

Static versus Dynamic Routing Revisited

As stated previously, static knowledge is administered manually. A network administrator enters it into the router's configuration. The administrator must manually update this static route entry whenever an internetwork topology change requires an update. Static knowledge may be private because by default, static routes are not conveyed to other routers as part of an update process. You may, however, program the router to share this knowledge.

Dynamic knowledge works differently. After the network administrator enters configuration commands to initiate dynamic routing, a routing process updates route knowledge automatically whenever new topology information is received from the internetwork. Changes in dynamic knowledge are exchanged between routers as part of the update process.

Static Route Example

Static routing has several useful applications when it reflects a network administrator's special knowledge about network topology. One such application is security. Dynamic routing tends to reveal everything known about an internetwork to sources outside it. For security reasons, it might be appropriate to conceal parts of an internetwork. Static routing allows an internetwork administrator to specify what is advertised about restricted partitions.

Another application is the accessibility of an internetwork partition by only one path. In such a case, a static route to the partition may be sufficient. This type of network is called a *stub network*. Configuring static routing to a stub network avoids the overhead

A **stub network** has only a single connection to a router.

FIGURE 3.18

Static routing entries can eliminate the need to allow route updates across the WAN link. (Reproduced with permission of Cisco Systems, Inc. Copyright © 2001 Cisco Systems, Inc. All rights reserved.)

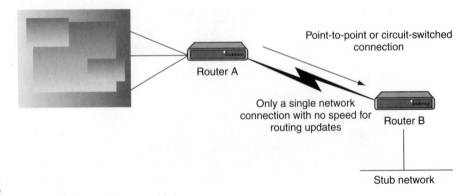

FIGURE 3.19

Dynamic routing enables routers to automatically use backup routes whenever necessary. (Reproduced with permission of Cisco Systems, Inc. Copyright © 2001 Cisco Systems, Inc. All rights reserved.)

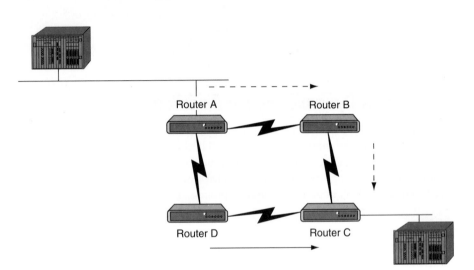

of dynamic routing. For example, in Figure 3.18, router A is configured with a static route to the remote stub network. There is no reason to allow periodic routing updates across the WAN link between router A and router B, as would occur with dynamic routing.

Dynamic Routing Topology Changes

The internetwork illustrated in Figure 3.19 adapts differently to topology changes depending on whether it uses statically or dynamically configured knowledge.

Static knowledge allows the routes to route a packet properly from network to network. In Figure 3.19, router A refers to its routing table and follows the static knowledge to relay the packet to router D. Router D does the same and relays the packet to router C. Router C delivers the packet to the destination host. But what happens if the path between routers A and D fails? Obviously, router A will not be able to relay the packet to router D with a static route. Until router A is manually reconfigured to relay packets by way of router B, communication with the destination host is impossible.

Dynamic knowledge offers more flexibility. According to the routing table generated by router A, a packet can reach its destination over the preferred route through router D. However, a second path to the destination is available by way of router B. When router A recognizes the link to router D is down, it adjusts its routing table making the path through router B the preferred path to the destination. The routers continue sending packets over this link. When the path between routers A and D is restored to service, router A can once again change its routing table to indicate a preference for the counterclockwise path through routers D and C to the destination network.

Dynamic Routing Operations

The success of dynamic routing depends on two basic router functions: maintenance of a routing table and timely distribution of knowledge in the form of routing updates to other routers.

Dynamic routing relies on a routing protocol to disseminate knowledge. A routing protocol defines the set of rules used by a router when it communicates with neighboring routers. A routing protocol describes:

- How updates are conveyed.
- What knowledge is conveyed.
- When to convey knowledge.
- How to locate recipients of the updates.

In Figure 3.20, router 1 uses IP's RIP to pass routing information from its routing table to router 2.

Routing Metrics

When a routing algorithm updates the routing table, its primary objective is to determine the best information to include in the table. Each routing algorithm interprets "best" in its own way. The algorithm generates a number, called the *metric*, for each path through the network. Typically, the smaller the metric, the better the path.

Metrics can be calculated based on a single characteristic of a path or by combining several characteristics. Routing protocols use the following metrics:

- *Hop count:* Refers to the number of routers a packet must go through to reach a destination. The lower the hop count, the better the path. Path length is used to indicate the sum of the hops to a destination.

Metric defines a standard of measurement utilized by router algorithms.

Hop count is a routing metric that refers to the number of routers a packet goes through to reach a destination.

FIGURE 3.20

Routing protocols maintain and distribute routing information. (Reproduced with permission of Cisco Systems, Inc. Copyright © 2001 Cisco Systems, Inc. All rights reserved.)

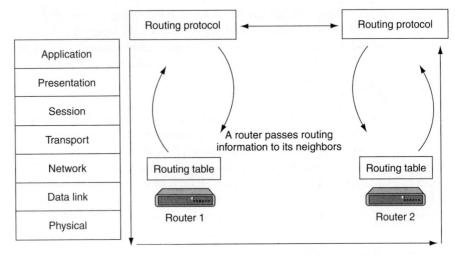

- Ticks: Used in conjunction with Novell IPX RIP to reflect delay. Each tick is one-eighteenth of a second.
- Path cost: The sum of the costs associated with each link to a destination. Costs are assigned automatically or manually to the process of crossing a network. Slower networks typically have a higher cost than faster networks. The lowest cost route is believed to be the fastest route available.

Bandwidth is the rating of a link's maximum throughput.

Delay is the time between the initiation of a transaction by a sender and the first response received.

Reliability reflects the propensity of network links to fail and the speed in which they are repaired.

- *Bandwidth:* The rating of a link's maximum throughput. Routing through links with greater bandwidth does not always provide the best routes. If a high speed link is busy, sending a packet through a slower link might be faster.
- *Delay:* Depends on many factors, including the bandwidth of network links, the length of queues at each router in the path, network congestion on links, and the physical distance to be traveled. Although contingent on a conglomeration of variables that change with internetwork conditions, delay is still a common and useful metric.
- Load: Dynamic factor that can be based on a variety of measures, including CPU use and packets processed per second. Monitoring these parameters on a continual basis can be resource intensive.
- *Reliability:* Reflects the propensity of network links to fail and the speed in which they are repaired. You can take multiple reliability factors into account when assigning reliability ratings. Reliability ratings are usually assigned by the network administrator but can be calculated dynamically by the protocol.
- Maximum transmission unit (MTU): The maximum message length in octets that is acceptable to all links on the path. It would be considered the fastest path to travel along a route that supports larger MTUs and allows maximum packet sizes to be used from end to end.

Although not used directly by the router, expense is another important metric influence. Some organizations might not care about performance as much as operating expenses. Even though the bandwidth is less and the delay is longer, sending packets over leased lines rather than through more expensive public lines may be preferable to some enterprises.

Convergence Time

It is virtually impossible for all routers in a network to detect a topology change simultaneously. In fact, depending on the routing protocol in use, as well as numerous other factors, a considerable time delay may pass before all routers in that network reach a consensus, or agreement, on what the new topology is. This delay is referred to as *convergence time*. Convergence time is not immediate. The only uncertainty is how much time is required for convergence to occur.

Some factors that can exacerbate the time delay inherent in convergence include the following:

- A router's distance (in hops) from the point of change
- The number of routers in the network using dynamic routing protocols
- Bandwidth and traffic load on communication links
- A router's load
- Traffic patterns via the topological change
- The routing protocol used

The effects of some of these factors can be minimized through careful network engineering. A network can be engineered to minimize the load on any given router or communication link. Other factors, such as the number of routers in the network, must be accepted as risks inherent in a network's design. It may be possible to engineer the network such that fewer routers need to converge. By using static routes to interconnect stubs to the network, you reduce the number of routers that must converge. This directly reduces convergence times. Given these factors, it becomes clear that two keys are vital in minimizing convergence time: selecting a routing protocol that can calculate routes efficiently and designing the network properly.

Route Calculation

Convergence is absolutely critical to a network's capability to respond to operational fluctuations. The key factor in convergence is communications among the routers in the network. Routing protocols are responsible for providing this function. Specifically, the routing protocols are designed to enable routers to share information about routes to the various destinations within the network.

One symptom of network instability that may arise is known as route flapping. *Route flapping* is the rapid vacillation between two or more routes. Flapping happens during a topology change. All the routers in the network must converge on a consensus of the new topology. Consequently, the routers begin sharing routing information. In an unstable network, a router (or routers) may be unable to decide on a route to a

Convergence time is the time delay needed for all routers to detect and agree on the change in the network topology.

Route flapping is the rapid vacillation between two or more routes.

destination. During convergence, a router may alter its primary route to any given destination as a result of the last received update. In complex but unstable networks with redundant routes, a router may find itself deciding on a different route to a given destination every time it receives an update. Each update nullifies the previous decision and triggers another update to the other routers. These other routers, in turn, adjust their own routing tables and generate new updates. This vicious cycle is known as flapping. It may be necessary to power down affected routers and slowly develop convergence in the network one router at a time.

Unfortunately, all routing protocols are not created equal. In fact, one of the best ways to assess the suitability of a routing protocol is to evaluate its capabilities to calculate routes and converge relative to other routing protocols. A routing protocol's convergence capability is a function of its capability to calculate routes. The efficacy of a routing protocol's route calculation is based on three factors:

- Whether the protocol calculates and stores multiple routes to each destination
- The manner in which routing updates are initiated
- The metrics used to calculate distances or costs

Storing Multiple Routes

Some routing protocols attempt to improve their operational efficiency by only recording a single route to each known destination. The drawback to this approach is that when a topology change occurs, each router must calculate a new route through the network for the impacted destinations.

Other protocols accept the processing overheads that accompany larger routing table sizes and store multiple routes to each destination. Under normal operating conditions, multiple routes enable the router to balance traffic loads across multiple links. If, or when, a topology change occurs, the routers already have alternative routes to the impacted destinations in their routing tables. Having an alternative route already mapped out does not necessarily accelerate the convergence process. It does, however, enable networks to sustain topology changes more gracefully.

Initiating Updates

Some protocols use the passage of time to initiate routing updates. Others are event driven. Event driven means that these router updates are initiated whenever a topological change is detected. Holding all other variables constant, event-driven updates result in shorter convergence times than timed updates.

Timed Updates

A timed update is a very simple mechanism. Time is decreased in a counter as it elapses. When a specified period of time has elapsed, an update is performed regardless of whether a topological change has occurred. This has two implications:

- Many updates will be performed unnecessarily. This wastes bandwidth and router resources.
- Convergence times can be needlessly inflated if route calculations are driven by the passing of time.

Event-Driven Updates

Event-driven updates are a much more sophisticated means of initiating routing updates. An event-driven update is initiated only when a change in the network's topology has been detected. Because a topology change causes the need for convergence, this is the more efficient approach. The update initiator can be selected by choosing a routing protocol—each protocol implements either one or the other. Therefore, this is the one factor that must be considered when selecting a routing protocol.

Routing Metrics

The routing protocol determines another important mechanism: its metrics. There is a wide disparity in terms of number and type of metrics used.

Quantity of Metrics

Simple routing protocols support as few as one or two routing metrics. More sophisticated protocols can support five or more metrics. It is safe to assume that the more metrics there are, the more varied and specific they are. Therefore, the greater the variety of available metrics, the greater the ability to tailor the network's operation to the particular needs of the network. For example, the simple distance vector protocols use a euphemistic metric: distance. In reality, that distance is not related at all to geographic mileage, much less to the physical cable mileage that separates source and destination machines. Instead, distance usually just counts the number of hops between those two points.

Link state protocols may afford the capability to calculate routes based on four factors:

- Traffic load
- Available bandwidth
- Propagation delay
- The network cost of a connection (which can be an estimate more so than the actual value)

Most of these factors are highly dynamic in a network—they vary by time of day, day of week, and so forth. As these metrics vary, so does the network's performance. Therefore, the intent of dynamic routing metrics is to allow optimal routing decisions to be made using the most current information available.

Static versus Dynamic Metrics

Some metrics are simplistic and static whereas others are highly sophisticated and dynamic. Static metrics usually offer the capability to customize their values when they are configured. After this is done, each value remains a constant until it is manually changed.

Dynamic protocols enable routing decisions to be made based on real-time information about the state of the network. These protocols are supported only by the more sophisticated link state or hybridized routing protocols.

Congestion: An Overview

Congestion occurs when network traffic is in excess of network capacity.

Congestion occurs when the amount of network traffic transmitted on a particular medium exceeds the bandwidth of that medium. Congestion leads to lost data packets and timeouts. The users of the network perceive the network to be slow, but they may not understand the cause of the slowness.

Temporary congestion can be expected in every network. Periodic congestion often occurs because of the "bursty" nature of today's network applications. Bursty traffic transitions from many packets per second to a few packets per second, based on the tasks being performed. Causes of chronic congestion must be identified and remedied. *Chronic congestion* is defined as congestion that occurs regularly or as a result of daily or regular use of the network rather than congestion that occurs only at heavy times. If chronic congestion causes packet loss, the network should be examined to determine the possible cause.

Chronic congestion is congestion that occurs regularly.

WAN links generally experience congestion more often because their available bandwidth is typically smaller than LAN bandwidth. When LAN-to-LAN traffic crosses a WAN link, it is most often the WAN link that is the bottleneck. A network analyzer can help determine the network's current bandwidth utilization. This information is typically presented as a percentage of the total possible bandwidth. An analyzer may indicate the average bandwidth utilization at 10 percent on a 10Base T network. This indicates that the average traffic level is 1 Mbps (10 percent of 10 Mbps). Where does all this traffic that causes congestion come from? Understanding the source of the traffic can enable you to reconfigure the network to control the traffic flow based on the type of traffic and its source and ultimate destination. Traffic sources in IP networks, IPX networks, and other multiprotocol networks are looked at next.

The **File Transfer Protocol** transfers files between networks.

Traffic in an IP Network

An IP network has many sources of data traffic and overhead traffic.

The **Trivial File Transfer Protocol** is a simplified version of FTP.

- User applications: Data traffic is usually generated by user applications. These applications initiate file transfers using the *File Transfer Protocol (FTP)* and *Trivial File Transfer Protocol (TFTP)*. Electronic mail is another common source of data traffic—mail uses *Simple Mail Transfer Protocol (SMTP)*.
- Routing protocol updates: Routing protocols send updates periodically or when routing information changes.
- Encapsulated protocol transport: Noncontiguous (or dissimilar) networks can be joined by encapsulating the network traffic in IP packets and sending that

The **Simple Mail Transfer Protocol** provided electronic mail services.

traffic across the IP network. If the two noncontiguous networks generate large amounts of traffic, slow links in the IP network could become congested.

- Broadcasts: Overhead traffic is generated by a variety of broadcast traffic. This can include Token Ring explorer packets, address resolution protocol (ARP) packets, RIP routing packets, or a workstation's dynamic host configuration protocol (DHCP) requests. Any application that needs to address the entire network will do so with a broadcast packet that is sent to every station within a shared medium, which can utilize a large portion of bandwidth. Depending on the applications and operating systems running on a particular network, the specifics and quantity of broadcasts may differ greatly. A healthy network is a network where broadcasts do not exceed 20 percent of total utilization.

Traffic in an IPX Network

An IPX network has many sources of data traffic and overhead traffic.

- User services: Users accessing NetWare servers and printers, especially over WAN links, can cause network congestion.
- Routing protocol updates: Periodic routing updates can add to network congestion.
- *Service Advertising Protocol (SAP)* announcements: SAP is the protocol used to announce network services such as file, print, and directory services. Much of the traffic in an IPX network is inherent to NetWare support of client services. SAP traffic is overhead but is required for announcements about service availability. SAP traffic can congest the WAN link. This congestion can result in loss of SAPs, which can cause intermittent service interruptions.
- Client/server *keepalive updates:* NetWare clients and servers use keepalive traffic (watchdog traffic) to verify that connections are active.

> The **Service Advertising Protocol** announces available network services.

> **Keepalive updates** are messages that verify connections are active.

Traffic in Other Multiprotocol Networks

A multiprotocol network has several protocol suites active at the same time. All user data traffic for the different protocols is active at the same time with many concurrent data transfers taking place. In addition, the overhead traffic for each protocol requires a portion of the bandwidth.

There is some underlying traffic on the media associated with the lower layers of the OSI reference model. All of the following require some portion of the medium's data carrying capacity:

- ARP to resolve logical-to-physical addressing issues except IPX because the host address is typically the MAC address already.
- Layer 2 keepalives to maintain connectivity.
- Tokens for accessibility on a Token Ring network.
- TTL updates that periodically indicate the remaining lifetime of distributed information.

Managing Traffic Congestion

Network congestion results from too much traffic at one time. To resolve congestion, the traffic must be either reduced or rescheduled. The following methods can be used to manage congestion:

- Filtering user and application traffic
- Filtering broadcast traffic placement of routers and choice of routing protocols
- Adjusting timers on periodic announcements and configuring routing protocols
- Providing static entries in routing tables
- Prioritizing traffic queuing techniques

Filtering User and Application Traffic

You can use standard and extended security access lists to filter user and application traffic. Traffic filters can keep some traffic from reaching critical links.

Filtering Broadcast Traffic

If broadcast traffic levels become excessive, they can be filtered in areas where they are not required. On IP networks, network segments can be managed and a better routing protocol can be selected. On IPX networks, SAP NetWare broadcasts can be filtered and RIP broadcasts can be managed.

Adjusting Timers on Periodic Announcements

Some periodic broadcasts, such as SAP packets, have configurable transmission timers to lengthen the interval between broadcasts. Lengthening the timers reduces the overall traffic load on the link. You can lengthen the time between SAP updates to minimize traffic load by the periodic SAP broadcasts.

Providing Static Entries in Routing Tables

Using static entries (entries that are locally defined, not dynamically learned) in a routing table can eliminate the need to advertise routes dynamically across that link. This technique effectively reduces congestion on WAN links.

Prioritizing Traffic

Traffic that uses a slow link, such as a WAN link, can be prioritized to ensure that critical applications do not time out. You can reorder the application's traffic in a first-come, first-served order to give some traffic preference. You can also control bandwidth on WAN links using queuing techniques, which enable you to specify where traffic is forwarded first through the router.

Hybrid Routing

So far, distance vector and link state routing protocols have been discussed. An emerging third type of protocol combines aspects of both and is referred to as balanced *hybrid routing*. The balanced hybrid routing protocol uses distance vectors with more accurate metrics to determine best paths to destination networks. However, a balanced hybrid differs from most distance vector protocols by using topology changes to trigger routing database updates.

The balanced hybrid routing type converges relatively quickly like link state protocols. However, it differs from these protocols by emphasizing economy in the use of required resources such as bandwidth, memory, and processor overhead. An example of a balanced hybrid protocol is OSI's IS-IS routing protocol.

> ● **Hybrid routing** uses distance vectors with more accurate metrics to determine best routes to destinations.

Basic Routing Processes

Even if a network uses distance vector or link state routing mechanisms, its routers must perform the same basic routing functions. The network layer must relate to and interface with various lower layers. Routers must be capable of seamlessly handling packets encapsulated into different lower level frames without changing the packet's layer 3 addressing.

LAN-to-LAN Routing

Figure 3.21 shows an example of network layer interfacing in LAN-to-LAN routing.

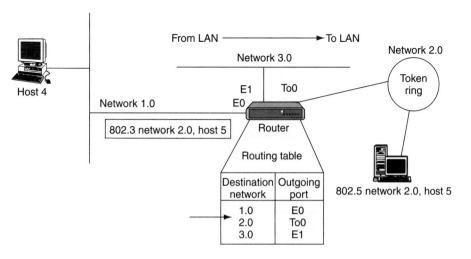

FIGURE 3.21

The router uses the destination network address contained in the packet to look up a route. (Reproduced with permission of Cisco Systems, Inc. Copyright © 2001 Cisco Systems, Inc. All rights reserved.)

In the example, packet traffic from source host 4 on Ethernet network 1.0 needs a path to destination host 5 on network 2.0. The LAN hosts depend on the router and its consistent network addressing to find the best path. When the router checks its router table entries, it discovers that the best path to destination network 2.0 uses outgoing port To0, the interface to a Token Ring LAN.

Although the lower layer framing must change as the router switches packet traffic from the Ethernet on network 1.0 to the Token Ring on network 2.0, the layer 3 addressing for source and destination remains the same. In Figure 3.21, the destination address remains network 2.0, host 5 despite the different lower layer encapsulations.

As an internetwork grows, the path taken by a packet might encounter several relay points and a variety of data link types beyond the LANs. For example, in Figure 3.22, a packet from the top workstation at address 1.3 must traverse three data links to reach the file server at address 2.4, shown on the bottom.

The routed communications follows these basic steps:

1. The workstation sends a packet to the file server by encapsulating the packet in a Token Ring frame addressed to router A at the data link layer and the file server at the network layer.

FIGURE 3.22

Routers maintain the end-to-end address information as they forward the packet. (Reproduced with permission of Cisco systems, Inc. Copyright © Cisco Systems, Inc. All rights reserved.)

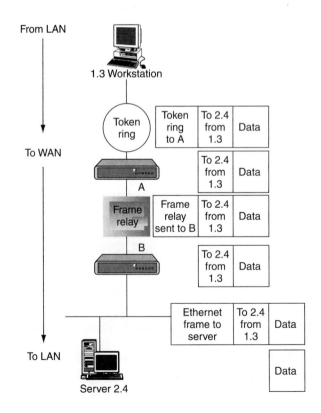

2. When router A receives the frame, it removes the packet from the Token Ring frame, encapsulates it in a frame relay frame, and forwards the frame to router B.
3. Router B removes the packet from the frame relay frame and forwards the packet to the file server in a newly created Ethernet frame.
4. When the file server at 2.4 receives the Ethernet frame, it extracts and passes the packet to the appropriate upper layer process.

The routers enable LAN-to-WAN packet flow by keeping the network layer source and destination addresses constant while encapsulating the packet at the interface to a data link that is appropriate for the next hop along the path.

Snapshot Routing

A fourth type of routing called *snapshot routing* provides an efficient and cost-effective routing solution in *dial on demand routing (DDR)*. Snapshot routing is a routing update mechanism that answers two key concerns:

- It eliminates the need for configuring and maintaining large static tables because dynamic routing protocols can now be used on DDR lines.
- It cuts down the overhead of these routing updates.

ISDN is the primary target for snapshot routing, but other media such as dedicated leased lines can also benefit from its reduction of periodic updates.

What Is Snapshot Routing?

Snapshot routing allows dynamic routing protocols to operate over DDR lines. Usually, routing broadcasts (routes and services) are filtered out on DDR interfaces and static definitions are configured instead.

With snapshot routing, normal updates are sent across the DDR interfaces for a short duration of time called an "active period." After this active period, the routers enter a quiet period during which the routing tables at both ends of the link are frozen. Snapshot routing is, therefore, a "triggering" mechanism that controls routing update exchange in DDR scenarios. Only in the active period do routers exchange dynamic updates. During the quiet period, no updates go through the link (up or down) and the routing information previously collected is kept unchanged (frozen) in the routing tables.

How Do Active Periods Alternate with Quiet Periods?

By default, each DDR connection (triggered by user data) starts an active period during which routers update their tables. Afterward, routers freeze the received information, enter the quiet period, and stop sending routing updates until the next data

Snapshot routing allows dynamic routing protocols to operate over DDR lines.

Dial on demand routing permits routing over ISDN lines.

connection or the expiration of the quiet period. If, during the quiet period, no connection has been triggered, the snapshot mechanism triggers a call to update its information at the end of the quiet period interval. Moreover, snapshot's principle is to let normal routing protocols update their routing tables. The routing protocol timers remain unchanged when snapshot routing is used. Therefore, in order for the received information to represent the network topology reliably, the active period needs to last long enough to let several routing updates come through the link. The active period's minimum duration is five minutes.

As an option, you can prevent snapshot routing from using data connections as active periods. Snapshot routing will then update its routing table regularly after each quiet period.

Snapshot routing can be used with all periodic route update protocols for all supported protocols on DDR lines. These protocols are:

- RIP for IP
- RTMP for AppleTalk
- RIP and SAP for IPX
- RTP for VINES

Why Use Snapshot Routing?

Snapshot routing brings you the following benefits:

- It allows you the use of dynamic routing protocols like RIP, IPX, RIP/SAP, VINES Real Time Transport Protocol (RTP), and Apple's Routing Table Maintenance Protocol (RTMP) in DDR environments. Information about network changes is spread automatically at the other end of DDR links.
- It allows you to get rid of static routes, and as such, it eases network administration. Administering static routes can be cumbersome. With snapshot routing, you have to configure snapshot and the routing protocols, which is not only easier than implementing static routes everywhere, but it also allows a seamless evolution of the network. No need to add, suppress, or modify static routes—with snapshot routing, the network adapts transparently as the topology changes. Snapshot routing's key advantages include easy evolution and manageability.
- Snapshot routing strictly controls the on demand costs. You specify the connection time used for exchanging routing updates.

Finally, as the network expands, snapshot routing will scale easily to the size of your network. Using static routes would not scale in terms of manageability of the network configuration. Snapshot routing, therefore, has better scaling properties than other solutions.

Summary

Routing is the act of moving information across an internetwork from an SA to a DA. Along the way, at least one intermediate node is encountered. Routing is often contrasted with bridging, which might seem to accomplish the same thing to the casual observer. The primary difference is that bridging occurs at layer 2, the data link layer of the OSI model, whereas routing occurs at layer 3, the network layer. The distinction of where each device transfers data provides routing and bridging with different information to use in the process of transmitting data from the SA to the DA. The two functions accomplish their tasks in different ways.

Routing has been covered in computer science literature for more than two decades, but routing did not achieve commercial popularity until the mid-1980s. Practically every company and every office has at least one router at its location. The primary reason for this time lag is that networks in the 1970s were fairly simple, homogeneous environments. Only recently has large-scale internetworking become popular.

Routing involves two basic activities: determining optimal routing paths and transporting information groups called packets through an internetwork. In the context of the routing process, transporting packets through an internetwork can be referred to as switching. Although switching is straightforward, path determination can be complex.

A scalable network is one that can be adjusted without major modification as time and resources require. Many of today's internetworks need to be scalable because they are experiencing phenomenal growth. The growth is primarily a result of the increasing demands for connectivity in business and telecommuting.

Scalable internetworks are described as networks that are experiencing constant growth. These networks must be flexible and expandable. The best managed scalable internetworks are designed to follow a hierarchical model of routing. A hierarchical model simplifies the management of the internetwork and allows for controlled growth without overlooking requirements. In building the network it is recommended to follow a three-layer hierarchical internetworking model. The layers are defined as follows:

- Core Layer: The core layer is the central internetwork for the entire enterprise and may include LAN and WAN backbone. The primary function of the core layer is to provide an optimized and reliable transport structure. Core routers provide services that optimize communication among routers at different sites or in different logical groupings. In addition, core routers provide maximum availability and reliability. Core routers should be able to maintain connectivity when LAN or WAN circuits fail. A fault-tolerant network design ensures that failures do not have a major impact on network connectivity. Core routers must be reliable because they carry information about all the routes in an internetwork. If one of the core routers fails, that router will affect routing on a larger scale than if an access router fails.

- Distribution Layer: The distribution layer represents the campus backbone. The primary function of the distribution layer is to provide access to various parts of the internetwork as well as access to services. Distribution routers control access to resources that are available at the core layer. Distribution routers must make efficient use of bandwidth. In addition, a distribution router must address the QoS needs for different protocols by implementing policy based traffic control to isolate backbone and local environments. Policy based traffic control enables you to prioritize traffic to ensure the best performance for the most time-critical and time-dependent applications. Distribution routers need to select the best path to different locations in order to make efficient use of bandwidth.

- Access Layer: The access layer provides access to corporate resources for a workgroup on a local segment. Access routers control traffic by localizing broadcasts and service requests to the access media. Access routers must also provide connectivity without compromising network integrity. The routers at the access point must be able to detect if a telecommuter dialing in is legitimate and to require minimal authentication steps for the telecommuter to gain access to the corporate LAN. Access routers are placed where security and filtering must be defined. The primary function of the access layer router is to reduce the amount of overhead by keeping unnecessary traffic out of the core of the network.

A hierarchy simplifies tasks such as addressing and device management. Using an addressing scheme that maps to the hierarchy reduces the need to redo the network addresses as a result of growth. Knowing the placement of devices in the hierarchy enables one to program all routers within one layer in a consistent manner because all routers must perform similar tasks.

There are several characteristics that classify internetworks as scalable. The following represent the key characteristics that should be considered when planning an internetwork design strategy:

- Reliable and available: Being reliable and available means being dependable and accessible 24 hours a day, 7 days a week (24×7). Failures need to be isolated and recovery must be invisible to the end user.

- Responsive: Responsiveness includes managing the QoS needs for the various protocols being utilized without affecting a response from the end user to the help desk. The internetwork must be able to respond to latency issues that might be common for one routing architecture but not for another.

- Efficient: Large internetworks must optimize their use of resources, especially bandwidth. Reducing the amount of overhead traffic such as unnecessary broadcasts, service location, and routing updates results in an increase in data throughput without increasing the cost of hardware or the need for additional WAN services.

- Adaptable: Adaptability includes accommodating disparate networks and interconnecting independent network clusters, as well as integrating legacy technologies.
- Accessible but secure: Accessibility and security must include the capability to enable connections into the internetwork using dedicated, dial up, and switched services while maintaining network integrity.

Chapter 4 will discuss the aspects of network management and network management tools such as SNMP V1 and V2, and RMON.

Review Questions

1. Explain routing.
2. What are the two basic routing activities?
3. What are the three layers that make up the network hierarchical model?
4. What are the characteristics of a scalable internetwork?
5. Explain a routing update.
6. What is a link state advertisement?
7. Define a routing domain.
8. What is the difference between an autonomous system and an area?
9. What are the three criteria to differentiate routing algorithms?
10. What are the design goals of routing algorithms?
11. What are some of the types of routing algorithms?
12. What is the difference between a routed protocol and a routing protocol?
13. What is the difference between static and dynamic routing?
14. What is the difference between a flat network and a hierarchical network?
15. What are the differences between distance vector and link state routing algorithms?
16. What is the counting to infinity problem?
17. Explain what two methods can be used to solve the counting to infinity problem.
18. What are hold-down timers?
19. What are triggered updates?
20. Name some distance vector and link state routing protocols.

Summary Questions

1. What are some of the techniques link state protocols use for correcting potential problems arising from LSP distribution?
2. Compare distance vector protocols with link state routing protocols.
3. What is multiprotocol routing?
4. What are the two factors that determine the success of dynamic routing protocols?

5. What is a routing metric?
6. List the typical routing metrics that routers use in their routing tables.
7. Explain convergence time.
8. What is route flapping?
9. Explain a timed update and an event-driven update.
10. What is chronic congestion?

Further Reading

Chappell, Laura. *Advanced Cisco Router Configuration*. Indianapolis: Cisco Press, 1999.

McQuerry, Steve. *Interconnecting Cisco Network Devices*. Indianapolis: Cisco Press, 2000.

Perlman, Radia. *Interconnections Bridges and Routers*. Reading: Addison-Wesley, 1992.

Teare, Diane. *Designing Cisco Networks*. Indianapolis: Cisco Press, 1999.

Network Management Fundamentals

Objectives

- Define network management.
- Explain the requirements necessary for network management.
- Discuss the OSI management functional areas.
- Explain a network management system (NMS) and the elements that comprise a network management system.
- Discuss network management architecture and explain distributed versus centralized network management.
- Explain proxies and how they relate to network management.
- Explain network monitoring and how it impacts network management.
- Discuss performance, fault, and accounting monitoring of a network management system.
- Explain network control and how it impacts a network management system.
- Discuss configuration control and security control and how they impact a network management system.

Key Terms

Introduction

Network management means different things to different people. In some cases, it involves a solitary network consultant monitoring network activity with an outdated protocol analyzer. In other cases, network management involves a distributed database, autopolling of network devices, and high-end workstations generating real-time graphical views of network topology changes and traffic.

For certain, networks have grown in importance and have become critical in the business world. Within a given corporation, the trend is toward larger, more complex networks supporting more applications and more users. As these networks grow in scale, two facts become evident:

- The network along with its associated resources and distributed applications becomes indispensable to the corporation.
- As a result of large, complex networks more things can go wrong, disabling the network or a portion of the network or degrading network performance to an unacceptable level.

A large and complex network cannot be assembled and managed by human effort alone. The complexity of such a network dictates the use of automated network management tools. If the network contains equipment from multiple vendors, the urgency for such tools and the difficulty in supplying them increase, creating additional problems for the network administrator.

As networked installations become larger, more complex, more heterogeneous, and cover more territory between central sites and remote sites, the cost of network management rises. To control costs, standardized tools are needed across a broad spectrum of product types that include end systems, bridges, routers, and telecommunications equipment that can be used in a mixed vendor environment. In response to this need, the *Simple Network Management Protocol (SNMP)* was developed to provide a tool for multivendor, interoperable network management.

SNMP refers to a set of standards for network management including a protocol, a database structure specification, and a set of data objects. SNMP was adopted as the standard for TCP/IP based Internets in 1989 and has enjoyed widespread popularity. In 1991, a supplement to SNMP referred to as *remote network monitoring (RMON)* was issued. RMON extends the capabilities of SNMP to include management of LANs as well as the devices attached to those networks. In 1993, an upgrade to SNMP known as *SNMP version 2 (SNMP v2)* was proposed, and a revision of SNMP v2 was formally adopted in 1996. SNMP v2 adds functional enhancements to SNMP and codifies the use of SNMP on OSI based networks. In addition, in 1996, RMON was extended with a revision known as RMON2.

In general, network management is a service that employs a variety of tools, applications, and devices to assist human network managers in monitoring and maintaining networks.

> The **Simple Network Management Protocol** provides a tool for multivendor, interoperable network management.

> **Remote monitoring** extends the capabilities of SNMP.

> **SNMP version 2** adds functional enhancements to SNMP.

Network Management Requirements

With any network management design it is best to begin by understanding the end user's requirements. The way to accomplish this task is to consider the features that are most important to the user. The following are the principal driving forces for justifying an investment in network management:

1. Controlling corporate strategic assets: Networks and network resources are increasingly vital for most organizations. Without effective control, these resources do not provide the necessary payback that corporate management requires.
2. Controlling complexity: The continued growth in the number of network components, end users, interfaces, protocols, and vendors threatens management with loss of control over what is connected to the network and how network resources are used.
3. Improving service: End users expect the same or improved service as the information and computing resources of the organization grow and become distributed.
4. Balancing various needs: The information and computing resources of an organization must provide a spectrum of end users with various applications at given levels of support with specific requirements in the areas of performance, availability, and security. Someone must be assigned to control resources and to balance the various needs that result.

5. Reducing downtime: As the network resources of a corporation become more important, minimum availability requirements approach 100 percent. In addition to proper fault-tolerant network design, network management has an indispensable role to play in ensuring high availability of its resources.

6. Controlling costs: Resource utilization must be monitored and controlled to enable essential end user needs to be satisfied with reasonable costs.

Background

The early 1980s saw tremendous expansion in the area of network deployment. As companies realized the cost benefits and productivity gains created by network technology, they began to add networks and expand existing networks almost as rapidly as new technologies and products were introduced. By the mid-1980s, certain companies were experiencing growing pains from deploying many different and sometimes incompatible network technologies.

The problems associated with network expansion affect daily network operation management and strategic network growth planning. Each new network technology requires its own set of experts. In the early 1980s, the staffing requirements alone for managing large, heterogeneous networks created a crisis for many corporations. An urgent need arose for automated network management including what is called network capacity planning integrated across diverse environments.

OSI Management Functional Areas

The ISO defined five key functional areas of network management. This functional classification was developed solely for the OSI environment. However, although developed for the OSI environment, these functional areas have gained broad acceptance by vendors of both standardized and proprietary network management systems. The five key functional areas defined for network management by the ISO are:

1. *Fault management:* The facilities that enable the detection, isolation, and correction of abnormal operation of the OSI environment.
2. *Accounting management:* The facilities that enable charges to be established for the use of managed objects and costs to be identified for the use of those managed objects.
3. *Configuration management:* The facilities that exercise control, identify, collect data from, and provide data to managed objects for the purpose of assisting in providing for the continuous operation of interconnection devices.
4. *Performance management:* The facilities needed to evaluate the behavior of managed objects and the effectiveness of communication activities.
5. *Security management:* The facilities that address those aspects of OSI security essential to operate OSI network management correctly and to protect managed objects.

Fault management attempts to ensure that network faults are detected and controlled.

Accounting management facilities collect network data relating to resource usage.

Configuration management facilities detect and determine the state of a network.

Performance management facilities analyze and control network performance.

Security management facilities control access to network resources.

Fault Management

To maintain the proper operation of a complex internetwork, a network manager must take care that systems as a whole and each essential component are in proper operational order. When a fault occurs, it is important for the network manager to perform the following tasks as quickly as possible:

- Determine where the fault is.
- Isolate the rest of the network from the failure so that the rest of the network can continue to function without any downtime or interference from the faulted segment of the network.
- Reconfigure or modify the network in such a way as to minimize the impact of the operation without the failed component.
- Repair or replace the failed component as quickly as possible to restore the network to its initial state before the failure.

Faults are to be distinguished from errors. A fault is an abnormal condition that requires management attention (or action) to repair whereas an error is a single event. A fault is usually indicated by the failure to operate correctly or by excessive errors. If a communication line is cut, no signals can get through from the source to the destination. A crimp in a cable may cause distortions so that there is a persistently high *bit error rate (BER)*. End users expect a fast and reliable problem resolution. Most end users will tolerate an occasional network outage. When these infrequent outages do occur, the end user expects to receive immediate notification and have the problem corrected promptly. To provide this level of fault resolution requires rapid and reliable fault detection as well as diagnostic management functions. The impact and duration of faults can be minimized by the use of fault-tolerant components and alternate communication links to give the network a degree of redundancy. After correcting a fault and restoring a system to its full operational state, the fault management service must ensure that the problem is truly resolved and that no new problems are introduced. This requirement is referred to as tracking and control. Fault management should have a minimal effect on network performance.

●
A **bit error rate**
is the ratio of bits
that contain errors.

Accounting Management

In many corporate networks, individual cost centers or departments, or even individual project accounts, are charged for the use of network services. These charges are internal accounting procedures rather than actual cash transfers from one department to another; nevertheless, they are important to the participating end users. Even if no internal charging is utilized, the network manager needs to track the use of network resources by end user or end user class for the following reasons:

- An end user or group of end users may be abusing their access privileges and burdening the network at the expense of other end users.

- End users may be making inefficient use of the network, especially bandwidth, and the network manager can assist in changing procedures to improve overall network performance.
- The network manager is in a better position to plan for anticipated network growth if end user activity is maintained and tracked.

The network manager needs to specify the kinds of accounting information to be recorded at various network elements, the desired interval between sending the recorded information to higher level management elements, and the algorithms to be used in calculating the charging. Accounting reports should be generated under network management control. To limit access to accounting information, the accounting department must provide the capability to verify the end user's authorization to access and manipulate that information.

Configuration Management

Modern data communication networks are composed of individual components and logical subsystems that can be configured to perform many different applications, for example, the device driver in an operating system. The same device can be configured to act either as a router or as an end system element or both. Once it is decided how a device is to be utilized, the configuration manager can choose the appropriate software, set of attributes, and parameters for that device.

Configuration management is concerned with initializing a network and gracefully shutting down part or all of the network. Configuration management is also concerned with maintaining, adding, and updating the relationships between each component and the status of each individual component during network operation. Start-up and shutdown operations on a network are the specific responsibilities of configuration management. It is often desirable for the operations on certain components to be performed unattended or during off-peak hours.

The network manager needs the capability to identify the components that comprise the network and to define the desired connectivity of these components. Those who regularly configure a network with the same or a similar set of resource parameters need ways to define and modify default parameters and to load these predefined sets of parameters into the specified network components. The network manager must be able to change the connectivity of network components when the end user needs the components to be changed. The reconfiguration of a network is often desired in response to performance evaluation or in support of network upgrade, fault recovery, or security checks.

End users often need or want to be notified of the status of network resources and components. Therefore, end users should be informed when changes in configuration occur. Configuration reports can be generated either on a regular basis or in response to a request for such a report. Before reconfiguration, end users often want to inquire about the upcoming status of network resources and their parameters. Usually, network managers want only authorized end users to manage and control network operation.

Performance Management

Modern data communication networks are composed of various components that must intercommunicate and share data and resources. In some cases, it is critical to the effectiveness of an application that the communication over the network is within certain performance guidelines.

Performance management of a computer network comprises two broad functional categories: monitoring and controlling. The *monitoring function* tracks activities on the network. The *controlling function* enables performance management to make adjustments to improve overall network performance. The following performance issues are of concern:

- What is the capacity level?
- Is there excessive traffic or overutilization?
- Has throughput been reduced to unacceptable limits?
- Are there any network bottlenecks?
- Is response time increasing or decreasing?

> The **monitoring function** tracks activities on the network.

> The **controlling function** enables performance management to improve network performance.

To deal with these concerns, the network manager must focus on some initial set of resources to be monitored for a period of time to assess performance levels. This includes associating appropriate metrics and parameters with relevant network resources as indicators of different levels of performance. One example would be what count of retransmissions on a transport connection is considered to be a performance problem requiring attention? Performance management, therefore, must monitor many resources to provide information in determining acceptable network operating levels. By collecting and analyzing the information and then using the results analysis as feedback to the prescribed set of parameters, the network manager can become more adept at recognizing situations indicative of present or impending performance degradation.

Before using a network for a particular application, an end user may want to know such things as the average and worst response times and the reliability of network services. Performance must be known in sufficient detail to assess specific end user queries. End users expect network services to be managed in a way that consistently affords their applications good response time.

Network managers need performance statistics to help them plan, manage, and maintain large internetworks. Performance statistics can be utilized to recognize potential bottlenecks in the network so that corrective action can be taken before they cause serious problems. For example, routing tables can be changed to balance or redistribute traffic load during times of peak usage or when a bottleneck is identified by a rapidly growing load in one specific area. Over the long term capacity planning based on such performance statistics can indicate the proper decisions to make with regard to increasing bandwidth or an expansion of leased lines in a particular area.

Security Management

Security management is concerned with managing information protection and access control facilities. The access control facilities include generating, distributing, and storing encryption keys. Passwords and other authorization or access control information must be maintained and distributed. Security management is also concerned with monitoring and controlling access to computer networks and to all or part of the network management information obtained from the network elements. Logs are an important security tool along with examination of audit records and security logs, as well as the enabling and disabling of these logging facilities.

Security management provides facilities for the protection of network resources and end user information. Network security facilities should be available for authorized users only. End users want to know that proper and effective security policies are in force and that the security facilities are secure.

Network Management System (NMS)

A network management system is a collection of tools for network monitoring and control.

A *network management system (NMS)* is a collection of tools for network monitoring and control that is integrated in two ways:

- The network management system contains a single operator interface with a powerful but user-friendly set of commands for performing most or all network management tasks.
- The NMS has a minimal amount of separate equipment. Most of the hardware and software required for network management is incorporated into the existing user equipment.

An NMS consists of incremental hardware and software additions implemented among existing network components. The software used in accomplishing the network management tasks resides in the host computers and communication processors such as front-end processors, terminal cluster controllers, bridges, and routers. An NMS is designed to view the entire internetwork as a unified architecture, with addresses and labels assigned to each specific point and specific values of each individual network element and link that is known to the system. The active elements of the network provide regular feedback of status information to the network control center.

A network management entity is a collection of software devoted to the network management task.

Network Management Configuration

Figure 4.1 suggests a typical NMS architecture.

Each network element contains a collection of software devoted to the network management task, referred to in the diagram as a *network management entity (NME)*. Each NME performs the following functions:

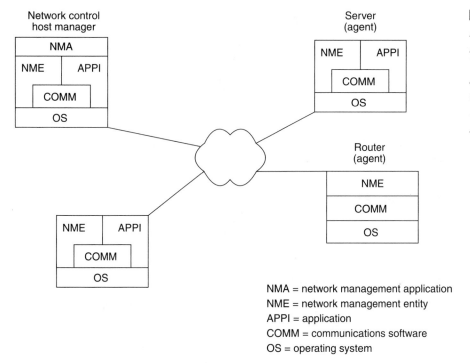

FIGURE 4.1
*Elements of an
NMS. (Stallings,
William; SNMP,
SNMP v2. Copy-
right © 1996,
1999 Pearson
Education.)*

- Collects statistics on communication and network related activities
- Stores statistics locally
- Responds to commands from the network control center, including commands to:
  1. Transmit collected statistics to network control center
  2. Change a parameter such as a hold-down timer
  3. Provide status information such as active links
  4. Generate artificial traffic to perform a test

At least one host in a network is designated as the network control host, or manag-ing agent. In addition to NME software, the network control host includes a collec-tion of software called the *network management application (NMA)*. The NMA includes an operator interface to allow an authorized user to manage the network. The NMA responds to user commands by displaying information and/or by issuing commands to NMEs throughout the network. This communication is carried out us-ing an application level network management protocol that employs the communi-cation architecture in the same manner as any other distributed application.

Other elements in the network that are part of the NMS include an NME that responds to requests from a manager agent. The NME in such managed systems is generally re-ferred to as an agent module or, simply, an agent. Agents are implemented in end systems that support end user applications as well as elements that provide a communications service such as front-end processors, cluster controllers, bridges, and routers.

The **network management application** in-cludes an operator interface to allow an authorized user to manage the network.

Some factors to consider:

1. Because the network management software relies on the host operating system and on the communications architecture, most offerings to date are designed for use on a single vendor's equipment.
2. The network control host communicates with and controls the NMEs in other systems.
3. For maintaining high availability of the network management function, two or more network control hosts are utilized. In normal operations, one of the centers is idle or simply collecting statistics, while the other is used for control. If the primary network control host fails, the backup system can be used.

Network Management Architecture

Most network management architectures use the same basic structure and set of relationships. End stations (managed devices), such as computer systems and other network devices, run software that enables them to send alerts when they recognize problems. A typical problem would be when one or more user determined thresholds are exceeded. Upon receiving these alerts, the management entities are programmed to react by executing one, several, or a group of actions, which may include operator notification, event logging, system shutdown, and automatic attempts at system repair.

Management entities also can poll end stations to check the values of certain variables. Polling can be automatic or user initiated, but agents in the managed devices respond to all polls. Agents are software modules that first compile information about the managed devices in which they reside, then store this information in a management database, and finally provide it proactively or reactively to management entities within NMSs via a network management protocol. Well-known network management protocols include the SNMP and *Common Management Information Protocol (CMIP)*. Management proxies are entities that provide management information on behalf of other entities. Figure 4.2 depicts a typical network management architecture.

●
Common Management Information Protocol is a network management protocol.

Distributed Network Management versus Centralized Network Management

The configuration illustrated in Figure 4.1 depicts a centralized network management strategy with a single network control center and possibly a standby center. *Centralized network management* is the strategy that mainframe vendors and information system executives have traditionally favored. A centralized NMS implies central control. In a mainframe dominated configuration, the centralized network management of resources makes sense because the key resources reside in a computer center and service is provided to remote users. Centralized network management also makes sense to managers

●
Centralized network management implies central control.

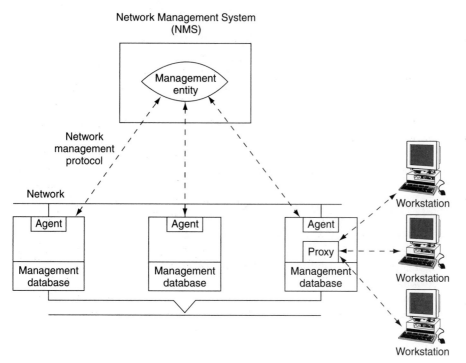

Network Management System
(NMS)

Management
entity

Network
management
protocol

Network

Agent

Management
database

Agent

Management
database

Agent

Proxy

Management
database

Workstation

Workstation

Workstation

FIGURE 4.2

A typical network management architecture maintains many relationships. (Reproduced with permission of Cisco Systems, Inc. Copyright © 2001 Cisco Systems, Inc. All rights reserved.)

responsible for the total information system assets of an organization. A centralized NMS enables the manager to maintain control over the entire configuration, balancing resources against needs and optimizing the overall utilization of resources.

Recently, the centralized computing model has given way to a distributed computing architecture. With applications shifted from data centers to remote departments, network management is also becoming distributed. The same factors come into play with *distributed network management* as with centralized network management:

- The proliferation of lower cost
- High power PCs and workstations
- The widespread use of departmental LANs
- The need for local control
- The optimization of distributed applications

A distributed management system replaces the single network control center with interoperable workstations located on LANs distributed throughout the enterprise. The distributed strategy allows department level managers, who must watch over downsized applications and PC LANs, the tools that are needed to maintain responsive

Distributed network management replaces the single network control center with interoperable workstations distributed throughout the system.

networks, systems, and applications for their local end users. To prevent disorder, a hierarchical architecture is used in addition to the following elements:

- Distributed management stations are given limited access for network monitoring and control, defined by the departmental resources they serve.
- One central workstation, along with a backup, has global access rights and the ability to manage all network resources. The centralized workstation can also interact with less enabled management stations to monitor and control their network operations.

Although maintaining the capacity for central control, the distributed approach offers a number of benefits to corporations:

1. Network management traffic overhead is minimized. Much of the network traffic is confined to the local segment.
2. Distributed management offers greater scalability. Adding additional management capability is a matter of deploying another inexpensive workstation at the desired location.
3. The use of multiple networked stations eliminates the single point of failure that exists with centralized network management systems.

Figure 4.3 illustrates the basic structure used for most distributed network management systems presently available in today's market. The management clients are found closest to the end users. The management clients allow the user access to management services and information and provide an easy-to-use *graphical user interface (GUI)*. Depending on access privileges, a client workstation may access one or more management servers. The management servers are the heart of the distributed system. Each management server supports a set of management applications and a *Management Information Base (MIB)*. The management servers also store common management data models and route management information to applications and clients. Those devices to be managed that share the same network management protocol as the management servers contain agent software and are managed directly by one or more management servers. For other devices, management servers can reach the resources only through a vendor-specific element manager, or proxy.

The flexibility and scalability of the distributed management system are evident as illustrated in Figure 4.3. As additional resources are added to the configuration, each is equipped with agent software or linked to a proxy. In a centralized system, this growth might eventually overwhelm a central station. In a distributed system, additional management servers and client workstations can be added to cope with the extra resources. The growth of the overall configuration will occur in a structured way such as adding an additional LAN with a number of attached PCs. The growth of the management system mirrors this underlying resource growth, with servers and clients added where the new resources are located.

- **A graphical user interface** uses pictorial as well as textual representations of data.

- A **Management Information Base** is a database of network management information.

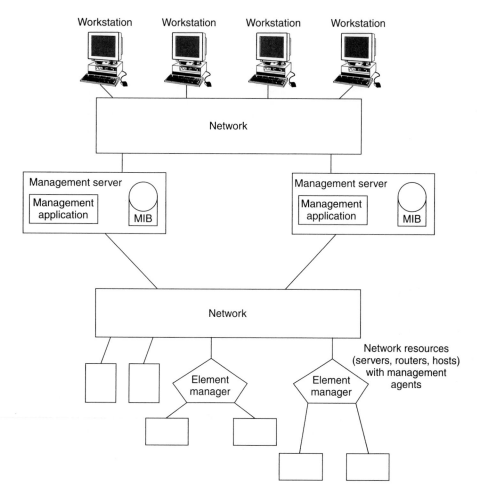

FIGURE 4.3
Typical distributed management system architecture. (Reproduced with permission of Cisco Systems, Inc. Copyright © 2001 Cisco Systems, Inc. All rights reserved.)

Proxies

The configuration of Figure 4.1 shows that each component of management interest includes an NME, with common network management software across all managers and agents. In an actual configuration it may not be practical or possible to have such an arrangement. The actual configuration could include any of the following different types of systems: older systems that may not support the current network management standards, small systems that would be unduly burdened by a full-blown NME implementation, or components such as modems and multiplexers that do not support additional software.

To handle these special cases, it is common to have one of the agents in the configuration serve as a *proxy* for one or more other elements. When an agent performs in a

A **proxy** is an entity that essentially stands in for another entity.

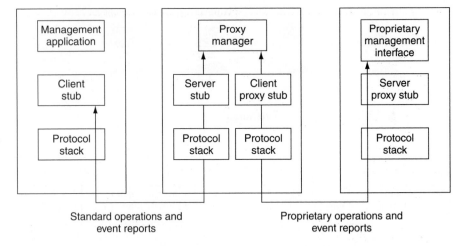

proxy role, the agent acts on behalf of one or more other elements. A network manager who wishes to obtain information or control the element communicates with the proxy agent. The proxy agent then translates the manager's request into a form for the target system and uses an appropriate network management protocol to communicate with the target system. Responses from the target system back to the proxy are translated and passed onto the manager.

Figure 4.4 illustrates a structured architecture that enables a management application to manage a proprietary resource through standard operations and event reports that are translated by the proxy system into proprietary operations and event reports. A proxy system that is engineered in this manner is using a *remote procedure call (RPC)* mechanism. The RPC mechanism is frequently associated with distributed systems software and provides a flexible and easy-to-use facility for supporting the proxy function.

The **remote procedure call** provides a flexible facility for supporting the proxy function.

Network Monitoring

The *network monitoring* portion of network management is primarily concerned with observing and analyzing the status and behavior of the end systems, intermediate systems, and subnetworks that create the configuration that is to be monitored and managed.

Network monitoring consists of three major design areas.

Network monitoring consists of three major design areas:

1. Access to monitored information: How to define monitoring information and how to get the monitored information from a resource to a manager.
2. Design of monitoring mechanisms: What is the best way to retrieve information from resources.
3. Application of monitored information: How the monitored information is best utilized in the various management functional areas.

Network monitoring encompasses the five functional areas: Fault, Accounting, Security, Configuration, and Performance Management. Before considering the design of a network monitoring system, it is best to discuss what type of information is of interest to a network monitor. Then one can consider the alternatives for configuring the network monitoring function.

Network Monitoring Information

Network monitoring information can be classified in three types:

- Static information: *Static information* is information that characterizes the current configuration and the elements in the current configuration. For example, these elements may include the number and identification of ports on a router. This type of information will change infrequently.
- Dynamic information: *Dynamic information* is related to events in the network in real-time operation. Examples of dynamic information would be a change of a state in a protocol machine or the transmission of a packet on the network.
- Statistical information: *Statistical information* can be derived from dynamic information. An example would be the average number of packets transmitted per unit of time by an ES.

A static database has two major components: a configuration database containing basic information about the computer and networking elements, and a sensor database containing information about sensors incorporated into the elements, which are used to retrieve real-time readings.

The dynamic database primarily is concerned with collecting information about the state of the various network elements and events that are detected by the sensors. The statistical database includes useful performance measures. Figure 4.5 illustrates the relationships among the three database components.

The nature of the monitored information has implications concerning the information retrieval and storage for purposes of monitoring. Static information is generated by the element that is being monitored. Thus, as an example, a router maintains its own configuration information. The router's information can be made available directly to a monitor if the element has the appropriate agent software. The information from the router can also be made available to a proxy that will make it available to a monitor.

Dynamic information is also collected and stored by the network element responsible for the underlying events. However, if a system is attached to a LAN, then much of its event activity can be observed by another system on the LAN. Remote monitoring refers to a device on a LAN that can observe and gather information concerning all of the traffic being transmitted on the LAN. For example, the total number of packets transmitted by an element or device on a LAN could be recorded by the device itself or by a remote monitor that is listening on the same LAN. Some dynamic information may be generated only by the element itself such as the current number of network level connections.

Static information characterizes the current configuration and the elements in the current configuration.

Dynamic information is related to events in the network in real-time operation.

Statistical information can be derived from dynamic information.

FIGURE 4.5

Organization of a Management Information Base. (Reproduced with permission of Cisco Systems, Inc. Copyright © 2001 Cisco Systems, Inc. All rights reserved.)

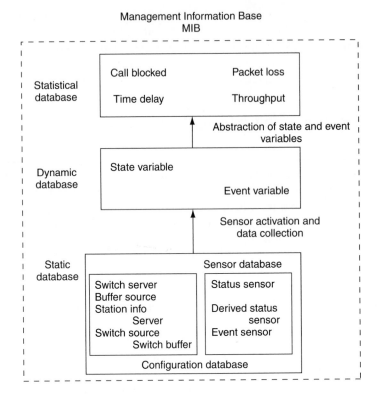

Statistical information may be generated by any system that has access to the underlying dynamic information. The statistical information could be generated toward the network monitor itself. This type of activity would require that all of the raw data must be transmitted to the monitor where the data would be analyzed and summarized. If the monitor does not need access to all of the raw data, then monitor processing time and network capacity could be saved if the system that contains the dynamic data does the summarization and sends the results to the monitor.

Network Monitoring Configurations

Figure 4.6 illustrates the architecture for network monitoring in functional terminology. Part (a) of Figure 4.6 illustrates the four major components of a network monitoring system described as follows:

- Monitoring application: The monitoring application component includes the functions of network monitoring that are visible to the user such as performance monitoring, fault monitoring, and accounting monitoring.
- *Manager function:* The manager module at the network monitor performs the basic monitoring function of retrieving information from other elements attached to the configuration.

The **manager function** performs the basic monitoring function of retrieving information.

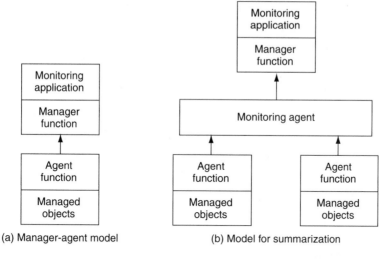

- *Agent function:* The agent module gathers and records management information for one or more network elements and communicates the information back to the monitor.
- *Managed objects:* This is the management information that represents resources and their activities.
- *Monitoring agent:* This additional module generates summaries and statistical analysis of management information. If remote from the manager, this module acts as an agent and communicates the summarization information to the manager.

These functional modules may be configured in any number of ways. The station that serves as the host for the monitoring application is itself a network element and is subject to monitoring. The network monitor generally includes agent software and a set of managed objects as illustrated in Figure 4.7 (a).

It is vital to monitor the status and behavior of the network monitor to ensure that the monitor continues to perform its function and to assess the load on itself and on the network. One key requirement is that the network management protocol be instrumented to record the amount of network management traffic into and out of the network monitor.

Figure 4.7 (b) illustrates the most common configuration for monitoring other network elements. This configuration requires that the manager and agent systems share the same network management protocol, MIB syntax, and semantics. Figure 4.7 (c) illustrates a network monitoring system that may also include one or more agents that monitor traffic on a network. These are often referred to as external or remote monitors. Finally, for network elements that do not share a common network management protocol with the network monitor a proxy agent is needed as illustrated in Figure 4.7 (d).

The **agent function** gathers and records management information for one or more network elements.

Managed objects represent resources and their activities.

The **monitoring agent** generates summaries and statistical analysis of management information.

FIGURE 4.7

Network monitoring configurations. (Reproduced with permission of Cisco Systems Inc. Copyright © 2001 Cisco Systems, Inc. All rights reserved.)

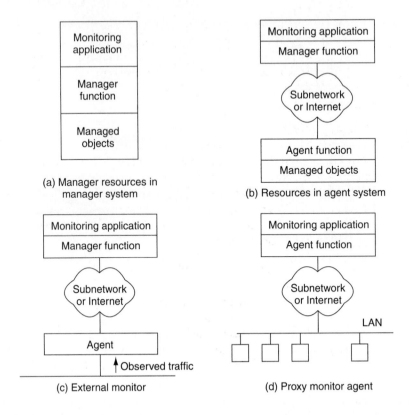

(a) Manager resources in manager system

(b) Resources in agent system

(c) External monitor

(d) Proxy monitor agent

Polling and Event Reporting

Information that is useful for network monitoring is retrieved and stored by agents and made available to one or more manager systems. Two techniques are used to make the agent information available to the manager: polling and event reporting.

Polling is a request and response interaction between a manager and an agent.

Polling is a request and response interaction between a manager and an agent. The manager can query any agent for which the manager has authorization and request the values of various information elements; the agent responds with information from its MIB. The request may be specific—listening to one or more named variables. A request may also be in the nature of a search—asking the agent to report information matching certain criteria or to supply the manager with information about the structure of the agent's MIB. A manager system may use polling to learn about the configuration that it is managing or to obtain an update of conditions periodically for a particular moment of an activity that may create an alarm condition. A manager system can use polling to investigate an area in detail after being alerted to a problem. Polling is also used to generate a report on behalf of a user and to respond to specific user queries.

Event reporting is initiated with the agent while the manager is in the role of a listener.

Event reporting is initiated with the agent while the manager is in the role of a listener waiting for incoming information. An agent may generate a report periodically to give the manager the current status of a particular event. The reporting period may be preconfigured or set by the manager. An agent may also generate a report when a significant event such as a change of state or an unusual event such as a fault occurs. Event reporting is useful for detecting problems as soon as they occur or real time. Event reporting is more efficient than polling for monitoring objects whose states or values change relatively infrequently.

Polling and event reporting are useful, and a network monitoring system will employ both methods. The relative emphasis placed on the two methods varies greatly in different systems. Telecommunication management systems have placed a high reliance on event reporting. The SNMP approach, in contrast, puts little reliance on event reporting. OSI systems management tends to fall somewhere between these extremes. Both SNMP and OSI systems management, as well as proprietary schemes, allow the user considerable latitude in determining the relative emphasis on the two approaches. The choice of emphasis depends on a number of factors:

- The amount of network traffic generated by each method
- Robustness in critical situations
- The time delay in notifying the network manager
- The amount of processing in managed devices
- The trade-offs of reliable versus unreliable transfer
- The network monitoring applications being supported
- The contingencies required in case a notifying device fails before sending a report

Performance Monitoring

An absolute prerequisite for the management of any communications network is the ability to measure the network's performance or performance management. A system or activity cannot be managed and controlled properly unless its performance can be monitored. The difficulty facing the network manager is the selection and use of appropriate indicators that can measure the network's performance in an efficient manner and give the necessary information needed to solve a fault. Among the many problems that may arise in selecting the proper performance indicators are the following:

- There are too many indicators in use.
- The meanings of most indicators are not yet clearly understood.
- Some indicators are introduced and supported by certain manufacturers only.
- Most indicators are not suitable for comparison with each other.
- Frequently, the indicators are accurately measured but incorrectly interpreted.
- In many cases, the calculations of indicators take too much time, and the final results can hardly be used for controlling the environment.

There are two types of categories in which indicators can be grouped: service oriented measures and efficiency oriented measures. The principal means of judging that a network is meeting its performance requirements is that specified service levels are maintained to the satisfaction of the users. Consequently, service oriented indicators are of the highest priority. The manager is also concerned with meeting these requirements at minimum cost, hence the need for efficiency oriented measures.

Availability is the percentage of time that a network system, component, or application is available for a user.

Availability

Availability can be expressed as the percentage of time that a network system, component, or application is available for a user. Depending on the application, high availability can be significant. For example, in an airline reservation system, a one-minute outage may cost a huge amount of money. In the banking industry, a one-hour outage may cost millions of dollars.

The **mean time between failure** is the length of time a user may expect a system to work before a fault occurs.

Availability is based on the reliability of each individual component that makes up the network. Reliability is the probability that a component will perform its specified function for a specified time period under specified conditions. Component failure is usually expressed by the *mean time between failure (MTBF)*. The availability A, may be expressed as

$$A = \frac{MTBF}{MTBF \times MTTR}$$

The **mean time to repair** is the average time required to return a failed system to service.

where *MTTR* is defined as the *mean time to repair* a system following a system failure. The availability of a system depends on the availability of its individual component parts plus the system organization. Some components may be redundant, such that the loss of a component can result in reduced capability but the system may still be able to operate. Figure 4.8 illustrates two simple configurations. In

FIGURE 4.8

Availability of serial and parallel connections. (Reproduced with permission of Cisco Systems Inc. Copyright © 2002 Cisco Systems, Inc. All rights reserved.)

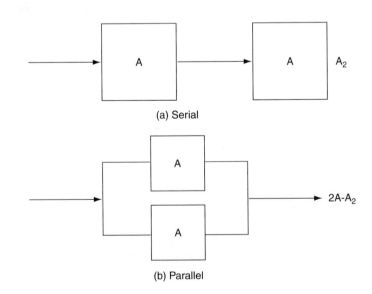

(a) Serial

(b) Parallel

part (a), two components are connected in a serial system. Both components must function properly for the system to be considered available. These devices could be two routers at opposite ends of a communication link. When two network components are connected serially, the availability of the combinations is A^2 if the availability of each component is A. If the availability of each router is 0.98, the availability of the link with two routers is $0.98 \times 0.98 = 0.96$. Figure 4.8 (b) illustrates two devices operating in parallel. This system could be two switches connecting to a core router; if one link fails the other is automatically utilized as a backup link. In this case, the dual link is unavailable only if both individual links are unavailable. If the availability of each link is 0.98, then the probability that one of them is unavailable is $1 - 0.98 = 0.02$. The probability that they are both unavailable is $0.02 \times 0.02 = 0.04$. Thus, the availability of the combined unit is $1 - 0.004 = 0.996$.

The availability analysis can become quite complex as the configurations become more complicated. Not only is the availability of the components taken into account but also the expected load on the system is factored into the equation. Consider a dual link system as illustrated in Figure 4.8 (b). The two links are being used to connect a multiplexer to a host system. Nonpeak periods account for 40 percent of requests for service, and during those periods either link can handle the traffic. During peak periods both links are required to handle the full load, but one link can handle 80 percent of the peak load. The availability of the system can be expressed as:

$$A_{1f} = \text{(capability when one link is up)} \times \Pr[1 \text{ link up}] + \text{(capability when two links are up)} \times \Pr[2 \text{ links up}]$$

where Pr means the probability of.

The probability that both links are up is A^2, where A is the availability of either link. The probability that exactly one multiplexer is up is $A(1- A) + (1- A) A = 2A - 2A^2$. Inserting a value for A of 0.9, then $\Pr[1 \text{ link up}] = 0.9 \times 0.9 = 0.81$, and $\Pr[2 \text{ links up}] = 0.18$. Recalling that one link is sufficient in a parallel system for nonpeak loads, the result of availability is:

$$A_{1f}\text{(nonpeak)} = (1.0)(0.18) + (1.0)(0.81) = 0.99$$

And, for peak periods,

$$A_{tf}\text{(peak)} = (0.8)(0.18) + (1.0)(0.81) = 0.954$$

Overall availability of the system is:

$$A_{1f} = 0.6 \times A_f\text{(peak)} + 0.4 \times A_f\text{(nonpeak)} = 0.9684$$

Therefore, on average the system can handle requests for service approximately 97 percent of the time whether peak or nonpeak hours.

Response Time

**Response time
is the time it takes
a system to react
to a given input.**

Response time is the time it takes for a system to react to a given input. In an interactive transaction, response time may be defined as the time between a user's last keystroke and the beginning result that is displayed by the computer. Different types of applications may require a slightly different definition for response time. In general, response time is the time it takes for a system to respond to a request to perform a particular task. The ideal requirement for the time period for response time should be a short interval. However, shorter response times invariably impose a greater cost. The cost factor can involve two sources:

1. Central Processing Unit (CPU) processing power: The faster the CPU, the shorter the response time. Increased processing power means increased cost.
2. Competing requirements from other service requests: As more requests are demanded from the CPU, providing rapid response time to some processes may penalize other processes that need CPU time.

The value of a given level of response time must be assessed versus the cost of achieving that response time. Design difficulties are faced when a response time of less than one second is required. There are six general ranges of response time:

1. Greater than 15 seconds: This rules out conversational interaction. For certain types of applications, certain types of users may be content to sit at a terminal for more than 15 seconds waiting for the answer to a single simple inquiry. For a busy person, waiting for more than 15 seconds seems intolerable. If such delays occur, the system should be designed so that the user can turn to other activities and request response time at some later time.
2. Greater than 4 seconds: These response time ranges are generally too long for a conversation requiring the operator to retain information in short-term memory for the computer operator. Such delays would be inhibiting in problem-solving activity and frustrating in data input activity or order entry activity. After a major closure, delays from 4 seconds to 15 seconds may be tolerable.
3. From 2 to 4 seconds: A delay longer than 2 seconds may be inhibiting to terminal operations demanding a high level of concentration. A wait of 2 to 4 seconds at a terminal can seem surprisingly long when users are absorbed and committed to complete what they are doing. A delay in this range may be acceptable after a minor closure has occurred.
4. Less than 2 seconds: When a terminal user has to remember information throughout several responses, the response time must be short. The more detailed information remembered, the greater the need for responses of less than 2 seconds. For elaborate terminal activities, 2 seconds represents an important response time limit.
5. Subsecond response time: Certain types of thought intensive work, especially with graphics applications, require short response times to maintain the user's interest and attention for long periods of time.

6. Decisecond response time: A response to pressing a key and seeing the character displayed on the screen or clicking a screen object with a mouse needs to be almost instantaneous—less than 0.1 second after the action takes place. Interaction with a mouse requires extremely fast interaction if the designer is to avoid the use of command syntax.

When a computer and a user interact at a pace that ensures that neither has to wait on the other, productivity increases significantly, the cost of work that is being completed on the computer therefore drops, and quality tends to improve. Years ago, it was commonly accepted that a relatively slow response time—up to two seconds—was acceptable for most interactive applications because the user was thinking about the next task that had to be performed. Today, productivity increases as rapid response times are achieved. The results reported on response time are based on an analysis of on-line transactions. A transaction consists of a user command from a terminal and the system's reply. The fundamental unit of work for on-line system users can be divided into two time sequences:

1. User response time: The time span between the moment a user receives a complete reply to one command and enters the next command, commonly referred to as think time.
2. System response time: The time span between the moment the user enters a command and the moment a complete response is displayed on the terminal.

An example of the effect of reduced response time would be a computer aided design (CAD) software program whereby each transaction by a user in a graphics intensive program alters in some way the graphic image being displayed on the screen. The rate of transactions increases as system response time falls and rises dramatically once system response time falls below one second. As the system response time falls, so does the user response time. This has to do with the effects of short-term memory and human attention span.

To measure response time, a number of elements need to be examined. Although it may be possible to directly measure the total response time in a given network environment, this factor alone is of little use in correcting problems or planning for the growth of the network. For these purposes, a detailed breakdown of response time is needed to identify bottlenecks and potential bottlenecks.

Figure 4.9 illustrates a typical networking situation and indicates the seven elements of response time common to most interactive applications. Each element is one step in the overall path an inquiry takes through a communications configuration, and each element contributes a portion of the overall response time.

- Inbound terminal delay: The delay in receiving an inquiry from the terminal to the communications line. There is no noticeable delay at the terminal itself, so the delay is directly dependent on the transmission rate from the terminal to controller. For example, if the data rate on the line is 9,600 bps = 300 characters per

FIGURE 4.9

Elements of response time. (Reproduced with permission of Cisco Systems, Inc. Copyright © 2001 Cisco Systems, Inc. All rights reserved.)

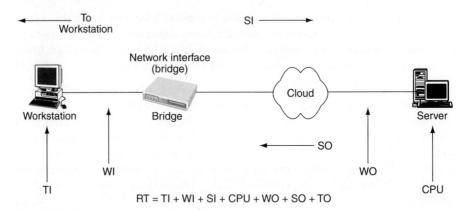

RT = TI + WI + SI + CPU + WO + SO + TO

RT = response time
TI = inbound terminal delay
WI = inbound queuing time
SI = inbound service time
CPU = CPU processor delay
WO = outbound queuing time
SO = outbound service time
TO = outbound terminal delay

second, the delay is 1/300 = 3.33 milliseconds per character. If the average message length is 100 characters, the delay will be 0.33 second.

- Inbound queuing time: The time required for processing by the controller device. The controller is dealing with input from a number of terminals as well as input from the network to be delivered to the terminals. An arriving message will be placed in a buffer to be served in turn. The busier the controller, the longer the delay for processing.
- Inbound service time: The time to transmit the communications link network or other communications facility from the controller to the host's front-end processor. This element is composed of a number of elements based on the structure of the communications facility. If the facility is a public packet-switched network, it must be treated as a single network element. If the facility is a private network such as a WAN or LAN, leased line, or other user configured facility, then a breakdown of this element will be needed for network control and planning.
- CPU processor delay: The time the front-end processor, the host processor, and the disk drives at the computer center spend preparing a reply to an inquiry. This element is usually outside the control of the network manager.
- Outbound queuing time: The time a reply spends at a port in the front-end processor waiting to be dispatched to the network or communications line. As with the controller, the front-end processor will have a queue of replies to be serviced, and the delay is greater as the number of replies waiting to be transmitted increases.

- Outbound service time: The time to transmit the communications facility from the host's front-end processor to the controller.
- Outbound terminal delay: The delay at the terminal itself. This is primarily a result of line speed.

Response time is relatively easy to measure and is one of the most important classes of information needed for network management.

Accuracy

Accurate transmission of data between user and host or between two hosts is essential for any network. Because of the built-in error correction mechanisms in protocols, such as the data link and transport protocols, accuracy is generally not a user concern. *Accuracy* is useful in monitoring the rate of errors that must be corrected. Accuracy may give an indication of an intermittently faulty line or the existence of a source of noise or interference that should be corrected.

Accuracy monitors the rate of errors that must be corrected.

Throughput

Throughput is an application oriented measure. Examples of throughput include:

- The number of transactions of a given type for a certain period of time
- The number of customer sessions for a given application during a certain period of time
- The number of calls for a circuit-switched environment

It is useful to track these measures over a period of time to get a feel for the demands that can be projected on bandwidth and the likely performance trouble spots.

Throughput is the rate of information arriving at a particular point in a network system.

Utilization

Utilization is more fine grained than throughput. *Utilization* determines the percentage of time that a resource is in use over a given period of time. The most important use of utilization is to search for potential bottlenecks and areas of congestion. This is important because response time usually increases exponentially as the utilization of a resource increases. As a result of this exponential behavior, congestion can quickly get out of control if it is not spotted early and corrected promptly.

Looking at a profile of resources that are in use at any given time period and that are idle, the network analyst may be able to find overcommitted or underutilized resources and consequently adjust the network accordingly. By redirecting traffic or changing the relative data rates of the various links, a closer balance can be achieved between planned load and actual load, thereby reducing the total required capacity and using resources more efficiently.

Utilization determines the percentage of time that a resource is in use over a given period of time.

Performance Monitoring Function

Performance monitoring encompasses three components: performance measurement, which is the actual gathering of statistics regarding network traffic and timing; performance analysis, which consists of software for reducing and presenting the data; and synthetic traffic generation, which permits the network to be observed under a controlled load environment.

Performance measurement is often accomplished through agent modules within the devices that are attached to the network such as hosts, routers, and bridges. These agents are in a position to observe the amount of traffic into and out of a node, the number of connections, and the traffic per connection, as well as other measures that provide a detailed picture of the behavior of that node. This measurement has the disadvantage of having to process resources within the node. Measurements of data packet size, packet type, packet interarrival time, channel acquisition delay, communication delay, and collision and transmission count can be used to answer a number of questions:

- Is traffic evenly distributed among network users or are there source destination pairs with unusually heavy traffic?
- What is the percentage of each type of packet? Are some packet types of unusually high frequency, indicating an error or an inefficient protocol?
- What is the distribution of data packet sizes?
- Are collisions a factor in receiving transmitted packets, indicating possible faulty hardware or protocols?

These areas are of interest to the network manager. Other questions can be answered by users concerning response time and throughput and determining how much growth the network can absorb before certain performance thresholds are crossed.

Fault Monitoring

The objective of fault monitoring is to identify faults as quickly as possible after they occur and to identify the cause of the fault so that remedial action may be taken. In a complex environment, locating and diagnosing faults can be difficult. There are specific problems associated with fault observation as indicated by the following:

- Unobservable faults: Certain faults are inherently unobservable locally. For example, the existence of a deadlock between cooperating distributed processes may not be observable locally. Other faults may not be observable because vendor equipment is not instrumented to record the occurrence of a fault.
- Partially observable faults: A node failure may be observable but the observation may be insufficient to pinpoint the problem. A node may not be responding because of the failure of some low level protocol in an attached device.
- Uncertainty in observation: Even when detailed observations of faults are possible, there may be uncertainty and even inconsistencies associated with the ob-

servations. A lack of response time from a remote device may mean that the device is locked up, the network is partitioned, congestion caused the response to be delayed, or the local timer is faulty.

Once faults are observed, it is necessary to isolate the fault to a particular component. The following is a list of problems that can arise:

- Multiple potential causes: When multiple technologies are involved the potential points of failure and the types of failure increase. This makes it harder to locate the source of a fault.
- Too many related observations: A single failure can affect many active communication paths. A failure in one layer of the communications architecture can cause degradations or failures in all dependent higher layers. A failure in a T1 line will be detected in the routers as a link failure and in the workstations as transport and application failures. Because a single failure may generate many secondary failures, the proliferation of fault monitoring data that can be generated in this manner can obscure the single underlying problem.
- Interference between diagnosis and local recovery procedures: Local recovery procedures may destroy important evidence concerning the nature of the fault, thereby disabling diagnosis.
- Absence of automated testing tools: Testing to isolate faults is difficult and costly to administer.

Fault Monitoring Functions

The first requirement of a fault monitoring system is that it must be able to detect and report faults. At a minimum, a fault monitoring agent will maintain a log of significant errors and events. These logs, or summaries, are available to authorized manager systems. A system that operates primarily by polling would rely on these logs. The fault monitoring agent has the capability to report errors independently to one or more managers. To avoid overloading the network, the criteria for issuing a fault report must be reasonably tight.

In addition to reporting known, existing faults, a good fault monitoring system will be able to anticipate faults. Generally, this involves setting up thresholds and issuing a report when a monitored variable crosses a threshold. For example, if the fraction of transmitted packets that suffers an error exceeds a certain value, this may indicate that a problem is developing along the communications path. If the threshold is set low enough, the network manager may be alerted in time to take action that can avoid a major failure in the system.

The fault monitoring system should also assist in isolating and diagnosing the fault. Examples of tests that a fault monitoring system should have at its command include:

- Connectivity test
- Data integrity test

- Protocol integrity test
- Data saturation test
- Connection saturation test
- Response time test
- Loopback test
- Function test
- Diagnostic test

An effective user interface is required more for fault monitoring than for other areas of network monitoring. In complex situations, faults will be isolated, diagnosed, and ultimately corrected only by the cooperative effort of human user and monitor software.

Accounting Monitoring

Accounting monitoring is primarily a matter of tracking user usage of network resources. The requirements for this function vary widely. In some environments, accounting may be quite general. An internal accounting system may be used only to assess the overall usage of resources and to determine what proportion of the cost of each shared resource should be allotted to each department. In other cases, particularly for systems that offer a public service or systems with only internal users, it is required that usage be broken down by account, by project, or even by individual user for the purposes of billing. The information gathered by the monitor system in this case must be more detailed and more accurate than that required for a general system.

Examples of resources that may be subject to accounting include the following:

- Communication facilities: LANs, WANs, leased lines, dial up lines, and PBX systems
- Computer hardware: Workstations and servers
- Software and systems: Applications and utility software in servers, a data center, and end user sites

For any given type of resource, accounting data are collected based on the requirements of the organization. The following communications related accounting data might be gathered and maintained on each user:

- User identification: Provided by the originator of a transaction or a service request
- Receiver: Identifies the network component to which a connection is made or attempted
- Number of packets: Count of data transmitted
- Security level: Identifies the transmission and processing priorities
- Time stamps: Associated with each principal transmission and processing event such as transaction start and stop times
- Network code status: Indicates the nature of any detected errors or malfunctions
- Resources used: Indicates which resources are invoked by this transaction or service agent

Network Control

The network control portion of network management is concerned with modifying parameters and causing actions to be taken by the end systems, intermediate systems, and subnetworks that comprise the configuration to be managed.

All of the five major functional areas of network management (performance, fault, accounting, configuration, and security) involve monitoring and control. Traditionally, the emphasis in the first three of these areas has been on monitoring, while the last two areas are more concerned with control.

Configuration Control

Configuration management is concerned with the initialization, maintenance, and shutdown of individual components and logical subsystems within the total configuration of network and communication resources of an installation. Configuration management can dictate the installation process by identifying and specifying the characteristics of the network components and resources that will constitute the network. Managed resources include identifiable physical resources such as a server or router and lower level objects such as a transport layer retransmission timer. Configuration management can specify initial or default values for attributes so that managed resources commence operation in desired states, possess the proper parameter values, and form the desired relationships with other network components.

While the network is operating, configuration management is responsible for monitoring the configuration and making changes in response to user commands or in response to other network management functions. If the performance monitoring function detects that response time is degrading because of an imbalance in traffic, configuration management may adjust the configuration to achieve a proper traffic level. If fault management detects and isolates a fault, configuration management may alter the configuration to bypass the fault.

Configuration management includes the following functions:

- Define configuration information
- Set and modify attribute values
- Define and modify relationships
- Initialize and terminate network operations
- Distribute software
- Examine values and relationships
- Report on configuration status

The final two items in the preceding list are configuration monitoring functions. Through a query-response interaction, a manager station may examine configuration information maintained by an agent station. Via an event report, an agent may report a change in status to a manager.

Define Configuration Information

Configuration management describes the nature and status of resources that are of interest to network management. The configuration information includes a specification of the resources under management and the attributes of those resources. Network resources include physical resources such as end systems ESs, routers, bridges, communication facilities and services, communication medium and modems, and logical resources such as timers, counters, and virtual circuits. Attributes include name, address identification number, states, operational characteristics, software version number, and release level. Configuration information may be structured in a number of different ways:

- As a simple structured list of data fields with each field containing a single value. SNMP utilizes this approach.
- As an object oriented database. One or more objects represent each element of interest to management. Each object contains attributes whose values reflect the characteristics of the represented element. An object may also contain behaviors, such as notifications to be issued if certain events relating to this element occur. The use of containment and inheritance relationships allows relationships among objects to be defined. OSI network management uses this approach.
- As a relational database. Individual fields in the database contain values that reflect characteristics of network elements. The structure of the database reflects the relationships among network elements.

Although this information is to be accessible to a manager station, it is generally stored near the resource in question, either in an agent node if the resource is part of that node, or in a proxy node if the node containing the resource does not support agent software. The network control function should enable the user to specify the range and type of values to which the specified resource attributes at a particular agent can be set. The range is a list of all possible states or the allowed upper and lower limits for parameters and attributes. The type of value allowable for an attribute may also be specified.

The network control function should be able to define new object types or data element types, depending on the database. Ideally, it should be possible to define these new objects on-line and to have such objects created at the appropriate agents and proxies. In virtually all systems, this function is performed off-line as part of configuring a network element, rather than being possible dynamically.

Set and Modify Attribute Values

The configuration control function should enable a manager station to remotely set and modify attribute values in agents and proxies. There are two limitations on this capability:

1. A manager must be authorized to make the modification of a particular attribute at a particular agent or proxy at a particular time. This is a security concern, which will be addressed later.

2. Some attributes reflect the reality at a resource and cannot by their nature be modified remotely. One item of information could be the number of physical ports on a router. Although each port may be enabled or disabled at any particular time, the actual number of ports can be changed only by a physical action at the router, not by a remote parameter setting action.

The modification of an attribute will change the configuration information at the agent or proxy. Generally, modifications fall into three categories:

1. Database update only: When a manager issues a modify command to an agent, one or more values in the agent's configuration database are changed if the operation succeeds. In some cases there is no other immediate response on behalf of the agent. A manager may change contact information such as the name and address of the person responsible for this resource. The agent responds by updating the appropriate data values and returning an acknowledgment to the manager.

2. Database update plus resource modification: In addition to updating values in the configuration database at an agent, a modify command can affect an underlying resource. If the state attribute of a physical port is set to be disabled, then the agent not only updates the state attribute but also disables the port so that it is no longer in use.

3. Database update plus action: In some network management systems there are no direct action commands available to managers. Rather, there are parameters in the database that when set cause the agent to initiate a certain action. A router might maintain a reinitialized parameter in its database. If an authorized manager sets this parameter to TRUE, the router would go through a reinitialization procedure, which would set the parameter to FALSE and reinitialize the router.

The user should be able to load predefined default attribute values such as default states, values, and operational characteristics of resources on a systemwide, individual node, or individual layer basis.

Define and Modify Relationships

A relationship describes an association, connection, or condition that exists between network resources or network components. Examples of relationships are a topology, hierarchy, physical or logical connection, or a management domain. A management domain is a set of resources that share a set of common management attributes or a set of common resources that share the same management authority.

Configuration management should allow on-line modification of resources without taking all or part of the network down. The user should be able to add, delete, and modify the relationships among network resources.

One example of the use of relationships is to manage the link layer connection between LAN nodes at the level of service access point (LLC) SAP (/) and of logical link control. An LLC connection can be initialized in one of two ways: First, the LLC protocol in one node can issue a connection request to another node, either in response to higher layer software or a terminal user command. These are referred to as switched connections. Second, a network manager station could set up a fixed, or permanent, LLC connection between two nodes. This type of connection setup would designate the SAP in each node that served as an endpoint for the connection. The manager software under operator command should be able to break a permanent or switched connection. Another useful feature allows a backup or alternate address to be designated in case the primary destination fails to respond to a connection request.

Initialize and Terminate Network Operations

Configuration management should include mechanisms that enable users to initialize and shut down a network or subnetwork operation. Initialization includes verification that all resource attributes and relationships have been properly set; notifying users of any resource, attribute, or relationship still needing to be configured; and validating users' initialization commands. For termination, mechanisms are needed to allow users to request retrieval of specified statistics, blocks, or status information before the termination procedures have been completed.

Distribute Software

Configuration management should provide the capability to distribute software throughout the configuration to ESs such as hosts, servers, and workstations and ISs such as bridges, routers, and application level gateways. This requires facilities to permit software loading requests, to transmit the specified versions of software, and to update the configuration tracking systems.

In addition to executable software, the software distribution function should also encompass tables and other data that drive the behavior of a node. Foremost in this category is the routing table used by bridges and routers. There may be accounting, performance, or security concerns that require management intervention into routing decisions that cannot be solved by mathematical algorithms alone. The user mechanisms need to examine, update, and manage different versions of software and routing information. Users should be able to specify the loading of different versions of software or routing tables based on particular conditions, such as error rates.

Security Control

The requirements of information security within an organization have undergone two major changes in the last several decades. Prior to the widespread use of data processing equipment, the security of an organization's valuable information was provided primarily by physical and administrative means. With the introduction of the computer, the need for authorized tools for protecting files and other information stored on the computer became evident. This is especially the case for a shared system, such as a time sharing system. The need is even more acute for systems that can be accessed over a public telephone or data network. The generic name for the collection of tools designed to protect data and thwart hackers is computer security.

The second major change that affects security is the introduction of distributed systems and the use of networks and communications facilities for carrying data between terminal users and between computer and computer. Network security measures are needed to protect data during their transmission.

The security management portion of network management deals with the provision of computer and network security for the resources under management, including the network management system itself.

Security Threats

To understand the types of threats that exist to security, a definition of security requirements is necessary. Computer and network security address three requirements:

- Secrecy: This requires that information in a computer system be accessible only for reading by authorized parties, including printing, displaying, and other forms of disclosure that reveal the existence of an object.
- Integrity: This requires that computer system assets can be modified only by authorized parties. The modification includes writing, changing, changing status, deleting, and creating.
- Availability: This requires that the computer system or network is best characterized by viewing the function of the computer system to be providing information. In general, there is a flow of information from a source, such as a file or region of main memory, to a destination, such as another file or user.

Figure 4.10 (a) illustrates the normal flow of data through the network from source to destination while Figure 4.10 (b–e) shows the four general categories of threats.

1. Interruption: An asset of the system has been destroyed or becomes unavailable or unusable. This is a threat to availability. Examples include the destruction of a piece of hardware such as a hard disk, the cutting of a line of communication, or the disabling of the file management system.
2. Interception: An unauthorized party gains access to an asset. This is a threat to secrecy. The unauthorized party could be a person, a program, or a computer. Examples include wiretapping to capture data in a network and the illicit copying of files or programs.
3. Modification: An unauthorized party not only gains access but also tampers with an asset. This is a threat to integrity. Examples include changing values

FIGURE 4.10

*Security threats.
(Reproduced with
permission of
Cisco Systems, Inc.
Copyright © 2001
Cisco Systems, Inc.
All rights reserved.)*

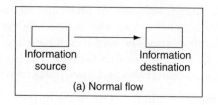

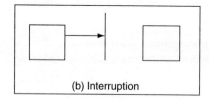

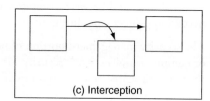

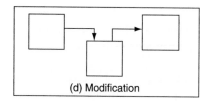

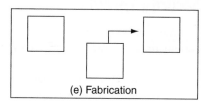

in a data file, altering a program so that it performs differently, and modifying the content of messages being transmitted in a network.

4. Fabrication: An unauthorized party inserts counterfeit objects into the system. This is also a threat to integrity. Examples include the insertion of spurious messages in a network or the addition of records to a file.

The assets of a computer system can be categorized as hardware, software, data, and communication lines and networks. Figure 4.11 indicates the nature of the threats each category of assets faces.

Threats to Hardware

The main threat to computer system hardware is in the area of availability. Hardware is the most vulnerable to attack and the least amenable to automated controls. Threats include accidental and deliberate damage to equipment as well as theft. The proliferation of personal computers and workstations and the increasing use of LANs increase the potential for losses in this area. Physical and administrative security measures are needed to deal with these threats.

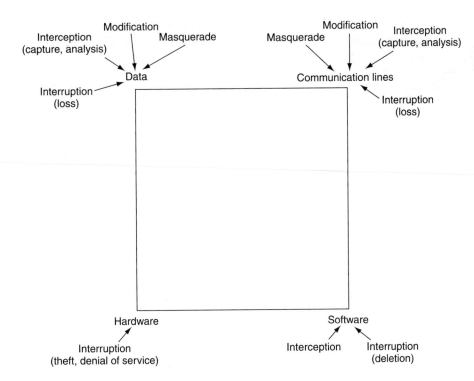

Threats to Software

The operating system, utilities, and application programs are what make computer system hardware useful to businesses and individuals. Several distinct threats need to be considered.

A key threat to software is availability. Software, especially application software, is surprisingly easy to delete. Software can also be altered or damaged to render it useless. Careful software configuration management, which includes making backups of the most recent version of software, can maintain high availability.

A more difficult problem to deal with is software modification that results in a program that still functions but that behaves differently than before. Computer viruses and related attacks fall into this category. A final problem is software secrecy. Although certain countermeasures are available, by and large the problem of unauthorized copying of software has not been solved.

Threats to Data

Typically, hardware and software security are concerns of computing center professionals or individual concerns of personal computer users. A more widespread problem is data security, which involves files and other forms of data controlled by individuals, groups, and business organizations. Security concerns with respect to data are broad, encompassing availability, secrecy, and integrity. In the case of availability, the concern is with the destruction of data files, which can occur either accidentally or maliciously.

The obvious concern with secrecy is the unauthorized reading of data files or databases. A less obvious secrecy threat involves the analysis of data which manifests itself in the use of so-called statistical databases that provide summary or aggregate information. The existence of aggregate information does not threaten the privacy of the individuals involved. As the use of statistical databases grows, the potential for disclosure of personal information increases. In essence, characteristics of constituent individuals may be identified through careful analysis.

Data integrity is a major concern in most installations. Modifications to data files can have consequences ranging from minor to disastrous.

Threats to Communication Lines and Networks

Communication systems are used to transmit data. The concerns of availability, security, and integrity that are relevant to data security apply as well to network security. In this context, threats are conveniently categorized as passive or active as illustrated in Figure 4.12.

Passive threats are in the nature of eavesdropping or monitoring the transmissions of an organization. The goal of the attacker is to obtain transmitted information. Two types of threats are release of message contents and traffic analysis.

Most observers clearly understand the threat of release of message contents. A telephone conversation, an electronic mail message, or a transferred file may contain sen-

FIGURE 4.12

Active and passive network security threats. (Reproduced with permission of Cisco Systems, Inc. Copyright © 2001 Cisco Systems, Inc. All rights reserved.)

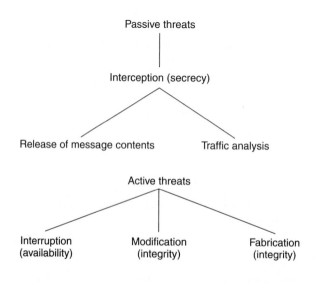

sitive or confidential information. The goal is to prevent the attacker from learning the contents of these transmissions.

The second passive threat, traffic analysis, is more subtle and often less applicable. *Encryption* is a common technique that can make the contents of messages or other information traffic incomprehensible so that the attackers, even if they captured the message, would be unable to extract the information. Passive threats are difficult to detect because they do not involve any alteration of the data. However, it is feasible to prevent these attacks from being successful. The emphasis in dealing with passive threats is on prevention and not detection.

The second major category of threats is active threats. Active threats involve some modification of the data stream or the creation of a false stream. Active threats can be divided into three subcategories, which are message stream modification, *denial of service (DoS)*, and masquerade.

Message stream modification means that some portion of a legitimate message is altered or that messages are delayed, replayed, or reordered to produce an unauthorized effect.

DoS prevents or inhibits the normal use or management of communications facilities. This attack may have a specific target such as an entity may suppress all messages directed to a particular destination. Another form of service denial is the disruption of an entire network, either by disabling the network or by overloading it with messages to degrade performance.

A masquerade takes place when one entity pretends to be a different entity. A masquerade attack usually includes one of the other two forms of active attack. Such an attack can take place by capturing and replaying an authentication sequence.

Active threats present the opposite characteristics of passive threats. Although passive attacks are difficult to detect, measures are available to prevent their success. On the other hand, it is quite difficult to prevent active attacks because this would require the physical protection of all communication facilities and paths at all times. Instead, the goal with respect to active attacks is to detect these attacks and to recover from any disruptions or delays caused by them.

Encryption is a technique that can make the contents of messages or other information traffic incomprehensible.

Denial of service prevents the normal use or management of communications facilities.

Threats to the NMS

Because network management is a set of applications and databases on various hardware platforms distributed throughout the configuration, all of the threats discussed earlier can be considered threats to the NMS. In addition, three security concerns specific to network management are possible:

1. User masquerade: A user who is not authorized to perform network management functions may attempt to access network management applications and information.
2. Network manager masquerade: A computer system may masquerade as a network manager station.

3. Interference with manager–agent interchange: One threat is the observation of manager–agent protocol traffic to extract sensitive management information. More damaging would be the modification of such traffic to disrupt the operation of the agent or the resources that the agent is managing.

Security Management Functions

The security facility of a system or network of systems consists of a set of security services and mechanisms. The functions of security management can be grouped into three categories:

- Maintain security information
- Control resource access service
- Control the encryption process

Maintain Security Information

As with other areas of network management, security management is based on the use of management information exchanges between managers and agents. The same sort of operations is employed for security management as for other areas of network management—the difference is only in the nature of the management information used. Examples of objects appropriate for security management include keys, authentication information, access right information, and operating parameters of security services and mechanisms.

Security management keeps account of activity, or attempted activity, with these security objects in order to detect and recover from attempted or successful security attacks. The following functions are related to the maintenance of security information:

- Event logging
- Monitoring security audit trails
- Monitoring usage and the users of security related resources
- Reporting security violations
- Receiving notification of security violations
- Maintaining and examining security logs
- Maintaining backup copies for all or part of the security related files
- Maintaining general network user profiles and usage profiles for specific resources to enable references for conformance to designated security profiles.

Control Resource Access Service

One of the central services of any security facility is access control. Access control involves authentication and authorization services and the actual decision to grant or refuse access to specific resources. The access control service is designed to protect a broad range of network resources. Those resources of particular concern for the network management function are the following:

- Security codes
- Source routing and route recording information
- Directories
- Routing tables
- Alarm threshold levels
- Accounting tables

Security management manages the access control service by maintaining general network user profiles and usage profiles for specific resources and by setting priorities for access. The security management function enables the user to create and delete security related objects, change their attributes or state, and affect the relationships between security objects.

Control the Encryption Process

Security management must be able to encrypt any exchanges between managers and agents as needed. In addition, security management should facilitate the use of encryption by other network entities. This function is also responsible for designating encryption algorithms and providing key distribution.

Summary

Network management means different things to different people. In some cases, it involves a solitary network consultant monitoring network activity with an outdated protocol analyzer. In other cases, network management involves a distributed database, autopolling of network devices, and high-end workstations generating real-time graphical views of network topology changes and traffic.

For certain, networks have grown in importance and have become critical in the business world. Within a given corporation, the trend is toward larger, more complex networks supporting more applications and more users. As these networks grow in scale, two facts become evident:

- The network along with its associated resources and distributed applications becomes indispensable to the corporation.
- As a result of large, complex networks more things can go wrong, disabling the network or a portion of the network or degrading network performance to an unacceptable level.

A large and complex network cannot be assembled and managed by human effort alone. The complexity of such a network dictates the use of automated network management tools. If the network contains equipment from multiple vendors, the urgency for such tools and the difficulty in supplying them increase, creating additional problems for the network administrator.

As networked installations become larger, more complex, more heterogeneous, and cover more territory between central sites and remote sites, the cost of network management rises. To control costs, standardized tools are needed across a broad spectrum of product types that include end systems, bridges, routers, and telecommunications equipment that can be used in a mixed vendor environment. In response to this need, the SNMP was developed to provide a tool for multivendor, interoperable network management.

SNMP refers to a set of standards for network management including a protocol, a database structure specification, and a set of data objects. SNMP was adopted as the standard for TCP/IP based Internets in 1989 and has enjoyed widespread popularity. In 1991, a supplement to SNMP referred to as RMON was issued. RMON extends the capabilities of SNMP to include management of LANs as well as the devices attached to those networks. In 1993, an upgrade to SNMP, known as SNMP version 2 (SNMP v2) was proposed and a revision of SNMP v2 was formally adopted in 1996. SNMP v2 adds functional enhancements to SNMP and codifies the use of SNMP on OSI based networks. In addition, in 1996, RMON was extended with a revision known as RMON2.

In general, network management is a service that employs a variety of tools, applications, and devices to assist human network managers in monitoring and maintaining networks. The correct term is ISO defined five key functional areas of network management. This functional classification was developed solely for the OSI environment. However, although developed for the OSI environment, these functional areas have gained broad acceptance by vendors of both standardized and proprietary network management systems. The five key functional areas defined for network management by the ISO are:

1. Fault management: The facilities that enable the detection, isolation, and correction of abnormal operation of the OSI environment.
2. Accounting management: The facilities that enable charges to be established for the use of managed objects and costs to be identified for the use of those managed objects.
3. Configuration management: The facilities that control, identify, collect data from, and provide data to managed objects for the purpose of assisting in providing for the continuous operation of interconnection devices.
4. Performance management: The facilities needed to evaluate the behavior of managed objects and the effectiveness of communication activities.
5. Security management: The facilities that address those aspects of OSI security essential to operate OSI network management correctly and to protect managed objects.

To maintain a complex internetwork, a network manager must take care that systems and essential components are in proper operational order. When a fault occurs, it is important for the network manager to perform the following tasks as quickly as possible:

- Determine where the fault is.
- Isolate the rest of the network from the failure so that the rest of the network

can continue to function without any downtime or interference from the faulted segment of the network.

- Reconfigure or modify the network in such a way as to minimize the impact of the operation without the failed component.
- Repair or replace the failed component as quickly as possible to restore the network to its initial state before the failure.

Faults are to be distinguished from errors. A fault is an abnormal condition that requires management attention (or action) to repair whereas an error is a single event. A fault is usually indicated by the failure to operate correctly or by excessive errors. If a communication line is cut, no signals can get through from the source to the destination. A crimp in a cable may cause distortions so that there is a persistently high BER. End users expect a fast and reliable problem resolution. Most end users will tolerate an occasional network outage. When these infrequent outages do occur, the end user expects to receive immediate notification and have the problem corrected promptly. To provide this level of fault resolution requires rapid and reliable fault detection as well as diagnostic management functions. The impact and duration of faults can be minimized by the use of fault-tolerant components and alternate communication links to give the network a degree of redundancy. After correcting a fault and restoring a system to its full operational state, the fault management service must ensure that the problem is truly resolved and that no new problems are introduced. This requirement is referred to as tracking and control. Fault management should have a minimal effect on network performance.

In many corporate networks, individual cost centers or departments, or even individual project accounts, are charged for the use of network services. These charges are internal accounting procedures rather than actual cash transfers from one department to another; nevertheless, they are important to the participating end users. Even if no internal charging is utilized, the network manager needs to be able to track the use of network resources by end user or end user class for the following reasons:

- An end user or group of end users may be abusing its access privileges and burdening the network at the expense of other end users.
- End users may be making inefficient use of the network, especially bandwidth, and the network manager can assist in changing procedures to improve overall network performance.
- The network manager is in a better position to plan for anticipated network growth if end user activity is maintained and tracked.

The network manager needs to specify the kinds of accounting information to be recorded at various network elements, the desired interval between sending the recorded information to higher level management elements, and the algorithms to be used in calculating the charging. Accounting reports should be generated under network management control. To limit access to accounting information, the accounting department must provide the capability to verify the end user's authorization to access and manipulate that information.

Today's data communication networks are composed of individual components and logical subsystems that can be configured to perform many different applications, for example, the device driver in an operating system. The same device can be configured to act either as a router or as an end system element or both. Once it is decided how a device is to be utilized, the configuration manager can choose the appropriate software, set of attributes, and parameters for that device.

Configuration management is concerned with initializing a network and gracefully shutting down part or all of the network. Configuration management is also concerned with maintaining, adding, and updating the relationships between each component and the status of each individual component during network operation. Start-up and shutdown operations on a network are the specific responsibilities of configuration management. It is often desirable for the operations on certain components to be performed unattended or during off-peak hours.

The network manager needs the capability to identify the components that comprise the network and to define the desired connectivity of these components. Those who regularly configure a network with the same or a similar set of resource parameters need ways to define and modify default parameters and to load these predefined sets of parameters into the specified network components. The network manager must be able to change the connectivity of network components when the end user needs the components to be changed. The reconfiguration of a network is often desired in response to performance evaluation or in support of network upgrade, fault recovery, or security checks.

End users often need or want to be notified of the status of network resources and components. Therefore, end users should be informed when changes in configuration occur. Configuration reports can be generated either on a regular basis or in response to a request for such a report. Before reconfiguration, end users often want to inquire about the upcoming status of network resources and their parameters. Usually, network managers want only authorized end users to manage and control network operation.

Modern data communication networks are composed of various components that must intercommunicate and share data and resources. In some cases, it is critical to the effectiveness of an application that the communication over the network is within certain performance guidelines.

Performance management of a computer network comprises two broad functional categories: monitoring and controlling. The monitoring function tracks activities on the network. The controlling function enables performance management to make adjustments to improve overall network performance. The following performance issues are of concern:

- What is the capacity level?
- Is there excessive traffic or overutilization?
- Has throughput been reduced to unacceptable limits?
- Are there any network bottlenecks?
- Is response time increasing or decreasing?

To deal with these concerns, the network manager must focus on some initial set of resources to be monitored for a period of time to assess performance levels. This includes associating appropriate metrics and parameters with relevant network resources as indicators of different levels of performance. One example would be what count of retransmissions on a transport connection is considered to be a performance problem requiring attention? Performance management, therefore, must monitor many resources to provide information in determining acceptable network operating levels. By collecting and analyzing the information and then using the results analysis as feedback to the prescribed set of parameters, the network manager can become more adept at recognizing situations indicative of present or impending performance degradation.

Before using a network for a particular application, an end user may want to know such things as the average and worst response times and the reliability of network services. Performance must be known in sufficient detail to assess specific end user queries. End users expect network services to be managed in a way that consistently affords their applications good response time.

Network managers need performance statistics to help them plan, manage, and maintain large internetworks. Performance statistics can be utilized to recognize potential bottlenecks in the network so that corrective action can be taken before they cause serious problems. For example, routing tables can be changed to balance or redistribute traffic load during times of peak usage or when a bottleneck is identified by a rapidly growing load in one specific area. Eventually, capacity planning based on such performance statistics can indicate the proper decisions to make with regard to increasing bandwidth or an expansion of leased lines in a particular area.

Security management is concerned with managing information protection and access control facilities. The access control facilities include generating, distributing, and storing encryption keys. Passwords and other authorization or access control information must be maintained and distributed. Security management is also concerned with monitoring and controlling access to computer networks and to all or part of the network management information obtained from the network elements. Logs are an important security tool along with examination of audit records and security logs, as well as the enabling and disabling of these logging facilities.

Security management provides facilities for the protection of network resources and end user information. Network security facilities should be available for authorized users only. End users want to know that proper and effective security policies are in force and that the security facilities are secure.

Network monitoring is the most fundamental aspect of automated network management. Although many network management systems do not include network control features because of a lack of security mechanisms, all network management systems include a network monitoring component.

The purpose of network monitoring is to gather information about the status and behavior of network elements. Information to be gathered includes static information, related to the configuration, dynamic information, related to events in the network, and statistical information, summarized from dynamic information. Each managed device in the network includes an agent module responsible for collecting local management information and transmitting it to one or more management stations. Each management station includes NMA software plus software for communicating with agents. Information may be collected actively, by means of polling by the management station, or passively, by means of event reporting by the agents.

Chapter 5 discusses SNMP protocol and RMON and shows how network management systems give you total control of the network from a remote distance.

Review Questions

1. What is network management?
2. What are the requirements for justifying a network management system?
3. List the five functional areas of OSI management.
4. Define each of the five areas.
5. What is an NMS?
6. What are the components that make up an NMS?
7. Contrast the differences between distributed versus centralized network management systems.
8. What are proxies and how do they affect an NMS?
9. Explain network monitoring.
10. What are the three design areas of network monitoring?
11. What are the three classifications of network monitoring?
12. What are the four main components of a network monitoring system?
13. Explain polling and event reporting.
14. What are some of the problems that may arise in selecting performance indicators for a network monitoring system?
15. Define availability.
16. How is a component failure expressed?
17. What is response time?
18. What are the elements that comprise response time?
19. Define accuracy.
20. What is the difference between throughput and utilization?

Summary Questions

1. What are some of the questions that can be answered by performance monitoring?
2. What is the objective of fault monitoring?
3. What are the problems that can be experienced with fault monitoring?
4. What is accounting monitoring?

5. Explain network control.
6. What is configuration control concerned with?
7. What are the functions of configuration management?
8. Define security control.
9. What are the threats to security control?
10. What are the four types of threats?
11. Explain passive and active threats.

Further Reading

Kaeo, Merike. *Designing Network Security.* Indianapolis: Cisco Press, 1999.

Thompson, James M. *Performance and Fault Management.* Indianapolis: Cisco Press, 2000.

Simple Network Management Protocol

Objectives

- Explain the origins of Simple Network Management Protocol (SNMP) in relation to TCP/IP.
- Explain the evolution of SNMP.
- Discuss the SNMP architecture.
- Discuss network management protocol architecture.
- Explain trap directed polling.
- Explain proxies and their role in network management.
- Discuss the structure of management information (SMI).
- Explain the management information base (MIB) structure.
- Explain ASN.1 and its role with SNMP.
- Discuss the basic encoding rules for ASN.1.
- Explain how community names are used in SNMP.
- Discuss lexicographic ordering and how it is utilized in SNMP.
- Explain the five messages utilized by the SNMP.
- Explain the limitations of the protocol.

Key Terms

abstract syntax notation one
(ASN.1), 183

Advanced Research Projects Agency
(ARPA), 170

Introduction

The **U.S. Department of Defense** has frequently funded communication protocol.

The **Advanced Research Projects Agency** is a research and development organization.

ARPANET is a packet-switching network established in 1969.

The term Simple Network Management Protocol (SNMP) is actually used to refer to a collection of specifications for network management that include the protocol itself, the definition of data structures, and their associated concepts. SNMP is an application layer protocol that facilitates the exchange of management information between network devices. SNMP is part of the TCP/IP suite. SNMP enables network administrators to manage network performance, find and solve network problems, and plan for network growth. Two versions of SNMP currently exist: SNMP version 1 (SNMP v1) and SNMP version 2 (SNMP v2). Both versions have a number of features in common, but SNMP v2 offers enhancements such as additional protocol operations. Standardization of SNMP version 3 (SNMP v3) is pending. The development of SNMP follows a similar historical pattern to the development of the TCP/IP suite.

TCP/IP Origins

The starting point for TCP/IP dates back approximately to 1969 when the *U.S. Department of Defense (DoD)* funded, through the *Advanced Research Projects Agency (ARPA),* the development of one of the first packet-switching networks, *ARPANET* or

Advanced Research Projects Agency network. ARPANET's purpose was to study technologies related to the sharing of computer resources and to develop these technologies into data networks useful for daily DoD requirements. As ARPANET evolved, it rapidly grew in size to accommodate hundreds of hosts and thousands of terminals. A major issue from the start was interoperability. As a result of terminals and hosts being manufactured from many different vendors, specialized software needed to be developed to support everything from file transfer to terminal-host interaction. The problem became even more astute when ARPANET evolved into the Internet, forming a collection of wide area and local area networks with ARPANET as the core.

To solve the interoperability problem, ARPANET researchers developed a standardized set of protocols, which by the late 1970s became known as the present TCP/IP protocol suite. The TCP/IP suite was standardized as the official *Internet Architecture Board (IAB)* standard and was issued as *request for comments (RFCs)*. Ultimately, the suite of protocols became military standards.

TCP/IP met the standard requirements of the DoD and became standard in DoD procurements. An interesting and unexpected development was the use of TCP/IP in nonmilitary applications. The growth of nonmilitary application usage of TCP/IP began to take off in the mid-1980s, when efforts were being made to develop an international consensus around OSI. Despite OSI, TCP/IP grew rapidly and is today the dominant standardized communications architecture. The TCP/IP suite is a mature, working set of protocols that provides interoperability and a high level of functionality. The international standards have been slow to develop and are still evolving. Only recently have the standards become commercially available. The complexity of the OSI protocol suite has made the implementation of conformant interoperable software more difficult than with TCP/IP.

TCP/IP and Network Management

As TCP/IP was being developed, little thought was given regarding network management. Initially, virtually all of the hosts and subnetworks attached to ARPANET were based in an environment that included system programmers and protocol designers working on some aspect of the ARPANET research. Consequently, management problems could be left to the protocol experts who could modify the network through the use of some basic tools.

Through the late 1970s no such management tools existed. However, network managers did have one tool at their disposal that was used effectively for network management—the *Internet Control Message Protocol (ICMP)*. ICMP provides a means for transferring control messages from routers and other hosts to a host, in order to provide feedback about problems occurring in a particular environment. ICMP is available on all devices that support the IP. The most useful feature of ICMP is the echo/echo reply message pair. These messages provide a mechanism for testing that communication is possible between two network entities. The recipient of an echo message is obligated to return the contents of that message in an echo reply

The **Internet Architecture Board** is a board of internetwork researchers who discuss issues pertinent to Internet architecture.

Request for comments is used as the primary means for communicating information about the Internet.

The **Internet Control Message Protocol** provides a means for transferring control messages from routers and other hosts to a host.

message. Another useful pair of messages are time stamp and time stamp reply, which provide a mechanism for sampling the delay characteristics of the network. These ICMP messages can be used in addition to various IP header options, such as source routing and record route, to develop simple powerful management tools. One common example that is widely used is the *packet internet groper (PING)* program. PING can perform a variety of functions by using ICMP plus some additional options such as the interval between requests and the number of times to send a request. PING can determine if a physical network device can be addressed by verifying that a network can be addressed and verifying the operation of a server on a host. The PING capability can be used to observe variations in round-trip times and in datagram loss rates, which can help isolate areas of congestion and points of failure.

● **PING** can perform a variety of functions such as testing the reachability of a network device.

Combined with some supplemental tools, the PING capability was a satisfactory solution to the network management requirement for many years. In the late 1980s the Internet growth became exponential and attention focused on developing more powerful network management capabilities. The number of hosts attached to the Internet exploded. The growth in size has been accompanied by a growth in complexity. There has been an equally rapid and exponential growth in the number of subnetworks that are part of the Internet and the number of distinct administrative domains. Thus, the number of different entities that have management responsibility for part of the Internet has also grown.

With the number of hosts and the number of individual networks multiplying by the thousands, it became evident that it was no longer possible to rely on a small cadre of network experts to solve management problems. What was needed was a standardized protocol with far more functionality than PING and yet at the same time one that could be easily learned and used by a variety of people with network management responsibilities. The starting point in providing specific network management tools was the *Simple Gateway Monitoring Protocol (SGMP)* formalized in November of 1987. SGMP provided a straightforward means for monitoring gateways. The need soon arose for a more general-purpose network management tool. This led to the development of three promising approaches:

● The **Simple Gateway Monitoring Protocol** provided a straightforward means for monitoring gateways.

- High Level Entity Management System (HEMS): HEMS was a generalization of the first network management protocol that was used in the Internet—the Host Monitoring Protocol (HMP).
- Simple Network Management Protocol (SNMP): SNMP was an enhanced version of the SGMP.
- CMIP over TCP/IP (CMOT): This was an attempt to incorporate to the maximum extent possible the protocol (*Common Management Information Protocol (CMIP)*, services, and database structure being standardized by ISO for network management.

● The **Common Management Information Protocol** was created and standardized by the ISO for network management.

In 1988, the IAB reviewed these three proposals and approved further development of SNMP as a short-term solution and CMOT as the long-range solution. Intuitively, the IAB felt that within a reasonable period of time TCP/IP installations would transition to OSI based protocols. There was a reluctance to invest a substantial effort in application level protocols and services on TCP/IP that might soon have to be abandoned. The strategy of the IAB was to meet immediate needs by developing SNMP quickly to provide basic management tools and support the development of an experience base for doing network management. HEMS was more capable than SNMP, but the extra effort on a possibly dying solution was unwarranted. If CMIP could be implemented to run on top of TCP/IP, then it might be possible to deploy CMOT even before the transition to OSI. When the time came to move to OSI, the network management aspect of the move would require minimal effort.

To further solidify this strategy, the IAB dictated that SNMP and CMOT use the same database of managed objects. Both protocols were to use the same set of monitoring and control variables, in the same formats, within any host, router, bridge, or other managed device. Only a single Structure of Management Information (SMI), which is the basic format for conventions for objects, and a single Management Information Base (MIB), which is the actual structure of the database, would be defined for both protocols. This singular identity of databases would greatly facilitate the transition: Only the protocol and supporting software would need to be changed; the actual database would be the same in format and content at the time of transition. However, it soon became apparent that this binding of the two protocols at the object level was impractical. In OSI network management, managed objects are seen as sophisticated entities with attributes, associated procedures, and notification capabilities, as well as other complex characteristics associated with object oriented technology. In contrast, SNMP is not designed to operate with such sophisticated concepts. The objects in SNMP are not really objects at all from the viewpoint of object oriented technology. Rather, objects in SNMP are simply variables with a few basic characteristics such as data type and read only or read-write attributes. As a result, the IAB relaxed its condition of a common SMI/MIB and allowed SNMP and CMOT to develop independently of each other, and in parallel.

Evolution of SNMP

With the constraints of OSI compatibility removed, the SNMP developers were able to move rapidly and make tremendous strides in the completion of the Simple Network Management Protocol. SNMP mirrored the history of TCP/IP. SNMP soon became widely available on vendor equipment and flourished within the Internet. SNMP became the standardized management protocol of choice for the general user. Just as TCP/IP has outlasted all predictions of its usefulness, so has SNMP proven that it will be around for the long haul. The widespread deployment of OSI network management continues to be delayed and the CMOT language effort has languished.

The basic SNMP protocol is now in widespread use. Virtually all major vendors of host computers, workstations, bridges, routers, and hubs offer basic SNMP. Work is progressing on the use of SNMP over OSI and other non-TCP/IP suites. In addition, enhancements to SNMP have been pursued in a number of directions. The most important of these initiatives is the development of a remote monitoring capability to SNMP called RMON (which will be discussed later in this chapter). The Remote Monitoring (RMON) specification defines additions to the basic SNMP MIB as well as the functions that exploit the RMON MIB. RMON gives the network manager the ability to monitor subnetworks as a whole, rather than just individual devices on the subnetwork. Both vendors and users view RMON as an essential extension to SNMP, and RMON, though relatively new, is already widely deployed.

In addition to RMON, other extensions to the basic SNMP MIB have been developed. Some of these are vendor-independent and have to deal with standardized network interfaces, such as Token Ring and FDDI. Others are vendor-specific, private extensions to the MIB and do not add any new technology or concepts to SNMP. There is a limit as to how far SNMP can be extended by defining new and more elaborate MIBs. As SNMP is applied to larger and more sophisticated networks, its deficiencies become more apparent. Those deficiencies, which will be discussed later, are incorporated into SNMP v2.

The SNMP Architecture

● A network **management station** monitors and controls network elements.

● A **management agent** performs the functions requested by the management stations.

Implicit in the SNMP architectural model is a collection of network management stations and network elements. Network *management stations* execute management applications that monitor and control network elements. Network elements are devices such as hosts, gateways, and terminal servers that have *management agents* responsible for performing the network management functions requested by the network management stations. SNMP communicates between the network management stations and the agents in the network elements. The model of network management that is used for SNMP is broken down into the following elements:

● Management station
● Management agent
● Management information base
● Network management protocol

Figure 5.1 illustrates the managed devices in an SNMP configured network.

The management station is a stand-alone device but may possess capabilities to be implemented on a shared system. The management station serves as the interface for the human network manager into the network management system. The management station should possess the following:

● A set of management applications for data analysis and fault recovery
● An interface by which the network manager may monitor and control the network
● The capability of translating the network manager's requirements into the actual monitoring and control of remote elements in the network
● A database of information extracted from the MIBs of all the managed entities in the network

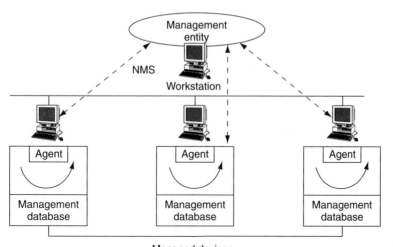

FIGURE 5.1
An SNMP managed network consists of managed devices, agents, and NMSs. (Stallings, William; SNMP, SNMP v2. Copyright © 1996, 1999 Pearson Education. All rights reserved.)

The other active element in the network management system is the management agent. Key platforms such as hosts, bridges, routers, and hubs may be equipped with SNMP agents so that they may be managed from a management station. The management agent responds to requests for information and actions from the management station and may asynchronously provide the management station with important information.

Network resources may be managed by representing them as objects. Each object is essentially a data variable that represents one aspect of the managed agent. The collection of objects is referred to as an MIB. The MIB functions as a collection of access points at the agent for the management station. These objects are standardized across systems of a particular class. For example, a common set of objects is used for the management of various routers. A management station can cause an action to take place at an agent or can change the configuration settings at an agent by modifying the value of specific variables.

The management station and agents are linked together by a network management protocol. The protocol used for the management of TCP/IP networks is SNMP, which includes the following key capabilities:

- *Get:* enables the management station to retrieve the value of objects at the agent
- *Set:* enables the management station to set the value of objects at the agent
- *Trap:* enables an agent to notify the management station of significant events

The standards do not specify the number of management stations or the ratio of management stations to agents. Generally, it is wise to have at least two systems capable of performing the management station function to allow for redundancy in case of failure. The other issue that arises is how many agents can a single management station handle. As long as SNMP remains simple in operation that number can be quite high.

Get enables the management station to retrieve the value of objects at the agent.

Set enables the management station to set the value of objects at the agent.

Trap enables an agent to notify the management station of significant events.

FIGURE 5.2
Configuration of
SNMP.
(Reproduced with
permission of Cisco
Systems, Inc.
Copyright © 2001
Cisco Systems, Inc.
All rights reserved.)

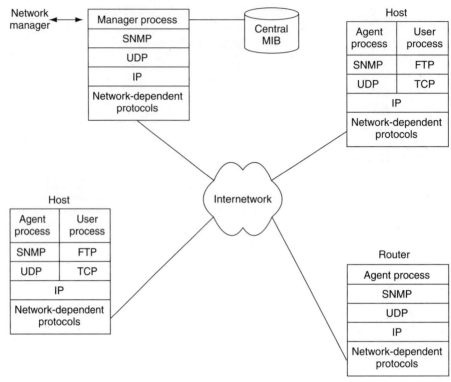

Network Management Protocol Architecture

The **User Datagram Protocol** exchanges datagrams without acknowledgments or guaranteed delivery.

SNMP was designed to be an application level protocol that is part of the TCP/IP suite. SNMP is intended to operate over the *User Datagram Protocol (UDP)*. Figure 5.2 illustrates the typical configuration of protocols for SNMP.

For a stand-alone management station, a manager process controls access to a central MIB at the management station and provides an interface to the network manager. The manager process achieves network management by using SNMP, which is implemented on top of UDP, IP, and the relevant network-dependent protocols such as Ethernet, FDDI, and X.25.

Each agent must also implement SNMP, UDP, and IP. In addition, an agent process interprets the SNMP messages and controls the agent's MIB. For an agent device that supports other applications such as FTP, both TCP and UDP are required. In Figure 5.2 the operational environment is composed of the portion of the host that contains the user process environment, which also encompasses FTP and other protocols such as TCP, IP, and network-dependent protocols. In addition, the operational environment also includes the router environment portion that contains IP and network-dependent protocols, as well as the internetwork itself. The operational environment refers to the

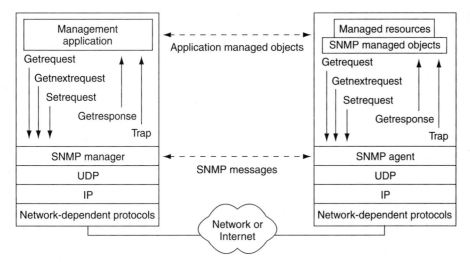

managed environment. Support is provided to the network management function by the host portion that contains the agent process consisting of SNMP and UDP; the manager station consisting of SNMP, UDP, IP, and network-dependent protocols; and the router process containing the agent process that consists of SNMP and UDP.

Figure 5.3 illustrates a closer look at the protocol context of SNMP. From a management station three types of SNMP messages are issued on behalf of a management application: GetRequest, GetNextRequest, and SetRequest. The first two are variations of the Get function. All three messages are acknowledged by the agent in the form of a GetResponse message, which is passed up to the management application. An agent may also issue a trap message in response to an event that affects the MIB and the underlying managed resources.

Because SNMP relies on UDP, which is a connectionless protocol, SNMP is itself connectionless. No ongoing connections are maintained between a management station and its agents. Instead, each exchange is a separate transaction between a management station and an agent.

Goals of the SNMP Architecture

The primary goal of SNMP explicitly minimizes the number and complexity of management functions realized by the agent. This goal is attractive in at least four respects:

1. The necessary development cost for management agent software necessary to support the protocol is accordingly reduced.
2. The degree of management function that is remotely supported is accordingly increased, thereby admitting the fullest use of Internet resources in the management task.

3. The degree of management function that is remotely supported is accordingly increased, thereby imposing the fewest possible restrictions on the form and sophistication of management tools.

4. Simplified sets of management functions are easily understood and used by developers of network management tools.

A second goal of the protocol is that the functional paradigm for monitoring and control be sufficiently extensible to accommodate additional and possibly unanticipated aspects of network operation and management.

A third goal is that the architecture be, as much as possible, independent of the architecture and mechanisms of particular hosts or gateways.

Trap-Directed Polling

If a management station is responsible for a large number of agents, and if each agent maintains a large number of objects, then it becomes impractical for the management station to regularly poll all agents for all of their readable object data. Instead, SNMP and the associated MIB are designed to encourage the manager to use a technique referred to as *trap-directed polling*.

Trap-directed polling can result in substantial savings of network capacity and agent processing time.

SNMP's strategy operates in the following manner. At initialization time and at infrequent intervals, a management station can poll all of the agents that it is aware of for key information, such as interface characteristics, and baseline performance statistics, such as the average number of packets sent and received over each interface for a given period of time. Once a baseline is established, the management station refrains from polling. Each agent is responsible for notifying the management station of any unusual event, such as the agent crashing and having to reboot; a communication link failure; or an overload condition as defined by the packet load crossing some predesigned threshold. These particular events are communicated in SNMP messages known as traps.

Once a management station is alerted to an abnormal condition, the management station may choose to take some action. At this point, the management station may direct polls to the agent reporting the event and to nearby agents in order to diagnose any problem and to gain more specific information about the abnormal condition.

Trap-directed polling can result in substantial savings of network capacity and agent processing time. The network is not made to carry management information that the management station does not need, and agents are not made to respond to frequent requests for uninteresting information.

Proxies

Using SNMP requires that all agents, as well as management stations, must support a common protocol suite such as UDP and IP. This requirement limits direct management to such devices and excludes other devices such as some bridges and modems

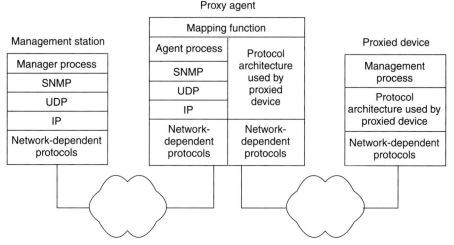

FIGURE 5.4

Proxy configuration. (Reproduced with permission of Cisco Systems, Inc. Copyright © 2001 Cisco Systems, Inc. All rights reserved.)

that do not support any part of the TCP/IP suite. There may be numerous small systems such as personal computers, workstations, and programmable controllers that do implement TCP/IP to support their applications; however, it is not desirable for them to add the additional burden of SNMP, agent logic, and MIB maintenance.

The concept of proxy was developed to accommodate devices that do not implement SNMP. This capability enables an SNMP agent to act as a proxy for one or more other devices, meaning the SNMP agent acts on behalf of the proxied devices. Figure 5.4 indicates the type of protocol architecture that is often involved in this type of arrangement. The management sends queries concerning a device to its proxy agent. The proxy agent converts each query into the management protocol that the device is using. When the agent receives a reply to a query, the agent passes that reply back to the management station. If an event notification from the device is transmitted to the proxy, the proxy sends the event notification on to the management station in the form of a trap message.

SNMP Management Information

As with any NMS, the foundation of a TCP/IP based NMS is a database containing information about the elements to be managed. The database in both TCP/IP and OSI environments is referred to as the management information base (MIB). Each resource to be managed is represented by an object. The MIB is a structured collection of these managed objects. In SNMP, the MIB is a database structure in the form of a tree. Each system such as a workstation, server, router, and bridge, which is part of the network or internetwork, maintains an MIB that reflects the status of the managed resources at that particular system. A network management entity can monitor the resources at that system by reading the values of objects in the MIB and controlling the resources at that system by modifying those values.

To serve the needs of the NMS, the MIB must meet certain criteria:

- The object used to represent a particular resource must be the same at each system. For example, consider information stored concerning the TCP entity at a system. The total number of connections opened over a period of time consists of active opens and passive opens. The MIB at the system could store any two of the three relevant values such as the number of active opens, number of passive opens, or the number of total opens. If different systems select different pairs for storage, it is difficult to write a simple protocol to access the required information. The MIB definition for TCP/IP specifies that the active and passive open counts is stored.
- A common scheme for representation must be used to support interoperability. This point is addressed by defining a structure of management information (SMI).

Structure of Management Information

The **Structure of Management Information** specifies the rules used to define managed objects in the MIB.

The *Structure of Management Information (SMI)*, which is defined in RFC 1155, contains the general framework in which an MIB can be defined and constructed. The SMI identifies the data types that can be used in the MIB and specifies how resources within the MIB are represented and named. The idea behind the SMI is to encourage simplicity and extensibility within the MIB. Consequently, the MIB can store only simple data types: scalars and two-dimensional arrays of scalars. SNMP can only retrieve scalars, including individual entries in a table. The SMI does not support the creation or retrieval of complex data structures. This concept is in contrast to that used with OSI management, which provides for complex data structures and retrieval modes to support greater functionality.

SMI avoids complex data types to simplify the task of implementation and to enhance interoperability. MIBs contain vendor created data types, and unless tight restrictions are placed on the definition of vendor data types interoperability will suffer.

To provide a standardized way of representing management information, the SMI must be able to do the following:

- Provide a standardized technique for defining the structure of a particular MIB.
- Provide a standardized technique for defining individual objects, including the syntax and the value of each object.
- Provide a standardized technique for encoding object values.

MIB Structure

All managed objects in the SNMP environment are arranged in a hierarchical or tree structure. Figure 5.5 shows where an organization would fit in the MIB structure, while Figure 5.6 illustrates the overall basic structure of the MIB. The leaf objects of

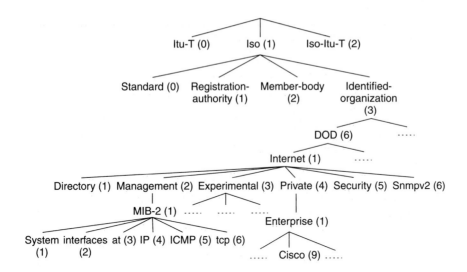

FIGURE 5.5

The MIB tree illustrates the various hierarchies assigned by different organizations. (Reproduced with permission of Cisco Systems, Inc. Copyright © 2001 Cisco Systems, Inc. All rights reserved.)

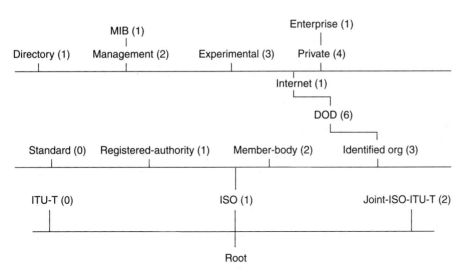

FIGURE 5.6

MIB tree. (Reproduced with permission of Cisco Systems, Inc. Copyright © 2001 Cisco Systems, Inc. All rights reserved.)

the tree are the actual managed objects, each of which represents some resource, activity, or related information that is to be managed. The tree structure itself defines a grouping of objects into logically related sets.

Associated with each type of object in an MIB is an identifier of the ASN.1 type OBJECT IDENTIFIER. The purpose of the identifier is to name the object. As a result of the value associated with the type OBJECT IDENTIFIER being hierarchical, the naming convention also serves to identify the structure of object types. The *object identifier* is a unique identifier for a particular object type. Its value consists of a se-

The **object identifier** is a unique identifier for a particular object type.

quence of integers. The set of defined objects has a tree structure, with the root of the tree being the object referring to the ASN.1 standard.

Beginning with the root of the object identifier tree, each object identifier component value identifies an arc in the tree. There are three nodes at the first level: iso, itu-t, and joint-iso-itu-t. Under the iso node, one subtree is for the use of other organizations, one of which is the U.S. Department of Defense (dod). RFC 1155 states that one subtree under dod will be allocated for administration by the IAB. The path to the IAB from the root of the tree reads as follows:

```
internet OBJECT IDENTIFIER ::= (iso (1) org (3) dod (6) 1)
```

Thus, the internet node has the object identifier value of 1.3.6.1. This value serves as the prefix for the nodes at the next lower level of the tree.

The SMI document defines four nodes under the internet node:

- Directory: reserved for future use with the OSI directory (X.500)
- Management (mgmt): used for objects defined in IAB approved documents
- Experimental: used to identify objects defined unilaterally
- Private: used to identify objects defined unilaterally

The mgmt subtree contains the definitions of management information bases that have been approved by the IAB. Presently there are two versions of the MIB that have been developed: MIB-1 and MIB-2. The second MIB is an extension of the first. MIB-1 and MIB-2 are provided with the same object identifier in the subtree because only one of the MIBs would be present in any configuration.

Additional objects can be defined for an MIB in one of three ways:

1. The MIB-2 subtree can be expanded or replaced by a completely new revision soon to be released as MIB-3. In order to expand MIB-2, a new subtree must be defined. An example of expanding the MIB-2 subtree would be the remote network monitoring MIB is defined as the sixteenth subtree under MIB-2.
2. An experimental MIB can be constructed for a particular application. Such objects may subsequently be moved to the mgmt subtree. Examples of experimental MIBs include the various transmission media MIBs that have been defined such as the IEEE 802.5 Token Ring LAN, which is defined in RFC 1227.
3. Private extensions can be added to the private subtree. An example is the MUX MIB defined in RFC 1227.

The private subtree currently has only one child element defined—the enterprise node. The private portion of the subtree is reserved to allow vendors to enhance the management of their devices and to share this information with other users and vendors who might need to interoperate with their systems. A branch within the enterprise subtree is allocated to each vendor who registers for an enterprise object identifier. Dividing the internet node into four subtrees provides a strong foundation for the evolution of MIBs. As vendors and other implementers experiment with new objects, they are gaining a good deal of practical experience before these objects are accepted

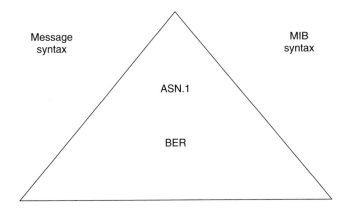

as part of the standardized mgmt specification. The MIB is useful for managing objects that fit within the standardized portion of the MIB and flexible enough to adapt to changes in technology and product offerings. This evolutionary character within SNMP mirrors that of the protocols within the TCP/IP suite. All of the protocols within the TCP/IP suite underwent extensive experimental use and debugging before being finalized as standard protocols.

SNMP and Abstract Syntax Notation One (ASN.1)

Abstract syntax notation one (ASN.1) illustrated by Figure 5.7 offers a standard platform-independent method of representing data across any internetwork. ASN.1 shows up in many places when different aspects are examined for SNMP.

The SMI requires that ASN.1 set the syntax for messages and for the objects in the MIBs. The ASN.1 basic encoding rules (BER) provide the transfer syntax (encoding) to transmit that abstract syntax. From another angle, ASN.1 defines the data types and the BER defines the way SNMP will serialize the data for transmitting.

ASN.1 covers platform independence in multiple areas by:

- Helping resolve character data type combinations like EBCDIC and ASCII
- Resolving big endian versus little endian incompatibilities
- Setting structure alignments on byte versus word boundaries
- Dealing with the size such as the number of octets/bytes of the data types
- Determining which bits are high order and which are low order

ASN.1 is a formal language developed and standardized by ITU-T (X.208) and ISO (ISO 8824). ASN.1 is important for the following reasons:

- ASN.1 can be used to define abstract syntaxes of application data.

Abstract syntax notation one offers a method of representing data across any internetwork.

FIGURE 5.8

The use of abstract and transfer syntaxes. (Reproduced with permission of Cisco Systems, Inc. Copyright © 2001 Cisco Systems, Inc. All rights reserved.)

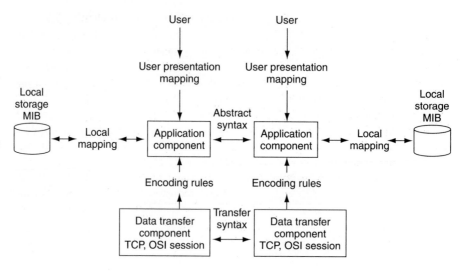

● ASN.1 is used to define the structure of application and presentation protocol data units.

● ASN.1 is used to define the management information database for SNMP and OSI systems management.

Figure 5.8 illustrates the underlying concepts of how ASN.1 is applied in the SNMP protocol model.

Key terms relevant to ASN.1 are:

● Abstract syntax: describes the generic structure of data independent of any encoding technique used to represent the data. The syntax allows data types to be defined and values of those types to be specified.

● *Data type:* a named set of values. A type may be simple, which is defined by specifying the set of its values, or structured, which is defined in terms of other types.

● *Encoding:* the complete sequence of octets used to represent a data value.

● Encoding rules: a specification of the mapping from one syntax to another. Specifically, encoding rules determine algorithmically any set of data values defined in an abstract syntax or the representation of those values in a transfer syntax.

● *Transfer syntax:* the way in which data are actually represented in terms of bit patterns while in transit between presentation entities.

Data type is a named set of values.

Encoding is a complete sequence of octets used to represent a data value.

Transfer syntax represents data in terms of bit patterns while in transit between presentation entities.

A communication architecture in an end system consists of two major components: the data transfer component and the application component. The data transfer component is concerned with the mechanisms for the transfer of data between end systems. In the TCP/IP suite, the data transfer component consists of

TCP or UDP and below. In the OSI architecture, the data transfer component consists of the session layer and below. The application component is the user of the data transfer component and is concerned with the end user's application. Again in the TCP/IP suite, the application component consists of an application such as SNMP, FTP, SMTP, or Telnet. In the OSI protocol suite, the application component consists of the application layer, which is comprised of a number of application service elements, and the presentation layer. As the boundary is crossed from the application to the data transfer component, there is a significant change in the way that data are viewed. The data transfer component receives the data from an application and is specified as the binary value of a sequence of octets. This value can be directly assembled into *service data units (SDUs)* for passing between layers and into *protocol data units (PDUs)* for passing between protocol entities within a layer. The application component is concerned with a user's view of the data. The user view of the data is one of a structured set of information such as text in a document, a personnel file, an integrated database, or the visual display of image information. The user is primarily concerned with the semantics of data. The application component must provide a representation of this data that can be converted to binary values, and the application component must also be concerned with the syntax of the data.

Service data units are units of information passed between layers.

Protocol data units are units of information passed between entities within a layer.

The approach to support application data is illustrated in Figure 5.8. The application component has information represented in an abstract syntax that deals with data types and data values. The abstract syntax formally specifies data independently from any specific representation. An abstract syntax has many similarities to the data type definition aspects of conventional programming languages such as Pascal and C. Application protocols describe their PDUs in terms of abstract syntax.

The abstract syntax is used for the exchange of information between application components in different systems. The exchange consists of application level PDUs, which contain control information and user data. Within a system, the information represented using an abstract syntax must be mapped into some form for presentation to the human user. In addition, the abstract syntax must be mapped into some local format for storage. Such a mapping is used in the case of the MIB. Elements within the MIB are defined using the abstract syntax. The abstract syntax notation is employed by a user to define the MIB, and the application must then convert the user definition to a form suitable for local storage.

The application component must also translate between the abstract syntax of the application and a transfer syntax that describes the data values in a binary form that is suitable for interaction with the data transfer component. An abstract syntax can include a data type of character while the transfer syntax could specify ASCII or EBCDIC encoding. The transfer syntax defines the representation of the data to be exchanged between data transfer components. The translation from abstract syntax to the transfer syntax is accomplished by means of encoding rules that specify the representation of each data value of each data type. This approach for the exchange of application data solves the two problems that relate to data representation in a distributed, heterogeneous environment:

- There is a common representation for the exchange of data between differing systems.
- Internal to a system, an application uses some particular representation of data. The abstract/transfer syntax scheme automatically resolves differences in representation between cooperating application entities.

ASN.1 Concepts

The basic building block of an ASN.1 specification is the module. ASN.1 is a language that can be used to define data structures. A structure definition is in the form of a named module. The name of the module can then be used to reference the structure. The module name can be used as an abstract syntax name and an application can pass this name to the presentation service to specify the abstract syntax of the APDUs that the application wishes to exchange with a peer application entity.

Modules have the basic form:

```
<modulereference> DEFINITIONS ::=
    BEGIN
    EXPORTS
    IMPORTS
    AssignmentList
End
```

The modulereference is a module name followed optionally by an object identifier to identify the module. The EXPORTS construct indicates which definitions in this module may be imported by other modules. The IMPORTS construct indicates which type and value definitions from other modules are to be imported into this module. Neither the IMPORTS nor EXPORTS constructs may be included unless the object identifier for the module is included. The AssignmentList consists of type assignments, value assignments, and macro definitions. Type and value assignments have the form

```
<name> ::= <description>
```

Lexical Conventions

ASN.1 structures, types, and values are expressed in a notation similar to that of a programming language. The following lexical conventions are followed:

1. Layout is not significant—multiple spaces and blank lines can be considered as a single space.

2. Comments are delimited by pairs of hyphens (- -) at the beginning of the comment and at the end of a comment.
3. Identifiers (names of values and fields), type references (names of types), and module names consist of upper- and lowercase letters, digits, and hyphens.
4. An identifier begins with a lowercase letter.
5. A type reference or a module name begins with an uppercase letter.
6. A built-in type consists of all capital letters. A built-in type is a commonly used type for which a standard notation is provided.

Abstract Data Types

ASN.1 is a notation for abstract data types and their values. A type can be viewed as a collection of values. The number of values that a type may take may be infinite. The type INTEGER has an infinite number of values.

Types can be classified into four major categories:

- Simple: These are atomic types with no components.
- Structured: A structured type has components.
- Tagged: These are types derived from other types.
- Other: This category includes the CHOICE and ANY types.

Every ASN.1 data type, with the exception of CHOICE and ANY, has an associated tag. The *tag* consists of a class name and a nonnegative integer tag number. There are four classes of data types, or four classes of tag:

- *UNIVERSAL:* Generally useful, application-independent types and construction mechanisms that are defined in the standard and are listed in Table 5.1.
- *Application-wide:* Data types that are relevant to a particular application. These are defined in other standards.
- *Context-specific:* Types that are also relevant to a particular application, but applicable in a limited context.
- Private: Types that are defined by users and not covered by any standard.

A data type is usually identified by its tag. ASN.1 types are the same if and only if their tag numbers are the same. UNIVERSAL 4 refers to OctetString, which is of class UNIVERSAL and has tag number 4 within the class. Tag types each fall into one of the tag classes. Under each class the decimal value of the tag type is unique and either primitive (P) or constructed (C). By combining the two class bits with the single format bit and the tag number to identify the tag type, the unique hexadecimal numbers are seen in the first column. The tag types identify the values that the value field of the tag carries when transmitting SNMP messages. The tag types also provide support for the MIB. The syntax for each object in the MIB uses ASN.1's naming to provide simplicity and extensibility.

The **tag** consists of a class name and a nonnegative integer number.

UNIVERSAL data types are application-independent types.

Application-wide data types are relevant to a particular application.

Context-specific data types are applicable in a limited context.

Table 5.1 ASN.1 Type Summary

Hex Value	Class	Format	Type
02	UNIVERSAL	P	Integer
04	UNIVERSAL	P	OctetString
05	UNIVERSAL	P	Null
06	UNIVERSAL	P	Object Identifier
30	UNIVERSAL	C	Sequence of
40	Application	P	IP Address
41	Application	P	Counter
42	Application	P	Gauge
43	Application	P	Time Ticks
44	Application	P	Opaque
A0	Context-specific	C	GetRequest PDU
A1	Context-specific	C	GetNextRequest PDU
A2	Context-specific	C	GetResponse PDU
A3	Context-specific	C	SetRequest PDU
A4	Context-specific	C	Trap

Universal Types

The UNIVERSAL class of ASN.1 consists of application-independent data types that are of general use. Within the UNIVERSAL class only the following data types are permitted to define MIB objects:

- Integer (UNIVERSAL 2)
- OctetString (UNIVERSAL 4)
- Null (UNIVERSAL 5)
- Object identifier (UNIVERSAL 6)
- Sequence, sequence-of (UNIVERSAL 16)

The first four are primitive types that are the basic building blocks of the other types of objects. The object identifier is a unique identifier of an object, consisting of a set of integers known as subidentifiers. The sequence is read from left to right and defines the location of the object in the MIB tree structure. For example, the object identifier for the object tcpConnTable is derived as:

```
iso   org   dod   internet   mgmt   mib-2   tcp   tcpConnTable
 1     3     6       1         2       1      6        13
```

The identifier would be written as 1.3.6.1.2.1.6.13. The last item in the preceding list consists of the constructor type sequence and sequence-of. These types are used to construct tables.

Application-Wide Types

The APPLICATION class of ASN.1 consists of data types that are relevant to a particular application. Each application including SNMP is responsible for defining its own APPLICATION data types. RFC 1155 lists a number of application-wide data types, and other types may be defined in future RFCs. The following types are defined:

- network address: This type is defined using the CHOICE construct to allow the selection of an address format from one of a number of protocol families. Currently, the only defined address is the IP Address.
- IP address: This is a 32-bit address using the format specified in IP.
- *counter:* This is a nonnegative integer that may be incremented but not decremented. A maximum value of $2^{32} - 1(4,294,967,295)$ is specified. When the counter reaches its maximum, it wraps around and starts increasing again from zero.
- *gauge:* This is a nonnegative integer that may increase or decrease, with a maximum value of $2^{32} - 1$. If the maximum value is reached, the gauge remains latched at that value until reset.
- *time ticks:* This nonnegative integer counts the time in hundredths of a second relative to a certain event. When an object type is defined in the MIB that uses this type, the definition of the object type identifies the referenced event.
- *opaque:* This type supports the capability to pass arbitrary data. The data are encoded as OCTET STRING for transmission. The data itself may be in any format defined by ASN.1 or another syntax.

The counter, also known as the rollover counter, is one of the most common types used in defining objects. Typical applications are to count the number of packets or octets sent or received. An alternative type of counter that the SMI designers considered is the latch counter, which sticks at its maximum value and must be reset. The latch counter was rejected because of a potential problem. Suppose that more than one management system is allowed access to a particular counter; that is, more than one management system can monitor a device. When a latch counter reaches its maximum and needs to be reset, there are two alternatives that can happen:

1. Designate one management system as responsible for latch reset. The problem with this approach is that if that system fails, the counter remains stuck at its latched value.
2. Allow any management system the authority to reset the counter when it is deemed appropriate. The problem here has to do with time lag involved in communication across a distributed system. Several systems may reset the same counter, resulting in lost information.

A **counter** is a nonnegative integer that may be incremented but not decremented.

A **gauge** is a nonnegative integer that may increase or decrease.

The **time ticks** type is a nonnegative integer that counts the time in hundredths of a second relative to a certain event.

The **opaque** type supports the capability to pass arbitrary data.

These difficulties are avoided with rollover counters. After a rollover counter has wrapped around several times, it is difficult for the management station to periodically poll the object to keep track of wraparounds. This should not have to be done very often because 32-bit counters are used.

A gauge is used to measure the current value of some entity, such as the current number of packets stored in a queue. A gauge can also be used to store the difference in the value of some entity from the start to the end of a time interval. This enables a gauge to be used to monitor the rate of change of the value of an entity.

The gauge type is referred to as a latched value. Once a gauge reaches its maximum value, the gauge will not roll over to zero. If the gauge represents a value that increases beyond the maximum, the gauge remains stuck at its maximum value. If the represented value subsequently falls below the gauge maximum, one of the following alternatives could be adopted:

1. Allow the gauge to decrease so that the gauge always has the same value as the modeled value so long as the modeled value remains in the range of the gauge.
2. Leave the gauge stuck at its maximum value until it is reset by management action.

There is no consensus on the correct interpretation. Latched counters do not necessarily apply to latched gauges. Adopting the second alternative in the preceding list creates a potential problem of multiple managers being allowed to reset the gauge similar to that of latched counters. A positive feature of choosing the second alternative is if the latched value is not immediately reset, it tells the management stations that some parameter has been exceeded, such as maximum queue size.

The time ticks type is a relevant timer: Time is measured relative to some event, such as start-up or reinitialization within the managed system. Although such values are unambiguous within the managed system, they cannot be directly compared to timer values in other systems. Using an absolute time value when utilizing the ASN.1 representation may be a problem because most systems running the TCP/IP suite do not support a time synchronization protocol. Thus, an absolute time type is impractical for SNMP.

One important type that is left out of the SNMP SMI is the threshold type. A threshold is used in the following manner: If the threshold value is crossed in either a positive or negative direction, depending on the definition of the threshold, an event is triggered and an event notification is sent to the management station. The SMI engineers feared that this capability could lead to event floods in which a managed system's threshold is repeatedly crossed and the system floods the network with numerous event notifications. A particularly deadly kind of event flood is one in which the event is triggered by congestion. The creation of an event flood as a result of congestion execrates the condition being reported. This problem has been corrected through the RMON MIB, which defines a form of threshold.

Defining Objects

A management information base consists of a set of objects. Each object has a type and a value. The object type defines a particular kind of managed object. The definition of an object type is therefore a syntactic description. An *object instance* is a particular instance of an object type that has been bound to a specific value.

ASN.1 includes a number of predefined universal types and a grammar for defining new types that are derived from existing types. One alternative for defining managed objects would be to define a new type called Object. Then, every object in the MIB would be of this type. This approach is technically possible, but it would result in unwieldy definitions because you must allow for a variety of value types including counters and gauges. In addition, the MIB supports the definition of two-dimensional tables, or arrays, of values. A general-purpose object type would have to include parameters that encompass all of these possibilities and alternatives.

Because managed objects may contain a variety of information to represent a variety of entities being managed, it would be better to define an open-ended set of new types, one for each general category of managed object. This could be done in ASN.1. The drawback to this alternative is the only restriction on the definition of a new managed object type is that the definition be written in ASN.1, which will cause considerable variation in the format of object definitions. This variety will make it more difficult for the user or implementor of an MIB to incorporate a variety of object types. The use of relatively unstructured object type definitions complicates the task of using SNMP for interoperable access to managed objects.

A more attractive alternative, and the one used by SNMP, is to use a macro to define a set of related types used to define managed objects. A *macro definition* gives the syntax of a set of related types, while a *macro instance* defines a specific type. The following are levels of definition for macros:

- Macro definition: defines the legal macro instances; specifies the syntax of a set of related types.
- Macro instance: an instance generated from a specific macro definition by supplying arguments for the parameters in the macro definition; specifies a particular type.
- Macro instance value: represents a specific entity with a specific value.

The macro used for the SNMP MIBs was initially defined in RFC 1155 (Structure of Management Information) and later expanded in RFC 1212 (Concise MIB Definitions). The RFC 1155 version is used for defining objects in MIB-1. The RFC 1212 version, which includes more information, is used for defining objects in MIB-2 and other recent additions to the MIB.

> An **object instance** is a particular instance of an object type that has been bound to a value.

> A **macro definition** gives the syntax of a set of related types.

> A **macro instance** defines a specific type.

Basic Encoding Rules (BER)

The **basic encoding rules** is a document that describes the rules for encoding data units described in the ASN.1 standard.

The *basic encoding rules (BER)* is an encoding specification developed and standardized by ITU-T (X.209) and ISO (ISO 8825). This document describes a method for encoding values of each ASN.1 type as a string of octets.

Encoding Structure

The **type-length-value** structure is used to encode any ASN.1 value.

The basic encoding rules define one or more ways to encode any ASN.1 value as an octet string. Figure 5.9 illustrates the *type-length-value (TLV)* structure. The encoding is based on the use of a TLV structure. Any ASN.1 value can be encoded as a triple with the following components:

- type: indicates the ASN.1 type, as well as the class of the type and whether the encoding is primitive or constructed.
- length: indicates the length of the actual value representation.
- value: represents the value of the ASN.1 type as a string of octets.

This structure is recursive. For any ASN.1 value that consists of one or more components, the "value" portion of its TLV encoding itself consists of one or more TLV structures. There are three methods for encoding an ASN.1 value:

1. primitive, definite-length encoding
2. constructed, definite-length encoding
3. constructed, indefinite-length encoding

The method chosen depends on the ASN.1 type of the value to be encoded and whether or not the length of the value is known based on the type.

Primitive, Definite-Length Encoding

The **primitive, definite-length encoding** method can be used for simple types and types derived from the simple types by implicit tagging.

The *primitive, definite-length encoding* method can be used for simple types and types derived from the simple types by implicit tagging. As illustrated in Figure 5.9, the BER format consists of three fields:

- identifier: This field encodes the tag (class and tag number) of the ASN.1 value. The first two bits indicate one of the four classes. The next bit is zero to indicate that this is a primitive encoding. The remaining five bits of the first octet can encode a tag number that distinguishes one data type from another within the designated class. For tags whose number is greater than or equal to 31, those five bits contain the binary value 11111, and the actual tag number is contained in the last seven bits of one or more additional octets. The first bit of each additional octet is set to 1, except for the last octet, which is set to 0.
- length: This field specifies the length in octets of the contents field. If the length is less than 128, the length field consists of a single octet beginning with a zero. If the length field is greater than 127, the first octet of the length field contains

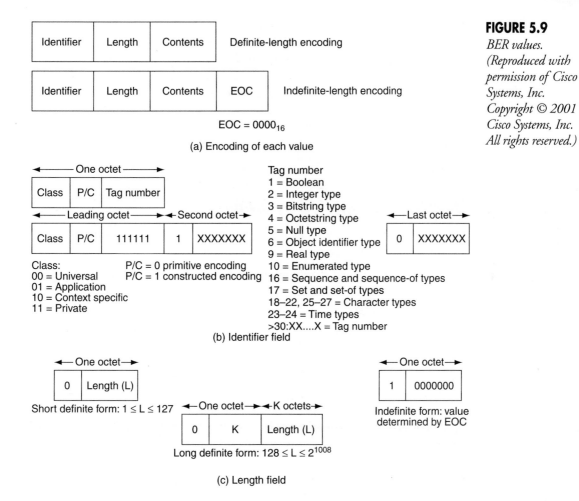

FIGURE 5.9

BER values.
(Reproduced with
permission of Cisco
Systems, Inc.
Copyright © 2001
Cisco Systems, Inc.
All rights reserved.)

a seven-bit integer that specifies the number of additional length octets; the additional octets contain the actual length of the contents field.

● contents: This field directly represents the ASN.1 value as a string of octets.

Constructed, Definite-Length Encoding

The *constructed, definite-length encoding* method can be used for simple string types, structured types (sequence, sequence-of, set, set-of), types derived from simple string types and structured types by implicit tagging, and any type defined by explicit tagging. This encoding method requires that the length value be known in advance. The BER format consists of three fields:

● identifier: same as for primitive, definite-length encoding, except the P/C bit is set to 1 to indicate that this is constructed encoding.

The **constructed, definite-length encoding** method can be used for simple string types, structured types, types derived from these types by implicit tagging, and any type defined by explicit tagging.

- length: same as for primitive, definite-length encoding.
- contents: contains the concatenation of the complete BER encodings (identifier, length, contents) of the component of the value; there are three possible cases:
    1. simple strings and types derived from simple strings by implicit tagging: the concatenation of the BER encodings of consecutive substrings of the value.
    2. structured types and types derived from structured types by implicit tagging: the concatenation of the BER encodings of components of the value.
    3. types defined by explicit tagging: the BER encoding of the underlying value.

Operations Supported by SNMP

The only operations that are supported in SNMP are the alteration and inspection of variables. Three general-purpose operations may be performed on scalar objects.

1. Get: A management station retrieves a scalar object value from a managed station.
2. Set: A management station updates a scalar object value in a managed station.
3. Trap: A managed station sends an unsolicited scalar object value to a management station.

It is not possible to change the structure of an MIB by adding or deleting object instances, such as adding or deleting a row of a table. Nor is it possible to issue commands for an action to be performed. Access is provided only to leaf objects in the object identifier tree. It is not possible to access an entire table or row of a table with one atomic action. These restrictions greatly simplify the implementation of SNMP and they also limit the capability of the network management system.

Communities and Community Names

Network management can be viewed as a distributed application. Like other distributed applications, network management involves the interaction of a number of application entities supported by an application protocol. In SNMP network management, the application entities are the management station applications and the managed station (agent) applications that use SNMP, which is the supporting protocol.

SNMP network management has several characteristics not typical of all distributed applications. The application involves a one-to-many relationship between a management station and a set of managed stations. The management station is able to get and set objects in the managed stations and is able to retrieve traps from the managed stations. From an operational or control point of view, the management station "manages" a number of managed stations. There may be a number of management stations, each of which manages all or a subset of the managed stations in the configuration. These subsets may overlap.

SNMP network management is also viewed as a one-to-many relationship between a managed station and a set of management stations. Each managed station controls its own local MIB and must be able to control the use of that MIB by a number of management stations. There are three aspects to this control:

1. authentication service: The managed station may wish to limit access to the MIB to authorize managed stations.
2. access policy: The managed station may wish to give different access privileges to different management stations.
3. proxy service: A managed station may act as a proxy to other managed stations. This may involve implementing the authentication service and/or access policy for the other managed systems on the proxy system.

All of these aspects relate to security concerns. In an environment in which responsibility for network components is split, managed systems need to protect themselves and their MIBs from unwanted and unauthorized access. SNMP provides only a primitive and limited capability for such security, which is referred to as community. An SNMP *community* is a relationship between an SNMP agent and a set of SNMP managers that defines authentication, access control, and proxy characteristics. The community concept is a local one, defined at the managed system. The managed system establishes one community for each desired combination of authentication, access control, and proxy characteristics. Each community is given a unique *community name*, and the management stations within that community are provided with and must employ the community name in all Get and Set operations. The agent may establish a number of communities with overlapping management station membership.

> An SNMP **community** is a relationship between an SNMP agent and a set of SNMP managers that defines authentication, access control, and proxy characteristics.

Because communities are defined locally at the agent, different agents may use the same name. This identity of names is irrelevant and does not indicate any similarity between the defined communities. A management station must keep track of the community name or names associated with each of the agents that the management station wishes to access.

> A **community name** is a unique name given to each community. The community name functions as a password.

Authentication Service

An authentication service is concerned with ensuring that a communication is authentic. In the case of an SNMP message, the function of an authentication service is to assure the recipient that the message that is received is from the source from which it claims to be. Every message Get request or Put request from a management station to an agent includes a community name. The community name functions as a password, and the message is assumed to be authentic if the sender knows the password.

As a result of this limited form of authentication many network managers are reluctant to allow anything other than network monitoring; that is, Get and Trap operations. Network control operates via a Set operation and is clearly a more sensitive area. The community name could be used to trigger an authentication procedure, with the name

functioning simply as an initial password screening device. The authentication procedure could involve encryption/decryption for more secure authentication functions.

Access Policy

By defining a community, an agent limits access to its MIB to a selected set of management stations. By the use of more than one community, the agent can provide different categories of MIB access to different management stations. There are presently
two aspects of access control:

- SNMP MIB view: a subset of the objects within an MIB. Different MIB
  views may be defined for each community. The set of objects in a view need
  not belong to a single subtree of the MIB.
- SNMP access mode: an element of the set [READ-ONLY, READ-WRITE].
  An access mode is defined for each community.

The combination of an MIB view and an access mode is referred to as an SNMP community profile. A community profile consists of a defined subset of the MIB at the
agent, plus an access mode for those objects. The SNMP access mode is applied uniformly to all objects in the MIB view. If the access mode READ-ONLY is selected, it
applies to all objects in the view and limits the management station's access to this
view to read-only operations.

Within a community profile, two separate access restrictions must be reconciled. The
definition of each MIB object includes an ACCESS clause. Table 5.2 shows the rules
for reconciling an object's ACCESS clause with the SNMP access mode imposed for
a particular view. The rules are very straightforward.

Table 5.2 Relationship between MIB ACCESS Category and SNMP Access Mode

MIB ACCESS Category	SNMP Access Mode	
	READ-ONLY	*READ-WRITE*
Read-only	Available for Get, Set, and Trap operations	
Read-write	Available for Get and Trap operations	Available for Get, Set, and Trap operations
Write-only	Available for Get and Trap operations, but the value is implementation-specific	Available for Get, Set, and Trap operations, but the value is implementation-specific for Get and Trap operations
Not accessible	Unavailable	

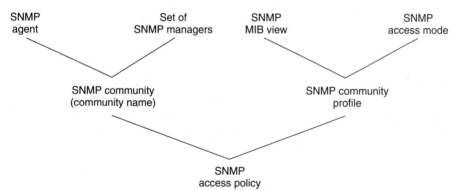

Even if an object is declared as write-only, it may be possible with SNMP to read that object, which is an implementation-specific function. Figure 5.10 illustrates the concept that a community profile is associated with each community defined by an agent. The combination of an SNMP community and an SNMP community profile is referred to as an SNMP access policy.

Proxy Service

The community concept is also useful in supporting the proxy service. A proxy is an SNMP agent that acts on behalf of other devices. The other devices are foreign, in that they do not support TCP/IP and SNMP. The proxied system may support SNMP but the proxy is used to minimize the interaction between the proxied device and network management systems.

For each device that the proxy system represents, the proxy device maintains an SNMP access policy. The proxy knows which MIB objects can be used to manage the proxied system and their access mode.

Instance Identification

Every object in an MIB has a unique object identifier, which is defined by the position of the object in the tree structured MIB. When an access is made to an MIB, via SNMP or some other means, the access is a specific instance of an object that is wanted, not an object type.

Columnar Objects

For objects that appear in tables, which are referred to as *columnar objects,* the object identifier alone does not suffice to identify the instance. There is one instance of each object for every row in the table. Therefore, some convention is needed by which a specific

Columnar objects are objects that appear in tables.

instance of an object within a table may be identified. If an object instance is referenced and not defined in the MIB, reference to object instances is achieved by a protocol-specific mechanism. Therefore, it is the responsibility of each management protocol adhering to the SMI to define the mechanism. This means that SNMP must define the referencing convention. SNMP actually defines two techniques for identifying a specific object instance: a serial-access technique and a random-access technique. The serial-access technique is based on a lexicographic ordering of objects in the MIB structure. The random-access technique works as follows. A table consists of a set of zero or more rows. Each row contains the same set of scalar object types, or columnar objects. Each columnar object has a unique object identifier that is the same in each row. The values of the INDEX objects of a table are used to distinguish one row from another. A combination of the object identifier for a columnar object and a set of values of the INDEX objects specifies a particular scalar object in a particular row of the table. The convention used in SNMP is to concatenate the scalar object identifier with the values of the INDEX objects, listed in the order in which the INDEX objects appear in the table definition.

As a simple example, let us consider the ifTable in the interfaces group where there is only one INDEX object, ifIndex, whose value is an integer in the range between 1 and the value of ifNumber, with each interface being assigned a unique number. Suppose you want to know the interface type of the second interface of a system. The object identifier of ifType is 1.3.6.1.2.1.2.2.1.3. The value of ifIndex of interest is 2. The instance identifier for the instance of ifType corresponding to the row containing a value of ifIndex of 2 is 1.3.6.1.2.1.2.2.1.3.2. The value of ifIndex has been added as the final subidentifier in the instance identifier.

The convention works as follows: Given an object whose object identifier is y, in a table with INDEX objects i1, i2 … iN, then the instance identifier for an object instance of y in a particular row is:

$$y.(i1).(i2). . ...(iN)$$

The remaining detail that must be solved is exactly how the value of an object instance is converted to one or more subidentifiers. RFC 1212, which defines the OBJECT-TYPE macro used for MIB-2, includes the following rules for each INDEX object instance:

- Integer-valued: A single subidentifier takes the integer value (valid only for nonnegative integers).
- String-valued, fixed-length: Each octet of the string is encoded as a separate subidentifier, for a total of n subidentifiers for a string of length n octets.
- String-valued, variable length: For a string of length n octets, the first subidentifier is n; this is followed by each octet of the string encoded as a separate subidentifier, for a total of n + 1 subidentifiers.
- Object-identifier-valued: For an object identifier with n subidentifiers, the first subidentifier is n; this is followed by the value of each subidentifier in order, for a total of n + 1 subidentifiers.
- IPAddress-valued: There are four subidentifiers in the familiar a.b.c.d notation.

The Problem of Ambiguous Row References

RFC 1212 defines the INDEX clause for the OBJECT-TYPE macro. The purpose of the INDEX clause is to list the objects whose object values will unambiguously distinguish a conceptual row. When the INDEX clause is applied to tables that were originally defined in MIB-I, unambiguous reference is not always possible. The IN-DEX object for ipRouteTable, in the ip group, is ipRouteDest. It is not always the case that only a single route will be stored for any given destination. Two rows will have the same value for ipRouteDest, and the instance identification scheme results in two or more object instances with the same instance identifier. To solve this problem an idea has been proposed to add yet another subidentifier to the instance identifier, under the control of the agent. When two or more rows have the same values of INDEX objects the agent designates one such row as primary and appends the subidentifier 1, designates another row as secondary and appends the subidentifier 2, and so on.

For example, to find the next hop of an entry in the ipRouteTable associated with a destination IP address of 89.1.1.42 the following rules would be followed: The desired instance identifier is ipRouteNextHop.89.1.1.42. If multiple rows have been assigned for the same destination and the manager is interested in the next hop along the primary route, then the instance identifier is ipRouteNextHop.89.1.1.42.1. This proposal has been rejected because it was deemed to complex. This technique would always have to be used for a particular table, whether or not it contained multiple rows with the same index, or the manager would somehow have to discover which references were ambiguous and required the additional subidentifier.

The strategy that has been adopted is to avoid future definition of tables that cannot be unambiguously referenced and to replace existing tables that suffer from such an ambiguity.

Conceptual Table and Row Objects

For table and row objects such as tcpConnTable and tcpConnEntry, no instance identifier is defined. This is because these are not leaf objects and therefore are not accessible by SNMP. In the MIB definition of these objects, their ACCESS characteristic is listed as "not-accessible."

Scalar Objects

In the case of *scalar objects,* there is no ambiguity between an object type and an instance of that object. There is only one object instance for each scalar type. For consistency with the convention for tabular objects and to distinguish between object type and an object instance, SNMP dictates that the instance identifier of a nontabular scalar object consists of its object identifier concatenated with 0.

In **scalar objects** there is no ambiguity between an object type and an instance of that object.

Lexicographic Ordering

An object identifier is a sequence of integers that reflects a hierarchical or tree structure of the objects in the MIB. Given the tree structure of an MIB, the object identifier for a particular object may be derived by tracing a path from the root to the object. Because object identifiers are sequences of integers, object identifiers exhibit a lexicographic ordering. Traversing the tree object identifiers in the MIB, provided the child nodes of a parent node are always depicted in ascending numerical order, can generate lexicographic ordering. The ascending numerical order extends to object instance identifiers, because an object instance identifier is also a sequence of integers. An ordering of object and instance identifiers is important because a network management station may not know the exact makeup of the MIB view that an agent presents to the network management station. The management station therefore needs some means of searching for and accessing objects without specifying them by name. With the use of lexicographic ordering, a management station can in effect traverse the structure of a MIB. At any point in the tree, the management station can supply an object or an object instance identifier and ask for the object instance that occurs next in the ordering.

Figure 5.11 illustrates how object instance identifiers can be seen as part of the hierarchical ordering of objects. The example shows the ipRouteTable in the MIB-2 ip group, as seen through an MIB view that restricts the table to just three entries whose values appear in the following table:

ipRouteDest	ipRouteMetric1	ipRouteNextHop
9.1.2.3	3	99.0.0.3
10.0.0.51	5	89.1.1.42
10.0.0.99	5	89.1.1.42

FIGURE 5.11

An example subtree of objects and object instances. (Reproduced with permission of Cisco Systems, Inc. Copyright © 2001 Cisco Systems, Inc. All rights reserved.)

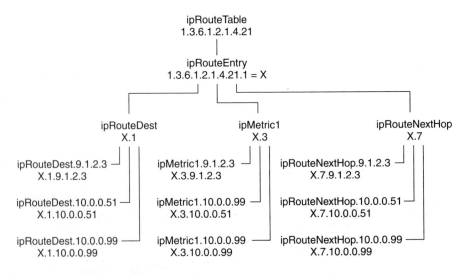

Table 5.3 Lexicographic Ordering of Objects and Object Instances in Figure 5.11

Object	Object Identifier	Next Object Instance Lexicographic Order
IpRouteTable	1.3.6.1.2.1.4.21	1.3.6.1.2.1.4.21.1.1.9.1.2.3
IpRouteEntry	1.3.6.1.2.1.4.21.1	1.3.6.1.2.1.4.21.1.1.9.1.2.3
IpRouteDest	1.3.6.1.2.1.4.21.1.1	1.3.6.1.2.1.4.21.1.1.9.1.2.3
IpRouteDest.9.1.2.3	1.3.6.1.2.1.4.21.1.1.9.1.2.3	1.3.6.1.2.1.4.21.1.1.10.0.0.51
IpRouteDest.10.0.0.51	1.3.6.1.2.1.4.21.1.1.10.0.0.51	1.3.6.1.2.1.4.21.1.1.10.0.0.99
IpRouteDest.10.0.0.99	1.3.6.1.2.1.4.21.1.1.10.0.0.99	1.3.6.1.2.1.4.21.1.3.9.1.2.3
IpRouteMetric1	1.3.6.1.2.1.4.21.1.3	1.3.6.1.2.1.4.21.1.3.9.1.2.3
IpRouteMetric1.9.1.2.3	1.3.6.1.2.1.4.21.1.3.9.1.2.3	1.3.6.1.2.1.4.21.1.3.10.0.0.51
IpRouteMetric1.10.0.0.51	1.3.6.1.2.1.4.21.1.3.10.0.0.51	1.3.6.1.2.1.4.21.1.3.10.0.0.99
IpRouteMetric1.10.0.0.99	1.3.6.1.2.1.4.21.1.3.10.0.0.99	1.3.6.1.2.1.4.21.1.7.9.1.2.3
IpRouteNextHop	1.3.6.1.2.1.4.21.1.7	1.3.6.1.2.1.4.21.1.7.9.1.2.3
IpRouteNextHop.9.1.2.3	1.3.6.1.2.1.4.21.1.7.9.1.2.3	1.3.6.1.2.1.4.21.1.7.10.0.0.51
IpRouteNextHop.10.0.0.51	1.3.6.1.2.1.4.21.1.7.10.0.0.51	1.3.6.1.2.1.4.21.1.7.10.0.0.99
IpRouteNextHop.10.0.0.99	1.3.6.1.2.1.4.21.1.7.10.0.0.99	1.3.6.1.2.1.4.21.1.1.x

The tree in Figure 5.11 is drawn to emphasize its logical interpretation as a two-dimensional table. The lexicographical ordering of the objects and object instances in the table can be seen as traversing the tree. The ordering is seen in Table 5.3.

SNMP Formats

With SNMP, information is exchanged between a management station and an agent in the form of an SNMP message. Each message includes a version number indicating the version of SNMP, a community name to be used for this exchange, and one of five types of protocol data units. This structure is illustrated in Figure 5.12 and the constituent fields are defined in Table 5.4. The GetRequest, GetNextRequest, and SetRequest PDUs have the same format as the GetResponse PDU, with the error-status and error-index fields always set to 0. This convention reduces by one the number of different PDU formats with which the SNMP must operate.

Although a PDU type field is depicted in Figure 5.12, no PDU type field is defined in the ASN.1 type. Because the basic encoding rules for ASN.1 use a type-length-value structure, the type field of a PDU appears as an artifact of the BER encoding rules of the PDU.

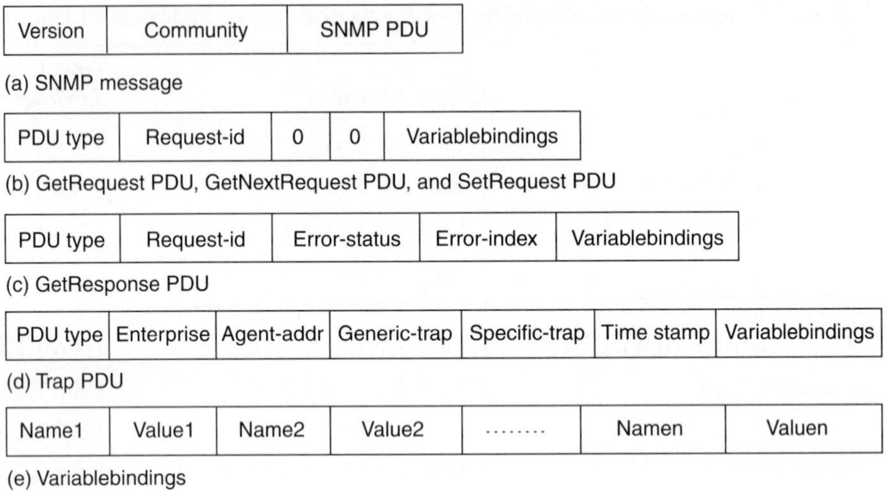

FIGURE 5.12
SNMP formats.
(Reproduced with
permission of Cisco
Systems, Inc.
Copyright © 2001
Cisco Systems, Inc.
All rights reserved.)

Version	Community	SNMP PDU

(a) SNMP message

PDU type	Request-id	0	0	Variablebindings

(b) GetRequest PDU, GetNextRequest PDU, and SetRequest PDU

PDU type	Request-id	Error-status	Error-index	Variablebindings

(c) GetResponse PDU

PDU type	Enterprise	Agent-addr	Generic-trap	Specific-trap	Time stamp	Variablebindings

(d) Trap PDU

Name1	Value1	Name2	Value2	....	Namen	Valuen

(e) Variablebindings

Table 5.4 SNMP Message Fields

Field	Description
Version	SNMP version (version 1 or version 2).
Community	A pairing of an SNMP agent with some arbitrary set of SNMP application entities. (The name of the community acts as a password to authenticate the SNMP message.)
request-id	Used to distinguish among outstanding requests by providing each request with a unique id.
error-status	Used to indicate that an exception occurred while processing a request; values are noError (0), tooBig (1), noSuchName (2), badvalue (3), readOnly (4), genErr (5).
error-index	When error-status is nonzero, may provide additional information by indicating which variable in a list caused the exception. (A variable is an instance of a managed object.)
variablebindings	A list of variable names and corresponding values. (In some cases, such as GetRequest PDU, the values are Null.)
enterprise	Type of object generating trap; based on sysObjectID.
agent-addr	Address of object generating trap.
generic-trap	Generic trap type; values are coldStart (0), warmStart (1), linkDown (2), linkUp (3), authenticationFailure (4), egpNeighborLoss (5), enterprise-Specific (6).
specific-trap	Specific trap code.
time-stamp	Time elapsed between the last reinitialization of the last network entity and the generation of the trap; contains the value of sysUpTime.

Transmission of an SNMP Message

In principle, an SNMP entity performs the following actions to transmit one of the four PDU types to another SNMP entity.

1. The PDU is constructed, using ASN.1 structure defined in RFC 1157.
2. This PDU is then passed to an authentication service, together with the source and destination transport addresses and a community name. The authentication service then performs any required transformations for this exchange, such as encryption or the inclusion of an authentication code, and returns the result.
3. The protocol entity then constructs a message consisting of a version field, the community name, and the result from step 2.
4. This new ASN.1 object is then encoded using the basic encoding rules and passed to the transport service.

In practice, authentication is not typically used.

Receipt of an SNMP Message

In principle, an SNMP entity performs the following actions upon reception of an SNMP message:

1. It does a basic syntax check of the message and discards the message if it fails to parse.
2. It verifies the version number and discards the message if there is a mismatch.
3. The protocol entity then passes the user name, the PDU portion of the message, and the source and destination transport addresses supplied by the transport service that delivered the message to an authentication service.
   - If authentication fails, the authentication service signals the SNMP protocol entity, which generates a trap and discards the message.
   - If authentication succeeds, the authentication service returns a PDU in the form of an ASN.1 object that conforms to the structure defined in RFC 1157.
4. The protocol entity does a basic syntax check of the PDU and discards the PDU if it fails to parse. Otherwise, using the named community, the appropriate SNMP access policy is selected and the PDU is processed accordingly.

In practice, the authentication service serves merely to verify that the community name authorizes the receipt of messages from the source SNMP entity.

Variable Bindings

All SNMP operations involve access to an object instance. It is possible in SNMP to group a number of operations of the same type (Get, Set, Trap) into a single message. If a management station wants to get the values of all the scalar objects in a particular group at a particular agent, it can send a single message requesting all values and get a single response listing all values. This technique can greatly reduce the communications burden of network management.

To implement multiple object exchanges, all of the SNMP PDUs include a variablebindings field. This field consists of a sequence of references to object instances, together with the value of those objects. Some PDUs are concerned only with the name of the object instance (such as Get operations). In this case, the receiving protocol entity ignores the value entries in the variablebindings field. RFC 1157 recommends that in such cases the sending protocol entity use the ASN.1 value NULL for the value portion of the variablebindings field.

GetRequest PDU

The GetRequest PDU is issued by an SNMP entity on behalf of a network management station application. The sending entity includes the following fields:

- PDU type: This indicates that this is a GetRequest PDU.
- request-id: The sending entity assigns numbers such that each outstanding request to the same agent is uniquely identified. The request-id enables the SNMP application to correlate incoming responses with outstanding requests. It also enables an SNMP entity to cope with duplicated PDUs generated by an unreliable transport service.
- Variablebindings: This lists the object instances whose values are requested.

The receiving SNMP entity responds to a GetRequest PDU with a GetResponse PDU containing the same request-id as illustrated in Figure 5.13. The GetRequest operation is atomic, either all of the values are retrieved or none of the values are retrieved. If the responding entity is able to provide values for all of the variables listed in the incoming variablebindings list, then the GetResponse PDU includes the variablebindings field, with a value supplied for each variable. If at least one of the variable values cannot be supplied, then no values are returned. The following error conditions can occur:

1. An object named in the variablebindings field may not match any object identifier in the relevant MIB view or a named object may be of an aggregate type and therefore not have an associated instance value. In either case, the responding entity returns a GetResponse PDU with an error-status of noSuchName and a value in the error-index field that is the index of the problem object in the variablebindings field. If the third variable listed in the incoming variablebindings field is not available for a Get operation, then the error-index field contains a 3.
2. The responding entity may be able to supply values for all of the variables in the list, but the size of the resulting GetResponse PDU may exceed a local limitation. In that case, the responding entity returns a GetResponse PDU with an error-status of tooBig.
3. The responding entity may not be able to supply a value for at least one of the objects for some other error-status of genErr and a value in the error-index field that is the index of the problem object in the variablebindings field.

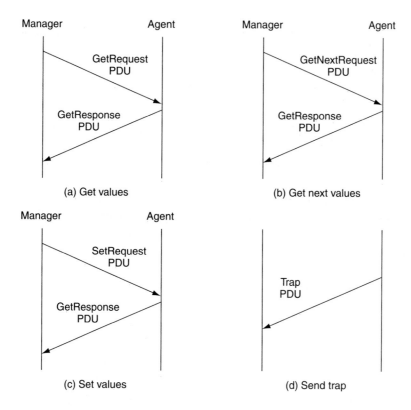

FIGURE 5.13
*SNMP PDU
sequences.
(Reproduced with
permission of Cisco
Systems, Inc.
Copyright © 2001
Cisco Systems, Inc.
All rights reserved.)*

SNMP allows for only the retrieval of leaf objects in the MIB tree. It is not possible to retrieve an entire row of a table, for example, ipRouteEntry, or to retrieve an entire table simply by referencing the table object, for example, ipRouteTable. A management station can retrieve an entire row of a table at a time simply by including each object instance of the table in the variablebindings list. For example, in Figure 5.11 the management station can retrieve the first row with

```
GetRequest (ipRouteDest.9.1.2.3, ipRouteMetric1.9.1.2.3,
ipRouteNextHop.9.1.2.3)
```

The rules for responding to a GetRequest PDU place a burden on the network management station to know how to use this PDU. If the network management station requires numerous values, then it is desirable to ask for a large number of values in a single PDU. If a response is not possible for even one of the objects, or if the response to all objects is too big for a single GetResponse PDU, then no information is returned. Request-ids are used to distinguish among outstanding requests. A GetResponse must include the same request-id value as the corresponding request PDU. No other use is made of this field in SNMP. The value of this field depends on how it is implemented in SNMP and how it is used by network management applications using SNMP.

GetNextRequest PDU

The GetNextRequest PDU is almost identical to the GetRequest PDU. The Get-NextRequest PDU has the same PDU exchange pattern as the GetRequest PDU (Figure 5–12b). The only difference is that in the GetRequest PDU, each variable in the variablebindings list refers to an object instance whose value is to be returned. In the GetNextRequest PDU, for each variable, the respondent is to return only the value of a simple object instance and not an aggregate object such as a subtree or a table. The Next in GetNextRequest refers to the next object instance in lexicographical order, not just the next object. Like GetRequest, GetNextRequest is atomic, either all requested values are returned or none of the values are returned. This minor difference between GetRequest and GetNextRequest has tremendous implications. It allows a network management station to discover the structure of an MIB view dynamically. It also provides an efficient mechanism for searching a table whose entries are unknown.

Retrieving a Simple Object Value

If a network management station wished to retrieve the values of all of the simple objects in the udp group from an agent, the management station could send a GetRequest PDU in the following form:

```
GetRequest (udpInDatagrams.0, udpNoPorts.0, udpInErrors.0,
udpOutDatagrams.0)
```

If the MIB view for this community at the agent supported all of these objects, then a GetResponse PDU would be returned with the values for all four objects:

```
GetResponse ((udpInDatagrams.0 = 100), (udpNoPorts.0 = 1),
(udpInErrors.0 = 2), (udpOutDatagrams.0 = 200))
```

where 100, 1, 2, and 200 are the correct values of the four object instances. If one of the objects was not supported, then a GetResponse PDU with an error code of noSuchName would be returned, but no values would be returned. To ensure getting all available values with the GetRequest PDU, the management station would have to issue four separate PDUs. The use of the GetNextRequest PDU used in the context GetNextRequest (udpInDatagrams, udpNoPorts, udpIn-Errors, udpOutdatagrams) will cause the agent to return the value of the lexicographically next object instance to each identifier in the list. Suppose that all four objects are supported. The object identifier for udpInDatagrams is 1.3.6.1.2.1.7.1. The next instance identifier in lexicographical order is udpIn-Datagrams.0, or 1.3.6.1.2.1.7.1.0. Similarly, the next instance identifier after udpNoPorts is udpNoPorts.0 and so on. If all values are available, then the agent returns a GetResponse PDU of the form:

```
GetResponse ((udpInDatagrams.0 = 100), (udpNoPorts.0 = 1),
(udpInErrors.0 = 2), (udpOutdatagrams.0 = 200))
```

which is the same as before. Suppose that udpNoPorts is not visible in this view, and the same GetNextRequest PDU is issued. The response is:

```
GetResponse ((udpInDatagrams.0 = 100), (udpInErrors.0 = 2),
(udpInErrors.0 = 2), (udpOutdatagrams.0 = 200).
```

The identifier for udpNoPorts.0, which is 1.3.6.1.2.1.7.2.0, is not a valid identifier in this MIB view. Therefore, the agent returns the value of the next object instance in order, which in this case is 1.3.6.1.2.1.7.3.0 = udpInErrors.0.

When an agent receives a GetRequest PDU with the object identifiers of a set of objects, the result includes the value of the requested object instances for all those object instances that are available. For those that are not, the next object instance value in order is returned. This is a more efficient way to retrieve a set of object values when some might be missing than the use of the GetRequest PDU.

Retrieving Unknown Objects

The rules for the use of the GetNextRequest PDU require that the agent retrieve the next object instance that occurs lexicographically after the identifier is supplied. There is no requirement that the supplied identifier actually represents an object or object instance. For example, returning back to the UDP group, because udpInDatagrams is a simple object there is no object whose identifier is udpInDatagrams.2 or 1.3.6.1.2.1.7.1.2. If this identifier is supplied to an agent in a GetNextRequest PDU, the agent simply looks for the next valid identifier. The agent does not check the validity of the supplied identifier. A value is returned for udpNoPorts.0 or 1.3.6.1.2.1.7.2.0. A management station can use the GetNextRequest PDU to probe an MIB view and discover its structure. If the management station issues a GetNextRequest (udp), the response will be the GetResponse (udpInDatagrams.0 = 100). The management station learns that the first supported object in this MIB view is udpInDatagrams, and it obtains the current value of that object at the same time.

Accessing Table Values

The GetNextRequest PDU can be used to search a table efficiently. Consider again the example in Figure 5.11. Recall that the table contains three rows with the following values:

ipRouteDest	ipRouteMetric1	ipRouteNextHop
9.1.2.3	3	99.0.0.3
10.0.0.51	5	89.1.1.42
10.0.0.99	5	89.1.1.42

The management station wishes to retrieve the entire table and does not currently know any of its contents or even the number of rows in the table. The management station can issue a GetNextRequest with the names of all of the columnar objects:

```
GetNextRequest (ipRouteDest, ipRouteMetric1, ipRouteNextHop)
```

The agent will respond with the values from the first row of the table:

```
GetResponse ((ipRouteDest.9.1.2.3 = 9.1.2.3),
(ipRouteMetric1.9.1.2.3 = 3), (ipRouteNextHop.9.1.2.3 =
99.0.0.3)).
```

The management station can then store those values and retrieve the second row with

```
GetNextRequest (ipRouteDest.9.1.2.3, ipRouteMetric1.9.1.2.3,
ipRouteNextHop.9.1.2.3).
```

The SNMP agent responds with:

```
GetResponse ((ipRouteDest.10.0.0.51 = 10.0.0.51),
ipRouteMetric1.10.0.0.51 = 5),
ipRouteNextHop.10.0.0.51 = 89.1.1.42)).
```

The following exchange occurs:

```
GetNextRequest (ipRouteDest.10.0.0.51,
ipRouteMetric1.10.0.0.51, ipRouteNextHop.10.0.0.51)
GetResponse ((ipRouteDest.10.0.0.99 = 10.0.0.99),
ipRouteMetric1.10.0.0.99 = 5), ipRouteNextHop.10.0.0.99 =
89.1.1.42)).
```

The management station does not know that this is the end of the table, thus it proceeds with

```
GetNextRequest (ipRouteDest.10.0.0.99,
ipRouteMetric1.10.0.0.99, ipRouteNextHop.10.0.0.99).
```

No further rows are in the table, so the agent responds with those objects that are next in the lexicographical ordering of objects in this MIB view:

```
GetResponse ((ipRouteMetric1.9.1.2.3 = 3),
ipRouteNextHop.9.1.2.3 = 99.0.0.3),
(ipNetToMediaIfIndex.1.3 = 1)).
```

The example assumes that the next object instance is the one shown in the third entry of the response. Because the object names in the response list do not match those in the request, this signals the management station that it has reached the end of the routing table.

SetRequest PDU

The SetRequest PDU is issued by an SNMP entity on behalf of a network management station application. The SetRequest PDU has the same exchange pattern (Figure 5.12c) and the same format as the GetRequest PDU (Figure 5.12b). The SetRequest PDU is used to write an object value rather than read one. The variablebindings list in the SetRequest PDU includes both object instance identifiers and a value to be assigned to each object instance listed.

The receiving SNMP entity responds to a SetRequest PDU with a GetResponse PDU containing the same request-id. The SetRequest operation is atomic; either all of the variables are updated or none of the variables are updated. If the responding entity is able to set values for all of the variables listed in the incoming variablebindings list, then the GetResponse PDU includes the variablebindings field with a value supplied for each variable. If at least one of the variable values cannot be supplied, then no values are returned, and no values are updated. The same error conditions used in the case of GetRequest may be returned such as noSuchName, tooBig, and genErr. One other error condition may be reported: badvalue. Badvalue is returned if the SetRequest contains at least one pairing of variable name and value that is inconsistent. The inconsistency could be in the type, length, or actual value of the supplied value.

Trap PDU

The Trap PDU is issued by an SNMP entity on behalf of a network management agent application. Trap PDU is used to provide the management station with an asynchronous notification of some significant event. The Trap PDU format is quite different from that of the other SNMP PDUs. The fields are:

- PDU type: indicating that this is a Trap PDU.
- enterprise: identifies the network management subsystem that generated the trap. (Its value is taken from sysObjectID in the System group.)
- agent-addr: the IP address of the object generating the trap.
- generic-trap: one of the predefined trap types.
- specific-trap: a code that indicates more specifically the nature of the trap.
- time-stamp: the time between the last reinitialization of the network entity that issued the trap and the generation of the trap.
- variablebindings: additional information relating to the trap (the significance of this field is implementation-specific).

The generic-trap field may take on one of seven values:

- coldStart (0): the sending SNMP entity is reinitializing itself such that the agent's configuration or the protocol entity implementation may be altered. Typically, this is an unexpected restart due to a crash or major fault.

- warmStart (1): the sending SNMP entity is reinitializing itself such that neither the agent's configuration nor the protocol entity implementation is altered. Typically, this is a routine restart.
- linkDown (2): signals a failure in one of the communications links of the agent. The first element in the variablebindings field is the name and value of the ifIndex instance for the referenced interface.
- linkUp (3): signals that one of the communication links of the agent has come up. The first element in the variablebindings field is the name and value of the ifIndex instance for the referenced interface.
- authenticationFailure (4): this signals that the sending protocol entity has received a protocol message that has failed authentication.
- egpNeighborLoss (5): this signals that an EGP neighbor for whom the sending protocol entity was an EGP peer has been marked down and the peer relationship no longer exists.
- enterpriseSpecific (6): signifies that the sending protocol entity recognizes that some enterprise-specific event has occurred. The specific-trap field indicates the type of trap.

Unlike the GetRequest, GetNextRequest, and SetRequest PDUs, the Trap PDU does not elicit a response from the other side (Figure 5.12d).

Limitations of SNMP

The user who relies on SNMP for network management needs to be aware of its limitations. Following is a list of the limitations for SNMP:

1. SNMP may not be suitable for the management of truly large networks because of the performance limitations of polling. When polling is done only at start-up time and in response to a trap, the management station may have an out-of-date view of the network. The management station will not be alerted to congestion problems in the network. With SNMP, you must send one packet to receive one packet of information. This type of polling results in large volumes of routine messages and yields problem response times that may be unacceptable.
2. SNMP is not well suited for retrieving large volumes of data, such as an entire routing table.
3. SNMP traps are unacknowledged. In a typical case where UDP/IP is used to deliver trap messages, the agent cannot be sure that a critical message has reached the management station.
4. The basic SNMP standard provides only trivial authentication. Basic SNMP is better suited for monitoring than control.
5. SNMP does not directly support imperative commands. The only way to trigger an event at an agent is indirectly, by setting an object value. This is a less flexible and less powerful scheme than one that would allow some sort of remote procedure call with parameters, conditions, status, and results to be reported.

6. The SNMP MIB model is limited and does not readily support applications that make sophisticated management queries based on object values or types.
7. SNMP does not support manager-to-manager communications. There is no mechanism that allows a management system to learn about the devices and networks managed by another management system.

Many of these deficiencies are addressed in SNMP v2.

Summary

The heart of the SNMP framework is the simple network management protocol itself. The protocol provides a straightforward, basic mechanism for the exchange of management information between manager and agent.

The basic unit of exchange is the message, which consists of an outer message wrapper and an inner protocol data unit (PDU). The message header includes a community name, which allows the agent to regulate access. For any given community name, the agent may limit access to a subset of objects in its MIB, known as an MIB view.

Five types of PDUs may be carried in an SNMP message. The GetRequest PDU, issued by a manager, includes a list of one or more object names for which values are requested. The GetNextRequest PDU also is issued by a manager and includes a list of one or more objects. In this case, for each object named, a value is to be returned for the object that is lexicographically next in the MIB. The SetRequest PDU is issued by a manager to request that the values of one or more objects be altered. For all three of these PDUs, the agent responds with a GetResponse PDU, which contains the values of the objects in question or an error-status explaining the failure of the operation. The final PDU is the Trap, which is issued by an agent to provide information to a manager concerning an event.

SNMP is designed to operate over the connectionless User Datagram Protocol (UDP). SNMP can be implemented to operate over a variety of transport level protocols.

Review Questions

1. What is SNMP and what is its main purpose?
2. What are the five elements that comprise the SNMP architecture?
3. Define Get, Set, and Trap.
4. What are the goals of the SNMP architecture?
5. What is trap-directed polling?
6. How are proxies used in network management?
7. What are the criteria that must be met by the MIB to serve the needs of the network management system?
8. To provide a standardized way of representing management information, what must SMI be capable of doing?

9. What are the four nodes defined by SMI under the internet node?
10. How does ASN.1cover platform independence in multiple areas?
11. Define abstract syntax, data type, encoding, and transfer syntax.
12. What is the basic building block of ASN.1?
13. What are the lexical conventions followed by ASN.1 structures?
14. What are the four classes of tag in ASN.1?
15. What is a gauge?
16. What is a counter?
17. What is opaque?
18. What are time ticks?
19. If the represented value falls below the gauge maximum, what are the two alternatives that could be adopted?
20. Define macro definition and macro instance.

Summary Questions

1. What are the three structures defined in the basic encoding rules?
2. Define type, length, and value.
3. What are the three methods for encoding an ASN.1 value?
4. What are the general-purpose operations that can be performed on scalar objects?
5. What is the purpose of the authentication service?
6. What are the two aspects of access control?
7. What is a proxy?
8. Define lexicographic ordering.
9. What are the five formats used by SNMP?
10. What is a PDU?
11. What are the steps in the transmission of an SNMP message?
12. What are the steps in the receipt of an SNMP message?
13. What are the error conditions that can exist in a GetRequest PDU?
14. What are the limitations of SNMP?

Further Reading

Stallings, William. *SNMP, SNMPv2, RMON*. Reading: Addison-Wesley, 1993.

Thompson, James M. *Performance and Fault Management*. Indianapolis: Cisco Press, 2000.

Systems Network Architecture

Objectives

- Explain how Systems Network Architecture (SNA) is used in today's corporate networks.
- Discuss SNA subarea networking.
- Explain Advanced Peer-to-Peer Networking (APPN).
- Explain SNA network configurations.
- Discuss the components that comprise an SNA network.
- Explain SNA network addressing, routes, virtual routes, and explicit routes.
- Define the seven layers that comprise the SNA architecture.
- Explain the various types of SNA sessions.
- Define the relationship between the OSI seven layer model and the SNA model.
- Describe the various SNA protocols.
- Explain the relationship between Synchronous Data Link Control (SDLC) and SNA.
- Explain how the Virtual Telecommunications Access Method (VTAM) and the Network Control Program (NCP) operate in the SNA environment.

Key Terms

3270 applications, 215

Advanced Communications Function (ACF), 231

Advanced Peer-to-Peer Networking (APPN), 215

Advanced Program-to-Program Computing (APPC), 215

bind, 252

boundary network node (BNN), 228

Class of Service (COS) Table, 238

Introduction

●
Systems Network Architecture is IBM's computer network architecture.

IBM networking today consists of essentially two separate architectures that branch from a common origin. Before contemporary networks existed, IBM's *Systems Network Architecture (SNA)* ruled the networking landscape; thus it is often referred to as traditional or legacy SNA.

With the rise of personal computers, workstations, and client/server computing, the need for a peer based networking strategy was addressed by IBM with the creation of *Advanced Peer-to-Peer Networking (APPN)* and *Advanced Program-to-Program Computing (APPC)*.

Although many of the legacy technologies associated with mainframe based SNA have been brought into APPN based networks, real differences exist.

Overview of Corporate Networks

Corporate networks have become the foundation for businesses. The idea of conducting business today without a corporate network leaves a company at a competitive disadvantage. The use of the Internet shows that networks not only have become the foundation for business but are quickly becoming a means of generating business. The network is now a vital part of conducting business—managing and growing this new corporate asset has evolved two schools of networking beliefs:

- Centralized control and management
- Decentralized control and management

Two different approaches to networking are hierarchical and peer.

Hierarchical Networks

The concept of a *hierarchical network* is that of a master and slave. The master dictates and controls while the slave listens and obeys. This top-down architecture centralizes processing and maintains a manageable structure. Hierarchical networks were first chosen for computer networks because of their predictive and deterministic qualities. The slaves—the computer terminals—cannot access data without the master—the centralized processor—recognizing their connection or establishing their connection to the network. As business became dependent on a network to connect workers in remote offices, a standard for hierarchical networking became a necessity. The standard that emerged from this requirement became the IBM Systems Network Architecture, a networking protocol.

The majority of large corporations rely in some manner on IBM mainframe computers to access their vital corporate data. Corporations use interactive applications executing on these mainframes to process the data into information that is useful to the business processes. Interactive applications deliver the information to office workers using computer terminals through network connections. As illustrated in Figure 6.1, the mainframe (master) sends information to the front end processor (slave), which then distributes it to the controllers, terminals, and printers. The traditional terminal type used for accessing information from IBM mainframe interactive applications is the IBM 3270 terminal. Applications using the IBM 3270 terminal for displaying and requesting information from the mainframe are collectively known as *3270 applications*. The 3270 applications grew in popularity because of their inherent predictive

Advanced Peer-to-Peer Networking enhances the original SNA.

Advanced Program-to-Program Computing is SNA system software that allows high speed communication between programs on different computers.

Hierarchical networking involves top-down architecture that centralizes processing and maintains a manageable structure.

Applications known as **3270 applications** use the IBM 3270 terminal for displaying and requesting information from the mainframe.

FIGURE 6.1

The hierarchical network model applied to SNA. (Reproduced with permission of Cisco Systems, Inc. Copyright © 2001 Cisco Systems, Inc. All rights reserved.)

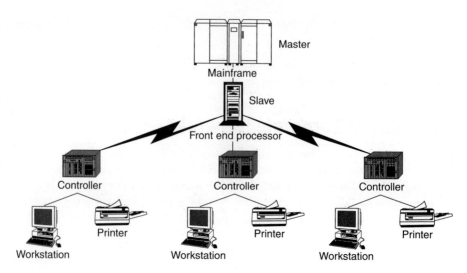

and deterministic behavior as predicated by SNA. This growth required a means to increase the size of the network while maintaining reliable delivery of information. SNA bases its architecture on static network knowledge and includes a networking scheme for large networks using the concept of subareas and static routes.

SNA Subarea Networking

SNA subarea routing asks the simple question: How do I get there from here? SNA defines the answer to this question in every SNA device responsible for delivering information to the 3270 terminal.

SNA *subarea networks* are composed of three types of devices:

- Mainframe
- Front end processor (FEP)
- Cluster controller

The *mainframe* in an SNA subarea network is the master, which controls all other resources in the network. The *front end processor (FEP)* cannot perform these duties without a directive from the mainframe. The FEP also assists in the delivery of information.

Cluster controllers connect the printers and 3270 terminals to the SNA network. The controllers themselves connect to the FEP as shown in Figure 6.2. The controllers cannot deliver information to the 3270 terminals until they receive permission from the mainframe.

Static or predefined routes in the mainframe and FEPs identify the path taken by information messages from the mainframe to the terminals. The assignment of subarea

---

Subarea networks are composed of three types of devices responsible for delivering information.

The **mainframe** controls all other resources in the network.

The **front end processor** assists in the delivery of information.

Cluster controllers connect the printers and 3270 terminals to the SNA network.

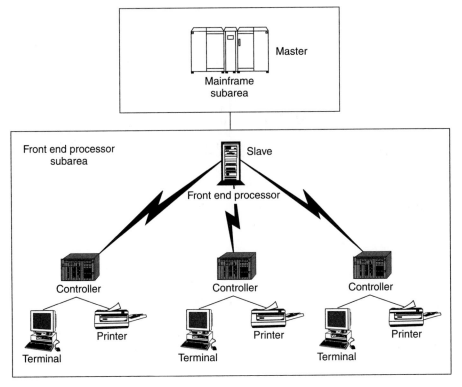

FIGURE 6.2
SNA subarea network components. (Reproduced with permission of Cisco Systems, Inc. Copyright © 2001 Cisco Systems, Inc. All rights reserved.)

numbers to the mainframe and FEPs is the foundation of SNA subarea networking. When a mainframe or FEP has a subarea number assigned, it becomes an SNA subarea. Forward and reverse routes in each subarea provide the mapping information to deliver messages.

Multiple routes through multiple subareas provide alternate paths for the delivery of messages. After a route is selected, the messages between the two subareas must traverse the same forward and reverse routes. Figure 6.3 illustrates the subarea and SNA route concepts. If a link or subarea fails during the delivery of messages, all connections through the disabled path will fail and need reestablishment in order to use a predefined alternate route. The larger the subarea network, the more possible routes there are to define. This leads to a complex static route definition with the potential for errors without dynamic rerouting of messages.

Implementation in a hierarchical network requires reinitializing at some point in time—the master, or FEPs—to keep the changes permanently. This disruption causes valuable downtime on the network. In the end, SNA's inherent need to know caused it to lose its place as the dominant networking architecture, which gave rise to reliable, yet dynamic, peer-to-peer networks.

FIGURE 6.3

SNA subareas and routing. (Reproduced with permission of Cisco Systems, Inc. Copyright © 2001 Cisco Systems, Inc. All rights reserved.)

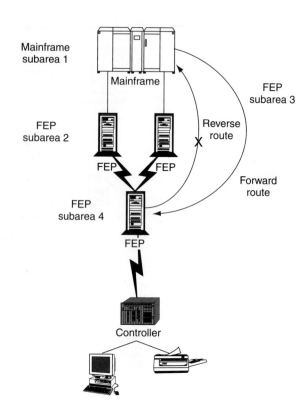

Peer Networks

Peer networking involves the ability to establish connections between two stations.

Peer networking involves the ability to establish connections between two stations independent of any other resource. In a peer network, there is no resource that controls the entire network. This independence has given rise to the delivery of messages in the form of dynamic routing.

Dynamic routing in peer networks, as compared with static routing in hierarchical networks, is a learned process versus a defined process. The network devices in a peer network communicate with each other to learn the topology of the network as illustrated in Figure 6.4. Network resources added or removed dynamically are learned almost instantaneously, allowing for uninterrupted network topology changes. There are three dominant peer network architectures in the majority of large corporations:

- Transmission Control Protocol/Internet Protocol (TCP/IP)
- Novell/Internet Packet EXchange (Novell/IPX)
- Advanced Peer-to-Peer Network (APPN)

Both TCP/IP and Novell/IPX deliver SNA messages in encapsulation techniques. APPN delivers SNA messages in native SNA format.

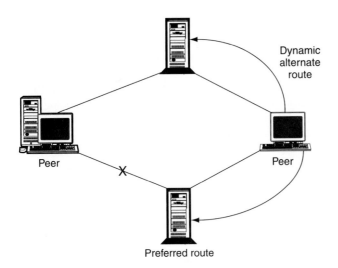

FIGURE 6.4

Peer network topology and routing. (Reproduced with permission of Cisco Systems, Inc. Copyright © 2001 Cisco Systems, Inc. All rights reserved.)

APPN Networking

APPN removes the hierarchical controls while maintaining predictive yet dynamic networks. APPN owes its roots, beginning in the early 1980s, to connecting small business networks based on IBM's System/36 (S/36) computers. The S/36 is the precursor platform to IBM's Advanced Series/400 (AS/400) computing platform.

Small companies with disparate physical offices required a network for connecting their S/36 or AS/400 computers. They lacked the technical staff to manage the infrastructure. APPN's ability to dynamically manage the changes in small business networks became an advantage, enabling APPN to be adapted to handle large corporation SNA networks as illustrated in Figure 6.5.

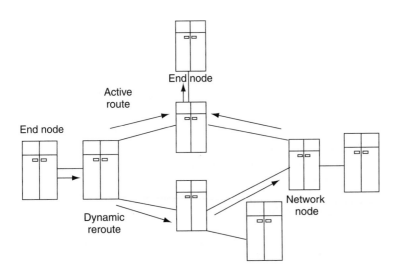

FIGURE 6.5

APPN network topology. (Reproduced with permission of Cisco Systems, Inc. Copyright © 2001 Cisco Systems, Inc. All rights reserved.)

network nodes provide the core of the networking services under APPN architecture.

End nodes are not capable of routing traffic and rely on an adjacent NN for services.

APPN networks revolve around applications communicating with other applications. The basis of this communication is the *network node (NN)* and the *end node (EN)*. For the most part, applications reside on APPN ENs. APPN NNs provide the core of the networking services under APPN architecture. NNs maintain directories, network resources, and the topology of the network. NNs communicate with each other, keeping themselves informed of network topology changes. Topology maps kept by the APPN NNs using High Performance Routing (HPR) allow the rerouting of application-to-application messages without disrupting the connection as found in hierarchical networks.

TCP/IP Networking

TCP/IP has its origins not in the business world but in the government and educational sectors. In the mid-1970s the now infamous ARPAnet, based on TCP/IP networking, fostered communication between military and educational research facilities. Initially, TCP/IP networks were used for remote file access, file transfer, and electronic mail. Today, TCP/IP networking and its dynamic abilities have proven to be an asset to corporate networks, allowing a structured yet rapid growth. Typically, a TCP/IP network consists of four networking resources as illustrated in Figure 6.6:

- IP network
- Host
- Bridge
- Router

An IP network is similar to the SNA subarea because network resources assigned to the same IP network are addressed by using the IP network number. An IP network maps to a physical networking media. Therefore, IP network numbers associate a logical network to a physical network.

TCP/IP networks were first based on local area networks (LANs). Limitations to LAN wiring restricted the number of network resources on the LAN. Increasing the size of LANs and, therefore, the IP network became a concern for early network engineers. The answer to this concern is bridging. Bridges extend the physical LAN limitations and thereby, theoretically, increase the number of resources assigned to an IP network. Multiple IP networks require a router to provide connectivity between them. Routers have the ability to connect multiple IP networks simultaneously, which transports messages between IP networks transparently. Routers contain network topology maps detailing the number of known IP networks and the best route between each network. The routers share information between them, keeping an up-to-date database dynamically.

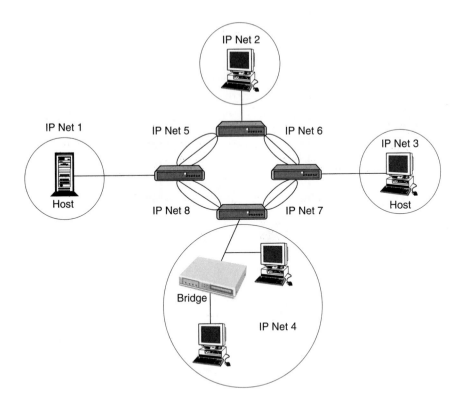

FIGURE 6.6
TCP/IP network topology. (Reproduced with permission of Cisco Systems, Inc. Copyright © 2001 Cisco Systems, Inc. All rights reserved.)

SNA Network Configurations

SNA is still the dominant networking architecture for vital corporate data accessed using IBM mainframe applications. Through the years, the typical IBM SNA network has evolved into various configurations supporting small to worldwide networks.

SNA uses the word domain as a means of describing the scope of resource control by the master or mainframe. In actuality, there is a software program executing on the mainframe called *Virtual Telecommunications Access Method (VTAM)*, which acts as the SNA master. VTAM's control over other SNA resources defines a domain.

Single Domain

In SNA, a *single domain* consists of a single VTAM that controls, or owns, all of the SNA resources attached to the network. These include the terminals, printers, controllers, lines, and FEPs.

Virtual Telecommunications Access Method is a software program that controls communication between logical units.

A **single domain** consists of a single VTAM that controls all of the SNA resources attached to the network.

FIGURE 6.7

*SNA single do-
main network.
(Reproduced with
permission of Cisco
Systems, Inc.
Copyright © 2001
Cisco Systems, Inc.
All rights reserved.)*

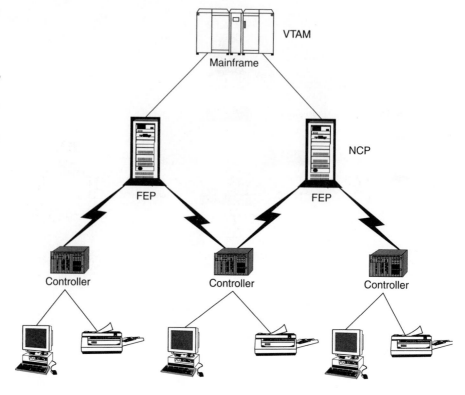

●
The **Network
Control
Program** routes
and controls the
flow of data be-
tween a communi-
cations controller
and other network
resources.

A single domain depicted in Figure 6.7 may comprise a single mainframe with one
or more FEPs directly or remotely attached through a subarea connection and all
of the controllers attached to each FEP. Each terminal or printer attached to a controller is owned by VTAM. Every communications line, controller, terminal, and
printer is defined to the FEP through the use of the *Network Control Program
(NCP)*. During VTAM's activation sequence of network resources, it becomes
aware of the resources attached to each FEP and, hence, determines the bounds of
its domain.

●
A **multidomain**
network consists of
more than one
VTAM controlling
network resources.

Multidomain

In an SNA *multidomain* network, there is more than one VTAM controlling network resources. Each VTAM is in communication with the others; however, they
do not share network topology information. Each VTAM may share control of the

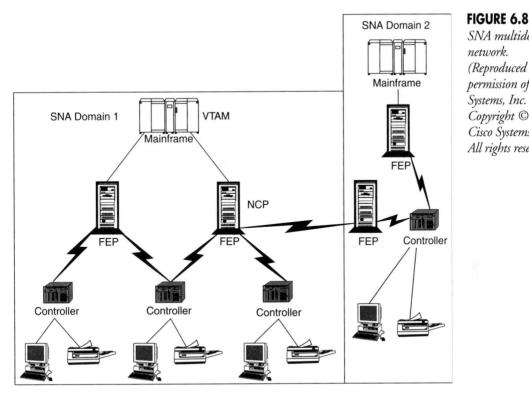

FEPs in an SNA network if the physical and logical connections are defined appropriately, as illustrated in Figure 6.8.

Multinetwork

At one point, though, the size of some SNA networks reached impractical limits. Many corporations created *multinetworks* (multiple independent networks) only to find out later that these independent networks needed communications between them. This requirement led to *Systems Network Interconnection (SNI)*. SNI connects two disparate, independent networks, allowing communication between the VTAM network resources as depicted in Figure 6.9. Communication between the networks happens through an NCP SNI gateway.

An SNI gateway is an FEP that shares its existence between two SNA networks. The distinction between SNA network resources is found in the addition of a network identifier (NETID) to the SNA resource name.

Multinetworks
are multiple independent networks
connected by SNI.

The **Systems
Network Interconnection**
connects multiple
SNA networks.

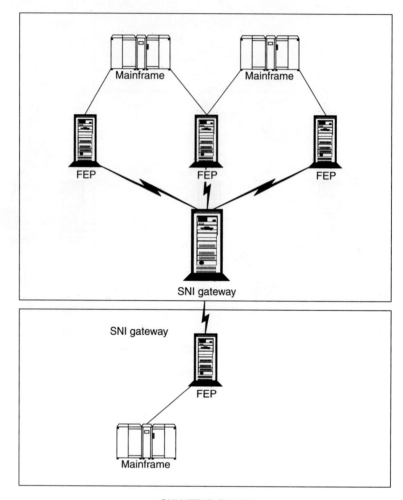

SNI NET ID EUNET

Point-to-Point

The simplest of SNA network configurations is the point-to-point connection. Point-to-point connections, as illustrated in Figure 6.10, provide remote locations with dedicated bandwidth to an FEP.

Point-to-point connections between FEPs are similar to a serial line backbone, which will be described next. Use of a multiplexer allows for enhanced connectivity solutions while still preserving bandwidth for the SNA resources.

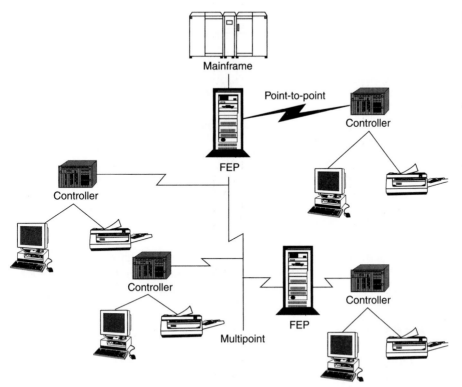

Multipoint

The evolution of SNA communications includes the requirement of having multiple locations connected to an FEP over a single line. Each location connected to the line is considered a point or drop on that line. Figure 6.10 shows a multipoint or multidrop line configuration.

The first incarnation of multipoint lines required that the same type of controllers attach to the line. The increasing cost efficiencies of corporations fostered a mixed multipoint line. Mixed multipoint lines allow different types of network resources to connect to, and share, the communications line. On a mixed multipoint line, both controllers and FEPs attach to the same communication line.

Serial Line Backbone

The simplest of the WAN backbone configurations is the serial line backbone illustrated in Figure 6.11.

FIGURE 6.11

Serial WAN back-bone topology. (Reproduced with permission of Cisco Systems, Inc. Copyright © 2001 Cisco Systems, Inc. All rights reserved.)

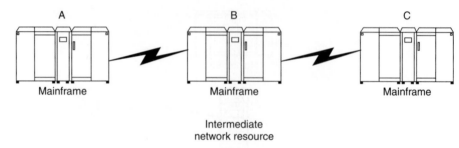

Serial lines are the basis of other backbone infrastructures. Serial lines connect two endpoints at full bandwidth speeds. Connections between two locations may require messages to traverse an intermediate network resource (INR). Although this type of communication works, it does not bode well for resources off the INR when large amounts of data are passing through the INR to a far endpoint. This concern leads to the idea of logical lines multiplexed over a physical line.

SNA LANs

SNA controllers and FEPs, as well as the mainframe, attach to LANs. The most popular method, depicted in Figure 6.12, has been controller connectivity over the LAN to an FEP.

The FEP attaches to the mainframe using channel communications links. LAN connectivity is typically Token Ring. Ethernet and Fiber Distributed Data Interchange (FDDI) are also used.

A **Token Ring interface coupler** is a controller through which an FEP connects to a Token Ring.

FEP connectivity to a Token Ring LAN uses a *Token Ring interface coupler (TIC)*. Controllers attached to the Token Ring use the network address of the TIC to connect to the FEP. SNA, being deterministic, is a nonroutable protocol that requires a LAN bridging technique for use on Token Ring LANs called source route bridging (SRB) (see Chapter 2). SRB embeds the description of physical paths to and from the TIC in the message. If any network resource described on the SRB path fails, the controller must reestablish connectivity to the FEP TIC to discover a new path through the network.

Frame Relay

The **Internet Engineering Task Force** is responsible for developing Internet standards.

A standard technique for sending SNA over frame relay is the *Internet Engineering Task Force (IETF)* Request For Comments (RFC) 1490. The RFC 1490 describes an encapsulation technique for placing SNA messages in frame relay messages and sending them through the network. IBM's FEP (IBM 3745) platform, using NCP version 7 or higher, enables the front end to participate directly in a frame relay network.

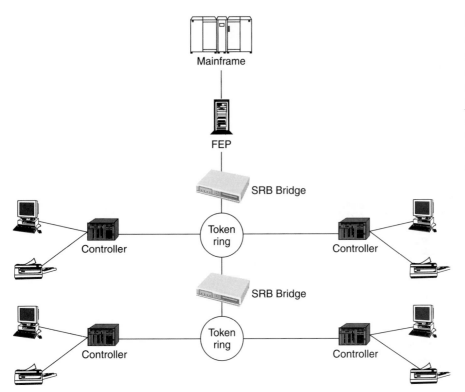

FIGURE 6.12

SNA connectivity over Token Ring LANs. (Reproduced with permission of Cisco Systems, Inc. Copyright © 2001 Cisco Systems, Inc. All rights reserved.)

There are two types of frame relay devices:

- Frame relay terminal equipment (FRTE)
- Frame relay switch equipment (FRSE)

A *frame relay terminal equipment (FRTE)* device is an end device on a frame relay network. It is the access device for remote network resources to send messages over the frame relay network to another FRTE. The IBM 3745 and the IBM controller, IBM 3174, participate in a frame relay network as FRTE devices. The IBM 3745 is also capable of acting as a *frame relay switch equipment (FRSE)* device in a frame relay network. As a switching device, FRTE devices attach to the IBM 3745, which in turn switches the messages from the FRTE to the appropriate far-end FRTE. As illustrated in Figure 6.13, an IBM 3174 FRTE connects to an IBM 3745 FRSE, which in turn has two virtual connections to the mainframe. Each virtual connection terminates in different IBM 3745s that are channel attached to the mainframe. This configuration provides alternative paths for the remote FRTE device with the ability to switch within the network should an intermediate FRSE experience an outage.

Frame relay terminal equipment devices are end devices on a frame relay network.

Frame relay switch equipment devices are switching devices in a frame relay network.

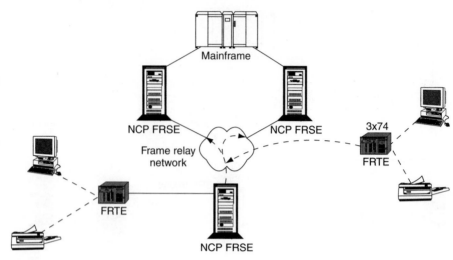

SNA Subarea Network

> The **intermediate network node** subarea participates in the forwarding of messages to subarea destinations.

The traditional SNA subarea network uses static routes called explicit routes (ERs). Each subarea in an SNA subarea network uses ERs to map to the physical communications line connecting the subareas. These subarea lines, or links, are defined as transmission groups. Multiple subarea links may be grouped together to form a single transmission group. Multiple ERs may be mapped to a transmission group.

> The **boundary network node** subarea performs polling and activation/deactivation procedures.

SNA subarea nodes perform *intermediate network node (INN)*, *boundary network node (BNN)*, or both functions as illustrated in Figure 6.14. INN subareas participate in the forwarding of messages to subarea destinations attached directly to the INN subarea or to subarea destinations attached directly to the INN subarea or to subareas beyond the directly attached subareas. BNN subareas also perform polling and activation/deactivation procedures with the directly attached controllers.

APPN Network

> **High performance routing** allows APPN networks to maintain SNA sessions.

APPN provides SNA with the dynamics of peer networks and the predictive deterministic qualities of traditional SNA subarea hierarchical networks. APPN networks have the ability to reroute SNA messages nondisruptively using a feature called *high performance routing (HPR)*. HPR allows APPN networks to maintain SNA sessions even

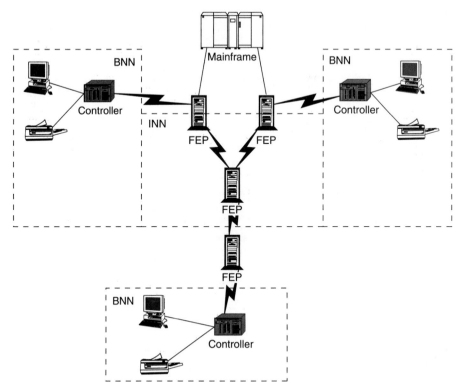

FIGURE 6.14

SNA network sub-area node functions example. (Reproduced with permission of Cisco Systems, Inc. Copyright © 2001 Cisco Systems, Inc. All rights reserved.)

when a network resource used for forwarding SNA messages fails. The APPN NNs sense this failure and reroute messages around the failing NN as illustrated in Figure 6.15. Each NN uses a type of service qualifier for each SNA message. ENs contain the applications used in an APPN network, and register the application name and the type of service the application expects with the NN attached to the EN. In APPN, applications establish a conversation between applications using an application name. The EN requests the NN to discover the EN location of the requested application. The ENs assist in setting up the application session. The use of application names makes APPN very flexible in allowing the movement of applications to different ENs dynamically.

APPN networks, like SNA networks, have NETIDs, or network names. Unique APPN networks are connected using an APPN border node (BN). An APPN BN is similar to an SNI gateway. The BN connects to both APPN networks and assumes an EN type of function registering applications found in one APPN network with the attached NN in the other APPN network. Different types of border nodes exist, providing different levels of routing and topology support. They are Peripheral Border Node (PBN) and Extended Border Node (EBN).

FIGURE 6.15
APPN internet-work example. (Reproduced with permission of Cisco Systems, Inc. Copyright © 2001 Cisco Systems, Inc. All rights reserved.)

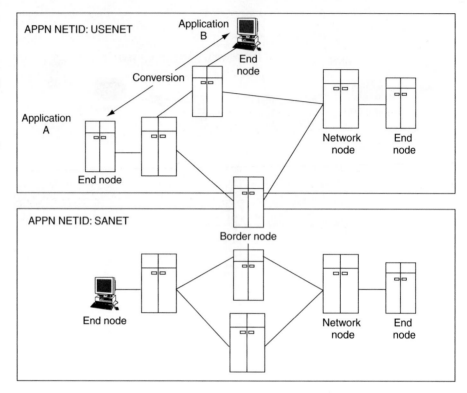

Foundation SNA Concepts

As IBM began to increase its computing power and reach into the business arena, it also had to increase the reach of its data processing. The answer to IBM's need to provide flexible communications from the main computing system out to the remote offices was Systems Network Architecture (SNA). SNA was born in September 1974. The objective of SNA was to further the batch processing into interactive terminal and print applications. This was accomplished by using a hierarchical structure that was composed of seven layers.

Some fundamental concepts to understanding SNA are as follows:

- SNA is a hierarchical network (master-slave).
- SNA uses predefined routes for transporting data between source and destination.
- ACF/VTAM is the master of an SNA network.
- SNA is the network architecture enabling communication between network addressable units.

Network Components

An SNA network is comprised of basic hardware and software components. In a classic SNA network, key hardware components would include the following:

- Mainframe CPU
- Communications controller and front end processor—37×5 (3725, 3745)
- Cluster controller—3×74 (3174, 3274)
- End user workstation/printer

The master-slave relationship was the mode in which this network operated. The master portion executed on the mainframe CPU, termed the access method, and the slaves were the rest of the network components. To allow for the increase of network functionality, a new hardware/software component was created to front end the mainframe processing and control the network; thus, the network control program was born.

To support the aforementioned network hardware list, the following software was developed: *Advanced Communications Function*/Virtual Telecommunications Access Method (VTAM) and Advanced Communications Function/Network Control Program (NCP).

Another vital piece of the IBM SNA strategy is the application software that communicates with the access method. The relationship created between these two software programs allows for each software product to specialize on core competencies. The access method takes care of the transport of the data, and the application manages data integrity. An excellent example of this genre of software is IBM's Customer Information Control System (CICS). Figure 6.16 portrays the quintessential SNA network.

Advanced Communications Function is a group of SNA products that provides distributed processing and resource sharing.

Nodes

The hardware and software categories define the physical and logical components of an SNA network. There are three types of nodes:

- Host subarea nodes
- Communications controller subarea nodes
- Peripheral nodes

Notice that Figure 6.17 uses the same diagram as Figure 6.16. The only difference is that it uses SNA nodal terminology instead of representing each component as its physical-logical function.

A host subarea node is the main engine for the network functionality. Its hardware platform is the mainframe CPU, and the software component is VTAM. The communications controller subarea node is the outboard engine that runs and dispatches network packets throughout the network. Its hardware platform is the communications FEP, and the software component is the NCP. The remainder of the network components comprise the peripheral nodes. This list includes cluster controllers, distributed processors, end user workstations, and printers.

FIGURE 6.16

*The SNA network.
(Reproduced with
permission of Cisco
Systems, Inc.
Copyright © 2001
Cisco Systems, Inc.
All rights reserved.)*

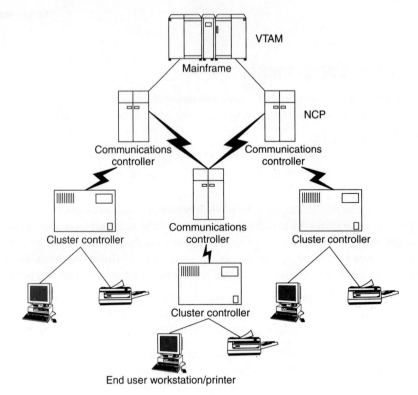

FIGURE 6.17

*SNA nodes.
(Reproduced with
permission of Cisco
Systems, Inc.
Copyright © 2001
Cisco Systems, Inc.
All rights reserved.)*

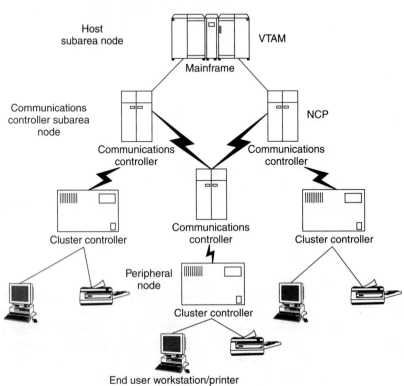

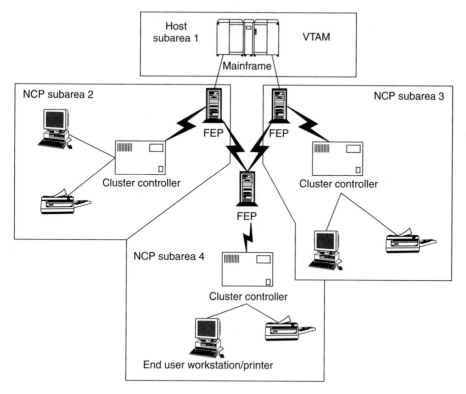

Subareas

In SNA, sections of the total network are broken down into smaller segments. The rule in creating a smaller section, or subarea, is that there must be a VTAM or NCP to manage the area. When defining and working with SNA networks, subarea numbers are present in many definitions ranging from VTAM to the AS/400 emulating as a peripheral node. Figure 6.18 shows a representation of subareas composed of VTAMs and NCPs.

Links

An SNA *link* is any connection used to join two subareas. The physical connection could be a fiber channel, the classic telecommunications circuit, or even satellite communications. There can be more than one link between subarea nodes. Even if there is only one physical connection, a good design includes multiple links to better segregate logical sessions for better data control and flow and to provide redundant routes in the event of a link failure.

A **link** is any connection used to join two subareas.

At the end of each side of the logical link is the link station. The link station transmits data over the link using data link control protocols. SNA supports the following data link protocols:

- System/390 data channel
- Synchronous Data Link Control (SDLC)
- Binary synchronous communication (BSC)
- S/S
- X.25

Network Addressable Units

Network Addressable Units are the endpoints for sessions.

The magic of computer networking is the cohesive manner in which disparate components communicate. The communication of these individual functions is accomplished under the designation of *Network Addressable Units (NAUs)*. NAUs are elements within an SNA network that are the endpoints for sessions. As their name suggests, NAUs must be addressable by the network using addresses that are unique within that network.

In SNA there are three types of NAUs:

- The logical unit (LU)
- The physical unit (PU)
- The System Services Control Point (SSCP)

Logical Units

The **logical unit** is the outermost access point into the network.

The *logical unit (LU)* is the outermost access point into the network. We commonly think of only a terminal or printer as an LU, but an application program is also an LU in the SNA network. There is a one-to-one relationship between a port in the network and the presence of the logical LU. The ultimate goal of the network is to establish a working relationship between the workstation and the application, thus creating an LU-LU session.

All LUs are not created equal. Some application programs can have multiple sessions simultaneously, which are called parallel LU-LU sessions. In a further classification of LUs, different capabilities are possible for each subtype:

- LU type 1: SNA Character String (SCS) printer
- LU type 2: 3270 Interactive Terminals
- LU type 3: Data Stream Compatible (DSC) printer
- LU type 6.2: Application programs

Physical Units

The **physical unit** presents the LU data to the link for transport.

These LU sessions are managed by a *physical unit (PU)*. A mainframe processor, communications controller, or cluster controller represents a PU. The job of a PU is to present the LU data to the link for transport. There is a one-to-many relationship between a PU and LUs.

Again, as with LUs, there are different classifications of PUs:

- PU type 1: Legacy distributed systems controller—S/3X
- PU type 2: Cluster controllers
- PU type 2.1: Advanced peer-to-peer networking nodes
- PU type 4: Communications controllers
- PU type 5: VTAMs

System Services Control Points

The final type is the *System Services Control Point (SSCP)*. The SSCP functionality lies within the access method at the mainframe, thus VTAM. The host subarea node is the only network component that can activate, control, or inactivate network resources. In SNA jargon, under a VTAM's control is called "in its domain." Small networks that have only one VTAM image are called single domain, and those that have more than one VTAM are called multidomain.

Network Addressing

Network addresses uniquely identify NAUs, links, and link stations in a subarea network. These addresses allow for the routing of messages between subareas. An SNA network address comprises two components: a subarea address and an element address.

The subarea address is a unique number defined and allocated during system generation and network activation. Consequently, these addresses have no significance to operators or users. Instead, SNA allows the definition and allocation of a name to a given NAU, which VTAM associates with the SNA network address. In this way, VTAM performs a function similar to that of a *domain naming system (DNS)*.

In recent times, the terminology has changed whereby the term subarea address is used to describe what was previously referred to as network addresses. This change has been brought about by the advent of different network addressing requirements needed in APPN.

Much the same as the TCP/IP addressing schema, underneath the shell of an LU name is a network/host numerical representation. SNA uses a subarea element number to identify each and every device in the network. SNA uses names to allow for an easier nomenclature, which enables users to operate more comfortably—again, another analogy to DNS under TCP/IP structure.

Building on the concept of a subarea, a VTAM or NCP, we can begin to describe the addressing in an SNA network. The subarea number for VTAM is defined by the HOSTSA parameter and the subarea number for NCP is determined by an NCP definition parameter, SUBAREA. The element addresses are then assigned from a pool of addresses that both VTAM and NCP maintain for their attached resources. SNA

The **System Services Control Point** is a focal point within an SNA network for managing network configurations and providing services for network and users.

A **domain naming system** translates names of network nodes into addresses.

networks are generated and defined in an off-line process. Unlike the network routers, programs are created and syntax checked prior to the load. The terminology used by SNA systems programmers is "to perform gen." During the generation process, the Network Definition Facility (NDF) of NCP manages the process of handing out a unique address for every element defined in that subarea. There are many more elements inside an SNA network than just LUs. Each SDLC line needs to be addressed as well as PU and Token Ring attached LINE, PU, and LU. After the generation of devices is completed, network changes are usually still required. Because the NCP cannot be reloaded without creating a disruptive environment, dynamic configuration is also available for many devices.

When resources are locally attached to the mainframe, VTAM handles all of the addressing assignments for those physical devices as well as all of the applications. SNA networks hand out these addresses at the time they load the network. This occurs when VTAM is started at the host and when a new copy of NCP is loaded into the FEPs. SNA addresses have a 31-bit address field. In the first portion of the address, 16 bits are used for the subarea number and the next 15 bits are used for the element. As a result, we can have 65,535 subareas and 32,768 elements within a given subarea. The introduction of VTAM Version 4 Release 1 (V4R1) enabled the use of 16 bits for the element address. This allows VTAM to address 65,535 elements within a given subarea. An enhancement to VTAM V4 has expanded the element address from 65,535 to 1.6 million elements per subarea.

Routes

Another aspect of SNA networking is the concept of routes. Routes are logical notions of physical connections in a network. These logical roadways are defined and used in such a manner as to provide quick response time and alternate pathways through the network. With the building blocks of priority and performance in place, SNA routes will enable terminal sessions to have a quicker response time than printers.

When connecting NCPs to each other, more than one connection is possible. If multiple links are used they are called parallel links. The parallel links are coupled together and are called a *transmission group (TG)*. A TG can be viewed as a single, logical, or composite link for the purposes of routing. Within subarea networks only PU 4 nodes define multiple TGs, whereas all other nodes define only single link TGs. To define these links to the systems, PATH statements are created. PATH statements are the definitions that tell the VTAMs and NCPs how to get to the next hop in the network. A PATH statement consists of physical and logical routes.

● In a **transmission group** the parallel links are coupled together.

● An **explicit route** is the physical connection between two subareas.

Explicit Routes

An *explicit route (ER)*, also known as a physical route, is the physical connection between two subareas. This connection can be a communications line or channel. This ER is mapped to a TG. You can have more than one ER in a specific TG. It would

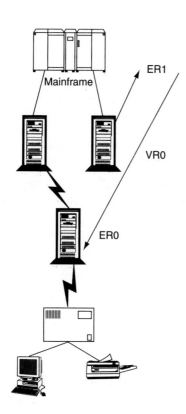

make sense to have two separate physical circuits compose one TG for nondisruptive backup in the event of a circuit failure as illustrated in Figure 6.19.

The following rules are involved when defining ERs:

● They must have a forward and reverse path.
● They must reverse the same set of subarea nodes and TGs.

Virtual Routes

A *virtual route (VR)*, also known as a logical route, is the logical connection between two subareas. A VR is mapped to an ER. One or more VRs can be mapped to an ER. This many-to-one relationship allows for greater flexibility in data flow control. The primary usage of a VR is to separate one type of traffic from another. A good example is the prioritization of terminal over printer traffic.

A **virtual route** is the logical connection between two subareas.

Class of Service Table

The **Class of Service Table** is used by subarea nodes to determine the optimal route to establish a given session.

The highways that provide multiple logical circuits over physical media and the ability to prioritize one type of data over another are now in place. The remaining question becomes: How is the type of data tagged? The answer is the *Class of Service (COS) Table.* The COStab is assembled at VTAM and has three classes that can be attached to the data. The three classes are up to the user to define but are typically high, medium, and low. Each session that is created has characteristics attached to it: one of them is the COS name.

In traditional subarea SNA, routes do not get dynamically distributed throughout the network. These routes can be thought of as static routes that are predefined and contain only the adjacent neighbor's information. Although there is no mechanism for the network itself to know a change in the path and propagate it automatically, a change can be made and a VTAM operator command can be issued to update all the particular tables.

Layers

IBM SNA model components map closely to the OSI reference model. The descriptions that follow outline the role of each SNA component in providing connectivity among SNA entities. Figure 6.20 illustrates the SNA model against the OSI model. Each layer is designed to be autonomous from the other yet still be able to commu-

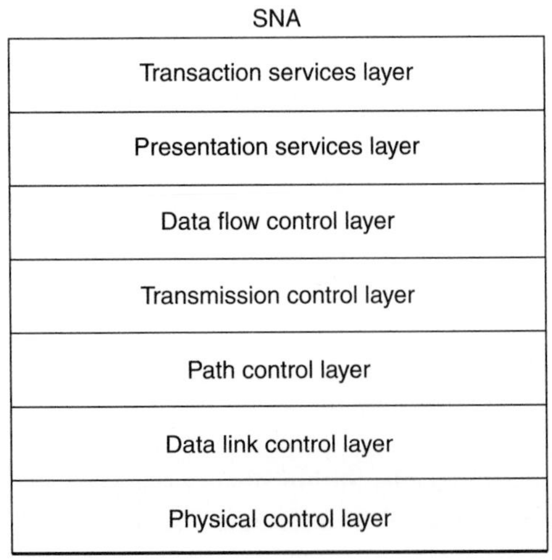
SNA
Transaction services layer
Presentation services layer
Data flow control layer
Transmission control layer
Path control layer
Data link control layer
Physical control layer

FIGURE 6.20
The SNA architectural model. (Reproduced with permission of Cisco Systems, Inc. Copyright © 2001 Cisco Systems, Inc. All rights reserved.)

nicate freely. The seven layers to the SNA protocol set are defined as follows and will be discussed from the bottom (layer 1) to the top (layer 7).

Physical Control Layer: The *physical control layer* of the model describes the physical interface for the transmission medium. The physical layer defines the electrical signaling characteristics.

Data Link Control Layer: In the second layer of SNA several protocols are defined including Synchronous Data Link Control (SDLC) protocol for hierarchical communication, and the Token Ring network communication protocol for LAN communication between peers. The *data link control layer* is also responsible for the workings of the software or firmware that handles the logical components of data communications. The scheduling and error recovery take place in the SDLC and mainframe OS/390 channel protocols.

Path Control Layer: All of the routing information and data movement is controlled under the notion of paths located in the *path control layer*. All of the different types of SNA sessions use these fundamental principles for transmitting data from source to destination. At the path control layer the aforementioned concepts of virtual or explicit routes and the idea of TGs are implemented. The path control layer has three subgroups:

- Transmission groups—provide connections between subarea nodes.
- Explicit routes—determine which TG will be chosen between subarea nodes.
- Virtual routes—provide the physical path the logical session will take as it works its way through the network.

Transmission Control Layer: The *transmission control layer* provides a reliable end-to-end connection service, as well as encrypting and decrypting services.

Data Flow Control Layer: All of the session flow between the LUs is handled at the *data flow control layer*. The packet of information that is sent out by the application is handed to VTAM for delivery. At this layer VTAM must organize related requests and responses into *request/response units (RUs)*. VTAM creates chains to link single packets of information together that are too big for one buffer size of data. VTAM uses brackets to maintain the application-to-LU session until all of the data have been received. The final product of the data is called a path information unit (PIU), which is illustrated in Figure 6.21.

Presentation Services Layer: The *presentation services layer* of the SNA model is where application programs define verbs for transaction programs. The verbs determine the appearance of the output of the data to a terminal, printer, or application. This layer also specifies data transformation algorithms that translate data from one format to another, coordinate resource sharing, and synchronize transaction operations.

The **physical control layer** defines the electrical signaling characteristics.

The **data link control layer** is responsible for the transmission of data over a particular physical link.

The **path control layer** controls all of the routing information and data movement.

The **transmission control layer** provides a reliable connection service.

The **data flow control layer** determines and manages data flow.

Request/response units are request and response messages exchanged between NAUs.

In the **presentation services layer** the application programs define verbs for transaction programs.

LH	TH	RH		RU	LT
Data link control	Path control	Services	Control	Application or SNA command	Data link control
		Transaction presentation	Data flow transmission		
		Basic information unit (BIU)			
	Path information unit (PIU)				
Basic link unit (BLU)					

The **transaction services layer** contains all of the rules for the end user to access the network and send requests to the presentation layer.

Transaction Services Layer: The *transaction services layer* contains all of the rules for the end user to access the network and send requests to the presentation layer. The functions of control and operation are also housed at this layer. SSCP-to-LU sessions are controlled here and are vital for the operation of the network. The SSCP-to-PU session is used to activate and deactivate links, load the same domain software, and assign dynamically created addresses.

In this layer, you also see the function of session establishment between two LUs. The SSCP-to-SSCP session and the SSCP-to-LU sessions are controlled out of this layer. User password validation, access authority, and session parameters are selected.

Management services are also performed in this layer. These services perform the monitoring, testing, tracing, and statistical recording for all of the myriad session partners.

Sessions

Using the SNA terminology, four permutations of SNA sessions can exist:

- SSCP to SSCP
- SSCP to PU
- SSCP to LU
- LU to LU

Figure 6.22 illustrates the four sessions.

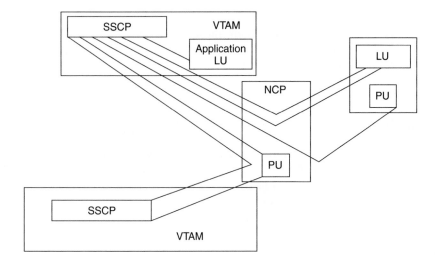

SSCP to SSCP

SSCP to SSCP is the Session SW Services Control Point of VTAM that communicates with another VTAM. This session is used to pass information from one VTAM to another, usually to set up LU-to-LU sessions in a cross domain environment. This SSCP-to-SSCP session will contain the same network or a different network, for example, an SNI session.

SSCP to PU

SSCP to PU is the conversation between a VTAM and a subordinate controller. This can be a large and smart communication controller such as an FEP37×5 or a smaller cluster controller (3×74). This session is established during the activation process of loading or starting the network. The main aim for this session is to act as a conduit for subordinate resources.

In the case of an FEP, the subordinate resources are large in number, consisting of an entire subarea's portion of network devices. For a small cluster controller it is probably just the LUs. Requests can be either the activation or deactivation of LUs or the passing of network management information.

SSCP to LU

The SSCP-to-LU session occurs during network activation or loading when the endpoint LU acknowledges the request. In the case of a terminal or printer, LU activation occurs when the remote device responds positively to the ACTLU request. As for an application scenario, this session is established when the Access Control Block, (ACB) has been activated. It is this all-important contact that allows the ultimate activation to occur—the joining of the end user to its desired application. Acknowledgment of this session occurs when the end user reaches the company logon screen. In VTAM jargon, the company logon screen is known as the USSMSG10 screen.

LU to LU

The LU-to-LU connection is most familiar and probably the most relevant to the MIS community at large. After a request is received from an LU to VTAM to start a session, VTAM will go through its routine to fulfill the request. VTAM will search for the device and the application (session establishment), present to each one the possible characteristics for the session (Bind), and ultimately perform the transport of data from the source to the destination.

Application LUs can have more than one session at a time, called parallel sessions. In many situations, the application is the *primary LU (PLU)* and the terminal/printer or the end user is the secondary LU (SLU).

The **primary LU** initiates a session with another LU.

Open Systems Interconnection and SNA

Thus far the discussion has focused on IBM's Systems Network Architecture, a proprietary model that has been designed for and used by IBM product sets. Standards committees have created an open model for all other market participants to share. One such model was developed by the International Organization for Standardization, called Open Systems Interconnection (OSI) (see Chapter 1).

SNA was designed to allow for the communication of IBM products. Although no direct correlation can be drawn between the layers, the functions provided by each are quite similar. Figure 6.23 outlines the two architectures and the SNA protocols depicting their close technical relationship.

SNA Protocols

The *Customer Information Control System (CICS)* supports transaction processing functions. CICS commands are used to build applications for local and remote SNA systems. CICS functions include file syntax translation, terminal to application communication, security, distributed file access, transaction tracking, recovery and reversal, storage management, and a restart feature.

The **Customer Information Control System** supports transaction processing functions.

The *Information Management System (IMS)* is a transaction processing product similar to CICS. IMS makes it easier to access and use databases. IMS is composed of two products: the IMS Transaction Manager and the IMS Database Manager. Using IMS, multiple applications can share databases, use message switching, and prioritize transaction scheduling. IMS also provides file syntax translation.

The **Information Management System** makes it easier to access and use databases.

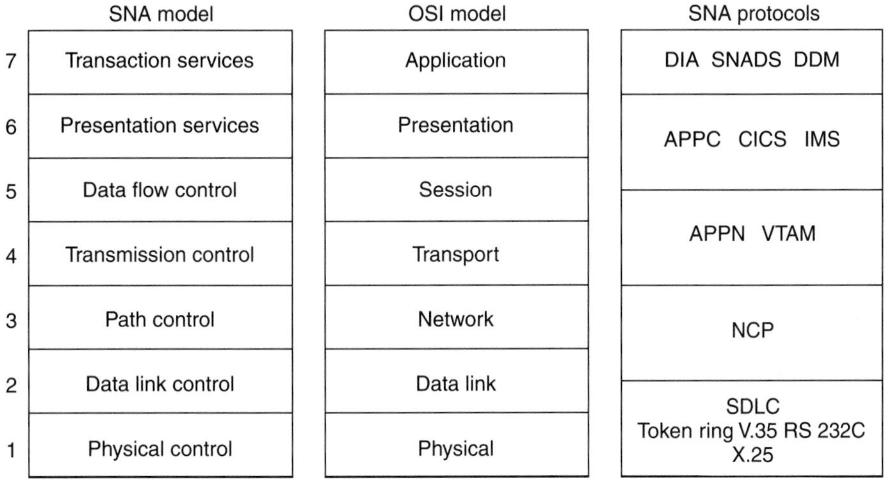

	SNA model		OSI model		SNA protocols
7	Transaction services		Application		DIA SNADS DDM
6	Presentation services		Presentation		APPC CICS IMS
5	Data flow control		Session		APPN VTAM
4	Transmission control		Transport		
3	Path control		Network		NCP
2	Data link control		Data link		SDLC
1	Physical control		Physical		Token ring V.35 RS 232C X.25

FIGURE 6.23
SNA and the OSI reference model. (Reproduced with permission of Cisco Systems, Inc. Copyright © 2001 Cisco Systems, Inc. All rights reserved.)

Distributed Data Management (DDM) provides transparent remote file access to SNA clients by using the operating system call interception service method (similar to NetWare shell).

SNA Distribution Services (SNADS) initiates store and forward transfers of messages/documents between distributed clients.

Document Interchange Architecture (DIA) defines the functions required for exchanging documents between different computer systems.

Advanced Program-to-Program Communications (APPC) is built on APPN to provide peer-to-peer communications without involving the mainframe host.

Figure 6.24 shows a comparison between the OSI layers and the SNA model architecture.

Downstream Physical Units

As it grew and spread its wings, SNA adapted to LAN protocols, namely Token Ring. The method of attaching 3×74 controllers to an SNA NCP is by means of a leased line. Newer and much faster Token Ring attached offerings are gateways and downstream PUs.

When users began to wire LANs on their premises, they also began to use this connection for connectivity to the mainframe (see Figure 6.25). One way to attach these systems to the mainframe was by using a gateway 3174.

The 3174 gateway Token Ring media access control (MAC) layer address is the destination MAC address for the non-gateway 3174 Token Ring attached controllers. The 3174 gateway is therefore providing an FEP function to the non-gateway 3174s (downstream PUs) and a single PU that represents all the downstream PUs to the FEP and the mainframe.

Distributed Data Management provides peer-to-peer communication and file sharing.

SNA Distribution Services initiates store and forward transfers of messages between distributed clients.

Document Interchange Architecture defines the functions required for exchanging documents between computer systems.

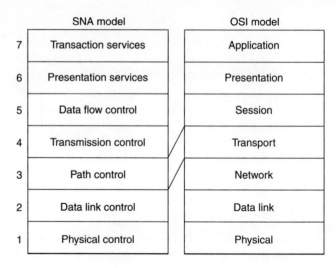

	SNA model	OSI model
7	Transaction services	Application
6	Presentation services	Presentation
5	Data flow control	Session
4	Transmission control	Transport
3	Path control	Network
2	Data link control	Data link
1	Physical control	Physical

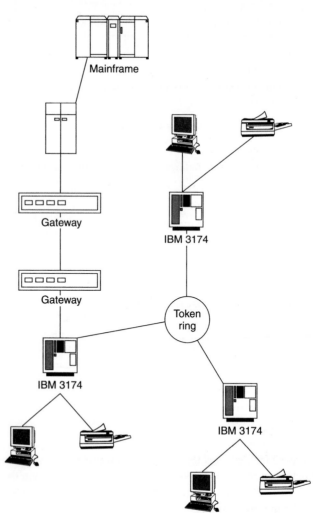

SNA/SDLC Frame Formats

Data passes over an SNA line through the second layer of the SNA seven layer model, the data link control layer. *Synchronous Data Link Control (SDLC)* is a very orderly and controlled method for sending information, checking that the information was received correctly, and retransmitting the information if an error occurred.

There are three types of frames that comprise an SDLC approach:

- The Information frame
- The Supervisory frame
- The Unnumbered frame

The frames break down in the following manner:

1. The Information frame (I-frame) passes SNA requests and responses.
2. The Supervisory frame acknowledges I-frames and reflects the status of the NAU as either ready to receive more data—receiver ready (RR)—or not ready to receive data—receiver not ready (RNR).
3. The Unnumbered frame passes SDLC commands used for data link management.

Message Unit Formats

In SNA there are three types of message units, and depending on where your data is along the path, their destination will dictate what that message unit is composed of:

1. Network Addressable Units (NAUs) use the Basic Information Unit (BIU) to pass information between other NAUs. The BIU is created by the LU and the LU attaches a Request Header (RH) to the Request Unit (RU). Only NAUs use Request Headers. When this has been assembled, the LU passes this on to the Path Control Layer for the additional routing information.
2. The Path Control Layer affixes a Transmission Header (TH) to the BIU. With the transmission header attached, the BIU is then promoted to the stage of a path Information Unit.
3. The Path Control Layer then passes this packet to the Data Link control mechanism for preparation to transmit the PIU.
4. The Data Link Layer adds a Link Header (LH) and a Link Trailer (LT) to the PIU.
5. Now, completely compiled, the packet of data has been formally assembled into an SDLC frame, also known as a Basic Link Unit (BLU).

Link Header

The *Link Header (LH)* contains three fields as illustrated in Figure 6.26:

- The flag
- The SDLC station address
- The Link Header control field

Synchronous Data Link Control is a method for sending, checking, and retransmitting information.

The **Link Header** contains three fields: the flag, the SDLC station address, and the Link Header control field.

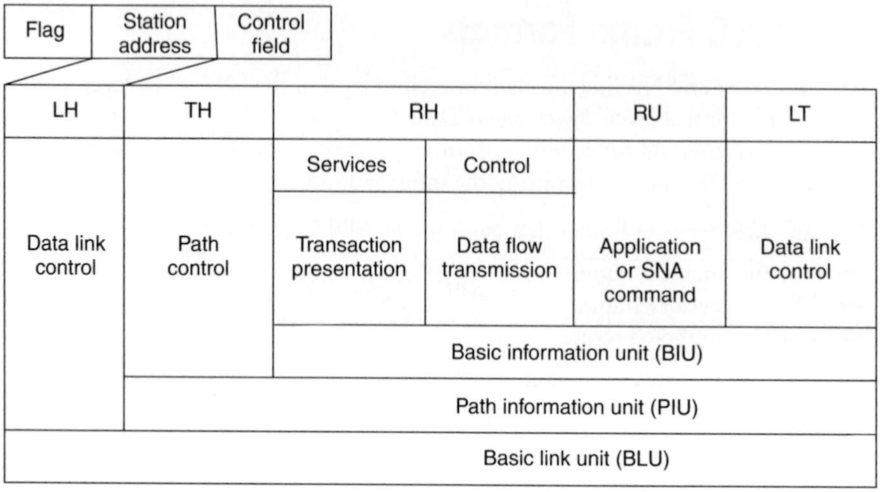

FIGURE 6.26
Link Header.
(Reproduced with
permission of Cisco
Systems, Inc.
Copyright © 2001
Cisco Systems, Inc.
All rights reserved.)

Flag	Station address	Control field

LH	TH	RH			RU	LT
		Services	Control			
Data link control	Path control	Transaction presentation	Data flow transmission		Application or SNA command	Data link control
		Basic information unit (BIU)				
	Path information unit (PIU)					
Basic link unit (BLU)						

The flag is always a hexadecimal value of 7E. This value notifies the data link layer that this is the start of a new SDLC frame.

The SDLC station address can be the address of a single station or a group of stations. In the case of a multipoint configuration, it denotes the destination of the frame. In a traditional NCP, it is the PU ADDR = parameter. The SDLC station address is always a hexadecimal address.

A broadcast function can also be used, (a value of FF) to alert all stations. In addition, a no stations address can be defined, which has a value of 00 and is a reserved address.

The Link Header control field is important because it describes the contents of the field. The field can be unnumbered, supervisory, or information. The unnumbered frame says that the SDLC link level command is being issued. The supervisory frame will convey the status of the state of the receiver, (RR or RNR) or if the prior frame was rejected. The I-frame keeps tally of the frames being sent and received.

Transmission Header

SNA supports a plethora of devices. To discern each device type from the next, SDLC has employed the use of five formats. The format identifiers (FID) are as follows:

Non-SNA FID	Type 0	Used for non-SNA traffic between adjacent nodes
PU 4 and PU 5 FID	Type 1	SNA traffic between subarea nodes
PU 4, PU 5, PU 2 FID	Type 2	SNA traffic between subarea nodes and PU 2.1 adjacent peripheral nodes

FID type	OAF	Sequence number field

LH	TH	RH		RU	LT
		Services	Control		
Data link control	Path control	Transaction presentation	Data flow transmission	Application or SNA command	Data link control
		Basic information unit (BIU)			
	Path information unit (PIU)				
Basic link unit (BLU)					

PU 4 and PU 1 FID	Type 3	SNA traffic between an NCP and a peripheral node
PU 4 and PU 4 FID	Type 4	SNA traffic between subarea nodes that support virtual route protocols
PU 4 and PU 5 FID	Type F	Specific SNA commands between subarea nodes that support virtual and explicit router protocols

The most commonly used FIDs are types 2 and 4. The *Transmission Header (TH)* contains the detail as to which FID type is being discussed. Figure 6.27 illustrates the format of the (TH).

The TH's purpose is to aid the path control level of the SNA seven layer model. The path control layer is responsible for transmitting data from the source to the destination. It is in this TH that the source and destination are denoted in the Origin Address Field (OAF) and the Destination Address Field (DAF). Also found in the TH is the sequence number of this PIU, the length of the Request Unit (RU), and whether it is the first, last, middle, or only RU in this transmission.

The **Transmission Header** aids the path control level of the SNA model.

		RH type Data type SDI BCI PI CEBI	DR ER PI	CDI EDI CEBI

LH	TH	RH		RU	LT
		Services	Control		
Data link control	Path control	Transaction presentation	Data flow transmission	Application or SNA command	Data link control
		Basic information unit (BIU)			
		Path information unit (PIU)			
		Basic link unit (BLU)			

BCI = Beginning Chain Indicator
CDI = Change Direction Indicator
CEBI = Conditional End Bracket Indicator
DR = Definite Response

EDI = Enciphered Data Indicator
ER = Exception Request/Response Indicator
PI = Pacing Indicator
SDI = Sense Data Indicator

For FID type 4s, there is additional information regarding explicit and virtual route information, origin and destination subarea, and element addresses.

Request/Response Header

Figure 6.28 illustrates the format of the *request/response header*. Directly following the TH is a three-byte header that informs us of the type of information being transmitted. The header type is indicated by the first bit, Bit 0. If Bit 0 is equal to 0, it is a request header. If Bit 0 is equal to 1, it is a response header.

The request header provides information to the PIU on how to control the session. Certain information that is provided includes whether or not you are running this session with definite response, pacing, bracketing, or its position in a chained packet.

The response header is used to provide appropriate information to the SDLC protocol. If there are negative responses because of an error in the transmission, the response header will contain related information concerning the reason for the error. A negative response to a request sets the sense indicator bit equal to 1, which indicates that SNA sense data will follow.

Request/Response Units

Request units (RUs) follow request headers. This unit can vary in length and may contain some end user data or an SNA command. The RUs can, in theory, be infinite in size. In practice, you would want to limit the size of the RU to the size of the buffer at the end device. In traditional SNA networks that were composed of 9.6 Kbps analog lines and 3×74 controllers, a value of 256 bytes was a good rule of thumb for the size of the buffer. In today's expanding high bandwidth networks, values of 256 bytes are not optimum.

Link Trailer

The final field in the SDLC format frame is the *link trailer (LT)*. The LT houses two fields as illustrated in Figure 6.29:

- The Frame Check Sequence
- The Link Trailer Flag

The *Frame Check Sequence (FCS)* is used to keep the cadence of the packets going in and out on the network. The transmitting link station executes an algorithm based

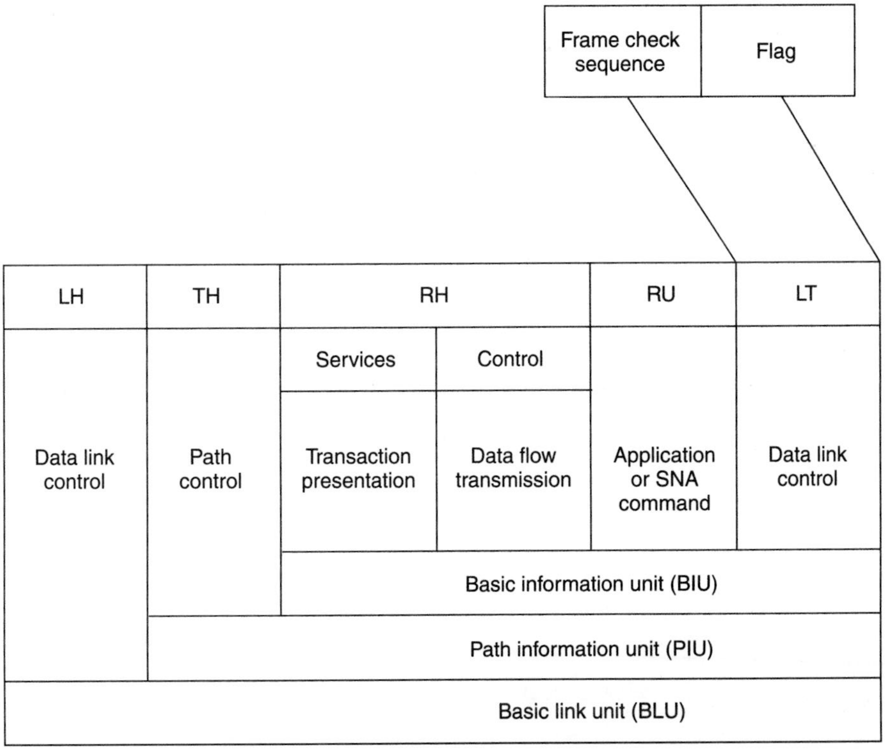

The **link trailer** is the final field in the SDLC format.

The **Frame Check Sequence** keeps the cadence of the packets going in and out on the network.

FIGURE 6.29
*Link trailer.
(Reproduced with permission of Cisco Systems, Inc.
Copyright © 2001 Cisco Systems, Inc.
All rights reserved.)*

LH	TH	RH		RU	LT
		Services	Control		
Data link control	Path control	Transaction presentation	Data flow transmission	Application or SNA command	Data link control
		Basic information unit (BIU)			
	Path information unit (PIU)				
Basic link unit (BLU)					

Cyclic Redundancy Checking is an error checking technique.

on *Cyclic Redundancy Checking (CRC)*. The data for the computation is the link header address field through the RU. The receiving link station performs a similar computation and checks its results against the FCS. If the results are incorrect, a transmission error is sent back to the originating link station and a retransmission will be scheduled.

The Link Trailer Flag indicates that the frame has ended and a new frame is expected. This too is depicted by a hexadecimal 7E.

IBM's Advanced Communications Function/Virtual Communications Access Method (VTAM)

Virtual Communication Access Method (VTAM) is a software package developed and supported by IBM that runs on a mainframe platform and is the cornerstone of an SNA network. VTAM directly controls the transmission of data to and from the network at large. VTAM has many complex abilities to manage, operate, and control the flow of data from application to end user.

Suppose there is an SSCP-to-LU session established with a remote terminal, and the user comes in and wants to log on to an application. VTAM will receive that request and try to establish an LU-to-LU session for that terminal as illustrated in Figure 6.30. The terminal as well as the application is an LU. The first thing VTAM does is to check to see if the application is available within its own control. In other words, to see if an SSCP-LU session is active. This session would then be termed as a same domain session. VTAM determines that it does have an active SSCP-LU session with the requested application. VTAM then passes the request over to the application, and after session start-up procedures, a session is established. At this point, VTAM is out of the flow of data. The data will flow between the application and the terminal until VTAM is in some way required to terminate the session.

Returning to the first example, suppose that the application was not residing on the same mainframe image as VTAM as illustrated in Figure 6.31. VTAM will check a

FIGURE 6.30

Same domain session logon example. (Reproduced with permission of Cisco Systems, Inc. Copyright © 2001 Cisco Systems, Inc. All rights reserved.)

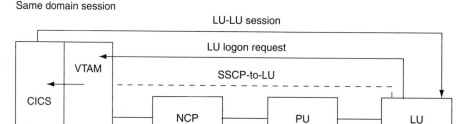

Same domain session

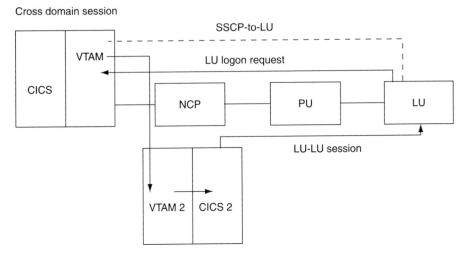

Cross domain session

FIGURE 6.31

Cross domain session establishment. (Reproduced with permission of Cisco Systems, Inc. Copyright © 2001 Cisco Systems, Inc. All rights reserved.)

table built to reference a list of LUs it knows about that are not within the scope or domain. These entries are entitled *Cross Domain Resources (CDRSC)*. They can be created by hard coded entries or are dynamically created when VTAM searches and discovers a resource. VTAM checks the CDRSC tables to see if the requested LU is known. If the resource is found, the LU representing the user who is trying to log on is presented to the other LU owning VTAM for session establishment, and the process begins.

If the LU resource is not found, VTAM will check with other VTAMs over SSCP-SSCP sessions for the LU resource. If no other VTAM has knowledge of the requested LU, the session request is denied.

Another function in VTAM that has a separate name is Cross Domain Resource Manager (CDRM). CDRMs handle the interaction between two VTAMs. If the other CDRM has knowledge, VTAM will:

● Pass along the resource for session establishment.
● Update its own CDRSC table to save a repetitious search for the next resource that wants that particular application.

What happens if VTAM has exhausted all of the VTAMs within its network? Figure 6.32 illustrates what happens next. VTAM will now go outside its network and check with its cross network connections. Similar to firewalling, autonomous SNA networks can communicate without the independence of being in the same network by using SNI.

VTAM will contact its SNI connected CDRMs in hopes of finding the destination resource. If it is found, the same procedure happens with session establishment commencing and updates to CDRSC tables for future reference.

If there were no such active resource in the network, VTAM would send a response back to the terminal indicating that the session could not be bound. That response is not as

●
Cross Domain Resources are entries in a table that reference a list of LUs.

FIGURE 6.32

VTAM cross network session establishment. (Reproduced with permission of Cisco Systems, Inc. Copyright © 2001 Cisco Systems, Inc. All rights reserved.)

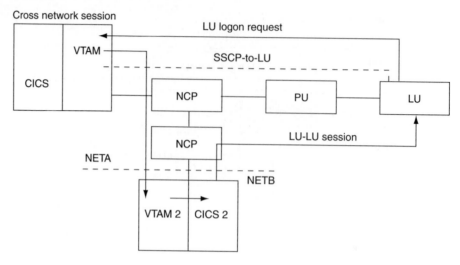

cryptic as it may first appear. In SNA networking after the two partner resources have been located, the next step is to establish the session. In order to do that, VTAM must present the terminal (SLU) to the application (PLU) for connection. VTAM must know the details about the capabilities of the resource. VTAM receives its information through the definitions. There is a specific reference in the LU definition to a VTAM LOG-MODTAB (LOGon MODe TABle) and entry. This table entry has a list of characteristics and features that comprise each and every resource in VTAM's network. VTAM will then gather the session information, which consists of buffer sizes, screen sizes, types of data streams, and definite/exception response mode to the destination resource.

A **bind** is a request to establish a session between the device and the application.

This presentation of the request of a session between the device and the application is called a *bind*. In legacy SNA, binds are either accepted or refused. In current APPN networks and with the advent of intelligent devices out in the field that are LU 6.2 capable, some binds can be negotiated.

Network Control Program (NCP)

IBM'S Network Control Program (NCP) runs outbound of the mainframe in the communications controller. Typically, in today's SNA network the traditional IBM 3745 is depicted as the hardware platform for this software. NCP is designed to handle the end station traffic and management of the device and act as a conduit to VTAM. Defined in NCP are all of the SDLC lines, Token Ring connections, adjacent subarea link stations, PUs, and LUs.

An NCP gen is created at the mainframe host and is loaded into the box during activation. Another less highlighted piece of IBM software actually performs any loading and/or dumping of the NCP; it is called Systems Support Program (SSP). NCP can be likened to the IOS router configuration as it pertains to form and function,

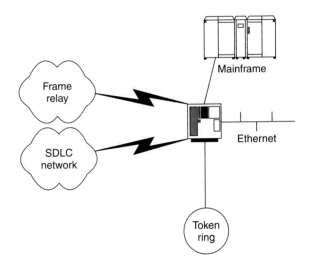

FIGURE 6.33

NCP in an SNA network. (Reproduced with permission of Cisco Systems, Inc. Copyright © 2001 Cisco Systems, Inc. All rights reserved.)

but not process. An NCP gen is assembled beforehand and has software checks before loading into the device. Figure 6.33 illustrates how NCP fits in the SNA network.

Summary

The IBM mainframe and SNA form the foundation of 90 percent of the Fortune 1000's information delivery to corporate end users and customers. At one time, IBM's mainframe and its Systems Network Architecture (SNA) owned the corporate networking arena. However, corporate America's embrace of TCP/IP and routers has dethroned IBM and SNA from its once dominant perch. However, this did not diminish the importance of delivering the information from mainframe computers with applications written in support of IBM's SNA.

IBM networking today consists of essentially two separate architectures that branch from a common origin. Before contemporary networks existed, IBM's Systems Network Architecture (SNA) ruled the networking landscape; thus it is often referred to as traditional or legacy SNA.

With the rise of personal computers, workstations, and client-server computing, the need for a peer based networking strategy was addressed by IBM with the creation of Advanced-Peer-to-Peer Networking (APPN) and Advanced Program-to-Program Communications (APPC).

Although many of the legacy technologies associated with mainframe based SNA have been brought into APPN based networks, real differences exist. This chapter has tried to cover some of the differences.

Chapter 7 will discuss the IPX and RIP protocols, which have replaced the SNA architecture.

Review Questions

1. How does the concept of hierarchical networks correlate to SNA?
2. What is SNA subarea networking?
3. What are the three devices that comprise the subarea network?
4. What is peer networking?
5. Give some examples of APPN networking.
6. What are ENs and NNs?
7. What are the four resources that comprise the TCP/IP network?
8. What is a single domain?
9. What is a multidomain?
10. What is SNI and what is its purpose?
11. What are the two types of frame relay devices and how do they operate in an SNA network?
12. What are the differences between virtual routes and explicit routes?
13. What are the fundamental concepts to understanding SNA?
14. What are the network components that comprise an SNA network?
15. What are links and what is their function in an SNA network?
16. What are NAUs?
17. What are the three types of NAUs?
18. How does network addressing operate in an SNA environment?
19. List the seven layers that comprise the SNA architectural model.
20. What are the different sessions that exist in SNA?

Summary Questions

1. What is an SSCP-to-SSCP session?
2. What is an SSCP-to-PU session?
3. What is an SSCP-to-LU session?
4. What is an LU-to-LU session?
5. How does SNA interoperate with OSI?
6. List the SNA protocols?
7. How are downstream physical units attached to an SNA network?
8. What are the three types of SDLC frames?
9. What are the three types of message units in SNA?
10. What is VTAM?

Further Reading

Sackett, George. *SNA Networking.* Indianapolis: Cisco Press, 1999.

NetWare Protocols

Objectives

- Explain why NetWare is still a dominant player in the marketplace.
- Discuss the various NetWare concepts.
- Explain the Internet Packet EXchange (IPX) Protocol.
- Explain how the Routing Information Protocol (RIP) operates.
- Explain how IPX addressing works and how to determine IPX addresses.
- Explain the functions of Sequenced Packet Exchange (SPX) Protocol and SPX II.
- Discuss the purpose of NetWare Core Protocol (NCP).
- Discuss how the NetWare Link Services Protocol (NLSP) operates and how it helps to solve the problems created by RIP.

Key Terms

Introduction

By far the most frequently installed workgroup client-service network operating system is Novell NetWare (although, as of this writing, Microsoft networking may have passed Novell). The NOS allows workstations and their associated file servers to exchange files, send and retrieve mail, and provide an interface to SNA terminal emulation and database programs, among other applications.

Novell NetWare's popularity grew because of its ability to support multiple manufacturers' network interface cards and many different types of access methods (Ethernet, Token Ring, ARCnet, and FDDI). Novell's install base primarily includes DOS based personal computers, but it also offers connectivity services to AppleTalk (which will be discussed in Chapter 8), UNIX, IBM SNA (see Chapter 6), and OS/2 environments. Because of its low cost during the ramp-up of the LAN environment in the early 1980s, the access method of ARCnet was very popular with NetWare environments. NetWare supported the ability to bridge packets to and from ARCnet, Ethernet, and Token Ring networks and, therefore gave users the ability to communicate to file servers and users regardless of the access method that was implemented. Because NetWare supported a native IPX router, NetWare can support the translation between varieties of access methods.

As Ethernet and Token Ring grew in popularity and became the access methods of choice, NetWare aided the migration to these networks from ARCnet. Furthermore, Novell supported almost any manufacturer of network interface cards (NICs). Novell developed a strong operating system from this environment and until a few years ago was the number one manufacturer of workgroup computing operating systems, claiming from 60 to 70 percent of the market share. Finally, Novell's design goal was to be the most competitive as well as to have the highest performance LAN operating system.

The key topics for discussion are (1) the LAN operating system of Internet Packet Exchange (IPX), (2) the WAN protocol operation of Routing Information Protocol (RIP), and (3) the user interface workstations and servers, including the workstation shell—the NetWare File Service Core Program—and the NetWare Core Protocol. As a

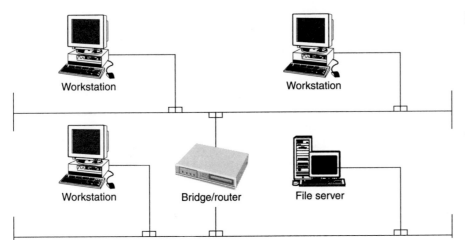

Workstation

Workstation

Workstation

Bridge/router

File server

user when you start and log in to the NetWare network, exactly what happens and how does the file server and workstation communicate over the LAN or an internetwork?

For users in the Novell environment there are two primary or user identifiable physical entities: a workstation and a server interconnected through a LAN as illustrated in Figure 7.1. The workstation is usually a personal computer with a DOS, OS/2, UNIX, or Apple operating system. The workstation makes requests for file and print services from an entity known as a file server. The file server runs a proprietary operating system known as NetWare Core Protocol (NCP). NCP has the job of servicing requests from the users' workstations and returning responses to these requests.

Concepts

Novell NetWare is a LAN workgroup network operating system that permits workstations and their servers to communicate. The entities of NetWare that allow this are:

- Access Protocols (Ethernet, Token Ring, ARCnet, ProNET-10, FDDI)
- Internet Packet Exchange (IPX)
- Routing Information Protocol (RIP)
- Service Access Protocol (SAP)
- Sequence Packet Exchange (SPX)
- NetWare Core Protocol (NCP)—run in the server and the workstation; in the workstation it is known as the shell.

Figure 7.2 shows the relationship between the OSI model and these NetWare processes.

The ability to send NetWare commands and user data across the network relies on a *Xerox Network System (XNS)* network layer Internet Datagram Protocol (IDP) derivative

●
**Xerox Net-
work System** is
a protocol suite
used by many net-
working companies
as their primary
transport protocol.

FIGURE 7.2

The OSI model and Novell NetWare. (Reproduced with permission of Cisco Systems, Inc. Copyright © 2001 Cisco Systems, Inc. All rights reserved.)

Application	NetWare core protocol (NCP)	Service advertising protocol (SAP)	Routing information protocol (RIP)
Presentation	NetWare core protocol (NCP)	Service advertising protocol (SAP)	Routing information protocol (RIP)
Session		NetBIOS	
Transport	Sequence packet exchange (SPX)		
Network	Internet packet exchange (IPX)		
Data link	Access protocols and wiring techniques Ethernet, token ring, ARCnet, coaxial cable, unshielded twisted pair.		
Physical	Access protocols and wiring techniques Ethernet, token ring, ARCnet, coaxial cable, unshielded twisted pair.		

Internet Packet Exchange is the protocol used for transferring data from servers to workstations.

protocol known as *Internet Packet Exchange (IPX)*. IPX is a network layer protocol implementation that allows data to be transferred across a local or a wide area network. IPX was derived using the IDP protocol of XNS. The protocols that allow users access to their file servers are known as NCP and the workstation shell program.

Internet Packet Exchange (IPX)

Before networks and the architecture of distributed computing, files that resided on a personal computer remained on the personal computer and were transferred to another computer by copying the file to a diskette and physically transporting that diskette to the other computer. This delivery system was either by human intervention or by addressing the diskette and having a mail service deliver the disk. Initial network protocols did allow for files to be exchanged between computers but it was limited to that. They did not allow the files to be downline loaded to the requesting station or to be seamlessly executed like the files that were located on a local hard disk once they were loaded to the requesting station.

With the advent of a network, data are still transported to another computer but the data to be transferred need to be formatted so that the network will understand what to do with the data. There are network delivery commands that need to be transferred between the users' workstations and their file servers so that the data are delivered to the proper place. Whether the data are user data or network commands, the data need to be formatted for transfer to the LAN. The IPX software is the network software used for this process. IPX provides the delivery service of data.

IPX is the interface between the file server operating software NCP/workstation shell (the Novell program that runs on the workstation) and the access protocols (Ethernet, Token Ring, or ARCnet). IPX will accept data from the workstation's shell or NCP and format the information for transmission onto the network. IPX will also accept data from the LAN and format it so it can be understood by the shell or NCP. IPX follows the XNS network layer protocol of Internet Datagram Protocol (XNS IDP). The IPX protocol was implemented for use on local and remote networks. Novell followed the XNS architecture and adapted it for use in their environment. IPX formats (provides addressing) data for network transmission and provides the delivery system for this data. NCP is the program that determines, on a DOS workstation, whether the data are destined for a network device or whether the data are destined for a local device (on the workstation).

There are two reasons for using IPX as the network level interface. First, because IPX follows the XNS protocol, and this architecture was specifically built to run on top of Ethernet, IPX was designed to run on LANs. IPX is the only networking protocol that was designed to run on top of Ethernet. Other protocols such as TCP/IP and AppleTalk were adapted to run on top of Ethernet. Because the architecture was already written, and IPX was an open specification, Novell simply had to implement the software to the architecture.

The second reason for using IPX as the network level interface is not only will IPX carry data but it can also easily route this data. This has many advantages. It allows NetWare to support multiple network architectures easily. Novell calls this bridging. Novell called it bridging because IPX bridged together multiple networks. In reality, it was routing. This allowed ARCnet, Ethernet, Token Ring, and StarLAN networks to exchange data between the different LANs transparently. It is true that bridges did not enter the commercial marketplace until 1985, but this is well after Novell had established itself as a major player in network operating systems. With the release of NetWare 3.x, Novell has changed the term and now properly refers to this ability as routing.

By implementing only the network layer stack and a user interface (the shell), the amount of RAM consumed in a user's workstation was very small, sometimes as small as 20K of RAM. This was very important, for the first generation of PCs was limited to 640K of RAM to run application software. Applications could not be loaded into upper memory. Later versions of PC operating systems allowed this. As the application software grew larger, the NetWare software remained the same and allowed larger applications to run.

IPX is a full implementation XNS IDP protocol with NetWare adapted features. With this in mind, IPX has a maximum packet size of 576 bytes. Novell does not strictly adhere to this when data are exchanged between two devices on a local LAN. The actual data size is 546 bytes because the IPX packet header consumes 30 bytes. IPX contains two entry points and two exit points. Data and commands are entered to and from either an application or the LAN. Data are transmitted from IPX from the same two points. It all depends on the direction of the data flow.

Because data are packetized and transmitted from the network layer software, data are delivered on a best-effort basis. The transport layer *Sequenced Packet Exchange (SPX)* is the protocol that provides reliable packet delivery. By implementing proprietary transport

Sequenced Packet Exchange is the protocol that provides reliable packet delivery.

layer software in NCP and simple transport software in the shell, NetWare did not have to implement full transport layer software in the protocol stack of the client workstation. This saved valuable RAM at a time when DOS workstations had a limitation of memory in which to run applications. By implementing a small, reliable protocol stack such as IPX, the speed at which stations communicated with each other increased, especially when the medium was upgraded from ARCnet to Ethernet. SPX was developed later and implemented into the shell. SPX is primarily used for peer-to-peer communications and utility programs, such as RCONSOLE, and SNA gateways. IPX provides many functions, which can be grouped into two categories: packet formatting and data delivery.

IPX Routing Architecture

The Data Delivery System

Packet Structure

Before developing a complete understanding of how data are transmitted and received on a network, some intermediate functions need to be addressed. The following text will address this. Any data that are to be transmitted on a LAN need to be formatted for transmission to the network. All data handed down to IPX from an upper layer protocol are encapsulated into an entity known as a packet. This process is similar to writing a letter. You write the letter, place the letter into an envelope, and attach an address to the receiver (the destination). On the envelope you place the address of the destination and your return address so the post office knows where to deliver the letter and the receiver knows where to send a response if needed. This is similar to the process IPX uses to format the data to deliver the data to the network.

Figure 7.3 illustrates the IPX packet structure. The proprietary packet structure of IPX contains the following fields:

Checksum is a method for checking the integrity of transmitted data.

Checksum: This field contains the *checksum* (16 bits) for the IPX packet. The checksum can be thought of as a fancy parity check. The objective of the checksum is to ensure that the bits transmitted are the same bits that are received. No bits in the packet were transposed during the transmission. The sending station will perform the checksum algorithm on the packet and place the result of the checksum in this field. The receiving station will also perform a checksum on the IPX portion of the packet and generate a checksum. That checksum is checked with the checksum in the packet. If there is a match, that packet is said to be good. If the two checksums do not match, that packet is said to contain an error and the packet will be discarded. Because this algorithm is performed at the data link layer (a CRC of 32 bits provides better accuracy than a 16-bit CRC), Novell has opted to disable this feature of IPX, considering it to be redundant and time consuming, resulting in unnecessary overhead. With IPX this field is set to FFFF to indicate that checksumming is turned off. There are some IPX routers that will use this field to check whether the packet is a Novell encapsulated IPX packet or a packet of another type.

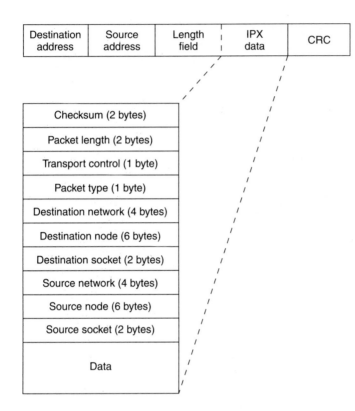

FIGURE 7.3
Novell proprietary IPX packet format. (Reproduced with permission of Cisco Systems, Inc. Copyright © 2001 Cisco Systems, Inc. All rights reserved.)

Length: The length field is used to indicate the total length of the IPX packet, including the IPX header checksum field. This means the length of the IPX header and data fields. The minimum length allowed is 30 bytes (the size of the IPX header fields) and the maximum number is 576 (indicating a maximum of 546 bytes for the total data field). For communications on a LAN, the number of bytes may be as high as the transmission medium allows: 1,500 bytes for Ethernet, 1,496 for IEEE 802.3 (including IEEE 802.2 headers), and 4,472 for 4-Mbps Token Ring. Under Novell's *packet burst mode,* large packets may be transferred between two stations residing on different LANs. This feature is available under NetWare 4.x.

Transport Control: The transport control field is initially set to 0 by the sending station. This field counts the number of *hops* (number of routers) the packet encountered along the way. Because the maximum number of routers a packet is allowed to traverse is 15 (a network 16 hops away is considered unreachable) the first 4 bits are not used. This is also used by routers that support Service Access Protocol (SAP) reporting and other file servers to indicate how far away a server providing certain services is from the recipient of the packet. When a packet type is transmitted onto the network, the sending station will set this field to 0. As the packet traverses each

Packet burst mode allows large packets to be transferred between two stations residing on different LANs.

Hops refer to the number of routers a data packet encounters along the way.

Table 7.1 Xerox Assigned Packet Types in Hex

Protocol	Packet Type (hex)
Unknown	0
Routing Information	1
Echo	2
Error	3
Packet Exchange Protocol (PEP) used for SAP	4
Sequenced Packet Exchange (SPX)	5
Experimental	10-1F
NetWare Core Protocol	11
NetBIOS	14

router if needed on its way to the destination, each router will increment this field by 1. The router that sets it to 16 will discard the packet.

Packet Type: The packet type field is used to indicate the type of data in the data field. This is the Xerox registration number for Novell NetWare. It identifies the XNS packet as a NetWare packet. Because IPX is a derivative of XNS's IDP protocol it follows the assigned types given by Xerox as defined in Table 7.1.

Destination Host: The destination host field is a 48-bit field that contains the physical address of the final (not any intermediate hosts such as routers that it may traverse on the way to the destination) destination network station. An analogy of this is the address displayed on a letter. Another analogy is the seven-digit number minus the area code on the phone system. If the physical addressing scheme does not use all 6 bytes then the address should be filled in using the least significant portion of the field first and the most significant portion should be set to 0s (users for ARCnet and ProNET-10). For Ethernet and Token Ring, it is the 48-bit physical address of the NIC. This address indicates the final ultimate destination. It does not indicate any physical address of any intermediate stops along the way.

The **socket** is an indicator of the process to be accessed on the destination station.

A **socket number** is an integer number assigned to a specific process running on a network station.

Destination Socket: The destination *socket* is a 16-bit field that is an indicator of the process to be accessed on the destination station. A *socket number* is an integer number assigned to a specific process running on a network station such as the file service that runs on a file server. Each and every service that runs on a file server will be assigned a socket number in hexadecimal. For example, if the file service is assigned a socket number of 0451, any workstation requesting this service must set this field to 0451 to be properly serviced by the file server. Because IPX follows the XNS standard the socket numbers defined in Table 7.2 are reserved. Other socket numbers that

Table 7.2 Assigned Socket Numbers in Hex

Registered with Xerox	0001-0BB8
User definable	0BB9 and higher

are reserved and that may not be used without the permission of Xerox Corporation are in the range of 1-BB8. A number other than this may be used dynamically.

Source Network: The source network is a 32-bit field that contains the network number of the source network. This indicates the network number from which the packet originated. A network number of 0 indicates that the physical network where the source resides is unknown. Any packet of this type received by a router will have the network number set by the router. When IPX is initialized on a workstation, it may obtain the network number by watching packets on the LAN and derive its number from there. It may also find its network number from the router. Network numbers are not assigned to the workstation.

Source Host: The source host is a 48-bit field that contains the physical address of the source host, which is the network station that submitted the packet. This field represents the host number from which the packet originated. Like the destination host field, if the physical address is less than 6 bytes long, the least significant portion of the field is set to the address and the most significant portion of the field is set to 0s (users for ARCnet and ProNet-10). Otherwise, it is set to the 48-bit address of the LAN interface card.

Source Socket: This 16-bit field contains the socket number of the process that submitted the packet. It is usually set to the number in the dynamic range (user definable range). Source, destination host and source, and destination network are self-explanatory. Sockets are a little more elusive.

Multiple processes may be running on a workstation, such as OS/2 and UNIX, and multiple processes will definitely be running on a file server. Sockets are the addresses that indicate the endpoint for communication. A unique socket number indicates which process running on the network station should receive the data. Sockets represent an application process running on a network workstation. There are two types of sockets: static and dynamic. Static sockets are reserved sockets that are assigned by the network protocol or application implementor, such as Novell, and cannot be used by any other process on the network. Dynamic sockets are assigned randomly and can be used by any process on the network.

For example, to access the file services of a server, IPX would fill in the destination and source network, the destination host number of the file server, and source host number of its workstation. The destination socket number would be set to 0451 in hex. This is a well-known socket number because it is static (it will never change). Novell has defined this particular socket number. The source socket is assigned by IPX at the source workstation and will be a dynamic socket number. IPX will pick the socket number from a range of 4,000 to 6,000 hex. The source socket is used by the destination as a socket number to reply to. It indicates the socket number that made the request. In this way, when the packet arrives at the server, the server will know that the packet is destined for this particular host. The server will also know the transmitting station is requesting something from the server (socket 0451). Deeper into the packet is a control code that indicates exactly what the transmitter of the packet wants such as creating a file, deleting a file, directory listing, or printing a file. Once the command is interpreted and processed, the server will return data to the transmitter of the packet. The server needs to know which endpoint of the workstation will receive this data (which process submitted the request). This is the purpose of the source socket number. The file server will format a packet, reverse the IPX header fields (source and destination headers), set the destination socket number to the number indicated in the received packet, of source socket number and transmit the packet (see Table 7.3).

The socket number (source or destination network number, source or destination host number, and source or destination port number) is the absolute address of any process on the network. With the combination of these fields, any process on any network on any network station can be found. IPX controls all socket numbering and processing.

Packet formatting is a function that transmits data across the network.

That is one of the functions provided by IPX, which is referred to as the *formatting of data into a packet* so that the data may be transmitted across the network. The next function of the IPX protocol is the ability to route packets directly to a workstation

Table 7.3 NetWare Assigned Port (Socket) Values

File Server	
0451h	NetWare Core Protocol
Router Static Sockets	
0452h	Service Access Protocol
0453h	Routing Information Protocol
Workstation Sockets	
4000h–6000h	Dynamically assigned sockets used for workstation interaction with file servers and other network communications
0455h	NetBIOS
0456h	Diagnostic packet

on the same LAN or to a network station on a remote LAN. The Novell implementation, which allows this function to occur, is referred to as routing.

Encapsulation at the Data Link Layer

Novell provides support for several *encapsulation* methods. This resulted from a long history of supporting many different vendors' network interface cards.

The six methods of data link encapsulation supported by Novell are:

1. Novell proprietary
2. IEEE 802.3 with IEEE 802.2
3. Ethernet
4. IEEE 802.3 with SNAP
5. Token Ring
6. ARCnet

Encapsulation is a method that is used when bridging dissimilar networks.

The six encapsulation methods of packets are illustrated in Figure 7.4. When installing a Novell network, the installer must choose between the encapsulation methods. It is necessary for the installer to select one. The software will not try to figure out the format. Once set, a transmitting station will format the packet and the receiver will read the packet according to the setting during the installation. Two communicating stations must use the same encapsulation type.

Some network installations prefer the Novell proprietary. This is basically the IEEE 802.3 MAC header encapsulation. Immediately following the length field will be the beginning of the IPX header, which will be set to FFFF, indicating that the checksumming is turned off. Some router vendors use this field to indicate that it is a Novell packet. IEEE 802.2 is not included in the packet. The reason for the entire packet formats is compatibility. During the ramp-up of Ethernet, different vendors supported different encapsulation techniques. The Ethernet packet header was the first encapsulation technique used with Ethernet networks. When IEEE 802.3 formally adopted CSMA/CD (Ethernet), they changed the packet format. Novell again changed the packet format to include this new type. With the IEEE 802.3 packet format, Novell supports both the IEEE 802.2 and the SNAP protocols. Novell decided to support its own packet format.

The installer of a Novell network may choose any of the preceding encapsulation methods. It all depends on the type of network interface cards that are installed in the network. In addition, Novell supports Token Ring encapsulation methods. The two methods supported are IEEE 802.5 with IEEE 802.2 SNAP. Figure 7.4 has some of the fields filled in. On the Ethernet frame, the Ethertype of 8137 identifies a Novell frame. On the IEEE 802.3 with IEEE 802.2, E0 is the SAP address assigned by the IEEE to Novell. On the SNAP packet, the organization of 000000 identifies an encapsulated Ethernet packet and 8137 is the Ethertype for Novell.

FIGURE 7.4

Encapsulation methods. (Reproduced with permission of Cisco Systems, Inc. Copyright © 2001 Cisco Systems, Inc. All rights reserved.)

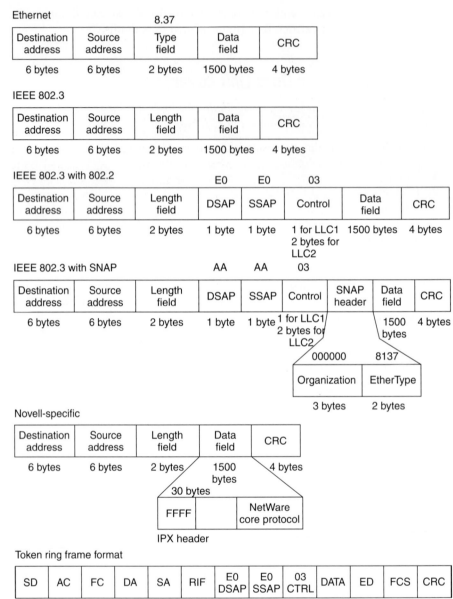

IPX Routing Functions

The routing function allows packets to be forwarded to different networks through a device known as a router. An analogy would be when a phone number such as 292-2769 is dialed in New Jersey, the local switching office knows by the first three digits (the local exchange number) that the number is local and should be routed to a destination on the local phone system. If necessary, the call is switched between exchange offices in the local area to its final destination. When a number such as 718-337-5678 is dialed, the local switching office knows the call is to be routed to a distant location. In this case the call is to be routed to an exchange in New York State. The aforementioned example shows how network IDs and routers work.

There are two available types of routers on a Novell network. First, Novell implements a routing function in their operating system. Previous to Novell 3.x, Novell documentation called their routers bridges. However, in reality they are routers and since NetWare 3.x Novell has officially changed the names to internal and external router. The internal router usually performs some other task or service, as well as the routing function. These tasks or services may be file and print services or a gateway service to SNA applications. The external router is a workstation such as a personal computer consisting of multiple network interface cards, and its sole function is to route packets. The external router performs no additional tasks or services.

Independent router manufacturers such as Wellfleet Communications, 3Com, and Cisco Systems can also participate in a Novell network, providing IPX routing functions. These independent router manufacturers are fully compliant with the Novell routing scheme. The independent router manufacturers usually provide multiprotocol routers that will route other types of packets such as TCP, AppleTalk, and DECnet as well as NetWare packets. The protocols are routed simultaneously in the same router. These are high performance routers specifically built to perform routing functions. They are not to be used as personal computers acting as routers.

To route a packet, routers will accept only packets directly addressed to them and will determine the best path on which to forward the packet. This involves multiple processes, which are explained next. Routing tables will be discussed first.

IPX Routing Tables

Routers need to know of all other available routers and therefore all other active networks on its internet. The IPX router keeps a complete listing of the networks listed by their network numbers. This is known as a *routing table*. Each router in a NetWare internetwork will contain a table similar to Table 7.4.

The entries in the routing table will let the router determine which path to forward the packet to Figure 7.5 illustrates a network depicting the aforementioned routing table. Network numbers are 32-bits in length. As shown in Figure 7.5, the networks are separated by special devices known as routers. The combination of all the networks together is called an internet or internetwork (see Chapter 1). Each of the routers needs to know of each network on the internet, and the Routing Information Protocol (IPX-RIP) process is the method for exchanging this information. Among other things, Novell's IPX protocol changes XNS IDP RIP implementation slightly to provide a timer, which are referred to as ticks. This provides RIP the ability to be a true distance vector routing protocol with true cost attributes, not just a hop count. *Ticks* are the amount

A **routing table** keeps a complete listing of the networks listed by their network numbers.

Ticks are primarily used to set timers on a NetWare workstation.

Table 7.4 IPX Routing Table

Network Number	Number of Hops to Network	Number of Ticks to Network	NIC	Intermediate Address of Forwarding Router	Net Status	Aging Timer
00000020	1	2	A	Local		0
00000030	1	2	B	Local		0
00000040	1	2	C	Local	R	0
00000050	2	3	B	02608C010203		1
00000060	2	3	A	02608C040506		2
00000070	3	4	A	02608C010304		2

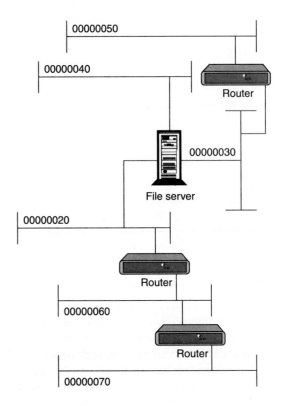

FIGURE 7.5

Network number assignment with routers. (Reproduced with permission of Cisco Systems, Inc. Copyright © 2001 Cisco Systems, Inc. All rights reserved.)

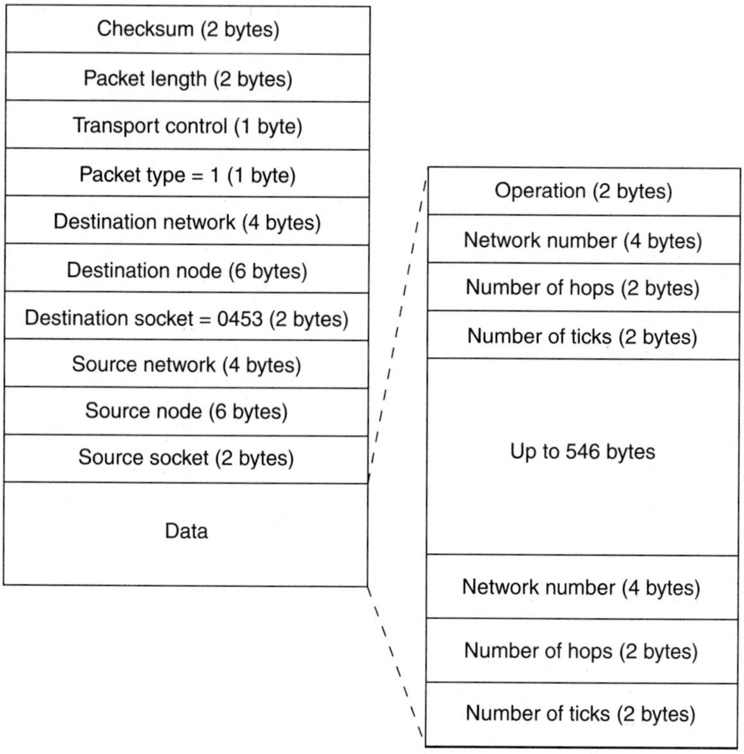

Checksum (2 bytes)
Packet length (2 bytes)
Transport control (1 byte)
Packet type = 1 (1 byte)
Destination network (4 bytes)
Destination node (6 bytes)
Destination socket = 0453 (2 bytes)
Source network (4 bytes)
Source node (6 bytes)
Source socket (2 bytes)
Data

Operation (2 bytes)
Network number (4 bytes)
Number of hops (2 bytes)
Number of ticks (2 bytes)
Up to 546 bytes
Network number (4 bytes)
Number of hops (2 bytes)
Number of ticks (2 bytes)

FIGURE 7.6
IPX RIP packet format. (Reproduced with permission of Cisco Systems, Inc. Copyright © 2001 Cisco Systems, Inc. All rights reserved.)

of time that is required to reach that path. Ticks are primarily used to set timers on a NetWare workstation. Table 7.4 is depicted in the PC file server routing table illustrated in Figure 7.5. The file server has networks 20, 30, and 40 directly connected.

The first entry of the router table, the network number, contains the network numbers that are in place on the internet. A router will exchange its routing table with other routers on the network. The actual entries in the table that are exchanged with other routers are network number, hops, and ticks. This information is transferred via an RIP packet (explained later) illustrated in Figure 7.6. The other entries pertain to the local router and are not distributed by each router. Every router will build its own entries. A router will receive these updates (routing tables) from other routers on the internet through a process known as RIP. From this received information, a router will build its own table and thus a picture of the internet.

The second entry shows the number of routers that must be traversed to reach this network. Anytime that a packet must traverse a router to reach a destination, the process of traversing the routers is referred to as a hop. Therefore, if a packet must cross over four routers to reach the final network, it is said that the network is 4 hops away. The term hops is also called a *metric*. Four hops is the same as a 4-metric count.

Metric is another name for the term *hops*.

The next entry is the tick counts. This number indicates an estimated time necessary to deliver a packet to a destination on that network. This time is based on the segment type. A tick is approximately $\frac{1}{18}$ of a second. This number is derived from an IBM type personal computer clock being ticked at 18 times a second. In actuality, there are 18.21 ticks in a personal computer clock for every second elapsed. At a minimum, this field will be set to a 1—18 would indicate 1 second.

For locally attached segments with more than 1 Mbps transmission speed (Ethernet and Token Ring), the NIC driver will assume a tick of 1. For serial network segments (X.25, synchronous lines of T1 and 64 Kbps, and asynchronous), the driver will periodically poll to determine this time delay. For a T1 circuit, the tick counter is usually 6 to 7 ticks per segment. Any changes in this time will be indicated to the router and propagated to other routers on the network. These numbers in the tables are cumulative, meaning that as each router broadcasts its routing table, this number will not be reset. It is the sum of all the paths' tick counts to reach a destination network.

The network in-
terface card
records the num-
ber from which the
network can be
reached.

The *network interface card (NIC)* entry field records the NIC number from which the network can be reached. It indicates the controller card from which the router received this reachability information. A Novell file server can hold four network interface cards or NICs. It is the same as a physical port number in a stand-alone router (not a personal computer or file server acting as a router). The intermediate address entry contains the physical node address of the router that can forward packets to each segment. If the network is directly attached, the entry will be empty. If the network to be reached requires the use of another router, this entry contains the physical address of the next router to send the packet to. The physical address is extracted from RIP updates (a router broadcasting table) sent by those routers. An entry in the NIC field would be valid only if the router were located in a PC. Otherwise, this field would indicate the router physical port number.

The net status entry indicates whether the network is considered reliable. The age entry is used to indicate how long it has been since a routing update has been made. This field is used to age-out (delete) entries for networks that have not been heard from in a certain amount of time. These timers follow the XNS specifications. This number can be in seconds or in minutes, depending on the manufacturer of the router.

In short, a routing table contains a listing of network numbers and an associated path, whether direct or indirect, to deliver the packet to its final destination network. With the exception of the next hop router address, the entries in the routing table do not contain any physical addresses of the network stations that reside on the internet. The only physical address in the table is that of another router to which a packet destined for a remote network may be addressed. Routers do not know which other end stations are on the networks they connect to. The final destination (physical address of the final destination) is embedded in the IPX header (the destination host). Once the router determines that the final destination network number is directly attached to the router, it will extract the destination host number from the IPX header and address the packet and deliver it to the directly attached network segment.

Routing Information Protocol (RIP)

Background

The *Routing Information Protocol (RIP)* is a distance vector routing protocol that uses hop count as its metric. RIP is widely used for routing traffic in the global Internet and is an *Interior Gateway Protocol (IGP),* which means that RIP performs routing within a single autonomous system. Exterior gateway protocols, such as *Border Gateway Protocol (BGP),* perform routing between different autonomous systems. The original incarnation of RIP was the Xerox protocol GWINFO. A later version, known as routed, shipped with Berkeley Standard Distribution (BSD) Unix in 1982. RIP itself evolved as an Internet routing protocol, and other protocol suites use modified versions of RIP. The AppleTalk *Routing Table Maintenance Protocol (RTMP)* and the Banyan VINES *Routing Table Protocol (RTP),* for example, are based on the Internet Protocol (IP) version of RIP. The latest enhancement to RIP is the RIP 2 specification, which allows more information to be included in RIP packets and provides a simple authentication mechanism.

IP RIP is formally defined in two documents: Request For Comments (RFC) 1058 and 1723. RFC 1058 issued in 1988 describes the first implementation of RIP, while RFC 1723 issued in 1994 updates RFC 1058. RFC 1058 enables RIP messages to carry more information and security features. To exchange their tables with other routers on the internet, IPX uses an algorithm known as RIP. The RIP algorithm is the most widely used routing algorithm in use today. Variations of this protocol exist on TCP/IP, AppleTalk, IPX, XNS, and a host of other proprietary XNS vendor implementations.

The functions of the RIP protocol are:

1. Allow the workstation to attain the fastest route to a network by broadcasting a route request packet, which will be answered by the routing software on the Novell file server or by a router supporting IPX RIP.
2. Allow routers to exchange information or update their internal routing tables.
3. Allow routers to respond to RIP requests from workstations.
4. Allow routers to become aware when a route path has changed.

Novell IPX Addressing

A Novell IPX address has 80 bits: 32 bits for the network number and 48 bits for the node number. The node number contains the MAC address of an interface. Novell IPX supports multiple logical networks on an individual interface; each network requires a single encapsulation type. Novell RIP is the default routing protocol on older NetWare products, while NetWare Link State Routing Protocol (NLSP) is the default routing protocol on NetWare 4.11 and higher. NetWare clients automatically discover available network services because Novell servers and routers announce the

---

The **Routing Information Protocol** is a distance vector routing protocol that uses hop count as its metric.

An **Interior Gateway Protocol** performs routing within a single autonomous system.

A **Border Gateway Protocol** performs routing between different autonomous systems.

The **Routing Table Maintenance Protocol** was derived from RIP.

The **Routing Table Protocol** is the VINES routing protocol based on RIP.

FIGURE 7.7
Each IPX interface has a unique 10-digit byte address. (Reproduced with permission of Cisco Systems, Inc. Copyright © 2001 Cisco Systems, Inc. All rights reserved.)

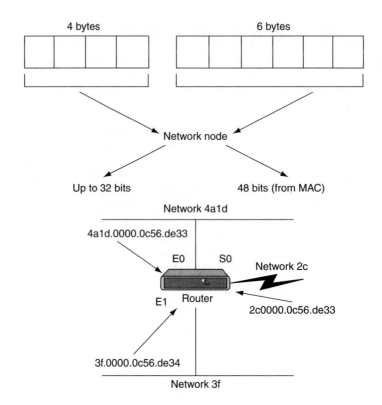

services using SAP broadcasts. The filtering of service advertisements is a critical issue in Novell networks. SAP traffic can become excessive and can severely impact bandwidth available for user data traffic. An example of one type of SAP advertisement is Get Nearest Server (GNS), which enables a client to locate the nearest server for login. Novell IPX addressing uses a two-part address: the network number and the node number.

The IPX network number can be up to 4 bytes (8 hexadecimal digits) in length. Usually, only the significant digits are listed. The network administrator assigns this number. Figure 7.7 illustrates the IPX network 4a1d. Other IPX networks shown are 2c and 3f. The IPX node number is 6 bytes (12 hexadecimal digits) in length. This number is usually the MAC address obtained from a network interface card. Figure 7.7 features the node 0000.0c56.de33. Another node address is 0000.0c56.de34.

● An **Address Resolution Protocol** is used to map an IP address to a MAC address.

Notice in Figure 7.7 that the same node number appears for both E0 and S0. Serial interfaces do not have MAC addresses, so the router created this node number for S0 by using the MAC address from E0. Each interface retains its own address. The use of the MAC address in the logical IPX address eliminates the need for an *Address Resolution Protocol (ARP)*.

How to Determine the IPX Address

You must use a valid network address when you configure a router. Because Novell NetWare networks are likely to be established already with IPX addresses, you can determine the existing IPX address from these already established networks. The IPX network address refers to the logical wire—all routers on the same wire must share the same IPX network address.

The first and recommended way to find out what address to use is to ask the NetWare administrator. Make sure that the NetWare administrator specifies the IPX network address for the same network as the NetWare file server (or other source of the address) specified by the NetWare administrator for that cabling system. If you cannot obtain an IPX address from the NetWare administrator, you can get the IPX address directly from a neighbor router. Following is a list of the several methods available:

- If the router is another Cisco router, you can use a show cdp neighbors detail command to view the required information.
- You can Telnet to the neighbor router, enter the appropriate mode, and then display the running configuration on the neighbor.
- If the neighbor router is not a Cisco router (for example, a NetWare PC based router or a NetWare file server), you might be able to attach or log in and use the NetWare config utility to determine the address.

On any router that is being configured to support IPX routing you must use the same IPX network address that already exists on that network. Alternately, if you have access to the server console, you can use the NetWare config command. The config command displays a window with the IPX address of the segment that the file server shares with the router.

Routing Updates

RIP sends routing update messages at regular intervals and when the network topology changes. When a router receives a routing update that includes changes to an entry, it updates its routing table to reflect the new route. The metric value for the path is increased by one, and the sender is indicated as the next hop. RIP routers maintain only the best route (the route with the lowest metric value) to a destination. After updating its routing table, the router immediately begins transmitting routing updates to inform other network routers of the change. These updates are sent independently of the regularly scheduled updates that RIP routers send.

Novell RIP is a distance vector routing protocol. RIP uses two metrics to make routing decisions: ticks (a time measure) and a hop count (a count of each router traversed). RIP checks its two distance vector metrics by first comparing the ticks for path alternatives. If two or more paths have the same tick value, RIP compares the hop count. If two or more paths have the same hop count, the router will use

FIGURE 7.8

RIP routers peri-odically broadcast updates of their routing tables. (Reproduced with permission of Cisco Systems, Inc. Copyright © 2001 Cisco Systems, Inc. All rights reserved.)

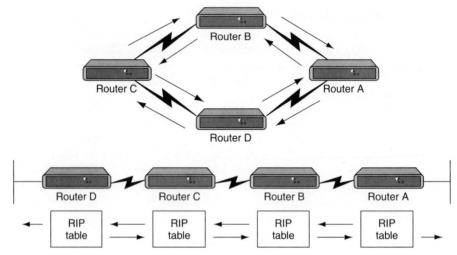

the age of the entry as the tiebreaker—the most recent entry in the tables will be preferred over the older entry. Each IPX router periodically broadcasts copies of its RIP routing table to its directly connected networks as illustrated in Figure 7.8. Upon receipt of these broadcasts, the neighbor IPX routers add distance vectors as required before broadcasting copies of their RIP tables to their other attached networks. A split-horizon algorithm prevents the neighbor from broadcasting RIP tables about IPX information back to the networks from which it received that information. RIP also uses an aging mechanism to handle conditions where an IPX router goes down without any explicit message to its neighbors. Periodic updates reset the aging timer.

Advertising is a router process in which routing updates are sent at specified intervals.

Routing table updates are sent at 60-second intervals. RIP uses User Datagram Protocol (UDP) data packets to exchange routing information. These routing updates occur every 30 seconds, and this process is termed *advertising*. If a router does not receive an update from another router for 180 seconds or more, it marks the routes served by the nonupdating router as being unusable. If there is no update after 240 seconds, the router removes all routing table entries for the nonupdating router. This update frequency can cause excessive overhead traffic on some internetworks.

NLSP, Novell's link-state routing protocol, is another alternative to RIP. NLSP is derived from the OSI Intermediate System-to-Intermediate System (IS-IS) protocol. NLSP will interoperate with RIP and SAP to erase the transition and provide backward compatibility with RIP internetworks that have no need for link-state routing. Many Novell customers want to reduce the excessive distance vector overhead packet traffic in RIP and SAP. Link-state routing requires less ongoing bandwidth, but link-state updates can also have problems, especially in large networks using a single area.

RIP Routing Metric

RIP uses a single metric (hop count) to measure the distance between the source and a destination network. Each hop in a path from source to destination is assigned a hop-count value, which is typically 1. When a router receives a routing update that contains a new or changed destination network entry, the router adds one to the metric value indicated in the update and enters the network in the routing table. The IP address is used as the next hop.

RIP prevents routing loops from continuing indefinitely by implementing a limit on the number of hops allowed in a path from the source to a destination. The maximum number of hops in a path is 15. If a router receives a routing update that contains a new or changed entry, and if increasing the metric value by one causes the metric to be infinity (i.e., 16), the network destination is considered unreachable.

RIP Stability Features

To adjust for rapid network topology changes, RIP specifies a number of stability features that are common to many routing protocols. RIP implements the split-horizon and hold-down mechanisms to prevent incorrect routing information from being propagated. In addition, the RIP hop-count limit prevents routing loops from continuing indefinitely.

RIP Timers

RIP uses numerous timers to regulate its performance. These include a routing update timer, a route time-out, and a route-flush timer. The routing update timer clocks the interval between periodic routing updates. Generally, this timer is set to 30 seconds, with a small random number of seconds added each time the timer is reset to prevent collisions. Each routing table entry has a route time-out timer associated with it. When the route time-out timer expires, the route is marked invalid but is retained in the table until the route-flush timer expires.

RIP Packet Formats

The following section focuses on the IP RIP and RIP version 2 packet formats illustrated in Figure 7.9. The illustration is followed by descriptions of the fields illustrated. The following descriptions summarize the IP RIP packet format fields as illustrated in Figure 7.9.

- Command: Indicates whether the packet is a request or a response. The request asks that a router send all or part of its routing table. The response can be an unsolicited regular routing update or a reply to a request. Responses

FIGURE 7.9

RIP packet formats. (Reproduced with permission of Cisco Systems, Inc. Copyright © 2001 Cisco Systems, Inc. All rights reserved.)

Command	Version number	Zero	Address family identifier	Zero	Address	Zero	Zero	Metric
1 byte	1 byte	2 bytes	2 bytes	2 bytes	4 bytes	4 bytes	4 bytes	4 bytes

IP RIP packet

Command	Version number	Unused	Address format identifier	Route tag	IP Address	Subnet mask	Next hop	Metric
1 byte	1 byte	1 byte	2 bytes	2 bytes	4 bytes	4 bytes	4 bytes	4 bytes

RIP Version 2 packet

contain routing table entries. Multiple RIP packets are used to convey information from large routing tables.

- Version Number: Specifies the RIP version used. This field can signal different potentially incompatible versions.
- Zero: Not used.
- Address Family Identifier (AFI): Specifies the address family used. RIP is designed to carry routing information for several different protocols. Each entry has an address family identifier to indicate the type of address being specified.
- Address: Specifies the IP address for the entry.
- Metric: Indicates how many internetwork hops (routers) have been traversed in the trip to the destination. This value is between 1 and 15 for a valid route or 16 for an unreachable route.

Up to 25 occurrences of the AFI, address, and metric fields are permitted in a single IP RIP packet. Up to 25 destinations can be listed in a single RIP packet. The following descriptions summarize the RIP 2 packet format fields as illustrated in Figure 7.9.

- Command: Indicates whether the packet is a request or a response. The request asks that a router send all or a part of its routing table. The response can be an unsolicited regular routing update or a reply to a request. Responses contain routing table entries. Multiple RIP packets are used to convey information from large routing tables.
- Version: Specifies the RIP version used. In an RIP packet implementing any of the RIP 2 fields or using authentication, this value is set to 2.
- Used: This value is set to zero.

- Address Family Identifier (AFI): Specifies the address family used. RIP is designed to carry routing information for several different protocols. Each entry has an address family identifier to indicate the type of address specified. The address family identifier for IP is 2. If the AFI for the first entry in the message is 0xFFF, the remainder of the entry contains authentication information. Currently, the only authentication type is a simple password.
- Route Tag: Provides a method for distinguishing between internal routes (learned by RIP) and external routes (learned from other protocols).
- IP Address: Specifies the IP address for the entry.
- Subnet Mask: Contains the subnet mask for the entry. If this field is zero, no subnet mask has been specified for the entry.
- Next Hop: Indicates the IP address of the next hop to which packets for the entry should be forwarded.
- Metric: Indicates how many internetwork hops (routers) have been traversed in the trip to the destination. This value is between 1 and 15 for a valid route or 16 for an unreachable route.

SAP: Supporting Service Advertisements

All the servers on NetWare internetworks can advertise their services and addresses. All versions of NetWare support SAP broadcasts to announce and locate registered network services, as illustrated in Figure 7.10. Adding, finding, and removing services on the internetwork are dynamic because of SAP advertisements. Each SAP service is an object type identified by a hexadecimal number. Examples are:

- 4 NetWare File Server
- 7 Print Server
- 278 Directory Server

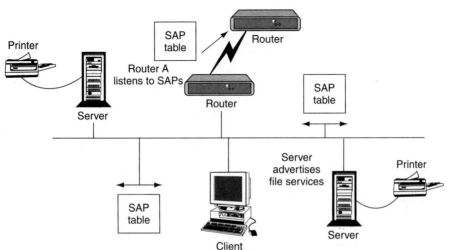

FIGURE 7.10

SAP packets advertise all NetWare network services. (Reproduced with permission of Cisco Systems, Inc. Copyright © 2001 Cisco Systems, Inc. All rights reserved.)

All servers and routers keep a complete list of the services available throughout the network in server information tables. Like RIP, SAP also uses an aging mechanism to identify and remove table entries that become invalid.

By default, service advertisements occur at 60-second intervals. However, although service advertisements might work well on a LAN, broadcasting services can require too much bandwidth to be acceptable on large internetworks or in internetworks linked on WAN serial connections.

Routers do not forward SAP broadcasts. Instead, each router builds its own SAP table and forwards the SAP table to other routers. By default, this occurs every 60 seconds. SAP advertisements can be filtered on input or output, or from a specific router:

- An IPX input: SAP filter allows the administrator to control services that are added to the router's SAP table from a specified interface.
- The IPX output: SAP filter allows the administrator to specify services included in SAP updates sent out to a specified interface.
- The IPX router: SAP filter statement is used to filter SAP messages received from a specified router on a specified interface.

The Get Nearest Server (GNS) Process

The NetWare client-server interaction begins when the client powers up and runs its client start-up programs. These programs use the client's network adapter on the LAN and initiate the connection sequence for the NetWare client software to use.

Get Nearest Server (GNS) is a broadcast that comes from a client using SAP. NetWare file servers respond with an SAP reply Give Nearest Server, as illustrated in Figure 7.11. From that point on, the client can log in to the target server, make a connection, set the packet size, and proceed to use server resources, provided the client is an authorized user of those resources.

● Get Nearest Server is a type of SAP advertisement that enables a client to locate the nearest server for login.

FIGURE 7.11
GNS is a broadcast from a client needing a server. (Reproduced with permission of Cisco Systems, Inc. Copyright © 2001 Cisco Systems, Inc. All rights reserved.)

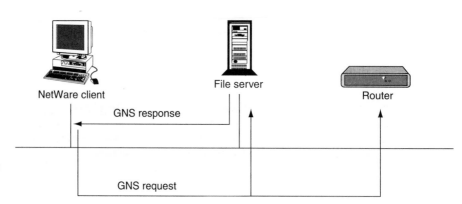

If a NetWare server is located on the segment, it will respond to the client request. If there are no NetWare servers on the local network, the router will respond to the GNS query with the address of the nearest server (or service) specified by the client. An administrator might want to filter the extent of GNS responses. To filter GNS responses, the administrator uses a GNS output filter to limit the SAP table listing of nearest or preferred servers that respond to the GNS broadcast.

Sequenced Packet Exchange (SPX)

The SPX protocol is a transport layer protocol that provides connection oriented services on top of the connectionless IPX protocol. SPX is used when a reliable virtual-circuit connection is needed between two stations. The SPX protocol takes care of flow control and sequencing issues to ensure that packets arrive in the right order. SPX also ensures that destination node buffers are not overrun with data that arrive too rapidly.

Prior to data transmission, SPX control packets are sent to establish a connection, and a connection ID is associated for that virtual circuit. This connection ID is used in all data transmissions. At the end of data transmission an explicit control packet is sent to break down the connection. SPX uses an acknowledgment scheme to make sure that messages arrive at the destination. Lost packets are re-sent, and sequencing is used to keep track of packets so that they arrive in the proper order and are not duplicated.

SPX uses a time-out algorithm to decide when a packet needs to be retransmitted. The time-out is dynamically adjusted based on the delay experienced in packet transmission. If a packet times out too early, its value is increased by 50 percent. This process can continue until a maximum time-out value is reached or the time-out value stabilizes. To verify that a session is still active when there is no data activity, SPX sends probe packets to verify the connection. The frequency of these probe packets can be controlled by settings in the NET.CFG file.

Interestingly, many SPX connections can use the same IPX socket as illustrated in Figure 7.12. This allows multiple connection IDs to be multiplexed and demultiplexed across the same IPX socket.

Figure 7.13 shows the SPX packet structure, and the following list describes the format fields as illustrated in Figure 7.13:

- The Connection Control field is used for regulating flow of data across the connection. The bit sequence 0001000, for example, is used as an end of message signal, and the bit sequence 01000000 indicates that an acknowledgment is requested.
- The Data Stream Type indicates the nature of the data contained in the SPX data field. It is used to identify the upper layer protocol to which the SPX data must be delivered. It serves a similar role to the Packet Type field in the IPX packet.
- The Source Connection ID and the Destination Connection ID are the virtual circuit numbers used to identify a session. These IDs are used to demultiplex separate virtual circuits on a single socket.

FIGURE 7.12

SPX connection IDs and sockets. (Reproduced with permission of Cisco Systems, Inc. Copyright © 2001 Cisco Systems, Inc. All rights reserved.)

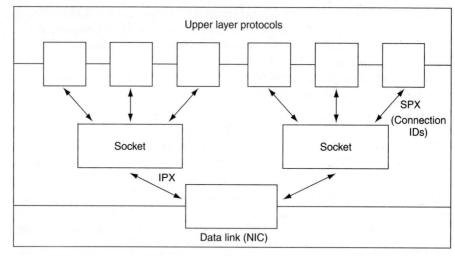

FIGURE 7.13

SPX and OSI. (Reproduced with permission of Cisco Systems, Inc. Copyright © 2001 Cisco Systems, Inc. All rights reserved.)

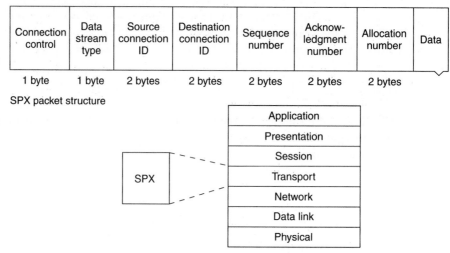

- The Sequence Number field numbers every packet sent. This is used by SPX to detect lost and out-of-sequence packets.
- The Acknowledgment Number field is used to indicate the next packet the receiver expects. It means that all packets prior to the acknowledgment value have been received correctly.
- The Allocation Number indicates how many free buffers the receiver has available on a connection. This value is used by the sender to pace the sending of data. The use of the allocation number helps avoid overwhelming the receiver with packets that do not have a corresponding buffer available to hold them.

The NetWare workstation does not usually use the SPX protocol. It uses the IPX protocol directly. Reliability of transmission is maintained by the NCP protocol. SPX is used to establish remote connections between the print server and remote printers. SPX also is used in NetWare SQL and remote connections to the NetWare file server through RCONSOLE.

Sequenced Packet Exchange II (SPX II)

Novell introduced SPX II in 1993 to provide improvements over the older SPX protocol in the following areas:

- Window flow control
- Larger packet sizes
- Improved negotiation of network options
- Safer method of closing connections

Novell cited the following reasons for providing a newer SPX protocol (SPX II):

- The poor performance of SPX when compared with IPX.
- The small packet size (maximum 576 bytes) used by SPX because of the absence of a facility to negotiate a larger packet size between two endpoints.
- Lack of proper flow control (the sending of additional data packets before the acknowledgment of a previous packet is sent).
- Lack of an orderly release mechanism, which ensures that data are not lost when a connection is closed.

Like SPX, SPX II is a connections oriented transport layer (OSI layer 4) protocol (see Figure 7.14). SPX II is designed to be backwards compatible with SPX. This

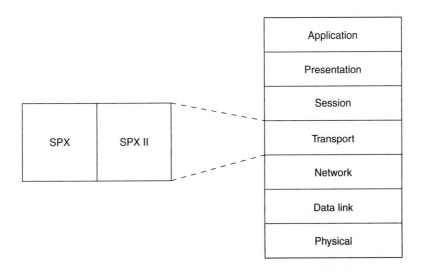

FIGURE 7.14
SPX II and OSI. (Reproduced with permission of Cisco Systems, Inc. Copyright © 2001 Cisco Systems, Inc. All rights reserved.)

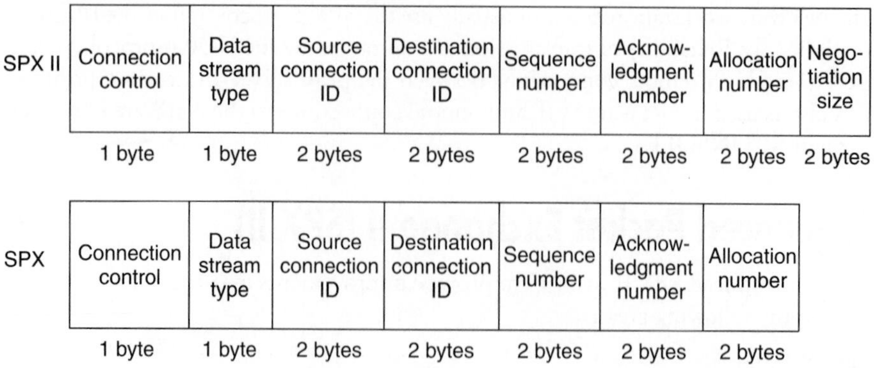

FIGURE 7.15

SPX and SPX II packet structures. (Reproduced with permission of Cisco Systems, Inc. Copyright © 2001 Cisco Systems, Inc. All rights reserved.)

SPX II	Connection control	Data stream type	Source connection ID	Destination connection ID	Sequence number	Acknow-ledgment number	Allocation number	Nego-tiation size
	1 byte	1 byte	2 bytes	2 bytes	2 bytes	2 bytes	2 bytes	2 bytes

SPX	Connection control	Data stream type	Source connection ID	Destination connection ID	Sequence number	Acknow-ledgment number	Allocation number
	1 byte	1 byte	2 bytes	2 bytes	2 bytes	2 bytes	2 bytes

means that SPX II is designed to recognize SPX packets as well as the enhanced SPX II packet. To achieve this goal, the SPX II packet format is designed to be a superset of the SPX packet structure. Figure 7.15 compares the packet structure of SPX II. As you can see, SPX II has an additional field called the Negotiation Size, which is 2 bytes long. The additional Negotiation Size field makes the SPX II header including the IPX header 14 bytes. Whereas the data portion of the SPX packet is limited to 534 bytes (576 bytes minus 12 bytes of SPX header and 30 bytes of IPX header), the data portion of the SPX II portion is as follows:

$$\text{Maximum media packet size} - \text{SPX II header size} - \text{IPX header size} = \text{maximum media packet size} - 44$$

Apart from the Negotiation Size, additional bits are defined for the first field of the SPX II packet called the Connection Control field. Figure 7.16 shows these additional bits, and Table 7.5 describes them.

SPX II uses the bit fields defined in Figure 7.16 to operate in an SPX compatible mode or the SPX II enhanced mode. If such a packet is received from a remote end, it indicates that the remote end supports SPX II. If the SPX II bit is not set in the Connection Control field, it indicates the older SPX packet.

You can switch from SPX to SPX II enhanced mode during connection establishment by the SPX protocol software. During connection establishment, the SPX protocol software can negotiate the maximum packet size to be used for the connection. To do this, set the NEG bit (bit 6) to 1 in the connection request.

The **NetWare Core Protocol** requests and replies to requests for file and print services.

NetWare Core Protocol (NCP)

The *NetWare Core Protocol (NCP)* is used to implement NetWare's file services, print services, name management services, file locking, synchronization, and bindery operations. A bindery refers to the internal database of network objects kept on the NetWare server.

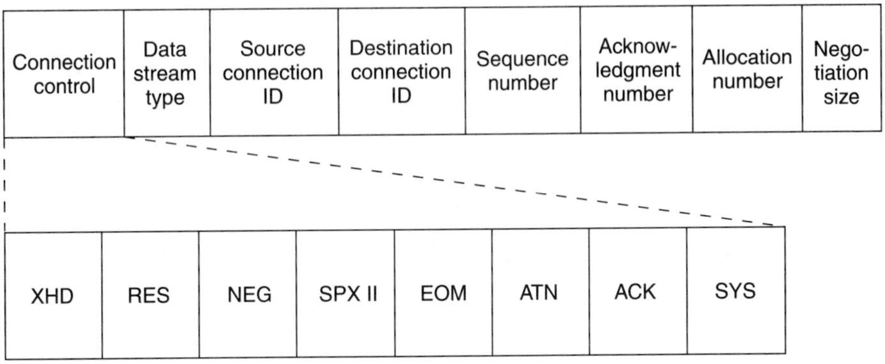

FIGURE 7.16
*SPX Connection
Control field.
(Reproduced with
permission of Cisco
Systems, Inc.
Copyright © 2001
Cisco Systems, Inc.
All rights reserved.)*

Table 7.5 SPX II Connection Control Field Descriptions

Field Abbreviation	Bit Position	Description
XHD	8	Reserved for extended header
RES	7	Reserved. Must be 0
NEG	6	Negotiate Size Request/Response
SPX II	5	Indicates an SPX II packet structure
EOM	4	End of message; indicates end of message
ATN	3	Gains the attention of the remote end
ACK	2	Requests that the remote end acknowledge the reception of the data packet
SYS	1	Indicates a system packet. System packets are used for controlling the SPX connection and are not sent to the application process. System packets are not numbered by the Sequence Number field.

NCP is implemented at the workstation and the NetWare server. On the workstation side, NCP is implemented in the NetWare shell and is limited to making requests for services to an NCP server. The NetWare server (NCP server) contains a full implementation of NCP that can execute or process requests for NCP services. NCP provides transparent remote file and print services to a NetWare client. These remote services appear to be local to the client.

NCP directly uses the IPX protocol, avoiding the use of SPX or Packet Exchange Protocol (PXP). This enables NCP to be more efficient because it avoids the protocol overhead of the SPX and PXP protocols. NCP provides its own mechanism for session control, error detection, and retransmission. Figure 7.17 shows an NCP packet structure.

FIGURE 7.17

NCP packet structure. (Reproduced with permission of Cisco Systems, Inc. Copyright © 2001 Cisco Systems, Inc. All rights reserved.)

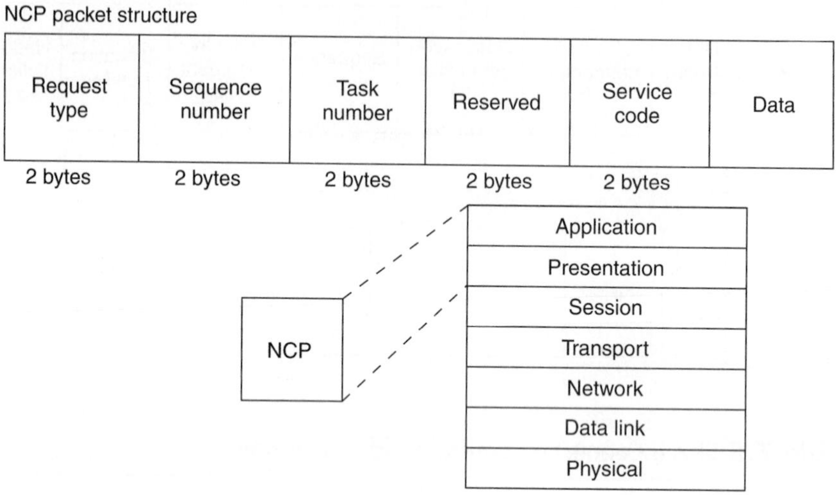

NCP packet structure

Request type	Sequence number	Task number	Reserved	Service code	Data
2 bytes	2 bytes	2 bytes	2 bytes	2 bytes	

NCP

- Application
- Presentation
- Session
- Transport
- Network
- Data link
- Physical

Create Service Connection, Negotiate Buffer Size, Logout, and **Get Server Date and Time** are NCP request types.

In Figure 7.17, the Request Type field indicates the type of NCP request. Examples of NCP request types are *Create Service Connection, Negotiate Buffer Size, Logout, Get Server Date and Time,* Get Station Number, and End of Job. The Sequence Number is used as a transaction ID field and identifies an NCP request and its corresponding response. The Service Code further identifies the service requested by the workstation.

Packet Burst Mode

The NCP numbers each request and reply packet with a sequence number. This sequence number is used as a transaction ID field to identify an NCP request and its corresponding response for a particular session. The session is identified by the connection number and is placed in every NCP transaction.

The NCP transaction models the client-server interaction between a workstation and a NetWare server quite well. This transaction introduces a new set of problems when NetWare servers are used in WANs. Typically, WAN link capacities today are in the range of tens of kilobits per second. This is quite small in relationship to megabits per second speed used in LANs, causing WANs to run at slower speeds than LANs. In addition, WANs have longer delays because they span longer distances. Using a single request and single response model as illustrated in Figure 7.18 means that the effective throughput of the transaction is as follows:

$$E = (Q + N*Pn)/(N*Td)$$

where E = effective throughput of the NCP transaction
Q = size of the request packet
N = number of reply packets
Pn = size of the Nth single reply packet
Td = round-trip delay

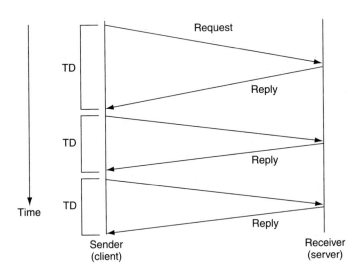

FIGURE 7.18

Single request/reply transactions. (Reproduced with permission of Cisco Systems, Inc. Copyright © 2001 Cisco Systems, Inc. All rights reserved.)

As can be seen in Figure 7.18, if the reply is larger than a single packet it has to be sent in a series of successive transactions, each of which takes additional time equal to the round-trip delay.

Many earlier NetWare routers have a limit of 512 bytes per packet, which means that if a 64 Kb file had to be transferred, 128 of the 512 bytes would have to be transmitted. To get an idea of what the throughputs are like, substitute numerical values in the preceding equation. Let us assume that the size of the request packet is 128 bytes, and the reply is a 1,000-byte packet. Also, assume that the round-trip delay on the link is 1 second, and the reply consists of 4 packets. Incorporating these values into the equation, you get the following:

$$E = (128 + 4{^*}1,000) / (4{^*}1) = 1,032 \text{ bytes/sec} = 8,256 \text{ bits/sec}$$

In packet burst mode, a single read reply could be sent as a series of successive packets that do not have to wait for an NCP acknowledgment of every message sent. Also, an NCP request can consist of a series of requests that do not have to wait to be acknowledged by a reply. Figure 7.19 shows a request and a three-packet reply using packet burst.

The effective throughput is now computed by the equation:

$$Ep = (Q + N{^*}Pn) / Td$$

Ep = effective throughput of the NCP transaction using packet burst
Q = size of the request packet
Pn = size of the Nth single reply packet
Td = round-trip delay
N = number of reply packets

FIGURE 7.19
*Single request/
multiple reply.
(Reproduced with
permission of Cisco
Systems, Inc.
Copyright © 2001
Cisco Systems, Inc.
All rights reserved.)*

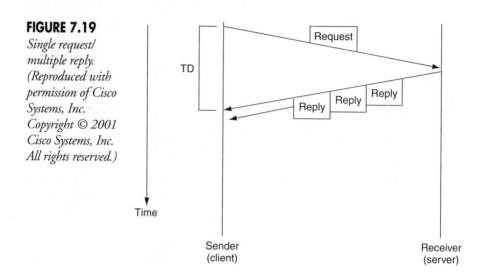

Using the preceding numerical example, you can calculate the effective throughput using packet burst as follows:

$$Ep = (128 + 4{}^{*}1{,}000) / 1 = 4{,}128 \text{ bytes/sec} = 33{,}024 \text{ bits/sec}$$

You can see that the effective throughput for packet burst in this example is 4 times that of the normal throughput. This is not surprising because dividing equation 2 by equation 1 reveals the following:

$$(3)\ Ep / E = N \text{ or}$$
$$(4)\ Ep = N{}^{*}E$$

Effective throughput of packet burst, therefore, is N times that of a normal NCP. Tests performed by Novell reveal that performance improvements of up to 300 percent can be achieved on a WAN. Packet burst can also improve performance in a LAN by up to 50 percent. Burst mode can also be used in situations in which a transaction consists of a multiple request/single reply sequence. Burst mode implements a dynamic window size algorithm and a dynamic time-out mechanism. The dynamic window size allows burst mode to adjust the number of frames that can be sent in burst mode. The dynamic time-out, also called transmission metering, adjusts itself to line quality, line bandwidth, and line delay.

NetWare Link Services Protocol is a link state protocol designed to overcome the limitations of distance vector based protocols.

NetWare Link Services Protocol (NLSP)

NetWare Link Services Protocol (NLSP) is a link state protocol designed to overcome the limitations of distance vector based protocols such as RIP and SAP. Before discussing NLSP it is useful to examine some of the limitations of RIP and SAP, and why Novell has a new protocol to accomplish routing and service advertising.

RIP Problems

To find the best route to a given destination NetWare servers and clients use RIP. One problem with RIP is that it takes a longer time to stabilize compared with link state protocols. The technical description of this problem is that RIP has a slower convergence. IPX routers exchange routing information using RIP at periodic intervals (60 seconds). The RIP message contains a list of all routes known to that router. In other words, each RIP message contains the entire routing table of a router. For large networks, the routing table tends to be large, and the RIP messages are proportionately large. A router, upon receiving a RIP message from its neighbor, recomputes the routes in its table based on the new information and sends the recomputed routing table on its next broadcast interval. It may take several broadcast intervals before the routers have a consistent view of the network. This problem is known as slow convergence. In addition to slow convergence, RIP has no authentication mechanism to prevent an intruder from broadcasting an incorrect RIP message. Other known problems that plague distance vector based methods are count-to-infinity problems and lack of capability to set up routing domains so that routing in a designated area or domain can be done without impacting other areas of the network.

The count-to-infinity problem is a classic problem associated with distance vector schemes. This problem arises when a router (router A) sends recomputed information about a route to a router (router B) from which it originally received the information, as illustrated in Figure 7.20. If a link connected to router A is broken, router A waits for its next broadcast interval to send information about unreachable destinations to its neighbor. Meanwhile, router B not knowing about unreachable destinations sends its routing table to router A before router A can send its broadcast. This routing table contains old information about the status of the links, including the link is down. Router A thinks that the unreachable destinations are reachable through router B and updates its table with incorrect information. Because router B is a neighbor, router A adds its distance vector from router B to the route path information received from router A. When router A broadcasts its incorrect routing table to B,

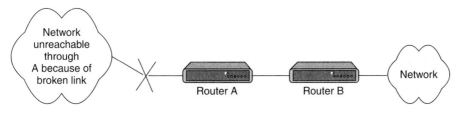

FIGURE 7.20

Scenario illustrating the count-to-infinity problem for distance vector protocols. (Reproduced with permission of Cisco Systems, Inc. Copyright © 2001 Cisco Systems, Inc. All rights reserved.)

router B adds its distance from router A to reach destinations through A. At each broadcast interval the distance metric to reach the unreachable destination grows until it reaches a value assumed to be infinity. In RIP, this value is 16 hops.

One solution to the count-to-infinity problem of RIP is *split-horizon,* and another is triggered update. In split-horizon, you do not report information about destinations from routers from which you originally received the information. In *triggered updates,* changed information is sent immediately without waiting for the next broadcast interval. Both RIP and SAP messages can consume a substantial fraction of the network bandwidth. In SAP, the service advertises the name of a service at periodic intervals of 60 seconds, and this too has the characteristics of a distance vector protocol. The destinations in the SAP protocol are the text strings that represent the name of advertised services. Both SAP and RIP are expensive in terms of the bandwidth they consume. In a LAN, the bandwidth used by SAP and RIP broadcasts is not significant, but it is significant on the slower WAN links. In NetWare 4.x the periodicity of broadcasts can be customized to reduce bandwidth requirements, but in general a better protocol is needed. For all the previously mentioned reasons, Novell created NLSP.

NLSP Messages

NLSP is designed to provide exchange of routing information between routers and the building of a routing table at each router that has a global view of the network. NLSP also can be used as a replacement for SAP. To provide backwards compatibility, NLSP can coexist and interoperate with existing RIP/SAP based networks. NLSP also can be expanded to provide hierarchical or area routing. Areas are defined using a <network number, network mask> pair. The network number and network mask are four bytes long. The network mask is used to define the area to which a router belongs. It can also be used for specifying multicast broadcasts for WAN networks.

Novell's NLSP implementation is based on ISO's IS-IS standard. The NLSP protocol has the following three types of messages: Hello, LSP, and SNP.

The NLSP *Hello messages* are used by NLSP routers to verify the up/down status of a router's link to its neighbor. The *Link State Protocol (LSP)* message is used to broadcast the identity of the router and the status of the links connected from it to the router's neighbors. The LSP lists link state information used to reach the router's neighbors. On a LAN that has several routers directly connected to it, the link state information reported by any router about the neighbor routers connected to the LAN is very similar. It therefore makes sense to have only one router, called the designated router (DR), send link state information rather than have all routers connected to the LAN send link state messages. In general, if N routers are on a LAN and each router sends link state messages, there would be N^*N messages on the LAN. Using a designated router, the LAN can be treated as a logical or pseudo-node on a network consisting of many LANs and WAN links. The designated router can be selected based

Split-horizon is a routing technique in which information about routes is prevented from exiting the router interface through which the information was received.

In **triggered updates**, changed information is sent immediately.

Hello messages are used by NLSP routers to verify the status of a router's link to its neighbor.

The **Link State Protocol** message broadcasts the identity of the router and the status of the links connected from it.

on a priority assigned to each router. This priority can be adjusted to influence the selection of a designated router. Assigning priorities to the DR is one improvement of NLSP over IS-IS. The LSP messages are used to compute the best route to a destination using Edsgar Djikstra's *shortest path first (SPF)* algorithm. The SPF algorithm is noteworthy because of the optimal route paths it yields for any router. SPF is run whenever the link state information for a router changes.

> ●
> The **shortest path first** algorithm is run whenever the link state information for a router changes.

Computation of the Forwarding Database in NLSP

NLSP uses the SPF algorithm to compute the forwarding database for a router. To compute the shortest path first from router R, perform the following steps:

Step 1. Place router R at the root of a tree and label it R(0). The number in parentheses represents the route path cost, which is 0 for router R to reach itself.
Step 2. For any new router, N, placed in the tree, examine the LSP messages. If for any neighbor X of N, the path from R to X through N is the best path so far to the neighbor X, put the router X on the tree with the label X (cost to N + cost from N to X).
Step 3. If alternative paths exist to a node X from router R, use the path with the least cost.

An example can clarify the previously stated rules for computing the routing table (also called forwarding database) based on the SPF algorithm. In Figure 7.21, it is required to compute the best routes from router B. The nodes in this figure represent routers, and the numbers on the arcs represent route path cost of the links. The cost of reaching nodes C, G, and A from B are 2, 3, and 5, respectively.

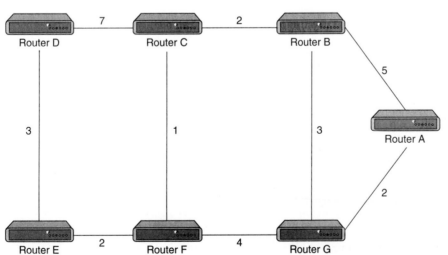

FIGURE 7.21

Example network illustrating the SPF algorithm. (Reproduced with permission of Cisco Systems, Inc. Copyright © 2001 Cisco Systems, Inc. All rights reserved.)

FIGURE 7.22

Example network continued. (Reproduced with permission of Cisco Systems, Inc. Copyright © 2001 Cisco Systems, Inc. All rights reserved.)

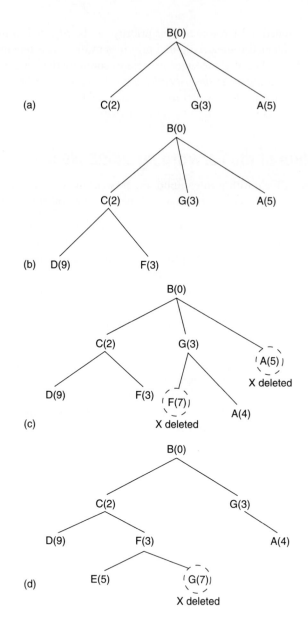

In Figure 7.21, node C is placed at the root of the tree, and the neighbor nodes reachable through C are placed next to it (see Figure 7.22a). In Figure 7.22b, nodes D and F are computed as follows:

$$\text{Cost to D} = \text{Cost to C} + \text{Cost from C to D} = 2 + 7 = 9$$
$$\text{Cost to F} = \text{Cost to C} + \text{Cost from C to F} = 2 + 1 = 3$$

FIGURE 7.22
(Continued)

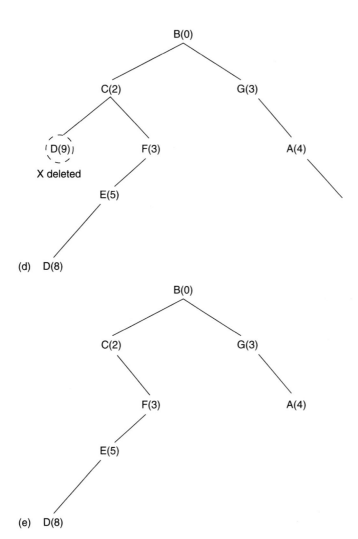

(d)

(e)

In Figure 7.22c, nodes F and A are shown reachable from node G. The costs to F and A are computed as follows:

$$\text{Cost to F} = \text{Cost to G} + \text{Cost from G to F} = 3 + 4 = 7$$
$$\text{Cost to A} = \text{Cost to G} + \text{Cost from G to A} = 3 + 1 = 4$$

The cost to F is 7 and is greater than the previously computed cost of 3 through node C, and therefore node F (7) is deleted from the tree. The cost to node A is 4, and this is less than the previously computed cost of 5 through node B. The previous node A (5) is therefore deleted from the tree.

Table 7.6 Forwarding Database of Router B

Destination	Forwarding Router	Cost
A	G	4
B	Not Applicable (self)	0
C	C	2
D	C	8
E	C	5
F	C	3

In Figure 7.22d, nodes E and G are shown reachable from node F. The costs to E and G are computed as follows:

$$\text{Cost to E} = \text{Cost to F} + \text{Cost from F to E} = 3 + 2 = 5$$
$$\text{Cost to G} = \text{Cost to F} + \text{Cost from F to G} = 3 + 4 = 7$$

The cost to G is 7 and is greater than the previously computed cost of 3 from node B, and therefore node G (7) is deleted from the tree. In Figure 7.22e, node D is shown reachable from node E. The cost to D is computed as follows:

$$\text{Cost to D} = \text{Cost to E} + \text{Cost from E to D} = 5 + 3 = 8$$

The cost to D is 8 and is less than the previously computed cost of 9 from node B, and therefore node D is deleted from the tree and is replaced with node D (8) under E (5).

The SPF algorithm yields the tree shown in Figure 7.22f, from which the forwarding database from C is computed as shown in Table 7.6.

Use of SNP in NLSP

The *Sequence Number Packet (SNP)* summarizes the information in the LSP database. Instead of sending the complete LSP database information, a summary of information about the LSP database is sent. The two types of SNP messages are partial and complete. Partial SNP messages are used to acknowledge specific LSP messages and Complete SNP (CSNP) messages summarize LSPs in a specified range. An example of PSNP usage is in a point-to-point link to acknowledge one or more LSP messages. For a LAN, the designated router can use CSNP to summarize the LSP database. If the CSNP indicates a discrepancy, you can request the missing information using PSNP. LSP messages can contain SAP information. The router closest to the service can place SAP information in the LSP message. The SAP information need not be transmitted periodically, and only one router transmits it. The link state protocol ensures that only changes in link state and service information are sent, and there is no

The **Sequence Number Packet** summarizes the information in the LSP database.

need for periodic broadcasts. As a precaution, NLSP forces a broadcast of service and link state information every two hours just in case of unforeseen problems caused by the loss of previous link state information.

Migrating to NLSP

To support existing RIP/SAP based networks, the NLSP can run on a router or Net-Ware server acting as a router at the boundary between an NLSP routing domain and an RIP/SAP domain as illustrated in Figure 7.23.

The routing information and service information acquired by the NLSP router can be translated into RIP/SAP information for networks based on these protocols. NLSP software is available for NetWare 3.11 and higher. If you are using a router built using ROUTGEN, you may consider upgrading to the NetWare Multiprotocol Router that has NLSP support. When converting from RIP to NLSP, you can convert gradually. If you convert WAN links to use NLSP, you see an immediate reduction in network bandwidth used by routing messages.

It is difficult to define an exact procedure for migrating to NLSP. The optimum strategy depends on the network topology, the speed at which you want to make the conversion, and the resources you have for troubleshooting problems that may arise because of the conversion. A general strategy for performing the conversion is outlined next.

The general strategy for converting to NLSP should be to convert the core of the WAN links to use NLSP for routing messages. Campuses (areas) that have LANs joining the core WAN infrastructure can continue using RIP when the WAN links are converted to using NLSP. Gradually, you can convert routes in each campus area to use NLSP. Assign addresses so that the campuses can be treated as a single area or multiple areas. Initially, you may not use hierarchical routing, but if you select your network addresses wisely, you can create areas and domains by changing only the network masks for a network address. If your network topology contains loops, convert all the routers within

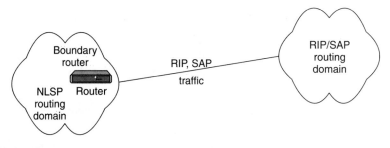

FIGURE 7.23

Coexistence of NLSP and RIP/SAP. (Reproduced with permission of Cisco Systems, Inc. Copyright © 2001 Cisco Systems, Inc. All rights reserved.)

the loop to use NLSP at the same time to avoid convergence problems associated with RIP and to reduce the network bandwidth used by RIP routing messages.

Mesh Networks and NLSP

Mesh networks allow the most flexible topology, as illustrated in Figure 7.24a, because they are characterized by redundant paths to reach a destination. Because of the redundant links, mesh networks lead to a more reliable design. However, if SAP and

FIGURE 7.24

Typical network configurations. (Reproduced with permission of Cisco Systems, Inc. Copyright © 2001 Cisco Systems, Inc. All rights reserved.)

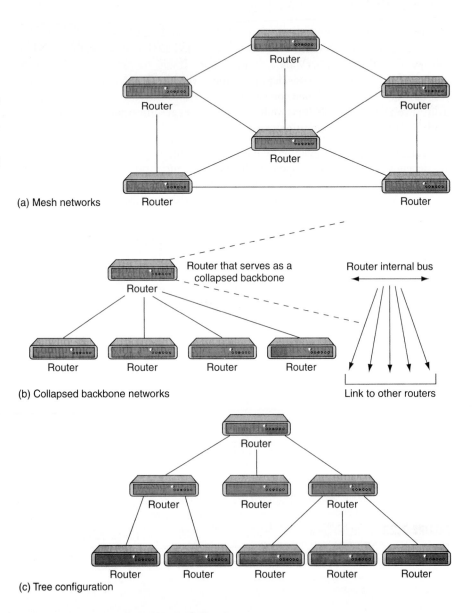

(a) Mesh networks

(b) Collapsed backbone networks

(c) Tree configuration

RIP are used in mesh networks, the broadcast traffic generated by these protocols can quickly overwhelm the network. For this reason, a collapsed backbone (Figure 7.24b) or tree configuration (Figure 7.24c) is often preferred to RIP/SAP based NetWare networks. NLSP makes the use of mesh networks in NetWare based networks more practical because NLSP does not rely on excessive use of broadcasts for its operation. NLSP reacts quickly to changes in the status of links in a mesh network.

Hierarchical Routing in NLSP

NLSP can be used to organize networks into hierarchical routing domains. Several networks can be connected by routers to form a routing area. Routers connecting networks within a routing area are called level 1 routers. Routing areas can be connected together to form a routing domain as illustrated in Figure 7.25.

The routers connecting routing areas within a routing domain are called level 2 routers. A single network administration authority such as the MIS department, *Network Operations Center (NOC)* for an organization, or a public carrier generally administers a routing domain. Routing domains administered by different organizations can be

The **Network Operations Center** is the organization responsible for maintaining a network.

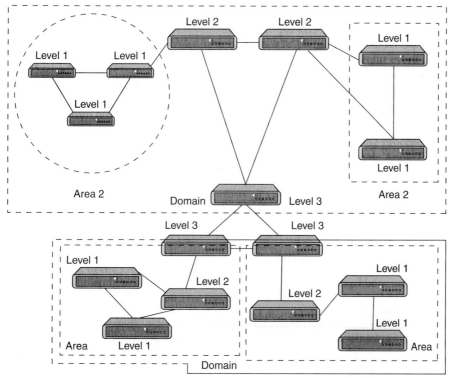

FIGURE 7.25

Hierarchical routing in NLSP. (Reproduced with permission of Cisco Systems, Inc. Copyright © 2001 Cisco Systems, Inc. All rights reserved.)

connected using level 3 routers. Figure 7.25 illustrates the different levels of routers. Level 2 routers also act as a level 1 area router within their own areas. Also, a level 3 router acts as a level 2 router within its domain. Use of level 2 routers reduces the amount of information that routers at each level need to know and process, and leads to improvements in the efficiency of the routing mechanism. For instance, level 1 routers need to store link state information in their areas only. They do not need to be aware of link state information for routers in other areas. To send traffic to other areas, a level 1 router needs to know its nearest level 2 router. Level 2 routers need to advertise the area addresses of the areas within their domains only. Similarly, level 3 routers need to advertise addresses of their domains only.

Managing Addresses for Hierarchical Routing Using NLSP

IPX uses a 32-bit address for each LAN segment. Additionally, NetWare 3.x and NetWare 4.x servers have unique internal network numbers. Routing areas and domains can be defined by network addresses and 32-bit network masks.

Consider a routing area that has a network address of D1127000 and a network mask of FFFFF000. Each F (hex) in the network mask represents a bit pattern of four 1s (1111). The network mask has the following meaning:

- A grouping of 1s in the network mask represents corresponding bits in the network address that refer to the network address of that area or domain.
- A grouping of 0s in the network mask represents corresponding bits in the network address that must be assigned to network segments with areas or domains.

Thus, a network mask of FFFFF000 applied to an area with a network address of D1127000 means that the address prefix D1127 is common to all network segments within the area. Each network segment within the area must have a network address unique in the last three hexadecimal digits because of their zero hexadecimal digits in the network mask. Figure 7.26 depicts examples of such network segments. With three hexadecimal digits assigned to a network segment in an area, 12-bit combinations can be assigned to network segments. This works out to be 4,096 network segments. The network mask should be selected to allow for growth in the network segments or segment areas.

Assigning Multiple Addresses for an NLSP Area

An NLSP area can have up to three different area addresses, each with a different network mask. NLSP treats the multiple addresses as synonyms for the routing area. Typically, the new addresses are added to represent alternate views of the network. NLSP allows routing areas to be organized without disrupting the network. For example, you could split a routing area into different areas by introducing new addresses to some of the routers and gradually phasing out the old addresses. When all routers in the area have separate area addresses, the area is split in two.

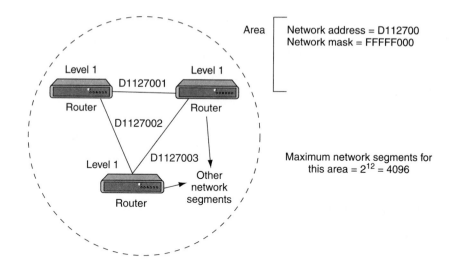

Area — Network address = D112700
Network mask = FFFFF000

Maximum network segments for
this area = 2^{12} = 4096

FIGURE 7.26

*Example of net-
work masks and
network addresses
for areas.
(Reproduced with
permission of Cisco
Systems, Inc.
Copyright © 2001
Cisco Systems, Inc.
All rights reserved.)*

Summary

In this chapter, you learned about native NetWare protocols. The protocols discussed were IPX, SPX, NetWare Core Protocol, RIP, SAP, and NLSP. Each of the protocols discussed has advantages and disadvantages for their use. RIP has been around for a long time and is still in use by many companies today. RIP poses several problems to network administrators, which have been overcome through the adoption of NLSP. However, before changing the network configuration from one protocol to another first make sure that the routers can accept the convergence and change in network topology. If not, you will be in for many sleepless nights until the problems are solved. Chapter 8 discusses another networking technology that has been around for a long time, AppleTalk.

Review Questions

1. What is the percentage of market share owned by Novell?
2. What type of network operating system is Novell?
3. What are the entities of NetWare?
4. What layer of the OSI model does IPX interact with?
5. What is IPX?
6. What are the three purposes of IPX?
7. What is the actual data size of an IPX packet?
8. Draw the packet structure of IPX.
9. What are sockets and how do they work?
10. What are the six encapsulation methods supported by Novell?

11. What does an IPX routing table mean?
12. What is a tick and a hop?
13. How many versions of RIP are in operation today?
14. What type of protocol is RIP?
15. How many bits comprise an IPX address?
16. How can a service advertisement be filtered?
17. How can you determine an IPX address?
18. What are the two methods used by RIP to make routing decisions?
19. How often does RIP update its routing tables?
20. What is meant by slow convergence?

Summary Questions

1. Where is NLSP derived from?
2. Draw a RIP version 1 and version 2 packet.
3. How often do service advertisements occur on a Novell network?
4. What is a Get Nearest Server process?
5. What layer of the OSI model does SPX interact with?
6. What protocol provides reliability of transmission for a Novell workstation?
7. What are the areas of improvement of SPX II over SPX?
8. What protocol is used to implement NetWare's file and print services?
9. Where is NCP implemented in a Novell network?
10. What protocol can improve both LAN and WAN performance?

AppleTalk Protocols

Objectives

- Discuss the goals of the AppleTalk networking scheme.
- Compare the AppleTalk protocol suite versus the OSI model.
- Explain the physical layer hardware of AppleTalk.
- Examine the media considerations of AppleTalk.
- Discuss the role of the Link Access Protocol (LAP) manager for LocalTalk.
- Explain AppleTalk addressing.
- Discuss the main network components for AppleTalk.
- Examine the differences between Phase 1 and Phase 2 of AppleTalk.
- Discuss AppleTalk routing.
- Examine each of the protocols that comprise the AppleTalk protocol suite.

Key Terms

Address Mapping Table (AMT), 319
AppleTalk Address Resolution Protocol (AARP), 311
AppleTalk Data Stream Protocol (ADSP), 347
AppleTalk Echo Protocol (AEP), 340
Apple File Protocol (AFP), 303
AppleTalk Session Protocol (ASP), 350

AppleTalk Transaction Protocol (ATP), 328
attention requests, 355
carrier sense (CS), 309
Carrier Sense with Multiple Access and Collision Avoidance (CSMA/CA), 308
Clear to Send, 309

Introduction

AppleTalk, a protocol suite developed by Apple Computer in the early 1980s, was developed in conjunction with the Macintosh computer. AppleTalk's purpose was to allow multiple users to share resources, such as files and printers. The devices that supply these resources are called servers, while the devices that make use of these resources such as a Macintosh computer are referred to as clients. Hence, AppleTalk is one of the early implementations of a distributed client-server networking system. This chapter provides an overview of AppleTalk's network architecture.

AppleTalk was designed with a transparent network interface. That is, the interaction between client computers and network servers requires little interaction from the user. In addition, the actual operations of the AppleTalk protocols are invisible to end users, who see only the result of these operations. Two versions of AppleTalk exist: AppleTalk Phase 1 and AppleTalk Phase 2.

Phase 1 supports a single physical network.

AppleTalk *Phase 1,* which is the first AppleTalk specification, was developed in the early 1980s strictly for use in local workgroups. Phase 1 therefore has two key limitations: its network segments can contain no more than 127 hosts and 127 servers, and it can support only nonextended networks. Extended and nonextended networks will be discussed later in detail.

AppleTalk *Phase 2,* which is the second enhanced AppleTalk implementation, was designed for use in larger internetworks. Phase 2 addresses the key limitations of AppleTalk Phase 1. In particular, Phase 2 allows any combination of 253 hosts or servers on a single AppleTalk network segment and supports both nonextended and extended networks.

Phase 2 supports multiple logical network.

The AppleTalk networking scheme was designed to be innovative but to adhere to standards wherever possible. Apple's main goal was to implement a network in every Macintosh inexpensively and seamlessly.

The key goals of the AppleTalk networking system are:

- Simplicity
- Plug and play
- Peer to peer
- Open architecture
- Seamless

When designing this system, the standard Macintosh user interface was not to be disturbed. Users should be able to implement an Apple network and not know that they are running on one. A user should be able to look at the screen and notice the familiar icons and window interfaces as if the Macintosh were operating locally. All exterior actions on a Macintosh were implemented in the chooser, and this is where the network extensions were also provided. The chooser is the program that allows peripheral access to the Apple operating system. This includes changing networks, printers, and other utility programs designed to change the default settings on the Apple Macintosh. The network menus should be as friendly to use as the operating system itself. The engineers accomplished this overwhelmingly.

Even the wiring scheme fits into these goals. Users should only need to plug in the network cable for the Apple operating system to be able to detect the network and work accordingly. The beginnings of AppleTalk gave us two entities to work with:

1. LocalTalk, which is the access method (physical and data link layers).
2. AppleTalk, more commonly known as AppleShare, which is the network operating system—OSI network through application layers.

The AppleTalk network operating system is an uncomplicated system and was an inexpensive solution at a time when other networking solutions were expensive. Macintosh computers have two large components that have given Apple computers a clear advantage in the personal computer marketplace: built-in networking and easy to use object oriented operating system.

The object oriented operating system provides users with computing capabilities based on objects. When working with the Apple personal computer, components of the operating system or application programs are accessed with icons that appear on the screen. Users utilize the mouse to access the icons, which provide an entrance into the operating system or to an application program.

AppleTalk and OSI

The AppleTalk network operating system consists of two entities: network protocols and hardware. AppleTalk protocols are arranged in layers, with each layer providing a service to another layer or to an application. The flow of data on an AppleTalk network will be illustrated.

Figure 8.1a depicts the layout of the AppleTalk protocols. Figure 8.1b shows the subset of the protocols. The network protocol is the software version of AppleTalk and the wiring scheme is the hardware component of AppleTalk. The hardware portion consists of the physical wiring, the connectors, and their physical interfaces. The hardware portion also includes the network access methods used with AppleTalk. Currently, AppleTalk can run over LocalTalk (Apple cabling and access method scheme) and the two other most popular access methods: Ethernet and Token Ring.

FIGURE 8.1a

AppleTalk and the OSI model.

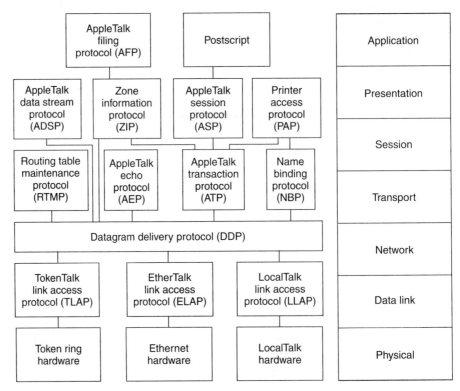

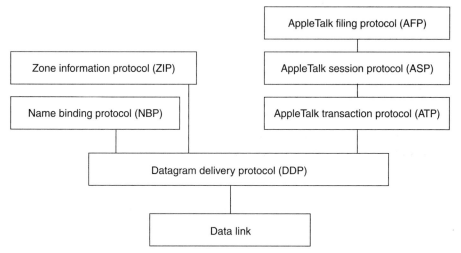

FIGURE 8.1b
*AppleTalk stack.
(Reproduced with
permission of Cisco
Systems, Inc.
Copyright © 2001
Cisco Systems, Inc.
All rights reserved.)*

The software portion of the AppleTalk stack consists of the OSI network layer through the application layers. The network layer has the ability to direct messages locally or through network extending devices such as routers. The stack also contains transaction oriented transport and session layer routines, which enable messages to be sent reliably through the network. Printing is also incorporated into the AppleTalk session layer. Finally, network workstations and servers may communicate through the network by the use of the *Apple File Protocol (AFP)* located at the application layer.

The Physical Layer—AppleTalk Hardware

AppleTalk can be used over many different types of media. This entity is comprised of the network controller and cabling systems that allow users to communicate with each other. The current media capable of handling the AppleTalk system include *LocalTalk* (for proprietary networking), *EtherTalk* (for use on Ethernet systems), *TokenTalk* (for use on Token Ring networks), and LANSTAR AppleTalk from Northern Telecom.

LocalTalk

Apple's low cost network implementation of LocalTalk consists of wiring (cable segments) and access methods to transmit data on the wiring. AppleTalk with LocalTalk is embedded into every Macintosh computer. The access method of LocalTalk is used only with Apple Macintosh computers and the related network devices like printers and modems. It is not used with any other network protocol.

Just as Ethernet defines the access methods that govern the transmission and reception of data on a special type of cable plant, LocalTalk accomplishes the same function and

●
Apple File Protocol provides access between the servers and clients.

●
LocalTalk is a protocol that is used for proprietary networking.

●
EtherTalk is a protocol that is used on Ethernet.

●
TokenTalk is a protocol that is used on Token Ring networks.

FIGURE 8.2

*LocalTalk devices:
(a) device connec-
tor, (b) bus cable,
(c) device connec-
tor, and (d) cable
extender.
(Reproduced with
permission of Cisco
Systems, Inc.
Copyright © 2001
Cisco Systems, Inc.
All rights reserved.)*

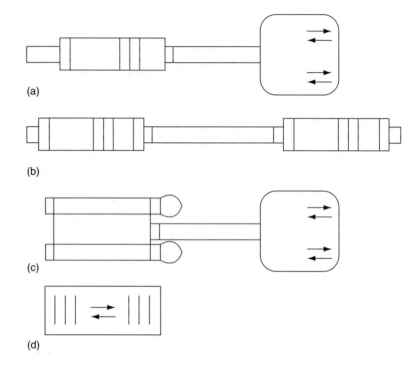

is an inexpensive way to connect workstations and their associated peripherals into a network. As shown in Figure 8.2, a LocalTalk network consists of the devices that connect AppleTalk network stations on a bus topology. The devices used are:

a. LocalTalk connector module shown as the two device connectors.
b. A 2-meter LocalTalk cable with locking connector shown as the bus cable.
c. A LocalTalk cable extender.
d. The DIN connector shells.

The device connector module is a small device with a cable attached to it that connects a network node to the LocalTalk cable. There are two types of connectors associated with it. An eight-pin round type connector called a *Deutsche Industrie Norm (DIN)* connector is used to connect to the Apple IIe, Apple IIGS, Macintosh Plus, Macintosh SE and Macintosh II computers, the LaserWriter II NT and LaserWriter II NTX printers, and the ImageWriter II printer with the LocalTalk option installed. The nine-pin rectangular connector is used on the Macintosh 128K, 512K, and Macintosh 512K enhanced computers and the LaserWriter and LaserWriter Plus printers.

A **Deutsche In-
dustrie Norm**
connector is used
in some Macintosh
and IBM PC-
compatible
computers.

LocalTalk cable is available in 10- and 25-meter lengths. Apple also sells a kit for custom-made lengths. Included in the kit are the connectors. Therefore, to run a longer cable between two network devices, a cable extender adapter is used to connect these cables to make longer runs of cable. In order for a network station to transmit and receive on the cable, a special device known as a transceiver is used. The transceiver is built into every Macintosh computer so that the user only has to connect the cable directly into the computer. It connects into the printer port in the back of the Macintosh. Printers participate in the AppleTalk scheme as autonomous devices and need not be directly connected to the Macintosh. Printers participate in the AppleTalk scheme as autonomous devices and need not be directly connected to the Macintosh. Apple specifications state that the longest single cable length is 300 meters. Other companies that have compatible wiring for AppleTalk have made modifications to this scheme and allow cable segments of up to 1,000 meters. This style of cabling is similar to the thin Ethernet cable scheme. To build the LocalTalk cable plant, it must be connected together. Figure 8.3 illustrates the concept of implementing a LocalTalk network. In order to build the LocalTalk cable plant, it must be connected together.

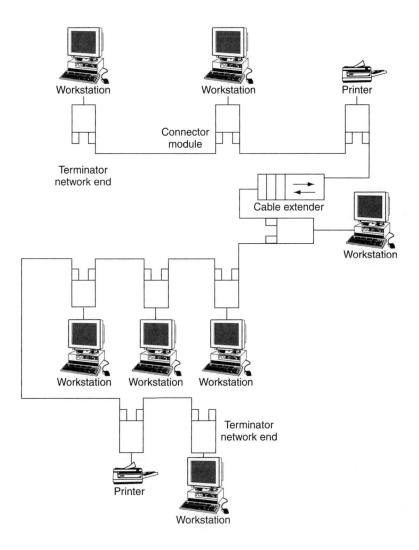

FIGURE 8.3

Apple LocalTalk network. (Reproduced with permission of Cisco Systems, Inc. Copyright © 2001 Cisco Systems, Inc. All rights reserved.)

Connecting the DIN connector to the printer port of the Apple computer connects devices to the LocalTalk network. The connector block at the other end of this connector has two RJ-11 plugs. The bus cable is connected into this and network stations are more or less concatenated.

Similar to Ethernet coaxial cable, the two end stations on the bus cable will have one connection to the bus and the other plug will be self-terminating. Only the two end stations will have the terminator in their block. All other stations will have the bus cable plugged into both connector blocks.

The speed on LocalTalk is 230.4 Kbps. This is much slower than the speed of Ethernet, which is 10 megabits per second, or Token Ring, which is 4 or 16 Mbps. The components used in the LocalTalk system were inexpensive, including the cabling system wire. This set limitations on the LocalTalk system that forced a low speed.

Media Considerations for AppleTalk

●

The **media attachment unit** performs physical layer functions.

With the capability of multimedia options for AppleTalk comes the need for a table to compare the three access methods. Because most Token Ring cables terminate in a central *media attachment unit (MAU),* the length of the lobe cable, which is the cable that connects the network station to the MAU, is usually 100 meters. Table 8.1 illustrates the comparison between the various access methods.

Data Link Functions

The data link functions consist of the following:

- Control panel software
- The LAP manager
- The AppleTalk Address Resolution Protocol (AARP)
- The Ethernet driver
- The Token Ring driver

The control panel: AppleTalk makes extensive use of the Macintosh (MAC) user interface, and part of the window system in the Apple user interface on the workstation is an icon called the control panel. In the control panel is the network control device package. The control panel uses the system folder to store all alternative AppleTalk connection files.

To allow connection to the LocalTalk bus, the user must use the Apple icon on the menu bar of the screen interface. Here, the operating system allows multiple choices into the Apple operating system. The user selects the chooser menu. In the chooser menu, icons represent each possible action on a network.

The user can select a network connection to use LocalTalk, EtherTalk, or TokenTalk. Only one physical interface to the network may be in use at a time. The network sta-

Table 8.1 LocalTalk and Ethernet Media Considerations

	LocalTalk	*Thick Ethernet*	*Thin Ethernet*	*LANSTAR AppleTalk*
Medium	Twisted pair	Coaxial	Coaxial	Twisted pair
Link Access Protocol	LLAP	IEEE 802.3	IEEE 802.3	LLAP
Transmission Rate	230.4 Kbps	10 Mbps	10 Mbps	2.56 Mbps
Maximum Length	1,000 ft	Segment: 1,640 ft	Segment: 656 ft	2,000 ft to star
		Network: 8,202 ft	Network: 3,281 ft	
Minimum Distance Between Nodes	No minimum	8.2 ft	1.5 ft	No minimum
Maximum Number of Nodes	32 Physical network: 1,023	Segment: 100 Physical network: 1,024	Segment: 30	1,344
Maximum Number of Active AppleTalk Nodes per Physical Cable Segment	32	Unlimited	Unlimited	1,344

Cable segment is a piece of cable not separated by a repeater device. Network segment is the total number of devices on all cable segments not separated by a network extending device such as a bridge or a router.

tion may be attached to both an Ethernet network and a LocalTalk network but can send and receive data only on one or the other, but not both. A user may attach to any network device on the active network.

Link Access Protocol (LAP) Manager for LocalTalk

The *Link Access Protocol (LAP) manager* entity is used to send and receive data over the selected media. The LAP manager is the data link layer for LocalTalk. For workstations that are connected to both LocalTalk and EtherTalk, the LAP manager sends packets to that network connection selected by the user in the control panel. For AppleTalk networks operating over a LocalTalk medium, the access method is *LocalTalk Link Access Protocol (LLAP)*.

One of LLAP's responsibilities is to handle access to the cable plant. The method is similar to that of Ethernet with one exception. The access method used by LocalTalk

The **LAP manager** is used to send and receive data over the selected media.

LocalTalk Link Access Protocol is the access method for AppleTalk networks operating over a LocalTalk medium.

Carrier Sense with Multiple Access and Collision Avoidance is the access method used by LocalTalk.

is *Carrier Sense with Multiple Access and Collision Avoidance* (CSMA/CA). Ethernet uses Collision Detection (CSMA/CD).

LLAP packet types include the following:

- LapENQ Inquiry packet used by the station to assign itself an address.
- LapACK A packet sent back in response to a lapENQ.
- LapRTS Sent to a destination station to indicate that a station has data for it.
- LapCTS Sent in response to a lapRTS to indicate to a source station that the destination station can accept data.

Figure 8.4 illustrates a LocalTalk and AppleTalk packet structure.

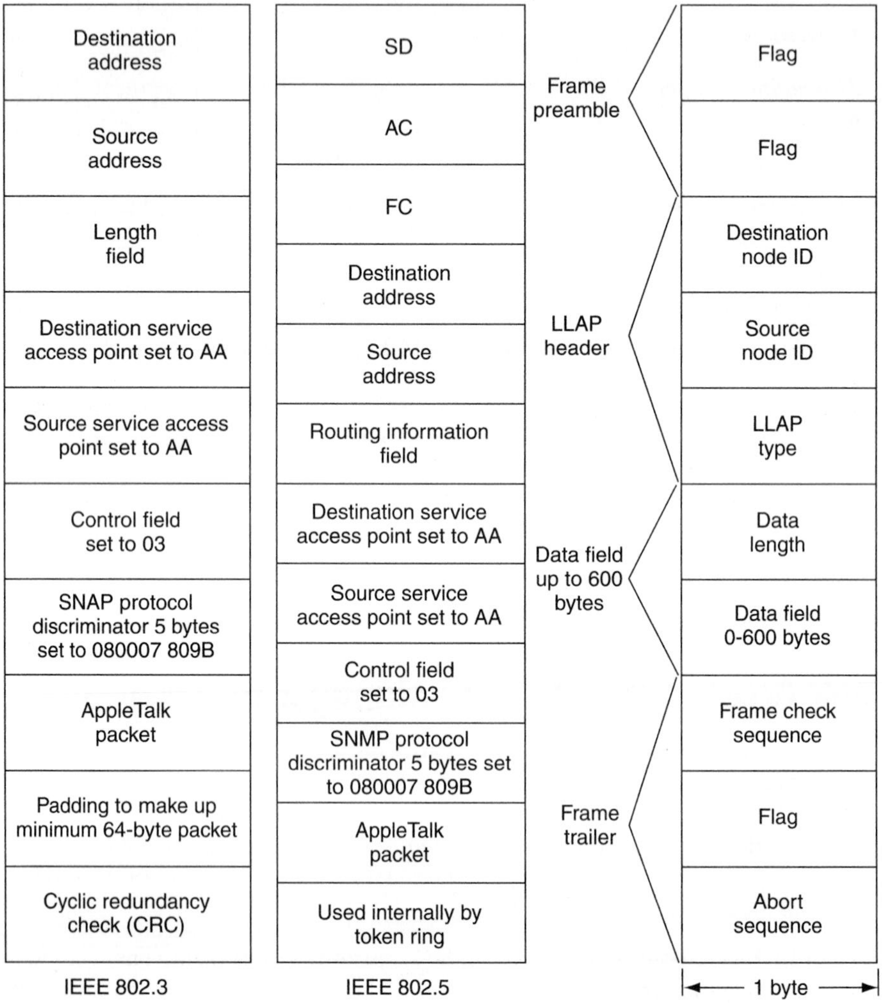

FIGURE 8.4
LocalTalk packet structure. (Reproduced with permission of Cisco Systems, Inc. Copyright © 2001 Cisco Systems, Inc. All rights reserved.)

LLAP Directed Transmissions (Station-to-Station Communications)

For station-to-station communications to take place, the source stations need to know the identity of the destination stations. Because there is only one cable plant and all stations need access to it, all active network stations compete for sole use of the cable plant for a limited amount of time. To gain access to the cable plant, one at a time, is the algorithm used in LLAP. LLAP uses Carrier Sense Multiple Access with Collision Avoidance (CSMA/CA).

In referring to Figure 8.5, LLAP first performs *carrier sense (CS)*. CS is the process of checking the cable plant to ensure that no one is currently using the cable to transmit. This is done by sensing electrical activity on the cable. If the cable is busy, the transmitter is requested to defer, which is the process of holding the data until the cable plant is clear. Once the line activity is quiet and has been idle for at least one InterDialogGap (IDG), which is 400 microseconds, it will wait an additional amount of random time. This random time is the result of the number of times it assumes a collision has occurred.

If the cable remains quiet throughout this whole time period, the network station will send a *Request to Send (RTS)* packet to the destination station. If the line becomes busy at any time during the initial interval, the whole process will be started over. The destination station must reply with a *Clear to Send (CTS)* packet to the originator within the InterFrame Gap (IFG) of 200 microseconds. When the originator receives the CTS packet it knows the destination is active and willing to accept packets. The source is now able to send data to the destination station. The source station must transmit the data within the IFG time period. If the destination needs to respond to the packet such as sending an ACK or a data response, it must repeat the foregoing procedure starting with an RTS packet. Once one station has transmitted any packet,

Carrier sense is the process of checking the cable plant to ensure that no one is using the cable to transmit.

A **Request to Send** packet requests a data transmission on a communications line.

A **Clear to Send** packet states that the destination is active and ready to receive packets.

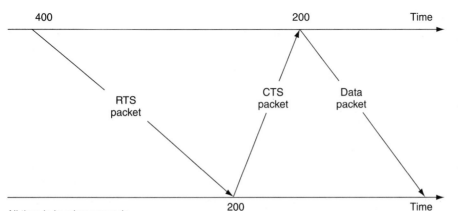

All time is in microseconds

FIGURE 8.5
AppleTalk with LocalTalk data transfer. (Reproduced with permission of Cisco Systems, Inc. Copyright © 2002 Cisco Systems, Inc All rights reserved.)

the cable is free to anyone to try and gain control over the cable. This is one of the reasons why the network operates at a slow rate. LocalTalk operates at 230.4 thousand bits per second or Kbps. This process is not completed when AppleTalk is running over Ethernet or Token Ring. When running over these access methods, AppleTalk follows their access methods. If a collision did occur, it should have occurred during the RTS-CTS handshake, and the corresponding LLAP control packet either the RTS or the CTS will be corrupted when received by one of the stations. The final result is that the CTS packet will never be received and the originating station will assume that a collision has occurred. The originating station must then back off and retry transmission. At this point, the originating station assumes that a collision has occurred because there is no circuitry defined in AppleTalk to determine that an actual collision has occurred. This reduces the cost of implementing AppleTalk with LocalTalk. The station will attempt 32 times to transmit a packet before notifying the upper layer software of its inability to do so.

LLAP Broadcast Transmissions

Broadcast transmissions are different from directed transmissions. Broadcast transmissions are intended for all stations on the local network. If there are 20 stations on the network, all 20 stations should receive a broadcast packet.

Broadcast frames will still wait for the cable to be quiet for at least one IDG. They will then wait an additional amount of random time. If the link is still quiet, the broadcast frame will send an RTS packet with the destination address set to 255 (FF in hex). If the line remains quiet for one IFG, the station will then transmit its data. It does not expect to receive any responses. This type of packet has many functions. A broadcast packet can be used to send a message to all stations at one time to find other stations on the network.

LLAP Packet Receptions

A network station will receive a packet if the packet's destination address in the packet received is the same as the receiving station's internal address and the receiving station finds no errors in the packet's frame check sequence (FCS).

A frame check sequence is an algorithm that is used to verify the data that was transmitted is the same data that was received. For simplicity sake, this algorithm is a complex parity checker. This data check guarantees the transmission of the packet to be 99.99 percent free of errors. The transmitting station computes the FCS upon building the packet. When the receiving station receives the packet, it will compute its own FCS. It will then compare its FCS with the FCS sent by the transmitting station. If there is a discrepancy between the two, the receiving station will discard the packet.

The LLAP of the receiving station will also drop a packet for other reasons such as too large or too small or the wrong type. It will handle this without interrupting the upper layer software.

AppleTalk Addressing

To communicate with another station requires more than sending RTS and CTS packets on the network. A packet must be addressed so that another station may receive this packet and decide if it is meant for that station.

AppleTalk was designed to run on top of LocalTalk. Unlike Ethernet or Token Ring, LocalTalk does not have an address burned into a prom of the LocalTalk hardware. With Ethernet and Token Ring, the manufacturer of the controller card usually assigns the MAC address of the controller card. During initialization, the network software that runs on these controller cards will read the address of the controller card and assign this as the address on the network. AppleTalk has an addressing scheme, but the address of a network workstation is decided by the network workstation's AppleTalk software during initialization of the network software. This type of address is also known as a protocol address. Originally, AppleTalk operating over LocalTalk had no physical addresses, only AppleTalk addresses.

As with any protocol that operates over a network, any attachment on the network must be identified so that other devices may communicate with it. Each device on a network is assigned a unique node address and a group network address. Because LocalTalk was devised as the hardware complement to AppleTalk, AppleTalk uses LocalTalk's addressing scheme. The address scheme will be used again when mapped to an Ethernet or Token Ring MAC layer address.

The AppleTalk addressing scheme was not changed during the migration process from AppleTalk to Ethernet and Token Ring. The AppleTalk address is combined with the Ethernet or Token Ring physical address. This addressing scheme requires the use of a new protocol known as *AppleTalk Address Resolution Protocol (AARP)*. The AARP protocol allows for a translation of the AppleTalk software address to Ethernet and Token Ring physical addresses.

Every network has many addresses. There will be one for each attachment such as workstations, routers, and servers to the network. On an AppleTalk network, each network station possesses a unique identity in the format of a numeric address. The assignment of this address to the network attachment is a dynamic process in AppleTalk. The address assignment process creates two advantages over hard coded addressing schemes, such as those used in Ethernet or Token Ring. First, there is less hardware required because the address is not fixed or burned into prom. Second, there is no central administration of IDs from vendor to vendor. Recall that in Ethernet and Token Ring the IEEE standards committee administers Ethernet and Token Ring network addressing. However, the dynamic node assignment can create problems when combined with network extending devices such as a bridge.

The **AppleTalk Address Resolution Protocol** maps a data link address to a network address.

The AppleTalk address will actually be two numbers: a 16-bit network address and an 8-bit node ID. A network address is similar to an area code in the phone system. The node ID is similar to the seven-digit phone number. The node ID ranges from 0 to 255 and the network number ranges from 0 to 65535. Each network attachment on the network will be identified by these two numbers together. For example, a single network station can be found on any AppleTalk internet if the network number and the node ID are known. Knowing the network number will track it down to a group of network stations, and knowing the node ID will single out a network station within that group. AppleTalk addressing takes the form of <network number, node ID> with the network number being 16 bits in length and the node ID being 8 bits in length. This address will then be mapped to a MAC address through the AARP process.

For AppleTalk Phase II there are two addressable types of networks: nonextended and extended networks. *Nonextended networks* are individual networks that contain one network number and one zone name. LocalTalk and AppleTalk Phase I/Ethernet are examples of nonextended networks. *Extended networks* are those individual networks that can contain more than one network number and multiple zone names. Extended networks were devised to allow more than 254 node IDs on a single physical network such as Ethernet and Token Ring.

There are some restrictions placed on the AppleTalk addresses. The network number of 0 indicates that the network number is unknown, and it specifies a local network to which the network station is attached. Packets containing a network address of 0 are meant for a network station on the local network segment. Network numbers in the range of FF00h through FFFEh are reserved. These numbers are used by network stations at start-up time to find their real network number from a router and at times when no router is available.

Node ID of 0 also has a special meaning. A packet with this address is destined for any router on the network as specified by the network part of the address. Packets utilizing this address will be routed throughout an internet and will be received by a router whose network address, not node ID, is included in the network ID field. A protocol such as the Name Binding Protocol (NBP) is one example of this type of addressing. It allows processes within routers to talk to each other without having to use up a node ID. Node ID 255 is reserved as a broadcast address. A packet containing this address is meant for all stations on the network. For AppleTalk Phase II, node ID 254 is reserved and may not be used.

With LocalTalk only, the available 254 node IDs are divided into two sections: those reserved for servers and those reserved for workstations. This eliminates the chance that a station was too busy to answer an inquiry packet. It also allows the separation of workstations from acquiring a server ID, which could be disastrous on an AppleTalk network. Because of the high speed and reliability of Ethernet and Token Ring, the EtherTalk and TokenTalk protocols do not implement the separation of node IDs. An AppleTalk node may acquire any address within those restric-

Nonextended networks are individual networks that contain one network number and one zone name.

Extended networks contain more than one network number and multiple zone names.

tions just mentioned. This is a dynamic process, meaning the number is randomly chosen by the AppleTalk protocol running on a network station. All stations on the network will participate in the selection of a node ID for a network station.

AppleTalk Network Components

AppleTalk networks are arranged hierarchically. Four basic components form the basis of an AppleTalk network: sockets, nodes, networks, and zones. Figure 8.6 illustrates the hierarchical organization of those components in an AppleTalk internetwork.

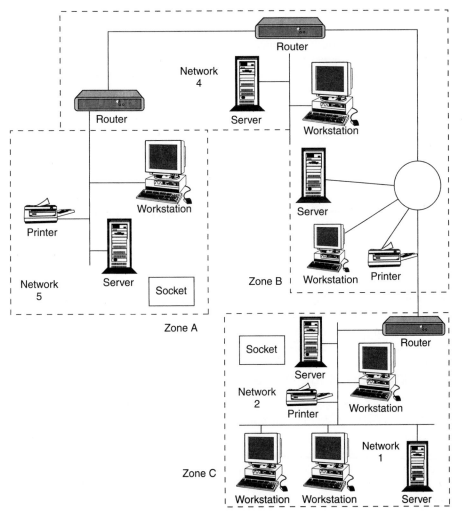

FIGURE 8.6

The AppleTalk internetwork consists of hierarchy components. (Reproduced with permission of Cisco Systems, Inc. Copyright © 2001 Cisco Systems, Inc. All rights reserved.)

Sockets

An AppleTalk *socket* is a unique, addressable location in an AppleTalk node. A *socket* is the logical point at which upper layer AppleTalk software processes and network layer *Datagram Delivery Protocol (DDP)* interact. These upper layer processes are known as socket clients. Socket clients own one or more sockets, which they use to send and receive datagrams. Sockets can be assigned statically or dynamically. Statically assigned sockets are reserved for use by certain protocols or other processes. Dynamically assigned sockets are assigned by DDP to socket clients upon request. An AppleTalk node can contain up to 254 different socket numbers.

> An AppleTalk **socket** is a unique, address-able location in an AppleTalk node.

> The **Datagram Delivery Proto-col** is responsible for the socket-to-socket delivery of datagrams over an AppleTalk internetwork.

Nodes

An AppleTalk *node* is a device that is connected to an AppleTalk network. This device might be a Macintosh computer, a printer, an IBM PC, a router, or some other similar device. Within each AppleTalk node exist numerous software processes called sockets. The function of these sockets is to identify the software processes operating on the device. Each node in an AppleTalk network belongs to a single network and a specific zone.

> A **node** is a de-vice that is con-nected to an AppleTalk network.

Networks

An AppleTalk *network* consists of a single logical cable and multiple attached nodes. The logical cable is composed of either a single physical cable or multiple physical cables interconnected by using bridges or routers. AppleTalk networks can be nonextended or extended.

> A **network** con-sists of a single logical cable and multiple attached nodes.

Zones

An AppleTalk *zone* is a logical group of nodes or networks that is defined when the network administrator configures the network. The nodes or networks need not be physically contiguous to belong to the same AppleTalk zone.

> A **zone** is a logi-cal group of nodes or networks that is defined when the network adminis-trator configures the network.

Nonextended Node ID Address Selection

On a nonextended network, server and workstation node IDs range from 1 to 254 (0 and 255 are reserved). Recall that a nonextended network is a physical cable plant that contains one network number and one or no zone name. Figure 8.7 illustrates a nonextended network. This type of network is assigned exactly one network ID and one zone name. An example of a nonextended network is LocalTalk or Phase I AppleTalk with Ethernet framing. When the network station initializes, it will assign itself a node ID. The station will then send out a special packet, known as an inquiry

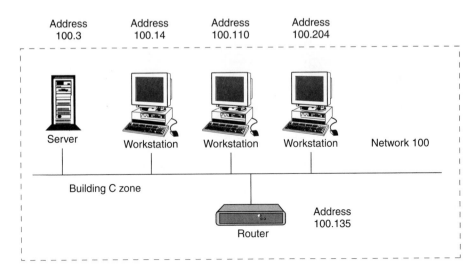

control packet containing this address, to see if any other station has already reserved this number for its use. If no response is received, the node will then use this number. If a response is received to this inquiry packet, the requesting node will randomly choose another number, submit another inquiry control packet, and wait for a response. It will continue this process until a node ID is acquired.

The node will then send out a request to the router to find out the 16-bit network number that has been assigned to the particular network. If a response is not received, the node will assume that no router is currently available and will use network ID of 0. If a router later becomes available, the node will switch to the new network number when it can. It may not switch immediately because previous connections may have already been established. If a response is received, the response packet will contain the network number assigned to that network and the node will then use that node ID.

Extended Network Node ID Selection

AppleTalk Phase II introduced a new network numbering scheme and a concept referred to as an extended network. An extended AppleTalk network is a physical network segment that can be assigned multiple network numbers, and, theoretically, may have up to 16 million network attachments on it. An extended network can afford this expansion, for each network station is assigned a combination of a 16-bit network ID and an 8-bit node ID. Network IDs are like area codes in the phone system. Network IDs usually, but not always, identify groups of nodes with common network IDs assigned to them.

Extended networks can also be assigned multiple zone names. With extended networks, there is one less node allowed. Node ID FE (decimal 254) is reserved; therefore, there are only 253 node IDs allowed per network ID. Implementation of

FIGURE 8.8

An extended net-work can be as-signed multiple network numbers. (Reproduced with permission of Cisco Systems, Inc. Copyright © 2001 Cisco Systems, Inc. All rights reserved.)

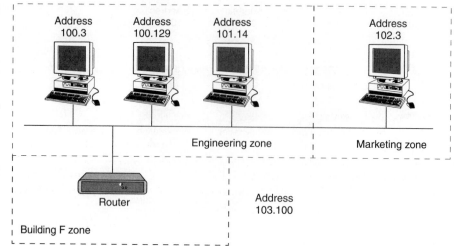

AppleTalk network IDs is different from other network ID implementations in that multiple network IDs are allowed on the same cable segment. Figure 8.8 illustrates an extended network. To allow AppleTalk to run on Ethernet and Token Ring networks, the aforementioned implementation of multiple network IDs on the same cable segment had to be taken into account because the maximum number of network attachments on an extended cable segment of Ethernet is 1,024. Token Ring still maintains the maximum number of 260. For those stations operating on an extended network, the network ID acquisition is a little different. First, the network station will assign a provisional node address to itself. This is assigned by the data link and its only purpose is to talk to a router. The start-up network ID is taken from the range of FF00h to FFEEh. This range is reserved for use with start-up stations and may not be permanently assigned to any network.

A unique twist to this acquisition process is that if the network station was previously started on the network, it will have reserved its previous node ID and network ID on its disk. This is referred to as hinting. When the network station starts up, it will try this address first. If this address is no longer valid, it will start from the beginning. Otherwise, the network station will send out a special packet to the router. The router will respond with a list of the valid network IDs for that network segment. If no router is available, the reserved start-up network number is used and will be corrected later when the router becomes available.

For those who are familiar with the manual filtering capabilities of bridges or routers, since node IDs and network IDs are dynamic, it is impossible to guess which network station is assigned to a network number because node IDs and network IDs are dynamic. The manual filtering of bridges or routers for AppleTalk is generally reserved to zone names.

AppleTalk Phase 1 and Phase 2

When AppleTalk Phase 1 was implemented, no more than 254 network stations could attach and be active on a single network cable segment. One network number and one zone name existed for each cable segment not separated by a router. Phase 1 also supported the Ethernet framing format.

With AppleTalk Phase 2, the network addressing was extended so that many network numbers could exist on a single cable plant. Phase 2 network numbers are 16-bits wide, allowing for the possibility of over 16 million network stations per network segment. Support for the Ethernet framing format was also changed to the IEEE 802.3 with SNAP headers framing format. AppleTalk Phase 2 allows 253 network station addresses per network number, one less than Phase 1. However, you may now have multiple network IDs on the same cable plant. AppleTalk Phase II also supports Token Ring networks using 802.2 SNAP frames. Most Apple network implementations today have switched over to AppleTalk Phase 2.

Recall that AppleTalk utilizes addresses to identify and locate devices on a network in a manner similar to the process utilized by such common protocols as TCP/IP and IPX. These addresses are assigned dynamically and are composed of three elements:

- Network number: a 16-bit value that identifies a specific AppleTalk network either nonextended or extended.
- Node number: an 8-bit value that identifies a particular AppleTalk node attached to the specified network.
- Socket number: an 8-bit number that identifies a specific socket running on a network node.

AppleTalk addresses are usually written as decimal values separated by a period. For example, 10.1.50 means Network 10, Node 1, Socket 50. This might also be represented as 10.1, socket 50. Figure 8.9 illustrates the AppleTalk network address format.

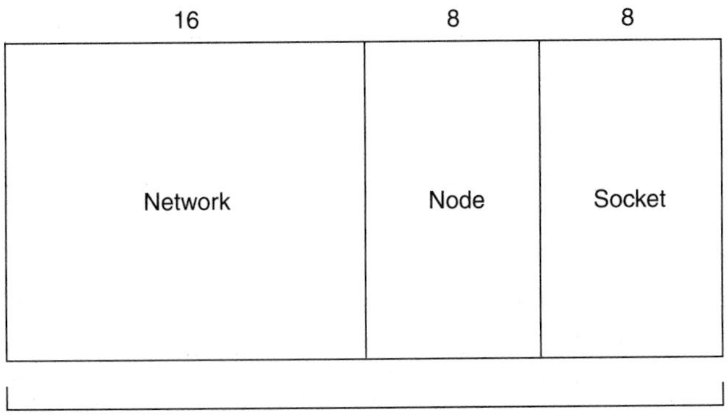

FIGURE 8.9

The AppleTalk network address consists of three distinct elements. (Reproduced with permission of Cisco Systems, Inc. Copyright © 2001 Cisco Systems, Inc. All rights reserved.)

The AppleTalk Address Resolution Protocol (AARP)

The previously mentioned AppleTalk addressing scheme was developed by Apple Computer and designed to work with LocalTalk. Since AppleTalk's inception, Ethernet and Token Ring have taken over as the preferred network implementations, especially since the price for controller cards has dropped considerably. LocalTalk is still used in similar network environments and by no means is a dead access method.

To have AppleTalk run on an Ethernet or Token Ring network, the AppleTalk addressing scheme must be mapped internally to conform to the 48-bit MAC address (also known as the hardware or physical address) used in Ethernet and Token Ring. AARP is the protocol that accomplishes this task.

AARP will use the following three types of packets:

1. Request packet: The request packet is used to find another node's AppleTalk address/MAC address. To send information to another station on a local network, the station must know its MAC address and its AppleTalk address. This packet is sent out to find the address.
2. Response packet: The response packet is used to respond to a node's request for an address mapping.
3. Probe packet: The probe packet is used to acquire an AppleTalk protocol address and to make sure that no one else on the local network is using this address. This packet is sent out up to 10 times, once every 200 ms (1/5 of a second), 10 times in 2 seconds.

For network identification all software and hardware protocol stacks operating within a network station are addressed with integer numbers. When AppleTalk and LocalTalk were first devised an 8-bit wide integer was selected for this purpose. Ethernet uses 48-bit addresses for network attachment identification. A software entity was devised to interpret the difference between the two forms of addressing. AARP is similar to TCP/IP's ARP protocol. AARP resides between the Link Access Protocol and the LAP manager. AARP performs the following functions:

1. Selecting a unique address for a client.
2. Mapping the protocol address to the specific physical hardware address.
3. Determining which packets are destined for a specific protocol.

The AARP process is only used when implementing AppleTalk on top of Ethernet or Token Ring. AARP is not implemented for LLAP (LocalTalk). The AppleTalk software address has not been replaced. A scheme has been devised to work with it in order to make it compatible with other network protocol architectures. The AARP process is similar to the TCP/IP process that maps the 32-bit TCP/IP addresses to the Ethernet or Token Ring MAC hardware address. Knowing how ARP operates will help you understand how AARP functions.

AARP Address Mappings

Each node that uses AARP maintains a table of address mappings. For example, Ethernet maintains a table of the AppleTalk protocol addresses and their associated IEEE 802.3 physical addresses. This table is referred to as the *Address Mapping Table (AMT)*. Instead of requesting the mapping each time a station needs to talk to another station on the network, the AMT will have a listing of all known network stations on its local network. The entries in the table include AppleTalk addresses and their corresponding MAC hardware addresses. When a network station needs to talk to another station on the network, it will first look up the mapping in the AMT. Only if it is not there will it send out a request packet. Once an entry in the AMT is accomplished, AARP maintains this table. After a certain time, AARP will age out and delete old addresses and update with new ones.

AARP receives all AARP request packets because they are sent out with a broadcast MAC destination address. AARP will discard the packet if it does not need to respond to the packet; however, it will check the sender's 48-bit hardware address and AppleTalk's 8-bit node address embedded into the AARP request packet. It will extract this information, use it in its initialization table, and then discard the packet. Table 8.2 shows an example of an AMT for a given network station. The table shows four entries. When a network station that contains this table would like to talk to another station, it must find the hardware address of the remote station. It will first consult this table. To communicate with another station over Ethernet or Token Ring, a network station must know the destination's AppleTalk and MAC addresses.

If the requesting station wanted to talk to station 16.3, it would look in this table and find the MAC address for 16.3. Then, it would build a packet and in the data link header for addressing would enter the MAC address of 02608c010101. If the entry for 16.3 were not in the table, AARP would build a request packet and transmit the packet to the network. Upon receiving a response, it would add the contents of the response packet to the AMT and then build a packet for 16.3.

The **Address Mapping Table** is a list of all known network stations on its local network.

Table 8.2 An AMT Table

Network	*Node ID Address*
16.3	02608c010101
16.4	02608c014567
16.90	02608c958671
17.20	02608c987654

AARP Node ID Assignment

When a network station initializes, AARP assigns a unique protocol address for each protocol stack running on this station. Either AARP can perform this or the client protocol stack can assign it and then inform AARP. For AARP to assign the address, it must accomplish three things:

1. Assign a tentative random address that is not already in the Address Mapping Table.
2. Broadcast a probe packet to determine if any other network station is using the newly assigned address.
3. If the address is not already in use, AARP permanently uses this number for the workstation.

If AARP receives a response to its probe packet, it will then try another number and start the whole algorithm over again until it finds an unused number.

When a network operating system submits data to the data link layer to be transmitted on the network, the protocol will supply the destination address. In AppleTalk, this will be supplied as a protocol address. This address will then be mapped to the corresponding MAC address for that particular destination station. This mapping is what AARP accomplishes.

Examining Received Packets

When AARP receives packets, it will operate only on AARP packets. All other packets are for the LAP. AARP does not provide any functionality other than AARP request, response, and probe packets.

Once AARP has provided the mapping of addresses, the data link protocol may then accomplish its work. The access protocol of Ethernet remains unchanged to operate on an AppleTalk network. AARP does not interfere with the operation of Ethernet or Token Ring.

LAP Manager for EtherTalk and TokenTalk

Ethernet was developed in Xerox's Palo Alto Research Center (PARC). Initially known as the experimental Ethernet, it was first utilized in 1976. In 1980, a cooperative effort by Digital, Intel, and Xerox led to a public document known as the "Blue Book" specification. The formal title was Ethernet version 1.0. In 1982, these three companies again converged and developed Ethernet version 2.0. This is the standard by which Ethernet operates today.

In the beginning, the components that made up an Ethernet network were extremely expensive and well over budget for many corporations. Because Ethernet is an open architecture and not proprietary many companies have jumped on the bandwagon and developed Ethernet products of their own. Because Ethernet is an open specification, all Ethernet products are compatible with each other. You may buy Ethernet products from one company and more Ethernet products from another company and the two will work with each other at least at the data link layer. Changes in cabling strategies and mass production of the chip sets have also led to price reductions in Ethernet. Currently, Ethernet is the most popular networking scheme.

The origins of Token Ring date back to 1969. It was made popular when IBM selected it as its networking scheme. Token Ring was adopted by the IEEE 802.5 committee in 1985, and it offered some advantages over the Ethernet scheme. Some of these advantages included a star-wired topology and embedded network management. The cost of Token Ring was extremely high at first. The price reductions throughout the last few years and its advantages over Ethernet have allowed this networking scheme to become very popular.

AppleTalk was derived as an alternative to this high cost of networking. AppleTalk, with LocalTalk, operates at all layers of the OSI model. It is a true peer-to-peer network and is built into every Macintosh computer. It is a simplex plug and play type of networking that easily allows users to access the services of a network. However, Ethernet and Token Ring are still the most popular networking schemes with many advantages over the LocalTalk access method, and as a result AppleTalk has been adopted to allow for this. To allow these networks compatibility with AppleTalk, the primary layer that is replaced is the LocalTalk data link layer and the way the LAP manager operates with it.

When an AppleTalk network station wishes to communicate with another AppleTalk network station, it must provide its data link with an AppleTalk protocol address, which consists of a 16-bit network number and an 8-bit node ID. When Ethernet and Token Ring are used as the medium, a major change is invoked. The address technique used with these access methods is 48 bits long. The AppleTalk protocol address must be translated into this format before a packet may be transmitted on an Ethernet or Token Ring network.

Extensions were made to AppleTalk's LAP in the form of *EtherTalk Link Access Protocol (ELAP)* and *TokenTalk Link Access Protocol (TLAP)* to accommodate Ethernet and Token Ring access methods. This LAP manager also uses a new function known as AppleTalk Address Resolution Protocol (AARP) to translate the 48-bit address to the AppleTalk addresses. ELAP and TLAP function according to their access methods of Ethernet (CSMA/CD) and Token Ring, respectively.

● **EtherTalk Link Access Protocol** accommodates Ethernet and Token Ring access methods.

Figure 8.10 illustrates the mapping of the AppleTalk media access methods to the bottom two layers of the OSI reference model.

● **TokenTalk Link Access Protocol** also functions according to its access method.

Theory

Simply stated, EtherTalk and TokenTalk are the Ethernet and Token Ring access methods that have AppleTalk running on top of them. The LocalTalk access method is stripped out and replaced with either Ethernet or Token Ring.

FIGURE 8.10

AppleTalk media access mapping to the bottom two layers of the OSI reference model. (Reproduced with permission of Cisco Systems, Inc. Copyright © 2001 Cisco Systems, Inc. All rights reserved.)

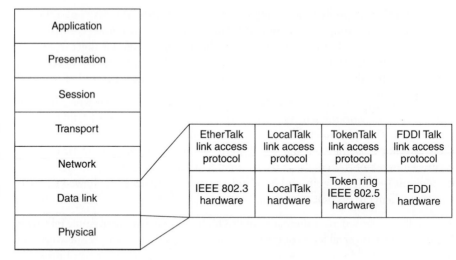

If you have an Ethernet network, EtherTalk will allow AppleTalk to run on that network. The same is true for TokenTalk. The major changes were made in the LAP manager. All layers above the network layer are still AppleTalk. Figure 8.11 illustrates that the one noticeable change in the attachments or devices is the printer being connected to an AppleTalk file server and not directly to the LocalTalk network. Ethernet and Token Ring interfaces are not widely used for direct attachment of Apple printers.

Token Ring packets are illustrated in Figure 8.4. Compare them with the AppleTalk and LocalTalk packet headers shown in Figure 8.4. As far as AppleTalk is concerned, it is running on top of LocalTalk. Only the LAP manager is changed.

FIGURE 8.11

AppleTalk on Ethernet. (Reproduced with permission of Cisco Systems, Inc. Copyright © 2001 Cisco Systems, Inc. All rights reserved.)

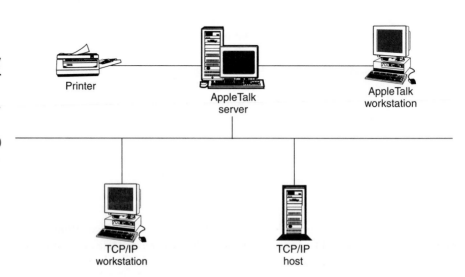

To allow AppleTalk Phase II to operate over Ethernet and Token Ring, the easiest method was through the use of the IEEE protocol specification 802.2 Type 1. The IEEE 802.2 data link specification consists of two types. Type 1 is a connectionless protocol, meaning that an established link between a source and destination network station does not have to exist before data transmission occurs. Type 2 is connection oriented, meaning the opposite of Type 1. EtherTalk and TokenTalk use IEEE 802.2 Type 1 packet formats for transmission over the network.

A subpart of that protocol is defined as Subnetwork Access Protocol (SNAP). The IEEE 802.2 specification replaced the concept of the type field with the concept of a Service Access Point (SAP). SAPs allow the distinction of a single protocol within a packet. It was determined when IEEE 802.2 was being written that many protocols might exist on a network station or on a network. When a packet is transmitted, it should be known which process of the protocol stack submitted the packet and to which process of the protocol stack the packet is destined. For example, you may have AppleTalk, XNS, and TCP/IP all running on the same network station. All three may send and receive packets from the same network connection. To determine which protocol stack any received packets are destined for, a SAP number is used. SNAP allowed these non-IEEE protocols to run on the IEEE 802.x data link. It allowed an easy port of existing LAN protocols.

Packet Formats for Ethernet and Token Ring

The IEEE 802.2 protocol also allowed for migration of existing packet types and network protocols to the IEEE packet type. Type 1 is the most commonly used on Ethernet and Token Ring when the IEEE 802.2 protocol is used. To allow for this, SNAP was invented. As illustrated in Figure 8.4, the *Source Service Access Point (SSAP)* is set to AAh and the *Destination Service Access Point (DSAP)* is also set to AAh. The control field is set to 03h to indicate an unnumbered information packet. The next 5 bytes represent the protocol discriminator, which describes the protocol family to which the packet belongs.

First, with 3 bytes of 0s in this field, the packet would be determined as an encapsulated Ethernet framed packet. This would tell the data link software that the next byte following the SNAP header is the type field of an Ethernet packet and to read the packet accordingly. The first 3 bytes could also read 08-00-07 in hex, which indicates that the frame is an encapsulated IEEE 802.3 framed packet with the following 2 bytes indicating a protocol ID field. The protocol discriminator allows proper translation when the frame is forwarded onto a network.

For example, if a frame traversed multiple media types such as Token Ring, FDDI, and Ethernet, the frame format would change for each type of media traversed. So, if the frame were received on a Token Ring port of a router, and it needed to be forwarded to an Ethernet network, the router would need to know which type of frame format to use on the Ethernet network—Ethernet version 2.0 or IEEE 802.3. This

The **Source Service Access Point** is the SAP of the network node designated in the source field of a packet.

The **Destination Service Access Point** is designated in the destination field of a packet.

is the purpose of the protocol discriminator. If the first 3 bytes were 00-00-00, then it would use Ethernet V2.0 frame format. If this field is set to a number other than 0, it will use the 802.3 frame format. The one exception to this rule is when you are bridging and not routing an AARP frame. Bridges employ a translation table to indicate special occurrences and will format this correctly. This is according to the IEEE 802.1h specification.

The SNAP address of 00-00-00-80-F3 is used to identify AARP packets. Following this is the AppleTalk data field, which will contain the AppleTalk OSI network layer protocol of DDP information. The last 2 bytes are the protocol ID, and it is set to 809B (hop) to indicate the packet is an AppleTalk packet.

TLAP packet formats are like ELAP packet formats, with the exception of the internal fields for the Token Ring controller and the source routing fields for packet routing with bridges. All IEEE 802.2 SNAP fields are the same.

Referring to Figure 8.4, the first few bytes pertain only to the data link layer of Token Ring. These bytes include the Access Control field and the Frame Control field. These fields indicate to the Token Ring data link controller how to handle the packet and which type of packet it is. The next byte is similar to an Ethernet packet address. It is the 48-bit physical address of the network station for which the packet is intended. The source address is the 48-bit physical address of the network station that transmitted the packet to the ring. The next fields are the routing information fields according to the source routing protocol defined by IBM and the IEEE 802.5 committee. Following the routing fields are the IEEE 802.2 SNAP headers of DSAP, SSAP, and control. The DSAP and SSAP fields contain AAh, and the control field will contain 03h. Following this is the protocol discriminator, which is set to 080007809B to indicate AppleTalk is the protocol for this packet.

The ELAP packet format is depicted in Figure 8.4. The first 6 bytes are the physical destination address of an Ethernet network station. The next 6 bytes are the physical address of the Ethernet station that transmitted the packet. The next field is the length field, which indicates to the data link the amount of data residing in the data field, excluding the pad characters.

What follows the length field is how AppleTalk easily resides on an Ethernet network. At byte 14 is the IEEE assigned SNAP DSAP header of AAh, and the next byte is the SNAP SSAP of AAh. AA is assigned by the IEEE to indicate that the packet is for the SNAP format, and all data link drivers should read the packet as such. Following the DSAP and SSAP bytes is the field to indicate the control. This byte indicates to the data link that it is a Type 1 (connectionless) IEEE 802.2 packet (03h). Following this is the SNAP protocol discriminator, which is set to 08-00-07-80-9B to indicate AppleTalk and the data link should read the packet according to AppleTalk protocol specifications. If that particular node is not running the AppleTalk protocols, the network station software will simply discard the packet. If it is running the AppleTalk protocols, it will accept and decipher the packet according to the AppleTalk specification.

Operation

When the AppleTalk protocol is started, ELAP or TLAP will ask AARP to assign a dynamic protocol address to it. All that was replaced here is the data link layer. AppleTalk above layer 2 remained the same. No changes were made to it. The data link layer changed to accommodate the new access technique of Ethernet or Token Ring.

Figure 8.12 depicts how three LAP managers may be installed but only one may be used at a time. The upper layer protocols may switch to any of the three protocols. The one small change to the LAP manager is in the form of AARP attached to the Ethernet and Token Ring protocols.

The previous sections discussed the data link and physical layers for AppleTalk. The software portion of AppleTalk begins at the network layer entity of the DDP, which will be discussed next.

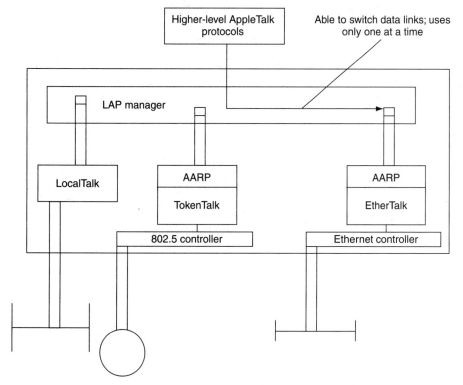

FIGURE 8.12

LAP manager. (Reproduced with permission of Cisco Systems, Inc. Copyright © 2001 Cisco Systems, Inc. All rights reserved.)

The AppleTalk Network Layer:
Datagram Delivery Protocol (DDP)

The DDP layer resides at the OSI network layer and allows data to flow between two or more communicating stations on the network. There are multiple entities that make up this layer:

- The Datagram Delivery Protocol (DDP)
- The Routing Table Maintenance Protocol (RTMP)
- The AppleTalk Echo Protocol (AEP)

There are many functions provided by the DDP, including data delivery, routing, and socket assignment.

The data link layer is nothing more than a car that carries passengers. It accepts passengers and will take those passengers to the destination they indicate. Although the data link delivers data given to it by the network layer, based on a node-to-node relationship, DDP establishes a concept referred to as sockets. When data is transmitted or received by a station, the packet must have someplace to attach to in the network code. This is the purpose of the socket. Because many processes may be running on a network workstation, the DDP must be able to identify which process the packet should be delivered to.

With DDP, communication between two stations is now on a socket-to-socket basis for data delivery and reception. All the communication is accomplished on a connectionless service, meaning the data is delivered to a process known as a socket, and once delivered, the receiving station does not acknowledge the originating station.

How Sockets Operate in AppleTalk Addressing

To communicate with another device, a network workstation will need an addressable endpoint to connect to. The initiator of the communication transfer must indicate the final destination—a software endpoint—for this data. The socket number tells the network station software to deliver the incoming packet to a specific process or application in the network station. This is the only purpose of the socket.

Socket numbers are addressable endpoints in any network station that represent an application program or another process running in that network station. There will be one unique socket number for each process that is running in a single network station. Because many processes may be running in a network station at any one time, each process must be uniquely identified with a socket number. If you know the socket number, you will know the process or application that owns it. Socket numbers are also used to identify the process that submitted the packet. If a file server receives a data packet and must respond to it, it needs to know the socket number in the source station that will receive this data.

All communications between a source and a destination station on the network will attach to each other through a socket. A socket is abstract to the user. It is not a physical device and is used by the networking software as an end connection point for data delivery. There are source and destination sockets. The source socket is from the originator of the connection and the destination socket indicates the final connection point—the end addressable point on the destination station. A source station will initiate communication with a source socket identified in the DDP header of a packet so that the destination station will know where to attach when a response is generated.

AppleTalk implements sockets somewhat differently than other network protocols. In other network protocols sockets are assigned as static (well-known) and dynamic sockets. AppleTalk well-known sockets are those sockets that are directly addressed to DDP only. There are no well-known sockets for applications that run on the internet. The well-known sockets are addressed in Table 8.3.

When a network station receives a packet that has a known socket, DDP will act on the packet. Only the DDP process uses these types of sockets. This is the difference between AppleTalk and other types of protocols for using sockets.

Other protocols such as TCP/IP and NetWare will assign static sockets for every known process in the network station. A committee will assign these socket numbers, and once a process is assigned this socket no other service may duplicate it. This well-known socket is universal. All applications that are written will be addressed to these socket numbers. AppleTalk allows its processes such as file service, mail service, and print service to ask DDP for a socket number, and it could be different every time the service is started on the network. Like the node ID assignment, socket numbers are also assigned dynamically. For example, the router table update process that runs in AppleTalk is statically assigned socket number 1. Any process that wishes to communicate to the router must identify the packet with a destination socket number 1.

Table 8.3 Socket Values

DDP Socket Value (hex)	Description
00h	Invalid
FFh	Invalid
01h	RTMP socket
02h	Names information socket (NIS)
04h	Echoer socket
06h	Zone information socket
80h–FEh	Dynamically assigned

FIGURE 8.13
DDP. (Reproduced with permission of Cisco Systems, Inc. Copyright © 2001 Cisco Systems, Inc. All rights reserved.)

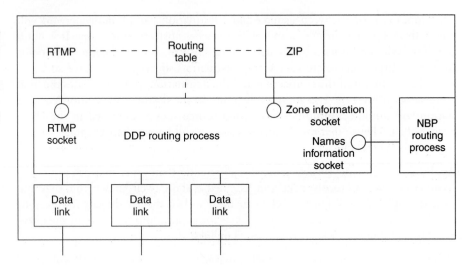

Figure 8.13 illustrates a listing of the router sockets and their associated applications. The only well-known sockets used by DDP are referenced in Table 8.4.

Dynamic sockets are assigned at process initiation. For example, when a process on a workstation initiates, it will request a socket number. It is the DDP layer that assigns the dynamic socket numbers. To connect to a process that is not directed for DDP, an application wishing to use the *AppleTalk Transaction Protocol (ATP)*, the transport layer protocol for AppleTalk, would use a locally dynamically assigned socket to connect to the destination station's DDP process. DDP would accept this packet and look into the type field in the DDP header. For ATP, this field would contain a value of 3. DDP type fields are depicted in Table 8.4. DDP will strip the

●
AppleTalk Transaction Protocol is the transport layer protocol for AppleTalk.

Table 8.4 DDP Type Fields

DDP Type Field Value	Description
00h	Invalid
01	RTMP response or data packet
02	NBP packet
03	ATP packet
04	AEP packet
05	RTMP request packet
06	ZIP packet
07	ADSP packet

DDP packet headers off the packet and pass the rest of the packet to the ATP process. ATP would then act on the packet according to the information in the ATP headers.

Valid socket numbers are numbered 01 to FEh and are grouped as follows:

- 01h to 7Fh: Statically assigned sockets
- 3h and 05h: Reserved for Apple Computer's use only
- 40h to 7Fh: Experimental use only (not used in released products)
- 80h to FEh: Dynamically assigned for node-to-node communications

Recall that an AppleTalk node is assigned a 16-bit network number and an 8-bit node number. Sockets form the final part of the addressing scheme used by Apple. A network number is assigned to each network segment. Each network station is dynamically assigned a unique node ID. Now, with the socket number ID, any process running on an AppleTalk internet can be identified no matter where the process is running. With this three-part addressing scheme, you can find the node, the network that the node lies on, and the exact process running on that node. It takes the form of <network number> <node number> <socket number>. This is called the internet socket address.

Now that sockets have been identified, DDP also has a type field in its packet to identify the process running on top of DDP that the packet is intended for.

DDP will accept the packet on the indicated socket number. It will, in turn, look at the type field to determine which process to hand the packet off to. For example, when DDP receives a packet from its data link, and the type field is 06, it will strip off the DDP headers and turn the rest of the packet over to the Zone Information Protocol (ZIP) for further processing. DDP's job is completed, and it returns to listening for packets or for interrupts from the higher level protocols for packet delivery.

When a process starts it will ask the transport layer protocol of ATP to assign it a socket number. ATP will pass this call to DDP, which will find an unused dynamic socket number and pass it back to ATP. ATP will then pass this socket number back to the calling process. The calling process will then pass the socket number to the *Name Binding Protocol (NBP),* which will bind this socket number to a name. DDP logs this to a table to ensure that it will not be used again because socket numbers are not allowed to be duplicated in the same network station. All processes on an AppleTalk network are available to users through names and not socket numbers. NBP provides this service to the users, but uses socket numbers to find users and services on the AppleTalk internet.

The second function of the DDP is routing. This is the ability to forward packets that are destined to remote networks. DDP provides a dynamic routing protocol, which is similar to the Routing Information Protocol that is found on Novell NetWare networks.

The **Name Binding Protocol** is an AppleTalk transport level protocol that binds the socket number to a name.

Routers, Routing Tables, and Maintenance

Another function of DDP is to allow networks to form an internetwork through the use of routers. Recall that an internet consists of a number of local LANs connected into an internet through special devices known as routers. These devices physically and logically link one, two, or more individual networks together. Thus, a network is transformed into an internet. In one of the DDP processes, network stations may submit their packets to the router in order to route to a destination on a different network (LAN). For DDP to accomplish this, it uses static sockets (socket 1) to deliver special messages known as routing table updates.

Router Description

A router is a special device that enables a packet destined for networks other than the local network they were transmitted on. Routers are usually separate devices on the network that contain at least two or more physical ports, each connected to a cable plant. On the back of the router is at least two or more DB-15 Ethernet DIX connectors. These connectors are the physical connection point of the routers.

By Apple's definition, AppleTalk routers are available in three forms: local, half, and backbone. Local routers are attached to local networks, for example, networks that are located geographically close together on the same floor or between multiple floors in a building. Local routers interconnect local LANs. This is usually multiple network segments with all segments having station attachment. Local routers are connected directly to the network. There is not an intermediate device between the two networks and the routers (see Figure 8.14). For example, Ethernet-to-Ethernet router is a local router. Figure 8.14 illustrates a local router operating in an AppleTalk network.

Each segment separated by a router will be assigned a network number on a range of network numbers. This is illustrated in Figure 8.14 with the middle network having a range of 200 to 300. This means that network numbers 200 through 300 are reserved for this segment. Note that the different network numbers are the two routers connected to this network. Their network numbers are different but they are assigned to the same cable segment. When assigning a range to a network, the network numbers should be contiguous.

Half-routers connect geographically distant networks.

Half-routers are used to connect geographically distant networks. This type of connection is usually done through the telephone system on what is referred to as leased lines. These are special lines that the phone company has conditioned to accept digital data. They are not standard voice lines because standard voice lines are too noisy to carry most high speed data. Typical data rates are 56 Kbps and T1 (1.544 Mbps), although now most telephone companies are using fiber and T3 (45 Mbps) lines. Half-routers are important for their hop count. Because a serial line separates the router, each router on each end of the serial line is considered a half-router. The two taken together are considered one router. Consequently, no network number is as-

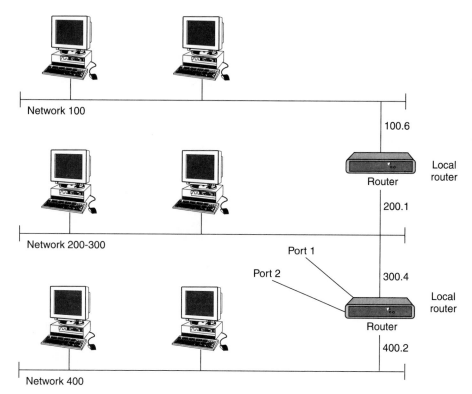

signed to the serial link, and a network separated by a serial line to another network is considered one hop away. All vendors do not support this method of routing. Some router vendors will assign a network number to the serial line and the two networks separated by a single line are considered two hops away.

Backbone routers connect AppleTalk network segments to a backbone network segment. Backbone network segments are networks connected to a backbone cable segment that is not AppleTalk. This type of network is illustrated in Figure 8.15. A backbone router is defined as a router that connects to a network having a higher throughput than the one to which it interconnects. Examples of these are FDDI backbones and 16 Mbps Token Ring.

Anytime that a packet must traverse a router to get to another network, the action is referred to as a hop. Throughout an AppleTalk internetwork, a packet may traverse no more than 15 routers and a distance of 16 hops is considered unreachable.

Operation

Routing of AppleTalk packets reduces the maximum size of a DDP packet to 586 bytes, even though the frame capacity of Ethernet and Token Ring is much higher.

FIGURE 8.15
LocalTalk to back-bone connection. (Reproduced with permission of Cisco Systems, Inc. Copyright © 2001 Cisco Systems, Inc. All rights reserved.)

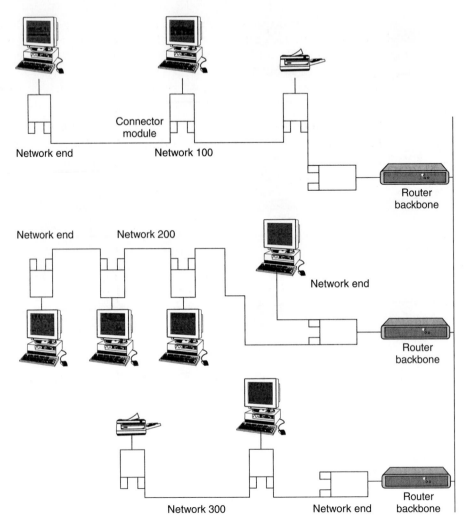

In an AppleTalk internet there are two types of devices for data. These are routing and nonrouting nodes. A routing node is a device whose main purpose is to route data for the internet. The other type of node is a nonrouting node, which is referred to as a user's workstation or a file/print server. Routers are devices that are usually autonomous from the other devices on the network. Their main function is to receive packets from other nodes on the network and forward them to their appropriate network.

When a network station has data to transmit, it will determine whether the packet is to be transmitted locally to another network station on the same LAN or to a network station that is remote—separated by a router. The network station accomplishes this by comparing the destination network number to its known network numbers. Recall that with

AppleTalk there can be no more than one network number assigned to a single network. Therefore, the destination network number cannot match any of the network numbers assigned to that network. If there is a match, the packet may be locally transmitted. If there is no match, the network station must employ a router to transmit the packet to its final destination. A router will discard a packet for an unknown destination network.

Simply stated, a router accepts packets directed to it, looks up the network address in a routing table, and forwards the packet to either a locally attached network or another router that will further route the packet to its destination. Each router will contain the following:

1. A data link handler (ELAP, TLAP, or LLAP)
2. A DDP routing process
3. A routing table
4. A process to update the routing table
5. A physical hardware connector known as a router port

The router ports may be connected to any of the previously mentioned router types, but no two active router ports may be attached to the same network cable. This is done with other routers to allow dynamic redundancy. You may connect two router ports to the same network as long as one of the ports is disabled. When the active router port fails, the disabled router port can then be made active, which allows manual redundancy. Each router port contains a port descriptor, which contains the following four fields:

1. A connection status flag: An indicator to distinguish between an AppleTalk port and another type port such as a serial link or a backbone network.
2. The port number: A number assigned to a physical port of a router. Used to identify a port in which to forward packets.
3. The port node ID: A router node ID for that port.
4. The port network number: The particular network number of the LAN connected to it.

When a port is connected to a serial link indicating a half-router, the port node address and port network number are not used. This is not true for routers that do not support the half-router function. When a port is connected to a backbone network, the port network number range is not used and the port node address is the address of the router on the backbone network.

Some router vendors treat AppleTalk routing like any other RIP. In other words, there are no such things as separation of backbone, half-router, and local router. A router is a router and network numbers will be assigned to each and every port of the router no matter what the connection is. AppleTalk packets are routed just like any other RIP packet in an RIP environment.

This is necessary to point out because routing of AppleTalk packets may be different from that of vendor to vendor. The concept of the backbone, half, and local routers is specific to the AppleTalk specifications. It is the recommended method by which AppleTalk is to be implemented.

The Router Table

To find other networks and their routers, AppleTalk routers use an algorithm similar to TCP/IP, XNS, and NetWare's IPX. It is known as a distance vector algorithm. The router maintains a table of network numbers (the vector) and the distance to the network number (distance, hop, or metric number).

For the router to know where to forward the packet, it must maintain a table that consists of network numbers and the routes to take in order to get there. Each entry in an AppleTalk router table contains three items: the port number for the destination network, the node ID of the next router, and the distance in hops to that destination network. The routing table consists of the following entries:

- Network range: This is the vector. This is a known network number that exists on the internet.
- Distance: This is the number of routers that must be traversed in order to reach the network number. This entry will have a 0 for locally attached networks.
- Port: The physical port on the router, which corresponds to the network number. If the hop count is 0, then this is the network number range assigned to that port. If the hop count is greater than 0, it is the port from which the network number was learned. Likewise, it indicates the port that the router will forward a packet to if the network number is so indicated in the packet.
- Status: Indicates the status of the path to that network.
- Next Router: If the network number is not locally attached to the router, this field indicates the next router in the path to the final destination network. If the network number is directly attached to the router, there will be no entry in this field. The router will forward the packet directly to that cable plant.

For the router to maintain its table, a process must be invoked to allow the routers to exchange data (their routing tables) for periodic updates. This allows the router to find shorter paths to a destination and to know when a new router is turned on or when a router has been disabled. Possibly, a new path must be taken to arrive at a particular network. The process that enables this maintenance is called the Routing Table Maintenance Protocol (RTMP).

A routing table is maintained in each of the routers that are on the AppleTalk internet. The table tells the router the network number and how to get there. Where did the router obtain this information? All router tables are constructed from information that comes from other routers on the network. In other words, each router will tell another router about its routing table. The protocol that maintains this is the RTMP protocol.

The **Routing Table Maintenance Protocol** allows routes to be dynamically discovered throughout the AppleTalk internet.

Routing Table Maintenance Program Protocol (RTMP)

The *Routing Table Maintenance Protocol (RTMP)* provides the logic to enable datagrams to be transmitted throughout an AppleTalk network through router ports. This protocol allows routes to be dynamically discovered throughout the AppleTalk internet. Devices, which are not routers such as a workstation, use part of this protocol known as the RTMP stub to find out their network numbers and the addresses of routers on their local network. Different DDP packet formats are illustrated in Figures 8.16b and c. Figure 8.16a depicts the general DDP packet header format. The type field would be filled in appropriately for each packet type of RTMP, NBP, ATP, AEP, ADSP, and ZIP. Figures 8.16b and c show the RTMP and NBP packet type. RTMP packets are further described in Figure 8.17. The hop field indicates how many routers a packet has traversed. Routing tables in the routers are exchanged between routers through the RTMP socket. By addressing the packet with this socket

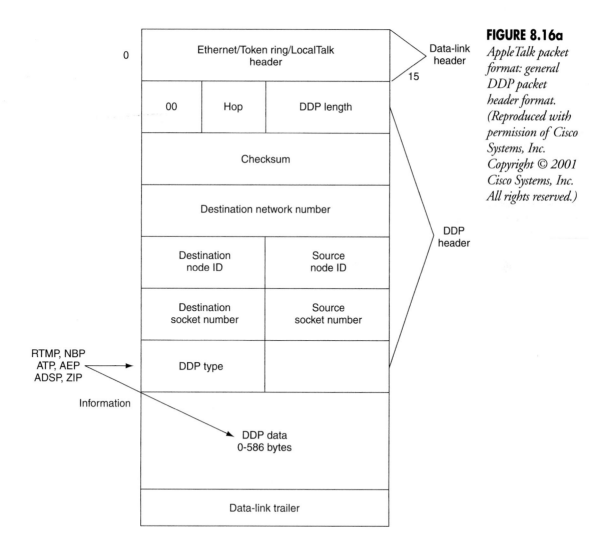

FIGURE 8.16a
AppleTalk packet format: general DDP packet header format. (Reproduced with permission of Cisco Systems, Inc. Copyright © 2001 Cisco Systems, Inc. All rights reserved.)

00	Hop	DDP length
Checksum		
Destination network number		
Source network number		
Destination node ID		Source node ID
Destination socket number		Source socket number
RTMP		

00	Hop	DDP length
Checksum		
Destination network number		
Source network number		
Destination node ID		Source node ID
Destination socket number		Source socket number
NBP tuple 1 NBP tuple n		

00	Hop	DDP length
Checksum		
Destination network number		
Source network number		
Destination node ID		Source node ID
Destination socket number		Source socket number
Echo		
Request or response		
Data to be or is being echoed up to 585 bytes		

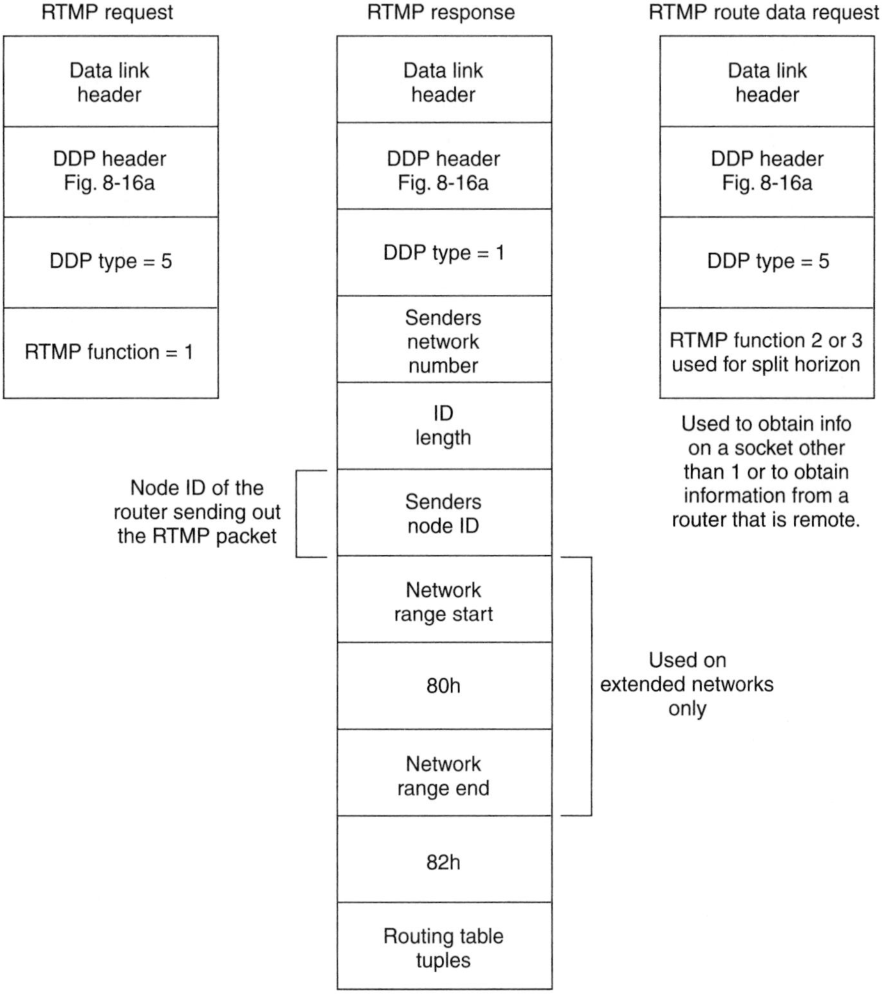

RTMP request	RTMP response	RTMP route data request
Data link header	Data link header	Data link header
DDP header Fig. 8-16a	DDP header Fig. 8-16a	DDP header Fig. 8-16a
DDP type = 5	DDP type = 1	DDP type = 5
RTMP function = 1	Senders network number	RTMP function 2 or 3 used for split horizon
	ID length	Used to obtain info on a socket other than 1 or to obtain information from a router that is remote.
	Senders node ID	
	Network range start	
	80h	Used on extended networks only
	Network range end	
	82h	
	Routing table tuples	

Node ID of the router sending out the RTMP packet

FIGURE 8.17

RTMP packets. (Reproduced with permission of Cisco Systems, Inc. Copyright © 2001 Cisco Systems, Inc. All rights reserved.)

number, DDP upon receipt of the packet will know exactly what the packet is and will interpret the packet without passing it on to another process. The DDP socket number of 1 indicates a routing update or request packet. Each RTMP packet includes a field called the routing tuple in the form of <network number, distance> for nonextended networks. Extended networks contain the header form <network number range start, distance, network number range stop, unused byte set to 82h>; then come routing tuples. In the routing tuple is the network number, which consumes 2 bytes, and the distance, which consumes 1 byte. The routing tuple contains a network number and the distance traveled in hops from the router to that network. When the router receives this packet, it will compare it to the entries already established in the table. New entries are added to the table and distances may be adjusted, depending on the information in the packet. There are two ways a router may receive its network number. One way is to configure the router as a seed router. The network administrator must manually assign the network numbers to each port on the router. Each port must be configured as a seed router or a nonseed router port. When a seed router initializes it will enter the network numbers assigned to its local ports into the table. Other routers on the network will obtain their network numbers from the seed router.

For a network that is served by multiple routers, the seed router will distribute the network numbers to the rest of the routers on that network. There is at least one seed router per network. There can be multiple seed routers per internet. The purpose of the seed router is to inform the other routers of their network numbers.

Nonseed routers will learn their network numbers and zone names through the seed routers. Configuring all routers as seed routers allows the network administrator to statically assign the network numbers to each of the router ports. The seed router will contain a list of network numbers on a per router port basis. The seed router can also contain a zone name listing on a per port basis.

Once a router has found its local port network numbers and entered them in its table, it will submit its router table to its locally attached network segments, being conscious of split horizon. This will inform other routers on those segments of the networks that the router knows about. Those routers will update their tables and then submit their tables to their locally attached cable segments. With this process, each router on the internet will eventually know about all network numbers and their associated routers. When a router updates its routing tables with a new or changed entry, the hop count is incremented by one more than the hop count number indicated in the table that it received. After this the router will transmit its table to the directly attached cable segments. This does not occur for entries in the table that are not new or are not changed from the previous update.

Nonrouting nodes such as user end stations are not expected to maintain routing tables like a router. A nonrouting node may acquire its network number and a router address (a router to which packets are submitted for forwarding data to another network) in one of two ways:

1. It may listen to the routing update messages sent out by routers on its local network. In these packets will be the local network number and the address of the router that sent the update.
2. It may also ask any router to respond by submitting a request for this information. The requesting node should receive a response packet that looks like a routing update packet but it contains no routing update entries. Instead, embedded in this packet will be the network number and address of the router that sent the response packet. Also, this packet is a directed response packet. It is not sent in broadcast mode.

The **Zone Infor-mation Protocol** enables network stations to be grouped together into a common zone name.

For nodes that are on extended networks, the end node will transmit a special packet known as the ZIP GetNetInfo request. This is a request to the router's *Zone Information Protocol (ZIP)*. ZIP is the protocol that enables network stations to be grouped together into a common zone name. Users needing connection to any network station will first choose a zone and then a network station. A router will respond to the packet with the network number range assigned to that zone. Normally, nodes on extended networks do not submit a routing request for network number information, but the AppleTalk specification does not disallow it.

The router that first responds to the request to an end node on either the extended or nonextended router is called the end nodes A-router. This is the router to which the nonrouting node such as a user's workstation will send its packets that must be routed. The entry in a nonrouting node for A-router may change if a different router responds to another routing request.

Aging Table Entries

In actuality, the end station's (nonrouting node) A-router will change each time a routing update is transmitted by a router. AppleTalk is also one of the few routing protocols that uses a hop count of 0. Any network number associated with a hop count of 0 is considered the local network number.

Each entry in a router table must be updated periodically to ensure that the path to a destination is still available. Otherwise, a route would stay in the table and may not be valid. To age out old entries, a timer is started on receipt of each routing tuple. If after a certain amount of time the validity timer (set to approximately 20 seconds) has expired without notification of a particular route through RTMP, the router will change the status of the select entries from good to suspect to bad and then finally it is deleted from the routing table. If at any time the router receives an update pertaining to that entry in the routing table, it will place the entry status back to good and start the timer process over again.

Aging of the A-router in a nonrouting node is set to 50 seconds. RTMP packets are depicted in Figure 8.17. There are three types. A request packet, a response packet, and a router data request. The request packet is used to obtain information from a router about a network or networks. In response to a request, a router will send the response packet. This will contain information about a network number, such as how many hops away the requested network is.

The route data request is used by a router to obtain information about another router, such as whether one router supports a protocol known as split-horizon. Split-horizon is the ability for a router to not announce information about a network on the same port from which it learned about the network. This means if a router learns about network 2 through port B, the router does not announce this information on port B. In the response packet notice the network range field. Any station on the network that reads this packet will know from this field what the range for its network is.

The router transmits RTMP routing update packets every 10 seconds. This means that a router will transmit its table to its locally backed cable segments every 10 seconds. RTMP is the most frequent of any of the protocols that use RIP as they are updating protocol. Novell's implementation of RIP is 60 seconds, while TCP/IP and XNS are every 30 seconds. Finally, AppleTalk routers do employ the split-horizon protocol. Routers find out about this from other routers by an indication in the RTMP Route Data Request Packet, which is a special type of routing information inquiry packet that does not require network number updates. Figure 8.18 illustrates a flowchart for router information flow.

AppleTalk Echo Protocol (AEP)

The *AppleTalk Echo Protocol (AEP)* resides as a transport layer protocol and allows any station to send a packet to another station, which will return the same packet, hence echo it. This allows stations to find other active stations on the network, but more important it allows a network station to determine packet round-trip delay times.

The AEP is used when a network workstation has found the destination station it wishes to communicate with and submits an echo packet to test connectivity. The station will submit an echo packet (DDP socket number 4) to the destination and will wait for a reply. When the reply does come, the station notes the time and submits the packet to establish a connection to that station. This can be used to establish timers for packet time-outs. The packet format for AEP is illustrated in Figure 8.16c.

The preceding sections discussed the delivery system for AppleTalk. The delivery system included the data link and network layers. All data are submitted to these layers for transmission on the network. The network and data link protocols do not care what the data are. The only job that these protocols perform is to provide a transportation service for the upper layer protocols. The protocols resident in the transport and session layers of the AppleTalk network model allow a session to be set up and maintained and ensure that the data are transmitted and received in a reliable fashion.

Names on AppleTalk

Throughout this section please refer to Figure 8.16a, the AppleTalk general DDP packet header format. Recall that the delivery system depends on certain numbers to deliver data. There is one physical address for every network station on the internet. There are network layer addresses so that the network layer is able to route the data over the internet if needed. Socket numbers are needed so that once data arrives at the final destination they are forwarded by DDP to the appropriate software process. As a result, there are many addresses used throughout the AppleTalk internet.

To eliminate the need for users to remember all the network addresses, node addresses, sockets, and so forth of the network and data link layers, a naming scheme has been devised. Because network addresses may change frequently as a result of dynamic node IDs, all services such as file, print, and mail on a network are assigned

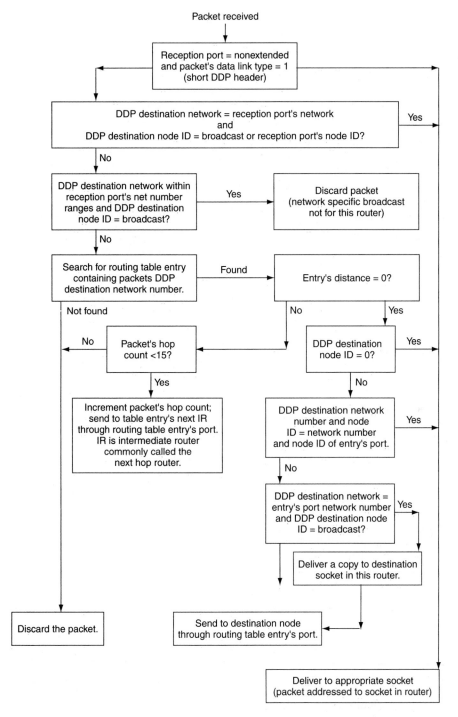

user definable or string names. These string names may not change frequently, so AppleTalk uses a name process to identify network stations on an internet. This naming process accommodates two things:

1. It is easier for a user on the internet to remember names and not numeric addresses.
2. It enables network stations to acquire different network addresses while retaining a static name.

This process is accomplished fully transparent to the users.

The network administrator usually assigns names on an AppleTalk network. Anyone needing access to these services cannot change the names that are being assigned. Names are used to request a service to be processed. This includes attaching to a file server or sending a print job to a network attached printer. All of these processes on the network are assigned numbers so that the network stations may communicate with each other using the string names that are typed into the network station. String names are only for users on the AppleTalk network. When network stations communicate with one another, they still use the full internet address of the network station and not the user defined name of the network station.

Defining the Names on AppleTalk

A **network visible entity** is any process on an AppleTalk internet that is accessible through a socket in DDP.

AppleTalk has a concept within a network referred to as *network visible entities (NVE)*. The user can see the actual physical devices. These devices include file and print servers and routers. These services are represented on the network as sockets, and each socket is assigned a numeric address. In this manner, a network station may logically attach to the service and not to the actual device. An NVE is any process on an AppleTalk internet that is accessible through a socket in DDP. NVEs are the application processes to which users attach from their network stations.

An entity can assign an entity name to itself. This name consists of three fields, and each may contain a maximum of 32 characters. Any of these fields may be user defined, meaning there is not a defined method for assigning the names. Once the names have been identified, and what constitutes a name has been defined, these names must be mapped to an internet address so that network stations on the network may attach and pass data. The protocols that handle names on the AppleTalk network are the Name Binding Protocol (NBP) and the Zone Information Protocol (ZIP).

The process to translate or map between numbers and names is the function invoked by NBP. NBP maintains a table that translates between character names and their corresponding full internet addresses. The name binding process is the process by which a network station will acquire the internet socket address of a destination network station by use of its string name. It converts the string name to the numeric address used internally by network stations. This process is accomplished completely transparent to the user. All of these name-to-address mappings are maintained in tables.

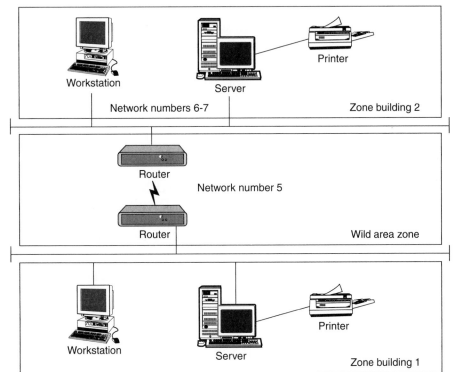

FIGURE 8.19
Simple AppleTalk zone network. (Reproduced with permission of Cisco Systems, Inc. Copyright © 2001 Cisco Systems, Inc. All rights reserved.)

The Zone Information Protocol maps network numbers and zone names. A zone is a logical grouping of network stations no matter where these network stations are located on the internetwork. Zones do not reflect on network addresses. Network stations grouped into a zone have many different network addresses. The grouping of zones is called the AppleTalk internet. Each network number requires at least one zone name. Network station placement in a zone is fairly liberal. Network stations may participate in one zone or many zones. Zones may cross networks via a router or routers. Zones may be spread through many different network numbers. Zone names originate in routers. The network administrator will assign the zone name and broadcast this name throughout the network. Routers maintain a listing of zone names and their associated network numbers through a table known as the *Zone Information Table (ZIT)*. A simple AppleTalk zone is illustrated in Figure 8.19.

Before any NVE can be accessed, the address of that entity must be obtained through a process referred to as name binding. Name binding is the mapping of a network name to its internet socket address (the network number, the node ID, and the socket, or port number). The NVE will start and request a socket number from ATP. ATP will return a socket number given to it by DDP and the NVE process will then determine its network and node addresses. The network ID, node ID, and socket

●
The Zone Information Table contains a listing of zone names and their associated network numbers.

number combine to form the internet socket address. The process will then notify NBP to bind the internet socket address to a name.

Each network station must maintain a names table, which contains an NVE name-to-internet address map for all entities in that particular network station. NBP does require the use of special nodes called name servers. NBP is distributed on all network stations on the AppleTalk internet. The process that maintains this table is available through the Names Information Socket. The Names Information Socket, which is static DDP socket number 2, is also responsible for accepting and servicing requests to look up names from within the network station and from the network.

A separate table is used called the names directory, which contains a distributed database listing of all the names tables on the internet. The names table contains only those internet socket address-to-name mappings of the NVEs on the local network station. The names directory contains a listing for NVEs on the internet.

NBP handles the names table and name lookup requests from client processes or requests from the network. Every network station on the AppleTalk internet has this process whether it is providing services for the network or uses the services of the internet. NBP is a distributed name service. This means that each network station acts as a name server. Use of a centralized name service means a single station on which all names reside. Centralized name servers are not usually found on an AppleTalk network, but they are allowed to exist on an AppleTalk network. The NBP process provides four types of services:

> The **name registration** process will register its name and socket number on its local node name table.

1. *Name registration:* Register its name and socket number on its local node name table.
2. Name deletion: Delete a name and corresponding socket.
3. Name lookup: Respond to a name registration.
4. Name confirmation: Affirm a name registration.

When any process starts on a network station it will register its name and associated socket number from the names table. When this process removes itself from the network station it will also ask NBP to remove its name and socket number from the names table. NBP will then place the NVE in the name directory. Name lookup occurs when a request for a name-to-internet address binding is required, and name confirmation occurs to check the validity of the current binding. Routers participate in this process through the use of zone mappings.

Name Registration

Any NVE working on the internet can place its name and corresponding internet socket address into its names table. This allows the entity to become visible on the network. First, the NBP process will check to see if its name is already in use. If it is, the attempt is aborted and the process will be notified. Otherwise, it is placed into the table.

The NBP process then places this name-to-address mapping in the names directory. Any entry into the names table is dynamic. It is reconstructed every time the network station is started. Using this process, every network user is registered on the internet. When the entity wants to delete itself from the names directory, it will do so by telling NBP to delete the entry.

On a single network, mapping a name involves three steps:

1. The requesting network station's NBP process broadcasts an NBP lookup packet addressed to the Name Information Socket number to the network. A lookup packet contains the name to be looked up.
2. Every active network station that has an NBP process will receive this packet and will perform a name lookup on its table not directory for a match to that name.
3. If a match is found a reply packet is generated and transmitted to the requesting socket. Included in this packet is the name's address.

If there is no response to this packet, the requesting network station assumes that the named entity does not exist on the local network and the name is added. This is name confirmation.

Two extra steps are added to find an NVE on an internet (those networks connected by a router). A router will pick up this request and will send out a broadcast lookup request for all networks in the requested zone. This process is done for zones and may stretch across multiple networks.

Zones

A zone is a group of networks that form an AppleTalk internet. ZIP is the protocol that maintains an internetwide map (a table) of zone-to-network names. NBP uses ZIP internet mappings to determine which networks belong to a given name. It is easier to address network stations based on their zone. An AppleTalk internet is a collection of network stations grouped together by zones. An extended network can have up to 255 zones in its zone list. An AppleTalk internet can theoretically have millions of zone names. A nonextended network can have at most one zone name.

The ZIP process uses routers to maintain the mapping of the internet network numbers to the zone names. ZIP also contains maintenance commands so that network stations can request the current mappings of the zone-to-internet network numbers. A router's participation is mandatory in an AppleTalk internet zone protocol. Just as a routing table is maintained in a router, a ZIT resides in each router on the AppleTalk internet. There is one ZIT for each physical router port on the router. Each table provides a list of the mappings between the zone name and internet addresses for every zone on the AppleTalk network. It looks like a routing table, but lookups are on the zone name instead of network numbers. This is similar to Novell's NetWare Service Advertisement Protocol.

The ZIT may consist of one zone name mapped to one network number or it may contain multiple zone names with multiple network numbers assigned to those zones. With the latter, it should be noted that zones could overlap each other. An example of three logical zones is depicted in Figure 8.19. An example of a zone overlap would be that the printer in zone building 2 could exist in zone building 1, except that it is still attached to the segment of network numbers 6 to 7. Not all routers support zone names on serial backbones as depicted in Figure 8.19. Some routers do provide support of half-routers, which allow point-to-point links such as the serial line between two routers, but there is no zone name on the point-to-point link.

To establish the zone table, the zone process in a router will update other routers through the ZIP socket. The requesting router will form a ZIP request packet, input into this packet a list of network numbers, and transmit it to the node A-router. This router will reply with the zone names that it knows about with the associated network numbers included. Because this process is accomplished over DDP the request contains a timer and, upon expiration of that timer, ZIP will retransmit a request.

To maintain a ZIT, ZIP monitors the router's routing table and not the ZIT for changes in the entries in the table. If a new network number has been found on the internet, ZIP will send out request packets to other routers in an attempt to find a zone name associated for the new network number. Therefore, ZIP maintains its zone name table by monitoring the routing table. When a network number is deleted from the routing table the network number will be deleted from all routing tables on all routers on the internet. ZIP will also delete its zone name entry for that network number. Any network station on the internet may request the mappings from ZIP by transmitting a ZIP request packet to the network. If the router does not know the zone name it will not reply.

For a user's workstation there are three special ZIP packets that enable it to operate properly with the ZIP. These are:

1. GetZoneList: Requests a list for all zones on the internet.
2. GetMyZone: Gets the zone name for the local network (nonextended networks only).
3. GetLocalZones: Obtains a listing of all the zones on the requester's network.

Just as a network station may keep its last known node ID stored, a network station upon initialization may have a previous zone name that it was using the last time the station was active on the network. If so, the network station upon initialization will broadcast a packet called the GetNetInfo to the network. In this request packet will be the last zone name it worked with or, if the zone name is unknown, the packet will not contain a zone name. Routers on this network will respond to this request with information on the network number range and a response to the zone name requested. If the zone name requested is okay, the network station will use this zone name. If not, the router will respond with the zone name used on that network.

To see the ZIP in action is as simple as signing onto the network and requesting connection to a service on the network. The Apple Mac chooser will bring up a listing of zone names for logon or to access a service.

Transport Layer Services: Reliable Delivery of Data

The previous sections discussed how stations identify themselves, gain access to a network, find zones, names, routers, and how to connect to an NVE via internet sockets. All of this allows communications to exist with data being delivered on a best-effort, connectionless method. With the connectionless delivery system there is no guarantee that the data were delivered or, if it was delivered, that the packets in a transferred data segment arrived at the destination in the same order in which they were sent.

Consider transferring a file that is 250 Kbytes long. This is larger than any network protocol could transfer in one packet. Therefore, many packets are used to transfer the data. If we used only a connectionless protocol, the data might arrive in the wrong order, or one packet out of a thousand packets that were transmitted might never arrive at its intended destination. A protocol that allows information to flow reliably in an AppleTalk network is called the AppleTalk Transaction Protocol (ATP).

Without transport layer services the application itself would have to provide the purpose of reliable delivery. To require every application to build this into their software is like reinventing the automobile. The network software should provide this for every network application, and the ATP is the protocol that provides this function.

AppleTalk provides two methods for data delivery: transaction based and data stream—the *AppleTalk Data Stream Protocol* (ADSP). Transaction based protocols are based on the request-response method commonly found on client workstation-to-server communication. Data stream protocols provide a full-duplex reliable flow of data between network stations. ATP is a transaction based transport layer protocol.

The protocols that make up this OSI transport layer in the AppleTalk model consist of the following:

1. AppleTalk Transaction Protocol (ATP)
2. Printer Access Protocol (PAP)
3. AppleTalk Session Protocol (ASP)
4. AppleTalk Data Stream Protocol (ADSP)

These protocols guarantee the delivery of data to its final destination. Two stations use ATP to submit a request to which some type of response is expected. The response is usually in the form of a status report or a result from the destination to the source of the request.

●
AppleTalk Data Stream Protocol provides a full-duplex reliable flow of data between network stations.

AppleTalk Transaction Protocol (ATP)

ATP as stated earlier is the protocol on AppleTalk that provides reliable delivery service for two communicating stations. There are three types of packets that are sent:

●
Transaction Request is a request initiated by the requester.

1. *Transaction Request (TReq):* a transaction request initiated by the requester
2. *Transaction Response (TResp):* the response to a transaction request
3. *Transaction Release (TRel):* releases the request from the responding ATP transaction list

There is a Transaction Identifier (TID) that assigns a connection number to each transaction. Each request and response between a source and destination is as simple as the three preceding commands taken as a whole. First, the requesting station submits a TReq. The destination station will respond with a TResp. The requesting station will then acknowledge the transaction response by sending it a TRel. This handshaking will continue until the session is released by the session layer (see Figure 8.20).

●
Transaction Response is the response to a transaction request.

●
Transaction Release releases the request from the responding ATP transaction list.

ATP provides adequate services for most sessions. Because most network protocols restrict the size of their packets on the medium, multiple packets may be needed for TResp. TReq packets are limited to a single packet. Multiple requests are sent as multiple TReqs. A TResp packet is allowed to be a sequential packet. When the requester receives all the TResp packets in response to a TReq, the request is said to be complete and the message is delivered as one complete message to the client of ATP. ATP will assemble the message as one complete message and then deliver it to a client of ATP. Every ATP packet has a sequence number in its header. This field is 8 bits wide. This does not mean that it has 255 sequence numbers. Each bit is a sequence number. Therefore, if bit 0 is turned on, this is sequence number 0; if bit 1 is turned on, this is sequence number 1 and so forth.

FIGURE 8.20

ATP transaction. (Reproduced with permission of Cisco Systems, Inc. Copyright © 2001 Cisco Systems, Inc. All rights reserved.)

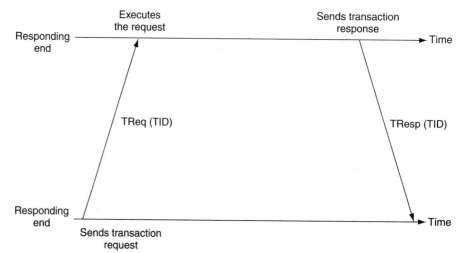

In a TReq packet, it will indicate to the receiver the number of buffers available on the requester. When the receiver receives this packet it will know how many response packets the requester is expecting. When this field is set by a TResp packet it is used as a sequence number (0 to 7). As stated before, each bit is not taken as a binary number; each bit represents an integer number in the range of 0 to 7. This will indicate the number of the packet in the sequence of the response packets. Therefore, the sequencer on the requester can place the incoming packets in the appropriate buffer. As each good packet is received, the requester makes a log of this. For each packet not received in sequence, the requester can make a request for a retransmission of this packet in the sequence. For example, if a TReq is transmitted and the sequence bitmap is set to 00111111, it is indicating to the destination that it has reserved six buffers for a response. The requester expects six packets of TResp in this response. As the TResp packets are sent back, the TResp packets will have a bitmap set to indicate the number of the packet that is being responded to. If the requester receives all packets except number 2 (00000010), it will send another TReq for the same information with its sequence bitmap set to 00000010. The destination should respond with that and only that TResp packet. The ATP data field and not the ATP header would be the session header and the application header.

Printer Access Protocol (PAP)

The *Printer Access Protocol* (PAP) uses the connection oriented transport. PAP is the protocol used to deliver data that a client workstation would send to a printing device on the network. This is the protocol that allows an application to write to the printer across the network. PAP uses the services of NBP and ATP to find the appropriate addresses of the receiving end and to write data to the destination.

PAP provides five basic functions described in the following four steps:

1. Opening and closing a connection to the destination
2. Transferring of data to the destination
3. Checking on the status of a print job
4. Filtering duplicate packets

PAP calls NBP to get the address of the server's listening socket for printer services (the SLS). Data can be passed only after this socket is known. After the connection is established, data are exchanged. There is a two-minute timer to determine if either end of the connection is closed. If the timer expires, the connection is closed. Either end of the connection may close the connection.

A workstation may determine the status of any print job. This may be executed with or without a connection being established. As for filtering duplicate packets, sequence numbers are assigned to each packet. A response from either end must include the original sequence number. Figure 8.21 illustrates the block diagram of the PAP. Notice how the printer has the AppleTalk protocol stack in the printer software. This architecture

Printer Access Protocol is used to deliver data that a client workstation would and to a printing device on the network.

FIGURE 8.21

Printer connections for AppleTalk. (Reproduced with permission of Cisco Systems, Inc. Copyright © 2001 Cisco Systems, Inc. All rights reserved.)

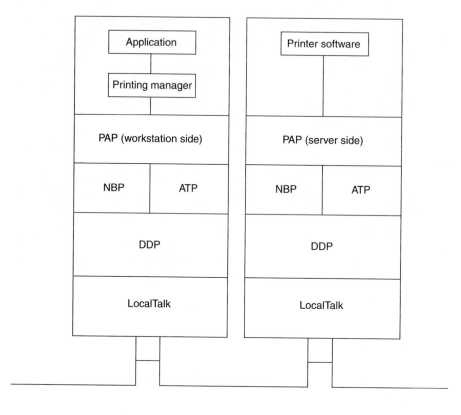

is shown for a printer connected to the LocalTalk network. For EtherTalk and To-kenTalk networks, the printer software is a spooler that resides in the Mac's printer port. The Mac uses AppleTalk to communicate between the Mac and the printer.

AppleTalk Session Protocol (ASP)

AppleTalk Session Protocol opens, maintains, and closes sessions and sequence requests from upper client software.

The *AppleTalk Session Protocol (ASP)* does not care about the delivery of data. ASP's sole purpose is very straightforward. ASP, like most session layer protocols, opens, maintains, and closes sessions and sequence requests from upper client software. ASP uses the lower layer protocol ATP to sequence these requests.

ASP does provide a method for passing commands between a network station and its connected service. ASP provides delivery of those commands without duplication and makes sure the commands are received in the same order in which they were originally submitted. This orderly retrieval of commands is then responded to as the results become available. Like other layers of the OSI model, the session layer ASP of the AppleTalk model communicates only with another ASP layer on another network station. ASP is based on a client-server model. Upon initiation, a service (NVE) on a server will ask ATP to assign a socket number to the service. ATP should respond to this request with a socket number received by DDP, and this number is the one used by the service on the server to let the network know about the service. The service will then notify NBP of the name and its socket number. This will be placed in the names table of that network station offering this service. This will allow any network station to find the service on the network. ASP will then provide a passive open connection on itself so that other stations on the internet may connect to it. This is referred to as the session listening socket (SLS).

There are three types of named sockets associated with AppleTalk that operate at the session layer:

1. *Session listening socket (SLS):* a service being offered listens on this socket.
2. *Workstation session socket (WSS):* identifies a socket in the requesting workstation when a connection attempt is being made to a service.
3. *Server session socket (SSS):* identifies to the workstation, once a connection is made, the socket number to which all future transactions are referred.

Figure 8.22 illustrates this interaction of sockets. The client side of ASP, upon seeking a connection to a remote resource, must find the full internet address of the remote service (the NVE).

ASP is said to have placed an active open connection request. In the active open connection request, ASP will find the full internet address of the remote service and give itself a socket number, which is the workstation session socket (WSS), and will then request a connection of the remote resource. The server should have open sockets residing on DDP and will accept the connection request on one of these sockets. The virtual circuit connection or session is then maintained between these two sockets—the WSS and the server socket, now known as the SSS. ASP builds upon ATP to provide a secondary level of transport service that is commonly required by client-workstation environments.

Establishing a Session

Two handshaking protocol steps must be performed before any session is established between a workstation and its file server.

1. The workstation and server must ask ASP to find out the maximum allowable command and reply sizes.
2. The workstation must find out the address of the SLS number by issuing a call to NBP.

The SLS is an addressable unit to which the workstation will send all requests. This socket is opened by ASP, and the file server will make this socket well known on the network by broadcasting this socket to known entities through the use of NBP. This is the socket number for which the file server, for example, is listening for connection requests.

The **session listening socket** is an addressable unit to which the workstation will send all requests.

The **workstation session socket** identifies a socket in the requesting workstation when a connection attempt is being made to a service.

The **server session socket** identifies to the workstation the socket number to which all future transactions are referred.

FIGURE 8.22

*Connection at-
tempts and sessions
using sockets.
(Reproduced with
permission of Cisco
Systems, Inc.
Copyright © 2001
Cisco Systems, Inc.
All rights reserved.)*

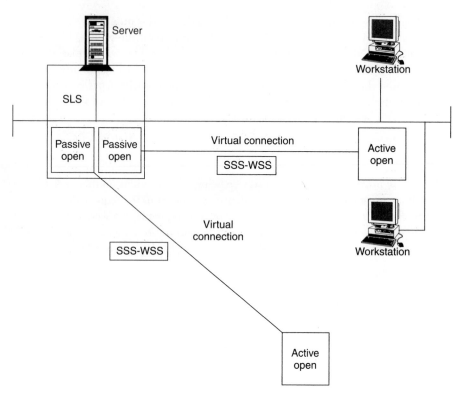

Refer to Figure 8.23.

1. The workstation will ask ASP to open a session with a particular service on a file server.
2. ASP will transmit a session request packet to the file server. This packet will contain the workstation's socket number to which the file server may respond. This WSS is an addressable unit in which the file server knows which process in the workstation to respond to.
3. If the server accepts the session request it will respond with the following:
   - Acceptance indicator: an entity within the packet to indicate to the workstation that the request is accepted.
   - Session identifier: an integer number that the server assigns to identify any particular session on the file server. Each unique workstation request is assigned an identifier.
   - Address of the server session socket: upon accepting a connection from a remote workstation, the server will respond to this request with a new socket number that will uniquely identify that session on the server.

The session is now considered established and all communication will continue between the two using the WSS, the connection ID, and the SSS.

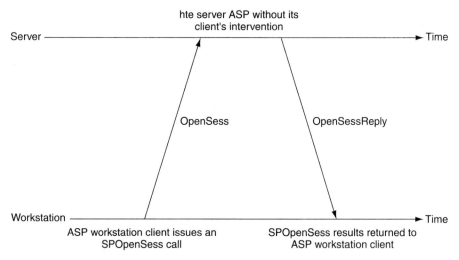

Server ────────────────────────────────────→ Time

hte server ASP without its
client's intervention

OpenSess   OpenSessReply

Workstation ──────────────────────────────→ Time

ASP workstation client issues an
SPOpenSess call

SPOpenSess results returned to
ASP workstation client

FIGURE 8.23

Session requests.
(Reproduced with
permission of Cisco
Systems, Inc.
Copyright © 2001
Cisco Systems, Inc.
All rights reserved.)

Session Maintenance

Once the session is established, ASP must maintain the session. There are maintenance commands specific to the management of any AppleTalk session. The three
maintenance commands are as follows:

1. Tickling
2. Command sequencing
3. Discarding duplicate requests

Once a session is established it will remain active until one end of the connection decides to quit the session. One other possible occurrence for session termination is
when the path has become unreliable or one end has become unreachable. There will
be times when the session will not be passing data or commands from one end of the
connection to the other. You may establish a connection, start up an application, and
then not enter any data for a while. When this occurs, each end of the circuit does not
know if any circumstance may have occurred that would make the session inoperable.

To find out if the other side is operating normally, AppleTalk employs a protocol known
as Tickling. *Tickling* is the operation whereby each network station involved in a connection will periodically send a packet to the other to ensure the other end is functioning properly. This is referred to as a keep-alive packet. If either end does not receive this
packet at certain time intervals, a timer will expire and the session will be disconnected.

Command Sequencing

If the same packet is received twice, discarding duplicate requests is the process by
which the second request is discarded. This may happen when the originating station
submits a packet and does not receive a response for it. The destination of this packet

Tickling is the
operation whereby
each network sta-
tion involved in a
connection will
send a packet to
the other to ensure
that the other end
is functioning
properly.

may have been busy or the packet may have been delayed while being forwarded through a busy router and the response packet was delayed. In any case, a timer will expire in the origination station and a packet will be retransmitted. Meanwhile, the server responds to the original packet and then receives the retransmission. The destination station must have a way of knowing that a particular packet was responded to. It does so with sequence numbers. Each incoming packet is checked for the sequence number. If a packet received on the file server corresponds to a response sequence number, the duplicate packet will be discarded.

Data Transfer

Once the connection is established and is operating properly, each end may send packets that will read or write requests to the respective sockets. These reads and writes are actually data transfers between the two network stations.

ASP handles all requests from the workstation and the server. ASP is an initiator of requests and a responder to any requests on the network. Any of these commands will translate into ATP requests and responses. ASP contains three formats.

1. Commands
2. Writes
3. Attention requests

As depicted in Figure 8.24a, ASP commands ask the file server to perform a function and to respond to these requests. Any command will translate into an ATP request of the SSS. Any command will translate into an ATP response. A write is performed in Figure 8.24b. ASP on the server issues a single write command with its appropriate transaction ID. The workstation responds, stating that it is okay to continue writing

FIGURE 8.24a

Data transfer. (Reproduced with permission of Cisco Systems, Inc. Copyright © 2001 Cisco Systems, Inc. All rights reserved.)

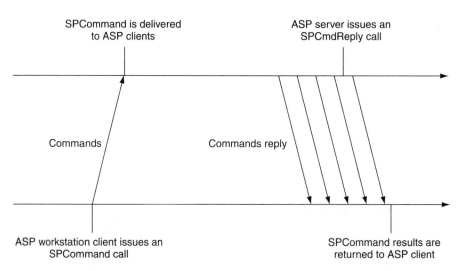

SPCommand is delivered to ASP clients

ASP server issues an SPCmdReply call

Commands

Commands reply

ASP workstation client issues an SPCommand call

SPCommand results are returned to ASP client

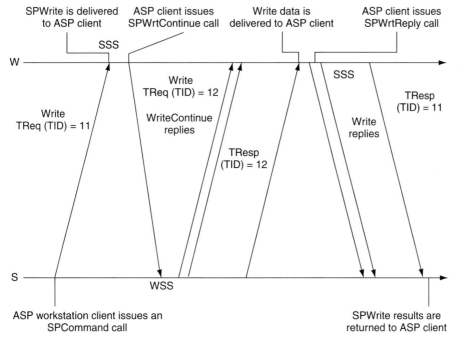

FIGURE 8.24b
ASP write. (Reproduced with permission of Cisco Systems, Inc. Copyright © 2001 Cisco Systems, Inc. All rights reserved.)

on TID 12, which the server will do. At a certain point in time, the workstation replies to the writes ending with a TResp for TID 12. After all the write replies are sent to the server, the workstation acknowledges the end by a TResp on TID 11 for the server's TID. If the workstation's command to its file server was to read part of a file, the response would be in multiple packets returned to the file server if the file was larger than the largest packet size allowed on the network. *Attention requests* are those requests that need immediate attention. No data will follow this packet.

● **Attention requests** are requests that need immediate attention.

Session End

Refer to Figure 8.25a and b. The workstation or the server may close an established session. The workstation may close a session by sending a close session command to the same socket on the server known as the SSS in which the server was first established. The server may close a session by sending a command to the workstation WSS in which the session was first established. After a session is closed, the ASP on both the server and the workstation must be notified so that each may delete the proper information from their session tables. All entries for those sessions in the tables are deleted.

FIGURE 8.25a

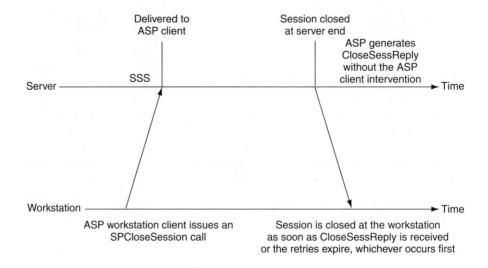

ASP workstation client issues an
SPCloseSession call

Session is closed at the workstation
as soon as CloseSessReply is received
or the retries expire, whichever occurs first

FIGURE 8.25B

*Server close.
(Reproduced with
permission of Cisco
Systems, Inc.
Copyright © 2001
Cisco Systems, Inc.
All rights reserved.)*

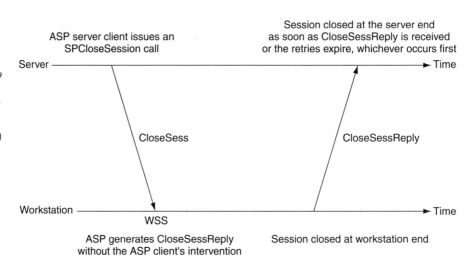

ASP generates CloseSessReply
without the ASP client's intervention

Session closed at workstation end

The previous sections conclude the discussion of the protocols needed for the AppleTalk network. These protocols accept data from the application layer and format them so that they may be transmitted or received on the network. A session may be started and maintained using the aforementioned protocols. Any application may be written using those protocols. The largest application for AppleTalk is the AppleTalk File Protocol (AFP). AFP allows network stations to share file and print services with each other.

AppleShare and the AppleTalk File Protocol (AFP)

For a workstation to access files and other services of another workstation the requesting station must invoke a high level service known as AppleTalk File Protocol. AFP is Apple's method of controlling server, volume, directory, file, and desktop calls between a client and workstation on an AppleTalk network. For those familiar with other protocols such as Microsoft's Server Message Block (SMB), Novell Netware Control Protocol (NCP), and Sun's Network File System (NFS), they are similar to Apple's AppleShare and AFP. Apple decided not to follow these implementations, for they were in various stages of completion and SMB was written for DOS and NFS was written for UNIX. Neither could implement all of Apple's file structure easily. So Apple created a new remote file system using a protocol called AFP. The name given to use this is AppleShare. AFP is a service that uses the AppleTalk protocols to allow network stations to share data over the internet.

There are entities within the AppleTalk network that must be known before fully understanding the AppleTalk file/print network. These entities are:

- File servers: A file server is a device on the network that acts as a repository for files. These files can include applications as well as user data. File servers are usually the most powerful network stations and contain a large capacity hard disk. Multiple file servers are usually found on any network and can be segregated to file server, data server, and print server.
- Volumes: A volume is the top level of the Apple directory structure. Volumes may be separated as an entire disk drive or may be many partitions of a disk drive.
- Directories and files: These entities are stored in each volume. In any Apple machine they are arranged in a branching-tree structure. Directories are not accessed as a file. Directories are an addressable holding area for files. They may branch into other directories because they carry a parent-child relationship.
- File forks: A file consists of two forks: a resource fork and a data fork. A resource fork contains system resources such as the icons and windows. A data fork contains the data in the file in unstructured sequence.

The protocol known as AppleShare allows an Apple computer to share its files with other users on the network. Every network station on the AppleTalk network may be set up to share files with any other station on the network. Or they may be dedicated file servers on an internet and all stations may have access to these servers provided they are allowed to do this. Each AppleShare folder has an owner and that owner will determine the access rights for that folder. The access rights are private, group, or all users. The ownership can be transferred to another owner.

Communication between a user's workstation and an AppleShare file server is accomplished using AFP. AFP operates as a client of ASP. AFP is an upper layer protocol to ASP. It will use ASP to allow a connection known as a session between a client and a file server. This session is a virtual connection that will allow protocol control information and user data to flow between the two network stations.

Prior to session establishment, a requesting station must know the address of the server it wishes to communicate with. Specifically, it must obtain its session listening socket or SLS. Servers will ask ATP for this. After this is accomplished, the server will use NBP to register the file server's name and type on the socket just received. For a workstation to find a server, the workstation will place a lookup call to NBP. NBP should return the address of all the NVEs within the zone specified by the workstation.

All the user has to do to connect to a file server is select Chooser under the Apple icon and select the AppleShare icon. Chooser will then ask the user to select a zone. Once the user selects a zone, the Chooser will ask the user to choose a file server. The user then logs into the server. Once the authentication is established, the workstation would establish a session to the server and communication would begin.

Summary

In this chapter you learned about the AppleTalk protocol suite. You learned that AppleTalk, a protocol suite developed by Apple Computer in the early 1980s, was developed in conjunction with the Macintosh computer. AppleTalk's purpose was to allow multiple users to share resources, such as files and printers. The devices that supply these resources are called servers, while the devices that make use of these resources such as a Macintosh computer are referred to as clients. Hence, AppleTalk is one of the early implementations of a distributed client-server networking system. This chapter provides an overview of AppleTalk's network architecture.

AppleTalk was designed with a transparent network interface. That is, the interaction between client computers and network servers requires little interaction from the user. In addition, the actual operations of the AppleTalk protocols are invisible to end users, who see only the result of these operations. Two versions of AppleTalk exist: AppleTalk Phase 1 and AppleTalk Phase 2. You also learned about AppleTalk routing and how AppleTalk interconnects to other protocol suites. Chapter 9 discusses Digital Network Architecture (DNA), also referred to as DECnet Phase IV, and the protocols for DNA. Chapter 9 also discusses the OSI protocols and their role in the Internet architecture.

Review Questions

1. What is the difference between AppleTalk Phase I and Phase II?
2. What are the key goals of the AppleTalk networking system?
3. What is LocalTalk?
4. What speed does LocalTalk operate at and why?
5. What is the medium for LocalTalk?

6. What is the LAP manager?
7. What is an AppleTalk address composed of?
8. What are the four basic components that comprise an AppleTalk network?
9. How many different socket numbers can be contained on an AppleTalk node?
10. Explain nonextended and extended node ID addressing.
11. How are AppleTalk addresses assigned?
12. What are the three types of packets utilized by the AARP protocol?
13. What is the AMT and what is its function?
14. What is the process for AARP to assign an AppleTalk address?
15. What are ELAP and TLAP?
16. What are the entities that make up the DDP?
17. What are sockets?
18. What is the function of the NBP?
19. What is the difference between half-routers and backbone routers?
20. What entries comprise an AppleTalk routing table?
21. What is RTMP and what function does it perform in an AppleTalk network?
22. What is the difference between seed routers and nonseed routers?

Summary Questions

1. How can a nonrouting node acquire its network number and routing address?
2. What are the three packet types in an RTMP packet?
3. What layer does the AEP operate at and what is its function?
4. Why are sockets needed in helping to resolve names in AppleTalk?
5. What two items does the AppleTalk naming process accommodate?
6. What is an NVE and what function does it perform in the naming process?
7. What is the function of the ZIP?
8. What type of name server is used in the AppleTalk NBP process?
9. What are the four types of services provided by the NBP naming process?
10. What are the three steps involved in mapping a name on a single AppleTalk network?
11. What is a zone?
12. What does the ZIP process use to map network numbers to zone names?
13. What is the ATP, where does it operate, and what is the function of the ATP?
14. What are the three packet types that provide reliable delivery service by ATP?
15. What is the PAP, what layer of the OSI does it operate at, and what is the function of the PAP?

DECnet Phase IV Digital Network Architecture

Objectives

- Explain the history of DECnet and discuss its relationship to the OSI model.
- Explain DECnet Phase IV routing.
- Discuss each of the DECnet layers individually and explain their functions and DECnet implementations.
- Discuss the various protocols that make up each layer.
- Discuss the DECnet addressing that is used in routing packets.
- Explain the concept of areas and designated router (DR) and how it is applied in DECnet routing.
- Examine the routing database.
- Explain the forwarding of data in a DECnet environment.

Key Terms

actual maximum path cost
(AMaxc), 370
actual maximum path length
(AMaxh), 370
adjacency, 367
areas, 374

block size (BlkSize), 387
Broadcast End Node Adjacency
(BEA), 368
circuit, 368
Command Terminal Protocol
(CTP), 404

Introduction

DECnet is a group of data communication products, including a protocol suite developed and supported by Digital Equipment Corporation (DEC), hereafter referred to as Digital. The first version of DECnet, released in 1975, allowed two directly attached PDP-11 minicomputers to communicate. In recent years Digital has included support for nonproprietary protocols, but DECnet remains the most important of Digital's network product offerings. This chapter provides a summary of the DECnet protocol suite, Digital's networking architectures, and the overall operation of DECnet traffic management. Figure 9.1 illustrates a DECnet internetwork with routers interconnecting two LANs that contain workstations and VAXs.

Several versions of DECnet have been released. The first allowed two directly attached minicomputers to communicate. Subsequent releases expanded the DECnet functionality by adding support for additional proprietary and standard protocols, while remaining compatible with the immediately preceding release. This means that the protocols are backward compatible. Currently, two versions of DECnet are in widespread use: DECnet Phase IV and DECnet/OSI.

DECnet Phase IV is the most widely implemented version of DECnet. However, DECnet/OSI is the most recent release. DECnet Phase IV is based on the Phase IV Digital Network Architecture (DNA) and supports proprietary Digital protocols and

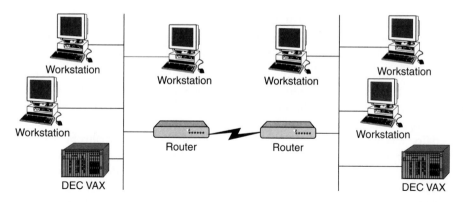

FIGURE 9.1

In a DECnet based internetwork, routers interconnect workstations and VAXs.
(Reproduced with permission of Cisco Systems, Inc. Copyright © 2001 Cisco Systems,
Inc. All rights reserved.)

other proprietary and standard protocols. DECnet Phase IV is backward compatible
with DECnet Phase III, the version that preceded it.

DECnet/OSI also called *DECnet Phase V* is backward compatible with DECnet
Phase IV and is the most recent version of DECnet. This version is based on the
DECnet/OSI DNA, protocols, and other proprietary and standard protocols.

History

DECnet Phase I was introduced in 1975 and ran on Digital's PDP-11 computers under the RSX operating system. Phase I allowed program-to-program communication,
such as file transfer and remote file management between two or more computers.
Phase I supported asynchronous, synchronous, and parallel communication devices.
Phase I was a simplex system compared with today's topologies and protocols, but it
did allow for most of the same capabilities of today's internets. At the time it was considered very powerful.

With the Phase I release and the next two subsequent releases, the PDP-11 computer was the network. All computers communicated to each other through the
PDP-11. Ethernet was not available until the release of DECnet Phase IV. The
data link for this network was a control message protocol known as Digital Data
Command Message Protocol (DDCMP), which operated on serial links between
the PDP-11s.

DECnet Phase II was introduced in 1978 and provided several enhanced features over
DECnet Phase I. Included in the Phase II release was support over more operating
systems: TOPS-20 and Digital Multiplex System (DMS). Also enhanced were the remote file management and file transfer capabilities. DECnet Phase II supported a

● **DECnet
Phase V**, also
called
DECnet/OSI, is the
most recent version
of DECnet.

● **DECnet Phase I**
allowed program-
to-program
communication.

maximum of 32 nodes on a single network. Most important with the release of Phase II was the mandatory change of the code. The code that provided DECnet was rewritten. With this release and all subsequent DECnet releases, Digital kept the end user interface the same. Only the underlying code changed.

DECnet Phase III was introduced in 1980. Included in the Phase III release was adaptive routing, which is the ability to find a link failure and route around it, and the support of 256 nodes. Remote terminal, known as Remote Virtual Terminal capability, was first introduced in the Phase III release. This functionality allowed end users to remotely log into remote processors as if the user were directly attached to that processor. Also included in the Phase III release was the addition of downline loading operations, which allowed a program to be loaded and run in another remote computer. Phase III was the first phase that supported IBM's SNA architecture.

DECnet Phase IV Digital Network Architecture (DNA)

DECnet Phase IV introduced the support for Ethernet and the expansion of the number of nodes supported.

The release that is by far the most popular and most often associated with Digital's products is called *DECnet Phase IV*. Phase IV was introduced in 1982. The Phase IV release introduced the support for Ethernet and the expansion of the number of nodes supported.

Because Digital was one of the originating companies to develop Ethernet into a standard they incorporated this into DECnet Phase IV. The second most noticeable improvement was the idea of area routing. Today, this routing technique is similar to OSI's IS-IS routing. The area routing technique remains the strongest link in the DECnet architecture.

The number of nodes that were supported in this release was increased to 64,449. The node total of 64,449 is split into the concept of areas and nodes. It is a 16-bit number with the first 6 bits assigned to the area and the remaining 10 bits allotted to the node number. Nodes are separated into distinct logical areas. The specification allows for 63 areas to be defined with 1,023 nodes in each of 63 separate areas. This was shortsighted on Digital's part. This may seem like a lot of addresses, but it disallows the building of large Digital internetworks that interoperate. With DECnet, even personal computers need to be assigned a full node address.

Digital's routing is different from all other protocol type routing in that the routers keep track of all the end nodes in the area. This greatly increases the size of a router's routing table. Separating nodes into areas not only reduces the traffic of updating routers but also decreases the size of routing tables per area. The network terminal concept introduced in Phase III was reinvented as network virtual terminal in Phase IV. Also introduced with this release was X.25 packet switching network support.

Future versions of the Phase IV release included support TCP/IP and Token Ring. both network TCP/IP and Token Ring protocols became very popular throughout the 1980s, but Digital did not allow these protocols to operate on DECnet Phase IV nodes. Around 1990, DEC finally delivered its version of TCP/IP for the Virtual

Memory System (VMS) operating system after years of allowing a company known as Wollongong to provide this functionality.

Previous to the direct support of TCP/IP on VMS, DEC did provide TCP/IP functionality for Ethernet through its Ultrix operating system. The Ultrix operating system was actually the Berkeley UNIX operating system, which was adapted to run on VAX computers.

DECnet Phase IV and OSI

The *Digital Network Architecture (DNA)* is a comprehensive layered network architecture that supports a large set of proprietary and standard protocols. The Phase IV DNA is similar to the architecture outlined by the OSI reference model. As with the OSI reference model, the Phase IV DNA utilizes a layered approach whereby specific layer functions provide services to protocol layers above it and depend on protocol layers below it. Unlike the OSI model, the Phase IV DNA is composed of eight layers. Figure 9.2 illustrates how the eight layers of the Phase IV

Digital Network Architecture is a layered network architecture that supports a large set of proprietary and standard protocols.

OSI reference model

| Application |
| Presentation |
| Session |
| Transport |
| Network |
| Data link |
| Physical |

DECnet phase IV DNA

| Network management | Network application |
| Session control |
| End communications |
| Routing |
| Data link |
| Physical |

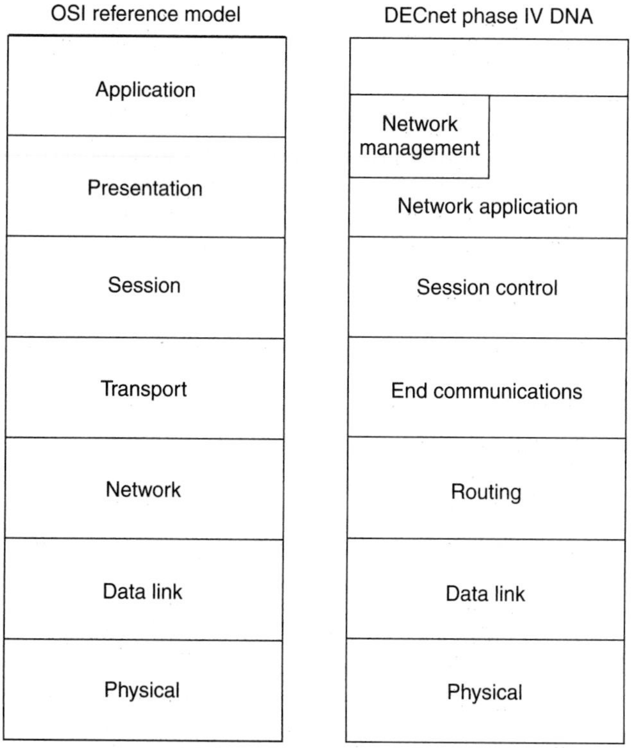

FIGURE 9.2

Phase IV consists of eight layers that map to the OSI layers. (Reproduced with permission of Cisco Systems, Inc. Copyright © 2001 Cisco Systems, Inc. All rights reserved.)

The **user layer** represents the user-network interface, supporting user services and programs.

The **network management layer** represents the user interface to network management information.

The **network application layer** provides various network applications, such as remote file access and virtual terminal access.

The **session control layer** manages logical link connections between end nodes.

The **end communications layer** handles flow control, segmentation, and reassembly functions.

The **routing layer** performs routing and other functions.

DNA relate to the OSI reference model. The *user layer* represents the user-network interface, supporting user services and programs with a communicating component. The user layer corresponds roughly to the OSI application layer. The *network management layer* represents the user interface to network management information. This layer interacts with all the lower layers of the DNA and corresponds roughly with the application layer. The *network application layer* provides various network applications, such as remote file access and virtual terminal access. This layer corresponds roughly to the OSI presentation and application layers. The *session control layer* manages logical link connections between end nodes and corresponds roughly to the OSI session layer. The *end communications layer* handles flow control, segmentation, and reassembly functions and corresponds roughly to the OSI transport layer. The *routing layer* performs routing and other functions and corresponds roughly to the OSI network layer. The data link layer manages physical network channels and corresponds to the OSI data link layer. The physical layer manages hardware interfaces and determines the electrical and mechanical functions of the physical media. This layer corresponds to the OSI physical layer. Digital includes support at the physical layer for:

1. X.21
2. EIA-232-D
3. ITU-T V.24/V.28
4. Ethernet

At the data link layer, DNA supports not only Ethernet IEEE 802.2/IEEE 802.3 but also a proprietary protocol known as Digital Data Communications Message Protocol (DDCMP). DDCMP is now a seldom used protocol with the introduction of Ethernet Phase IV support. Most companies have switched to Ethernet LAN and WAN for their Digital minicomputers. Figures 9.3 and 9.4 review the physical and data-link DNA layers. Figure 9.5 depicts the network routing layer. Digital differs from other protocol implementations at the network layer. The network layer is where Digital becomes proprietary, meaning that it is not an open protocol. Not just anyone can build this software protocol suite because Digital holds the patents and copyrights to the architecture.

DECnet/OSI DNA Implementations

The DECnet/OSI DNA defines a layered model that implements three protocol suites: OSI, DECnet, and TCP/IP. The OSI implementation of DECnet/OSI conforms to the seven layer OSI reference model and supports many of the standard OSI protocols. The Digital implementation of the DECnet/OSI supports the lower layer TCP/IP protocols and enables the transmission of DECnet traffic over TCP transport protocols. Figure 9.6 illustrates the three DECnet/OSI implementations.

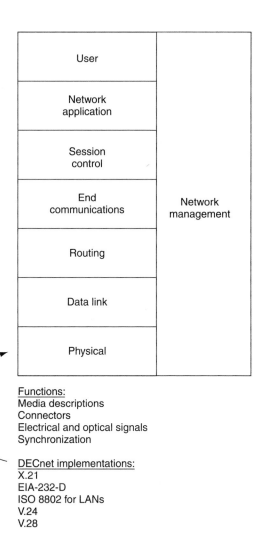

FIGURE 9.3

The physical layer. (Reproduced with permission of Cisco Systems, Inc. Copyright © 2001 Cisco Systems, Inc. All rights reserved.)

The Routing Layer: DECnet Phase IV Routing

Before studying the routing methods, some specific terms will be used throughout the discussion. The following list of terms is to be used as a reference:

● *Adjacency:* According to the DECnet specification, it is a [circuit, node ID] pair. For example, an Ethernet LAN with n attached nodes (network stations) is considered n − 1 adjacencies and does not include itself for that router on that Ethernet. Basically, an adjacency is a node and its associated data link connection. This could also be a synchronous connection to a router provided by a phone line.

FIGURE 9.4

The data link layer. (Reproduced with permission of Cisco Systems, Inc. Copyright © 2001 Cisco Systems, Inc. All rights reserved.)

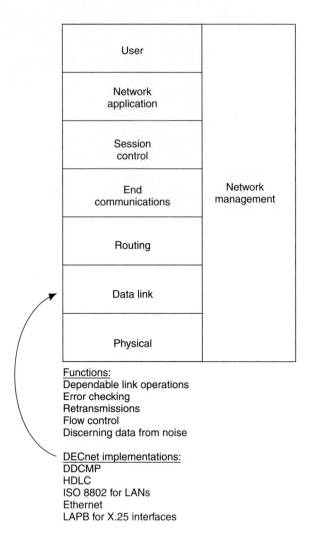

Functions:
Dependable link operations
Error checking
Retransmissions
Flow control
Discerning data from noise

DECnet implementations:
DDCMP
HDLC
ISO 8802 for LANs
Ethernet
LAPB for X.25 interfaces

- *Broadcast End node Adjacency (BEA):* A router connected to the same Ethernet as this node.
- *Circuit:* An Ethernet network (not internet) or a point-to-point link.
- *Connectivity algorithm:* The algorithm in the decision process whose function is to maintain path lengths (or hops). The routing function.
- *Cost:* An integer number assigned to a router interface. This represents the cost of the port for routing.

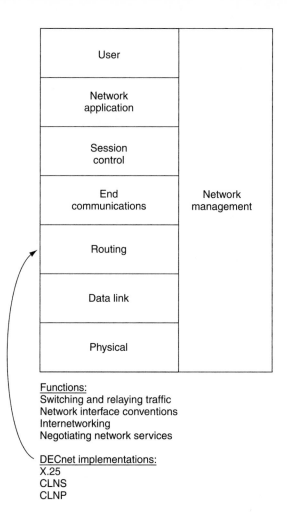

FIGURE 9.5

The network layer. (Reproduced with permission of Cisco Systems, Inc. Copyright © 2001 Cisco Systems, Inc. All rights reserved.)

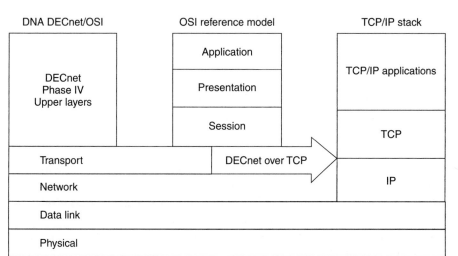

FIGURE 9.6

The OSI, DECnet, and TCP/IP are all supported by DECnet/OSI DNA. (Reproduced with permission of Cisco Systems, Inc. Copyright © 2001 Cisco Systems, Inc. All rights reserved.)

- *Designated router (DR):* The router on the Ethernet segment chosen to perform additional duties beyond that of a normal router. This includes informing end nodes on the circuit of the existence and identity of the Ethernet routers. Also used for the intra-area routing for end nodes.
- *End node:* A node that cannot route.
- *Node:* A device on the network. A router, bridge, minicomputer, and personal computer are examples.
- *Hop:* The logical distance between two adjacent nodes. Nodes that are directly connected are said to be one hop away.
- *NBEA:* The number of broadcasting end node adjacencies.
- *NBRA:* The number of broadcasting router adjacencies.
- *NN:* The maximum node number in an area. Maximum value: less than 1,024.
- *NA:* A level 2 router parameter only, the maximum area number. Maximum value: less than 64.
- *Maxh:* Maximum hops possible in a path to a reachable node in an area value. Maximum value: less than or equal to 30.
- *Maxc:* The maximum path cost in an area. Maximum value: less than or equal to 63.
- *AMaxh:* Pertaining to level 2 routers only, it is the actual maximum path length to any area. Maximum value: less than or equal to 30.
- *AMaxc:* Actual maximum path cost to any area. Maximum value: less than or equal to 1,022.
- *Path:* The route a packet takes from the source to the destination node. There are three types of paths:
    1. End node to end node (direct routing)
    2. End node through level 1 routers in the same area
    3. End node through level 2 routers to an end node destination in another area
- *Path cost:* The sum of the circuit costs along a path between two nodes.
- *Path length:* The number of hops between two nodes.
- *Point-to-point link:* The link between two routers, usually a remote serial line connecting two routers.
- *Traffic assignment algorithm:* The algorithm in the decision process that calculates the path costs in the routing database.

Functions of DECnet Routing

DECnet provides a first glimpse into hierarchical routing. Unlike other dynamic routing methods, routing tables are not distributed to all nodes on the entire internet. Routing tables are held to the local area routers with end nodes reporting their status to their local routers.

The routing method of DECnet divides the DECnet internet into areas. There are 63 areas that are allowed in a DECnet network. Each area may have up to 1,023 addressable nodes in one area. Routers that provide routing functions between other routers in a local area are referred to as level 1 routers. Routers that provide routing between two areas are referred to as level 2 routers. Upon initialization, end nodes report their status to the router with a MAC destination multicast address of all routers. This is called the end node hello packet. All level 1 routers will pick up these packets and build a level 1 database table of all known end nodes for the circuit they are attached to. These end node multicasts will not be multicast across the router. This means that a router that receives an end node hello will consume the packet. The router does not forward this packet. Instead, level 1 routers will transmit their database tables consisting of all end nodes to other level 1 routers. All level 1 routers will be updated with all known end nodes in a particular area. Level 1 routers also maintain the address of the nearest level 2 router. To reduce the memory required to keep track of all end stations, DECnet routing is split into distinct areas. Level 1 routers track only the state of the end nodes in their area. Level 2 routers maintain a database of all known level 2 routers in the entire DECnet internet. Level 2 routers must perform level 1 functionality, including the receiving of all local level 1 updates for their area. They do provide the additional ability to route between areas.

Table 9.1 summarizes the definitions pertaining to the services and components used in DECnet routing.

Table 9.1 Services and Components in DECnet Routing

DECnet Routing Layer Functions	
Service	Component
Packet paths	Determines path for packets if more than one path exists.
Topology changes	Alternate paths are used if a node or circuit fails; routing modules are changed to reflect changes.
Packet forwarding	Forwards packet to end communication layer at destination node or the next node if packet is not destined for the local node.
Node visits	Limits number of nodes that a packet can visit.
Buffer management	Manages buffers at nodes.
Packet return	Returns packets to end communication layer if packets are addressed to unreachable nodes if requested by end communication layer.
Data link monitoring	Monitors errors detected by the data-link layer.
Statistics	Gathers event data for network management layer.
Node verification	If requested by the network management layer, exchanges passwords with adjacent node.

Addressing

The addressing scheme used by DECnet Phase IV is based on area and node numbers. The total address is 16 bits long with the area number being the first 6 bits and the node number being the final 10 bits. To carry this addressing scheme from Phase III over to Phase IV, which supports Ethernet, an algorithm was derived to produce the conversion. Recall that an Ethernet address is 48 bits long. The algorithm works as follows. Although other protocols use a translator function to map a protocol address to a MAC address, DECnet literally maps its protocol address directly to the MAC address. A DECnet address is used in the DECnet packet header, but the translation of it to a MAC address overrides the burnt-in MAC address of the LAN controller. A table is not maintained to map a DECnet address to a MAC address.

```
Conversion Example:
Example: DECnet address 6.9 (area 6, node number 9).
Put into binary with 6 bits for the area number and
10 bits for the node ID:
000110.0000001001
Split into two 8-bit bytes gives:
00011000 00001001
Swap the two bytes:
00001001 00011000
Convert to hex:
0918
```

Adding the HIORD (Digital's constant high-order bytes AA 00 04 00), explained next, gives AA 00 04 00 09 18 as the 48-bit physical address for a node residing in area 6 and a node ID of 9. Allowing for 6 bits of area address and 10 bits for node ID gives a total of 65,535 possible addresses in a DECnet environment. Figure 9.7 illustrates another example.

Address Constraints

The following addresses are reserved by Digital:

High-order bytes are the 4 bytes placed on the beginning of a 48-bit converted DECnet address.

- *HIORD* (*high-order bytes*): The 4 bytes placed on the beginning of a 48-bit converted DECnet address. This is used after the address conversion is accomplished to complete the 48-bit DECnet address. These bytes are AA 00 04 00 xx xx.
- All routers: This is the multicast address for packets that send out information pertaining to all the routers:

```
AB 00 00 03 00 00
```

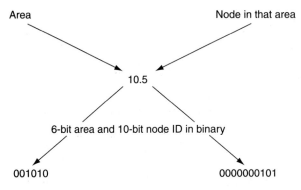

Area   Node in that area

10.5

6-bit area and 10-bit node ID in binary

001010   0000000101

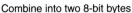

Combine into two 8-bit bytes

00101000   00000101

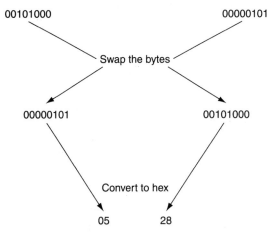

Swap the bytes

00000101   00101000

Convert to hex

05   28

Add the HIORD (4 hi-order bytes assigned
by DEC which are AA 00 04 00)

Gives the following 48-bit physical address
AA 00 04 00 05 28

FIGURE 9.7

MAC address conversion for a DECnet address. (Reproduced with permission of Cisco Systems, Inc. Copyright © 2001 Cisco systems, Inc. All rights reserved.)

● All Endnodes: This is a multicast address used by packets that contain information for all end nodes in an Ethernet segment, a circuit AB 00 00 04 00 00.

The protocol Ethernet type field should be set to 60-03, which indicates a DECnet packet. DECnet uses the Ethernet frame format. DEC supports SNAP for Token Ring packets.

Areas

A DECnet network is split into 63 definable *areas*. To accomplish area routing there are two types of routers: level 1 and level 2. *Level 1 routers* route data within a single area and keep track of all nodes that are in its area. Level 1 routers do not care about nodes that are outside their area. When communication takes place between two different areas, the level 1 routers will send data to a level 2 router.

Level 2 routers route traffic between two areas. Level 2 routers keep track of the least-cost path, not necessarily the fastest, to each area in the internetwork as well as the state of any nodes in the area. Figure 9.8 shows three areas: 10, 8, and 4. Level 1 routers forward data within their own area. They will not forward data outside their area. This is the function of the level 2 router. If a data packet is destined for another area, the level 1 router will forward the packet to its nearest level 2 router. A level 2 router automatically assumes the function of a level 1 router in that area. In turn, this router will ensure the packet makes it to the final area. Then a level 1 router will forward the data to the final destination node. Level 2 routers perform both level 1 and level 2 routing functions in the same router.

Finally, Figure 9.8 shows level 2 routers being connected through serial lines. This is not always the case. Level 2 routers may connect two segments of Ethernet cable. Each Ethernet segment would be defined as a different area. Unless there are multiple routers defined with a different area number connected to the same cable segment, there is only one area per cable segment.

The Routing Database

There are two types of nodes in DECnet areas:

1. End nodes are stations that do not have a routing capability. This means that they cannot receive and forward messages that are intended for other nodes. An end node can send data packets to another adjacent node such as on the same LAN, whether that is an end node or router. They cannot route a packet for another end node or router. Examples of end nodes are personal computers or VAX minicomputers with routing disabled.

2. A full function node, which is a routing node, can send packets to any node on the network. This includes sending messages through a router to local nodes as well as nodes that are in a different area. The computer equipment that Digital manufactures and sells includes the ability to act not only as a host but also as a host that can provide network functionality and user application functionality. The discussion will also include functionality of the network routing architecture. It will be discussed in terms of a device that provides routing

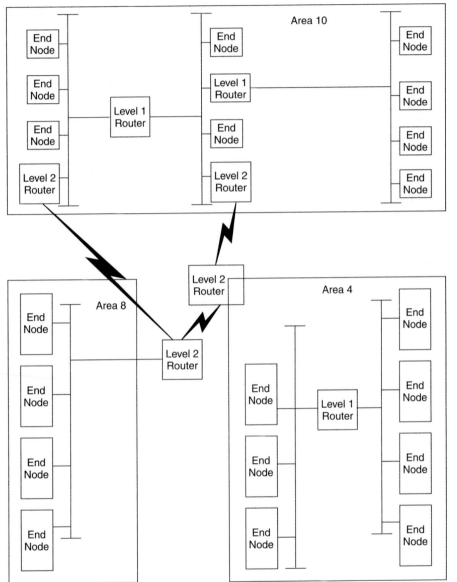

FIGURE 9.8

DECnet areas. Level 2 routers can be Ethernet-to-Ethernet as well as Ethernet-to-serial connections as illustrated. (Reproduced with permission of Cisco Systems, Inc. Copyright © 2001 Cisco Systems, Inc. All rights reserved.)

capabilities only and not the dual functionality of a host providing application functionality plus the network routing functions. To explain the function of routing the network will have two entities: an end node and a single device referred to as a router. A VAX mini and an external router can provide full function routing.

Declarations

Before defining the function of the DECnet routers, some declarations should be made to make the routing function of DECnet more understandable.

- Router database tables are constructed from updated routing messages received from adjacent nodes. These adjacent nodes could be other level 1 or level 2 routers or end nodes in the local area, which is the same area number that is configured by the router. Because a DECnet internet is divided into areas there are two types of tables: level 1 and level 2 tables.
- The function of level 1 routers is to store a distance measured by cost and hop information. This distance is from the router to any destination node within the local area. Level 1 routers receive database updates from adjacent level 1 routing nodes. This means that once a level 1 router builds a table containing a listing of end nodes that the router knows about, it will send this information to other level 1 routers referred to as adjacencies on its network. Therefore, the level 1 routing database table is updated periodically by the update routing message received from adjacent level 1 routing nodes. Included in this message is the router's listing of known active end nodes in the unique area.
- To update other level 1 routers, a routing node sends update routing messages to its adjacent routing nodes with cost/hop information. All level 1 routers will transmit this information to all other level 1 routers. This will allow all level 1 table information to be eventually propagated to all routing nodes. Level 1 routers will not propagate their information to other areas. This information is propagated to routers in their own area.
- Level 1 routers can function on Ethernet broadcast circuits or point-to-point, router-to-router through a serial line.
- Level 2 routers perform level 1 routing functions as well as level 2 routing. Level 2 routers store cost and hop values from the local area to any other destination area. The main function of the level 2 routers is to route information to other areas on the DECnet internet. Level 2 routers receive database updates periodically from adjacent level 2 routing nodes.
- Level 2 routers can function on Ethernet broadcast circuits and point-to-point circuits, router-to-router across a serial line.
- DECnet routers are adaptive to topology changes. If the path to a destination fails, the DECnet routing algorithm will dynamically choose an alternative

path if available. If at any time a physical line in the network goes down, all paths affected by that would recalculate circuit. Any time these algorithms re-execute, the contents of the databases are revealed to all other adjacent nodes on each respective network.

● Protocol types that are adaptive to different circuits include X.25, DDCMP, serial links, and Ethernet.

● Event-driven updates mean that other routers are updated immediately to any changes in the network that would affect the routing of a packet.

● Periodic updates are timed intervals at which routing updates test and hello messages will be sent. These timers are settable.

● An unknown address or network unreachable packet will be returned if requested.

● The number of nodes a packet has visited is tracked to keep it from endlessly looping in the network.

● Node verification-password protection is performed.

● Information is gathered for network management purposes.

DECnet Routing

DECnet routing occurs at the routing layer of the DNA in DECnet Phase IV and at the network layer of the OSI model in DECnet/OSI. The routing implementations are similar in both cases. The *DECnet Routing Protocol (DRP)* implements DECnet Phase IV routing. It is a relatively simple and efficient protocol whose primary function is to provide optimal path determination through a DECnet Phase IV network. Figure 9.9 provides a sample DECnet network to illustrate how the routing function is performed in a DECnet Phase IV network.

●
The **DECnet Routing Protocol** Determines the optimal path through a DECnet Phace IV network.

DECnet routing decisions are based on cost—an arbitrary measure assigned by network administrators to be used in comparing various paths through an internetwork environment. Costs typically are based on hop count or media bandwidth compared with other media bandwidth and other measures. The lower the cost, the better the path. When network faults occur, the DRP uses cost values to recalculate the best paths to each destination.

DECnet/OSI routing is implemented by the standard OSI routing protocols (ISO 8473, ISO 9542, and ISO 10589) and by DRP.

Forwarding of Packets

Recall that DECnet uses not only adaptive routing but also area routing. This means that DECnet is hierarchical in its routing. There is more than one layer in the routing as depicted in Figure 9.8. The routing techniques used by DECnet are not the same as those used in most networks today. When a source station is trying

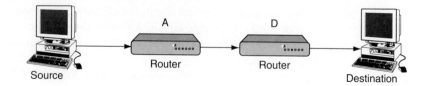

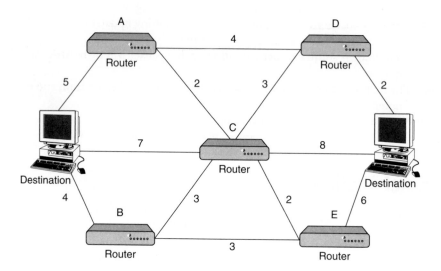

to send data to a destination station, the path it takes to transmit this data is called the path length and is measured by the number of hops. A path between the source and destination station may never exceed the maximum number of hops for the network.

Although DECnet uses hops, DECnet is not based on a distance vector algorithm. DECnet routes are based on a different cost factor. In other words, what is the cost to reach another destination on the network? The path with the lowest cost will be chosen as the path a packet will take. This allows realistic multiple paths to a destination to exist. The network manager may also assign a cost to each circuit on each station. The cost is an arbitrary number that will be used to determine the best path for data transfer. The total cost is the cumulative number that is between a source and destination station. The router will pick a path for data to travel based on the least cost involved. Figure 9.10 illustrates this concept.

The path to node D can take many different routes. Before data are sent, node A must decide which is the lowest cost to node D. As illustrated in Figure 9.10, A to B to C to D offers the lowest cost, so it is chosen as the path. Notice that option 2 has more hops than option 1. Hops are used in DECnet to determine when a packet

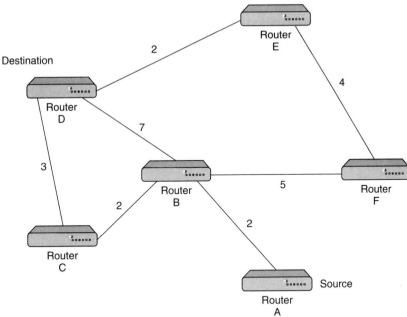

FIGURE 9.10

DECnet least-cost routing. (Reproduced with permission of Cisco Systems, Inc. Copyright © 2001 Cisco Systems, Inc. All rights reserved.)

1. A to B to D = 9 path cost
2. A to B to C to D = 7 path cost
3. A to B to F to E to D = 13 path cost

Therefore choose route 2

is looping. Assigning a real cost is a much more efficient way of routing. Suppose the metric was based on hops. In this case, option 1 would have been taken. But if the path from B to D is a 9,600-baud synchronous line and the paths from B to C and C to D are T1 serial lines, the hop algorithm will select the 9,600-baud line because it has the lowest hop count. Assigning costs allows the network administrator to effect the best path. One exception to cost based routing is when a node can be reached in two different ways and each path has the same cost. The router will arbitrarily select the route.

Cost numbers for circuits are arbitrarily assigned by a network administrator. There is not a strict standard in choosing them. The DECnet architecture does have an algorithm for assigning them. The general thought is the slower the circuit or link, the higher the cost than a T1 line. Table 9.2 shows the cost number most usually assigned.

Depending on the type of level of a router node, each routing node in a DECnet network maintains at least one database. On a level 1 router, the database contains entries on path length and path cost to every node in its area. It also maintains an entry for the whereabouts of a level 2 router in its area. Level 2 routers add a second database

Table 9.2 Possible Cost Parameters: Based on Line Speed

Speed	Cost	Speed	Cost
100 Mb/s	1	64 Kb/s	14
16 Mb/s	2	56 Kb/s	15
4 Mb/s	3	38.4 Kb/s	16
10 Mb/s	5	32 Mb/s	17
1.54 Mb/s	7	19.2 Kb/s	18
1.25 Mb/s	8	9.6 Kb/s	19
833 Kb/s	9	7.2 Kb/s	20
625 Kb/s	10	4.8 Kb/s	21
420 Kb/s	11	2.4 Kb/s	22
230.4 Kb/s	12	1.2 Kb/s	25
125 Kb/s	13		

known as an area routing table, which determines the least-cost path to the other area routers in the network not only the local area.

Node reachability explains that a node can be accessed if the computed cost and the hops it takes to get there are not exceeded by the number that is configured for that router. A lot of parameters may be configured for a DECnet router. This allows efficient utilization of memory in the router. If the DECnet network is small and has only a few nodes in a few areas, there is no need to configure the parameters with large entries. The memory will never be used. Most DECnet routers will allocate memory space based on the configuration parameters. It is important to know the DECnet topology before configuring a DECnet router. The memory space will not be allocated dynamically based on what the router finds on the network. In a DECnet environment, routers are not necessarily separate boxes that perform routing functions only. Routers may be contained in a micro VAX, a VAX, or any other node on the network.

Routing the Data

There are two types of messages that are transferred between two communicating stations, specifically at the End-to-End Communications layer, which corresponds to the OSI transport layer. The process of routing will add a route header to the packet that will enable the routers to forward the packet to its final destination based on the information that is in the route header portion of the packet. Control messages are exchanged between the routers, which initialize, maintain, and monitor the status between the routers or end nodes.

Figure 9.11a depicts the end station Hellos from nodes 5.3, 5.2, and 5.1, respectively. These are multicast packets that will be received by the 5.4 router. The 5.4 router is simply a level 1 router because it has no other area to communicate with. It will transmit level 1 routing updates from its Ethernet port.

Figure 9.11b shows two types of update packets that the routing function uses. One is for Ethernet and the other is for non-Ethernet or serial lines. The routers in Figure 9.11b are level 2 routers. As depicted in Figure 9.11b, the packet types used on broadcast networks such as Ethernet are:

1. Ethernet end node Hello messages: picked up by level 1 and level 2 routers.
2. Ethernet router Hello messages: sent to all routers on the Ethernet LAN by the router.
3. Level 1 routing message: sent by the level 1 router and contains the end node database.
4. Level 2 routing message: sent by the level 2 router and contains level 2 routing table information.

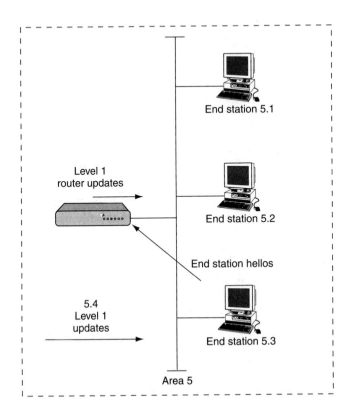

FIGURE 9.11a

Level 1 routing updates. (Reproduced with permission of Cisco Systems, Inc. Copyright © 2001 Cisco Systems, Inc. All rights reserved.)

FIGURE 9.11b

Level 1 routing updates continued. (Reproduced with permission of Cisco Systems, Inc. Copyright © 2001 Cisco Systems, Inc. All rights reserved.)

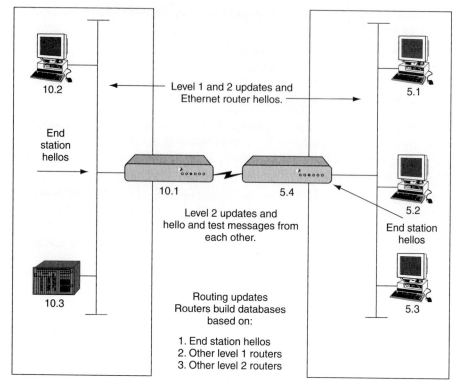

The packets that are transmitted between the routers on the serial lines and non-broadcast networks use the following routing messages only:

1. Initialization message
2. Verification message
3. Hello and Test message
4. Level 1 routing message
5. Level 2 routing message

The definitions of the types of routing control messages used by DECnet Phase IV routers are:

1. The routing message contains information that is used for updating the routing database of an adjacent node. Inside this message are the path cost and path length values for specific destinations.
2. The Ethernet router Hello message is used to initialize and monitor routers on an Ethernet circuit. This is a multicast packet and contains a list of all routers on the Ethernet circuit that the sending node has recently received via the Ethernet router Hello messages. This packet is multicast periodically to all other stations, routers, and end nodes alike on the same Ethernet circuit. This is done so that all routers are updated with the most current status of any other routers on the circuit. By

transmitting this message, the DR is also selected on that Ethernet circuit. Once the DR is selected, it will remain so until it is taken off-line for whatever reason. One purpose of the DR is to assist end stations in discovering that they are on the same circuit. This is accomplished by setting a special intra-Ethernet bit in a packet that it will forward from one end node to the other. This enables the end nodes on the same Ethernet segment to communicate without using the router.

Specifically, the Ethernet router Hello messages contain the sending router's ID, a timer called T3 (which is the timer for sending periodic Hello messages), and the sending router's priority. This message will also contain a listing of other routers the transmitting router has heard from. The list of other routers will contain their node number and their priority. A new router will be added to a router's table provided the two variables, which are the number of routers and the number of broadcasting routing adjacencies, are not exceeded. A router will broadcast this to all level 1 routers that contain a destination address of AB-00-00-03-00-00 and it will broadcast this to all end stations AB-00-00-04-00-00. A router will not declare itself the designated router until a certain time has passed, which is the DECnet default time of 5 seconds.

3. End nodes on an Ethernet circuit use the Ethernet end node Hello message to allow routers to initialize and monitor end nodes. This packet is transmitted as a multicast packet, and is transmitted by each active end node on an Ethernet circuit to allow all routers to find out about all active end nodes on an Ethernet circuit. The routers will use this to monitor the status of the end nodes on their Ethernet. The router will make an entry in its table for this end node. Once a router receives this packet, it should receive Hellos from this adjacency at certain timer periods. If it does not, it will delete this entry from its table and generate an event that the adjacency is down. This period of waiting is set to three times the router's T3 timer, the Hello timer for the router.

4. The Hello and Test message tests for the operational status of an adjacency. This type of packet is used on non-Ethernet (nonbroadcast) circuits and is used when no messages are being transmitted across that line. For example, on a serial line interconnecting two DECnet routers, there may be times when no messages are transmitted on that line. Periodically, when no valid messages are available to transmit, this message is transmitted so that the opposite end will not think the circuit is down. If this or any other message is not received within a certain time, the routing layer will consider the circuit to be down.

5. An Initialization message appears when a router is initializing a non-Ethernet circuit. Contained in this message is information on the type of node, the required verification, maximum message size, and the routing version.

6. A Verification message is sent when a node is initializing and a node verification must accompany the initialization.

Updating the Routers

Routing messages are propagated through the DECnet internet. Any node on the network can send out a routing message to an adjacent routing node. When the adjacent routing node receives that message it will compare the routing information in the message to the existing routing database table. If the information in the routing message contains new information not already in the routing table, the router updates its table and a new routing message is generated with the new information and sent to all other adjacent routers on the internet. In turn, those routers will update and send it to their adjacent nodes, and so on. This process is referred to as propagating the information.

If there are multiple areas on an internet, indicating level 2 routers, routers must discard messages that do not pertain to them. For example, when a level 1 router receives an end node or a router Hello message, the first check accomplished is that the incoming message's ID (area.node ID) is the same as the router's ID. The areas should match. If the areas do not match, the packet is discarded. This means that a level 1 router will not keep track of end nodes in other areas. A level 1 router will keep end node tables only for end nodes that have the same area ID as the router's. Level 2 routers must keep a table of adjacencies to other level 2 routers besides the adjacencies to its level 1 routers and end nodes in its own single area. Level 2 routers will discard any Ethernet end node Hello messages it may receive from other areas. When it does receive a level 2 update message, it will include that router's ID in its update table. The router will also include that ID in its next Ethernet router Hello message. A level 2 update will contain an area number and an associated hop and cost number to get to an area. These routing messages will be broadcast using the all-router's broadcast AA-00-00-03-00-00 in the destination address field of the MAC header.

When a router broadcasts its information to the network, and there is more information in the table than the Ethernet allows for a maximum size packet, multiple packets will be transmitted. This is similar to other routing protocols.

Routing Operation

As illustrated in Figure 9.12, the routing layer is actually divided into two sublayers:

1. The routing initialization sublayer performs only the initialization procedures. These procedures include initialization of the data link layer. This includes setting up the drivers and controlling the Ethernet, X.25, and DDCMP. This is the layer that the routing layer of DECnet must communicate with.
2. The routing control sublayer performs the actual routing, congestion control, and packet lifetime control. This layer controls five different processes as well as the routing database.

The routing control sublayer is made up of the following five processes:

1. Decision process: based on a connectivity algorithm that maintains path lengths and another algorithm that maintains path costs, this process will select a route for a received packet. Those algorithms are executed when the router receives a routing message, not a routable data message. The forwarding database tables will then be updated based on the outcome of the invoked

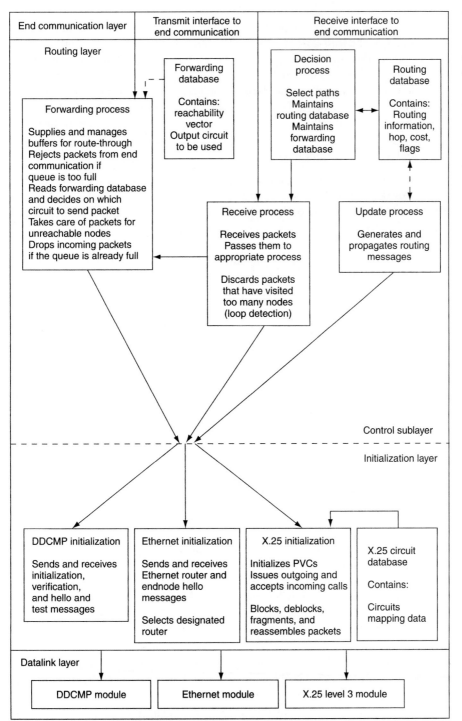

FIGURE 9.12

Routing layer components and their functions. (Reproduced with permission of Cisco Systems, Inc. Copyright © 2001 Cisco Systems, Inc. All rights reserved.)

End communication layer | **Transmit interface to end communication** | **Receive interface to end communication**

Routing layer

Forwarding database

Contains:
reachability
vector
Output circuit
to be used

Decision process

Select paths
Maintains
routing database
Maintains
forwarding
database

Routing database

Contains:
Routing
information,
hop, cost,
flags

Forwarding process

Supplies and manages
buffers for route-through
Rejects packets from end
communication if
queue is too full
Reads forwarding database
and decides on which
circuit to send packet
Takes care of packets for
unreachable nodes
Drops incoming packets
if the queue is already full

Receive process

Receives packets
Passes them to
appropriate process

Discards packets
that have visited
too many nodes
(loop detection)

Update process

Generates and
propagates routing
messages

Control sublayer

Initialization layer

DDCMP initialization

Sends and receives
initialization,
verification,
and hello and
test messages

Ethernet initialization

Sends and receives
Ethernet router and
endnode hello
messages

Selects designated
router

X.25 initialization

Initializes PVCs
Issues outgoing and
accepts incoming calls

Blocks, deblocks,
fragments, and
reassembles packets

X.25 circuit database

Contains:

Circuits
mapping data

Datalink layer

DDCMP module | **Ethernet module** | **X.25 level 3 module**

algorithms. The decision process will select the least-cost path to another level 1 router or the least-cost path to another area router (level 2 routing).

2. Update: this process is responsible for building and propagating the routing messages. These routing messages contain the path cost and path length for all destinations. Based on the decision process, it will transmit these messages when required to do so. This process will also monitor the circuits that it knows about by periodically sending routing messages to adjacent nodes. If the router is a level 1 router, it will transmit level 1 messages. However, if it is a level 2 router it will transmit level 1 and level 2 messages.

- Level 1 routing packets are sent to adjacent routers within their home area.
- Level 2 packets are sent to other level 2 routers.
- Level 1 routing packets contain information on all nodes in their home area.
- Level 2 routing packets contain information about all areas.
- Packets containing routing information are event-driven with periodic backup.

3. Forwarding process: this process looks in a table to find a circuit path to forward a packet to. If this path cannot be found, it is this process that will return the packet to the sender or will discard the packet depending on the option bits set in the packet route header. This process also manages all the buffers required to support the tables.

4. Receive process: this process inspects the packet's route header. Based on the packet type, this process gives the packet to another process for handling. Routing messages are given to the decision process, Hello messages are given to the node listener process, and packets that are not destined for the router are given to the forwarding process.

5. Congestion control: uses a function known as transmit management. This process handles the buffers, which are blocks of RAM memory set aside for storing or queuing information until it can be sent. If this number is exceeded, the router is allowed to discard packets to prevent the buffer from overflowing.

Packet lifetime control prevents packets from endlessly looping the network by discarding packets that have visited too many routers. This includes the loop detector, the node listener, and the node talker. The loop detector keeps track of how many times a packet has visited the node. It will remove the packet when it has visited a node a certain number of times. The node listener is the process that tracks the number of adjacencies. It determines when a node has been heard from and if the identity of an adjacent node has changed. This is the process that determines whether an adjacency is declared down. In combination with the node listener is the node talker process that provides for Hello packets to be transmitted. It places an artificial load on the adjacency so failures can be detected.

Initialization and circuit monitor is the means by which the router will obtain the identity of neighboring nodes. For Ethernet circuits, the Ethernet router Hello and the Ethernet end node Hello message perform these functions.

Forwarding of Data in a DECnet Environment

Data starts at the application layer of any network station and flow downward though the OSI model toward the physical layer. As the data pass through each of the layers, additional information referred to as header information is added to the beginning of the packet. The header information does not change the original information, which is the data from the application layer in the packet. The header information is merely control information so that any node that receives this packet will know whether or not to accept it and how to process it. If the packet is not intended for that node, the header information in the packet will contain information on how to handle the packet.

A router must check its routing table to determine the path for a packet to take to reach the final destination. If the destination is in the local area, the router will send the packet directly to the destination node or to another router in that area to forward the packet to the destination node. If the packet is for the node in a different area, the router will forward the packet to a level 2 router. That router will send it to another level 2 router in that area. The level 2 router in the other area will forward the packet to the destination node in the same manner as a level 1 router.

Table 9.3 depicts a typical level 1 and level 2 routing database table of a level 2 router. This would be router A, which is a level 2 router forwarding database in Figure 9.13.

From Table 9.3 notice that the number of reachable areas from this router is 3, the number of reachable nodes is 4 for areas 1 and 2, and for area 11 the number of adjacent nodes is 2. The table may look a little odd. In the level 1 database there is information to two different areas. Level 1 routers are not supposed to keep track of different area node IDs, only the end node IDs for their own area. This is a multiport router that contains more than just two ports. Therefore, it can assign area and node IDs on a per port basis. Consequently, this node is also a level 1 router for both area 1 and area 11. This router must keep track of the nodes in both areas that it connects to. In actuality, the router would probably keep two separate tables, one for each area. The separate routing tables are shown in the same table here for simplicity.

Definitions

Level 1 Table Descriptions

- *Forwarding destination of the packet (node and port):* This contains the DECnet router address and the port number on the router, assuming the router has multiple ports, for possible packet destinations.
- *Next hop:* The DECnet address (area and node address) that this router must send the packet to for it to be routed to its final destination.
- *Cost:* This is a user configurable number that the network administrator will assign to the circuit. Packets are routed based on the path with the lowest cost.
- *Hops:* This is the number of hops (the number of routers) that a packet must traverse before reaching the destination.
- *BlkSize (block size):* This is the maximum size of a packet that can be sent to a destination node.

Block size is the maximum size of a packet that can be sent to a destination node.

Table 9.3 Level 1 and Level 2 Routing Tables

LEVEL 1 INFORMATION

AREA 1

Node	Port	Next Hop	Cost	Hops	BlkSize	Priority	Timer
1	1	1.1	3	1	1500	40	—
2	1	1.1	3	1	1500	40	40
15	1	1.1	3	1	1500	50	40
4	2	1.90	3	1	1500	40	30
6	2	1.90	3	1	600	40	40
9.0	2	1.90	3	1	1500	40	—

AREA 11

Node	Port	Next Hop	Cost	Hops	BlkSize	Priority	Timer
2	3	11.2	3	1	1500	40	40
30	3	11.2	3	1	1500	40	90
15	3	11.2	3	1	1500		
1	3	11.2	3	1	1500		

LEVEL 2 INFORMATION

Area	Port	Next Hop	Cost	Hops	Time
1	Local	—	—	—	—
11	3	11.2	3	1	60
50	3	11.2	6	2	40

Priority is the individual priority of a router.

- *Priority:* The individual priority of that router. This is used to determine a designated router. The priority is also used in the following scenario: If the number of the router's parameter is exceeded, then the update of a routing database is as follows:

  1. The router with the lowest priority will be deleted from the database unless the new router has a lower priority than any other router in the database.
  2. If there are duplicate priority numbers, then the router with the lowest ID (48-bit Ethernet address) will be deleted from the database or the new router will not be added if its Ethernet address is the lowest. Ethernet addresses are not duplicated. There will be no decisions after this.

- *Time:* This is a parameter of how long this entry will remain in the table. This parameter is based on the configurable timer known as the Hello time.

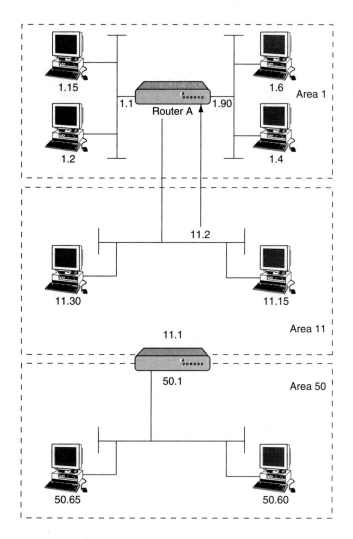

FIGURE 9.13

Multiple DECnet areas. (Reproduced with permission of Cisco Systems, Inc. Copyright © 2001 Cisco Systems, Inc. All rights reserved.)

Level 2 Table Descriptions

- *Destination of the packet (area and port)*: The DECnet area number, not node number, and the port number of the router to direct the packet to.
- *Next hop:* The DECnet address (area and node address) of the router to which a router will send the packet so that it may be routed to its final destination.
- *Cost.* Same as for level 1 table.
- *Hops.* Same as for level 1 table.
- *Time.* Same as for level 1 table.

It is important to note that a router has the possibility of routing a packet through multiple routes. The decision on which route to take is based solely on the cost associated with a path to the final destination.

End Node Packet Delivery

When an end node would like to communicate with another end node on the same Ethernet, it may do so under the following conditions:

- If the end node has the destination station's address in its cache table, it may communicate directly with the end node. Otherwise if the destination station's address is not in the table, it must send the directed packet to the designated router (see Figure 9.12). The designated router will then perform a lookup to see if the designated end node is active.
- If the destination station's address is in the designated router's table, it will forward the packet to the end node with the intra-area bit in the route header set to a 1 to indicate that this packet is destined for the local LAN. The destination station upon receipt of that packet will then send some type of response packet back to the originating station adding that station's address to its local cache. The designated router will not be used for further communication as indicated by the number 3 in Figure 9.14.
- The originating station, upon receipt of the destination station's response, will add the destination station's address into its cache and then communicate with it directly with no additional help from the designated router. The end node cache table may be aged out when the station indicated in the cache table has not been heard from for a specified amount of time.
- If the end node does not have the address of the designated router, it will try to send the packet to the destination anyway.

A flow chart for DECnet routing is illustrated in Figure 9.15. As illustrated in Figure 9.16a two areas are depicted on the same segment: area 5 and area 10. The only stipulation is that there must be two full-function nodes or routers on the

FIGURE 9.14

End node communication. (Reproduced with permission of Cisco Systems, Inc. Copyright © 2001 Cisco Systems, Inc. All rights reserved.)

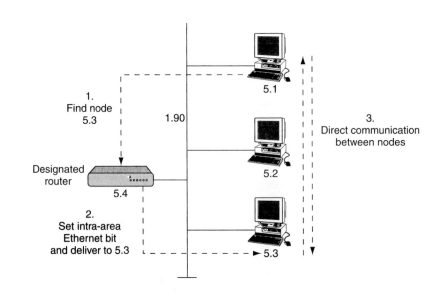

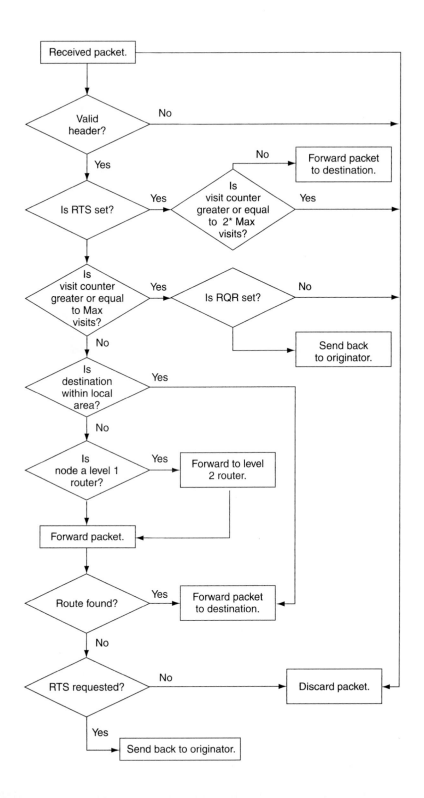

391

FIGURE 9.16a

Multiple areas per segment. (Reproduced with permission of Cisco Systems, Inc. Copyright © 2001 Cisco Systems, Inc. All rights reserved.)

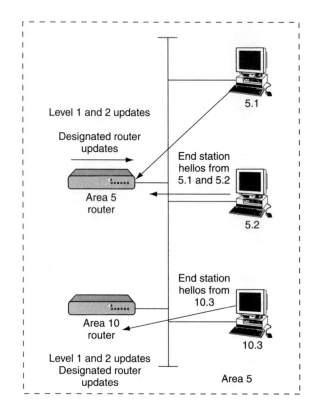

same segment of the Ethernet cable. Each of these routing nodes will have a different area. All of the rules that were stated previously remain the same (Ethernet router Hellos, level 1 updates, level 2 updates, and so forth). Each router controls the nodes in its own area. A single router port may not support two areas. Figure 9.16b shows a typical DECnet internet. It shows level 1 and level 2 routers, designated routers, end stations, and full-function nodes.

End Communications Layer: The DNA Transport Layer

The **Network Services Protocol** is a DECnet transport layer protocol.

The end communications layer (the transport layer on the OSI model) is used for reliable data transfer between two network stations on the network, regardless of where the two reside on the network. Reliable data transfer is accomplished by establishing a session, assigning sequence numbers between the source and destination stations, and then transmitting data over the virtual circuit. The DECnet transport layer protocol that accomplishes this service is called the *Network Services Protocol (NSP)*. It

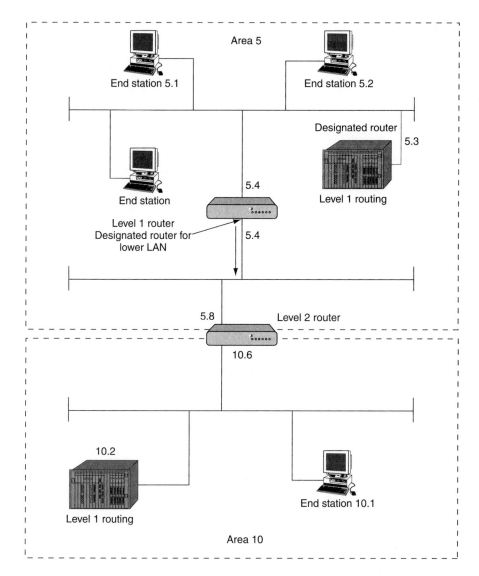

provides this as a service to the upper layer protocols. Figure 9.17 summarizes the functions of the DNA transport layer.

The NSP provides the following functions:

- Creates, maintains, and terminates logical links (virtual circuits)
- Guarantees the delivery of data and control messages in sequence to a specified destination by means of an error control mechanism
- Transfers data into and out of buffers

FIGURE 9.17

Functions of the transport layer. (Reproduced with permission of Cisco Systems, Inc. Copyright © 2001 Cisco Systems, Inc. All rights reserved.)

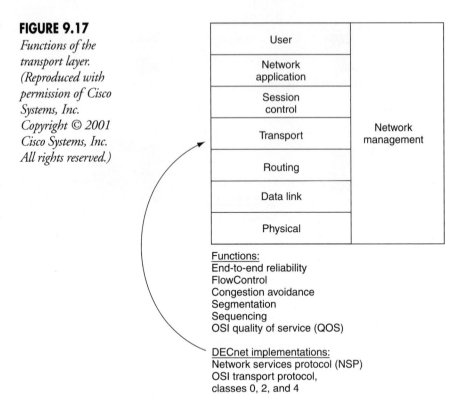

• Fragments data into segments and automatically puts them back together again at the destination
• Provides flow control
• Provides error control

To provide for logical link service, flow control, error control, and other functions, there are three types of NSP messages:

1. Data
2. Acknowledgement
3. Control

Table 9.4 expands on the functions of each type.

A logical link is a full-duplex logical channel that data may pass over. This is also known as a virtual circuit. The session layer of NSP only requests these connections; it is the responsibility of the end communications layer to perform the work. These processes are completely transparent to the user. NSP sets up a connection, manages it by providing sequencing and error control, and then terminates the link when requested to do so.

Table 9.4 NSP Messages

Type	Message	Description
Data	Data Segment	Carries a portion of a Session Control message. This has been passed to Session Control from higher DNA layers and Session Control has added its own control information.
Data	Interrupt	Carries urgent data, originating from higher DNA layers. It also may contain an optional Data Segment acknowledgement.
	Data Request	Carries data flow control information and optionally a Data Segment acknowledgement also called a Link Service message.
	Interrupt Request	Carries interrupt flow control information and optionally a Data Segment acknowledgement.
Acknowledgement	Data Acknowledgement	Acknowledges receipt of either a Connect Confirm message or one or more Data Segment messages and optionally another data message.
	Other Data Acknowledgement	Acknowledges receipt of one or more Interrupt, Data Request, or Interrupt Request messages and optionally a Data Segment message.
	Connect	Acknowledges receipt of a Connect Initiate message.
Control	Connect Initiate And Retransmitted	Carries a logical link Connect request from a Session Control module.
	Connect Confirm	Carries a logical link Connect acceptance from a Session Control module.
	Disconnect Initiate	Carries logical link Connect rejection or Disconnect request from a Session Control module.
	No resources	Sent when a Connect Initiate message is received and there are no resources to establish a new logical link (also called a Disconnect Confirm message).
	No link	Sent when a message is received for a nonexistent link
	No operation	Does nothing.

There may be several logical links between two stations or to many stations. Figure 9.18 shows the typical exchange between two end stations A and B having a logical link. The exchange of information data between nodes A and B can be carried over two types of channels:

1. Data channel: used to carry application data.
2. Other data subchannel: used to carry Interrupt messages, Data Request messages, and Interrupt Request messages.

FIGURE 9.18

Typical message exchange. (Reproduced with permission of Cisco Systems, Inc. Copyright © 2001 Cisco Systems, Inc. All rights reserved.)

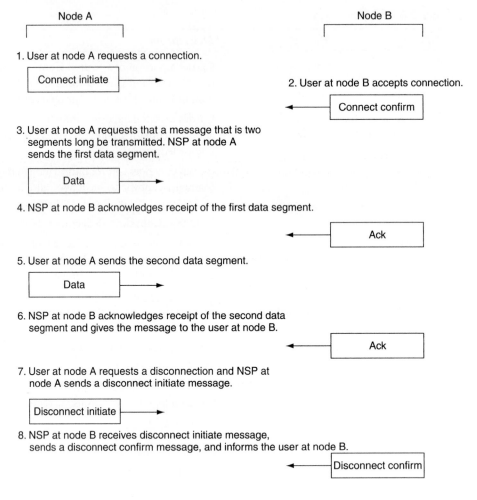

Node A   Node B

1. User at node A requests a connection.

Connect initiate →

2. User at node B accepts connection.

← Connect confirm

3. User at node A requests that a message that is two segments long be transmitted. NSP at node A sends the first data segment.

Data →

4. NSP at node B acknowledges receipt of the first data segment.

← Ack

5. User at node A sends the second data segment.

Data →

6. NSP at node B acknowledges receipt of the second data segment and gives the message to the user at node B.

← Ack

7. User at node A requests a disconnection and NSP at node A sends a disconnect initiate message.

Disconnect initiate →

8. NSP at node B receives disconnect initiate message, sends a disconnect confirm message, and informs the user at node B.

← Disconnect confirm

The transport layer also fragments data into a size that the routing layer will carry. There is a maximum number of bytes the routing layer will handle, and it is up to NSP to fragment the data and give it to the routing layer. Each fragment will contain a special number and other control information in the Data Segment message, which the receiving NSP layer will have to interpret in order to understand how to put the data back together again in the order to which it was sent. Only data that are transmitted over the data channel are fragmented.

To provide for error control, each end of the virtual link must acknowledge data that it received. Data that are received too far out of sequence will be discarded and a negative acknowledgement will be sent to the originator. Any station sending data will not discard them until it has received a good acknowledgement from the recipient of the data.

Flow Control

NSP provides flow control to ensure that memory is not overrun (buffer overflow). Both types of data (normal and interrupt data) can be flow controlled. When the logical link is first established, both sides of the link tell each other how the flow of data should be handled. This is based on two types:

1. No flow control.
2. The receiver will send the transmitter the number of data segments it can accept.

In addition to these rules, each receiving end of the link may at any time tell the opposite side the transmitter to stop sending data until further notice. To provide for efficiency any data type, whether it is a control data type or an application data segment, may contain the positive acknowledgement being returned for previously received data. This means that there is not necessarily a separate ACK packet for data that are being acknowledged.

The Session Control Layer

The session control layer protocol resides at the fifth layer of the OSI model (see Figure 9.19). The session control layer provides the following functions:

● *Mapping of node names to node addresses:* This maintains a node name-mapping table that provides a translation between a node name and a node address or its

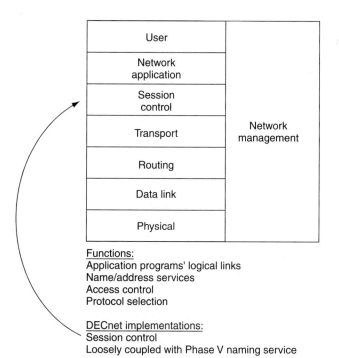

FIGURE 9.19

Functions of the session control layer.

adjacency. This allows the session layer to select the destination node address or channel number for outgoing connect requests, and it allows the session control layer to properly identify the node that is making the connect request.

● *Identifying end users:* This process determines if a process exists as requested by an incoming connection request.

● *Activating or creating processes:* The session control layer may start up a process or activate an existing process to handle an incoming connect request.

● *Validating incoming connect request:* Performs a validation sequence to find out if the incoming request should be processed.

These functions are divided into five actions:

1. *Requests a logical link between itself and a remote node:* If the application desires a logical link between itself and a remote node, it will request the session layer for this. It will identify the destination node address or a channel number for the end communications layer and issue a connect request to this layer. It will then start an outgoing connection timer. If this timer expires before hearing from the destination accept or reject, it will cause a disconnect to be reported back to the requesting application layer.

2. *Accepts or rejects a connect from a remote node:* The end communications layer will tell the session control layer of a connection request from a remote destination node. The session control layer will check the incoming packet for source and destination end user processes socket or port numbers. With this it will identify, create, or activate the destination address with a destination name in its table. Then, it will deliver any end user data to the destination application process. It will also validate any access control information.

3. *Sends and receives data across a valid logical link:* The session control layer will pass any data between the application layer to the end communications layer to be delivered to the network.

4. *Disconnects or aborts an existing logical link:* Upon notification from the application process, it will disconnect the session between two communicating nodes. Also, it will accept disconnect requests from the destination node and will deliver this to the application.

5. *Monitors the logical link.*

Figure 9.20 illustrates the relationships between the user processes and the session control and end communications layers.

Network Application Layer

●

The **Data Access Protocol** enables file transfer between nodes on the DECnet network.

DECnet defines the following modules at the network application layer, also illustrated in Figure 9.21:

1. The *Data Access Protocol (DAP)* is the remote file access protocol. It enables file transfer between nodes on the DECnet network.

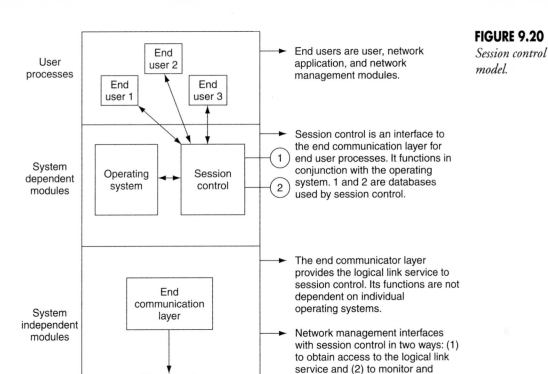

FIGURE 9.20

*Session control
model.*

User
processes

End
user 2

End
user 1

End
user 3

End users are user, network
application, and network
management modules.

System
dependent
modules

Operating
system

Session
control

1

2

Session control is an interface to
the end communication layer for
end user processes. It functions in
conjunction with the operating
system. 1 and 2 are databases
used by session control.

System
independent
modules

End
communication
layer

The network

The end communicator layer
provides the logical link service to
session control. Its functions are not
dependent on individual
operating systems.

Network management interfaces
with session control in two ways: (1)
to obtain access to the logical link
service and (2) to monitor and
control session control operations.

FIGURE 9.21

*Functions of the
network applica-
tion layer.*

User	
Network application	
Session control	Network management
Transport	
Routing	
Data link	
Physical	

Functions:
Filetransfer
Virtual terminals
E-Mail
SNA interconnects
Time servers

DECnet implementations:
Data access protocol
Network virtual terminal
Message router
SNA gateway access
Digitial time service
Distributed queuing service

The **Network Virtual Terminal Protocol** allows remote terminals to act as if they were local to the processor.

2. The *Network Virtual Terminal Protocol (NVT)* is the protocol that allows remote terminals to act as if they were local to the processor. It is the ability to remote to another host and act as if that terminal were locally attached.
3. The X.25 Gateway Access Protocol is the protocol that allows DECnet to interoperate over an X.25 link.
4. The SNA Access Protocol allows a DECnet network to interoperate with IBM's SNA.
5. The Loopback Mirror Protocol tests logical links.

Data Access Protocol (DAP)

The Data Access Protocol (DAP) is an application level protocol. DAP permits remote file access within DNA environments. The following is a listing of the functions of the DAP protocol:

1. Supports heterogeneous file systems.
2. Retrieves a file from an input device such as a disk file, a card reader, or a terminal.
3. Stores a file on an output device such as a magnetic tape, a line printer, or a terminal.
4. Transfers files between nodes.
5. Supports deletion and renaming of remote files.
6. Lists directories of remote files.
7. Recovers from transient errors and reports fatal errors to the user.
8. Allows multiple data streams to be sent over a logical link.
9. Submits and executes remote command files.
10. Permits sequential, random, and indexed access of records.
11. Supports sequential, relative, and indexed file organizations.
12. Supports wildcard file specifications for sequential file retrieval, file deletion, file renaming, and command file execution.
13. Permits an optional file checksum to ensure file integrity.

DAP Operation

The DAP process, by the source and destination communicating station, exchanges a series of messages. The initiation message contains information about operating a file system and buffer size. These are negotiation parameters. Table 9.5 indicates these messages. Figure 9.22 shows a node-to-node file transfer using DAP, and Figure 9.23 shows a file transfer over a DECnet network.

Network Virtual Terminal (NVT)

As illustrated in Figure 9.24, there are times when a user connected to one host may need a connection to another host on the network. Figure 9.24 illustrates the overall

Table 9.5 DAP Messages

Message	Function
Configuration	Exchanges system capability and configuration information between DAP speaking processes. Sent immediately after a logical link is established, this message contains information about the operating system, the file system, protocol version, and buffering capability.
Attributes	Provides information on how data are structured in the file being accessed. The message contains information on file organization, data type, format, record attributes, record length, size, and device characteristics.
Access	Specifies the file name and type of access requested: read, write, etc.
Control	Sends control information to a file system and establishes data streams.
Continue Transfer	Allows recovery from errors. Used for retry, skip, and abort after an error is reported.
Acknowledge	Acknowledges access commands and control messages used to establish data streams.
Access Complete	Denotes termination of access.
Data	Transfers file data over the logical link.
Status	Returns status and information on error conditions.
Key Definition	Specifies key definitions for indexed files.
Attributes Extension Allocation	Specifies the character of the allocation when creating or explicitly extending a file.
Attributes Etcetra Summary	Returns summary information about a file.
Attributes Etcetra Date and Time	Specifies time related information about a file.
Protection	Specifies file protection codes.
Attributes Etcetra Name	Sends name information when renaming a file or obtaining file directory data.

process for this. The NVT protocol allows terminal communications to exist remotely as if the terminal were directly attached to the remote computer.

The main protocol that is used for NVT is called the *Terminal Communication Protocol (TCP)*. It is the lower of the two sublayers. The other protocol is the Command Terminal Protocol (CTP). The TCP of the NVT is also referred to as the foundation

The **Terminal Communication Protocol** is the main protocol used for NVT.

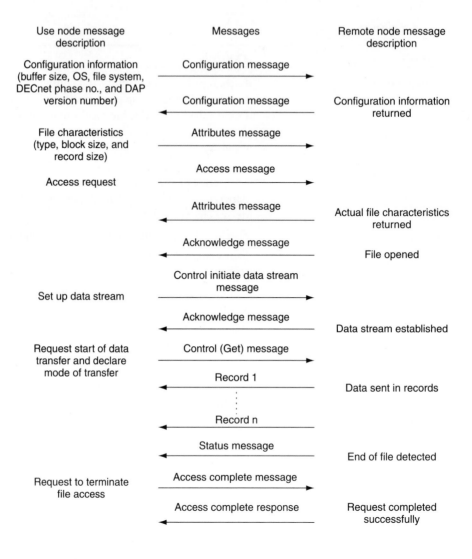

FIGURE 9.22

DAP message exchange sequential file retrieval.

layer. TCP's main goal is to establish and disconnect terminal sessions between applications and terminals. TCP is actually extending the session control layer to establish and disconnect sessions between end points or applications that are specific to terminals. The end point in the host is called the portal and the end point in the terminal end OS is called the logical terminal. A portal is a remote terminal identifier in the host, and a connection binds that identifier to an actual terminal. An NVT connection is called a binding.

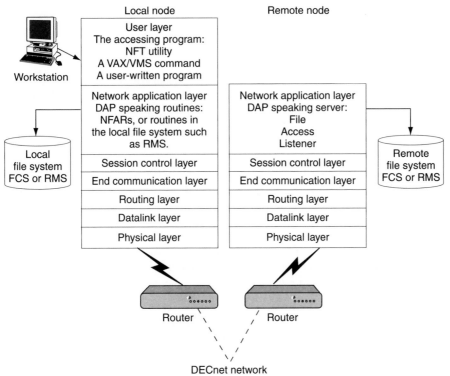

FIGURE 9.23
File transfer across a network.

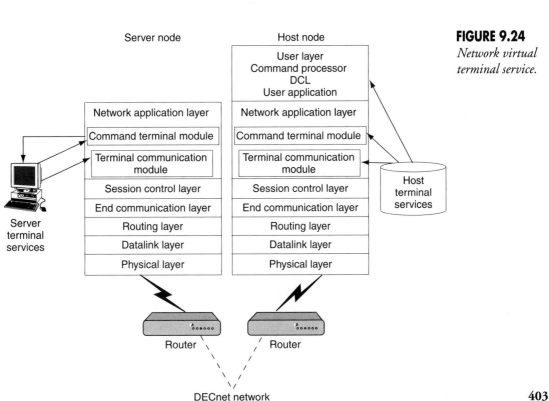

FIGURE 9.24
Network virtual terminal service.

The **Command Terminal Protocol** offers a set of functions oriented toward command line input.

The Command Terminal Protocol (CTP)

The *Command Terminal Protocol (CTP)* is the second sublayer of NVT. It offers a set of functions mainly oriented toward command line input. After a connection between a host and network terminal is made, it is this layer that provides control over the network.

Operation of NVT

With a terminal session between a terminal and a host the active terminal will request a binding to the destination host. This is illustrated in Figure 9.24. The terminal management module in the server system requests a binding to the host system. To accomplish this it will invoke the services of the terminal communication services function. Most DECnet users will know this as the SET HOST < Hostname> command entered at the terminal. The SET HOST command will allow a user at one host to connect to another host over the network. That is, a terminal connected to host A may now connect to remote host B over the network.

The terminal communication service tries to connect to the remote system by requesting a logical link from the session control layer. At the destination host, the incoming logical link request is accepted and allocates a portal. This is the beginning of the binding. The host module should accept the logical link. Once the logical link has been established, the Bind Request is sent and if accepted, a Bind Accept message will be sent back to the requester. This binding has now been established.

Now that the binding has been formed, the host will enter into command mode because the CTP will now interact on the connection. This action will take place in the first terminal request from the login process. Both ends of the connection will now enter into the command terminal protocol and will remain with this protocol until the end of the connection. Requests and responses will then take place as data are transferred across the link using the command terminal protocol.

The application program in the host system will send terminal service requests to the host operating system terminal services. The host terminal services issue corresponding requests to the host protocol module of a logout of the user from the host. Host terminal services issue corresponding requests to the host protocol module. The server protocol module reproduces those requests remotely and reissues them to the server terminal services.

Termination of this link usually comes from the application program running in the host system. This will happen by a logout by the host application. With this, the host terminal communication services will send an Unbind request. The server will respond by releasing the link also. Finally, the host disconnects the logical link. Figure 9.25 shows the highest layer of the DNA model, which is the user layer. This layer consists of the users or user applications that were written or are being used in a networked environment. This could be specifically written network management, database, or a program that is not part of the DNA model but uses DNA as part of its operation. Finally, Figure 9.26 shows the network management layer used by DNA. It will not be explained further here, but the figure shows the management functions of DNA.

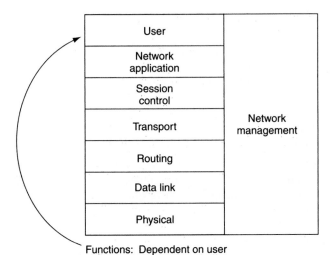

FIGURE 9.25
*Functions of the
user layer.*

Functions: Dependent on user

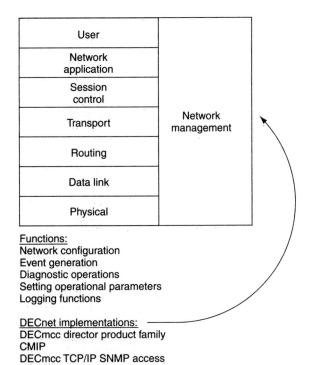

FIGURE 9.26
*Functions of the
network manage-
ment layer.*

Functions:
Network configuration
Event generation
Diagnostic operations
Setting operational parameters
Logging functions

DECnet implementations:
DECmcc director product family
CMIP
DECmcc TCP/IP SNMP access
DECmcc extended LAN manager
Token VIEW plus

Summary

In this chapter the history and backdrop of DECnet architecture were examined. The relationship between DECnet and OSI was also discussed. You learned about the routing layer and how routing and addressing are carried out in a DECnet network. You also learned about the various layers that comprise the DECnet architecture and the protocols that comprise each layer. Although DECnet is not widely used today in the corporate world, there are still companies that incorporate this legacy system in their network architecture. Some of the concepts of the DECnet architecture that have been incorporated into today's networking architectures include shortest path first and the concept of designated routers and areas. Chapter 10 discusses the protocols that comprise the OSI protocol suite.

Review Questions

1. When was DECnet Phase I introduced?
2. What is the most popular release of the DECnet architecture and why?
3. Define the following terms:
   a. adjacency
   b. designated router
   c. hop
   d. end node
   e. path length
4. What are the functions of the physical layer and what are its DECnet implementations?
5. What are the functions of the data link layer and what are its DECnet implementations?
6. What are the functions of the network layer and what are its DECnet implementations?
7. How is the DECnet address comprised?
8. What is an area?
9. How does a DECnet router forward packets?
10. What are the packet types used on Ethernet broadcast networks?
11. What are the packets that are transmitted on serial lines or nonbroadcast networks?
12. What are the five processes of the routing control sublayer?
13. What are the level 1 routing table descriptions?
14. What are the level 2 routing table descriptions?
15. What are the functions of the end communications layer and what are its DECnet implementations?
16. What is NSP and what are the functions provided by NSP?
17. What are the three types of NSP messages?
18. How does NSP provide flow control?

19. What are the functions of the session control layer and what are its DECnet implementations?
20. What are the five actions that divide the functions of the session control layer?

Summary Questions

1. What are the functions and DECnet implementations of the network application layer?
2. What is DAP and what are the functions of the DAP?
3. What is NVT and how does it operate?
4. What are the two sublayers of the NVT?
5. What are the functions and DECnet implementations of the network management layer?
6. What are the two types of channels that provide the exchange of information data?
7. What type of routing does DECnet perform?
8. What are the two types of nodes that comprise the routing database?
9. What are the address constraints in DECnet routing?
10. How many nodes are supported by DECnet Phase III?

Open Systems Interconnection Protocols

Objectives

- Explain how the Open Systems Interconnection (OSI) protocol suite developed.
- Define and explain the OSI routing components.
- Explain the routing algorithm used by the OSI protocol suite.
- Describe the method of addressing used by the OSI protocol suite.
- Explain how the transport and session layers work with each other to ensure connection establishment.

Key Terms

Introduction

The OSI protocol encompasses all seven layers of the OSI model, which encompasses many different protocols. At the data link layer, the protocol has been established to operate over HDLC, X.25, FDDI, IEEE 802.3, IEEE 802.4, and IEEE 802.5. The OSI architecture differs from other protocols starting at the network layer, which is where this chapter begins its discussion.

The intent of this chapter is to give the reader an introduction to the basic concepts and terminology of the OSI protocol suite. It will cover at a simplistic level the routing, transport, and application layers of OSI. The specifications listed on the following pages will give more information on this topic.

The Open Systems Interconnection (OSI) protocol suite is comprised of numerous standard protocols that are based on the OSI reference model. These protocols are part of an international program to develop data networking protocols and other standards that facilitate multivendor equipment interoperability. The OSI program grew out of the need for international networking standards and is designed to facilitate communication between hardware and software systems despite differences in underlying architectures.

All OSI standards and other international standards are generated by the International Standards Organization (ISO), the International Telecommunications Union (ITU), the Institute of Electrical and Electronics Engineers (IEEE), and other various committees. An International Standards document is prefixed by the letters ISO followed by a code number that indicates its origin such as (ISOxxxx). The first proposal generated is called a *Draft Proposal (DP)*. Next, the DP will become an *Agreed Draft Standard (DIS)*, but the code number will always remain the same. As agenda are published they will have the designations *PDAD*, *DAD*, and *AD*—again the code number will remain the same.

The ITU-T is headquartered in Geneva, Switzerland, and the ITU reports to the United Nations Organization. The principal members of the ITU are the *Post, Telegraph, and Telephone* authorities *(PTT)* of the member countries. The ITU is responsible for the wide area aspects of national and international communications, publishing quadrennial recommendations. Each new book publishes recommendations. The series prefixed by X relate to data network services. The data following the number, for example X.25 (1980), refers to the edition recommendation or update.

The following are the important ISO standards and specifications for the OSI protocol.

The **Draft Proposal** is an ISO standards document that is the first proposal generated.

An **Agreed Draft Standard** is a development step that represents near final status on a specification.

The **Post, Telegraph, and Telephone** authorities are the principal members of the ITU.

Physical Layer

ITU-T X.21:	15-pin physical connection specification for circuit-switched networks.
ITU-T X.21 bis:	25-pin connection similar to EIA/TIA RS-232-C.

Data Link layer

ISO 4335/7809:	High level data link control specifications (HDLC).
ISO 8802.2:	Local area logical link control (LLC) specifications.
ISO 8802.3:	(IEEE 802.3) Ethernet standard.
ISO 8802.4:	(IEEE 802.4) token bus standard.
ISO 8802.5:	(IEEE 802.5) Token Ring standard.

Network Layer

ISO 8473:	Network layer protocol and addressing specification for connectionless network service.
ISO 8208:	Network layer protocol specification for connection oriented service based on ITU-T X.25 specifications.
ITU-T X.25:	Specifications for connecting data terminal equipment to packet-switched networks.
ITU-T X.21:	Specifications for accessing circuit-switched networks.

Transport Layer

ISO 8072:	OSI transport layer service definitions.
ISO 8073:	OSI transport layer protocol specifications.

Session Layer

ISO 8326:	OSI session layer service definitions, including transport classes 0, 1, 2, 3, and 4.
ISO 8327:	OSI session layer protocol specifications.

Presentation Layer

ISO 8822/23/24:	Presentation layer specifications.
ISO 8649/8650:	Common-Application and Service Elements (CASE) specifications and protocols.

Application Layer

X.400:	OSI application layer specification for electronic message handling (electronic mail).
FTAM:	OSI application layer specification for file transfer and access method.
VTP:	OSI application layer specification for virtual terminal protocol.
JTM:	Job transfer and manipulation standard—similar to a *remote job entry* (RJE) function.

A **remote job entry** is an application that is batch oriented as opposed to interactive.

OSI Networking Protocols

Figure 10.1 illustrates the entire OSI protocol suite and its relationship to the layers of the OSI reference model. Each component is discussed briefly in this chapter.

OSI Physical and Data Link Layers

The OSI protocol suite supports numerous standard media access protocols at the physical and data link layers. The wide variety of media access protocols supported in the OSI protocol suite allows other protocol suites to exist seamlessly alongside OSI on the same network media. Supported media access protocols include IEEE 802.2 LLC, IEEE 802.3, Token Ring/IEEE 802.5, Fiber Distributed Data Interface (FDDI), and X.25.

OSI Network Layer

The OSI protocol suite specifies two routing protocols at the network layer: End System-to-Intermediate System (ES-IS) and Intermediate System-to-Intermediate

OSI reference model

OSI protocol suite

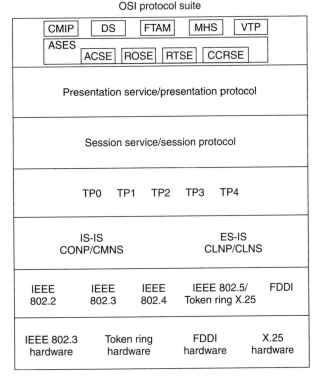

FIGURE 10.1

The OSI protocol suite maps to all layers of the OSI reference model. (Reproduced with permission of Cisco Systems, Inc. Copyright © 2001 Cisco Systems, Inc. All rights reserved.)

System (IS-IS). In addition, the OSI suite implements two types of network services: connectionless service and connection oriented service.

OSI Connectionless Network Service

OSI connectionless network service is implemented by using the *Connectionless Network Protocol (CLNP)* and *Connectionless Network Service (CLNS)*. CLNP and CLNS are described in the ISO 8473 standard.

CLNP is an OSI network layer protocol that carries upper layer data and error indications over connectionless links. CLNP provides the interface between the CLNS and upper layers.

In addition, CLNS provides best-effort delivery, which means that no guarantee exists that data will be lost, corrupted, misordered, or duplicated. CLNS relies on transport layer protocols to perform error detection and correction.

Connectionless Network Protocol is a network layer protocol that carries upper layer data over connectionless links.

Connectionless Network Service relies on transport layer protocols to perform error detection and correction.

OSI Connection Oriented Network Service

Connection Oriented Network Protocol carries upper layer data and error indications over connection oriented links.

OSI connection oriented network service is implemented by using the *Connection Oriented Network Protocol (CONP)* and *Connection Mode Network Service (CMNS)*.

CONP is an OSI network layer protocol that carries upper layer data and error indications over connection oriented links. CONP is based on the X.25 Packet-Layer Protocol (PLP) and is described in the ISO 8208 standard, X.25 Packet-Layer Protocol for DTE. CONP provides the interface between CMNS and the upper layers. It is a network layer service that acts as the interface between the transport layer and CONP and is described in the ISO 8878 standard.

Connection Mode Network Service establishes paths between communicating transport layer entities.

CMNS performs functions related to the explicit establishment of paths between communicating transport layer entities. These functions include connection setup, maintenance, and termination. CMNS also provides a mechanism for requesting a specific quality of service (QoS). This contrasts with CLNS.

Network Layer Addressing

OSI network layer addressing is implemented by using two types of hierarchical addresses: network service access point addresses and network entity titles.

A **network service access point** is a conceptual point on the boundary between the network and the transport layers.

A *network service access point (NSAP)* is a conceptual point on the boundary between the network and the transport layers. The NSAP is the location at which OSI network services are provided to the transport layer. Each transport layer entity is assigned a single NSAP, which are individual addresses in an OSI internetwork using NSAP addresses.

Figure 10.2 illustrates the format of the OSI NSAP address, which identifies individual NSAPs.

FIGURE 10.2

The OSI NSAP address is assigned to each transport layer entity. (Reproduced with permission of Cisco Systems, Inc. Copyright © 2001 Cisco Systems, Inc. All rights reserved.)

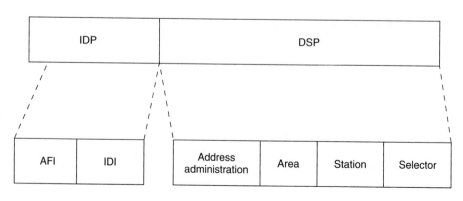

Two NSAP address fields exist: the *Initial Domain Part (IDP)* and the *Domain Specific Part (DSP)*. The IDP field is divided into two parts: the *Authority Format Identifier (AFI)* and the *Initial Domain Identifier (IDI)*. The AFI provides information about the structure and content of the IDI and the DSP fields such as whether the IDI is of variable length and whether the DSP uses decimal or binary notation. The IDI specifies the entity that can assign values to the DSP portion of the NSAP address.

The DSP is subdivided into four parts by the authority responsible for its administration. The Address Administration fields allow the further administration of addressing by adding a second authority identifier and by delegating address information to subauthorities. The Area field identifies the specific area within a domain and is used for routing purposes. The Station field identifies a specific station within an area and also is used for routing purposes. The Selector field provides the specific n-selector within a station and, much like the other fields, is used for routing purposes. The reserved n-selector 00 identifies the address as a network entity title (NET).

End-System NSAPs

An OSI end system (ES) often has multiple NSAP addresses, one for each transport entity that it contains. If this is the case, the NSAP address for each transport entity usually differs only in the last byte called the n-selector. Figure 10.3 illustrates the relationship between a transport entity, the NSAP, and the network service.

The **Initial Domain Part** contains an authority format identifier and a domain identifier.

The **Domain Specific Part** contains an area identifier, a station identifier, and a selector byte.

The **Authority Format Identifier** provides information about the structure and content of the IDI and the DSP fields.

The **Initial Domain Identifier** specifies the entity that can assign values to the DSP portion of the NSAP address.

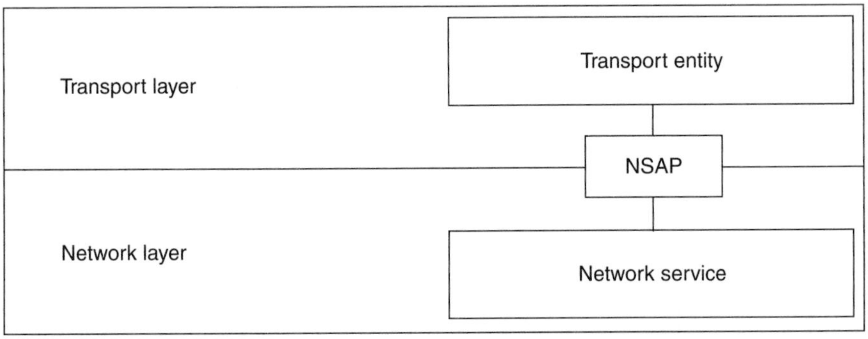

FIGURE 10.3

The NSAP provides a linkage between a transport entity and a network service.
(Reproduced with permission of Cisco Systems, Inc. Copyright © 2001 Cisco Systems, Inc. All rights reserved.)

The **network entity title** is a network address used in CLNS based networks.

A network entity title (NET) is used to identify the network layer of a system without associating that system with a specific transport-layer entity as an NSAP address does. NETs are useful for addressing intermediate systems (ISs), such as routers that do not interface with the transport layer. An IS can have a single NET or multiple NETs if it participates in multiple areas or domains.

OSI Protocols Transport Layer

The OSI protocol suite implements two types of services at the transport layer: connection oriented transport service and connectionless transport service.

Five connection oriented transport layer protocols exist in the OSI suite, ranging from Transport Protocol Class 0 through Transport Protocol Class 4. Connectionless transport service is supported only by Transport Protocol Class 4.

Transport Protocol Class 0 (TP0): the simplest OSI transport protocol performs segmentation and reassembly functions. TP0 requires connection oriented network service.

Transport Protocol Class 0 performs segmentation and reassembly functions.

Transport Protocol Class 1 (TP1): performs segmentation and reassembly and offers basic error recovery. TP1 sequences protocol data units (PDUs) and will retransmit PDUs or reinitiate the connection if an excessive number of PDUs are unacknowledged. TP1 requires connection oriented network service.

Transport Protocol Class 1 offers basic error recovery.

Transport Protocol Class 2 (TP2): performs segmentation and reassembly, as well as multiplexing and demultiplexing data streams over a single virtual circuit. TP2 requires connection oriented network service.

Transport Protocol Class 2 multiplexes and demultiplexes data streams over a single virtual circuit.

Transport Protocol Class 3 (TP3): offers basic error recovery and performs segmentation and reassembly, in addition to multiplexing and demultiplexing data streams over a single virtual circuit. TP3 also sequences PDUs and retransmits them or reinitiates the connection if an excessive number are unacknowledged. TP3 requires connection oriented network service.

Transport Protocol Class 3 offers basic error recovery and performs segmentation and reassembly.

Transport Protocol Class 4 (TP4): TP4 offers basic error recovery, performs segmentation and reassembly, and supplies multiplexing and demultiplexing of data streams over a single virtual circuit. TP4 provides reliable transport service and functions with either connection oriented or connectionless network service. It is based on the Transmission Control Protocol (TCP) in the Internet Protocols suite and is the only OSI protocol class that supports connectionless network service.

Transport Protocol Class 4 provides reliable transport service and functions.

OSI Protocols Session Layer

The session layer implementation of the OSI protocol suite consists of a session protocol and a session service. The session protocol allows session service users (SS-users) to communicate with the session service. An SS-user is an entity that requests the services of the session layer. Such requests are made at Session Service Access Points (SSAPs), and

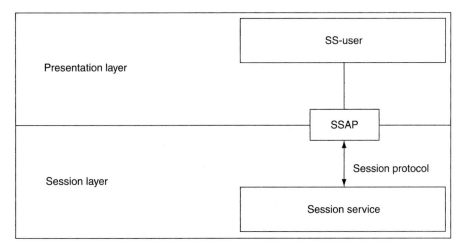

SS-users are uniquely identified by using an SSAP address. Figure 10.4 shows the relationship between the SS-user, the SSAP, the session protocol, and the session service.

Session service provides four basic services to SS-users. First, it establishes and terminates connections between SS-users and synchronizes the data exchange between
them. Second, it performs various negotiations for the use of session layer tokens,
which must be possessed by the SS-user to begin communicating. Third, it inserts
synchronization points in transmitted data that allow the session to be recovered in
the event of errors or interruptions. Finally, it allows SS-users to interrupt a session
and resume the session at a specific point.

Session service is defined in the ISO 8326 standard and in the ITU-T X.215 recommendation. The session protocol is defined in the ISO 8327 standard and in the
ITU-T X.225 recommendation. A connectionless version of the session protocol is
specified in the ISO 9548 standard.

OSI Protocols Presentation Layer

The presentation layer implementation of the OSI protocol suite consists of a presentation protocol and a presentation service. The presentation protocol allows
presentation service users (PS-users) to communicate with the presentation service.
A PS-user is an entity that requests the services of the presentation layer. Such requests
are made at Presentation Service Access Points (PSAPs). PS-users are uniquely identified by using PSAP addresses.

Presentation service negotiates transfer syntax and translates data to and from the transfer syntax for PS-users, which represent data using different syntaxes. The presentation

service is used by two PS-users to agree upon the transfer syntax that will be used. When a transfer syntax is agreed on, presentation service entities must translate the data from the PS-user to the correct transfer syntax.

The OSI presentation layer service is defined in the ISO 8822 standard and in the ITU-T X.216 recommendation. The OSI presentation protocol is defined in the ISO 8823 standard and in the ITU-T X.226 recommendation. A connectionless version of the presentation protocol is specified in the ISO 9576 standard.

OSI Protocols Application Layer

The application layer implementation of the OSI protocol suite consists of various application entities. An application entity is composed of the user element and the *application service element (ASE)*.

The **application service element** provides interfaces to the lower OSI layers.

The user element is the part of an application entity that provides services to user elements and, therefore, to application processes. ASEs also provide interfaces to the lower OSI layers. Figure 10.5 portrays the composition of a single application process

FIGURE 10.5

An application process relies on the PSAP and presentation service. (Reproduced with permission of Cisco Systems, Inc. Copyright © 2001 Cisco Systems, Inc. All rights reserved.)

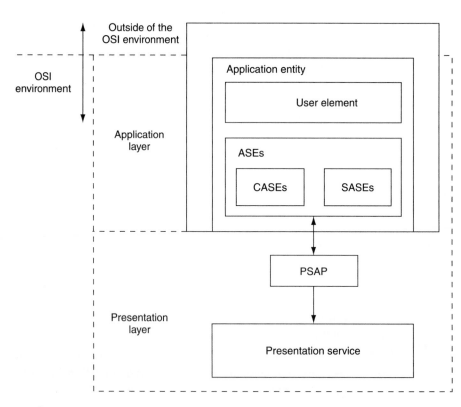

composed of the application entity, the user element, and the ASEs as well as its relationship to the PSAP and presentation service.

ASEs fall into one of the following classifications: Common-Application Service Elements (CASEs) and Specific-Application Service Elements (SASEs). Both of these might be present in a single application entity.

Common-Application Service Elements (CASEs)

Common-Application Service Elements (CASEs) are ASEs that provide services used by a wide variety of application processes. In many cases, multiple CASEs are used by a single application entity. The following four CASEs are defined in the OSI specification:

- Association Control Service Element (ACSE): Creates associations between two application entities in preparation for application-to-application communication.
- Remote Operations Service Element (ROSE): Implements a request-reply mechanism that permits various remote operations across an application association established by the ACSE.
- Reliable Transfer Service Element (RTSE): Allows ASEs to reliably transfer messages while preserving the transparency of complex lower layer facilities.
- Commitment, Concurrence, and Recovery Service Elements (CCRSE): Coordinates dialogues between multiple application entities.

Common-Application Service Elements provide services used by a wide variety of application processes.

Specific-Application Service Elements (SASEs)

Specific-Application Service Elements (SASEs) are ASEs that provide services used only by a specific application process such as file transfer, database access, and order-entry, among others.

Specific-Application Service Elements provide services used only by a specific application process.

OSI Protocols Application Processes

An application process is the element of an application that provides the interface between the application itself and the OSI application layer. Some of the standard OSI application processes include the following:

- Common Management Information Protocol (CMIP): Performs network management functions, allowing the exchange of management information between ESs and management stations. CMIP is specified in the ITU-T X.700 recommendation and is functionally similar to the Simple Network Management Protocol (SNMP).
- Directory Services (DS): Serves as a distributed directory that is used for node identification and addressing in OSI internetworks. DS is specified in the ITU-T X.500 recommendation.

- *File Transfer, Access, and Management (FTAM)*: Provides file transfer service and distributed file access facilities.
- Message Handling System (MHS): Provides a transport mechanism for electronic messaging applications and other applications by using store-and-forward services.
- Virtual Terminal Protocol (VTP): Provides terminal emulation that allows a computer system to appear to a remote ES as if it were a directly attached terminal.

Open Systems Interconnection (OSI) Routing Protocol

Background

The International Standards Organization (ISO) (it is known as both) developed a complete suite of routing protocols for use in the OSI protocol suite. These include Intermediate System-to-Intermediate System (IS-IS), End System-to-Intermediate System (ES-IS), and Interdomain Routing Protocol (IDRP).

IS-IS is based on work originally done at Digital Equipment Corporation for DECnet/OSI (DECnet Phase V). IS-IS originally was developed to route in ISO Connectionless Network Protocol networks. A version has since been created that supports both CLNP and Internet Protocol (IP) networks. This version is referred to as Integrated IS-IS and it has also been referenced as Dual IS-IS.

OSI routing protocols are summarized in several ISO documents including ISO 10589, which defines IS-IS. The American National Standards Institute (ANSI) X3S3.3 network and transport layers committee was the motivating force behind ISO standardization of IS-IS. Other ISO documents include ISO 9542, which defines ES-IS, and ISO 10747, which defines IDRP.

OSI Routing

There are five major components of the OSI routing architecture:

1. *End system (ES)*: An end system is considered to be some type of individual computing device such as a personal computer, a host system, or minicomputer. End systems usually need to know only the destination and the intermediate system to which they can transmit a message in order to have it forwarded to its final destination. These are similar to nodes in the DECnet hierarchical routing system.

2. *Intermediate system (IS)*: An intermediate system is usually a router. A router is an intelligent device that forwards data through a best path to reach its final destination. A router will submit packets directly to a node if that is where the destination is or directly to another router in the path to the final destination.

   Intermediate systems differ from the end systems in that they must contain the intelligence to route packets. They also must obtain information about the network on which they reside. This information consists of the various paths that make up the entire network to which they are connected.

3. *Routing domain (RD)*: A group of intermediate systems that operate a particular routing protocol such that the schematic of the overall network is the same for each IS. These domains are represented in two forms: hierarchical or flat.

4. *Hierarchical domain* is a domain that has more than one level. IS-IS specifically has two levels. These levels are known as Level 1 and Level 2.

5. *Flat domains* contain one level. There is now a distinction between ISs in this type of domain.

One other type of domain is called the *administrative domain*. This type of domain is structured so that each domain is a group of ISs that are owned by the same organization and therefore are administered as a single entity. This administration is simple in that it involves picking a protocol and defining the interfaces between the domains. The two domains—routing and administrative—are viewed as a single entity and the two are commonly known as the domain.

The routing algorithm used with OSI is called a link state algorithm (LSA). The LSA commonly associated with the TCP/IP protocol is Open Shortest Path First (OSPF). The link state algorithm is vastly different from the more common routing algorithms that are used with most protocols today. This algorithm is referred to as a distance vector algorithm (DVA) and it is found on TCP/IP, XNS, AppleTalk, IPX (Novell), and other protocols. Link state algorithms offer many benefits in contrast to the distance vector algorithm. Some of these advantages are faster convergence time by reducing the amount of time that routing loops may form and lower bandwidth consumption because only bad links are reported and not the whole routing table.

When routers were introduced to the commercial data communication marketplace, they used a static method of updating themselves. This meant that the routing tables had to be updated manually. The first protocol that allowed dynamic routing table updates was known as the Routing Information Protocol (RIP), and it is based on a distance vector algorithm. Distance vector algorithms produce routing tables based on geographical distance (the metric). If a destination is four hops away, the packet to be forwarded must traverse four routers before reaching the destination. This metric is key to the distance vector algorithm structure and operation. This metric could also be based on line bandwidth, line speed, and line delay. The distance vector algorithm is very simple and easy to implement. Most router implementations of distance vector algorithms simply allow the network administrator to turn on the protocol and allow the algorithm to determine the routes on the network. Because the distance vector algorithm's routing tables are based on the single distance metric, the algorithm is said to be flat, meaning it has only one logical peer.

Distance vector algorithms can be placed on a hierarchical level, but this requires manual configuration. Network administrators must manually place topological information into the routers, information such as how the routers are connected to one another and the status of each link. Network administrators must think about their

A **routing domain** is represented in two forms: hierarchical or flat.

A **hierarchical domain** has more than one level.

A **flat domain** contains one level.

The **administrative domain** is a group of domains administered as a single entity.

network in terms of line speed and so forth in order to place policy based routing into their network routers. This means that certain routers are trusted routers and they will update other routers on the network. However, this is all manual configuration.

When the routing protocol RIP is used, the protocol of TCP/IP usually comes to mind. RIP was developed by Xerox Network Systems (XNS). The protocol was placed in the public domain and was released by the UNIX operating system known as Berkeley UNIX. The wide use of the Berkeley UNIX operating system gave RIP its popularity. This was the first dynamic routing update protocol that allowed network administrators to simply assign metrics (hops) to an interface and let the RIP protocol handle the rest. No longer did network administrators have to statically enter a network number and the number of hops the routers had to traverse. RIP would produce this for you.

This had many advantages. Think of a network that involved 50 routers, which is not uncommon. When one router went down for any reason, all the routers had to mark any paths that used that router as being down. In a static environment, this meant going to each router and manually entering this into the routing table. With an up-date protocol, this path would be automatically marked as being down on the next routing update.

RIP was a breakthrough but it also brought many problems, with convergence and scalability being the largest. When a route goes down, it takes a certain amount of time before all the routers can be notified of this. Routing updates are sent from each router, but they are not synchronized. This means that routers send their routing up-dates at various times depending on when their timers expired.

Link state algorithms overcome the convergence and scalability problems experienced with distance vector algorithms. When a router is disabled or a new router is added to the network, having one router as the trusted router will detect the change and it will inform all the other routers of this change. In a link state algorithm internet, only one router will be given this responsibility. In distance vector algorithm internets, all routers are considered trusted and therefore all routers will eventually update each other (see Figure 10.6).

By dividing an internet into areas, loops are avoided while still allowing for multiple paths and, therefore, redundancy. The area process also helps the routing table up-dates when a change occurs. With a hierarchical internet, a change will affect only the particular nodes that are in that area. The updates are not passed to those nodes that are not considered to be trusted. That is, those routers that do not need to know about a change will not be updated, and the network will still operate efficiently. With this type of network, overhead is drastically reduced.

OSI Networking Terminology

The world of OSI networking uses some specific terminology such as end system, which refers to any nonrouting network nodes, and intermediate system, which refers to a router. These terms form the basis of the ES-IS and IS-IS OSI protocols. The ES-

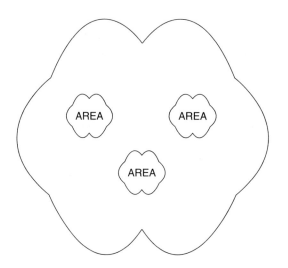

IS protocol enables ESs and ISs to discover each other. The IS-IS protocol provides routing between ISs. Other important OSI networking terms include area, domain, Level 1 routing, and Level 2 routing. An area is a group of contiguous networks and attached hosts that is specified to be an area by a network administrator or manager. A *domain* is a collection of connected area. Routing domains provide full connectivity to all end systems within them. Level 1 routing is routing within a Level 1 area, while Level 2 routing is routing between Level 1 areas.

A **domain** is a collection of connected area.

IS-IS Protocol

The IS-IS routing protocol is the routing algorithm used for OSI, and it is based on the link state algorithm architecture. IS-IS supports a two-level hierarchical structure within a domain (see Figures 10.7 and 10.8).

Routers are divided into Level 1 and Level 2 routers. Level 1 routers provide routing within an area. This type of routing consists of communication between ESs and Level 1 ISs, between Level 1 ISs, and between Level 1 and Level 2 ISs. Level 2 ISs, as a group, represent the backbone network. The backbone area provides routing between the areas, but within one domain.

As Figure 10.8 shows, the internet is divided into areas. This has many advantages. First, information about changes in the internet that affect only one nonbackbone area is limited to that area and to the Level 2 routers that are connected to the Level 1 routers in that area. This information will not be transmitted to any other routers on the internet.

This provides faster convergence because only the routers that need to know about a change are updated. It also provides fewer lost sessions and better response times. Because link state algorithm architected internets converge faster, bandwidth is better utilized on the whole internet.

FIGURE 10.7

Areas, routers, and end nodes. (Reproduced with permission of Cisco Systems, Inc. Copyright © 2001 Cisco Systems, Inc. All rights reserved.)

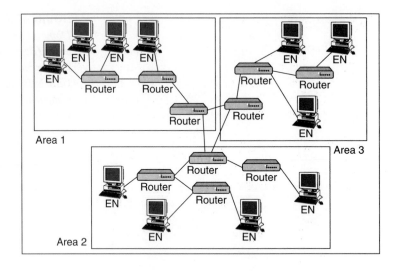

FIGURE 10.8

Level 1 and Level 2 routers. (Reproduced with permission of Cisco Systems, Inc. Copyright © 2001 Cisco Systems, Inc. All rights reserved.)

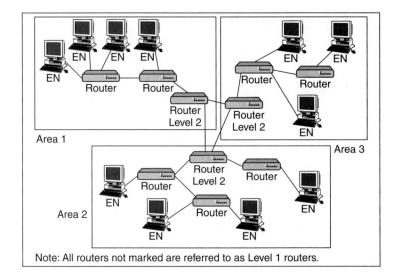

Each IS on a link state algorithm internet will contain an image of the whole internet topology. By contrasting this image with other ISs in the area, discrepancies can be resolved quickly. Also, link state algorithm routers usually implement some type of authentication procedure that is designed to ensure that only ISs that have been properly configured and are supposed to be operating on the network can gain access to the network.

OSI Addressing

The OSI addressing scheme is one of the most complex addressing schemes when compared with the other protocols. Instead of identifying a host with a number, the address can be variable in length and contain many parts. The address scheme was designed for virtually every different type of data communications environment possible. The basic unit of the OSI address is known as the NSAP.

The address scheme is truly global and hierarchical. It is hierarchical in that there are multiple authorities. The highest authority is the standards body known as the ISO, which can create up to 90 authorities. Each of these 90 authorities can create thousands of subauthorities. These subauthorities may, in turn, create their own subauthorities. Currently there are seven authorities recognized by ISO:

- Local: A local OSI network. A network that is not attached to any other public OSI internetwork.
- X.121: An authority that uses the X.25 addressing scheme.
- IO DCC: Consists of subnetworks corresponding to countries or a sponsored country (one not yet participating in ISO).
- F.69: Telex authority.
- E.163: Public switched telephone networks.
- E.164: ISDN
- ISO 6523-ICD: International Code Designator, a four-digit code according to the ISO 6523 standard.

Recall from Figure 10.2 that the OSI addressing scheme consists of four parts:

1. IDP: Contains the AFI and the IDI.
2. AFI: Contains a two-digit value between 0 and 99. This is used to describe the IDI format and the syntax of the DSP. In addition, the AFI identifies the authority responsible for allocating the values of the IDI. Table 10.1 contains a comparison of OSI addressing and TCP/IP addressing.
3. IDI: The IDI specifies the addressing domain. Table 10.2 contains a summary of the IDI formats and contents.

Table 10.1 Summary of IDI Formats and Contents

	Address Length	*Address Range*	*Address Format*
TCP/IP	32 bits	2^{30} nodes	network number.host
OSI	16-160 bits	2^{152} nodes	AFI/IDI/DSP

Table 10.2 The OSI Address

IDP		DSP			
AFI	*IDI*	*Organization ID*	*Subnet ID*	*MAC address*	*N-Sel*
47	0005	0032	1234	53184427	1

4. DSP: The DSP contains the address determined by the network authority. It is an address below the second level of the addressing hierarchy. It can contain addresses of end user systems on an individual network. The DSP identifies the NSAP to the final subnetwork point of attachment (SNPA). The SNPA is the point at which the network forwards the data to the underlying network for delivery to the network node.

Domains

In an OSI internetwork, the network is split into domains. Just as in TCP/IP, a large internet may be split into subnets. In OSI, domains are used to administer the large network address space. Also, in OSI subnetworks are used to identify a physical network. For example, Ethernet and Token Ring are considered subnetworks in the OSI scheme. The domains are areas of an internetwork that are administered by an addressing authority. The addressing authority ensures that all addresses within the domain are unique. A domain is allowed to cover multiple subnetworks.

●

The **Government OSI Profile** is a U.S. government specification for OSI protocols.

An example would be the U.S. government use of this addressing format for its implementation of OSI known as *Government OSI Profile (GOSIP)*. The AFI value will be 47 in hexadecimal. The IDI value will be 0005. This is the value that is established by the National Institute for Standards and Technology (NIST). The IDI value of 0005 signifies that the address was assigned by the addressing authority of NIST. This is for the entire federal government. The IDI value of 0006 is for the Department of Defense (DoD).

The DSP field will be divided into four subfields. The ORG ID is a value assigned by NIST to signify an individual government organization. The SUBNET ID is used by the organization to identify a subnetwork within the organization's subdomain. The END SYSTEM ID can be used in any way that the subnetwork administrator decides. More than likely, this will be the 48-bit MAC address that is the Ethernet or Token Ring physical address, but it does not have to be. The NSAP Selector (N-Sel) subfield identifies a higher layer transport entity. For example, a value of 1 in this field identifies the ISO Transport Protocol.

A publication produced by NIST gives the following address as an example:

```
47 00 05 00 32 12 34 53 18 44 27 01
```

which represents:

$$AFI = 47$$
$$IDI = 0005$$
$$ORG\ ID = 0032$$
$$SUBNET\ ID = 1234$$
$$END\ SYSTEM\ ID = 53184427$$
$$NSAP\ Selector = 1$$

This address is illustrated in Table 10.2. Figures 10.9 and 10.10 show the network layer routing service.

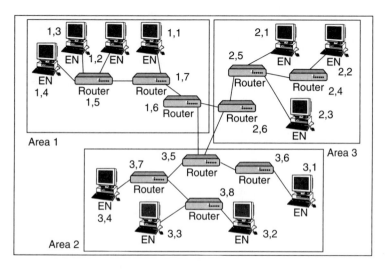

FIGURE 10.9

Routing opera-tions within and between areas. (Reproduced with permission of Cisco Systems, Inc. Copyright © 2001 Cisco Systems, Inc. All rights reserved.)

Entry in routing table at EN 3,3:
Destination = 1,3; next node = 3,8

Entry in routing table at R 3,8:
Destination = 1,3; next node = 3,7

Entry in routing table at R 3,7:
Destination = 1,3; next node = 3,5

Entry in routing table at R 3,5:
Destination = 1,3; next node = 3,8

Entry in routing table at R 1,6:
Destination = 1,3; next node = 1,6

Entry in routing table at R 1,7:
Destination = 1,3; next node = 1,5

Entry in routing table at R 1,5:
Destination = 1,3; directly attached

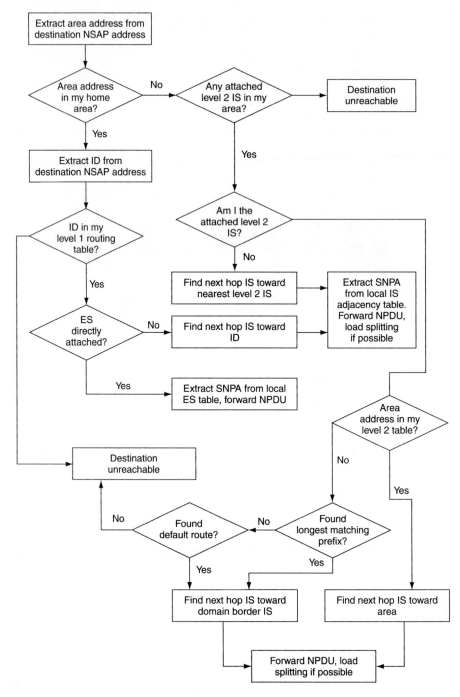

FIGURE 10.10

Flowchart illustrating OSI router's packet handling. (Reproduced with permission of Cisco Systems, Inc. Copyright © 2001 Cisco Systems, Inc. All rights reserved.)

Transport Layer: Connection Mode Transport Service

When a connection attempt is made at one end of the link and the other end of the link accepts the connection, a session is established. Messages are then sent at one end of this session and should be received at the other end of the session. Messages should be received in the same order they were sent. Any messages that are in error or are sent out of sequence should be re-sent. If this does not happen, the session will be terminated. The connection establishment is usually made by the transport layer receiving a request from the session layer. The transport layer will transmit a session request packet to the intended destination. During this phase the two stations will negotiate parameters, which will be supported, and any alternative classes. If the destination accepts this connection packet, it will transmit a connection confirm message to the originator. In this message the transport class that was selected from the incoming connection attempt packet will be specified.

After this is set up, data may flow across the session. The OSI transport layer supports full duplex communication, which means that each end may send and receive data simultaneously. There are special types of packets that transport data. The actual data transfer is accomplished by the network layer. As stated before, this layer may be *connection oriented network service (CONS)* or connectionless network service (CLNS). During this transfer data may be transmitted that bypass normal flow control procedures. Transport Classes 2 and 4 support this while Class 0 does not. During the transfer of data, Class 2 and Class 4 assign send sequence numbers to the packets. This sequence assignment operation is like most transport layer protocols in that the receiver will send an acknowledgment to the sender, which is dependent on the received sequence number. This acknowledgment number indicates exactly how many sequence numbers have been successfully received. Class 0 depends on the network layer to provide this type of reliability. These sequence numbers indicate in which order the transmitter sent the data.

To provide for flow control on the circuit, the OSI transport protocol uses a system called credits. The credits determine how many messages the destination is able to accept. A source station can only transmit data up to the amount of the credits. The destination, upon receiving the data, may then return credits to the sender. The sender may then send more data to the destination. If it does not receive any more credits, it cannot send any more data. It must wait. There are many methods that can be used in this scheme. The protocol can be relaxed to state that even if we have only enough room to accept one message, the destination may send more credits back to the source in anticipation that memory will be freed up by the time the next series of messages arrives. Other systems will adhere strictly to the credit scheme in allotting only those credits for which there is room.

Class 4 can use two types of flow control fields. These are the normal and the extended flow control fields. In the *normal flow control* there are 7-bit sequence numbers and 4-bit credit fields. *Extended flow control* has 31-bit sequence fields and 16-bit

● **Connection oriented network service** is an OSI protocol for packet-switched networks.

● The **normal flow control** has 7-bit sequence numbers and 4-bit credit fields.

● The **extended flow control** has 31-bit sequence fields and 16-bit credit fields.

credit fields. The reason for this is that in a high speed network such as FDDI, there lies the possibility that a sequence number may wrap to the starting sequence number before a previous sequence number has left the internet. This would allow two different packets, both with the same sequence number, to be on the network.

One of the unique things placed in the OSI transport layer Class 4 is the congestion avoidance algorithm. The congestion avoidance algorithm allows a network station to indicate in a transmitted packet that the network station was experiencing network congestion. This works in conjunction with the network layer. It allows the network layer to inform the transport layer of congestion possibilities and allows the transport layer to adjust its flow control window to reduce the amount of packets on the network. When the congestion is deemed to be reduced, the flow control window can be increased. The transport layer can also reduce its flow control window when it determines that the network layer has lost a packet. In this case, the transport layer will reduce its window to 1, allowing only the retransmitted message to be sent. It will then increase the window by 1 each time a packet is acknowledged.

In contrast to other network protocols, the transport layer has the responsibility of fragmenting session layer data. If the transport layer receives a message that is too large to be transmitted, it will fragment the data segment into multiple packets and transmit the multiple packets. As stated before, Classes 2 and 4 allow multiple transport connections to one network connection. When a connection is established, the transport layer will assign a 16-bit connection number to the connection. Each side will assign its own connection numbers and inform the remote end of the connection.

When all data has been transmitted and the connection is to be terminated, either side may accomplish this. Normally, the user will decide that a connection is no longer needed and will disconnect the session. This is a simple process because the side that requests the disconnect transmits a disconnect request to the remote side of the connection. Upon receiving this disconnect request, the remote side will transmit a packet confirming the disconnection.

Session Layer

The session control layer is the layer that resides between the application and the transport service. It is the application interface to the network. It is responsible for organizing communication between two applications. It is also responsible for the management of the communication between the two applications. Connection control is concerned with session establishment, maintaining the session, and terminating the session. In doing this, the session layer works directly with the transport layer to ask for a connection to be established. The session layer provides many services beyond that of the connection control facilities found in the transport layer. When this layer receives an inbound connection request it will extract certain information such as access control information, which includes a password and validation of the password.

The session layer also provides for a proxy entity to be connected, in that a user on one system may access the services of another system without the account control information. This can be seen as a guest account. It will also determine whether the requested application exists on the node. If the application is found, it will issue an error message back to the originating node.

Data is sent to and from an application using the session layer. The session layer will provide the buffering for this data. The sending of data includes methods for supporting three types of data:

1. Message interface: Allows users to send and receive messages of any desired size.
2. Segment interface: Allows users to send and receive messages using messages of predetermined size. This size is usually set to the allowable transport data unit size.
3. Stream interface: Data is viewed as a stream of bytes with an end of message make inserted.

Another duty of the session layer is the monitoring of the connection. This entails monitoring of the transport connection and will force a disconnect if it is known that the transport layer has detected a probable network disconnect or when the transport layer does not receive a response to a connection attempt.

To disconnect a session, the application will request the release of a connection at the transport layer. Before this happens, all data will be transmitted and then the connection will be terminated. The connection can be immediately disconnected if the application layer requests a connection abort and not a connection disconnect.

Session layer address resolution provides name-to-address resolution. In this, the protocol will maintain cached tables that keep track of objects such as applications and nodes, which not only reside in the local station but also reside on the network.

Applications

File Transfer, Access and Management, and X.400

This is the equivalent of the FTP program in the protocol of TCP/IP. It is a file transfer mechanism that promotes a master-slave relationship. The station requesting the connection is called the initiator and the station accepting the connection is called the responder.

There are five types of documents that may be transferred:

- FTAM-1 unstructured text
- FTAM-2 sequential text
- FTAM-3 unstructured text
- FTAM-4 sequential binary
- FTAM-5 simple hierarchical file

Just as with the transport service, there are five classes defined by FTAM:

1. **Transfer class:** This class allows the transfer of a file or parts of a file using a simplex underlying protocol. This allows a user to copy a file from a remote station or to copy a file to the remote station. It can allow a user to move a file. This allows a user to copy a file, but the original is deleted after the move is successful.
2. **Management class:** This allows files to be manipulated in that they can be renamed, deleted, and their attributes can be read. In some cases, the file attributes may also be manipulated. These fields cannot be transferred across the network.
3. **Transfer and management class:** Combines all the services provided in the two previous classes.
4. **Access class:** Allows a user to read and write to the file.
5. **Unconstrained class:** Allows an OSI designed to choose the functions to be implemented. There are a vast amount of operations that may be performed with an FTAM implementation. Not all of these functions are going to be supported in all the vendor FTAM's implementations. Therefore, FTAM functions will interoperate between different user systems, but the user's interface may be different.

It is true that FTAM allows remote file access on a network. The standard mandates the underlying protocol. It defines the semantics of the protocols but, unfortunately, it does not define the programming interface. Because of this the user interface to this application may be different from system to system.

X.400

The **X.400** system is the messaging system used to exchange messages between two stations on the network.

The **user agent** collects the information found in the message and sends it to an MTA.

The **message transfer agent** transmits the message to the intended recipients indicated on the message header.

The last application to be discussed is probably the first OSI application that was commercially available even before the OSI standard was finalized. The *X.400* system is the messaging system used to exchange messages (mail) between two stations on the network. This is based on a store-and-forward method. This means that a message is created and then sent to a service that will store the message. In proper time, this service will attempt to deliver the message to the proper recipient (forward the message). There are two entities in this protocol. They are the user agent (UA) and the message transfer agent (MTA).

- *User agent (UA)*: This is how the users access the X.400 system. The UA collects the information found in the message and sends it to a message transfer agent (MTA).
- *Message transfer agent (MTA)*: This system transmits the message to the intended recipients indicated on the message header. This system also returns notifications to a message originator.

In order to deliver the message, the user agent must be available. There are times when a UA is not available and the message must be stored for a period of time. X.400 allows this in the message stores. It is here that a message will be stored until the recipient can accept the message. The message store acts on behalf of the user agent and the message is placed on an X.400 disk drive. This is usually located on the same system as the MTA that is serving that user agent. All messages for that user will be placed in the same location. At certain time intervals, the X.400 system will attempt to deliver the message. If, after a number of times, it cannot deliver the message, it will inform the originator that it could not deliver it.

The messages that a user may send are a simple format, similar to writing a letter. At the top of the message is a header that contains the user's name, the intended user's name, and miscellaneous items such as a carbon copy. The message is addressed using a naming scheme, which contains name (proper and surname), group name, and organization name. Based on this, a user's message is completed and will be routed to the destination. A message will be transferred from message agent to message agent until the final user agent is found. The message will then be delivered to the user agent for delivery to the final user.

The message transfer system can implement two types of receipt notifications. These are delivery and nondelivery. When a message has been deemed not deliverable, the MTS will generate a nondelivery message back to the originator. On the other hand, a message originator can indicate that when the user creates the message a return receipt is requested. Upon the destination receiving the message, a notification will be generated that it was delivered.

End System-to-Intermediate System (ES-IS)

End System-to-Intermediate System (ES-IS) is an OSI protocol that defines how end systems (hosts) and intermediate systems (routers) learn about each other, a process known as configuration. Configuration must happen before routing between ESs can occur.

ES-IS is more of a discovery protocol than a routing protocol. It distinguishes between three different types of subnetworks: point-to-point subnetworks, broadcast networks, and general topology subnetworks. Point-to-point subnetworks, such as WAN serial links, provide a point-to-point link between two systems. Broadcast subnetworks, such as Ethernet and IEEE 802.3, direct a single physical message to all nodes on the subnetwork. General topology subnetworks, such as X.25, support an arbitrary number of systems. Unlike broadcast subnetworks, however, the cost of an n-way transmission scales directly with the subnetwork size on a general topology subnetwork. Figure 10.11 illustrates the three types of subnetworks.

ES-IS Configuration

ES-IS configuration is the process whereby ESs and ISs discover each other's configuration that routing between ESs can occur. ES-IS configuration information is transmitted at regular intervals through two types of messages: ES hello messages

FIGURE 10.11

*ES-IS can be de-
ployed in point-to-
point, broadcast,
and general topol-
ogy subnetworks.
(Reproduced with
permission of
Cisco Systems, Inc.
Copyright ©
2001 Cisco Sys-
tems, Inc. All
rights reserved.)*

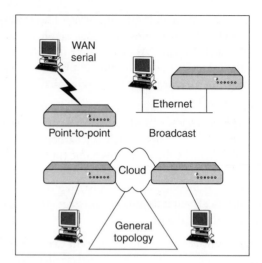

(ESHs) and IS hello messages (ISHs). ESHs are generated by ESs and sent to every
IS on the subnetwork. ISHs are generated by ISs and sent to all ESs on the subnetwork. These hello messages primarily are intended to convey the subnetwork and network layer addresses of the systems that generate them. Where possible, ES-IS
attempts to send configuration information simultaneously to many systems. On
broadcast subnetworks, ES-IS hello messages are sent to all ISs through a special multicast address that designates all end systems. When operating on a general topology
subnetwork, ES-IS generally does not transmit configuration information because of
the high cost of multicast transmissions.

ES-IS Addressing Information

The ES-IS configuration protocol conveys both OSI network layer and OSI subnetwork addresses. OSI network layer addresses identify either the NSAP, which is the
interface between OSI layer 3 and layer 4, or the NET, which is the network layer entity in an OSI IS. OSI subnetwork addresses or SNPAs are the points at which an ES
or IS is physically attached to a subnetwork. The SNPA address uniquely identifies
each system attached to the subnetwork. In an Ethernet network, for example, the
SNPA is the 48-bit Media Access Control (MAC) address. Part of the configuration
information transmitted by ES-IS is the NSAP-to-SNPA or NET-to-SNPA mapping.

OSI Routing Operation

Each ES lives in a particular area. OSI routing begins when the ESs discover the nearest IS by listening to ISH packets. When an ES wants to send a packet to another ES,
it sends the packet to one of the ISs on its directly attached network. The router then

looks up the destination address and forwards the packet along the best route. If the destination ES is on the same subnetwork, the local IS will know this from listening to ESHs and will forward the packet appropriately. The IS also might provide a redirect (RD) message to the source to tell it that a more direct route is available. If the destination address is an ES on another subnetwork in the same area, the IS will know the correct route and will forward the packet appropriately. If the destination address is an ES in another area, the Level 1 IS sends the packets to the nearest Level 2 IS. Forwarding through Level 2 ISs continues until the packets reach a Level 2 IS in the destination area. Within the destination area, ISs forward the packets along the best path until the destination ES is reached.

Link state update messages help ISs learn about the network topology. First, each IS generates an update specifying the ESs and ISs to which it is connected, as well as the associated metrics. The update is then sent to all neighboring ISs, which forward (flood) it to their neighbors, and so on. Sequence numbers terminate the flood and distinguish old updates from new ones. Using these sequence numbers updates the topology of the network. When the topology changes new updates are sent.

IS-IS Metrics

IS-IS uses a single required default metric with a maximum path value of 1,024. The metric is arbitrary and typically is assigned by a network administrator. Any single link can have a maximum value of 64, and path links are calculated by summing link values. Maximum metric values were set at these levels to provide the granularity to support various link types while at the same time ensuring that the shortest path algorithm used for route computation will be reasonably efficient. IS-IS also defines three optional metrics (costs): delay, expense, and error. The delay cost metric reflects the amount of delay on the link. The expense cost metric reflects communications cost associated with using the link. The error cost metric reflects the error rate of the link. IS-IS maintains a mapping of these four metrics to the quality of service (QoS) option in the CLNP packet header. IS-IS uses these mappings to compute routes through the internetwork.

IS-IS Packets Consist of Eight Fields

IS-IS uses three packet formats: IS-IS hello packets, link state packets (LSPs), and sequence number packets (SNPs). Each of the three IS-IS packets has a complex format with the following three different logical parts. The first part consists of an 8-byte fixed header shared by all three packet types. The second part is a packet-type-specific portion with a fixed format. The third part is also a packet-type-specific but of variable length. Figure 10.12 illustrates the logical format of IS-IS packets. Figure 10.13 shows the common header fields of the IS-IS packets.

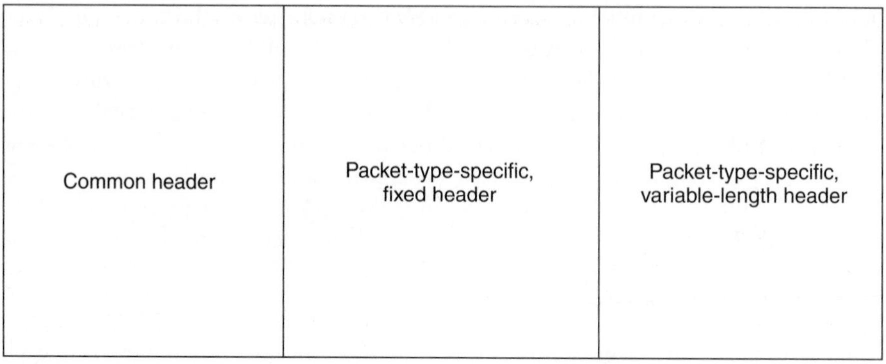

FIGURE 10.12

IS-IS packets consist of eight fields. (Reproduced with permission of Cisco Systems, Inc. Copyright © 2001 Cisco Systems, Inc. All rights reserved.)

Common header	Packet-type-specific, fixed header	Packet-type-specific, variable-length header

FIGURE 10.13

IS-IS packet fields. (Reproduced with permission of Cisco Systems, Inc. Copyright © 2001 Cisco Systems, Inc. All rights reserved.)

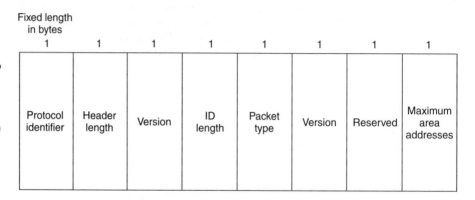

The following descriptions summarize the fields illustrated in Figure 10.13:

- Protocol identifier: Identifies the IS-IS protocol and contains the constant 131.
- Header length: Contains the fixed header length. The length always is equal to 8 bytes but is included so that IS-IS packets do not differ significantly from CLNP packets.
- Version: Contains a value of 1 in the current IS-IS specification.
- ID length: Specifies the size of the ID portion of an NSAP address. If the field contains a value between 1 and 8 inclusive, the ID portion of an NSAP address is that number of bytes. If the field contains a value of zero, the ID portion of an NSAP address is 6 bytes. If the field contains a value of 255 (all ones), the ID portion of an NSAP address is zero bytes.
- Packet type: Specifies the type of IS-IS packet (hello, LSP, or SNP).
- Version: Repeats after the packet type field.
- Reserved: Is ignored by the receiver and is equal to 0.
- Maximum area addresses: Specifies the number of addresses permitted in this area.

Following the common header, each packet type has a different additional fixed portion followed by a variable portion.

Integrated IS-IS

Integrated IS-IS is a version of the OSI IS-IS routing protocol that uses a single routing algorithm to support more network layer protocols than just CLNP. Integrated IS-IS is sometimes called Dual IS-IS, named after a version designed for IP and CLNP networks. Several fields are added to IS-IS packets to allow IS-IS to support additional network layers. These fields inform routers about the reachability of network addresses from other protocol suites and other information required by a specific protocol suite. Integrated IS-IS implementations send only one set of routing updates, which is more efficient than two separate implementations.

Integrated IS-IS represents one of two ways of supporting multiple network layer protocols in a router. The other is the ships-in-the-night approach. Ships-in-the-night routing advocates the use of a completely separate and distinct routing protocol for each network protocol so that the multiple routing protocols essentially exist independently. Essentially, the different types of routing information pass like ships in the night. Integrated routing has the capability to route multiple network layer protocols through tables calculated by a single routing protocol, thus saving some router resources. Integrated IS-IS uses this approach.

Interdomain Routing Protocol (IDRP)

The *Interdomain Routing Protocol (IDRP)* is an OSI protocol that specifies how routers communicate with routers in different domains. IDRP is designed to operate seamlessly with CLNP, ES-IS, and IS-IS. IDRP is based on the Border Gateway Protocol (BGP), an interdomain routing protocol (see Chapter 12) that originated in the IP community. IDRP features include the following:

- Support for CLNP quality of service (QoS).
- Loop suppression by keeping track of all RDs traversed by a route.
- Reduction of route information and processing by using confederations and the compression of RD path information.
- Security by using cryptographic signatures on a per packet basis.
- Route servers.

IDRP Terminology

IDRP introduces several environment-specific terms. These include *border intermediate system (BIS)*, routing domain (RD), *routing domain identifier (RDI)*, routing information base (RIB), and confederation. A BIS is an IS that participates in interdomain routing and, as such, uses IDRP. An RD is a group of ESs and ISs that

Interdomain Routing Protocol is an OSI protocol that specifies how routers communicate with routers in different domains.

A **border intermediate system** is an IS that participates in interdomain routing.

A **routing domain identifier** is a unique RD identifier.

FIGURE 10.14

*Domains commu-
nicate via border
intermediate sys-
tems (BISs).
(Reproduced with
permission of Cisco
Systems, Inc.
Copyright © 2001
Cisco Systems, Inc.
All rights reserved.)*

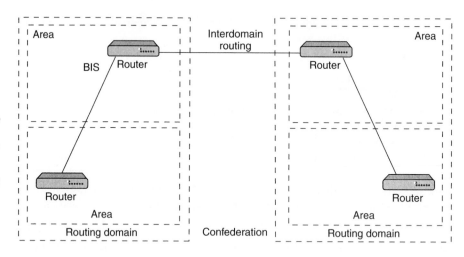

operate under the same set of administrative rules and share a common routing plan.
An RDI is a unique RD identifier. A RIB is a routing database used by IDRP that is
built by each BIS from information received from within the RD and from other
BISs. A RIB contains the set of routes chosen for use by a particular BIS. A

● A **confederation**
is a group of RDs
that appears to RDs
outside the confed-
eration as a
single RD.

confederation is a group of RDs that appears to RDs outside the confederation as a
single RD. The confederation's topology is not visible to RDs outside the confederation. Confederations must be nested within one another and help reduce network
traffic by acting as internetwork firewalls. Figure 10.14 illustrates the relationship between IDRP entities.

IDRP Routing

An IDRP route is a sequence of RDIs, some of which can be confederations. Each BIS
is configured to know the RD and the confederations to which it belongs. It learns
about other BISs, RDs, and confederations through information exchanges with each
neighbor. As with distance vector routing, routes to a particular destination accumulate
outward from the destination. Only routes that satisfy a BIS's local policies and have
been selected for use will be passed on to other BISs. Route recalculation is partial and
occurs when one of three events occurs: an incremental routing update with new routes
is received, a BIS neighbor goes down, or a BIS neighbor comes up.

Summary

In this chapter you learned about the OSI protocol suite. The OSI protocol encompasses all seven layers of the OSI model, which encompasses many different
protocols. At the data link layer, the protocol has been established to operate over

HDLC, X.25, FDDI, IEEE 802.3, IEEE 802.4, and IEEE 802.5. The Open Systems Interconnection (OSI) protocol suite is comprised of numerous standard protocols that are based on the OSI reference model. These protocols are part of an international program to develop data networking protocols and other standards that facilitate multivendor equipment interoperability. The OSI program grew out of the need for international networking standards and is designed to facilitate communication between hardware and software systems despite differences in underlying architectures. The International Organization for Standardization (ISO) developed a complete suite of routing protocols for use in the OSI protocol suite. These include Intermediate System-to-Intermediate System (IS-IS), End System-to-Intermediate System (ES-IS), and Interdomain Routing Protocol (IDRP).

IS-IS is based on work originally done at Digital Equipment Corporation for DECnet/OSI (DECnet Phase V). IS-IS originally was developed to route in ISO Connectionless Network Protocol (CLNP) networks. A version has since been created that supports both CLNP and Internet Protocol (IP) networks. This version is referred to as Integrated IS-IS and it has also been referenced as Dual IS-IS.

In Chapters 11 and 12 you will learn about interdomain routing. Chapter 11 discusses Open Shortest Path First (OSPF), and Chapter 12 discusses Border Gateway Protocol (BGP).

Review Questions

1. How many layers of the OSI reference model does the OSI protocol encompass?
2. What organization standardizes the OSI standards?
3. What are the three major components that comprise the OSI routing architecture?
4. What is an end system?
5. Give an example of an IS and how an IS operates.
6. What is a routing domain?
7. What are the two types of routing domains and define each?
8. What is the routing algorithm used with OSI?
9. What are the main differences between link state routing algorithms and distance vector algorithms?
10. What protocol is used as the routing algorithm for OSI and how does it work?
11. How is the OSI addressing scheme characterized?
12. What are the seven authorities recognized by ISO?
13. What are the four components that make up the OSI addressing scheme?
14. What is a domain and give examples?
15. Contrast TCP/IP addressing with OSI addressing.
16. List the fields that comprise an OSI address.
17. Place the following address in its appropriate fields: 47 00 05 00 32 12 34 53 18 44 27 01.

18. What are the five classes that comprise the transport layer?
19. What does the end system ID consist of?
20. What are the two main fields that make up an OSI address?

Summary Questions

1. What is the function of Class 0 in the transport layer?
2. Which class is responsible for basic recovery?
3. Which class is considered the multiplexing class?
4. What does CONS stand for?
5. How does the transport layer handle flow control in the OSI protocol suite?
6. What are the responsibilities of the session layer in the OSI protocol suite?
7. What are the three types of data transfer supported by the session layer?
8. How does the session layer monitor the connection?
9. What are the five classes defined in FTAM?
10. What are the two entities that make up X.400 messaging?

Interdomain Routing Basics Part 1: Open Shortest Path First Routing Protocol

Objectives

- Explain the time line of Open Shortest Path First (OSPF) development and the problems with the Routing Information Protocol (RIP), which necessitated the need for the development of OSPF.
- Explain the functional requirements of OSPF.
- Discuss the design decisions behind OSPF.
- Explain the differences between distance vector and link state routing protocols.
- Discuss the concept of Designated Router and Backup Designated Router and how elections are completed.
- Discuss the role of link state advertisements (LSAs) in the OSPF network.
- Explain how to identify an LSA instance.
- Define the link state database.
- Explain communication between OSPF routers.
- Discuss reliable flooding of link state advertisements.
- Explain the hierarchical routing concepts in OSPF.
- Discuss how to configure OSPF areas.

Key Terms

Introduction

Open Shortest Path First is a link state routing protocol developed for IP networks.

Open Shortest Path First (OSPF) is a routing protocol developed for IP networks by the Interior Gateway Protocol (IGP) working group of the Internet Engineering Task Force (IETF). The working group was formed in 1988 to design an IGP based on the Shortest Path First (SPF) algorithm for use in the Internet. Similar to the Interior Gateway Routing Protocol (IGRP), which will be discussed later, OSPF was created as a result of the mid-1980s when RIP was increasingly unable to serve large heterogeneous internetworks.

OSPF was derived from several research efforts including Bolt, Beranek, Newman's (BBN's) SPF algorithm developed in 1978 for the ARPANET (a landmark packet-switching network developed in the early 1970s by BBN); Dr. Radia Perlman's research on fault-tolerant broadcasting of routing information (1988); BBN's work on area routing (1986); and an early version of OSI's Intermediate System-to-Intermediate System routing protocol.

OSPF has two primary characteristics. The first is that the protocol is open, which means that its specification is in the public domain. The OSPF specification is published as Request for Comments (RFC) 1247. The second principal characteristic is that OSPF is based on the SPF algorithm, which is sometimes referred to as the Dijkstra algorithm—named for the person credited with its creation.

OSPF is a link state routing protocol that calls for the sending of link state advertisements (LSAs) to all other routers within the same hierarchical area. Information on attached interfaces, metrics used, and other variables is included in OSPF LSAs. As OSPF routers accumulate link state information they use the SPF algorithm to calculate the shortest path to each node.

As a link state routing protocol, OSPF contrasts with RIP, which is a distance vector routing protocol. Routers running the distance vector algorithm send all or a portion of their routing tables in routing update messages to their neighbors.

Functional Requirements

To understand the initial goals of the OSPF Working Group you need to understand the composition of the Internet in 1987. At that time the Internet was largely an academic and research network funded by the U.S. government. At the core of the Internet, the NFSNET backbone and its regional networks had replaced the ARPANET. Most of the Internet used static routing. The *Autonomous Systems* employing dynamic routing used RIP, while the *Exterior Gateway Protocol (EGP)* was being used between Autonomous Systems.

Both of these protocols were having problems. As the size of Autonomous Systems grew and as the size of the Internet routing tables increased, the amount of network bandwidth consumed by RIP updates was increasing, and route-convergence times were becoming unacceptable as the number of routing changes also increased. EGP's update sizes were also increasing. The topological restrictions imposed by EGP— which was a tree topology using the term "reachability protocol" instead of routing protocol—were rapidly becoming unmanageable.

It was decided to tackle the problem of producing an RIP replacement. The reasons for selecting to solve the RIP problem instead of the EGP problem were twofold. First, the RIP problem seemed easier to solve because RIP was a local scale type of protocol as opposed to EGP, which was a global protocol that had to run over the entire Internet. Second, an RIP replacement would have wider applicability because of its use in the Internet and in commercial TCP/IP networks or intranets. Tackling the EGP problem was left to the designers of BGP, which will be discussed in the next chapter. Figure 11.1 illustrates the historical time line of OSPF development.

Thus, the major requirement was to produce an intra-AS-routing protocol also called an Interior Gateway Protocol (IGP) that was more efficient than RIP. It was desired that the protocol consumed fewer network resources than RIP and converged faster than RIP when the network topology changed. Examples of network topology changes are the failure of communication links, router interfaces, or entire routers. After the failure, the best paths to certain destinations change. It takes time for any routing protocol to find the new best routes or paths, and the paths used in the meantime are sometimes suboptimal or even nonfunctional. The process of finding the new path is called convergence.

An **Autonomous System** is a collection of networks under a common administration.

The **Exterior Gateway Protocol** exchanges routing information between Autonomous Systems.

FIGURE 11.1

Timeline of OSPF development. (Reproduced with permission of Cisco Systems, Inc. Copyright © 2001 Cisco Systems, Inc. All rights reserved.)

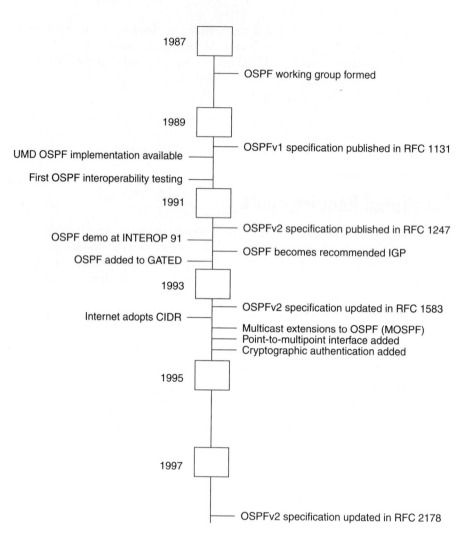

1987 — OSPF working group formed

1989 — OSPFv1 specification published in RFC 1131

UMD OSPF implementation available —

First OSPF interoperability testing —

1991 — OSPFv2 specification published in RFC 1247

OSPF demo at INTEROP 91 — — OSPF becomes recommended IGP

OSPF added to GATED —

1993 — OSPFv2 specification updated in RFC 1583

Internet adopts CIDR — — Multicast extensions to OSPF (MOSPF)
— Point-to-multipoint interface added
— Cryptographic authentication added

1995

1997

— OSPFv2 specification updated in RFC 2178

Other initial functional requirements for the OSPF protocol included the following:

● A more descriptive routing metric; RIP uses hop count as its routing metric, and path cost is allowed to range from 1–15 only. This created two problems for network administrators. First, it limited the diameters of their networks to 15 router hops. Second, administrators could not take into account such factors as bandwidth and/or delay when configuring their routing systems. The OSPF Working Group settled on a configurable link metric whose values ranged between 1 and 65,535 with no limitations on total-path cost. This design removes network diameter limitations and allows for configurations as illustrated in Figure 11.2,

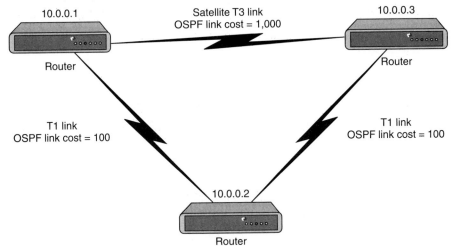

where a two-hop T1 path connection with an OSPF cost of 200 can be preferred over a direct satellite connection with an OSPF cost of 1,000.

- Equal cost multipath: OSPF had to be able to discover multiple best paths to a given destination, when they exist. However, it was not mandated how routers should use these multiple best paths for forwarding data packets. It did not seem necessary to standardize the choice of which of the multiple paths to use for a given packet. There are many strategies: round robin on a packet-by-packet basis, a hash on the source addresses, and so on. However, routers with different strategies can be mixed freely and a working network will still result.

- Routing hierarchy: OSPF had to be able to support very large routing domains, on the order of many thousands of routers. The only known way to scale routing to such a size is by introducing hierarchy. The OSPF hierarchy was implemented via a two level area routing scheme.

- Separate internal and external routes: Autonomous Systems running RIP were having trouble knowing which information to trust. You always want to trust information gained firsthand about your own internal routing domain over external routing information that has been injected into your domain. RIP did not distinguish between the two types of information. In OSPF, external routing information is labeled and is explicitly overridden by any internal routing information.

- Support of more flexible subnetting schemes: When OSPF was first being designed, the Internet was still using Class A, B, and C addresses. IP subnetting did exist, but subnets were always allocated by dividing a given Class A, B, and C address into equal sized pieces. However, it was felt that a more efficient use of address space was going to be needed in the future, so it was a requirement for the OSPF designers to have OSPF capable of routing to arbitrary [address, mask] combinations. In particular, OSPF had to be able to accommodate variable-length

Classless inter-domain routing allows routers to group routes together to cut down on the quantity of routing information.

Type of Service routing influenced the handling of the IP packets including the route that the packet would take through the Internet.

subnet masks (VLSMs), which will be discussed in Chapter 14, whereby a Class A, B, or C address could be carved into unequal sized subnets. This requirement anticipated *classless internet domain routing (CIDR)*. OSPF would have to be adjusted slightly to accommodate CIDR completely.

- Security: OSPF had to be able to administratively control which routers joined an OSPF domain. A common problem with RIP is that anyone can bring up a bogus RIP router advertising the default route or any other route, thereby disrupting routing. By authenticating received OSPF packets, a router would have to be given the correct key before it could join the OSPF routing domain. This requirement led to reserve space in OSPF packets for authentication data. However, the development of nontrivial authentication algorithms for OSPF would have to wait for several years.

- Type of Service routing: OSPF had to be able to calculate separate routes for IP *Type of Service (TOS) routing*. TCP/IP supports five classes of TOS: normal service, minimized monetary cost, maximize reliability, maximize throughput, and minimize delay. The idea behind TOS is that IP packets can be labeled with a particular TOS, which would then influence the handling of the packets, including possibly the route that the packet would take through the Internet called TOS based routing.

OSPF supported TOS based routing from the beginning, allowing the assignment of separate link metrics and building separate routing tables for each TOS. For example, in Figure 11.2, a second metric can be assigned for the maximize bandwidth TOS, allowing the satellite link to be preferred for file transfer traffic at the same time that the link is avoided by all other traffic types. TOS support was made optional, giving rise to many arguments about what should happen to a packet when a path minimizing monetary cost does not exist but a normal service path does: Should the packet be discarded, or should it be forwarded along the normal service path instead.

All of these efforts and arguments were really for naught. Although several implementations for OSPF TOS based routing were developed, TOS based routing has never been deployed in the Internet. This was due to a lack of real need for TOS and also because the hosts do not label packets with TOS because routers do not act on TOS, and the routers do not act on TOS because the packets are not labeled.

Backup Designated Router

When running over broadcast and NBMA segments, OSPF relies on the Designated Router. All LSAs flooded over a broadcast or NBMA subnet go through the subnet's Designated Router, which is responsible for reporting the subnet's local topology within a network LSA. This action leads to a robustness problem. When the Designated Router fails, neither LSAs nor user data can be forwarded over the subnet until a new Designated Router is established.

Even inadvertently replacing the current Designated Router with another causes some disruption, as all routers must synchronize with the new Designated Router and a network LSA is flooded to all OSPF routers, causing all routers to rerun their routing calculations. To make sure that a switch of Designated Router happens only on failures, the Designated Router election was designed so that routers newly added to the subnet always defer to the existing Designated Router.

To ensure that the switchover occurs as quickly as possible on failures the concept of *Backup Designated Router* was introduced. The Backup Designated Router also is elected and then prequalified to take over as Designated Router all while the current Designated Router keeps flooding over the subnet, going even before the Designated Router's failure is detected. As soon as the failure is discovered, the Backup Designated Router is promoted to Designated Router, because all other routers on the subnet have already synchronized with the Backup Designated Router. Thus, the switchover is relatively painless.

Designated Router Election

The Designated Router election process works as follows, using data transmitted in Hello packets. The first OSPF router on an IP subnet always becomes the Designated Router. When a second router is added, it becomes the Backup Designated Router. Additional routers that are added to the segment defer to the existing Designated Router and Backup Designated Router. The only time the identity of the Designated Router or Backup Designated Router changes is when the existing Designated Router or Backup Designated Router fails.

In the event of failure of a Designated Router or Backup Designated Router, there is an orderly changeover. Each OSPF router has a configured Router Priority value, a value between 0 and 127 inclusive. When the Designated Router fails the Backup Designated Router is promoted to Designated Router, and of the remaining routers the one having the highest Router Priority becomes the Backup Designated Router, using the routers' OSPF *Router Priority ID* to break ties. Then the new Backup Designated Router has to start the time-consuming procedure of initial database synchronization or Database Exchange with all other routers on the subnet.

Because being Designated Router or Backup Designated Router on a subnet consumes additional resources, you may want to prevent certain routers from assuming these roles. This is accomplished by assigning those routers a Router Priority of 0. One other time that Router Priority comes into play is in partitioned subnets. When a subnet is partitioned into two pieces, two Designated Routers are elected, one for each partition. When the partition heals, the Designated Router with the highest Router Priority assumes the role of Designated Router for the healed subnet.

OSPF Basics

OSPF belongs to the general category of routing protocols called link state protocols. In the last 15 years, link state protocols have become very popular alternatives to the more traditional distance vector algorithms. The link state routing algorithm has replaced the

The **Backup Designated Router** is elected and prequalified to take over as Designated Router as soon as a failure is discovered.

The **Router Priority ID** is a value between 0 and 127 that is given to each OSPF router. It is used to break ties when deciding the Backup Designated Router.

distance vector algorithm because it was starting to show signs of wear and tear. The goal of ARPANET was to find the least-delay paths through the network. The previous distance vector algorithm was taking a long time to converge on correct paths when the network topology changed, and it was generating a lot of control traffic in the process. The new link state routing protocol was extensively instrumented and monitored, and it was capable of meeting the stringent performance goals of consuming less than 1 percent of link bandwidth for control traffic and less than 2 percent of switch CPU for routing calculations, while responding to network changes in less than 100 milliseconds.

At the core of every link state routing protocol is a distributed, replicated database. This database describes the routing topology—the collection of routers in the routing domain and how they are interconnected. Each router in the routing domain is responsible for describing its local piece of the routing topology in link state advertisements (LSAs). These LSAs are then reliably distributed to all the other routers in the routing domain in a process called reliable flooding. Taken together, the collection of LSAs generated by all of the other routers is called the *link state database*. The link state routing protocol's flooding algorithm ensures that each router has an identical link state database, except during brief periods of convergence. Using the link state database as input, each router calculates its IP routing table, enabling the correct forwarding of IP data traffic.

The **link state database** is the collection of LSAs generated by all of the other routers.

An OSPF Example

The original Internet switches were interconnected via point-to-point leased lines. To this day all link state protocols operate in a very similar fashion, given such an environment. Consequently, the basic concepts of OSPF will be illustrated using a collection of routers interconnected by point-to-point links. Figure 11.3 depicts a specific example of a network topology and will be referred to throughout this chapter.

FIGURE 11.3

Point-to-point network topology. (Reproduced with permission of Cisco Systems, Inc. Copyright © 2001 Cisco Systems, Inc. All rights reserved.)

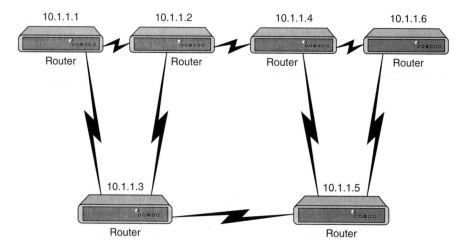

Suppose that this network has been considered up and operational for a long period of time. All of the six routers pictured will then have identical link state databases that describe a complete map of the network. Looking at this database, any of the six can tell how many other routers are in the network (5), how many interfaces router 10.1.1.4 has (3), whether a link connects to 10.1.1.2 and 10.1.1.4 (yes), and so on. The database also gives a cost for each link although this cost is not depicted in Figure 11.3. Let us assume that the cost of each link is 1.

From the database, each router has calculated the shortest paths to all other routers. For example, router 10.1.1.1 calculates that it has two equal-cost shortest paths to 10.1.1.6, one through 10.1.1.2 and 10.1.1.4 and the other through 10.1.1.3 and 10.1.1.5. Because they all have the same database, any router can calculate the routing table of any other—this useful property is taken advantage of by the Multicast Extensions to OSPF.

When the network is in a steady state, that is, no routers or links are going in or out of service, the only OSPF routing traffic is periodic Hello packets between neighboring OSPF routers and occasional refresh pieces of the link state database. *Hello packets* are usually sent every 10 seconds, and failure to receive Hellos from a neighbor tells the router of a problem in its connected link or neighboring router. Every 30 minutes a router refloods the pieces of the link state database that it is responsible for, just in case those pieces have been lost from or corrupted in one of the other router's databases. Suppose that the link between routers 10.1.1.2 and 10.1.1.4 fails. The physical or data link protocol in router 10.1.1.2 has links to routers 10.1.1.1 and 10.1.1.3, but it no longer has a link to 10.1.1.4. Router 10.1.1.2 will start the flooding of its new router-LSA by sending the LSA to routers 10.1.1.1 and 10.1.1.3. Router 10.1.1.3 will then continue the flooding process by sending the LSA to router 10.1.1.5, and so on. As soon as each router receives router 10.1.1.2's new router-LSA, the router recalculates its shortest path to 10.1.1.6, which is the one going through 10.1.1.3 and 10.1.1.5.

> **Hello packets** are used by routers for neighbor discovery and recovery.

In this example, each router has been identified by an IP address, and this is also how OSPF routers commonly identify one another. Each OSPF router has an OSPF Router ID. The *Router ID* is a 32-bit number that uniquely identifies the router within the OSPF routing domain. Although not required by the OSPF specification, the OSPF Router ID in practice is assigned to be one of the router's IP addresses.

> The **Router ID** is a 32-bit number that uniquely identifies the router within the OSPF routing domain.

Link State Advertisements (LSAs)

Each OSPF router originates one or more *link state advertisements (LSAs)* to describe its local part of the routing domain. Taken together, the LSAs form the link state database, which is used as input to the routing calculations. To provide organization to the database and to enable the orderly updating and removal of LSAs, each LSA must provide some bookkeeping and topological information. All OSPF LSAs start with a 20-byte common header, illustrated in Figure 11.4, which carries this bookkeeping information. These bookkeeping functions are described in the following sections.

> **Link state advertisements** form the link state database.

FIGURE 11.4

The LSA header. (Reproduced with permission of Cisco Systems, Inc. Copyright © 2001 Cisco Systems, Inc. All rights reserved.)

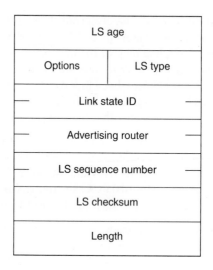

LS age	
Options	LS type
Link state ID	
Advertising router	
LS sequence number	
LS checksum	
Length	

Identifying LSAs

An OSPF link state database might consist of many thousands of LSAs. Individual LSAs must be distinguished during flooding and the various routing calculations. Three fields found in the common LSA header identify OSPF LSAs: LS Type, Link State ID, and Advertising Router.

LS Type Field

The LS Type field (link state type field) broadly classifies LSAs according to their functions. Five LS Types are defined by the base OSPF specification. These five link state types are listed in Table 11.1. LSAs with LS Type equal to 1 are called router-LSAs. Each router originates a single router-LSA to describe its set of active interfaces and neighbors. In a routing domain consisting solely of routers interconnected by point-to-point links, the link state database consists only of router-LSAs.

LSAs with LS Type equal to 2 are called network-LSAs. Each network LSA describes a network segment, such as a broadcast or nonbroadcast multiaccess network, along with the identity of the network's currently attached routers.

LSAs with LS Type equal to 3 (network summary LSAs, and LSAs with LS Type equal to 4 (Autonomous System Border Router [ASBR] summary LSAs), and 5 (Autonomous System external LSAs) are used to implement hierarchical routing within OSPF. One way of extending the OSPF protocol is to add new LS Types. Two LS Types have been added to those defined by the base specification. LSAs with LS Type equal to 6 are called group-membership LSAs and are used to indicate the location of multicast group members in MOSPF. LSAs with LS Type equal

Table 11.1 Types of Link State Advertisement

LSA Type	Name	Description
1	Router link entry (record) (O-OSPF)	Generated by each router for each area to which it belongs. It describes the state of the router's link to the area. These are only flooded within a particular area. The link status and cost are two of the descriptors provided.
2	Network link entry (O-OSPF)	Generated by the designated driver in multiaccess networks. It describes the set of routers attached to a particular network. Flooded within the area that contains the network only.
3 or 4	Summary link entry (IA-OSPF Inter area)	Originated by ABRs. Describes the links between the ABR and the internal routers of a local area. These entries are flooded throughout the backbone area to the other ABRs. Type 3 describes routes to networks within the local area and are sent to the backbone area. Type 4 describes reachability to ASBRs. These link entries are not flooded through totally stubby areas.
5	Autonomous System external link entry (E1-OSPF external Type-1; E2-OSPF external Type-2)	Originated by the ABSR. Describes routes to destinations external to the autonomous system. Flooded throughout an OSPF autonomous system except for stub and totally stubby areas.

to 7 are used in OSPF NSSA (not-so-stubby areas) to import a limited set of external information. In addition, LS Type 8 has been proposed—the external attributes LSAs to carry BGP path information across an OSPF routing domain in lieu of internal BGP.

OSPF routers are not required to store or forward LSAs with unknown LS Type. When OSPF is extended by adding new LS Types, this rule is maintained through the use of the Options field. Option bits are added supporting this particular new LS Type and are exchanged between routers in database description packets. By looking at the Options field that is advertised by its neighbor, a router then knows which LSAs to forward and which LSAs to keep to itself.

Link State ID Field

The Link State ID field uniquely distinguishes an LSA that a router originates from all other self-originated LSAs of the same LS Type. For compactness and convenience, the Link State ID often carries addressing information also. For example, the Link State ID of an *AS-external-LSA* is equal to the IP address of the externally reachable IP network being imported into the OSPF routing domain.

An **As-external-LSA** is used to implement hierarchical routing within OSPF.

Advertising Router Field

The *Advertising Router field* is set to the originating router's OSPF Router ID. A router can easily identify its self-originated LSAs as those LSAs whose Advertising Router is set to the router's own Router ID. Routers are allowed to update or to delete only self-originated LSAs.

Knowing which router has originated a particular LSA tells the calculating router whether the LSA should be used in the routing calculation and, if so, how it should be used. For example, *network-summary-LSAs* are used in the routing calculation only when their Advertising Router is reachable. When they are used the cost to the destination network is the sum of the cost to the Advertising Router and the cost advertised in the LSA.

Identifying LSA Instances

When a router wishes to update one of the LSAs it is originating, it must have some way to indicate to the other routers that this new LSA instance is more up to date and therefore should replace any existing instances of the LSA. In OSPF LSAs, the LS Sequence Number in the common LSA header is incremented when a new LSA instance is created. The LSA instance having the larger LS Sequence Number is considered to be more recent. If the LS Sequence Numbers are the same, the LS Age and the LS Checksum fields of the LSA are compared by the router before it declares that the two instances are identical.

LS Sequence Number Field

When a router has two instances of a particular LSA, it detects which instance is more recent by comparing the instance LS Sequence Numbers. The instance with the larger LS Sequence Number is more recent. The meaning of larger depends on the organization of the sequence number space as illustrated in Figure 11.5.

The original link state algorithm used a *circular sequence number space*. By carefully controlling the rate at which new LSA instances were created, all possible sequence numbers for a given LSA were constrained at any one time to lie in half-circle or semicircle. The largest sequence number was then the sequence number appearing on the counterclockwise edge of the semicircle. However, this scheme was not robust in the face of errors. In a possible network failure using this method of LS sequence numbering, bit errors can switch memory causing the accidental introduction of three instances of having an LSA with sequence numbers S_1, S_2, S_3 that are not constrained to a semicircle as illustrated in Figure 11.5. The switches can no longer determine which LSA instance was most recent because $S_1 < S_2 < S_3 < S_1$. As a result, the three LSA instances were continually flooded throughout the network, with each instance replacing the others in turn. This can continue until the entire network is power cycled. This type of problem is similar to a network virus. Once introduced by a switch, such bad data spread to all other switches and become extremely difficult to eradicate.

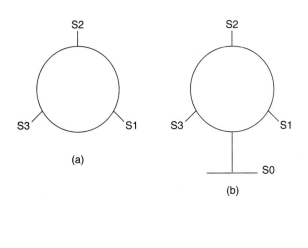

FIGURE 11.5
Various LS sequence number space organizations. (Reproduced with permission of Cisco Systems, Inc. Copyright © 2001 Cisco Systems, Inc. All rights reserved.)

To help avoid this type of network failure, called the ARPANET sequence bug, a *lollipop-shaped sequence number space* was introduced. In the lollipop-shaped sequence numbering scheme, each LSA is initially originated with the smallest sequence number S_0, which is part of the lollipop's handle. The sequence space then increments until it enters the circular part of the space. Version 1 of the OSPF protocol used the lollipop-shaped sequence space. Although the lollipop-shaped organization provides better protection against the ARPANET sequence bug, because sequence numbers falling into the lollipop handle never create ambiguities, the three sequence numbers S_1, S_2, and S_3 still cause a problem.

OSPFv2 uses a *linear sequence space*, which completely prevents the ARPANET sequence bug. PSPF LS Sequence Numbers are assigned 32-bit values. The first time an OSPF router originates a given LSA, it sets the LSA's LS Sequence Number to the smallest negative value. Thus S_0, which is referred to as the Initial Sequence Number, has a value of 0×80000001. Subsequently, each time the router updates the LSA, it increments the LSA's LS Sequence Number by 1. An LSA's LS Sequence Number monotonically increases until the maximum positive value is reached, which is referred to as S_{Max} or MaxSequenceNumber with a value of $0 \times 7fffffff$. At this point, when the router wishes to update the LSA it must start again with the initial sequence value of S_0, rolling over the sequence space. However, to get the other routers to accept this new LSA instance as the most recent, the router must first delete the LSA instance with sequence number S_{Max} from the routing domain before flooding the new instance with sequence number S_0. Because OSPF routers are not allowed to update their self-originated LSAs more than once every 5 seconds, in the absence of errors in either hardware or software implementations, a 32-bit sequence space will take more than 1,000 years to roll over!

A **lollipop-shaped sequence space** provides better protection against the ARPANET sequence bug.

A **linear sequence space** completely prevents the ARPANET sequence bug.

Besides using a linear sequence space in LSAs, OSPF has other features that guard against problems similar to the ARPANET sequence bug. First, all OSPF LSAs contain a checksum so that data corruption within an LSA is detected. Second, OSPF requires the LS Age field of all LSAs to be incremented at each hop during flooding, which eventually breaks any flooding loop by causing a looping LSA's LS Age field to reach the value MaxAge.

Verifying LSA Contents

An LSA may become corrupted during flooding or while being held in a router's memory. Corrupted LSAs can create havoc, possibly leading to incorrect routing calculations, black holes, or looping data packets. To detect data corruption, redundant information is added to the LSA as a checksum or parity check. In OSPF this function is provided by the *LS Checksum field* in the LSA header.

The **LS Check-sum field** is used to detect data corruption.

LS Checksum Field

Each OSPF LSA is checksummed to detect data corruption within the LSA header and contents. The checksum is calculated originally by the router that originates the LSA and then is carried with the LSA as it is flooded throughout the routing domain and stored within the link state database. A router verifies the checksum of an LSA received from a neighboring router during flooding; corrupted LSAs will be discarded by the router in hopes that the retransmitted LSA from the neighbor will be uncorrupted. A router also periodically verifies the checksums of all the LSAs in its link state database, guarding against its own hardware and software errors. Detection of such internal errors will generally cause the router's OSPF processing to reinitialize.

After an LSA instance is originated, its checksum is never altered. For this reason the checksum excludes the LSA's LS Age field, which is modified in flooding. This means that OSPF does not detect corruption of the LS Age field. Corruption of the LS Age field will in general cause no more harm than speeding up the rate of LSA originations.

The checksum is implemented by using the Fletcher checksum algorithm, which is also used in the OSI network and transport layers. The reason for using the Fletcher checksum in LSAs is that it is easy to calculate and it catches patterns of data corruption different from the standard Internet ones-complement checksum used by OSPF's protocol packets. During flooding, LSAs are carried in OSPF Link State Update packets. If for some reason data corruption fails to be detected by the Update packet's ones-complement checksum, Fletcher may still detect the corruption within individual LSAs.

The OSPF protocol also uses the LSA checksum as an efficient way to determine whether two instances of the same LSA both having the same LS Checksum fields and relatively the same age, also have the same contents, and are considered the same instance. This is a probabilistic comparison—there is no absolute insurance that when the checksums are equal so are the contents, although it seems to be highly likely in this context having the same LSA, same sequence number, and relatively the same LS age. Even if the assumption were wrong, OSPF would recover automatically whenever the LSA was refreshed within 30 minutes.

Removing LSAs from the Distributed Database

Under normal circumstances, every LSA in the link state database is updated at least once every 30 minutes. If an LSA has not been updated after an hour, the LSA is assumed to be no longer valid and is removed from the database. The LS Age field in the LSA header indicates the length of time since the LSA was last updated.

LS Age field

The *LS Age field* indicates the number of seconds since the LSA was originated. Under normal circumstances, the LS Age field ranges from 0 to 30 minutes. If the age of an LSA reaches 30 minutes, the originating router will refresh the LSA by flooding a new instance of the LSA, incrementing the LS sequence number and setting the LS age to 0 again.

If the originating router has failed, the age of the LSA continues to increase until the value of MaxAge (1 hour) is reached. At that time the LSA is deleted from the database. One hour is the maximum value that the LS Age field can ever attain. To ensure that all routers remove the LSA more or less at the same time, without depending on a synchronized clock, the LSA is reflooded at that time. All other routers will then remove their database copies on seeing the MaxAge LSA being flooded.

After an LSA's originating router has failed, it can take as long as an hour for the LSA to be removed from other routers' link state databases. Such an LSA is certainly advertising out-of-date information. OSPF guarantees that the LSA will not interfere with the routing table calculation. OSPF accomplishes this by requiring that a link be advertised by the routers at both ends of the link before using the link in the routing calculation.

OSPF also has a procedure, entitled *premature aging*, for deleting an LSA from the routing domain without waiting for its LS Age to reach MaxAge. Sometimes a router wishes to delete an LSA instead of updating its contents. By using premature aging, the router deletes the LSA from the distributed database by setting the LSA's LS Age field to MaxAge and reflooding the LSA. To avoid possible thrashing situations, a router is allowed to prematurely age only those LSAs that the router itself originated. If a router crashes or is removed from service without prematurely aging its self-originated LSAs, the said LSAs will remain in other routers' link state databases for up to an hour while they age out naturally.

The age of an LSA is also examined in order to implement other OSPF functions such as rate limiting the amount of OSPF flooding. The uses of the LS Age field are shown in Table 11.2.

The **LS Age field** indicates the number of seconds since the LSA was originated.

Premature aging is used to delete an LSA from the routing domain without waiting for its LS Age to reach MaxAge.

Table 11.2 Actions Taken by OSPF Routers Based on LS Age Fields

Constant	Value	Action of OSPF Router
MinLSArrival	1 second	Maximum rate at which a router will accept updates of any given LSA via flooding.
MinLSInterval	5 seconds	Maximum rate at which a router can update an LSA.
CheckAge	5 minutes	Rate at which a router verifies the checksum of an LSA contained in its database.
MaxAge Diff	15 minutes	When two LSA instances differ by more than 15 minutes they are considered to be separate instances, and the one with the smaller LS Age is accepted as more recent.
LSRefreshTime	30 minutes	A router must refresh any self-originated LSA whose age reaches the value of 30 minutes.
MaxAge	1 hour	When the age of an LSA reaches 1 hour, the LSA is removed from the database.

Other LSA Header Fields

Options

The Options field in the OSPF LSA header can indicate that an LSA deserves special handling during flooding or routing calculations. In addition to the LSA header, the Options field can appear in the OSPF Hello and Database Description packets. The Options field is 1 byte in length. In the base OSPF protocol specification, only two Option bits were defined. With the addition of OSPF protocol extensions, such as MOSPF, NSSA areas, and Demand Circuit extensions, five of the eight Option bits now have defined meanings.

Length

The Length field contains the length, in bytes of the LSA, counting both LSA header and contents. Because the Length field is 16 bits long an LSA can range in size from 20 bytes, which is the size of the LSA header, to more than 65,000 bytes. One cannot go all the way to 65,000 bytes because the LSA must eventually be transported within an IP packet that is itself restricted to 65,535 bytes in length. All OSPF LSAs are small, with the only LSA type that is likely to exceed a few hundred bytes being the router-LSA. Router-LSAs can be large when a router has many different interfaces, but they are still unlikely to exceed a few thousand bytes in length.

A Sample LSA: The Router-LSA

In the network environment of routers interconnected via serial lines, OSPF uses only a single LSA type: the router-LSA. Each OSPF router originates a single router-LSA, which reports the router's active interfaces, IP addresses, and neighbors. Figure 11.6 examines the router-LSA originated by router 10.1.1.1. An explanation of Figure 11.6 follows. All serial lines in the figure are unnumbered: The router's serial line interfaces have not been assigned IP addresses nor have IP subnets been assigned to the serial lines. Each router has been assigned an IP address that is not attached to any particular interface—the routers are labeled with these addresses. Following the usual convention, OSPF Router IDs have been assigned equal to the routers' IP addresses, although this is not strictly required by the OSPF specification. Finally, each interface is labeled with a pair of numbers. The first number is the interface's MIB-II IfIndex value, and the second is the OSPF output cost that has been assigned to the interface. Figure 11.7 displays the router-LSA that router 10.1.1.1 would originate after full OSPF relationships to its neighbors have been established.

Consider the setting of the fields in the standard LSA header. All LSAs are originated with LS Age set to 0. The Options field describes various optional capabilities supported by the router. The Link State ID of a router-LSA is set to the router's OSPF Router ID. The LS sequence number of the LSA is 0×80000006, indicating that five instances of the router-LSA have previously been originated. Next in the router-LSA is a byte labeled Router Type (Figure 11.7), describing the router's role in hierarchical routing as well as any special roles that the router might have in multicast routing. The router-LSA then indicates that three connections or links are being reported. The first two links are point-to-point connections to neighboring

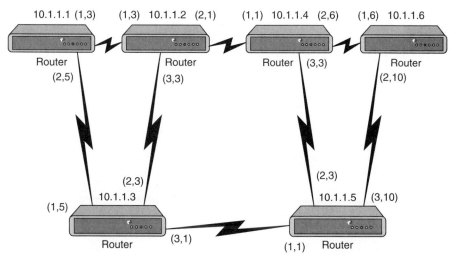

FIGURE 11.6

Point-to-point network topology. (Reproduced with permission of Cisco Systems, Inc. Copyright © 2001 Cisco Systems, Inc. All rights reserved.)

FIGURE 11.7

Router 10.1.1.1's router-LSA. (Reproduced with permission of Cisco Systems, Inc. Copyright © 2001 Cisco Systems, Inc. All rights reserved.)

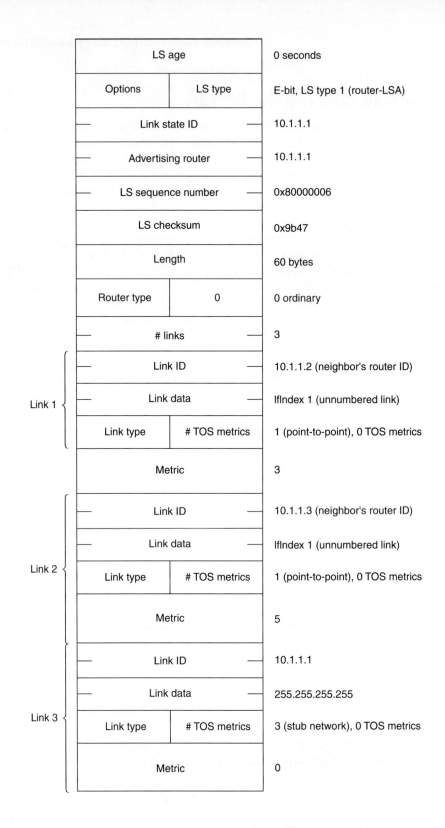

458

routers as indicated by the Link Type field. The Link ID of a point-to-point connection is the neighboring router's OSPF Router ID. Because these connections are unnumbered the Link Data field indicates the corresponding router interface's MIB-II IfIndex.

Each link also contains a Metric field. This field, ranging from 1 to 65,535, indicates the relative cost of sending data packets over the link. The larger the cost, the less likely the data will be routed over the link. The person setting up the network configures metrics. The metric can mean anything such as delay, dollar cost of sending traffic over the link, and so on. Adding metrics should be meaningful because OSPF calculates the cost of the path to be the sum of the cost of the path's component links. This means that delay would be a fine metric and so would transmission time for the link. Link bandwidth would probably not be a good metric because adding bandwidth along a path is not very meaningful. Figure 11.6 uses the metric weighted hop count. One starts by assigning each link a metric of 1 and then increases the cost of the less preferred links until the least-cost paths flow over the more desirable links.

Metrics can be asymmetric. One does not have to assign equal metrics to both sides of a link. Assigning equal metrics to both sides of a link is the norm, as has been done in Figure 11.6. In the original OSPF specification separate metrics could be assigned for each TOS, although this feature was removed due to a lack of deployment of TOS based routing. For backward compatibility the router-LSA still has space for TOS metrics, even though in practice the # TOS metrics for each link is always set to 0.

The last link in router 10.1.1.1's router-LSA advertises the router's own IP address. The address is advertised as a stub network connection. Stub networks are allowed to source or sink IP packets, but they do not carry transit traffic. The Link ID field of a stub network connection is advertised as the network's IP address, with the Link Data field as its IP mask. Connections to stub networks are allowed to advertise a Metric of 0.

The router-LSA is the one type of LSA that can get quite large. All of a router's interfaces must be advertised in a single router-LSA. The fixed part of a router-LSA is 24 bytes in length, with each advertised connection adding an additional 12 bytes. The router-LSA in Figure 11.7 is only $24 + 3^*12 = 60$ bytes long, but a router-LSA for a router with 100 interfaces would be 1,224 bytes long.

The Link State Database

The collection of all OSPF LSAs is called the link state database. Each OSPF router has an identical link state database. Link state databases are exchanged between neighboring routers soon after the routers have discovered each other; after that, the link state database is synchronized through a procedure called reliable flooding.

The link state database gives a complete description of the network: the routers and network segments and how they are interconnected. Starting with a link state database, one can draw a complete map of the network. This property can serve

Table 11.3 Link State Database for Figure 11.6

LS Type	Link State ID	Advertising Router	LS Checksum	LS Sequence No.	LSAge
Router-LSA	10.1.1.1	10.1.1.1	0x9b47	0x80000006	0
Router-LSA	10.1.1.2	10.1.1.2	0x219e	0x80000007	1,618
Router-LSA	10.1.1.3	10.1.1.3	0x6b53	0x80000003	1,712
Router-LSA	10.1.1.4	10.1.1.4	0xe39a	0x8000003a	20
Router-LSA	10.1.1.5	10.1.1.5	0xd2a6	0x80000038	18
Router-LSA	10.1.1.6	10.1.1.6	0x05c3	0x80000005	1,680

as a powerful debugging tool. By examining a single router's database, one immediately observes the state of all other routers in the network. OSPF is not a network monitoring protocol; it is a routing protocol. In performance of the routing function, a router uses the link state database as the input to the router's routing calculation.

Table 11.3 shows the link state database for the network in Figure 11.6. Each row in the table represents a single LSA. Each LSA is identified by the contents of the table's first three columns; the second three columns identify the LSA's instance.

One can quickly determine in OSPF whether two routers indeed have synchronized databases. First, determine that the two routers have the same number of LSAs in their link state databases. Second, the sums of their LSAs' LS Checksum fields are equal. When using the LS Checksums to compare LSA contents, comparing the sum of the LS Checksums to determine link state database synchronization is probabilistic in nature but useful in practice. For example, according to the database in Table 11.3, all routers should have six LSAs in their link state database, and the sum of the LSA's LS Checksums should be 0x2e43b.

Looking at the OSPF database, one can also immediately tell which parts of the network are changing the most. This would be the parts that are described by LSAs whose LS Sequence Numbers are changing the most and whose LS Age field never gets very large. In Table 11.3, for example, you can see that the router-LSAs originated by routers 10.1.1.4 and 10.1.1.5 are changing much more rapidly than LSAs originated by other routers. Looking at the network diagram in Figure 11.6, one can see that the point-to-point connection between 10.1.1.4 and 10.1.1.5 could possibly go up and down frequently. This information can be determined by looking at any router's link state database because all routers have exactly the same information.

In our sample network of routers connected via serial lines, the link state database is very simple and uniform, consisting of only router-LSAs. However, in the presence of network segments other than point-to-point links or when hierarchical routing is employed or when OSPF extensions such as MOSPF are deployed, other LSA types are included in the link state databases.

Communicating Between OSPF Routers: OSPF Packets

Like most IP protocols, OSPF routers communicate in terms of protocol packets. OSPF runs directly over the IP network layer, doing without the services of UDP or TCP, which are used by RIP and BGP, respectively. When a router receives an IP packet with an IP protocol number equal to 89, the router knows that the packet contains OSPF data. Stripping off the IP header, the router finds an OSPF packet. One particular type of OSPF packet, together with its IP encapsulation, is depicted in Figure 11.8.

OSPF uses services of the IP header as follows:

- Because most OSPF packets travel only a single hop, namely between neighboring routers or peers, the TTL in the IP packet is almost always set to 1. This keeps broken or misconfigured routers from mistakenly forwarding OSPF packets. The one exception to setting the TTL to 1 comes in certain configurations of OSPF hierarchical routing.
- The Destination IP address in the IP header is always set to the neighbor's IP address or to one of the OSPF multicast addresses AllSPFRouters (224.0.0.5) or AllDRouters (224.0.0.6). One of these two multicast addresses is used.
- An OSPF router uses IP fragmentation/reassembly when it has to send a packet that is larger than a network segment's MTU. However, most of the time a router can avoid sending such a large packet by sending an equivalent set of smaller OSPF packets instead. This is sometimes called semantic fragmentation. For example, IP fragmentation cannot be avoided when a router is flooding an LSA that is itself larger or close to larger than the network segment's MTU. There is no semantic fragmentation for OSPF LSAs. IP fragmentation is also unavoidable when the router has so many neighbors on a broadcast segment that the size of the Hello packet exceeds the segment's MTU.
- A router sends all OSPF packets with IP precedence of Internetwork Control in hope that this setting will cause OSPF packets to be preferred over data packets, although in practice it seldom does.

All OSPF packets begin with a standard 24-byte header, which provides the following functions:

- An OSPF packet type field: There are five separate types of OSPF protocol packets: Hello packets (type = 1), which are used to discover and maintain neighbor relationships; Database Description packets (type = 2); Link State Request packets (type = 3); Link State Update packets (type = 4); and Link State Acknowledgement packets (type = 5); all used in link state database synchronization.

FIGURE 11.8

An OSPF Hello packet. (Repro - duced with permis- sion of Cisco Systems, Inc. Copyright © 2001 Cisco Systems, Inc. All rights reserved.)

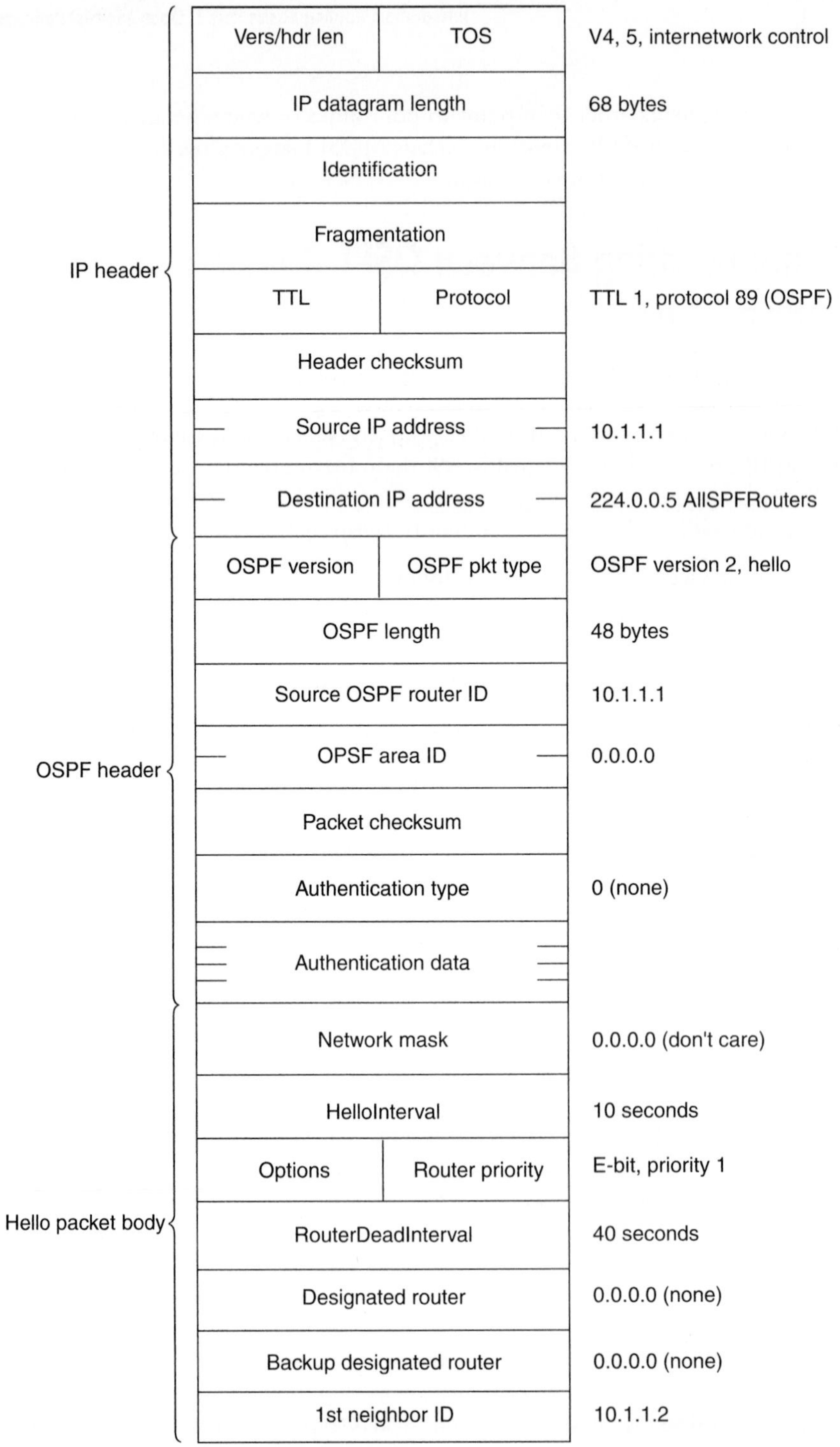

Vers/hdr len	TOS	V4, 5, internetwork control
IP datagram length		68 bytes
Identification		
Fragmentation		
TTL	Protocol	TTL 1, protocol 89 (OSPF)
Header checksum		
Source IP address		10.1.1.1
Destination IP address		224.0.0.5 AllSPFRouters
OSPF version	OSPF pkt type	OSPF version 2, hello
OSPF length		48 bytes
Source OSPF router ID		10.1.1.1
OPSF area ID		0.0.0.0
Packet checksum		
Authentication type		0 (none)
Authentication data		
Network mask		0.0.0.0 (don't care)
HelloInterval		10 seconds
Options	Router priority	E-bit, priority 1
RouterDeadInterval		40 seconds
Designated router		0.0.0.0 (none)
Backup designated router		0.0.0.0 (none)
1st neighbor ID		10.1.1.2

IP header

OSPF header

Hello packet body

- The OSPF Router ID of the sender: The receiving router can tell which OSPF router the packet came from.
- A packet checksum: This allows the receiving router to determine whether the packet has been damaged in transit, and if so, the packet is discarded.
- Authentication fields: For security, these fields allow the receiving router to verify that the packet was indeed sent by the OSPF router whose Router ID appears in the header and that the packet's contents have not been modified by a third party.
- An OSPF Area ID: This enables the receiving router to associate the packet to the proper level of OSPF hierarchy and to ensure that the OSPF hierarchy has been configured consistently.

Neighbor Discovery and Maintenance

All routing protocols provide a way for a router to discover and maintain neighbor relationships, also sometimes called peer relationships. A router's neighbors, or peers, are those routers with which the router will directly exchange routing information. In OSPF, a router discovers neighbors by periodically sending OSPF Hello packets out all of its interfaces. By default, a router sends Hellos out an interface every 10 seconds, although this interval is configurable as the OSPF parameter HelloInterval. A router learns the existence of a neighboring router when it receives the neighbor's OSPF Hello in turn.

The part of the OSPF protocol responsible for sending and receiving Hello packets is called OSPF's Hello protocol. The transmission and reception of Hello packets also enables a router to detect the failure of one of its neighbors. If enough time elapses, specified as the OSPF configurable parameter RouterDeadInterval, whose default value is 40 seconds without the router's receiving a Hello from a neighbor, the router stops advertising the connection to the router and starts routing data packets around the failure. In most cases, the failure of the neighbor connection should be noticed much earlier by the data link protocol. Detecting neighbor failures in a timely fashion is crucial to OSPF protocol performance. The time to detect neighbor failures dominates convergence time because the rest of the OSPF protocol machinery, including flooding updated LSAs and redoing the routing table calculations required to route packets around the failure, takes at most a few seconds.

Besides enabling discovery and maintenance of OSPF neighbors, OSPF's Hello protocol also establishes that the neighboring routers are consistent in the following ways:

- The Hello protocol ensures that each neighbor can send packets to and receive packets from each other. The Hello protocol ensures that the link between neighbors is bidirectional. Unless the link is bidirectional, OSPF routers will not forward data packets over the link. This procedure prevents a routing failure that can be caused by unidirectional links in other protocols, which is most easily demonstrated using RIP. Suppose that RIP routers A and B are connected and that packets can go from B to A but not vice versa. B may then send routing updates for network N to A, causing A to install B as the next hop for N. However, whenever A forwards a packet destined for network N, the packet will go into a black hole.

● The Hello protocol ensures that the neighboring routers agree on the HelloInterval and RouterDeadInterval parameters. This ensures that each router sends Hellos quickly enough so that losing an occasional Hello packet will not mistakenly bring the link down.

The OSPF Hello protocol performs additional duties on other types of network segments. Also, the OSPF Hello protocol is used to detect and negotiate certain OSPF extensions. Figure 11.8 shows the Hello packet that router 10.1.1.1 would be sending to router 10.1.1.2 in Figure 11.3 after the connection has been up for some time. By including router 10.1.1.2 in the list of neighbors recently heard from, router 10.1.1.1 indicates that it is receiving router 10.1.1.2's Hellos. On receiving this Hello packet, router 10.1.1.2 will then know that the link is bidirectional. The fields Network Mask, Router Priority, Designated Router, and Backup Designated Router are used only when the neighbors are connected by a broadcast or NBMA network segment. The Options field is used to negotiate optional behavior between the neighbors.

Database Synchronization

Database synchronization in a link state protocol is crucial. As long as the database remains synchronized, a link state protocol's routing calculations ensure correct and loop-free routing. It is no surprise, then, that most of the protocol machinery in a link state protocol exists simply to ensure and to maintain synchronization of the link state database.

Database synchronization takes two forms in a link state protocol. First, when two neighbors start communicating they must synchronize their databases before forwarding data traffic over their shared link or routing loops may ensue. Second, there is the continual database resynchronization that must occur as new LSAs are introduced and distributed among routers. The mechanism that achieves this resynchronization is called reliable flooding.

Reliable Flooding

Reliable flooding is the procedure by which updated LSAs are sent throughout the routing domain.

As LSAs are updated with new information, they are sent throughout the routing domain by a procedure called *reliable flooding*. The flooding procedure starts when a router wishes to update one of its self-originated LSAs. This may be because the router's local state has changed. For example, one of the router's interfaces may have become inoperational, causing the router to reoriginate its router-LSA. Or the router may wish to delete one of its self-originated LSAs, in which case it sets the LSA's LS Age field to MaxAge. In any case, the router then floods the LSA, packaging the LSA within a Link State Update packet out of all of its interfaces.

When one of the router's neighbors receives the Link State Update packet, the neighbor examines each of the LSAs contained within the update. For each LSA that is uncorrupted, of known LS type, and more recent than the neighbor's own database copy, the neighbor installs that LSA in its link state database, sends an acknowledgement to the router, repackages the LSA within a new Link State Update packet, and sends it out all interfaces except the one that received the LSA in the first place. This procedure then iterates until all routers have the updated LSA. To achieve reliability, a router will periodically retransmit an LSA sent to a neighbor until the neighbor acknowledges receipt of the LSA by sending a Link State Acknowledgement packet listing the updated LSA's link state header.

Figure 11.9 shows an example of OSPF flooding. The example starts with router 10.1.1.3 updating its router-LSA and flooding the LSA to its neighbors at Time T1. The example then makes the simplifying assumptions that each router spends an equal time processing the update, that transmission time of the update is equal on all links, and that all of the updates are received intact. After a number of iterations less than or equal to the diameter of the network, in this example Time T3, all routers have the updated LSA. At Time T2, router 10.1.1.4 seems to receive the update from both 10.1.1.2 and 10.1.1.5 simultaneously. It is assumed that the update from 10.1.1.5 is processed first, thus at Time T3, 10.1.1.4 sends an update to 10.1.1.2 only. After Time T3, acknowledgements are sent. In OSPF, routers generally delay acknowledgements of received LSAs in hope that they can fit more LSA acknowledgements into a single Link State acknowledgement packet, which reduces routing protocol processing and bandwidth consumption. Not all updates require explicit acknowledgement. When updates cross with each neighbor sending the other the same update as in the case of routers 10.1.1 and 10.1.1.2 at Time T2, the received update is taken as an *implicit acknowledgement*, and no corresponding Link State Acknowledgement packet is required to be sent. In fact, as a result of this property, in any link direction, one update or acknowledgement is sent but never both. The preceding example has described flooding in a network of routers connected via serial lines. When different network segment types such as broadcast networks are present or when some routers run optional extensions to the OSPF protocol, flooding can get more complicated.

Flooding Robustness

OSPF's reliable flooding scheme is robust in the face of errors. By robust it is meant that even when transmission errors, link and/or router failures occur, the network continues to function correctly; link-state databases continue to be synchronized, and the amount of routing traffic remains at an acceptable level. OSPF's flooding achieves this robustness because of the following features:

- Flooding could be restricted to a minimal collection of links interconnecting all routers, which is called spanning tree. As long as the links in the spanning tree remain operational, router databases would remain synchronized. However, OSPF does not use a spanning tree; it floods over all of the links. As a result, the

> When updates cross with each neighbor sending the other the same update, the received update is known as an **implicit acknowledgement**.

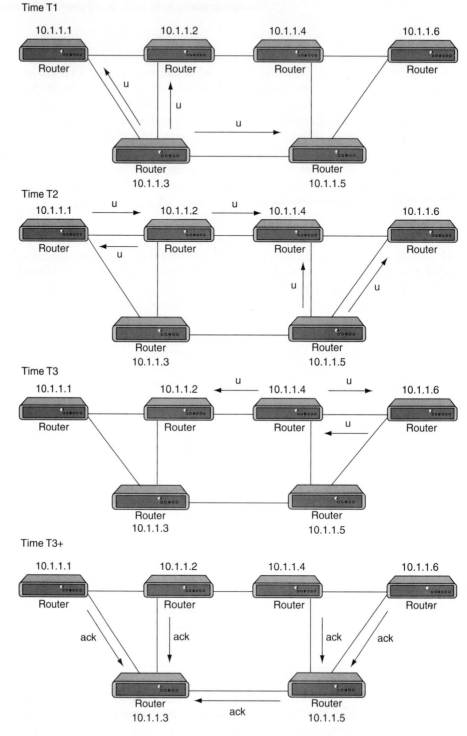

FIGURE 11.9

Reliable flooding, starting at 10.1.1.3, where u is a link state update packet and ack is a link state acknowledgement packet. (Reproduced with permission of Cisco Systems, Inc. Copyright © 2001 Cisco Systems, Inc. All rights reserved.)

failure of any link does not significantly disrupt database synchronization, as LSA updates simultaneously flow on alternate paths around link failure.

- Because of software errors a router might accidentally delete one or more LSAs from its database. To ensure that the router eventually regains database synchronization with the rest of the OSPF routing domain, OSPF mandates that all originators of LSAs refresh their LSAs every 30 minutes, incrementing the sequence numbers of the LSAs and reflooding them throughout the routing domain.

- To detect corruption of LSAs as they are flooded, each LSA contains a checksum field. An LSA's checksum is originally calculated by the LSA's originator and then becomes a permanent part of the LSA. When a router receives an LSA during flooding, it verifies the LSA's checksum. A checksum failure indicates that the LSA has been corrupted. Corrupted LSAs are discarded and not acknowledged in hope that the retransmitted LSA will be valid.

- Errors in implementation might lead to situations in which the routers disagree on which LSA instance is more recent, possibly causing flooding loops. An example of this was the problem caused by the ARPANET routing protocol's circular sequence space. To guard against this kind of problem, an OSPF router increments an LSA's LS Age field when placing the LSA into a Link State Update packet. This action breaks flooding loops because a looping LSA's LS Age field will eventually hit the MaxAge, at which time the LSA will be discarded. This action is similar to the breaking of IP forwarding loops through the use of the IP header's TTL field.

- To guard against a rapidly changing network element, for example, a link between routers that is continually going up and down causing excessive amount of control traffic, OSPF imposes rate limits on LSA origination—any particular LSA can be updated at most once every 5 seconds (MinLSInterval).

- To guard against routers that are updating their LSAs at too high a rate, an OSPF router will refuse to accept a flooded LSA if the current database copy was received less than 1 second ago. For example, suppose that router 10.1.1.1 in Figure 11.3 disregards the MinLSInterval limit and begins updating its router-LSA 100 times a second. Rather than flood the entire network with these excessive updates router 10.1.1.1's neighbors, 10.1.1.2 and 10.1.1.3, will discard most of the updates from 10.1.1., allowing only 1 update in 100 to escape into the network at large. This action localizes the disruption to 10.1.1.1's immediate neighbors.

Given that (a) when an LSA is flooded either the LSA or an acknowledgement for the LSA is sent over every link, but not both, (b) each LSA is guaranteed to be flooded at least every 30 minutes—the refresh time—and (c) the most an LSA can be flooded is once every 5 seconds, one can calculate the minimum and maximum amount of link bandwidth consumed by flooding traffic. For example, in Figure 11.3 the sum of LSA sizes is 336 bytes, 24 bytes for the fixed portion of each router-LSA and 12 bytes for each link direction. This yields a maximum link bandwidth consumption by flooding of 537 bits/second and a minimum consumption of 1.5 bits/second. Use of the demand circuit extensions to OSPF can further reduce the amount of link bandwidth consumed by flooding, at the expense of reduced robustness.

Routing Calculations

An IP router does not use the link state database when forwarding IP datagrams. Instead, the router uses its IP routing table. Just how does an OSPF router produce an IP routing table from its OSPF link state database?

The link state database describes the routers and the links that interconnect them. However, only the links that are appropriate for forwarding have been included in the database. Inoperational links, unidirectional links detected by the Hello protocol, and links between routers whose databases are not yet synchronized have been omitted from the database.

The link state database also indicates the cost of transmitting packets on the various links. What are the units of this link cost? OSPF leaves that question unanswered. Link cost is configurable by the network administrator. There are only three restrictions on link cost. First, the cost of each link must be in the range of 1 to 65,535. Second, the more preferred links should have a smaller cost. And third, it must make sense to add link costs because the cost of a path is set to equal to the sum of the cost of its constituent links. For example, link delay makes a perfectly good metric. So does hop count. If you set the cost of each link to 1, the cost of a path is the number of links in the path.

The link state database for the network of Figure 11.3 is shown in Figure 11.10. The metrics used in this example are also called weighted hop count. Small integers are assigned to each link, with more preferred links getting smaller values than others. Link costs have been assigned symmetrically in this example, although they do not have to be. Each router can advertise a different cost for the router's own link direction. If costs are assigned asymmetrically, paths will generally be asymmetric, with packets in one direction taking a given path and responses traveling a different path. IP works

FIGURE 11.10

Link state database. (Reproduced with permission of Cisco Systems, Inc. Copyright © 2001 Cisco Systems, Inc. All rights reserved.)

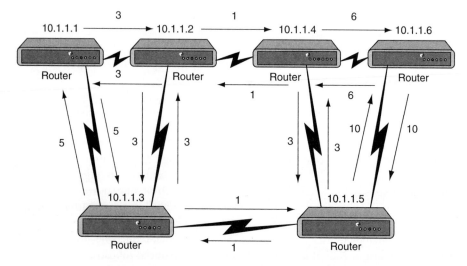

fine in the face of such asymmetric paths, although many network administrators dislike asymmetry because it makes routing problems more difficult to debug.

Each link in Figure 11.10 is advertised by two routers, namely those at either end of the link. For example, the link between 10.1.1.2 and 10.1.1.4 is advertised by both routers in their router-LSAs, each with a cost of 1. To avoid using stale information in the routing calculation, OSPF requires that both routers advertise the link before it can be used by the routing calculation. For example, suppose that router 10.1.1.4 in Figure 11.10 fails. Router 10.1.1.4's router-LSA advertising the now inoperational links to routers 10.1.1.2, 10.1.1.5, and 10.1.1.6 will persist in the link state database for up to an hour. But since these routers will no longer be advertising their halves of the link to 10.1.1.4, data traffic will be successfully rerouted around the failure.

The link state database in Figure 11.10 reconstitutes the network map in the form of a directed graph. The cost of a path in this graph is the sum of the path's component link costs. For example, there is a path of cost 4 from router 10.1.1.3 to 10.1.1.4, going through router 10.1.1.5. On the graph, a router can run any algorithm that calculates shortest paths. An OSPF router typically uses Dijkstra's *Shortest Path First (SPF) algorithm* for this purpose.

Dijkstra's algorithm is a simple algorithm that efficiently calculates all at once the shortest paths to all destinations. The algorithm incrementally calculates a tree of shortest paths. It begins with the calculating router adding itself to the tree. All of the router's neighbors are then added to a *candidate list*, with costs equal to the cost of the links from the router to the neighbors. The router on the candidate list with the smallest cost is then added to the shortest-path tree, and that router's neighbors are then examined for inclusion in or modification of the candidate list. The algorithm then iterates until the candidate list is empty.

Table 11.4 shows Dijkstra's algorithm as run by router 10.1.1.3 in Figure 11.10. At each iteration another router is added to the tree of shortest paths. While on the candidate list, each destination is listed with its current cost and next hops. These values are then copied to the routing table when the destination is moved from the candidate list to the shortest-path tree. The boldface type in Table 11.4 indicates those places where a destination is first added to the candidate list or when its values on the candidate list are modified. For example, in iteration 4, destination 10.1.1.6's entry on the candidate list is modified when a shorter path is discovered by going through router 10.1.1.4.

As you can see by going through the example, Dijkstra's algorithm ends up examining each link in the network once. When examining the link, the algorithm may place a destination onto the candidate list or modify a destination's entry on the candidate list. This operation also requires a sort of the candidate list because the router always wants to know which destination on the candidate list has the smallest cost. Because the sort is a $0(\log(n))$ operation and the size of the candidate list never exceeds the number of destinations, the performance of Dijkstra's algorithm is $0(l^*\log(n))$, where l is the number of links in the network and n is the number of destinations in our example equal to the number of routers.

The **Shortest Path First algorithm** is a routing algorithm sometimes called *Dijkstra's algorithm*.

The **candidate list** is a list containing all of the router's neighbors which includes costs equal to the cost of the links from the router to the neighbors.

Table 11.4 Dijkstra Calculation Performed by Router 10.1.1.3

Iteration	Destination Added to Shortest-Path Tree	Candidate List Destination (cost, next hops)
1	10.1.1.3	**10.1.1.5 (1, 10.1.1.5)** **10.1.1.2 (3, 10.1.1.2)** **10.1.1.1 (5, 10.1.1.1)**
2	10.1.1.5	10.1.1.2 (3, 10.1.1.2) **10.1.1.1 (4, 10.1.1.5)** 10.1.1.1 (5, 10.1.1.1) **10.1.1.6 (11, 10.1.1.5)**
3	10.1.1.2	**10.1.1.4 (4, 10.1.1.5, 10.1.1.2)** 10.1.1.1 (5, 10.1.1.1) 10.1.1.6 (11, 10.1.1.5)
4	10.1.1.4	10.1.1.1 (5, 10.1.1.1) **10.1.1.6 (10, 10.1.1.5, 10.1.1.2)**
5	10.1.1.1	10.1.1.6 (10, 10.1.1.5, 10.1.1.2)
6	10.1.1.6	Empty

The resulting shortest paths calculated by router 10.1.1.3 in Figure 11.10 are displayed in Figure 11.11. Multiple shortest paths have been found to 10.1.14, allowing load balancing of traffic to that destination called equal-cost multipath. An OSPF router knows the entire path to each destination. However, in IP's hop-by-hop routing paradigm only the first hop is needed for each destination. Extracting this

FIGURE 11.11

Shortest paths, as calculated by router 10.1.1.3. (Reproduced with permission of Cisco Systems, Inc. Copyright © 2001 Cisco Systems, Inc. All rights reserved.)

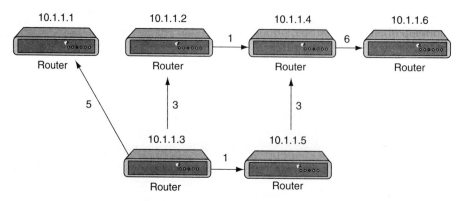

Table 11.5 Router 10.1.1.3's IP Routing Table

Destination	Next Hop(s)	Cost
10.1.1.1	10.1.1.1	5
10.1.1.2	10.1.1.2	3
10.1.1.4	10.1.1.2 10.1.1.5	4
10.1.1.5	10.1.1.5	1
10.1.1.6	10.1.1.2 10.1.1.5	10

information from the collection of shortest paths yields the IP routing table, as displayed in Table 11.5.

The routing calculation is the simplest part of OSPF. However, in the presence of network segments other than point-to-point links or when OSPF hierarchical routing or some of the OSPF extensions are used, the routing calculations get a little more complicated.

Hierarchical Routing in OSPF

Hierarchical routing is a technique commonly used when building large networks. As a network grows, so do the resource requirements for the network's management and control functions. In a TCP/IP network, resource consumption includes:

- Router memory consumed by routing tables and other routing protocol data. In an OSPF network this other data would include the OSPF link state database.
- Router computing resources used in calculating routing tables and other routing protocol functions. In an OSPF network these resources include the CPU required to calculate shortest-path trees.
- Link bandwidth, used in distributing routing data. In an OSPF network this includes the bandwidth consumed by OSPF's database synchronization procedures.

It is inevitable that resource requirements grow as the network grows. But the question is, How quickly do resource requirements need to grow? Looking at the routing table size as an example (employing the OSPF protocol as described earlier) one sees that the routing table size increases linearly, specifically one for one as the number of TCP/IP segments grows. This is called *flat routing* as each router in the network is aware of the existence and specific addresses belonging to each and every network segment. However, by employing hierarchical routing, the growth rate of the routing

Hierarchical routing is a technique commonly used when building large networks.

In **flat routing** each router in the network is aware of the existence and specific addresses belonging to each and every network segment.

FIGURE 11.12

Linear versus log-arithmic growth.
(Reproduced with permission of Cisco Systems, Inc. Copyright © 2001 Cisco Systems, Inc. All rights reserved.)

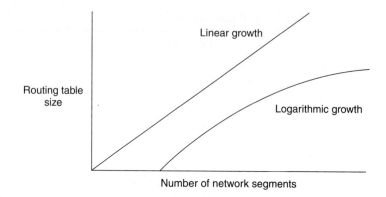

table can be slowed down to the order of the logarithm of the number of segments, written as $O(\log(n))$, where n is the number of network segments. Figure 11.12 illustrates the difference between linear and logarithmic routing table growth.

In hierarchical routing, an internet is partitioned into pieces, which in turn are grouped recursively into levels. At the lowest level, inside one of the lowest level partitions, routing is flat, with all routers knowing about all network segments within the partition. The routers have only sketchy information about other partitions. When forwarding a packet addressed to a remote destination, the routers rely on the higher levels of hierarchical routing to navigate the internet, eventually locating the partition containing the destination address.

Figure 11.13 shows an internet organized into a three level hierarchy. All addresses come from the address range 10.0.0.0/8. There are three second level partitions, with the lower left containing the 10.1.0.0/16 addresses, the lower right 10.2.0.0/16, and the upper partition 10.3.0.0/16. The nine first level partitions contain even more specific addresses, with the lower-left first level partition containing the addresses 10.1.1.0/24, and so on.

Suppose that an IP packet is sent by host 10.1.1.6 to the destination 10.3.3.5. The packet first appears in the first level partition 10.1.1.0/24. Because the destination is not in the range 10.1.1.0/24 it is handed to second level routing. At this point, the packet is forwarded to the correct second level partition 10.3.0.0/16 and then by second level routing in the 10.1.0.0/16 partition and then similarly handed to third level routing. At this point, the packet is forwarded to the correct second level partition 10.3.3.0/24, whereupon first level routing delivers the packet to the destination. This pattern of forwarding the packet up the hierarchy from first level to second to third and then back down again is what gives hierarchical routing its name.

Hierarchical routing reduces routing table size. Suppose that 16 network segments are in every first level partition in Figure 11.13. Examine the routing from the perspective of a router in the first level partition labeled 10.1.1. If flat routing were deployed throughout the figure, the router would have a routing table consisting of $9^{*}16 = 144$ entries, one for every network segment. However, because of the three level hierarchy, the router has 16 entries for the local segments within 10.1.1.0/24

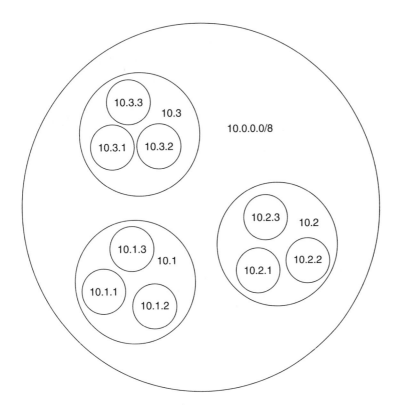

FIGURE 11.13

An internet employing hierarchical routing. (Reproduced with permission of Cisco Systems, Inc. Copyright © 2001 Cisco Systems, Inc. All rights reserved.)

and additional entries for 10.1.2.0/24, 10.1.3.0/24, and 10.2.0.0/16 for a total of 20 routing table entries, a marked reduction from 144.

There is often a trade-off involved in the routing table size reduction and reduction in other resources that can be accomplished with hierarchical routing. Namely, the information reduction can lead to suboptimal forwarding. Although packets are still forwarded to their destinations, the packets may take a longer path than can be found in a flat routing system. The worldwide Internet employs hierarchical routing to keep the routing table sizes and the memory and CPU demands on its routers to a manageable level. IP subnetting and CIDR addressing are tools used to implement the Internet's routing hierarchy. IP's 32-bit addressing generally limits the number of hierarchical levels to approximately four, although with Ipv6's 128-bit addressing more levels of hierarchy may be possible in the future.

OSPF implements a two level hierarchical routing scheme through the deployment of OSPF areas, which will be explained next. Furthermore, OSPF allows an internet to be split into additional levels by incorporating two levels of external routing information into the OSPF routing domain.

Hierarchical routing protocols are difficult to design. Most of the protocol bugs found in OSPF over the years have been encountered and repaired since OSPF's original design.

OSPF Areas

OSPF supports a two level hierarchical routing scheme through the use of OSPF *areas*. Each OSPF area is identified by a 32-bit Area ID and consists of a collection of network segments interconnected by routers. Each area has its own link state database, consisting of router-LSAs and network-LSAs describing how the area's routers and network segments are interconnected. Routing within the area is flat, with each router knowing exactly which network segments are contained within the area. However, detailed knowledge of the area's topology is hidden from all other areas—the area's router-LSAs and network-LSAs are not flooded beyond the area's borders.

Routers attached to two or more areas are called *area border routers (ABRs)*. Area border routers leak IP addressing information from one area to another in OSPF summary-LSAs. This enables routers in the interior of an area to dynamically discover destinations in other areas, which are the so-called interarea destinations, and to pick the best area border router when forwarding data packets to these destinations.

Figure 11.14 illustrates an example of an OSPF area configuration having four OSPF areas. Area 0.0.0.0's link state database consists of four router-LSAs. Area 0.0.0.1's link

FIGURE 11.14

A sample OSPF area configuration. (Reproduced with permission of Cisco Systems, Inc. Copyright © 2001 Cisco Systems, Inc. All rights reserved.)

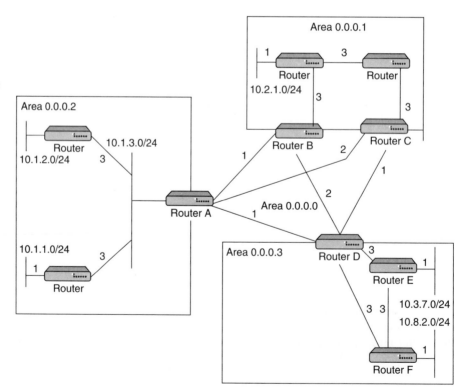

state database has four router-LSAs. Area 0.0.0.2's link state database has three router-LSAs and a network-LSA. Area 0.0.0.3's link state database has three router-LSAs. Because all of area 0.0.0.1's addresses fall into the range 10.2.0.0/16, routers B and C can be configured to aggregate area 0.0.0.1's addresses by originating a single summary-LSA with destination of 10.2.0.0/16. Likewise router A can be configured to aggregate area 0.0.0.2's addresses by advertising a single summary-LSA with destination equal to 10.1.0.0/16. However, area 0.0.0.3's addresses do not aggregate, so router D will end up originating two summary-LSAs, one for 10.3.7.0/24 and one for 10.8.0/24.

Figure 11.15 displays in detail the summary-LSA that router B uses to leak area 0.0.0.1's addressing information into area 0.0.0.0. Router B has been configured to aggregate all of area 0.0.0.1's addresses into a single advertisement for the prefix 10.2.0.0/16. The Link State ID for the summary-LSA is the prefix address 10.2.0.0. The prefix mask 255.255.0.0 is included in the body of the summary-LSA. Also included in the summary-LSA is the cost from the advertising router (router B) to the prefix. Since 10.2.0.0/16 is an aggregation, the cost in this case is set to the cost from router B to the most distant component of 10.2.0.0, in this case, 10.2.2.0/24, at a cost of 7. Area 0.0.0.1's addresses are distributed throughout area 0.0.0.0, as the summary-LSA is flooded throughout area 0.0.0.0. From area 0.0.0.0, the addresses are advertised into other areas.

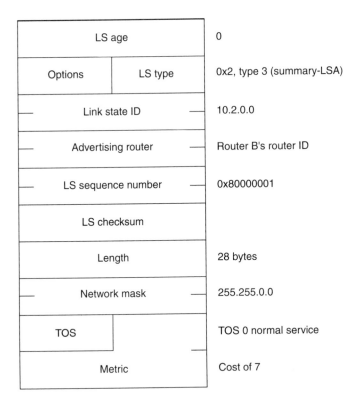

FIGURE 11.15

Summary-LSA advertised by router B into area 0.0.0.0. (Reproduced with permission of Cisco Systems, Inc. Copyright © 2001 Cisco Systems, Inc. All rights reserved.)

Splitting an OSPF routing domain into areas reduces OSPF's demands for router and network resources. Because the area's link state database contains only router-LSAs and network-LSAs for the area's own routers and networks the size of the link state database is reduced, along with the amount of flooding traffic necessary to synchronize the database. If aggregation is employed at area boundaries, routing table size is also reduced. The cost of the shortest-path calculation is $0(i^* \log(n))$, where I is the number of router interfaces, and n is the number of routers. Thus, as the routing domain is split into areas, the cost of the shortest-path calculation decreases. Splitting the routing domain adds some amount of summary-LSAs to the database and routing calculations. However, summary-LSAs are smaller than router-LSAs, and the routing calculations involving summary-LSAs are cheaper than the shortest-path calculation. In fact, the routing calculations for all summary-LSAs in an area are like the processing of a single RIP packet.

In 1991, the Working Group recommended that the size of an OSPF area be limited to 200 routers, based solely on the cost of the shortest-path calculation. However, this estimate is probably outdated because router CPU speeds have increased considerably since then. In reality, maximum area size is implementation specific. Some vendors of OSPF routers are building areas of 500 routers, whereas others recommend that the number of routers in an area be limited to 50.

In addition to allowing one to build much larger OSPF networks, OSPF areas provide the following functionality:

- Increased robustness: The effects of router and/or link failures within a single area are dampened external to the area. At most, a small number of summary-LSAs are modified in other areas, and when aggregation is employed, possibly nothing will change in the other areas at all.
- Routing protection: OSPF always prefers paths within an area (intra-area paths) over paths that cross area boundaries. This means that routing within an area is protected from routing instabilities or misconfiguration in other areas. For example, suppose that a corporation runs OSPF and assigns the engineering department as one area and the marketing department as another. Then, even if the marketing department mistakenly uses some of the engineering address prefixes, communication within engineering will continue to function correctly.
- Hidden prefixes: One can configure prefixes so that they will not be advertised to other areas. This capability allows one or more subnets to be hidden from the rest of the routing domain, which may be necessary for policy reasons: The subnets may contain servers that should be accessed only by clients within the same area.

Area Organization

The **backbone area** is a special area in an OSPF routing domain.

When an OSPF routing domain is split into areas, all areas are required to attach directly to a special area called the OSPF *backbone area*. The backbone area always has an Area ID 0.0.0.0. In the sample area configuration of Figure 11.14, areas 0.0.0.1, 0.0.0.2, and 0.0.0.3 attach directly to area 0.0.0.0 via the area border routers A, B, C, and D.

The exchange of routing information between areas is essentially a distance vector algorithm. Using router D in Figure 11.14 as an example, the exchange of routing information includes the following steps:

1. The area border routers A through D advertise the addresses of their directly connected areas by originating summary-LSAs into the backbone.
2. Router D receives all the summary-LSAs through flooding.
3. For any given destination router D examines all summary-LSAs advertising that destination, using the best summary-LSA to create a routing table entry for the destination, and then readvertises the destination into its attached area 0.0.0.3 in summary-LSAs of its own.

In particular, for the destination 10.2.0.0/16, router D sees two summary-LSAs, one from router C and one from router B, each advertising a cost of 7. Router D then selects the summary-LSA from C as better because the total cost through C is smaller. Router D then installs a routing table entry for 10.2.0.0/16 with a cost of 8: the cost advertised in router C's summary-LSA plus the cost from D to C. Finally, router D originates a summary-LSA for 10.2.0.0/16, with a cost of 8 into area 0.0.0.3 so that area 0.0.0.3's routers will know how to reach 10.2.0.0/16.

The similarities between the distribution of area routing information in OSPF and the operation of the canonical distance vector algorithm, RIP, are given in Table 11.6. The use of distance vector mechanisms for exchanging routing information between areas is the reason for requiring all areas to attach directly to the OSPF backbone. The larger the number of redundant paths in a network, the worse a distance vector algorithm's convergence properties. Requiring all areas to attach directly to the backbone limits the topology for interarea routing exchange to a simple hub-and-spoke topology as illustrated in Figure 11.16, which eliminates redundant paths and is not subject to distance vector convergence problems such as counting to infinity.

Table 11.6 Distribution of Area Routing Information Using Distance Vector Mechanisms

Area Routing Function	Analogous RIP Function
Originate summary-LSAs for directly attached areas into backbone.	Send directly attached nets in RIP packets.
Receive summary-LSAs via flooding.	Receive RIP packets from neighbors.
Add cost in summary-LSA to distance to summary-LSA originator.	Add 1 to cost of each received route.
Choose best summary-LSA.	Choose best route advertised by neighbor.
Originate own summary-LSAs into directly attached areas.	Advertise updated routing table in RIP packets sent to neighbors.

FIGURE 11.16
*Interarea routing
exchange in the
sample OSPF of
Figure 11.14.
(Reproduced with
permission of
Cisco Systems, Inc.
Copyright
© 2001 Cisco
Systems, Inc. All
rights reserved.)*

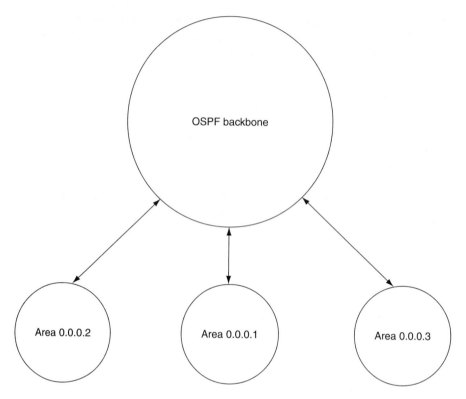

However, requiring direct connection of all areas to the OSPF backbone does not mean requiring physical connectivity to the backbone. The loss of information that enables OSPF area routing to scale can also lead to the selection of less efficient paths. If router D forwards a packet to the IP destination 10.2.1.20, it will forward the packet to router C instead of along the shorter path through router B. This behavior is due to the fact that router D does not even realize the existence, much less the location, of the network segment 10.2.1.0/24, and so forwards using the aggregated routing table entry of 10.2.0.0/16 instead.

Virtual Links

OSPF requires that all areas attach directly to the backbone area but not that the attachment be physical. One can take any physical arrangement of areas and attach them logically to the backbone through the use of OSPF virtual links. For example, suppose that the organization whose network is pictured in Figure 11.14 purchases two smaller companies and adds their networks as separate OSPF areas, as pictured in Figure 11.17. The two new areas do not attach physically to the backbone, so two virtual links are configured through area 0.0.0.3, the first having as end points

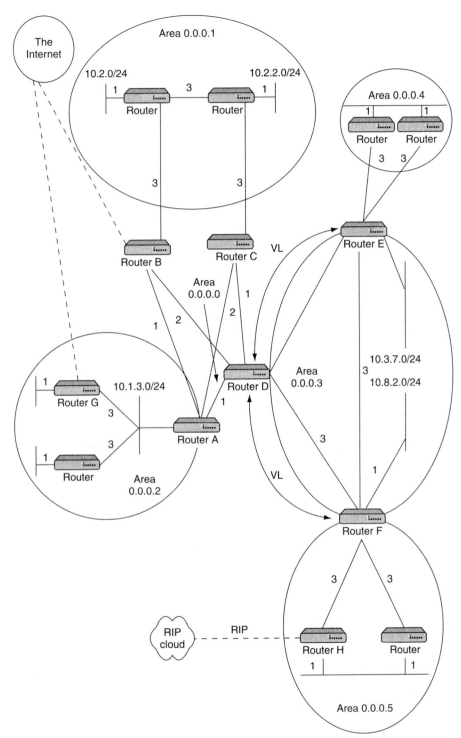

479

routers D and E, the second having as end points routers D and F. Virtual links allow summary-LSAs to be tunneled across nonbackbone areas, maintaining the desired hub-and-spoke topology for interarea routing exchange. In Figure 11.17, router A receives the summary-LSA from router E, after the summary-LSA has been tunneled across area 0.0.0.3. To evaluate the relative cost of the summary-LSA, router A sums the cost of the backbone path to router D, the cost of the virtual link to router E, and the cost advertised in router E's summary-LSA. In this fashion, the cost of the virtual link to router E, and the cost advertised in router E's summary-LSA. In this fashion, the virtual link acts like a point-to-point link that has been added to the backbone.

However, although the exchange of routing topology continues to follow a simple hub-and-spoke topology, the forwarding of data packets does not. The OSPF routing calculations for virtual links have a built-in shortcut calculation, allowing data packets to avoid the backbone area when the backbone is not on the shortest path. Using Figure 11.17 again as an example, routing information from area 0.0.0.4 goes to the backbone before being redistributed to area 0.0.0.5, but data traffic from area 0.0.0.4 to 0.0.0.5 simply traverses area 0.0.0.3, avoiding the backbone altogether.

Unfortunately, many administrators find it difficult to decide when and where to configure virtual links. It is possible to design algorithms so that the routers themselves can dynamically establish virtual links. In the future, these algorithms may relieve network administrators of the burden of configuring virtual links.

Incorporating External Routing Information

The entire Internet is not run as a single OSPF domain. Many routing protocols are in use in the Internet simultaneously such as OSPF, RIP, BGP, IGRP, and IS-IS to name a few. Generally, on the edge of an OSPF routing domain you will find routers that run on one or more of these routing protocols, in addition to OSPF. It is the job of these routers, called AS boundary routers (ASBRs), to import the routing information learned from the other routing protocols into the OSPF routing domain. This behavior allows routers internal to the routing domain to pick the best exit router when routing to destinations outside the OSPF routing domain, just as summary-LSAs allow routers within an OSPF area to pick up the right area exit when forwarding to interarea destinations.

Information learned from other routing protocols and for destinations outside of the OSPF routing domain is called external routing information, or external routes. External routes are imported into the OSPF routing domain in AS-external-LSAs originated by the AS boundary routers. Each AS-external-LSA advertises a single prefix.

Consider, for example, the network in Figure 11.17. Routers B and G are running BGP sessions to learn of destinations in the Internet at large. Tens of thousands of routes may be learned in this way. To date, the default-free core of the Internet carries more than 40,000 routes. However, probably not all of these routes would be imported into the OSPF domain. Routers B and G would import only those routes where the choice of B or G was important—when either exit would do, default routes imported by B and G

Table 11.7 OSPF's Four Level Routing Hierarchy

Level	Description
1	Intra-area routing
2	Interarea routing
3	External Type 1 metrics
4	External Type 2 metrics

would suffice. Still it would not be unusual for B and G to originate several thousand AS-external-LSAs into the routing domain. In addition, router H is exchanging RIP information with an isolated collection of RIP routers. Router H then imports each prefix that it learns from the RIP routers in an AS-external-LSA.

Paths internal to the OSPF routing domain are always preferred over external routes. External routes can also be imported into the OSPF domain at two separate levels, depending on metric type. This gives a four level routing hierarchy, as illustrated in Table 11.7.

Paths that stay within one level are always preferred over paths that must traverse the next level. For example, in the network in Figure 11.17, router H may import its RIP routes into OSPF as external Type 1 metrics, and routers B and G import their routing information as external Type 2 metrics. Then the routing preferences introduced by the OSPF hierarchy in Figure 11.17 are as follows, from most preferred to least preferred: (1) routing within any given OSPF area, (2) routing within the OSPF routing domain itself, (3) routing within the OSPF domain and RIP cloud, taken together, and (4) routing within the Internet as a whole.

Besides establishing two different routing levels, external Type 1 and Type 2 metrics have different semantics. The use of Type 1 metrics assumes that in the path from OSPF router to destination, the internal component (path to the ABSR advertising the AS-external-LSA) and external component (cost described by external Type 1 metric) are of the same order. For example, if the OSPF routing domain used hop count as its metric, setting each interface cost to 1 and RIP routes were imported as external Type 1 metrics, the combined OSPF and RIP system would operate more or less seamlessly, selecting paths based on minimum hop count even when they cross the OSPF-to-RIP boundary. In contrast, external Type 2 metrics assume that the external part of the path (cost given by the external Type 2 metric) is always more significant than the internal cost to the AS boundary router. This would be the case when BGP routes were imported as external Type 2 metrics, with the metric set to the BGP route's AS path length no matter what the cost to the advertising AS boundary router the whole OSPF routing domain is still only a single AS.

Figure 11.18 displays an AS-external-LSA. This AS-external-LSA assumes that router B has learned the prefix 8.0.0.0/8 through BGP, with an AS path length of 12, and that

FIGURE 11.18

An AS-external-LSA. (Reproduced with permission of Cisco Systems, Inc. Copyright © 2001 Cisco Systems, Inc. All rights reserved.)

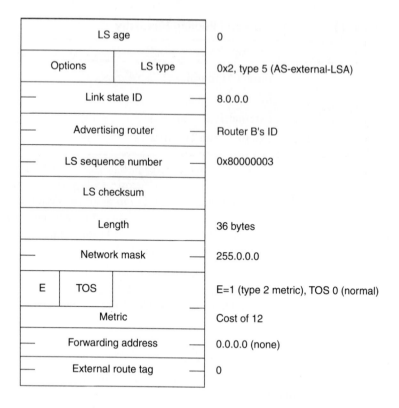

AS path length is being used as the OSPF external Type 2 metric. Note that the format of the AS-external-LSA is very similar to the summary-LSA: The Link State ID for both is the address prefix of the route being advertised, and both LSAs contain the network mask and route cost in the body of the LSA.

There are two fields in the AS-external-LSA of Figure 11.18 that have not been mentioned. The Forwarding Address field in this particular AS-external-LSA has been set to 0.0.0.0 to indicate that traffic destined for 8.0.0.0/8 should be forwarded to router B, the originator of the AS-external-LSA. However, by specifying another router's IP address in the Forwarding Address field, router B can have traffic forwarded to another router instead. This feature is used to prevent extra hops at the edge of the routing domain and would be done automatically by router B when necessary.

The External Route Tag field is not used by OSPF itself but instead is used to convey information between the routing protocols being run at the edge of the OSPF routing domain (BGP and RIP in Figure 11.17). For example, in the BGP–OSPF interaction specified, the external route tag is set when importing external routes on one edge of the routing domain to give routers on the other side of the routing domain information as to whether, and if so, how they should export this routing information to other Autonomous Systems. The external route tag is also used by the external-attributes-LSA as an alternative to BGP.

Interaction with Areas

How is external routing information conveyed across area borders in OSPF? One way this could have been done in OSPF was to reoriginate the AS-external-LSAs at area borders, just as OSPF does with summary-LSAs for interarea routes. However, this would have been expensive in terms of database size. When there are multiple area border routers for a given area, multiple AS-external-LSAs would have to be originated for each original AS-external-LSA: one origination per area border router. And within the area border routers, the situation would have been even worse, with each area border router holding a slightly different version of each AS-external-LSA for each attached area. OSPF takes a different tactic simply by flooding AS-external-LSAs across area borders. For example, the AS-external-LSA in Figure 11.18 is flooded throughout all areas in Figure 11.17; all routers in the network then hold this exact AS-external-LSA in their link state databases.

In particular, router H in area 0.0.0.5 has the AS-external-LSA in its link state database. However, to make use of this information router H must know the location of the originator of the AS-external-LSA, in this case, the ASBR router B. For this reason, OSPF advertises the location of ASBR's from area to area, using Type 4 summary-LSAs also called ASBR-summary-LSAs. The ASBR-summary-LSA that the area border router F originates into the area 0.0.0.5 in order to advertise the location of ASBR B is illustrated in Figure 11.19. The ASBR-summary-LSA is the OSPF Router ID of the ASBR whose location is being advertised. The concept of format, origination, and processing of ASBR-summary-LSAs is identical to that of summary-LSAs.

The AS-external-LSA is the only type of OSPF LSA, other than the proposed external-attributes-LSA, that is flooded throughout the entire OSPF routing domain. OSPF AS-external-LSAs have AS flooding scope, whereas router-LSAs, network-LSAs, and summary-LSAs, which are not flooded across area borders, have area flooding scope.

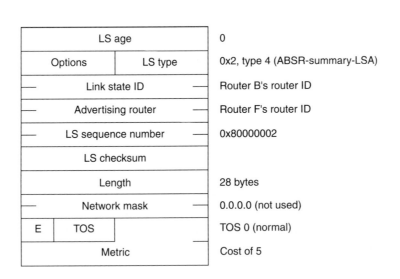

LS age		0
Options	LS type	0x2, type 4 (ABSR-summary-LSA)
— Link state ID —		Router B's router ID
— Advertising router —		Router F's router ID
— LS sequence number —		0x80000002
LS checksum		
Length		28 bytes
— Network mask —		0.0.0.0 (not used)
E	TOS	TOS 0 (normal)
Metric		Cost of 5

FIGURE 11.19

ASBR-summary-LSA originated by router F into area 0.0.0.5. (Reproduced with permission of Cisco Systems, Inc. Copyright © 2001 Cisco Systems, Inc. All rights reserved.)

There can be thousands of external routes imported into an OSPF routing domain in AS-external-LSAs, forming a large part of an OSPF area's link state database. For this reason, additional OSPF area types have been defined that restrict the amount of external routing information within an area, thereby limiting the resources that OSPF consumes in the area's routers and links, at the expense of reduced functionality.

OSPF Area Types

Normal OSPF areas have some desirable properties. Normal areas can be placed anywhere within the OSPF routing domain, although possibly requiring configuration of virtual links. Normal areas calculate efficient, although not always optimal, interarea and external routes through the use of summary-LSAs, ASBR-summary-LSAs, and AS-external-LSAs. And normal areas support ASBRs, directly importing external routing information from other routing protocols and then distributing this information to the other areas.

However, this support requires processing and bandwidth resources that may not be available everywhere. To have areas with smaller routers and to have low bandwidth links, OSPF forgoes some of these desirable properties and introduces two restricted area types: stub areas and NSSAs.

Stub Areas

Stub areas contain routers that have limited resources, especially when it comes to router memory.

Of all OSPF area types, stub areas consume the least resources. *Stub areas* (part of the original OSPF design) were designed to contain routers that had limited resources, especially when it came to router memory.

To conserve router memory, the link state database in stub areas is kept as small as possible. AS-external-LSAs are not flooded into stub areas; instead, routing to external destinations within stub areas is based simply on default routes originated by a stub area's ABRs. As a result, ASBRs cannot be supported within stub areas. To further reduce the size of the link state database, origination of summary-LSAs into stub areas is optional. Interarea routing within stub areas can also follow the default route. Without AS-external-LSAs and summary-LSAs stub areas can also follow the default route. Without AS-external-LSAs and summary-LSAs stub areas cannot support virtual links either and so must lie on the edge of an OSPF routing domain.

As a result of these restrictions, not all areas can become stub areas. For example, in Figure 11.17 only areas 0.0.0.1 and 0.0.0.4 can be configured as stub areas. The backbone area can never be configured as a stub, and areas 0.0.0.2 and 0.0.0.5 support ASBRs while area 0.0.0.3 needs to support virtual links.

Even if you can configure an OSPF area as a stub area, you may not want to. The lack of AS-external-LSAs and possibly summary-LSAs means that routing to external and possibly interarea destinations can take less efficient paths than in regular areas. The trade-off for the improved scaling properties of hierarchical routing is the possibility of suboptimal routes. With the even better scaling properties of stub areas comes the possibility of even more suboptimal routes.

Not-So-Stubby Areas (NSSAs)

Not-so-stubby areas (NSSAs) were defined as an extension to stub areas. Although most of the stub area restrictions such as preventing the flooding of AS-external-LSAs into the area and not allowing configuration of virtual links through the area were deemed acceptable by the NSSA designers, they wanted the ability to import a small amount of external routing information into the NSSA for later distribution into the rest of the OSPF routing domain.

A typical example of an NSSA is area 0.0.0.5 in Figure 11.17. You want to import routes learned from the RIP cloud into area 0.0.0.5 and then to distribute these routes throughout the rest of the OSPF routing domain. However, area 0.0.0.5 does not need the collection of AS-external-LSAs imported by routers B and G as a result of their BGP sessions. Instead, routing in area 0.0.0.5 to these BGP learned destinations can be handled by a single default route pointing at router F.

External routing information is imported into an NSSA by using Type-7-LSAs (LS Type = 7). These LSAs have area flooding scope and are translated at the NSSA boundary into AS-external-LSAs that allow the external routing information to be flooded to other areas. The NSSA border serves as a one-way filter for external information, with external information flowing from NSSA to other areas but not vice versa.

●
Not-so-stubby areas are extensions to stub areas.

Summary

In this chapter you learned about interdomain routing. The first aspect of interdomain routing is Open Shortest Path First (OSPF). OSPF has two primary characteristics. The first is that the protocol is open, which means that its specification is in the public domain. The OSPF specification is published as Request for Comments (RFC) 1247. The second principal characteristic is that OSPF is based on the SPF algorithm, which is sometimes referred to as the Dijkstra algorithm— named for the person credited with its creation.

OSPF is a link state routing protocol that calls for the sending of link state advertisements (LSAs) to all other routers within the same hierarchical area. Information on attached interfaces, metrics used, and other variables is included in OSPF LSAs. As OSPF routers accumulate link state information, they use the SPF algorithm to calculate the shortest path to each node.

As a link state routing protocol, OSPF contrasts with RIP, which is a distance vector routing protocol. Routers running the distance vector algorithm send all or a portion of their routing tables in routing update messages to their neighbors.

This chapter also discussed OSPF operations including establishing adjacencies, router advertisement, area configuration, and different area types.

In Chapter 12 "Interdomain Routing Basics Part II," you will learn about the companion protocol of OSPF: Border Gateway Protocol (BGP).

Review Questions

1. When did the development of OSPF begin?
2. What are the functional requirements of OSPF?
3. What were some of the problems being experienced by RIP?
4. What is meant by type of service routing?
5. What are the differences between link state protocols and distance vector protocols?
6. What is the Designated Router?
7. What is the Backup Designated Router?
8. What type of protocol is OSPF?
9. What was the first link state algorithm?
10. Explain what happens when a link fails between two routers.
11. What are LSAs and what are their functions?
12. Draw a diagram showing the fields of the LSA header.
13. How are LSA instances identified?
14. What are the three types of LS Sequence Number Space organizations?
15. How are LSA contents verified?
16. How are LSAs removed from the database?
17. What comprises the link state database?
18. How are link state databases synchronized?
19. How can one tell which parts of the network are changing the most in an OSPF database?
20. What services of the IP header are used by OSPF?

Summary Questions

1. What are the fields that comprise the OSPF header?
2. How does reliable flooding operate?
3. Describe the election process for Designated Router and Backup Designated Router.
4. Define the differences between hierarchical routing and flat routing.
5. What type of routing is OSPF?
6. What is an area in OSPF?
7. What are ABRs?
8. What are the types of LSAs?
9. What is the backbone area?
10. What is a virtual link?

Interdomain Routing Basics Part II: Border Gateway Protocol

Objectives

- Explain how today's Internet is organized.
- Discuss how routing takes place in the Internet.
- Define the differences between distance vector protocols and link state protocols with relationship to the Internet.
- Examine how autonomous systems (ASs) are applied to routing in the Internet and what their role is.
- Discuss different types of routing connections that are used in today's Internet.
- Explain how Border Gateway Protocol (BGP) works.
- Examine the BGP header format and the types of messages generated by the protocol.
- Discuss the different states that are deployed by the protocol.
- Explain peering sessions.
- Discuss how synchronization within an AS takes place.

Key Terms

Introduction

Routing involves two basic activities: determination of optimal routing paths and the transport of packets through an internetwork. The transport of packets through an internetwork is relatively straightforward. Path determination can be very complex. One protocol that addresses the task of path determination in today's networks is the Border Gateway Protocol (BGP).

BGP performs interdomain routing in Transmission Control Protocol/Internet Protocol (TCP/IP) networks. BGP is an Exterior Gateway Protocol (EGP), which means that it performs routing between multiple autonomous systems or domains and exchanges routing and reachability information with other BGP systems.

BGP was developed to replace its predecessor, the now obsolete EGP, as the standard exterior gateway routing protocol used in the global Internet. BGP solves serious problems with EGP and scales to Internet growth more efficiently.

Figure 12.1 illustrates core routers using BGP to route traffic between autonomous systems.

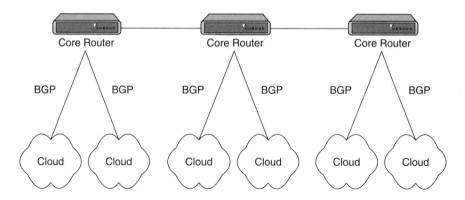

The Internet Today

The decommissioning of NFSNET (National Science Foundation) was done in specific stages to ensure continuous connectivity to institutions and government agencies that used to be connected to the regional networks. Today's Internet structure is a move from a core network such as NFSNET to a more distributed architecture operated by commercial providers such as Sprint, BBN, and others connected via major network exchange points. Figure 12.2 illustrates the general form of the Internet today.

The contemporary Internet is a collection of providers who have connection points called *points of presence (POPs)* over multiple regions. The Internet is a collection of POPs and the way its POPs are interconnected form a provider's network. Customers are connected to providers via the POPs. Customers of providers can be providers themselves. Providers who have POPs throughout the United States are called national providers.

Providers who cover specific regions (regional providers) connect themselves to other providers at one or multiple points. To enable customers of one provider to reach customers of another provider, Network Access Points (NAPs) are defined as interconnection points. The term Internet service provider (ISP) is usually used when referring to anyone who provides service, whether directly to end users or to other providers. The term network service provider (NSP) is usually restricted to providers who have NSF funding to manage the Network Access Points such as Sprint, Ameritech, and MFS. The term NSP, however, is also used more loosely to refer to any provider who connects to all the NAPs.

The Internet is a collection of autonomous systems (ASs) that define the administrative authority and the routing policies of different organizations. Autonomous systems run *Interior Gateway Protocols (IGPs),* such as RIP, IGRP, EIGRP, OSPF, and IS-IS, within their boundaries and interconnect via a BGP.

●
point of presence is a physical access point to a long-distance carrier interchange.

●
An **Interior Gateway Protocol** is an Internet protocol used to exchange routing information within an autonomous system.

FIGURE 12.2

The general structure of today's Internet. (Reproduced with permission of Cisco Systems, Inc. Copyright © 2001 Cisco Systems, Inc. All rights reserved.)

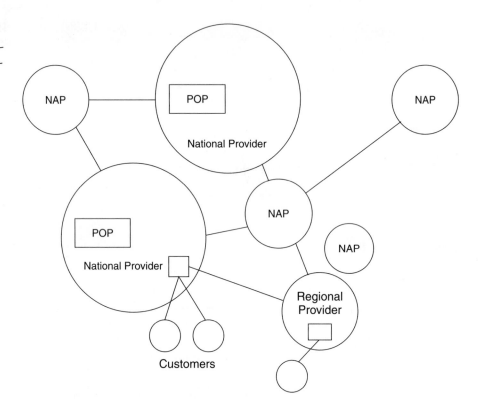

●

Routers direct traffic between hosts.

Routers are devices that direct traffic between hosts. Routers build routing tables that contain their collected information on all the best paths to all the destinations they know how to reach. They both announce and receive route information to and from other routers. This information goes into the routing tables.

Overview of Routers and Routing

Routers develop a hop-by-hop mechanism by keeping track of next hop information that enables a data packet to find its destination through the network. A router that does not have a direct physical connection to the destination checks its routing table and forwards the packet to another next hop router that is closer to that destination. The process repeats until the traffic finds it way through the network to its final destination.

EGPs, such as BGP, were introduced because IGPs do not scale in networks that go beyond the enterprise level. IGPs were never designed for the purpose of global internetworking because they do not have the necessary hooks to segregate enterprises into different administrations that are technically and politically independent from one another.

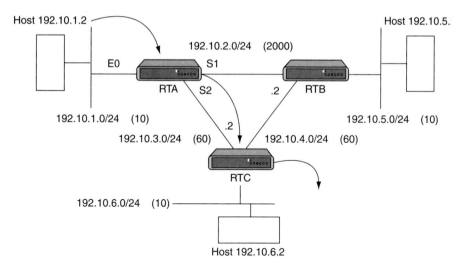

FIGURE 12.3

Basic routing behavior. (Reproduced with permission of Cisco Systems, Inc. Copyright © 2001 Cisco Systems, Inc. All rights reserved.)

RTA IP Routing Table (RIP)			RTA IP Routing Table (OSPF)		
Destination	Next Hop	Hop Count	Destination	Next Hop	Hop Count
192.10.1.0	Connected (E0)	—	192.10.1.0	Connected (E0)	—
192.10.2.0	Connected (S1)	—	192.10.2.0	Connected (S1)	—
192.10.3.0	Connected (S2)	—	192.10.3.0	Connected (S2)	—
192.10.4.0	192.10.2.2 (S1)	1	192.10.4.0	192.10.3.2 (S2)	120
	192.10.3.2 (S2)	1	192.10.5.0	192.10.3.2 (S2)	130
192.10.5.0	192.10.2.2 (S1)	1	192.10.6.0	192.10.3.2 (S2)	70
192.10.6.0	192.10.3.2 (S2)	1			

Figure 12.3 describes three routers (RTA, RTB, and RTC) connecting three local area networks (192.10.1.0, 192.10.5.0, and 192.10.6.0) via serial links. Each serial link is represented by its own network number, which results in three additional networks, 192.10.2.0, 192.10.3.0, and 192.10.4.0. Each network has a metric associated with it indicating the level of overhead (cost) of transmitting traffic on that particular link. For example, the link between RTA and RTB has a cost of 2,000, which is much higher than the cost of the link between RTA and RTC, which is 60. In practice, the link between RTA and RTB is a 56 Kbps link with much bigger delays than the T1 link between RTA and RTC and the T1 link between RTC and RTB combined.

Routers RTA, RTB, and RTC would exchange network information via some interior gateway protocol and build their respective IP routing tables. Figure 12.3 shows examples of RTA's IP routing table for two different scenarios: the routers are exchanging routing information via RIP in one scenario and OSPF in another. As an example of how traffic is passed between end stations, if host 192.10.1.2 is trying to reach host 192.10.6.2 it will first send the traffic to RTA. RTA will look in its IP routing table for any network that matches this destination and find that network 192.10.6.0 is reachable via next hop 192.10.3.2 (RTC) out Serial line 2 (S2). RTC would receive the traffic and would try to look for the destination in its IP routing table (not shown). RTC would discover that the host is directly connected to its Ethernet 0 interface (E0) and would send the traffic to 192.10.6.2.

In the preceding example, the routing is the same whether RTA is using the RIP or OSPF scenario. RIP and OSPF, however, fall into different categories of IGP protocols, namely distance vector protocols and link state protocols, respectively. In Figure 12.3 the results might be different depending on whether you are looking at the RIP or OSPF scenario. At this point it is useful to review the characteristics of both IGP protocol categories to see how protocols generally have evolved to meet increasingly sophisticated routing demands.

Distance Vector Protocols

Distance vector protocols such as RIP version 1 were mainly designed for small network topologies. The term distance vector derives from the fact that the protocol includes in its routing updates a vector of distances (hop counts). By using hop counts, distance vector protocols do not factor into the routing equation the overhead of sending information over a particular link. Low speed links are treated equally or sometimes preferred over a high speed link, depending on the calculated hop count in reaching a destination. This would lead to suboptimal and inefficient routing behaviors.

Consider, for example, the RTA routing tables illustrated in Figure 12.3. In the RIP case, RTA has listed the direct link between RTA and RTB to reach network 192.10.5.0. RTA prefers this link because it requires just one hop via RTB versus two hops via RTC and then RTB. But the preferred route is inefficient because the total cost of the routing path via RTC and then RTB (60 + 60 = 120) is much less than the cost of crossing the RTA-RTB link (2,000).

Another issue with hop counts is the count to infinity restriction: Distance Vector Protocols have a finite limit of hops (15) after which a route is considered unreachable. This would restrict the propagation of routing updates and would cause problems for large networks.

The reliance on hop counts is one deficiency of distance vector protocols; another deficiency is the way in which the routing information gets exchanged. Distance vector algorithms work on the concept that routers exchange all the network numbers they can reach via periodic broadcasts of the entire routing table. In large networks, the routing table exchanged between routers becomes very large and very hard to maintain, leading to slower convergence.

Convergence refers to the point in time at which the entire network becomes updated to the fact that a particular route has appeared or disappeared. Distance vector protocols work on the basis of periodic updates and hold-down timers. If a route is not received in a certain amount of time, the route goes into a hold-down state and gets aged out of the routing table. The hold-down and aging process translates into minutes in convergence time before the whole network detects that a route has disappeared. The delay between a route's becoming unavailable and its aging out of the routing tables can result in routing loops and black holes.

Another major drawback of distance vector protocols is their classful nature and their lack of support of *variable length subnet masks (VLSMs)* or classless interdomain routing (CIDR). Distance vector protocols do not exchange mask information in their routing updates. A router that receives a routing update on a certain interface will apply to this update its locally defined subnet mask. This would lead to confusion, in case the interface belongs to a network that is variably subnetted, and misinterpretation of the received routing update.

Finally, distance vector networks are considered to be flat. They present a lack of hierarchy, which translates into a lack of aggregation. This flat nature has made distance vector protocols incapable of scaling to larger and more efficient enterprise networks. RIP version 2 has added support for VLSM and CIDR, but it still carries most of the other deficiencies that its predecessor, RIP version 1, has.

variable length subnet mask is the ability to specify a different subnet mask for the same network number on different subnets.

Link State Protocols

Link state protocols such as OSPF are more advanced routing protocols that have addressed the deficiencies of distance vector protocols. Link state protocols work on the basis that routers exchange information elements, called link states, which carry information about links and nodes. This means that routers running link state protocols do not exchange routing tables. Each router inside a domain will have enough bits and pieces of the big puzzle that it can run a shortest path algorithm and build its own routing table.

The following are some of the benefits that link state protocols provide over distance vector protocols:

- No hop count: No limits on the number of hops a route can take. Link state protocols work on the basis of metrics rather than hop counts. As an example of a link state protocol's reliance on metrics rather than hop count, turn again to the RTA routing tables illustrated in Figure 12.3. In the OSPF case, RTA has picked the optimal path to reach RTB by factoring in the cost of the links. Its routing table lists the next hop of 192.10.3.2 (RTC) to reach 192.10.5.0 (RTB). This is in contrast to the RIP scenario, which resulted in a suboptimal path.

- Bandwidth representation: Link bandwidth and delays are factored in when calculating the shortest path to a certain destination. This leads to better load balancing based on actual link cost rather than hop count.
- Better convergence: Link and node changes are flooded into the domain via link state updates. All routers in the domain will immediately update their routing tables.
- Support for VLSM and CIDR: Link state protocols exchange mask information as part of the information elements that are flooded in the domain. As a result, networks with variable length masks can be easily identified.
- Better hierarchy: Whereas distance vector networks are flat networks, link state protocols divide the domain into different levels and areas. This hierarchical approach provides better control over network instabilities and a better mechanism to summarize routing updates across areas, specifically by lumping multiple contiguous routing updates into supersets of routing updates called aggregates.

Even though link state algorithms have provided better routing scalability, which enables them to be used in bigger and more complex topologies, they still should be restricted to interior routing. Link state protocols by themselves cannot provide a global connectivity solution required for Internet interdomain routing. In very large networks and in case of route fluctuation caused by link instabilities, link state retransmission and recomputation will become too large for any router to handle.

Segregating the World into Administrations

Exterior routing protocols were created to control the expansion of routing tables and to provide a more structured view of the Internet by segregating routing domains into separate administrations in which each has its own independent routing policies.

During the early days of the Internet an EGP was used. The NFSNET used EGP to exchange reachability information between the backbone and the regional networks. Although the use of EGP was widely deployed, its topology restrictions and inefficiency in dealing with routing loops and setting routing policies created a need for a new and more robust protocol. Currently, BGP4 is the de facto standard for Internet routing. BGP4 is an advanced exterior protocol that is providing the Internet with a controlled and loop-free topology.

Static Routing, Default Routing, and Dynamic Routing

Static routing refers to destinations being listed manually in the router.

Before introducing and looking at the basic ways in which autonomous systems can be connected to ISPs, there is a need to establish some basic terminology and concepts of routing. *Static routing* refers to destinations being listed manually in the router. Network reachability in this case is not dependent on the existence and state of the network itself. Whether a destination is up or down, the static routes would remain in the routing table, and traffic would still be sent toward that destination. *Default routing* refers to a last resort outlet—traffic to destinations that are unknown to the router will be sent to that default outlet. Default routing is the easiest form of routing for a domain connected to a single exit point.

Dynamic routing refers to routes being learned via an internal or external routing protocol. Network reachability is dependent on the existence and state of the network. If a destination is down, the route would disappear from the routing table, and traffic would not be sent toward that destination.

These three routing approaches are possibilities for all the AS configurations but usually there is an optimal approach. When illustrating autonomous systems throughout this discussion it will be considered whether static, dynamic, default, or some combination of these is optimal. Also under consideration is whether interior or exterior routing protocols are appropriate. Remember that static and default routing are not your enemy. The most stable but not so flexible configurations are the ones based on static routing. Trying to force dynamic routing on situations that do not really need it is just a waste of bandwidth, effort, and money.

Autonomous Systems

An *autonomous system (AS)* is a set of routers having a single routing policy, running under a single technical administration. The AS could be a collection of IGPs working together to provide interior routing. To the outside world, the whole AS is viewed as one single entity. Each AS has an identifying number that is assigned to it by an Internet Registry or provider. Routing information between ASs is exchanged via an exterior gateway protocol such as BGP4, as illustrated in Figure 12.4.

What is gained by segregating the world into administrations is the capability to have one large network—in the sense that the Internet could have been one huge OSPF network divided into smaller and more manageable networks. These smaller networks, called ASs,

Default routing refers to a last resort outlet.

Dynamic routing refers to routes being learned via an internal or external routing protocol.

An **autonomous system** is a set of routers having a single routing policy.

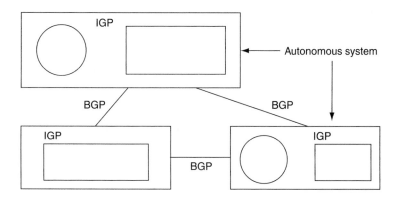

FIGURE 12.4

General illustration of AS relationships. (Reproduced with permission of Cisco Systems, Inc. Copyright © 2001 Cisco Systems, Inc. All rights reserved.)

FIGURE 12.5

Single-homed stub AS. (Reproduced with permission of Cisco Systems, Inc. Copyright © 2001 Cisco Systems, Inc. All rights reserved.)

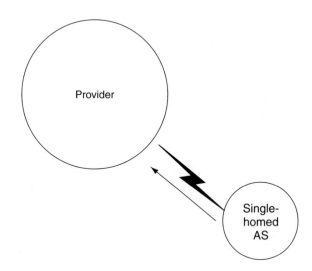

can now have their own set of rules and policies that will uniquely distinguish them from other ASs. Each AS can now run its own set of IGPs independent of IGPs in other ASs.

Stub ASs

● **A stub AS** reaches networks outside its domain via a single exit point.

An AS is considered to be stub when it reaches networks outside its domain via a single exit point. These ASs are also called single-homed with respect to another provider. Figure 12.5 illustrates a single-homed or *stub AS*.

A single-homed AS does not really have to learn Internet routes from its provider. Because there is a single way out all traffic can default to the provider. There are different methods for the provider to list the customer's subnets as static entries in its router. The provider would then advertise these static entries toward the Internet. This method would scale very well if the customer's routes can be represented by a small set of aggregate routes. When the customer has too many discontiguous subnets, listing all these subnets via static routes becomes inefficient.

Alternatively, IGPs can be used for the purpose of advertising the customer's networks. An IGP can be used between the customer and the provider for the customer to advertise its routes. This has all the benefits of dynamic routing where network information and changes are dynamically sent to the provider.

The third method by which the ISP can learn and advertise the customer's routes is to use BGP between the customer and the provider. In the stub AS situation, it is hard to get a registered AS number from the InterNIC because the customer's routing policies are an extension of the policies of the provider. Instead, the provider will give the customer an AS number from the private pool of ASs (65412-65535).

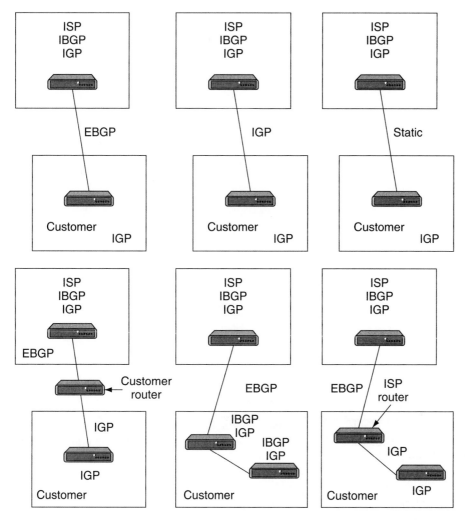

Quite a few combinations of protocols can be used between the ISP and the customer. Figure 12.6 illustrates some of the possible configurations, taking just stub ASs as an example. Providers might extend customer routers to their POPs or providers might extend their routers to the customer's network. Not every situation requires that a customer run BGP with its provider.

Multihomed Nontransit AS

A *multihomed nontransit AS* has more than one exit point to the outside world and does not allow transit traffic to go through it. An AS can be multihomed to a single

A **multihomed nontransit AS** has more than one exit point to the outside world and does not allow transit traffic to go through it.

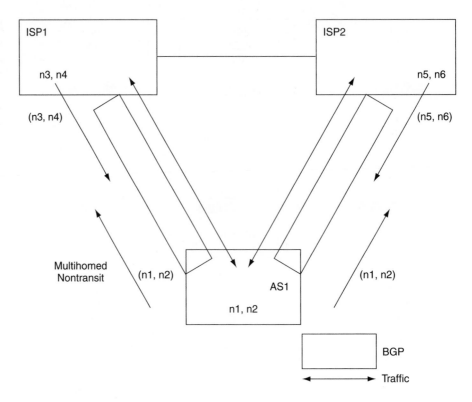

provider or multiple providers. Transit traffic is any traffic that has a source and destination outside the AS. Figure 12.7 illustrates an AS (AS1) that is nontransit and multihomed to two providers, ISP1 and ISP2.

A nontransit AS would only advertise its own routes and would not advertise routes that it learned from other ASs. This ensures that traffic for any destination that does not belong to the AS would not be directed to the AS.

In Figure 12.7, AS1 is learning about routes n3 and n4 via ISP1 and routes n5 and n6 via ISP2. AS1 is only advertising its local routes (n1, n2) and is not passing to ISP2 routes it learned from ISP1 or to ISP1 routes it learned from ISP2. This way, AS1 will not open itself to outside traffic, such as ISP1 trying to reach n5 or n6 and ISP2 trying to reach n3 and n4 via AS1. Of course, ISP1 or ISP2 can force their traffic to be directed to AS1 via default or static routing. As a precaution against this, AS1 could filter any traffic coming toward it with a destination not belonging to AS1.

Multihomed nontransit ASs do not really need to run BGP4 with their providers, although it is recommended and most of the time required by the provider.

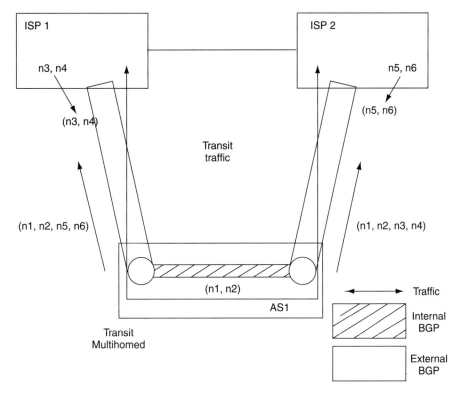

FIGURE 12.8

Multihomed transit AS using BGP internally and externally. (Reproduced with permission of Cisco Systems, Inc. Copyright © 2001 Cisco Systems, Inc. All rights reserved.)

Multihomed Transit AS

A *multihomed transit AS* has more than one connection to the outside world and can still be used for traffic by other ASs as illustrated in Figure 12.8. Transit traffic relative to the multihomed AS is any traffic with an origin and destination that does not belong to the AS.

Even though BGP4 is an exterior gateway protocol, it can still be used inside an AS as a pipe to exchange BGP updates. BGP connections inside an autonomous system are called *Internal BGP (IBGP),* whereas BGP connections between autonomous systems are called *External BGP (EBGP).* Routers that are running IBGP are called transit routers when they carry the transit traffic going through the AS. Routers that run EBGP with other ASs are usually called border routers.

A transit AS would advertise to one AS route that it learned from another AS. This way, the transit AS would open itself to traffic that does not belong to it. Multihomed transit ASs are advised to use BGP4 for their connections to other ASs and also internally to shield their internal nontransit routers from Internet routes. Not all routers inside a

A **multihomed transit AS** has more than one connection to the outside world and can still be used for traffic by other ASs.

Connections inside an autonomous system are called **Internal BGP**.

Connections between autonomous systems are called **External BGP**.

domain need to run BGP; internal nontransit routers could run default routing to the BGP routers, which alleviates the number of routes the internal nontransit routers must carry.

Figure 12.8 illustrates a multihomed transit autonomous system, AS1, connected to two different providers, ISP1 and ISP2, and in turn advertising all that it learned, including its local routes, to ISP1 and ISP2. In this case, ISP1 could use AS1 as a transit AS to reach networks n5 and n6, and ISP2 could use AS1 to reach networks n3 and n4.

Border Gateway Protocol Version 4

BGP went through different phases and improvements from its earlier version, BGP1, in 1989 to today's version, BGP4, which was deployed in 1993. BGP4 is the first version that handles aggregation (CIDR) and supernetting.

BGP imposes no restrictions on the underlying Internet topology. It assumes that routing within an autonomous system is done via an intra-autonomous system routing protocol. Intra means routing within an entity, and inter means between entities. BGP constructs a graph of autonomous systems based on the information exchanged between BGP neighbors. This directed graph environment is sometimes referred to as a tree. As far as BGP is concerned, the whole Internet is a graph of ASs, with each AS identified by an AS number. Connections between two ASs form a path and the collection of path information forms a route to reach a specific destination. BGP ensures that loop-free interdomain routing is maintained. Figure 12.9 illustrates this general path tree concept.

FIGURE 12.9

Example AS_path tree. (Reproduced with permission of Cisco Systems, Inc. Copyright © 2001 Cisco Systems, Inc. All rights reserved.)

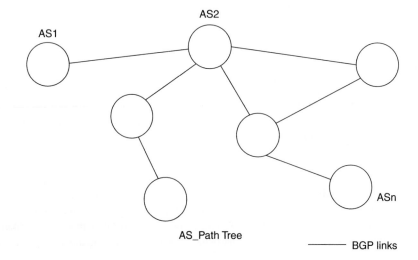

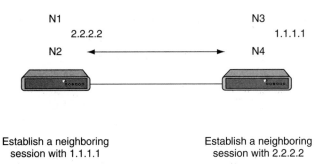

FIGURE 12.10
BGP routers become neighbors. (Reproduced with permission of Cisco Systems, Inc. Copyright © 2001 Cisco Systems, Inc. All rights reserved.)

How BGP Operates

BGP is a path vector protocol used to carry routing information between autonomous systems. The term *path vector* comes from the fact that BGP routing information carries a sequence of AS numbers that indicates the path a route has traversed. BGP uses TCP as its transport protocol. This ensures that all the transport reliability such as retransmission is taken care of by TCP and does not need to be implemented in BGP itself.

Two BGP routers form a transport protocol connection between each other. These routers are called *neighbors or peers*. Figure 12.10 illustrates this relationship.

Peer routers exchange multiple messages to open and confirm the connection parameters, such as the BGP version running between the two peers, for example, version 3 for BGP3 and version 4 for BGP4. In case of any disagreement between the peers, notification errors are sent and the peer connection does not get established. Initially, all candidate BGP routes are exchanged, as illustrated by Figure 12.11.

Incremental updates are sent as network information changes. The incremental update approach has shown an enormous improvement as far as CPU overhead and bandwidth allocation compared with periodic updates used by previous protocols,

> The **path vector** refers to the fact that BGP routing information carries a sequence of AS numbers that indicates the path a route has traversed.

> **neighbors or peers** are routers that form a transport protocol connection between each other.

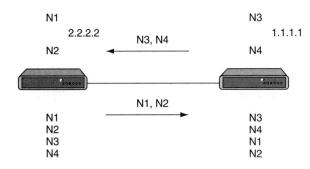

FIGURE 12.11
Exchanging all routing updates. (Reproduced with permission of Cisco Systems, Inc. Copyright © 2001 Cisco Systems, Inc. All rights reserved.)

FIGURE 12.12

N1 goes down; partial update sent. (Reproduced with permission of Cisco Systems, Inc. Copyright © 2001 Cisco Systems, Inc. All rights reserved.)

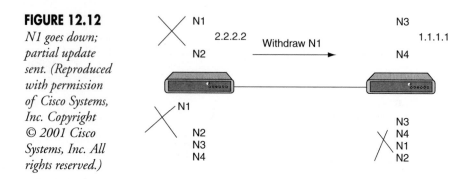

● **UPDATE messages** advertise routes between a pair of BGP routers.

such as EGP. Routes are advertised between a pair of BGP routers in UPDATE messages. The *UPDATE message* contains, among other things, a list of <length, prefix> tuples that indicate the list of destinations reachable via each system. The UPDATE message also contains the path attributes, which include such information as the degree of preference for a particular route.

In case of information changes, such as a route being unreachable or having a better path, BGP informs its neighbors by withdrawing the invalid routes and injecting new routing information. As illustrated in Figure 12.12, withdrawn routes are part of the UPDATE message. These are the routes no longer available for use. Figure 12.13 illustrates a steady state situation: If no routing changes occur, the routers exchange only KEEPALIVE packets.

● **KEEPALIVE messages** are sent between BGP neighbors to ensure that the connection is kept alive.

KEEPALIVE messages are sent periodically between BGP neighbors to ensure that the connection is kept alive. KEEPALIVE packets (19 bytes each) should not cause any strain on the router CPU or link bandwidth because they consume a minimal bandwidth about 2.5 bits/sec for a periodic rate of 60 seconds.

BGP keeps a table version number to keep track of the instance of the BGP routing table. If the table changes, BGP will increment the table version. A table version that is incrementing rapidly is usually an indication of instabilities in the network.

FIGURE 12.13

Steady state; N1 is still down. (Reproduced with permission of Cisco Systems, Inc. Copyright © 2001 Cisco Systems, Inc. All rights reserved.)

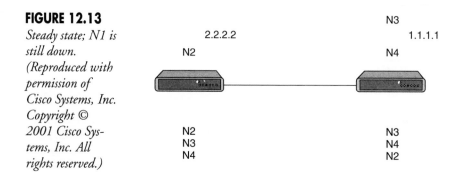

BGP Message Header Format

The BGP message header format is a 16-byte marker field followed by a 2-byte length field and a 1-byte type field. Figure 12.14 illustrates the basic format of the BGP message header. There may or may not be a data portion following the header, depending on the message type. KEEPALIVE messages, for example, consist of the message header only with no following data.

The marker field is used to either authenticate incoming BGP messages or detect loss of synchronization between two BGP peers. The marker field can have two formats:

- If the type of the message is OPEN or if the *OPEN message* has no authentication information, the marker field must be all ones.
- Otherwise, the marker field will be computed based on part of the authentication mechanism used.

The length indicates the total BGP message length including the header. The smallest BGP message is no less than 19 bytes (16 + 2 + 1) and no greater than 4,096. The type indicates the message type from the following possibilities:

- OPEN
- UPDATE
- NOTIFICATION
- KEEPALIVE

BGP Neighbor Negotiation

One of the basic steps of the BGP protocol is establishing neighbors between BGP peers. Without successful completion of this step, no exchange of updates will ever take effect. Neighbor negotiation is based on the successful completion of a TCP transport connection, the successful processing of the OPEN message, and periodic detection of the KEEPALIVE messages.

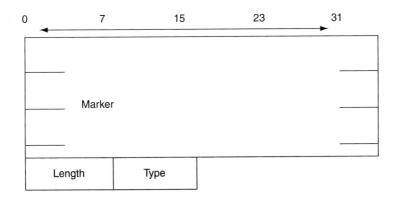

FIGURE 12.14

BGP message header format. (Reproduced with permission of Cisco Systems, Inc. Copyright © 2001 Cisco Systems, Inc. All rights reserved.)

FIGURE 12.15
OPEN message format. (Reproduced with permission of Cisco Systems, Inc. Copyright © 2001 Cisco Systems, Inc. All rights reserved.)

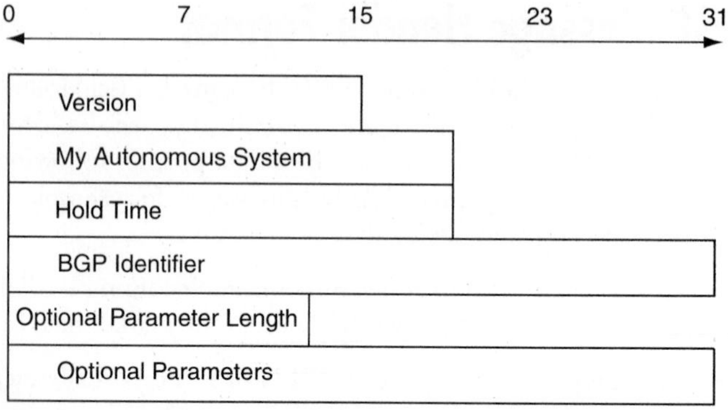

0	7	15	23	31

Version

My Autonomous System

Hold Time

BGP Identifier

Optional Parameter Length

Optional Parameters

OPEN Message Format

The **OPEN message format** consists of six fields: Version, My Autonomous System, Hold Time, BGP Identifier, Optional Parameters, and Optional Parameter Length.

Figure 12.15 illustrates the *OPEN message format.* The descriptions that follow summarize each of its fields:

- Version: A 1-byte unsigned integer that indicates the version of the BGP protocol, such as BGP3 or BGP4. During the neighbor negotiation, BGP peers agree on a BGP version number. BGP peers will try to negotiate the highest common version that they both support. Setting the version statically is usually used when the version of the BGP peers is already known.
- My Autonomous System: A 2-byte field that indicates the AS number of the BGP router.
- Hold Time: The maximum amount of time in seconds that may elapse between the receipt of successive KEEPALIVE or UPDATE messages. The hold timer is a counter that increments from zero to the hold time value. Receipt of a KEEPALIVE or UPDATE message causes the hold timer to be reset to zero. If the hold time for a particular neighbor is exceeded, the neighbor would be considered dead. The hold time is a 2-byte integer.

   The BGP router negotiates with its neighbor to set the hold time at whichever value is lower—its own hold time or its neighbor's. The hold time could be 0, in which case the hold time and the KEEPALIVE timers are never reset—that is, these timers never expire and the connection is considered to be always up. If not set to zero, the minimum recommended hold time is three seconds.
- BGP Identifier: A 4-byte unsigned integer that indicates the sender's ID. This is usually the router ID (RID), which is calculated as the highest IP address on the router or the highest loopback address at BGP session startup. Loopback address is the IP address of a virtual software interface that is considered to be up at all times, irrespective of the state of any physical interface.
- Optional Parameter Length: This is a 1-byte unsigned integer that indicates the total length in bytes of the Optional Parameters field. A length value of 0 indicates that no Optional Parameters are present.

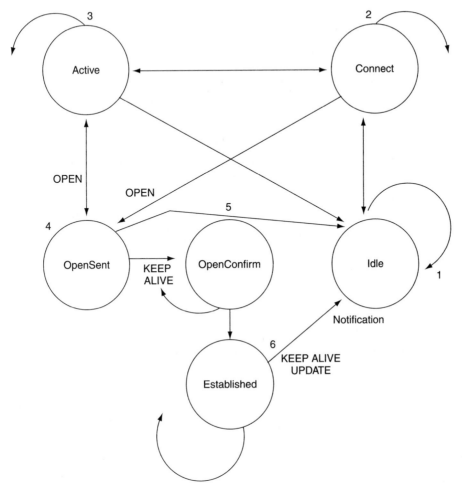

FIGURE 12.16
BGP neighbor negotiation finite state machine.

● Optional Parameters: This is a variable length field that indicates a list of optional parameters used in BGP neighbor session negotiation. This field is represented by the triplet <Parameter Type, Parameter Length, Parameter Value> with lengths of 1 byte, 1 byte, and variable length, respectively. An example of optional parameters is the authentication information parameter (type1), which is used to authenticate the session with a BGP peer.

Finite State Machine Perspective

BGP neighbor negotiation proceeds through different stages before the connection is fully established. Figure 12.16 illustrates a simplified *finite state machine (FSM)* that highlights the major events in the process with an indication of messages

●
The **finite state machine** is a computer system with a defined set of possible states and defined transitions from state to state.

(OPEN, KEEPALIVE, NOTIFICATION) sent to the peer in the transition from one state to the other.

The following key states in the FSM example illustrated in Figure 12.16 are summarized:

1. *Idle:* This is the first stage of the connection. BGP is waiting for a Start event, which is normally initiated by an operator. A Start event is usually caused by an administrator establishing a BGP session through router configuration or resetting an already existing session. After the Start event BGP initializes its resources, resets a connect retry timer, initiates a TCP transport connection, and starts listening for a connection that may be initiated by a remote peer. BGP then transitions to a Connect state. In case of errors, BGP falls back to the Idle state.

2. *Connect:* BGP is waiting for the transport protocol connection to be completed. If the TCP transport connection is successful, the state transitions to OpenSent (this is where the OPEN message is sent). If the connect retry timer expires, the state will remain in the connect stage, the timer will be reset, and a transport connection will be initiated. In case of any other event (initiated by the system or operator), the state will go back to Idle.

3. *Active:* BGP is trying to acquire a peer by initiating a transport protocol connection. If it is successful, it will transition to OpenSent (an OPEN message is sent). If the connect retry timer expires, BGP will restart the connect timer and fall back to the Connect state. Also, BGP is still listening for a connection that may be initiated from another peer. The state may go back to Idle in case of other events, such as a Stop event initiated by the system or the operator.

   In general, a neighbor state that is flip-flopping between Connect and Active is an indication that something is wrong with the TCP transport connection not taking effect. It could be because of many TCP retransmissions or the inability of a neighbor to reach the IP address of its peer.

4. *OpenSent:* BGP is waiting for an OPEN message from its peer. The OPEN message is checked for correctness. In case of errors, such as a bad version number or an unacceptable AS, the system sends an error NOTIFICATION message and goes back to Idle. If there are no errors, BGP starts sending KEEPALIVE messages and resets the KEEPALIVE timer. At this stage the hold time is 0, and the hold timer and the KEEPALIVE timer are not restarted.

   At the OpenSent state, the BGP will recognize, by comparing its AS number to the AS number of its peer, whether the peer belongs to the same AS (Internal BGP) or to a different AS (External BGP).

   When a TCP transport disconnect is detected, the state will fall back to Active. For any other errors, such as an expiration of the hold timer, the BGP will send a NOTIFICATION message with the corresponding error code and will fall back to the Idle state. Also, in response to a Stop event initiated by a system or operator, the state will fall back to Idle.

5. *OpenConfirm:* BGP waits for a KEEPALIVE or NOTIFICATION message. If a KEEPALIVE is received, the state will go to Established, and the neighbor nego-

tiation is complete. If the system receives an UPDATE or KEEPALIVE message, it restarts the hold timer assuming the negotiated hold timer is not set to 0. If a NOTIFICATION message is received, the state falls back to Idle. The system will send periodic KEEPALIVE messages at the rate set by the KEEPALIVE timer. In case of any transport disconnect notification or in response to any Stop event, the state will fall back to Idle. In response to any other event, the system will send a NOTIFICATION message with an FSM error code and will go back to Idle.

6. *Established:* This is the final stage in the neighbor negotiation. At this stage BGP starts exchanging UPDATE packets with its peers. Assuming that it is nonzero, the hold timer is restarted at the receipt of an UPDATE or KEEPALIVE message. If the system receives any NOTIFICATION message— if some error has occurred—the state will fall back to idle.

> The UPDATE messages are checked for errors, such as missing attributes or duplicate attributes. If errors are found a NOTIFICATION is sent to the peer, and the state will fall back to Idle. In case the hold timer expires, or a disconnect notification is received from the transport protocol, or a Stop event is received, or in response to any other event, the system will fall back to Idle.

NOTIFICATION Message

The previous examination of the FSM illustrated that many opportunities exist among the various states for error detection. A *NOTIFICATION message* is always sent whenever an error is detected, after which the peer connection is closed. Network administrators will need to evaluate these NOTIFICATION messages to determine the specific nature of errors that emerge in the routing protocol. Figure 12.17 illustrates the general message format.

The NOTIFICATION message is composed of the Error code (1 byte), Error subcode (1 byte), and a Data field (variable). The Error code indicates the type of the notification, the Error subcode provides more specific information about the nature of the error, and the Data field contains data relevant to the error such as a bad

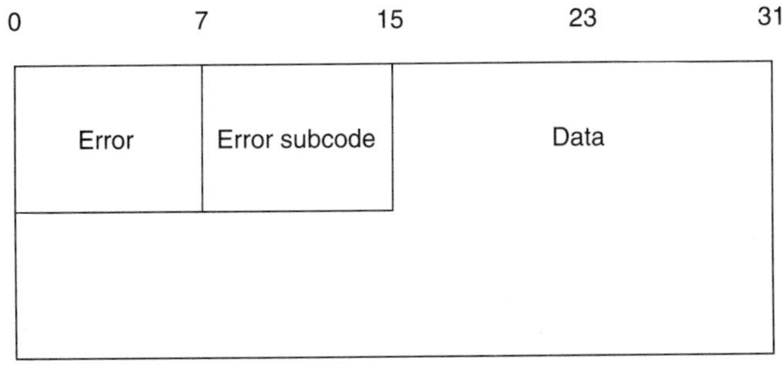

> A **NOTIFICATION message** is always sent whenever an error is detected.

FIGURE 12.17
NOTIFICATION message format.

Table 12.1 Possible BGP Error Codes

Error Code	Error Subcode
1-Message Header Error	1- Connection Not Synchronized
	2- Bad Message Length
	3- Bad Message Type
2-OPEN Message Error	1- Unsupported Version Number
	2- Bad Peer AS
	3- Bad BGP Identifier
	4- Unsupported Optional Parameter
	5- Authentication Failure
	6- Unacceptable Hold Time
3-UPDATE Message Error	1- Malformed Attribute List
	2- Unrecognized Well-Known Attribute
	3- Missing Well-Known Attribute
	4- Attribute Flags Error
	5- Attribute Length Error
	6- Invalid Origin Attribute
	7- AS Routing Loop
	8- Invalid NEXT_HOP Attribute
	9- Optional Attribute Error
	10- Invalid Network Field
	11- Malformed AS_path
4-Hold Timer Expired	NOT applicable
5-Finite State Machine Error for errors detected by the FSM	NOT applicable
6-Cease for fatal errors besides the ones already listed	NOT applicable

header, an illegal AS number, and so on. Table 12.1 lists possible errors and their subcodes.

KEEPALIVE Message

KEEPALIVE messages are periodic messages exchanged between peers to determine whether peers are reachable. The hold timer is the maximum amount of time that

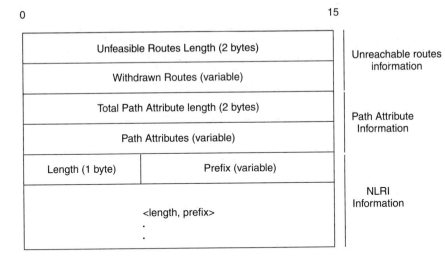

FIGURE 12.18
BGP routing up-date.

may elapse between the receipt of successive KEEPALIVE or UPDATE messages. The KEEPALIVE messages are sent at a rate that ensures that the hold time will not expire (the session is considered alive). A recommended rate is one-third of the hold time interval. If the hold time interval is zero, periodic KEEPALIVE messages will not be sent. The KEEPALIVE message is a 19-byte BGP message header with no data following it.

UPDATE Message and Routing Information

Central to the BGP protocol is the concept of the routing updates. Routing updates contain all the necessary information that BGP uses to construct a loop-free picture of the Internet. The following are the basic blocks of an UPDATE message:

- Network Layer Reachability Information (NLRI)
- Path attributes
- Unreachable routes

Figure 12.18 illustrates these components in the context of an UPDATE message format.

The NLRI is an indication in the form of an IP prefix route of the networks being advertised. The path attribute list provides BGP with the capabilities of detecting routing loops and the flexibility to enforce local and global routing policies. An example of the BGP path attributes is the AS_path attribute, which is a sequence of AS numbers a route has traversed before reaching the BGP router. For example, AS3 in Figure 12.19 is receiving BGP updates from AS2 indicating that network 10.10.1.0/24 (NLRI) is reachable via two hops, first AS2 and then AS1. Based on this information, AS3 will be able to direct its traffic to 10.10.1.0/24.

FIGURE 12.19
BGP routing up-date example.

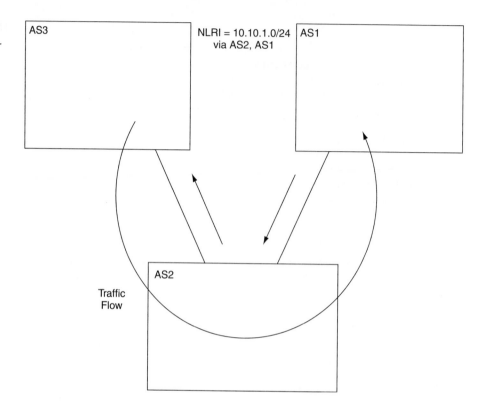

The third part of the UPDATE message is a list of routes that have become *unreachable* or WITHDRAWN. With the example illustrated in Figure 12.19, if 10.10.1.0/24 is no longer reachable or experiences a change in its attribute information, BGP can withdraw the route that it advertises by sending an UPDATE message that lists the new network information or that the network is unreachable.

Network Layer Reachability Information

BGP4 provides a new set of mechanisms for supporting classless interdomain routing (CIDR). The concept of CIDR is a move from the traditional IP classes (A, B, C) toward a concept of IP prefixes. The IP prefix is an IP network address with an indication of the number of bits (left to right) that constitute the network number. The *Network Layer Reachability Information (NLRI)* is the mechanism by which BGP supports classless routing. The NLRI is the part of the BGP routing update that lists the set of destinations about which BGP is trying to inform its other BGP neighbors. The NLRI consists of multiple instances of the 2-tuples <length, prefix>, where length is the number of masking bits that a particular prefix has.

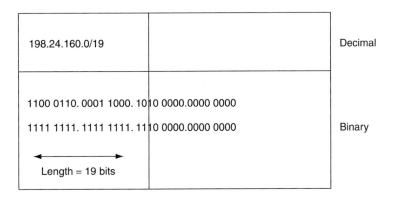

FIGURE 12.20
NLRI example.

Figure 12.20 illustrates the NLRI <19, 198.24.160.0>. The prefix is 198.24.160.0, and the length is a 19-bit mask (counting from the far left of the prefix).

Withdrawn Routes

Withdrawn routes provide a list of routing updates that are not feasible or are no longer in service and need to be withdrawn (removed) from the BGP routing tables. The withdrawn routes have the same format as the NLRI: an IP address and the number of bits in the IP address counting from left, as illustrated in Figure 12.21. Withdrawn routes are also represented by the tuple <length, prefix>. A tuple of the form <18, 192.213.134.0> indicates a route to be withdrawn of the form 192.213.134.0 255.255.192.0 or 192.213.134.0/18 in the CIDR format.

The Unfeasible Routes Length is the length in bytes of the total withdrawn routes. An UPDATE message can list multiple routes to be withdrawn at the same time or no routes to be withdrawn. An Unfeasible Routes Length of 0 indicates that no routes are to be withdrawn. On the other hand, an UPDATE message can advertise at most one route, which can be described by multiple path attributes. An UPDATE message that has no NLRI or Path Attribute information is used to advertise only routes to be withdrawn from service.

Withdrawn routes provide a list of routing updates that are not feasible or are no longer in service and need to be withdrawn from the BGP routing tables.

Length (1 byte)
Prefix (variable)

FIGURE 12.21
General form of the Withdrawn Routes field.

FIGURE 12.22
Path attribute type format.

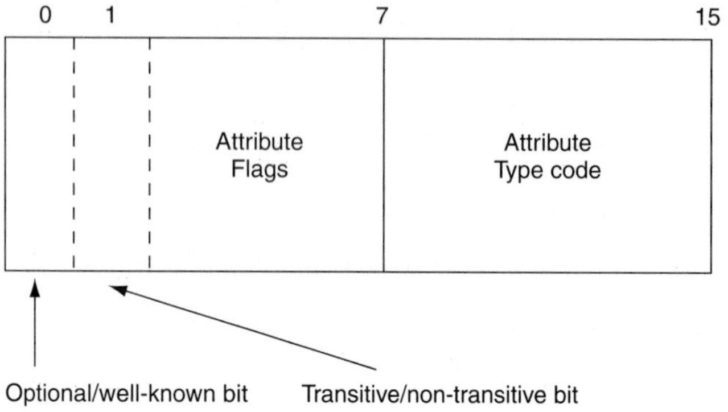

Path Attributes are a set of parameters used to keep track of route-specific information.

Path Attributes

The BGP *path attributes* are a set of parameters used to keep track of route-specific information such as path information, degree of preference of a route, next hop value of a route, and aggregation information. These parameters are used in the BGP filtering and route decision process. Every UPDATE message has a variable length sequence of path attributes. A path attribute is a triple of the form <attribute type, attribute length, attribute value> of variable length. The attribute type is a 2-byte field that consists of a 1-byte attribute flag and a 1-byte attribute type code. Figure 12.22 illustrates the general form of the Path Attribute Type field.

Path attributes fall under four categories: well-known mandatory, well-known discretionary, optional transitive, and optional nontransitive. These four categories are described by the first two bits of the Path Attribute Flags field.

The first bit of the Flags field indicates whether the attribute is optional (1) or well-known (0). The second bit indicates whether the optional attribute is transitive (1) or nontransitive (0). Well-known attributes are always transitive (second bit is always 1). The third bit indicates whether the information in the optional transitive attribute is partial (1) or complete (0). The fourth bit defines whether the attribute length is 1 byte (0) or 2 bytes (1). The other four bits in the Flags field are always set to 0.

A **well-known mandatory** attribute must be recognized by all BGP implementations.

A **well-known discretionary** attribute may or may not be sent in the BGP UPDATE message.

The following descriptions elaborate on the significance of each attribute category:

- *Well-known mandatory:* An attribute that has to exist in the BGP UPDATE packet. It must be recognized by all BGP implementations. If a well-known attribute is missing, a notification error will be generated. This is to make sure that all BGP implementations agree on a standard set of attributes. An example of a well-known mandatory attribute is the AS_path attribute.
- *Well-known discretionary:* An attribute that is recognized by all BGP implementations, but may or may not be sent in the BGP UPDATE message. An example of a well-known discretionary attribute is the LOCAL_PREF.

In addition to the well-known attributes, a path can contain one or more optional attributes. Optional attributes are not required to be supported by all BGP implementations. Optional attributes can be transitive or nontransitive.

- *Optional transitive:* In case an optional attribute is not recognized by the BGP implementation, that implementation would look for a transitive flag to see whether it is set for that particular attribute. If the flag is set the attribute is transitive, and the BGP implementation should accept the attribute and pass it along to other BGP speakers.
- *Optional nontransitive:* When an optional attribute is not recognized and the transitive flag is not set, the attribute is nontransitive and should be quietly ignored and not passed along to other BGP peers.

Building BGP Peer Sessions

Although BGP is meant to be used between autonomous systems to provide an interdomain loop-free topology, BGP can be used within an AS as a pipe between border routers running external BGP to other ASs. Recall that a neighbor connection, also called a *peer connection,* can be established between two routers within the same AS, in which case BGP is called internal BGP (IBGP). A peer connection can also be established between two routers in different ASs. BGP is then called external BGP (EBGP). Figure 12.23 contrasts these environments.

Upon neighbor session establishment and during the OPEN message exchange negotiation, peer routers compare AS numbers and determine whether they are peers in the same AS or in different ASs. The difference between EBGP and IBGP manifests itself in how each peer would process the routing updates coming from the other peer and in the way different BGP attributes are carried on external versus internal links. The neighbor negotiation process is mainly the same for internal and external neighbors as far as building the TCP connection at the transport level. It is essential to have IP connectivity between the two neighbors for the transport session to take place. IP connectivity has to be achieved via a protocol different from BGP; otherwise, the session will be in a race condition. In a race condition neighbors can reach one another via some IGP, the BGP session gets established, and the BGP updates get exchanged. The IGP connection goes away for some reason, but still the BGP TCP session is up because neighbors can still reach each other via BGP. Eventually, the session will go down because the BGP session cannot depend on BGP itself for neighbor reachability.

An IGP or static route can be configured to achieve IP connectivity. In essence, a ping packet containing a source IP address (the IP address of one BGP peer) and a destination IP address (the IP address of the second peer) must succeed for a transport session to initiate.

An **optional transitive** attribute is an optional attribute that is not recognized but the transitive flag is set.

An **optional nontransitive** attribute is an optional attribute that is not recognized and the transitive flag is not set.

A **peer connection** can be established between two routers in different ASs.

FIGURE 12.23

Internal and external BGP implementations.

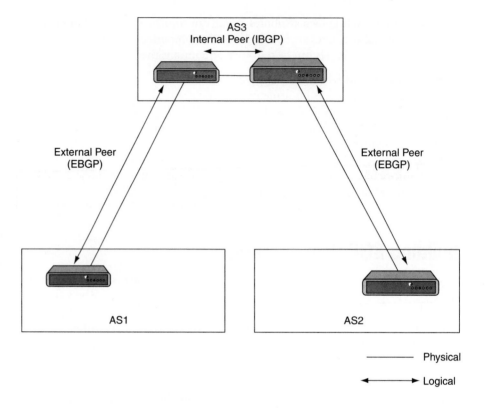

Physical versus Logical Connections

External BGP neighbors have a restriction on being physically connected. BGP drops any updates from its external BGP peer if the peer is not connected. However, some situations arise where external neighbors cannot be on the same physical segment. Such neighbors are logically but not physically connected. An example would be running BGP between external neighbors across non-BGP routers. In this situation there is an extra solution to override this restriction. BGP would require some extra configuration to indicate that its external peer is not physically attached.

EBGP multihop refers to EBGP peers that are not directly connected.

EBGP peers that are not directly connected are referred to as *EBGP multihop*. In Figure 12.24, RT2 is not able to run BGP, but RT1 and RT3 are. Thus, external neighbors RT1 and RT3 are logically connected and peer with one another via EBGP multihop.

On the other hand, neighbors within the same autonomous system (internal neighbors) have no restrictions whatsoever on whether the peer router is physically connected. As long as there is IP connectivity between the two neighbors, BGP requires no extra configurations. In Figure 12.24, RT1 and RT4 are logically but not physi-

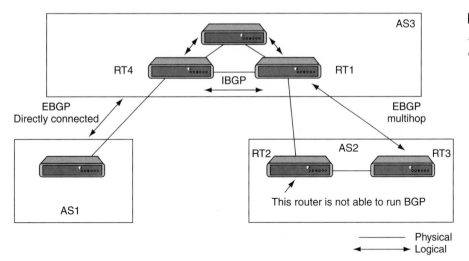

FIGURE 12.24
*EBGP multihop
environment.*

cally connected. Because both are in the same AS, no additional configuration is needed for them to run IBGP.

Obtaining an IP Address

The neighbor's IP address could be the address of any of the router's interfaces. The stability of the neighbor connection will rely on the stability of the IP address you choose. If the IP address belongs to an Ethernet card that has some hardware problems and is shutting down every few minutes, the neighbor connection and the stability of the routing updates will suffer. There is a *loopback interface* that is actually a virtual interface that is supposed to be up at all times. Tying the neighbor connection to a loopback interface will make sure that the session is not dependent on any hardware interface that might be problematic.

Adding loopback interfaces is not necessary in every situation. If external BGP neighbors are directly connected and the IP addresses of the directly connected segment are used for the neighbor negotiation, a loopback address is of no added value. If the physical link between the two peers is problematic, then the session will break with or without loopback.

●
A **loopback in-
terface** is actu-
ally a virtual
interface that is
supposed to be up
at all times.

Authenticating the BGP Session

The BGP message header allows authentication. Authentication is the measure of precaution against hackers who might present themselves as one of your BGP peers and feed you wrong routing information. Authentication between two BGP peers gives the capability to validate the session between you and your neighbor by using a combination of passwords and keys upon which you both agree. A neighbor that tries to establish a session without the use of these specific passwords and keys will not be permitted. The authentication feature uses the *Message Digest Algorithm version 5 (MD5).*

BGP Continuity Inside an AS

To avoid creating routing loops inside the AS, BGP does not advertise to internal BGP peer routes that are learned via other IBGP peers. Thus, it is important to maintain a full IBGP mesh within the AS. Every BGP router in the AS has to build a BGP session with all other BGP routers inside the AS. Figure 12.25 illustrates one of the common mistakes administrators make when setting BGP routing inside the AS.

In the example illustrated in Figure 12.25, an ISP has three POPs in San Jose, San Francisco, and Los Angeles. Each POP has multiple non-BGP routers and a BGP border router running EBGP with other ASs. The administrator sets an IBGP connection between the San Jose border router and the San Francisco border router. Another IBGP connection is set between the SF border router and the LA border router. In this configuration, EBGP routes learned via SJ will be given to SF, EBGP routes learned via SF are given to SJ and LA, and EBGP routes learned via LA are given to SF. Routing in this picture is not complete; EBGP routes learned via SJ will not be given to LA, and EBGP routes learned via LA will not be given to SJ. This is because the SF router will not pass on IBGP routes between SJ and LA. What is needed is an additional IBGP connection between SJ and LA shown via the dotted line. This situation can be handled better by using the concept of route reflectors, which is an option that scales much better in cases where the AS has a large number of IBGP peers.

FIGURE 12.25

Common BGP continuity mistake.

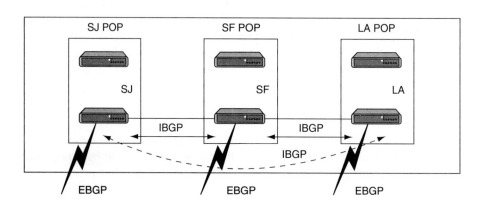

Synchronization Within an AS

BGP must be *synchronized* with IGP in such a way that it waits until the IGP has propagated routing information across your autonomous system before advertising transit routes to other ASs. It is important that your autonomous system be consistent about the routes that it advertises. If, for example, your BGP were to advertise a route before all routers in your AS had learned about the route through the IGP, your AS could receive traffic that some routers cannot yet route.

Whenever a router receives an update about a destination from an IBGP peer, the router tries to verify internal reachability for that destination before advertising it to other EBGP peers. The router would do so by checking for the existence of this destination in the IGP. This would give an indication whether non-BGP routers can deliver traffic to that destination; the router will announce it to other EBGP peers. Otherwise, the router will treat the route as not being synchronized with the IGP and would not advertise it. Consider the situation illustrated in Figure 12.26; ISP1 and ISP2 are using ISP3 as a transit AS. ISP3 has multiple routers in its AS and is running BGP only on the border routers. Even though RTB and RTD are carrying transit traffic, ISP3 has not configured BGP on these routers. ISP3 is running some IGP inside the AS for internal connectivity.

Assume that ISP1 is advertising route 192.213.1.1/24 to ISP3. Because RTA and RTC are running IBGP, RTA will propagate the route to RTC. Note that other routers besides RTA and RTC are not running BGP and have no knowledge so far of the existence of route 192.213.1.1/24. In the situation illustrated in Figure 12.26, if RTC advertises the route to ISP2, traffic toward the destination 192.213.1.1/24 will start flowing toward RTC. RTC will do a recursive lookup in its IP routing table and will direct the traffic toward the next hop RTB. RTB, having no visibility to the BGP routes, will drop the traffic because it has no knowledge

synchronization is the establishment of common timing between sender and receiver.

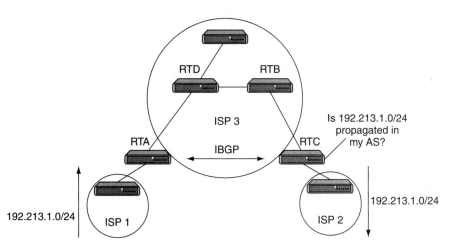

FIGURE 12.26
BGP route synchronization.

of the destination. This has happened because there is no synchronization between BGP and IGP.

The BGP rule states that a BGP router should not advertise to external neighbors destinations learned from inside BGP neighbors unless those destinations are also known via IGP. If a router knows about these destinations via IGP it assumes that the route has already been propagated inside the AS, and internal reachability is guaranteed.

The consequence of injecting BGP routes inside an AS is costly. Redistributing routes from BGP into the IGP will result in major overhead on the internal routers, which might not be equipped to handle that many routes. Routing can easily be accomplished by having internal non-BGP routers default to one of the BGP routers. Of course, this will result in routing suboptimality because there is no guarantee for shortest path for each route, but this cost is minimal compared with maintaining thousands of routes inside the AS.

The BGP Routing Process Simplified

BGP is a fairly simple protocol, which is why it is so flexible. Routes are exchanged between BGP peers via UPDATE messages. BGP routers receive the UPDATE messages, run some policies or filters over the updates, and then pass on the routes to other BGP peers. BGP keeps track of all BGP updates in a BGP routing table separate from the IP routing table. In case multiple routes to the same destination exist, BGP does not flood its peers with all those routes; rather, it picks the best route and sends it. In addition to passing along routes from peers, a BGP router may originate routing updates to advertise networks that belong to its own autonomous system. Valid local routes originated in the system, and the best routes learned from BGP peers, are then installed in the IP routing table. The IP routing table is used for the final routing decision. To model the BGP process imagine each BGP speaker having different pools of routes and different policy engines applied to the routes. The model would involve the following components:

- A pool of routes that the router receives from its peers.
- An Input Policy Engine that can filter the routes or manipulate their attributes.
- A decision process that decides which routes the router itself will use.
- A pool of routes that the router itself uses.
- An Output Policy Engine that can filter the routes or manipulate their attributes.
- A pool of routes that the router advertises to other peers.

Figure 12.27 illustrates this model. The subsequent discussion provides more details about each component.

Routes Received from Peers

BGP receives routes from external or internal peers. Depending on what is configured in the Input Policy engines, some or all of these routes will make it into the router's BGP table.

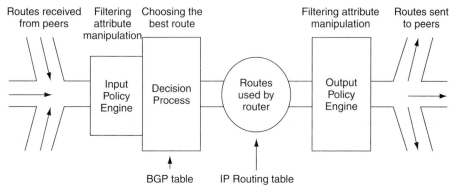

FIGURE 12.27
Routing process overview.

Input Policy Engine

The *Input Policy Engine* handles route filtering and attribute manipulation. Filtering is done based on different parameters such as IP prefixes, AS_path information, and attribute information. BGP also uses the Input Policy Engine to manipulate the path attributes to influence its own decision process and hence affect what routes it will actually use to reach a certain destination. If, for example, BGP chooses to filter a certain network number coming from a peer, it is an indication that BGP does not want to reach that network via that peer. Or, if BGP gives a certain route a better local preference it is an indication that BGP would like to prefer this route over other routes.

> The **Input Policy Engine** handles route filtering and attribute manipulation.

The Decision Process

BGP goes through a decision process to decide which routes it wants to use to reach a certain destination. The *decision process* is based on the routes that made it into the router after the Input Policy Engine was applied. The decision process is performed on the routes in the BGP routing table. The decision process looks at all the available routes for the same destination, compares the different attributes associated with each route, and chooses one best route.

> The **decision process** is performed on the routes in the BGP routing table.

Routes Used by the Router

The best routes as identified by the decision process are what the router itself uses and are candidates to be advertised to other peers and also to be placed in the IP routing table. In addition to routes passed on from other peers, the router originates updates about the networks inside its autonomous system. This is how an AS injects its routes into the outside world.

Output Policy Engine

The *Output Policy Engine* is the same engine as the Input Policy Engine but applied on the output side. Routes used by the router (the best routes) in addition to routes that the router generates locally are given to this engine for processing. The engine

> The **Output Policy Engine** is the same as the Input Policy Engine but applied on the output side.

FIGURE 12.28

Example routing environment.

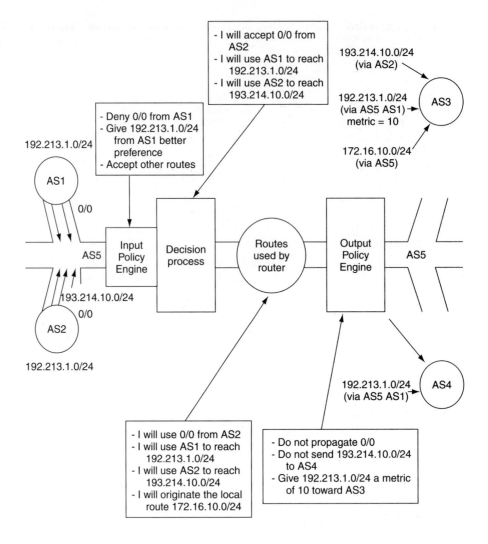

might apply filters and might change some of the attributes such as AS_path or metric before sending the update.

The Output Policy Engine also differentiates between internal and external peers; for example, routes learned from internal peers cannot be passed on to internal peers.

Routes Advertised to Peers

This is the set of routes that made it through the Output Engine and are advertised to the BGP peers—internal or external.

Example Routing Environment: Figure 12.28 illustrates routing in an example environment. In the figure, AS5 is receiving routes from both AS1 and AS2 and is originating its own routes 172.16.10.0/24. To simplify, consider just the flow of updates in one direction from left to right. By applying the engine model to AS5, the following results occur:

Routes received from peers (these are the routes coming from AS1 and AS2):

- 192.213.1.0/24 via AS1.
- 0/0 (this is a default route) via AS1.
- 193.214.10.0/24 via AS2.
- 0/0 (this is a default route) via AS2.
- 192.213.1.0/24 via AS2.

Input Policy Engine:

- Do not accept default route 0/0 from AS1.
- Give route 192.213.1.0/24 coming from AS1 better preference than route 192.213.1.0/24 coming from AS2.
- Accept all other routes (this will accept 193.214.10.0/24).

The decision process:

- Because 192.213.1.0/24 has better preference via AS1, I will reach 192.213.1.0/24 via AS1.
- I will reach 193.214.10.0/24 via AS2.
- I will accept 0/0 via AS2.

Routes used by the router:

- I will use 0/0 as default from AS2.
- I can reach 192.213.1.0/24 via AS1.
- I can reach 193.214.10.0/24 via AS2.
- Network 172.16.10.0/24 is one of my local networks that I want to advertise.

Output Policy Engine:

- Do not propagate the default route 0/0.
- Do not advertise 193.214.10.0/24 to AS4.
- Give 192.213.1.0/24 a metric of 10 when sent to AS3.

Routes advertised to peers:

- Toward AS3:
  - 192.213.1.0/24 via (AS5 AS1) (this means, first AS5 then AS1) with a metric of 10.
  - 172.16.10.0/24 (via AS5).
  - 193.214.10.0/24 (via AS5 AS2).

- Toward AS4:
  - 192.213.1.0/24 (via AS5 AS1).
  - 172.16.10.0/24 (via AS5).

Summary

The Border Gateway Protocol (BGP) has defined the basis of routing architectures in the Internet. The segregation of networks into autonomous systems has logically defined the administrative and political borders between organizations. Interior Gateway Protocols (IGPs) can now run independently of each other, but still interconnect via BGP to provide global routing.

BGP as a protocol presents some basic elements of routing that are flexible enough to allow total control from the administrator's perspective. The power of BGP lies in its attributes and filtering techniques, which are beyond the scope of this book. The intent here is to give a working knowledge of the protocol and how the protocol operates in the Internet. But as a means of exploring the subject further, attributes are simply parameters that can be modified to affect the BGP decision process. Route filtering can be done on a prefix level or a path level. A combination of filtering and attribute manipulation can achieve the optimal routing behavior.

Chapter 13 discusses two proprietary protocols, IGRP and EIGRP, and how they can be used to optimize routing.

Review Questions

1. Give a basic illustration showing how the Internet is set up today.
2. What is a POP and where is it located in the Internet?
3. Describe an NAP and define its role in the Internet hierarchy.
4. What is the Internet?
5. What are some protocols that are run by autonomous systems?
6. What are routers?
7. Define EGP.
8. What are distance vector protocols and what are some of the problems they have in routing?
9. What are link state protocols and how do they improve routing?
10. What is an autonomous system?
11. Define the differences between static, default, and dynamic routing.
12. What is a stub autonomous system?
13. Where is a multihomed nontransit AS used?
14. A single-homed AS does not really have to learn Internet routes from its provider. Why?
15. Where is a multihomed transit AS used in Internet routing?
16. What are BGP connections inside an AS called?
17. What are BGP connections outside the AS called?
18. How does BGP view the Internet?
19. What type of protocol is BGP?
20. Routers that run BGP between each other and form a transport connection are called?

Summary Questions

1. What happens in case of disagreement between peers?
2. How do peer routers exchange multiple messages?
3. How are routes advertised between a pair of BGP routers?
4. Illustrate the BGP header format.
5. What are KEEPALIVE messages?
6. What are the four messages generated by BGP?
7. What are the key states in the finite state machine?
8. What does a neighbor state that is flip-flopping between Connect and Active indicate?
9. What is the NLRI?
10. How are WITHDRAWN routes handled?

Advanced IP Routing

Objectives

- Discuss issues facing future addressing.
- Explain the benefits of hierarchical addressing.
- Explain how hierarchical addressing allows efficient allocation of addresses and reduced number of routing table entries.
- Discuss Variable Length Subnet Masks (VLSMs).
- Explain classful and classless routing updates.
- Examine how to calculate VLSMs.
- Discuss how VLSMs provide the capability to include more than one subnet mask within a network and the capability to subnet an already subnetted address.
- Explain route summarization.
- Examine how to summarize addresses in a VLSM designed network.
- Explain the concept of route redistribution.

Key Terms

Introduction

After a brief overview of TCP/IP including IP addressing and subnetting we will begin our discussion on advanced IP routing.

The Internet protocols are the world's most popular open-system (nonproprietary) protocol suite because they can be used to communicate across any set of interconnected networks and are equally well suited for LAN and WAN communications. The Internet protocols consist of a suite of communication protocols, of which the two best known are the Transmission Control Protocol (TCP) and the Internet Protocol (IP). The Internet protocol suite not only includes lower layer protocols such as TCP and IP, but also specifies common applications such as electronic mail, terminal emulation, and file transfer.

●
The **Defense Advanced Research Projects Agency** was a U.S. government agency that funded research for and experimentation with the Internet.

Internet protocols were first developed in the mid-1970s when the *Defense Advanced Research Projects Agency (DARPA)* became interested in establishing a packet-switched network that would facilitate communication between dissimilar computer systems at research institutions. With the goal of heterogeneous connectivity in mind, DARPA funded research by Stanford University and Bolt, Beranek, and Newman (BBN). The result of this development effort was the Internet protocol suite, completed in the late 1970s.

TCP/IP later was included with Berkeley Software Distribution (BSD) UNIX and has since become the foundation on which the Internet and the World Wide Web (WWW) are based. Documentation of the Internet protocols including new or revised protocols and policies are specified in technical reports called Request For Comments (RFCs), which are published and then reviewed and analyzed by the Internet community. Protocol refinements are published in the new RFCs. To illustrate the scope of the Internet protocols, Figure 13.1 maps many of the protocols of the Internet protocol suite and their corresponding OSI layers.

Internet Protocol (IP)

The Internet Protocol (IP) is a network layer (layer 3) protocol that contains addressing information and some control information that enables packets to be routed. IP is documented in RFC 791 and is the primary network layer protocol in the Internet protocol suite. Along with the Transmission Control Protocol (TCP), IP rep-

OSI Reference Model

Internet Protocol Suite

Application	FTP, Telnet, SMTP, SNMP	NFS
Presentation		XDR
Session		RPC
Transport	TCP, UDP	
Network	Routing Protocols IP	ICMP
Data Link	ARP, RARP	
Physical	Not Specified	

FIGURE 13.1

Internet protocols span the complete range of OSI model layers. (Reproduced with permission of Cisco Systems, Inc. Copyright © 2001 Cisco Systems, Inc. All rights reserved.)

resents the heart of the Internet protocols. IP has two primary responsibilities: providing connectionless, best-effort delivery of datagrams through an internetwork; and providing fragmentation and reassembly of datagrams to support data links with different maximum transmission unit (MTU) sizes.

IP Packet Format

An IP packet contains several types of information as illustrated in Figure 13.2.

The following describes the IP packet fields illustrated in Figure 13.2:

- Version: Indicates the version of IP currently being used.
- IP Header Length (IHL): Indicates the datagram header length in 32 bit words.
- Type-of-Service: Specifies how an upper layer protocol would like a current datagram to be handled, and assigns datagrams various levels of importance.
- Total Length: Specifies the length, in bytes, of the entire IP packet including the data and header.
- Identification: Contains an integer that identifies the current datagram. This field is used to help piece together datagram fragments.
- Flags: Consists of a 3-bit field of which the low order (least significant) bits control fragmentation. The low order bit specifies whether the packet can be fragmented. The middle bit specifies whether the packet is the last fragment in a series of fragmented packets. The third or high order bit is not used.
- Fragment Offset: Indicates the position of the fragment's data relative to the beginning of the data in the original datagram, which allows the destination IP process to reconstruct the original datagram.

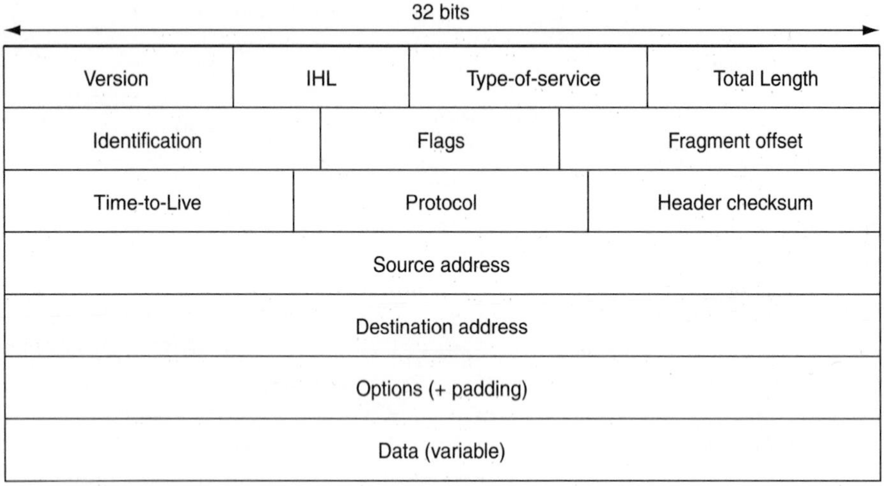

FIGURE 13.2

Fourteen fields comprise an IP packet. (Reproduced with permission of Cisco Systems, Inc. Copyright © 2001 Cisco Systems, Inc. All rights reserved.)

32 bits

Version	IHL	Type-of-service	Total Length
Identification		Flags	Fragment offset
Time-to-Live		Protocol	Header checksum
Source address			
Destination address			
Options (+ padding)			
Data (variable)			

- Time-to-Live: Maintains a counter that gradually decrements down to zero, at which point the datagram is discarded. This keeps packets from looping endlessly.
- Protocol: Indicates which upper layer protocol receives incoming packets after IP processing is complete.
- Header Checksum: Helps ensure IP header integrity.
- Source Address: Specifies the sending node.
- Destination Address: Specifies the receiving node.
- Options: Allows IP to support various options such as security.
- Data: Contains upper layer information.

IP Addressing

As with any other network layer protocol, the IP addressing scheme is intregal to the process of routing IP datagrams through an internetwork. Each IP address has specific components and follows a basic format. These IP addresses can be subdivided and used to create addresses for subnetworks.

Each host on a TCP/IP network is assigned a unique 32-bit logical address that is divided into two main parts: the network number and the host number. The network number identifies a network and must be assigned by the Internet Network Information Center (InterNIC) if the network is to be part of the Internet. An Internet service provider (ISP) can obtain blocks of network addresses from the InterNIC and can itself assign address space as necessary. The host number identifies a host on a network and is assigned by the local network administrator.

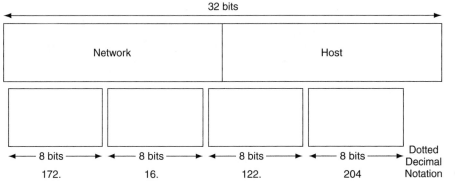

IP Address Format

The 32-bit IP address is grouped eight bits at a time, separated by dots, and represented in decimal format known as dotted decimal notation. Each bit in the octet has a binary weight 128, 64, 32, 16, 8, 4, 2, 1. The minimum value for an octet is 0, and the maximum value for an octet is 255. An IP address consists of 4 octets (1 octet = 8 bits) or a total of 32 bits. The value in each octet ranges from 0 to 255 decimal or 00000000 − 11111111. Here is how binary octets convert to decimal:

```
1    1    1    1    1    1    1    1
128  64   32   16   8    4    2    1
```

Now here is a sample octet conversion:

```
0    1    0    0    0    0    0    1
0    64   0    0    0    0    0    1  (0 + 64 + 0 + 0 + 0 + 0 + 0 + 1 = 65)
```

And this is a sample address representation (4 octets):

```
10          1.         23.        19        (decimal)
00001010.   00000001.  00010111.  00010011  (binary)
```

Figure 13.3 illustrates the basic format of an IP address.

IP Address Classes

The 4 octets are broken down to provide an addressing scheme that can accommodate large and small networks. IP addressing supports five different address classes: A, B, C, D, and E. Only classes A, B, and C are available for commercial use. The left-most (high order) bits indicate the network class. One address is reserved for the broadcast and one address is reserved for the network. In a Class A

Table 13.1 Reference Information About the Five IP Address Classes

IP Address Class	Format	Purpose	High Order Bit(s)	Address Range	No. Bits Network/ Host	Max. Hosts
A	N.H.H.H	Few large organizations	0	1.0.0.0 to 126.0.0.0	7/24	16,581,375
B	N.N.H.H	Medium size organizations	1,0	128.1.0.0 to 191.254.0.0	14/16	65,543
C	N.N.N.H	Relatively small organizations	1,1,0	192.0.1.0 to 223.255.254.0	22/8	245
D	N/A	Multicast groups	1,1,1,0	224.0.0.0 to 239.255.255.255	N/A (not available for commercial use)	N/A
E	N/A	Experimental	1,1,1,1	240.0.0.0 to 254.255.255.255	N/A	N/A

address such as 10.1.23.19, the first octet is the network portion, so the Class A example has a major network address of 10. Octets 2, 3, and 4, the next 24, bits are for the network administrator to divide into subnets and hosts as they see fit. Class A addresses are used for networks that have more than 65,536 hosts, actually up to 16,581,375 hosts.

In a Class B address such as 172.16.19.48, the first two octets are the network portion, so in the Class B example the major network address is 172.16. Octets 3 and 4, the next 16 bits, are for local subnets and hosts. Class B addresses are used for networks that have between 256 and 65,536 hosts.

In a Class C address such as 193.18.9.10, the first three octets are the network portion. The Class C example has a major network address of 193.18.9. Octet 4 (8 bits) is for local subnets and hosts, which is perfect for networks with less than 256 hosts.

Table 13.1 provides reference information about the five IP address classes.

Figure 13.4 illustrates the format of the commercial IP address class. Note the high order bits in each class. The class of addresses can be determined easily by examining the first octet of the address and mapping that value to a class range in Table 13.2.

In an IP address of 172.31.1.2, for example, the first octet is 172. Because 172 falls between 128 and 192, 172.31.1.2 is a Class B address. Table 13.2 summarizes the range of possible values for the first octet of each address class.

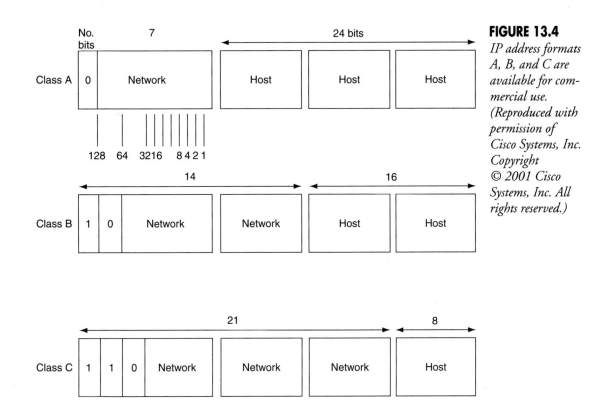

Table 13.2 A Range of Possible Values Exists for the First Octet of Each Address Class

Address Class	*First Octet in Decimal*	*High Order Bits*
Class A	1 to 126	0
Class B	128 to 191	10
Class C	192 to 223	110
Class D	224 to 239	1110
Class E	240 to 254	1111

IP Subnet Addressing

In order to use your addresses, you need to understand subnetting. Subnetting allows you to create multiple logical networks that exist within a single Class A, B, or C network. If you do not subnet, you will only be able to use one network from your Class A, B, or C network. Unless you have been assigned many major networks, you really need to subnet.

IP networks can be divided into smaller networks called subnetworks or subnets. Subnetting provides the network administrator with several benefits including extra flexibility, more efficient use of network addresses, and the capability to contain broadcast traffic because a broadcast will not cross a router. Subnets are under local administration. As such, the outside world sees an organization as a single network and has no detailed knowledge of the organization's internal structure.

A given network address can be broken up into many subnetworks. For example, 172.16.1.0, 172.16.2.0, 172.16.3.0, and 172.16.4.0 are all subnets within network 172.16.0.0. All 0s in the host portion of an address specifies the entire network. Each data link on a network must be a unique subnet, with every node on that link being a member of the same subnet. For serial interfaces you will need one subnet for every circuit, or wire—both ends of the serial connection will be in the same subnet.

IP Subnet Mask

A **subnet mask** is defined for each IP address.

A subnet mask is defined for each IP address. The *subnet mask* identifies which portion of the 4 octets is used to identify the data link, with the remaining bits identifying the node. If you want no subnetting, strictly use the default masks 255 followed by the number, 0 − wildcard:

```
Class A: 255.0.0.0
Class B: 255.255.0.0
Class C: 255.255.255.0
```

Borrowing bits from the host field and designating them as the subnet field create a subnet address. The number of borrowed bits varies and is specified by the subnet mask. Figure 13.5 shows how bits are borrowed from the host address field to create the subnet address field.

Subnet masks use the same format and representation technique as IP addresses. The subnet mask, however, has binary 1s in all bits specifying the network and subnetwork fields, and binary 0s in all bits specifying the host field. Figure 13.6 illustrates a sample subnet mask.

Subnet mask bits should come from the high order (left-most) bits of the host field as Table 13.3 illustrates. Details of Class B and C subnet mask types follow. Class A addresses are not discussed because they generally are subnetted on an 8-bit boundary.

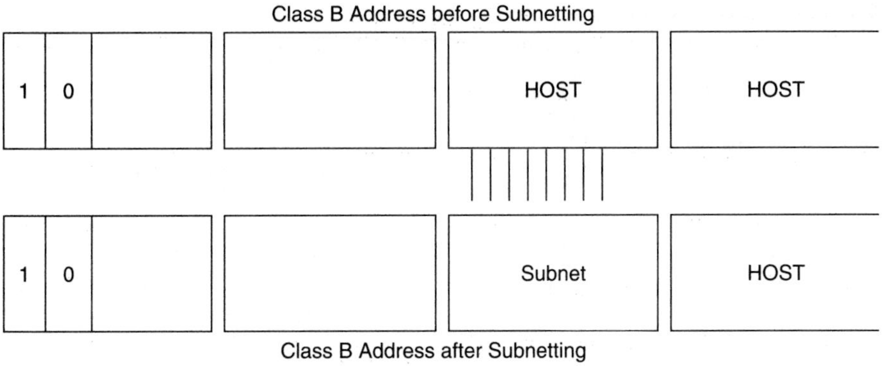

FIGURE 13.5

Bits are borrowed from the host address field to create the subnet address field. (Reproduced with permission of Cisco Systems, Inc. Copyright © 2001 Cisco Systems, Inc. All rights reserved.)

Class B Address before Subnetting

Class B Address after Subnetting

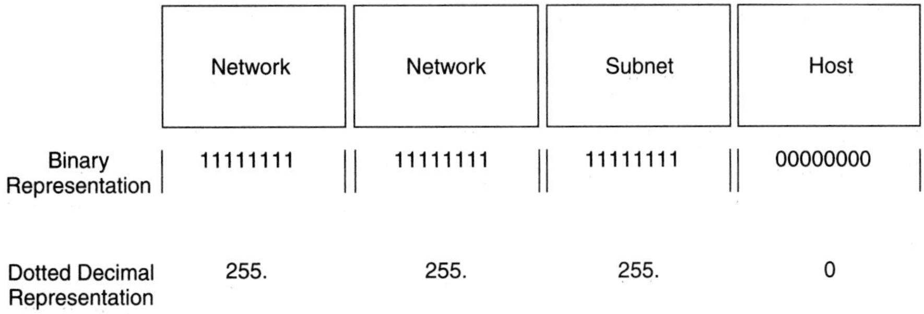

FIGURE 13.6

A sample subnet mask consists of all binary 1s and 0s. (Reproduced with permission of Cisco Systems, Inc. Copyright © 2001 Cisco Systems, Inc. All rights reserved.)

Table 13.3 Subnet mask bits come from the high order bits of the host field.

128	64	32	16	8	4	2	1	
1	0	0	0	0	0	0	0	= 128
1	1	0	0	0	0	0	0	= 192
1	1	1	0	0	0	0	0	= 224
1	1	1	1	0	0	0	0	= 240
1	1	1	1	1	0	0	0	= 248
1	1	1	1	1	1	0	0	= 252
1	1	1	1	1	1	1	0	= 254
1	1	1	1	1	1	1	1	= 255

Table 13.4 Class B Subnetting Reference Chart

Number of Bits	Subnet Mask	Number of Subnets	Number of Hosts
2	255.255.192.0	2	16,382
3	255.255.224.0	6	8,190
4	255.255.240.0	14	4,094
5	255.255.248.0	30	2,046
6	255.255.252.0	62	1,022
7	255.255.254.0	126	510
8	255.255.255.0	254	254
9	255.255.255.128	510	126
10	255.255.255.192	1,022	62
11	255.255.255.224	2,046	30
12	255.255.255.240	4,094	14
13	255.255.255.248	8,190	6
14	255.255.255.252	16,382	2

Various types of subnet masks exist for Class B and Class C subnets. The default subnet mask for a Class B address that has no subnetting is 255.255.0.0, while the subnet mask for a Class B address 171.16.0.0 that specifies 8 bits of subnetting is 255.255.255.0. The reason for this is that 8 bits of subnetting or $2^8 - 2$ (1 for the network address and 1 for the broadcast address) = 254 subnets possible, with $2^8 - 2$ = 254 hosts per subnet. The subnet mask for a Class C address 192.168.2.0 that specifies 5 bits of subnetting is 255.255.255.248. With 5 bits available for subnetting, $2^5 - 2$ = 30 subnets possible, with $2^3 - 2$ = 6 hosts per subnet.

The reference charts shown in Table 13.4 and Table 13.5 can be used when planning Class B and C networks to determine the required number of subnets and hosts and the appropriate subnet mask.

How Subnet Masks Are Used to Determine the Network Number

The router performs a set process to determine the network or, more specifically, the subnetwork address. First, the router extracts the IP destination address from the incoming packet and retrieves the internal subnet mask. The router then performs a logical AND operation to obtain the network number. This causes the host portion of the IP destination address to be removed while the destination network

Table 13.5 Class C Subnetting Reference Chart

Number of Bits	Subnet Mask	Number of Subnets	Number of Hosts
2	255.255.255.192	2	62
3	255.255.255.224	6	30
4	255.255.255.240	14	14
5	255.255.255.248	30	6
6	255.255.255.252	62	2

number remains. The router then looks up the destination network number and matches it with an outgoing interface. Finally, it forwards the frame to the destination IP address. Specifics regarding the logical AND operation are discussed in the next section.

Logical AND Operation

Three basic rules govern logically ANDing two binary numbers. First, 1 ANDed with 1 yields 1. Second, 1 ANDed with 0 yields 0. Finally, 0 ANDed with 0 yields 0. The truth table provided in Table 13.6 illustrates the rules for logical AND operations. Two simple guidelines exist for remembering logical AND operations: Logically ANDing a 1 with a 1 yields the original value and logically ANDing a 0 with any number yields 0.

Figure 13.7 illustrates that when a logical AND of the destination IP address and the subnet mask is performed the subnetwork number remains, which the router uses to forward the packet.

Let us use these two addresses for some examples: 171.68.3.3 and 171.68.2.3. If the subnet mask is 255.255.255.0, the first 24 bits are masked, so the router compares the first 3 octets of the two addresses. Because the masked bits are not the same, the router knows that these addresses belong to different subnets. If the subnet mask is 255.255.0.0, the first 16 bits are masked, so the router compares the first 2 octets of

Table 13.6 Rules for Logical AND Operations

Input	Input	Output
1	1	1
1	0	0
0	1	0
0	0	0

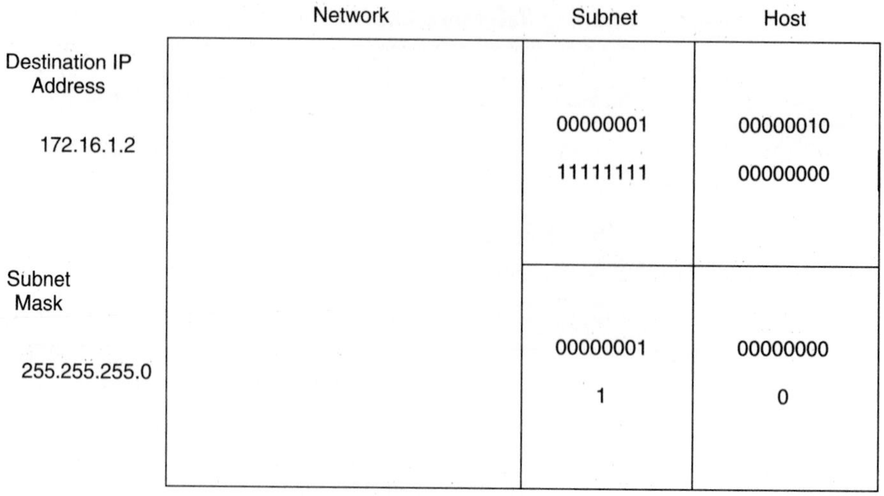

FIGURE 13.7

Applying a logical AND of the destination IP address and the subnet mask produces the subnetwork number. (Reproduced with permission of Cisco Systems, Inc. Copyright © 2001 Cisco Systems, Inc. All rights reserved.)

	Network	Subnet	Host
Destination IP Address 172.16.1.2		00000001 11111111	00000010 00000000
Subnet Mask 255.255.255.0		00000001 1	00000000 0

the two addresses. Because the masked bits are the same, the router knows that these addresses belong to the same subnet.

Nodes and routers use the mask to identify the data link on which the address resides. For instance, imagine that San Francisco proper is a class B network, and think of the streets as subnets. Each street must have a unique name. How would the post office deliver a letter or find the correct destination if there were two Lombard Streets? Each house number can be thought of as a unique identifier for that street. The house numbers themselves can be duplicated on other streets: 33 Market Street is not the same as 33 Van Ness Avenue.

```
San Francisco.Lombard.33
172.          68.   3.   3
San Francisco.Market.33
171.          68.   2.   3
```

Let us compare our sample addresses 171.68.3.3 and 171.68.2.3 against the subnet mask 255.255.240.0. We need to compare the binary representation of the third octet of the mask with the binary representation of the third octets of the addresses. To do this, we will perform a logical AND operation on the corresponding bits in each octet. The masked bits are those turned on or 1 in the mask. Because the masked bits in both addresses are the same, the router knows that these addresses belong to the same subnet.

Example 1: Class B

Let us use a Class B address to illustrate how subnetting works. Let us say you were assigned the Class B address 172.16 from the NIC. First, determine how many subnets you need and how many nodes per subnet you need to define. A typical and easy to use Class B subnet mask would be 8 bits. Because the third octet is the first free octet for Class B, you will start there. Therefore, an 8-bit subnet mask would be 255.255.255.0. This means it would have 254 subnets available and 254 addresses for nodes per subnet. Again, there are only 254 subnets available instead of 256 (0 − 255) because you should not use subnet 0 or a subnet consisting of all 1s. An all 1s subnet mask is also your broadcast address. Subnet 0 is also not recommended.

Example 2: Class B

Now let us take this example: You have just assigned an interface the address 172.16.10.50 with a mask of 255.255.255.0. What subnet is it in? First, represent the bits in binary for Class B. Start with the third octet because octets 1 and 2 are fixed.

```
SUBNET      HOST
00001010    00110010  (address representation—10.50)
11111111    00000000  (subnet mask representation—255.0)
00001010    00000000  (results of logical AND—subnet 10) 10
```

This address is in subnet 10 (172.16.0.0). Valid addresses for subnet 10 would be 172.16.10.1 through 172.16.10.254. Address 172.16.10.255 is the broadcast address for this subnet. According to the standard, any host ID consisting of all 1s is reserved for broadcast.

Example 3: Class B

Let us say you have a need for more subnets than 254. Remember, 254 is the maximum number of subnets in a single octet. Sticking with our Class B address, let us configure an 11-bit subnet. This means we will use all 8 bits from the third octet and the first three bits from the fourth octet. The subnet mask is now 255.255.255.224 (128 + 64 − 32 = 224). Now you need to find out what subnet the following address is in: 172.16.10.170 255.255.255.224. First, denote the address in binary representation just using octets 3 and 4 for a Class B address like this:

```
00001010    10101010  (address representation—10.70)
11111111    11100000  (subnet mask representation—255.0)
00001010    10100000  (results of logical AND—subnet 10) 10 160
```

The address is in subnet 172.16.10.160. The valid addresses for this subnet are 172.16.10.161 through 172.16.10.190 and .191 is the broadcast address. As soon as you hit 10.192, the bits in the subnet change and you move into the subnet 10.192.

Example 4: Class B

Let us take an example where the mask is shorter than 1 octet. Now we want only a few subnets, but we will need many hosts per subnet. We will use a 3-bit subnet mask.

Now we have 172.16.65.170 255.255.224.0 because the mask is now the first three bits of the third octet. What subnet is this address in?

```
01000001    10101010  (address representation—65.170)
11100000    00000000  (subnet mask representation—224.0)
01000000    00000000  (results of logical AND—subnet 64) 64
```

The subnet is 172.16.64.0. The range of addresses that would fall into subnet 64 would be 172.16.64.1 through 172.16.95.254 with 172.16.95.255 as the broadcast address. The next subnet would be 172.16.96.0. Class A and Class C map out exactly as Class B. The only differences are at which octet subnetting starts and how many octets you can use for subnetting.

Example 5: Class C

Suppose the NIC assigned the address 192.1.10.0. You will need to use the fourth octet for your subnetting needs. Let us use a 4-bit subnet mask and map out the following address: 192.1.10.200 255.255.255.240.

```
11001000    (address representation for 200)
11110000    (subnet mask representation for 240)
11000000    (results of logical AND—128 + 64 = 192)
```

Address 192.1.10.200 is on subnet 192. The valid range of addresses in this subnet would be 192.1.10.192 through 192.1.10.206 with .207 as the broadcast address. The next subnet would be .208.

Keeping the same subnet mask, you can choose different addresses to be in different subnets. For instance, address 192.1.10.17 255.255.255.240 is in subnet 16 and therefore has another unique subnet address, with valid addresses in the range of 192.1.10.17 through 192.1.10.30.

IP Address Issues and Solutions

Now that we have defined IP addressing and subnetting, let us look at some more complex issues in Internet routing. When IP was first defined in 1981, it was designed as a 32-bit number that had two components: a network address and a node (host) address. Classes of addresses were also defined: classes A, B, and C and later classes D and E. Since then, the growth of the Internet has been incredible. Following are two addressing issues that have resulted from this explosion:

- IP address exhaustion: This is largely a result of the random allocation of IP addresses by the NIC. It is also due to the fact that not all IP classes are suitable for a typical network topology.
- Routing table growth and manageability: One source indicates that in 1990 only about 5,000 routes needed to be tracked in order to use the Internet. By

1995, this number had grown to 35,000 routes. In addition to the exponential growth of the Internet, the random assignment of IP addresses throughout the world has also contributed to the exponential growth of routing tables.

IPv6 the next generation IP responds to these problems by introducing a 128-bit address. In the meantime, RFCs have been introduced that enable the current IP addressing scheme to be organized in a hierarchical manner. One particularly effective way of combating these problems is using addressing hierarchies in setting up the corporate network.

Using Addressing Hierarchies

What is *hierarchical addressing,* and why do you want to implement it? Perhaps the best known addressing hierarchy is the telephone network. The telephone network uses a hierarchical numbering scheme that includes country codes, area codes, and local exchange numbers as illustrated in Figure 13.8.

To call Aunt Judy in Alexandria, Virginia, from San Jose, California, dial the area code, 703, the Alexandria prefix, 555, and then Aunt Judy's local number, 1212. The central office first looks up the number 703 and determines that it is not in its local area. The central office immediately routes the call to a central office in Alexandria. The San Jose central office does not know where 555-1212 is, nor does it have to. It only needs to know the area codes, which summarize the local telephone numbers within an area. If there were no hierarchical structure, every central office would need to have every telephone number, worldwide, in its locator table. With a simple hierarchical addressing scheme, the central office uses country codes and area codes to determine how to route a call to a destination. A summary number (address) represents

Hierarchical addressing enables efficient allocation of addresses and reduced number of routing table entries.

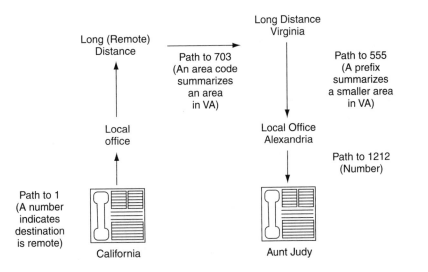

FIGURE 13.8

Telephone number hierarchy. (Reproduced with permission of Cisco Systems, Inc. Copyright © 2001 Cisco Systems, Inc. All rights reserved.)

a group of numbers. For example, an area code, such as 408, is a summary number for the San Jose area. That is, if you dial 408 from anywhere in the United States, and then a seven-digit telephone number, the central office will route the call to a San Jose central office. This is the kind of addressing strategy that as a network administrator you should implement in your own internetwork.

The benefits of hierarchical addressing are twofold:

● Efficient allocation of addresses: Hierarchical addressing enables one to optimize the use of the available addresses because you group them contiguously. With random address assignment you may end up wasting groups of addresses because of addressing conflicts.
● Reduced number of routing table entries: Whether it is with your Internet routers or your internal routers, you should try to keep your routing tables as small as possible by using route summarization. Route summarization is a way of having a single IP address represent a collection of IP addresses when you employ a hierarchical addressing plan. By summarizing routes, you can keep your routing table entries manageable, which means the following:
   ● More efficient routing
   ● Reduced number of CPU cycles when recalculating a routing table or when sorting through the routing table entries to find a match
   ● Reduced router memory requirements

Slowing IP Address Depletion

Since the 1980s several solutions have been developed to slow the depletion of IP addresses and to reduce the number of Internet route table entries by enabling more hierarchical layers in an IP address. The solutions are as follows:

● *Subnet masking:* RFC 950 (1985); 1812 (1995): Developed to add another level of hierarchy to an IP address. This additional level allows extending the number of network addresses derived from a single IP address.
● Variable Length Subnet Masks: RFC 1918 (1996): Developed to allow the network designer to utilize multiple address schemes within a given class of address. This strategy can be used only when it is supported by the routing protocol.
● Address Allocation for Private Networks: RFC 1918 (1996): Developed for organizations that do not need much access to the Internet. The only reason to have an NIC assigned IP address is to interconnect to the Internet. Any and all companies can use the privately assigned IP addresses within the organization, rather than using an NIC assigned IP address unnecessarily.
● *Network Address Translation:* RFC 1631 (1994): Developed for those companies that use private addressing or use non-NIC assigned IP addresses. This strategy enables an organization to access the Internet with an NIC assigned address without having to reassign the private or illegal addresses that are already in place.

Subnet masking allows extending the number of network addresses derived from a single IP address.

Private addressing was developed for organizations that do not need much access to the Internet.

Network address translation was developed for those companies that use private addressing or use non-NIC assigned IP addresses.

● Classless Interdomain Routing (CIDR): RFC 1518 and 1519 (1993): This is another method used for and developed for ISPs. This strategy suggests that the remaining IP addresses be allocated to ISPs in contiguous blocks, with geography being a consideration.

Variable Length Subnet Masks

Overview

Variable Length Subnet Masks (VLSMs) provide the capability to include more than one subnet mask within a class based address, and the capability to subnet an already subnetted network address. VLSMs do this by using a portion of the host address space as a subnet address. The term variable is used because the subnet address field can be variable in length, such as 2 bits, 3 bits, or 4 bits, as opposed to using a full byte for the subnet. These capabilities offer the following two benefits:

● Even more efficient use of IP addresses: Without the use of VLSMs companies are locked into implementing a single subnet within an NIC number in the entire network. With VLSMs you can create a subnet with only two hosts, which is ideal for serial links.

For example, consider that the 172.16.0.0/16 network address is divided into subnets using 172.16.0.0/24 masking and one of the subnetworks in this range, 172.16.14.0/24, is further divided into smaller subnets with the 172.16.14.0/27 masking as illustrated in Figure 13.9. These subnets range from 172.16.14.4 to 172.16.14.252. In Figure 13.9, one of the smaller subnets is further divided with the 172.16.14.128/30 subnet to be used on the WAN links.

> **Variable Length Subnet Masks** provide the capability to include more than one subnet mask within a class based address, and the capability to subnet an already subnetted network address.

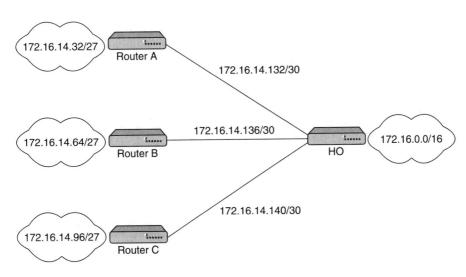

FIGURE 13.9

Subnet 172.16.14.0/24 is divided into smaller subnets. (Reproduced with permission of Cisco Systems, Inc. Copyright © 2001 Cisco Systems, Inc. All rights reserved.)

● Greater capability to use route summarization: VLSMs allow more hierarchical levels within your addressing plan, and thus allow better route summarization within routing tables. In Figure 13.9, for example, subnet 172.16.0.0/24 summarizes subnet 172.16.14.0. Subnet 172.16.0.0/24 includes all the addresses that are further subnetted, using VLSMs from subnets 172.16.14.0/27 and 172.16.14.128/30.

Classless and Classful Updates

Classless networks are not constrained by the Class A, B, and C designations that indicate the boundary for network and host portions.

VLSMs can be used when the routing protocol sends a subnet mask along with each network address. The protocols that support subnet mask information include RIP2, OSPF, Enhanced IGRP, BGP, and IS-IS. Networks running these protocols are called *classless* networks because they are not constrained by the Class A, B, and C designations that indicate the boundary for network and host portions. A prefix identifies the number of bits used for the network and host portions. This prefix accompanies all routing exchanges. RIP1 and IGRP do not support VLSMs. RIP1 and IGRP networks support only one subnet per network address because routing updates do not have a subnet masks field. As a result, upon receiving a packet, the router does one of the following to determine the network portion of the destination address:

● If the routing update information about the same network number is configured on the receiving interface, the router applies the subnet mask that is configured on the receiving interface.
● If the router receives information about a network address that is not the same as the one configured on the receiving interface, it will apply the default by class subnet mask, which is why RIP1 networks are referred to as *classful* networks. RIP1 route updates do not have a subnet mask field.

Classful networks are traditional networks that stay within their class boundaries and follow conventional addressing rules.

In Figure 13.10, for example, the RIP network router B is attached to network 172.5.1.0/24. Therefore, if router B learns about any network that is also a subnet of the 172.5.0.0 network, it will apply the subnet mask configured on its receiving interface (/24).

But notice how router C, which is attached to the 192.168.5.0/24 subnet, handles network address 172.5.2.0. Rather than assigning the network address the correct subnet mask (/24), it applies the default classful subnet mask for a Class B network address 172.5.0.0. It is impossible in this kind of environment to further subnet already subnetted IP addresses without causing confusion. Instead, VLSMs can be used only when the routing protocol sends subnet mask information along with the network address.

Calculating VLSMs

VLSMs allow you to subnet an already subnetted address. For example, consider that you need to assign subnetted address 172.6.32.0/20 to a network that has 10 hosts

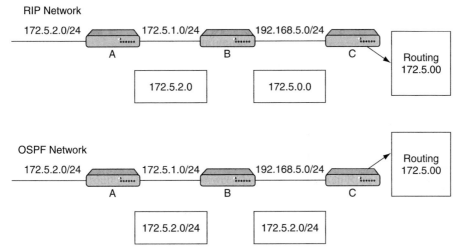

FIGURE 13.10

RIP1 routers cannot exchange network mask information along with route information like RIP2 and OSPF routers can. (Reproduced with permission of Cisco Systems, Inc. Copyright © 2001 Cisco Systems, Inc. All rights reserved.)

allowing 12 bits for the host portion. With this address, however, you get over 4,000 (2^2 2 − 4,094) host addresses, so you would be wasting about 4,000 IP addresses. With VLSM, you can further subnet address 172.6.32.0/20 to give you more network addresses and fewer hosts per network, which would provide more room to grow your network. If, for example, you subnet 172.6.32.0/20 to 172.6.32.0/26, you can now have up to 62 (2^6 − 2) subnetworks, each of which could support up to 62 (2^6 − 2) hosts. To further subnet 172.6.32.0/20 to 172.6.32.0/26 and gain five more network addresses, perform the following steps:

Step 1. Write 172.6.32.0 in binary form.
Step 2. Draw a vertical line between the twentieth and twenty-first bits, as illustrated in Figure 13.10.
Step 3. Draw a vertical line between the twenty-sixth and twenty-seventh bits, as shown in Figure 13.10.
Step 4. Calculate the five network addresses from lowest to highest in value, as shown in Figure 13.11. If necessary refer to the IP address calculation tables in Tables 13.7 and 13.8.

In Table 13.7, the # Bits column indicates how many bits have been taken from the host address bits from the classful address. For example, the first line indicated that two bits have been masked off from the host address portion to be used as subnet bits. This creates a subnet value of 11000000 in the third byte, which has the decimal equivalent of 192.

The Mask column indicates the mask value in decimal once the subnet bits have been masked off. For example, 255.255.192.0 is the mask for a Class B address that is subnetted using 2 bits for the subnet portion. The Effective Subnets and Effective Hosts

FIGURE 13.11

Calculating VLSMs may require binary conversion. (Reproduced with permission of Cisco Systems, Inc. Copyright © 2001 Cisco Systems Inc. All rights reserved.)

Subnetted Address: 172.6.32.0/20

In Binary: 10101100.00000110.00100000.00000000

VLSM Address: 172.6.32.0/26

In Binary: 10101100.00000110.001│ 00000.00 │ 000000

	Network		Subnet	VLSM Subnet	Host
1st Subnet	10101100.00000110		.0010	0000.01	00000000 = 172.6.32.64
2nd Subnet	172.	6	.0010	0000.10	00000000 = 172.6.32.128
3rd Subnet	172.	6	.0010	0000.11	00000000 = 172.6.32.192
4th Subnet	172.	6	.0010	0001.00	00000000 = 172.6.33.0
5th Subnet	172.	6	.0010	0001.01	00000000 = 172.6.33.64

Table 13.7 Subnet options using a Class B address that typically provides three bytes for the network portion and one byte for the host portion

# Bits	Mask	Effective Subnets	Effective Hosts	VLSM
2	255.255.192.0	2	16,382	/18
3	255.255.224.0	6	8,190	/19
4	255.255.240.0	14	4,094	/20
5	255.255.248.0	30	2,046	/21
6	255.255.252.0	62	1,022	/22
7	255.255.254.0	126	510	/23
8	255.255.255.0	254	254	/24
9	255.255.255.128	510	126	/25
10	255.255.255.192	1,022	62	/26
11	255.255.255.224	2,046	30	/27
12	255.255.255.240	4,094	14	/28
13	255.255.255.248	8,190	6	/29
14	255.255.255.252	16,382	2	/30

columns indicate how many different network and host numbers can be created with the bits masked off for the network portion and the remaining host portion. Network and host numbers using all 1s or all 0s are not counted in these numbers.

Table 13.8 defines the subnet options using a Class C address that typically provides three bytes for the network portion and one byte for the host portion.

Table 13.8 Class C IP Address Quantities

# Bits	Mask	Effective Subnets	Effective Hosts
2	255.255.255.192	3	62
3	255.255.255.224	7	30
4	255.255.255.240	15	14
5	255.255.255.248	31	6
6	255.255.255.252	63	2

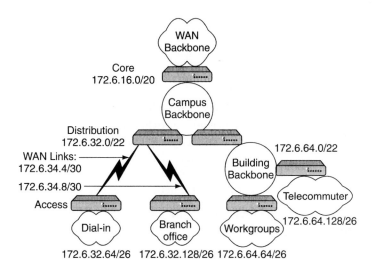

FIGURE 13.12
*A VLSM network.
(Reproduced with
permission of
Cisco Systems, Inc.
Copyright
© 2001 Cisco
Systems, Inc. All
rights reserved.)*

VLSM Example

Figure 13.12 illustrates a VLSM example used to optimize the number of possible addresses available for a network. Because point-to-point serial lines require only two host addresses you want to use a subnetted address that will not waste scarce subnet numbers.

Figure 13.12 illustrates where the addresses can be applied, depending on the network layer and the number of hosts anticipated at each layer. For example, the WAN links use Class B addresses with a prefix of /30. This prefix allows for 16,382 subnets $(2^{14} - 2)$ and only two hosts $(2^2 - 2)$, just enough hosts for a point-to-point connection between a pair of routers.

What Is Route Summarization?

This section discusses what route summarization is and how VLSMs maximize the use of route summarization.

FIGURE 13.13

Route summa-rization reduces router B's routing table size. (Repro-duced with per-mission of Cisco Systems, Inc. Copyright © 2001 Cisco Systems, Inc. All rights reserved.)

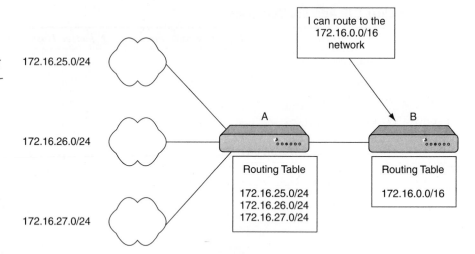

● **Route summa-rization**, also called **route ag-gregation** or **supernetting,** reduces the num-ber of routes that a router must main-tain because it represents a series of network num-bers as a single summary address.

● In **contiguous networks** the addresses are in sequential blocks in powers of two.

In large networks hundreds or even thousands of network addresses can exist. In these environments some routers may become overwhelmed. *Route summarization,* also called *route aggregation* or *supernetting,* reduces the number of routes that a router must maintain because it represents a series of network numbers as a single summary address. In Figure 13.13, for example, you can either send three routing update entries or summarize the address into a single network number 172.16.0.0/16.

Another advantage to using route summarization in large, complex networks is that it can isolate topology changes from other routers. That is, if a specific link in the 172.16.27.0/24 domain was intermittently failing, the summary route would not change, so no router external to the domain would need to keep modifying its routing table because of this problematic activity.

Route summarization is most effective within a subnetted environment when the network addresses are in *contiguous* (sequential) blocks in powers of two. For example, consider these two addresses:

● 130.129.0.0
● 130.192.0.0

Both addresses have nine matching bits in the beginning. If you were going to add more network addresses after you have used all numbers possible with these nine bits matching, you can dip into the next bit (the tenth bit) to define another group of addresses.

Routing protocols summarize or aggregate routes based on shared network numbers within the network. RIP2, OSPF, and Enhanced IGRP support route summarization based on subnet addresses, including VLSM addressing. Summarization is described in RFC 1518: An Architecture for IP Address Allocation with CIDR.

172.108.168.0 =	10101100.01101100.10101	000	.00000000
172.108.169.0 =	172.	108	.10101	001	.0
172.108.170.0 =	172.	108	.10101	010	.0
172.108.171.0 =	172.	108	.10101	011	.0
172.108.172.0 =	172.	108	.10101	100	.0
172.108.173.0 =	172.	108	.10101	101	.0

Number of Common Bits = 21   Noncommon
Summary: 172.108.168.0/21   Bits = 11

FIGURE 13.14

Find the common subnet bits to summarize routes. (Reproduced with permission of Cisco Systems, Inc. Copyright © 2001 Cisco Systems, Inc. All rights reserved.)

Summarizing Within an Octet

Figure 13.14 illustrates a summary route based on a full octet: 172.16.25.0/24, 172.16.26.0/24, and 172.16.27.0/24 could be summarized into 172.16.0.0/16. What if a router received updates for the following routes? How would the routes be summarized? Consider the following list of network addresses:

- 172.108.168.0
- 172.108.169.0
- 172.108.170.0
- 172.108.171.0
- 172.108.172.0
- 172.108.173.0

To determine the summary route, the router looks for the most number of high order bits that match. Referring to the list of IP addresses in Figure 13.14, the best summary route is 172.108.168.0/21.

To allow the router to aggregate the most number of IP addresses into a single route summary, your IP addressing plan should be hierarchical in nature. This approach is particularly important when using VLSMs.

In addition, you can summarize when the count is a power of two. The starting octet must be a multiple of the count. For example, you can summarize 8 bits starting with a multiple of 8, or 16 bits starting with a multiple of 16.

Summarizing Addresses in a VLSM Designed Network

A VLSM design allows a maximum use of IP addresses, as well as more efficient routing update communication when using hierarchical IP addressing. In Figure 13.15, route summarization occurs at the following two levels:

FIGURE 13.15

Summarization can be performed by multiple routers along a path. (Reproduced with permission of Cisco Systems, Inc. Copyright © 2001 Cisco Systems, Inc. All rights reserved.)

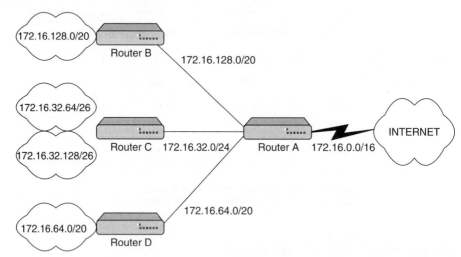

- Router C summarizes two routing updates from networks 172.16.32.64/26 and 172.16.32.128/26 into a single update, 172.16.32.0/24.
- Router A receives three different routing updates, but summarizes them into a single routing update before propagating it to the Internet.

Route Summarization Implementation Considerations

Route summarization reduces memory use on routers and routing protocol network traffic. Requirements for summarization to work correctly are as follows:

- Multiple IP addresses must share the same high order bits.
- Routing tables and protocols must base their routing decisions on a 32-bit IP address and prefix length that can be up to 32 bits.
- Routing protocols must carry the prefix length (subnet mask) with the 32-bit IP address.

For example, consider the binary equivalent of 172.21.134.0/16 and 172.21.138.0/16:

```
10101100.00010101.10001000.00000000
10101100.00010101.10001100.00000000
```

You can count the common high order bits at 21 and then summarize these routes as 172.21.134.0/21. The first 21 bits are common between the two addresses. The number 21 is used as the prefix.

Route Summarization in Routers

This section discusses the generalities of how routers handle route summarization. Details about how route summarization operates with a specific protocol should be referred to that protocol because every protocol summarizes routes a little differently.

Routers manage route summarization in two ways:

- Sending route summaries: Routing information advertised out an interface is automatically summarized at major classful network address boundaries only for some protocols such as RIP. A classful network address is a traditional address that has defined standard boundaries. For example, the address 12.0.0.0 is a classful address using the first byte for the network portion 12.0.0.0/8. Specifically, this automatic summarization occurs for those routes whose classful network address differs from the major network address of the interface to which the advertisement is being sent. For protocols such as OSPF, you must configure summarization.

    Route summarization is not always a good solution. You would not want to use route summarization if you needed to advertise all networks across a boundary, such as when you have noncontiguous networks. Protocols such as RIP2 allow you to disable automatic summarization.

- Selecting routes from route summaries: If more than one entry in the routing table matches a particular destination, the longest prefix match in the routing table is used. This is known as most specific route. Several routes may get close to one destination, but the most specific one will always be chosen. For example, if a routing table has different paths to 172.168.0.0/16 and to 172.168.5.0/24, packets addressed to 172.168.5.99 would be routed through 172.168.5.0/24 because that address has the longest prefix match.

Summarizing Routes in a Noncontiguous Network

Classful routing protocols summarize automatically at network boundaries. This behavior, which cannot be changed with RIP, has the following important results:

- Subnets are not advertised to a different major network.
- *Noncontiguous* subnets are not visible to each other.

In Figure 13.16, the 172.16.5.0 255.255.255.0 and 172.16.6.0 255.255.255.0 subnetworks are not advertised by RIP because RIP cannot advertise subnets. Therefore,

In **noncontiguous networks** the subnets are not visible to each other.

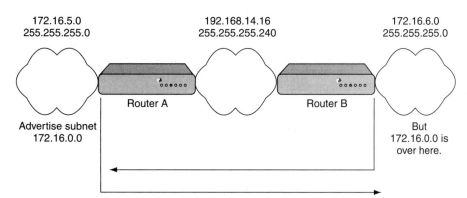

172.16.5.0
255.255.255.0

192.168.14.16
255.255.255.240

172.16.6.0
255.255.255.0

Router A

Router B

Advertise subnet
172.16.0.0

But
172.16.0.0 is
over here.

FIGURE 13.16
*RIP1 does not advertise subnets.
(Reproduced with permission of Cisco Systems, Inc. Copyright © 2001 Cisco Systems, Inc. All rights reserved.)*

each router advertises 172.16.0.0, which leads to confusion when routing across network 172.168.14.0 because this network receives routes about 172.16.0.0 from two different directions, so it cannot make a correct routing decision. RIP2 and OSPF resolve this situation when summarization is not used because the routes could be advertised with their actual subnet masks.

Advertisements are configurable with OSPF. There is an additional feature that routers offer which is called IP unnumbered that permits noncontiguous or nonsequential subnets to be separated by an unnumbered link. An unnumbered link does not have an address assigned to it.

IP Unnumbered

The IP unnumbered feature allows you to enable IP processing on a serial interface without assigning it an explicit IP address. The IP unnumbered interface can borrow the IP address of another interface already configured on the router, thereby conserving network address space.

On a router every interface connecting to a network segment must belong to a unique subnet. Directly connected routers have interfaces connecting to the same network segment and are assigned IP addresses from the same subnet. If a router needs to send data to a network that is not directly connected, it looks in the routing table and forwards the packet to the next directly connected hop toward the destination. If there is no route in the routing table, the router forwards the packet to the gateway of last resort. When a router that is directly connected to the final destination receives the packet, it delivers the packet directly to the end host.

The IP routing table contains either subnet routes or major network routes. Each route has one or more directly attached next hop addresses. Subnet routes are aggregated or summarized by default at major network boundaries in order to reduce size of the routing table.

Let us consider assigning IP addresses to the interface of a router using a Class B network that had been subnetted using 8 bits of subnetting. Every interface requires a unique subnet. Although each point-to-point serial connection has only two end points to address, if an entire subnet is assigned to each serial interface, there are then 254 available addresses for each interface where only two addresses are needed. If IP unnumbered is configured on each serial interface, address space is saved because the address of a LAN interface is borrowed and used as the source address for routing updates and packets sourced from the serial interface. In this way, address space is conserved. IP unnumbered only makes sense for point-to-point links.

A router receiving a routing update installs the source address of the update as the next hop in its routing table. Normally, the next hop is a directly connected network node. This is no longer the case if we use IP unnumbered because each serial interface borrowed their IP address from a different LAN interface, each in a different sub-net and possibly in a different major network. When IP unnumbered is configured, routes learned through the IP unnumbered interface have the interface as the next hop instead of the source address of the routing update. Thus, an invalid next hop address problem is avoided because the source of the routing update comes from a next hop that is not directly connected.

Other Addressing Considerations

This section discusses the use of private IP addresses, which are addresses that are not allowed on the public Internet because they are either reserved addresses or previously assigned. It also discusses network address translation, which is translation from one IP address to another address.

Using Private Addressing

Some organizations never need to connect to the Internet or any other external IP net-work. In other situations, some organizations may have only a few hosts or networks that never need to make connections external to their own network. For example, if the arrival and departure display monitors in a large airport are individually addressable via TCP/IP, most likely these displays need not be directly accessible from other networks.

In these cases you can use private addresses, as defined in RFC 1918: Address Allo-cation for Private Internets. This RFC specifies the following IP addresses as private:

- Class A: 10.0.0.0 to 10.255.255.255
- Class B: 172.16.0.0 to 172.31.255.255
- Class C: 192.168.0.0 to 192.168.255.255

Implementation Considerations

If it is decided that the entity will use these private addresses, you do not need to co-ordinate them with the Internet registry because they will never be broadcast exter-nal networks. You should do some planning before randomly assigning addresses. Some implementation considerations are as follows:

- Determine which hosts do not need to have network layer connectivity to the outside. These hosts are considered private hosts. Private hosts can communi-cate with all other hosts within your network, both public and private, but they cannot have direct connectivity to external hosts.
- Routers that connect to external networks should be set up with the appropri-ate packet and routing filters at both ends of the link to prevent the leaking of the private IP addresses. You should also filter any private networks from in-bound routing information to prevent ambiguous routing situations that can occur if routes to the private address space point outside the network.

FIGURE 13.17

A NAT router connects a private network to the Internet. (Reproduced with permission of Cisco Systems, Inc. Copyright © 2001 Cisco Systems, Inc. All rights reserved.)

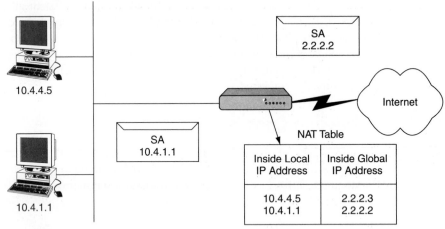

SA = Source Address

• Changing a host from private to public will require changing its address, and in most cases, its physical connectivity. In locations where such changes can be foreseen, you might want to configure separate physical media for public and private subnets to make these changes easier.

Understand that private addresses can connect to external hosts through a network address translation (NAT) capable device or a proxy device.

Accessing the Internet Using Private Addresses

If a host configured with a private IP address needed to access the Internet or other external hosts, its IP address would need to be reconfigured, and most likely the host device would need to be moved physically to a network that used a public IP address. Reconfiguring and reconnecting an entire network, building, or corporation can be a very costly venture in both time and resources. To avoid having to renumber all hosts RFC 1631, The IP Network Address Translator (NAT), was defined.

A NAT router or host is placed on the border of a stub domain and an internetwork that has a single connection to the Internet (referred to as the outside network).

The NAT router translates the internal local addresses into globally unique IP addresses before sending packets to the outside network as illustrated in Figure 13.17.

NAT takes advantage of the fact that relatively few hosts in a stub domain communicate outside of the domain at any given time. Because most of the hosts do not communicate outside of their stub domain, only a subset of the IP addresses in a stub domain must be translated into globally unique IP addresses when outside communication is necessary. NAT can also be used when you need to modify your internal addresses because you change ISPs. Rather than renumber your networks, use NAT to translate the appropriate addresses.

One disadvantage of using NAT is with network management. In order to track NAT activity, you need two network management hosts on either side of the NAT router because the SNMP IP address table does not go through the NAT router correctly.

Translating Inside Local Addresses

Figure 13.18 illustrates one of several NAT capabilities: the capability to translate addresses from inside your network to destinations outside of your network.

The steps shown in Figure 13.18 are defined as follows:

Step 1. User at host 10.4.1.1 opens a connection to host B.
Step 2. The first packet that the router receives from 10.4.1.1 causes the router to check its NAT table. If a translation is found because it has been statically configured, the router continues to step 3. If no translation is found, the router determines that address 10.4.1.1 must be translated. The router allocates a new address and sets up a translation of the inside local address 10.4.1.1 to a legal global address from the dynamic address pool. This type of translation entry is referred to as a simple entry.
Step 3. The router replaces the inside local IP address 10.4.1.1 with the selected inside global address 2.2.2.2 and forwards the packet.

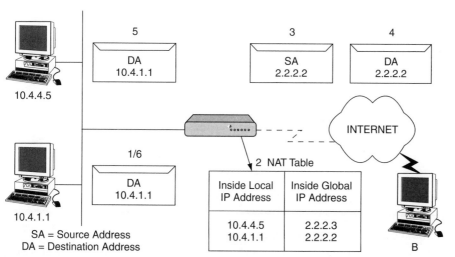

FIGURE 13.18

The private network address of 10.4.0.0 is never seen on the Internet. (Reproduced with permission of Cisco Systems, Inc. Copyright © 2001 Cisco Systems, Inc. All rights reserved.)

Step 4. Host B receives the packet and responds to 10.4.1.1 using the inside global IP address 2.2.2.2.

Step 5. When the router receives the packet with the inside global IP address of 2.2.2.2, the router performs a NAT table lookup using the inside global address as the reference. The router then translates the address back to 10.4.1.1 and forwards the packet to the host.

Step 6. Host 10.4.1.1 receives the packet and continues the conversation. For each packet, the router performs steps 2 through 5.

What Is Redistribution?

Route redistribution is the capability for boundary routers connecting different autonomous systems to exchange and advertise routing information received from one autonomous system to another autonomous system.

Routers allow internetworks using different routing protocols (referred to as autonomous systems) to exchange routing information through a feature called route redistribution. *Route redistribution* is defined as the capability for boundary routers connecting different autonomous systems to exchange and advertise routing information received from one autonomous system to the other autonomous system.

Within each autonomous system the internal routers IGRP and Enhanced IGRP (EIGRP), as in the case of Figure 13.19, have complete knowledge about all subnets that make up each network. The router interconnecting both autonomous systems is called an autonomous system boundary router (ASBR) and has both IGRP and EIGRP process active. The ASBR is responsible for advertising routes learned from one autonomous system into the other autonomous system.

In Figure 13.19, network 192.168.5.0 is known via the S0 interface. The routing table for the router in AS 300 contains routes such as 192.168.5.0 and 172.16.0.0 that are summarized at network boundaries. These routes are indicated by the D for Enhanced IGRP and EX for an external route that was learned from redistribution.

FIGURE 13.19

Redistribution enables routes to be learned from another routing protocol. (Reproduced with permission of Cisco Systems, Inc. Copyright © 2001 Cisco Systems, Inc. All rights reserved.)

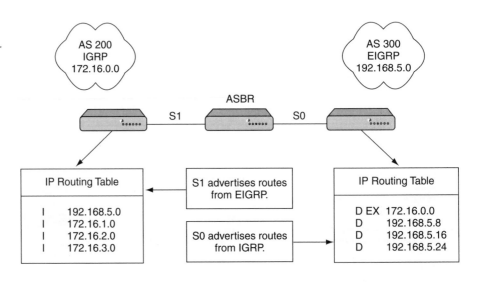

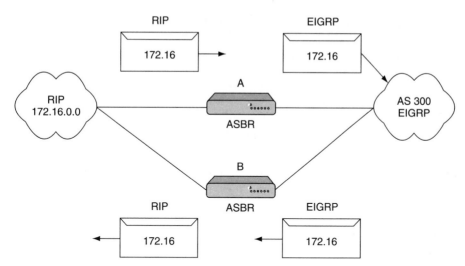

Redistribution Implementation Considerations

Redistribution, although powerful, increases the complexity and potential for routing confusion, so it should be used only when absolutely necessary. The key issues that arise when using redistribution are as follows:

- Routing feedback (loops): Depending on how you employ redistribution, routers can send routing information received from one autonomous system back into the autonomous system as illustrated in Figure 13.20.

    The feedback is similar to the split horizon problem that occurs in distance vector technologies. In Figure 13.20, information about network 172.16 crosses ASBR A to the EIGRP net. Because the internetwork has a loop, the same information propagates back to the RIP network through ASBR B.

- Incompatible routing information: Because each routing protocol uses different metrics to determine the best path, for example, RIP uses hop counts and OSPF uses cost, path selection using the redistributed route information may not be optimal. Because the metric information about a route cannot be translated exactly into a different protocol, the path a router chooses may not be the best.

- Inconsistent convergence time: Different routing protocols converge at different rates. For example, RIP converges slower than EIGRP, so if a link goes down, the EIGRP network will learn about it before the RIP network.

Selecting the Best Path

Most routing protocols have metric structures and algorithms that are not compatible with other protocols. In a network where multiple routing protocols are present,

Table 13.9 Administrative Distance for Routed Protocols

Route Source	Default Distance
Connected Interface	0
Static Route	1
EIGRP Summary Route	5
External BGP	20
Internal Enhanced IGRP	90
IGRP	100
OSPF	110
IS-IS	115
RIP	120
EGP	140
External Enhanced IGRP	170
Internal BGP	200
Unknown	255

the exchange of route information and the capability to select the best path across multiple protocols are critical.

The **administrative distance** is a rating that defines the believability of a routing protocol.

When routers learn two or more different routes to the same destination from two different routing protocols, they can select the best path by using the *administrative distance,* which defines the believability of a routing protocol. Each routing protocol is prioritized in order of most to least believable by using the value of administrative distance. This criterion is the first a router uses to determine which routing protocol to believe if two protocols provide route information for the same destination. The more believable protocol is selected, even when it advertises a suboptimal route.

What Protocol to Believe?

Table 13.9 lists the believability (administrative distance) of the protocols supported by the router.

The smaller the administrative distance, the more reliable the protocol. For example, if the router received a route to network 10.2.2.0 from IGRP and then received a route to the same network from OSPF, IGRP is more believable, so the IGRP version of the route would be added to the routing table.

When using redistribution, there may be a need occasionally to modify the administrative distance of a protocol so that it has preference. If you want the router to select

RIP learned routes rather than IGRP learned routes to the same destination, then you must increase the administrative distance for IGRP.

After an ASBR selects the routing protocol to which to listen, it must be able to translate the metric of the received route from the source routing protocol into the other routing protocol. If an ASBR receives a RIP route, for example, it will have a hop count as a metric. To redistribute it into OSPF, the hop count must be translated into a cost value. The cost value you define during configuration is referred to as the *seed* or default metric. After the seed metric is established, the metric will increment normally within the autonomous system.

Some precautions should be taken when a loop exists between two autonomous systems, as illustrated in Figure 13.20. Consider setting the default metric on the incoming redistributed route to something higher than what currently exists in the receiving protocol. This way you get some degree of automatic protection from routing loops. Suppose a routing loop exists where a routing protocol is getting preferred routes that it originally sourced. If those routes are in at a low metric, they can be preferred over the original route. So, imagine router Z in a RIP domain knows that network A, a route native to the RIP domain of which you are a member, is currently at a cost of three hops from the destination. Now suppose that the route to A has been redistributed to some foreign protocol and is now coming back into the RIP domain from the outside and router Z is only one hop away from the point of redistribution. If the redistributing router sets the metric for this new route to network A to one, router Z's cost to A is now only two hops and it now points toward the redistribution router. If the redistribution router sets the metric to four hops, for example, then all will be well even though technically a routing loop is occurring. Setting the incoming default metric low may create a black hole for routes.

> The **seed** or default metric is the cost value defined during configuration.

Redistribution Guidelines

At a high level, it is recommended that you consider employing the following guidelines when using redistribution:

- Be familiar with your network: This is the overriding recommendation. There are many ways to implement redistribution, so knowing your network will enable you to make the best decision. Use an analyzer to track your data flows.
- Do not overlap routing protocols: Do not run two different protocols in the same internetwork. Rather, have distinct boundaries between networks that use different protocols.
- One-way redistribution: To avoid routing loops and having problems with varying convergence time, allow routes to be exchanged in only one direction, not both directions. In the other direction, you should consider using a default route. This is important for cases when you have multiple points of redistribution. If you have only one ASBR, full two-way relationships are recommended.

● Two-way redistribution: In case of only one point of redistribution (one ASBR only), there are no potential problems. If you have multiple ASBRs and if you must allow two-way redistribution, enable a mechanism to reduce the chances of routing loops. Examples are default routes, route filters, and modification of the metrics advertised. With these types of mechanisms you can reduce the chances of routes imported from one autonomous system being reinjected into the same autonomous system as new route information.

Summary

In this chapter you learned how subnet masks, VLSMs, private addressing, and network address translation can enable more efficient use of IP addresses.

You learned that hierarchical addressing allows efficient allocation of addresses and reduced number of routing table entries. VLSMs, specifically, provide the capability to include more than one subnet mask within a network and the capability to subnet an already subnetted network address. Proper IP addressing is required to ensure the most efficient network communications system.

Chapter 14 discusses the data link protocols and their associated hardware.

Review Questions

1. Before implementing private addressing what should you do?
    a.
    b.
    c.
2. What are the private IP addresses, as defined by RFC 1918?
    a.
    b.
    c.
3. Provide one example of when NAT can be used.
4. Which IP address class has the fewest hosts per segment?
5. Which RFC defines NAT?
6. Which of the following is a Class A network address?
    a. 128.4.5.6
    b. 127.4.5.0
    c. 127.0.0.0
    d. 127.8.0.0
7. Which of the following is a Class B host address?
    a. 230.0.0.0
    b. 130.4.5.6
    c. 230.4.5.9
    d. 30.4.5.6

8. What are the two primary responsibilities of the IP protocol?
9. Which field in the IP header packet keeps packets from looping endlessly throughout the internetwork?
10. What is the unique bit pattern assigned each IP host?
11. How is a Class A address composed?
12. How is a Class B address composed?
13. How is a Class C address composed?
14. How many subnets are needed for serial interfaces?
15. What is the default subnet mask for a Class A network?
16. What is the default subnet mask for a Class B network?
17. What is the default subnet mask for a Class C network?
18. List the rules for the logical AND operation and construct the truth table for the logical AND operation.
19. How do subnet masks determine the network number?
20. Under what circumstances would you use VLSM?

Summary Questions

1. Explain the process of route summarization.
2. When would you use route summarization?
3. Using the network in Figure 13.21, indicate where route summarization can occur, and what the summarized address would be, by completing the following table.

Router C Route Table Entries	Routes That Can Be Advertised to Router D from Router C
172.16.1.64/28	
172.16.1.80/28	
172.16.1.96/28	
172.16.1.192/28	
172.16.1.2.208/28	
172.16.1.112/28	

4. What is IP unnumbered?
5. When would IP unnumbered be used?
6. What is route redistribution?
7. What are some guidelines to consider when using route redistribution?
8. How is one type of protocol considered more believable than another in route redistribution?
9. What is administrative distance?
10. What are some issues to be considered when implementing route redistribution?

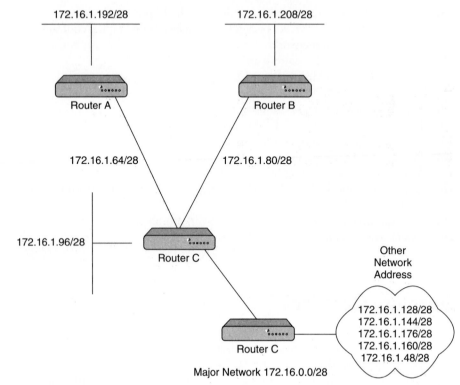

FIGURE 13.21

Where can route summarization occur on this network? (Reproduced with permission of Cisco Systems, Inc. Copyright © 2001 Cisco Systems, Inc. All rights reserved.)

172.16.1.192/28

172.16.1.208/28

Router A

Router B

172.16.1.64/28

172.16.1.80/28

172.16.1.96/28

Router C

Other
Network
Address

172.16.1.128/28
172.16.1.144/28
172.16.1.176/28
172.16.1.160/28
172.16.1.48/28

Router C

Major Network 172.16.0.0/28

The Data Link Protocols

Objectives

- Discuss the different types of data link protocols.
- Explain the differences between synchronous and asynchronous systems.
- Discuss the functions of timers at the data link layer.
- Examine the Automatic Request for Repeat (ARQ) Flow Control Protocol.
- Explain the difference between positive and negative acknowledgements.
- Discuss the types of host configurations.
- Explain polling and the different types of polling.
- Examine the types of transmission impairments that occur on the line and the different types of error checking that take place at the data link layer.
- Explain how the Binary Synchronous Control Protocol (BISYNC) operates.
- Examine how High Level Data Link Control (HDLC) operates.

Key Terms

Introduction

The transfer of data across the communications link must flow in a controlled, orderly manner. Because communication links experience distortions such as noise, a method must be provided to deal with periodic errors that occur. The data communications system must provide each station on the link with the capability to send data to another station, and the sending station must be assured that the data arrive error free at the receiving station. The sending and receiving stations must maintain complete accountability for all traffic. In the event the data are distorted, the receiver site must be able to direct the originator to resend the erroneous frame or correct the errors.

Data Link Controls operate at layer 2 providing control mechanisms for data link functions.

Data Link Controls (DLCs), also called line protocols, provide these services. They manage the flow of data across the communications path or link. Their functions are limited to the individual link. Link control is responsible only for the traffic between adjacent nodes or stations on a line. Once the data are transmitted to the adjacent node and an acknowledgement of the transmission is returned to the transmission site, the link task is complete for that particular transmission.

DLCs typically consist of a combination of software and hardware. The DLC provides the following functions:

- Synchronizing (logically, not physically) the sender and receiver through the use of flags/SYN characters.
- Controlling the flow of data to prevent the sender from sending too fast.
- Detecting and recovering from errors between two points on the link.
- Maintaining awareness of link conditions such as distinguishing between data and control and determining the identity of the communicating stations.

The data link layer rests above the physical layer in the OSI model. Generally, the data link protocol is medium independent and relies on the physical layer to deal with

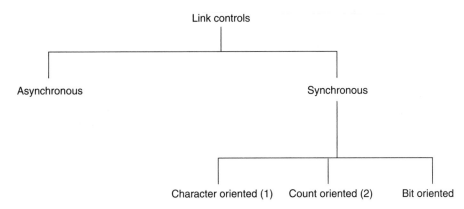

FIGURE 14.1
Formats of data link protocols. (Reproduced with permission of Cisco Systems, Inc. Copyright © 2001 Cisco Systems, Inc. All rights reserved.)

(1) Also called byte oriented

(2) Also called block oriented

the specific media, such as wire and radio, and the physical signals, such as electrical current, laser, and infrared.

Hundreds of different link protocols are used by the data communications industry. We will focus on the main family of High Level Data Link Control (HDLC) protocols. Figure 14.1 provides a simple explanation for our discussion.

Asynchronous Line Protocols

Asynchronous DLCs place timing bits around each character in the user data stream as illustrated in Figure 14.2. The start bit precedes the data and is used to notify the receiving site that data are on the path (start bit detection). The line is in an idle condition prior to the arrival of the start bit and remains in the idle state until a start bit is transmitted. The start signal initiates mechanisms in the receiving device for sampling, counting, and receiving the bits of the data stream. The data bits are represented as the mark signal (binary 1) and the space signal (binary 0). The user data bits are placed in a temporary storage area, such as a register or buffer, and are later moved into the terminal or computer memory for further processing.

Stop bits, consisting of one or more mark signals, provide the mechanism to notify the receiver that all bits of the character have arrived. Following the stop bits the signal returns to an idle level, thus guaranteeing that the next character will begin with a 1 to 0 transition. Even if the character is all 0s, the stop bit returns the link to a high or idle level.

This method is called *asynchronous transmission* because of the absence of continuous synchronization between the sender and receiver. The process allows a data character

Asynchronous DLCs place timing bits around each character in the user data stream.

Asynchronous transmission is the absence of continuous synchronization between the sender and receiver.

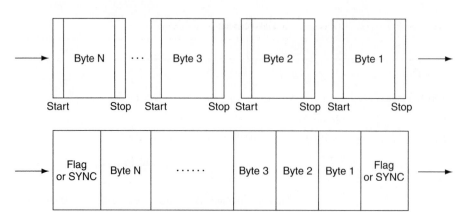

FIGURE 14.2

Asynchronous and synchronous transmissions. (Reproduced with permission of Cisco Systems, Inc. Copyright © 2001 Cisco Systems, Inc. All rights reserved.)

to be transmitted at any time without regard to any previous timing signal—the timing signal is part of the data signal. Asynchronous transmission is commonly found in machines or terminals such as teletypes and low speed computer terminals. The vast majority of personal computers use asynchronous transmission because its value lies in its simplicity.

Synchronous Line Protocols

> **Synchronous DLCs** place a preamble and a postamble bit pattern around the user data.

Synchronous DLCs do not surround each character with start/stop bits, but place a preamble and a postamble bit pattern around the user data. These bit patterns are usually called a SYN (synchronization) character, an EOT (end of transmission) character, or simply a flag. They are used to notify the receiver when user data are arriving and when the last user data have arrived.

It should be emphasized that data communications systems require two types of synchronization:

- At the physical level: to keep the transmitter and receiver clocks synchronized.
- At the link level: to distinguish user data from flags and other control fields.

Character Oriented Protocols

> **Character oriented protocols** rely on a specific code set to interpret the control fields.

The *character oriented synchronous data link controls* were developed in the 1960s and are still widely used today. The binary synchronous control (BSC or BISYNC) family are character oriented systems. These protocols rely on a specific code set such as ASCII and EBCDIC to interpret the control fields; thus, they are code dependent. Table 14.1 summarizes the functions of the most common character oriented control codes.

If machines on a link use different codes, one for ASCII and one with EBCDIC for example, the user must deal with two variants of the link protocol, and some type of

Table 14.1 Typical Character Oriented Control Codes

Character	Function
SYN	Synchronous idle (keeps channel active)
PAD	Frame pad (time fill between transmission)
DLE	Data link escape (used to achieve code transparency)
ENQ	Enquiry (used with polls/selects and bids)
SOH	Start of heading
ITB	End of intermediate block
ETB	End of transmission block
ETX	End of text
EOT	End of transmission (puts line in control mode)
BCC	Block check count
ACK0	Acknowledges even-numbered blocks
ACK1	Acknowledges odd-numbered blocks
WACK	Wait before transmitting

code conversion is required before the machines can communicate with each other. The problem is illustrated in Figure 14.3a.

Let us assume stations A and B use STX and ETX to represent start of user text and end of user text, respectively. It is evident that the two devices cannot communicate because the ASCII and EBCDIC codes use different bit sequences. Consequently, code conversion must be performed by one of the machines.

It is also possible that a code recognized as control could be created by the user application process. For instance, assume in Figure 14.3b that a user program creates a bit sequence, which is the same as the ETX (end of text) control code. The receiving station, upon encountering the ETX inside the user data, would erroneously assume the end of the transmission is signified by the user generated ETX. The control would accept the ETX as a protocol control and attempt to perform an error check on an incomplete frame, which would result in an error.

Obviously, control codes must be excluded from the user text field. Character protocols address the problem with the DLE control code. This code is placed in front of the control codes such as STX, ETX, ITB, and SOH to identify these characters as valid line control characters as depicted in Figure 14.3c. The simplest means to achieve code transparency is to use DLE.STX or DLE.SOH to signify the beginning of noncontrol data (user data) and DEL.ETX, DLE.ETB, or DLE.ITB to signify the end of user data. The DLE is not placed in front of user generated data. Consequently, if bit patterns resembling any of these control characters are created in the

FIGURE 14.3

*Character ori-
ented protocols.
(Reproduced with
permission of
Cisco Systems, Inc.
Copyright
© 2001 Cisco
Systems, Inc. All
rights reserved.)*

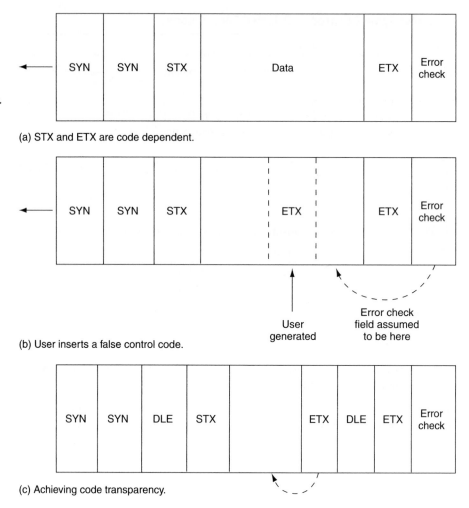

(a) STX and ETX are code dependent.

(b) User inserts a false control code.

(c) Achieving code transparency.

user text and encountered by the receiving station, the receiving station assumes that
they are valid user data because the DLE does not precede the character in question.

The DLE presents a special problem if it is generated by the end user application
process because it could be recognized as a control code. Character oriented protocols handle this situation by inserting a DLE next to a data DLE character. The receiver discards the first DLE of two successive DLEs and accepts the second DLE as
valid user data.

The character oriented protocols have dominated the vendors' synchronous line protocol products since the mid-1960s. While still widely used, they are being replaced
by count and bit protocols.

Count Oriented Protocols

In the 1970s, *count oriented protocols,* also called block protocols, were developed to address the code dependency problem. These systems exhibit one principal advantage over character oriented protocols in that they have a more effective means of handling user data transparency: They simply insert a count field at the transmitting station. This field specifies the length of the user data field. As a consequence, the receiver need not examine the user data field contents, but need only count the incoming bytes as specified by the count field.

The count oriented protocols are really a combination of character oriented protocols and bit oriented protocols. Figure 14.4 shows that certain control fields are code dependent and others are code transparent.

Count oriented protocols may encounter problems when the signals are transmitted across a digital link. A digital system may delete a digital frame on a link to recover clocks and resynchronize. It may also insert timing and control data into the transmission. A count protocol loses its receive counter in such a situation, and the protocol must recover the lost data with retransmissions.

Count oriented protocols have a more effective means of handling user data transparency: they simply insert a count field at the transmitting station.

Bit Oriented Protocols

Bit oriented data link control protocols were developed in the 1970s and are now quite prevalent throughout the industry. They form the basis for most of the new link level systems in use today.

Bit protocols do not rely on a specific code for line control. Individual bits within an octet are set to effect control functions. An 8-bit flag pattern of 01111110 is generated at the beginning and end of the frame to enable the receiver to identify the beginning and end of a transmission.

Bit oriented protocols form the basis for most of the new link level systems in use today.

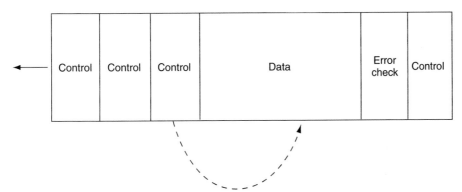

FIGURE 14.4
Count oriented protocols. (Reproduced with permission of Cisco Systems, Inc. Copyright © 2001 Cisco Systems, Inc. All rights reserved.)

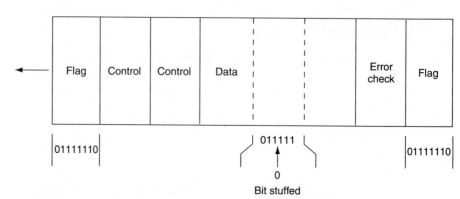

There will be occasions when a flaglike sequence, 01111110, is inserted into the user data stream by the application process. To prevent this occurrence, the transmitting site inserts a 0 bit after it encounters five continuous 1s anywhere between the opening and the closing flag of the frame. This technique is called bit stuffing and is similar in function to the DLE in character protocols and the count field in count protocols. As the frame is stuffed, it is transmitted across the link to the receiver. Figure 14.5 reviews the process. The receiver continuously monitors the bit stream. After it receives a 0 bit with five continuous 1 bits following, the receiver inspects the next bit. If it is a 0 bit, it pulls this bit out; it unstuffs the bit. In this manner, the system achieves code and data transparency. The protocol is not concerned about any particular bit code inside the data stream. Its main concern is to keep the flags unique.

Bit oriented protocols have virtually taken over the market for the newer products and offerings.

Comparison of Synchronous Protocols

Table 14.2 provides a comparison of character oriented, count oriented, and bit oriented protocols. It is obvious that no clear distinction exists between all the attributes of the protocols. The count oriented protocol is actually a hybrid of the other two protocols.

Controlling Traffic on the Link

Asynchronous Systems

Traffic control on an asynchronous link may use the same techniques as a synchronous link. For example, some systems encapsulate an asynchronous data stream into

Table 14.2 Comparison of Bit Protocols

Attribute	Character	Count	Bit
Start Framing	SYN SYN	SYN SYN	Flag
Stop Framing	Characters	Count Field	Flag
Retransmissions	Stop and Wait	Go-back-N or selective Repeat	Go-back-N or selective Repeat
Window Size	1	Various (255)	Various (7–127)
Frame Formats	Several	Few	1
Line Mode	HDX	HDX or FDX	HDX or FDX
Text Transparency	DLE code	Count Field	Bit Stuffing
Traffic Flow	TTD or WACK	Sliding Window	Sliding Window
Line Control	Full character	Full Character and Bits	Bits

the information field of a synchronous frame and transmit the frame with the synchronous protocol. Two methods that are widely used for controlling asynchronous traffic are (a) request to send/clear to send and (b) XON/XOFF.

Request to send/clear to send (RTS/CTS) (see Appendix C for further explanations) is considered a low level approach to protocols and data communications. Nonetheless, it is widely used because of its relationship to and dependence upon the widely used physical interface, Electronic Industries Association (EIA) RS-232C.

The use of EIA RS-232C to effect communications between DTEs is most common in a local environment because RS-232C is inherently a short-distance interface, typically constraining the channel to no greater than 19,200 feet. Devices can control the communications between each other by raising and lowering the RTS/CTS signals on the circuits (pins 4 and 5, respectively). A common implementation of this technique is found in the attachment of a terminal to a simple multiplexer. The terminal requests use of the channel by raising its RTS circuit (pin 4). The multiplexer responds to the request by raising the CTS circuit (pin 5). The terminal then sends its data to the multiplexer through the transmit data circuit (pin 2) and the multiplexer receives the data on its receive circuit (pin 3).

Another widely used technique is XON/XOFF. XON is an American National Standards Institute (ANSI/A5) transmission character. The XON is usually implemented by Device Control 1 (DC1). The XOFF, also an ANSI/A5 character, is represented by DC3. Peripheral devices such as printers, graphics terminals, or plotters can use the XON/XOFF approach to control incoming traffic. The master station, typically a computer, sends data to the remote peripheral site, which prints or graphs the data onto an output media. Because the plotter or printer may be slow relative to the transmission speed of the channel and the transmission speed of the transmitting computer, its buffer may become full. Consequently, to prevent overflow it transmits back to the computer an XOFF signal, which means stop transmitting or transmit off. The signals can be transmitted across an RS-232C connection, twisted-pair, or any type of media. As the buffers empty, the peripheral device transmits an XON to resume the data transfer.

Synchronous Systems

Regardless of the synchronous format used (character, count, or bit) the communications processes between stations are similar. The following processes are utilized by synchronous communications systems:

- Flow control of traffic between the machines
- Sequencing and accounting of traffic
- Actions to be taken in the event of error detection

Figure 14.6 provides an illustration of *synchronous link operations.* This example illustrates several important points about data link (level 2) communications. It shows DTE A is to transmit data to DTE B. The transmission goes through an intermediate point, a host computer located at station C. The most common approach is to pass the data from site to site until it finally reaches its destination. One important aspect of the process is in event 2 where C, the host computer, sends an acknowledgement of the data received to DTE A. This acknowledgement means the host computer has checked for the possible errors occurring during the transmission of the frame, and as best as it can determine, the data have been received without corruption. It so indicates by transmitting another frame on the return path—indicating acceptance.

Synchronous transmission does not require use of start and stop bits to indicate beginning of transmission thus reducing transmission overhead.

FIGURE 14.6

Link operations. (Reproduced with permission of Cisco Systems, Inc. Copyright © 2001 Cisco Systems, Inc. All rights reserved.)

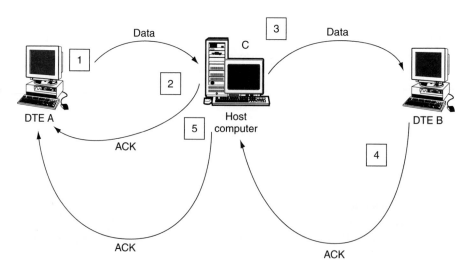

The data communications industry uses two terms to describe the event 2 response. The term ACK denotes a *positive acknowledgement;* the term NAK represents a *negative acknowledgement.* A NAK usually occurs because the signal is distorted due to faulty conditions on the channel such as noise. The frame sent to DTE A in event 2 will be either an ACK or a NAK. In the event of an error in the transmission, DTE A must receive a NAK so it can retransmit the data. It is also essential that the processes shown in events 1 and 2 are completed before event 3 occurs. If the host computer immediately transmitted the data to DTE B before performing the error check, DTE B could possibly receive erroneous data.

If DTE A receives an ACK in event 2, it assumes the data have been received correctly at DTE B, and the communications system at site A can purge these data from its queue. Continuing the process in events 3 and 4, assume that an ACK is returned from DTE B to the host computer. The end user at DTE A may assume through event 2 that the data arrived at the host computer. A false sense of security could result because event 2 only indicates that the data arrived safely at the host computer. If the data are lost between the host computer and DTE B sites, the DTE A user assumes no problem exists. This scenario provides no provision for an end-to-end acknowledgement. If an end user wishes to have absolute assurance that the data arrived at the remote site, event 5 is required. Upon receiving event 4 at the host computer, the host computer sends another acceptance (ACK) to DTE A.

The level 2 data link protocols do not provide end-to-end accountability through multiple links. Some systems provide this service at level 3, the network layer. However, the OSI model intends end-to-end accountability to be provided by the transport layer (level 4).

> A **positive acknowledgement** means the host computer has checked for possible errors and found the data to be error free.

> A **negative acknowledgement** means erroneous data could be received.

Functions of Timers

Many link protocols use timers in conjunction with logic states to verify that an event occurs within a prescribed time. When a transmitting station transmits a frame onto the channel, it starts a timer and enters a wait state. The value of the timer usually called T1 is set to expire if the receiving station does not respond to the transmitted frame within a set period of time. Upon expiration of the timer, one to n retransmissions are attempted, each with the timer T1 reset, until a response is received or until the link protocol's maximum number of retries is met. In this case, recovery is performed by higher level protocols or by manual intervention and troubleshooting efforts. The retry parameter is usually designated as parameter N2.

The T1 timer just described is designated as the acknowledgement timer. Its value depends on (a) round-trip propagation delay of the signal, which is usually a small value except for very long and very high speed circuits; (b) the processing time at the receiver including the queuing time of the frame; (c) the transmission time of the

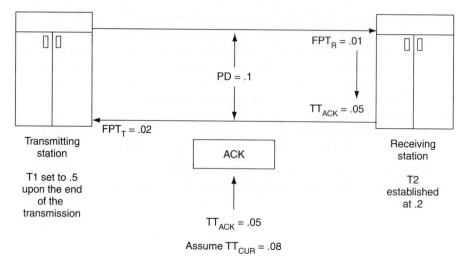

acknowledging frame; and (d) possible queue and processing time at the transmitter when it receives the acknowledgement frame.

The receiving station may use a parameter (T2) in conjunction with T1. T2's value is set to ensure that an acknowledgement frame is sent to the transmitting station before the T1 at the transmitter expires. This action precludes the transmitter from resending frames unnecessarily.

To illustrate the use of T1 and T2, the following algorithms describe a lower bound on T1 and an upper bound on T2. T1 is started at the end of the transmission of a frame.

$$T1_T = \geq T2_R + PD + FPT_R + TT_{CUR} + TT_{ACK} + FPT_T$$

$$T2_R = \leq T1_T - PD - FPT_R - TT_{CUR} - TT_{ACK} - FPT_T$$

Where T is the transmitter, R is the receiver, PD is the round-trip propagation delay, FPT is the frame processing time, TT_{ACK} is the admission time of the acknowledgement frame, and TT_{CUR} is the time to complete the transmission of the ongoing frames that are already in the transmit queue and cannot be pushed down into the queue. It can also describe, if relevant to a particular protocol, the queue time at the receiver before the acknowledgement frame is sent and T1 stopped.

Figure 14.7 illustrates the use of the T1 timer and the T2 parameter. For the values shown in the figure, the T1 and T2 values are set properly:

$$T1_T \text{ or } .5 = \geq .2 + .1 + .08 + .05 + .02$$
$$.5 = \geq .46$$
$$T2_R \text{ or } .2 = \leq .5 - .1 - .01 - .08 - .05 - .02$$
$$.2 = \leq .24$$

The number of timers varies, depending on the type of protocol and the designer's approach to link management. Some other commonly used timers are:

- Poll timer (also called P bit): defines the time interval during which a polling station such as a station requesting a frame from another station should expect to receive a response.
- NAK timer (also called a reject or selective reject timer): defines the time interval during which a rejection station should expect a reply to its reject frame.
- Link setup timer: defines the time interval during which a transmitting station should expect a reply to its link setup command frame.

The timing functions may be implemented by a number of individual timers, and the protocol designer or implementer is responsible for determining how the timers are set and restarted.

Automatic Request for Repeat (ARQ) Flow Control Protocol

When a station transmits a frame, it places a send sequence number in a control field. The receiving station uses this number to determine if it has received all other preceding frames with lower numbers. It also uses the number to determine its response. For example, after it receives a frame with send sequence number = 3, it responds with an ACK with a receive sequence number = 4, which signifies it accepts all frames up to and including 3 and expects 4 to be the send sequence number of the next frame. The send sequence number is identified as N(S) and the receive sequence number is identified as N(R).

Half-duplex (HDX) protocols only need to use two numbers for sequencing because they can have only one frame outstanding at a time. Most HDX protocols use the binary numbers 0 and 1 alternately. Full duplex FDX protocols typically use a greater range of sequence numbers because many frames may be outstanding at a time.

The term *automatic request for repeat (ARQ)* describes the process by which a receiving station requests a retransmission. As an example, the reception of a NAK with receive sequence number = 5 indicates that frame 5 is in error and must be retransmitted. The process is automatic, without human intervention. A stop-and-wait ARQ describes a similar half-duplex protocol that waits for an ACK or NAK before sending another frame.

> **Automatic request for repeat** is the process by which a receiving station requests a retransmission.

Inclusive Acknowledgement

One advantage of continuous ARQ is called *inclusive acknowledgement*. For example, a receiver might send an ACK of 5 and ACKs of 1, 2, 3, and 4 are not transmitted. The ACK of 5 means the station received and acknowledges everything up to and

> **Inclusive acknowledgement** is an advantage of continuous ARQ.

including 4; the next frame expected should have a 5 in its send sequence field. It is evident from this simple example that continuous ARQ protocols with inclusive acknowledgement can considerably reduce the overhead involved in the ACKs. In this example, one ACK acknowledges 4 frames considerably better than the stop-and-wait systems, in which an ACK is required for every transmission.

Stop-and-Wait ARQ Flow Control Protocol

Stop-and-Wait ARQ includes retransmission of data in case of lost or damaged frames.

Stop-and-Wait ARQ is a form of stop-and-wait flow control extended to include retransmission of data in case of lost or damaged frames. For retransmission to work, four features are added to the basic flow control mechanism:

- The sending device keeps a copy of the last frame transmitted until it receives an acknowledgement for that frame. Keeping a copy allows the sender to retransmit lost or damaged frames until they are received correctly.
- For identification purposes, both data frames and ACK frames are numbered alternately 0 and 1. A data 1 frame is acknowledged by an ACK 1 frame, indicating that the receiver has gotten data 1 and is now expecting data 0. This numbering allows identification of data frames in case of duplicate transmission, which is important in the case of lost acknowledgements.
- If an error is discovered in a data frame, indicating that it has been corrupted in transit, a NAK frame is returned. NAK frames, which are not numbered, tell the sender to retransmit the last frame sent. Stop-and-Wait ARQ requires that the sender wait until it receives an acknowledgement for the last frame transmitted before it transmits the next one. When the sending device receives a NAK, it resends the frame transmitted after the last acknowledgement, regardless of number.
- The sending device is equipped with a timer. If an expected acknowledgement is not received within an allocated time period, the sender assumes that the last data frame was lost in transit and sends it again.

When the receiver discovers that a frame contains an error, the receiver returns a NAK frame and the sender retransmits the last frame. For example, in Figure 14.8 the sender transmits a data frame: data 1. The receiver returns an ACK 1, indicating that data 1 arrived undamaged and it is now expecting data 0. The sender transmits its next frame: data 0. It arrives undamaged, and the receiver returns ACK 0. The sender transmits its next frame: data 1. The receiver discovers an error in data 1 and returns a NAK. The sender retransmits data 1. This time data 1 arrives intact, and the receiver returns ACK 1.

Any of the three frame types can be lost in transit. For a lost data frame, the sender is equipped with a timer that starts every time a data frame is transmitted. If the frame never makes it to the receiver, the receiver can never acknowledge it either positively or negatively. The sending device waits for an ACK or NAK frame until its timer goes off, at which point it tries again. It retransmits the last data frame, restarts its timer, and waits for an acknowledgement.

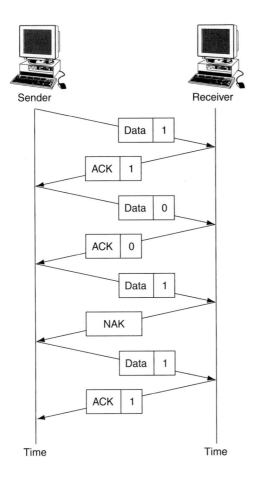

FIGURE 14.8

Stop-and-Wait ARQ damaged frame. (Reproduced with permission of Cisco Systems, Inc. Copyright © 2001 Cisco Systems, Inc. All rights reserved.)

In the case of a lost acknowledgement, the data frame has made it to the receiver and been found to be either acceptable or not acceptable. But the ACK or NAK frame returned by the receiver is lost in transit. The sending device waits until its timer goes off, then retransmits the data frame. The receiver checks the number of the new data frame. If the lost frame was a NAK, the receiver accepts the new copy and returns the appropriate ACK, assuming the copy arrives undamaged. If the lost frame was an ACK, the receiver recognizes the new copy as a duplicate, acknowledges its receipt, then discards it and waits for the next time.

Piggybacking

The newer line protocols permit the inclusion of the N(S) and the N(R) field in the same frame. This technique, called *piggybacking,* allows the protocol to piggyback an ACK (the N[R] value) onto an information frame sequenced by the N(S) value. As an

Piggybacking is the process of carrying acknowledgements within a data packet to save network bandwidth.

example, assume a station sends a frame with N(R) = 5 and N(S) = 1. The N(R) = 5 means all frames up to a number 4 are acknowledged and the N(S) = 1 means station B is sending a user information in this frame with a sequence of 1.

Sliding Windows Flow Control Protocol

Continuous ARQ devices use the concept of transmit and receive windows to aid in link management operations. A window is established on each link to provide a reservation of resources at both stations. These resources may be the allocation of specific computer resources or the reservation of buffer space for the transmitting device. In most systems, the window is established. For example, if stations A and B are to communicate with each other, station A reserves a window for B, and B reserves a window for A. The windowing concept is necessary to full duplex protocols because they entail a continuous flow of frames into the receiving site without the intermittent stop-and-wait acknowledgements. Consequently, the receiver must have a sufficient allocation of memory to handle the continuous incoming traffic. It is clear that window size is a function of buffer space and the magnitude of sequence numbers.

The windows at the transmitting and receiving site are controlled by state variables, which is another name for counter. The transmitting site maintains a send state variable (V[S]). It is the sequence number of the next frame to be transmitted. The receiving site maintains a receive state variable (V[R]), which contains the number that is expected to be in the sequence number of the next frame. The V(S) is incremented with each frame transmitted and placed in the send sequence field (N[S]) in the frame.

Upon receiving the frame, the receiving site checks for a transmission error. It also compares the send sequence number N(S) with its V(R). If the frame is acceptable, it increments V(R) by one, places it into a receive sequence number field N(R) in an ACK frame and sends it to the original transmitting site to complete the accountability for the transmission. If an error is detected or if the V(R) does not match the sending sequence number in the frame, a NAK with the receiving sequence number N(R) containing the value V(R) is sent to the original transmitting site. This V(R) value informs the transmitting DTE of the next frame that it is expected to send. The transmitter must then reset its V(S) and retransmit the frame whose sequence matches the value of N(R).

A useful feature of the sliding window scheme is the ability of the receiving station to restrict the flow of data from the transmitting station by withholding the acknowledgement frames. This action prevents the transmitter from opening its windows and reusing its send sequence numbers values until the same send sequence numbers have been acknowledged. A station can be completely throttled if it receives no ACKs from the receiver.

Table 14.3 Functions of State Variables and Sequence Numbers

Functions
Flow Control of Frames (Windows)
Detect Lost Frames
Detect Out-of-Sequence Frames
Detect Erroneous Frames

Uses
N(S) Sequence number of transmitted frame
N(R) Sequence number of the acknowledged frame(s). Acknowledges all frames up to N(R) − 1
V(S) Variable containing sequence number of next frame to be transmitted
V(R) Variable containing expected value of the sequence number N(S) in the next transmitting frame

Many data link controls use the numbers 0 through 7 for V(S), V(R), and the sequence numbers in the frame. Once the state variables are incremented through 7 the numbers are reused, beginning with 0. Because the numbers are reused the stations must not be allowed to send a frame with a sequence number that has not yet been acknowledged. For example, the protocol must wait for frame number 6 to be acknowledged before it uses a V(S) of 6 again. The use of 0–7 permits seven frames to be outstanding before the window is closed. Even though 0–7 gives eight sequence numbers, the V(R) contains the value of the next expected frame, which limits the actual outstanding frames to 7. Table 14.3 gives a brief review of state variables and sequence numbers.

Other Considerations

Three important goals of a line protocol are:

- To obtain high throughput
- To obtain fast response time
- To minimize the logic required at the transmitting and receiving sites to account for traffic such as ACKs and NAKs.

The transmit window is an integral tool in meeting these goals. One of the primary functions of the window is to ensure that by the time all the permissible frames have been transmitted, at least one frame has been acknowledged. In this manner, the window is kept open and the line is continuously active. The timers are the key to effective line utilization and window management.

A very large window permits continuous transmission regardless of the speed of the link and the size of the frames because the transmitter does not have to wait for acknowledgement from the receiver. Although this is true, a larger window size also means that the transmitter must maintain a large queue to store those frames that have not been acknowledged by the receiver.

The concepts of continuous ARQ are relatively simple, yet it should be realized that with a large communications facility the host computer or front-end processor is tasked with efficient transmission, data flow, and response time between itself and all the secondary sites attached to it. The primary host must maintain a window for every station and manage the traffic to and from each station on an individual basis.

Examples of Continuous ARQ Protocol Operations

Figure 14.9 provides an example of operations on a data link, which will allow us to tie together some of the concepts in the previous discussions. Station A sends four frames in succession to station B. Station A increments the send state variable V(S) with each transmission and places its value in the N(S) field of each frame. Consequently, the four user data frames would have the N(S) fields as illustrated in the figure. The illustration shows that station A's send state variable is incremented to the next frame to be transmitted and station B's receive state variable V(R) is incremented upon receiving an error-free frame. Thus, the receive state variable should always equal the value of the next expected frame N(S) field. Let us expand this idea further by moving to event 10 and seeing the effect of the receipt of the N(R) field on station A.

In event 10, station A receives a frame from station B with N(R) = 4. Notice that station A's V(S) of 4 equals the incoming N(R) = 4. Consequently station A knows that all preceding traffic has been received without problems because its V(S) equals the incoming N(R) value. The N(R) value is the value of the next expected frame, so it should equal the V(S), which is the value of the next frame to be transmitted. The process is elegantly simple. Flow control and traffic accountability are maintained by the continuous checking, incrementing, and rechecking of the state variables and sequence numbers.

What happens in the event of an error? Errors are dealt with through ARQ. User data are transmitted in frames with an error-checking field created by the transmitting site and checked by the receiving site. This field allows the receiver to detect transmission errors and request retransmission of erroneous frames. The transmitter automatically repeats these frames upon request from the data receiver.

In event 18, station A sends frame 6. However, station B detects an error in the frame. Due to error, station B does not increment its receive state variable. Rather, it inserts the current value of this variable into the N(R) field and sends it back to station A with a NAK control code or a bit set to indicate a negative acknowledgement in event 20. Station A then adjusts its state variables accordingly and retransmits the erroneous frame. In event 21, station A resets its send state variable to the received N(R) value of 6 and retransmits frame 6 in event 22. In event 23, the channel state variables and sequence numbers are once again synchronized, and the error has been resolved.

Event	V (S)		V (R)
1	0		0
2	0	N (S) = 0 ──────────────►	0
3	1		1
4	1	N (S) = 1 ──────────────►	1
5	2		2
6	2	N (S) = 2 ──────────────►	2
7	3		3
8	3	N (S) = 3 ──────────────►	3
9	4		4
10	4	◄──── N (R) = 4	4
11	4	N (S) = 4 ──────────────►	4
12	5		5
13	5	◄──── N (R) = 5, N (S) = 0	5	Note 1
14	5		5
15	5	N (S) = 5 ──────────────►	5
16	6		6
17	6	◄──── N (R) = 6	6
18	6	N (S) = 6 ──────────────►	6	Note 2
19	7		6
20	7	◄──── NAK, N (R) = 6	6
21	6		6
22	6	N (S) = 6 ──────────────►	6
23	7		7

FIGURE 14.9

Sequencing frames on the link. (Reproduced with permission of Cisco Systems, Inc. Copyright © 2001 Cisco Systems, Inc. All rights reserved.)

Note 1: An example of piggybacking.

Note 2: An error is detected.

Examples of Negative Acknowledgements and Retransmissions

To return to the scenario in Figure 14.9, event 20 can be implemented in a number of ways:

● Implicit reject
● Selective reject (SREJ)
● Reject (REJ) or Go-Back-N
● Selective reject-reject (SREJ-REJ)

Implicit reject uses the N(R) value to acknowledge all preceding frames and request the retransmission of the frame whose N(S) value equals the value in N(R). This technique works well enough on half-duplex links but should not be used on full duplex systems. Because a full duplex protocol permits simultaneous two-way transmission an N(R) value could be interpreted erroneously as either an ACK or a NAK. For example, if station A receives a frame with N(R) = 4 from station B and the station just sent a frame with N(S) = 4, station A does not know if B is NAKing or ACKing frame 4. On a half-duplex, stop-and-wait link, the N(R) = 4 would clearly mean that frame 4 is expected next. Explicit NAKs are really preferable—they include selective reject, reject, and selective reject-reject.

Selective reject (SREJ) requires that only the erroneous transmission be retransmitted. Reject (REJ) requires that not only the erroneous transmission be repeated but also all succeeding frames be retransmitted such as all frames behind the erroneous frame. Selective reject and reject are illustrated in Figure 14.10. Both techniques have advantages and disadvantages. Selective reject provides better line utilization because the erroneous frame is the only retransmission. However, as illustrated in Figure 14.10b, site B must hold frames 3, 4, and 5 to await the retransmission of frame 2. Upon its arrival, frame 2 must be inserted into the proper sequence before the data is passed to the end user application. The holding of frames can consume precious buffer space especially if the user device has limited memory available and several active links.

Reject (also called Go-Back-N) is a simpler technique. Once an erroneous frame is detected, the receiving station discards all subsequent frames in the session until it receives the correct retransmission. Reject requires no frame queuing and frame resequencing at the receiver. However, its throughput is not as high as selective reject because it requires the retransmission of frames that may not be in error. The reject ARQ approach is ineffective on systems with long round-trip delays and high data rates. One principal disadvantage of selective reject is the requirement that only one selective reject frame can be outstanding at a time. For example, assume site A has transmitted frames 0, 1, 2, 3, 4, and 5, and site B responds with SREJ with N(R) = 2. This response frame acknowledges 0 and 1 and requests the retransmission of 2. However, let us suppose another SREJ frame was sent by site B before the first SREJ condition cleared. As the frame flow in Figure 14.11 shows, multiple SREJs contradict the idea of N(R) value acknowledging the previous frames.

In our example, site A does not know if the second SREJ of N(R) = 4 acknowledges the preceding frames because it contradicts the previous SREJ of N(R) = 2. Again, the requirement for only one SREJ to be outstanding eliminates any ambiguity, but if the frame error rate is high and its occurrence exceeds the round-trip propagation delay between sites A and B, then the single SREJ convention can reduce the throughput on the channel. This problem stems from the fact that the channel may go idle while the stations wait for the first SREJ to clear. In our previous example, the SREJ N(R) = 4 could not be sent until site B had received its response to its first SREJ N(R) = 2. This effect is especially evident on circuits that experience long delays.

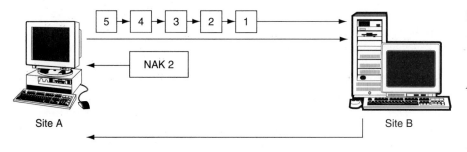

(a) Frames 1 through 5 transmitted with an error in frame 2

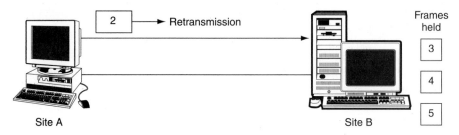

(b) Selective reject

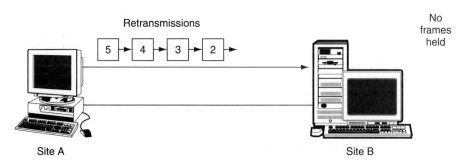

(c) Reject

One relief to the SREJ dilemma is to use the SREJ $N(R) = X$ to only refuse (NAK) the $N(S) = X$ frame. That is, the SREJ has no acknowledgement functions. In our frame flow diagram, site A would assume that SREJ $N(R) = 2$ and SREJ $N(R) = 4$ only NAK frames 2 and 4. Consequently, channel efficiency is not reduced significantly on error-prone channels or channels with long propagation delays.

However, if the SREJ has no positive acknowledgement as seen in our previous example, how is frame 3 acknowledged if SREJ $N(R) = 4$ does not inclusively

FIGURE 14.11

The problem with multiple selective rejects. (Reproduced with permission of Cisco Systems, Inc. Copyright © 2001 Cisco Systems, Inc. All rights reserved.)

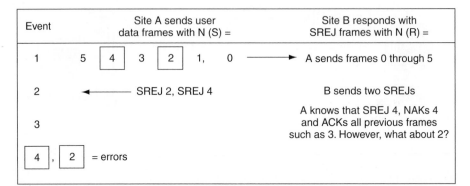

acknowledge preceding frames? The acknowledgement is through one of three alternatives. The first alternative is that a subsequent REJ can be used to acknowledge previous frames. A second alternative can use a regular user data frame with the N(R) value acknowledging preceding frames.

The last approach, selective reject-reject (SREJ-REJ), has been proposed as an alternative to the other two techniques. SREJ-REJ performs like SREJ except that once an error is detected, it waits to verify the next frame as correct before sending the SREJ. If the receiver detects the loss of two contiguous data frames, it sends a REJ instead and discards all subsequently received frames until the lost frame is received correctly. Also, if another frame error is detected prior to recovery after the first erroneous frame, the receiver discards frames received after the second erroneous frame until the first erroneous frame is recovered. Then, a REJ is issued to recover the second frame and the other subsequent discarded frames. SREJ-REJ is depicted in the frame flow diagram in Figure 14.12.

FIGURE 14.12

Using a combination of selective reject and reject. (Reproduced with permission of Cisco Systems, Inc. Copyright © 2001 Cisco Systems, Inc. All rights reserved.)

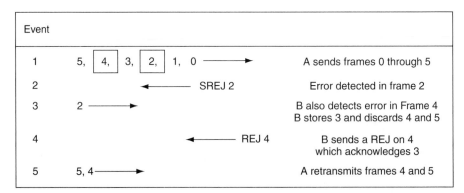

Host Configurations

On any communications channel there are two options for attaching hosts: point-to-point and multipoint.

Point-to-point connections have one host at the end of a communications link and another host at the remote end of the link. There is little contention for use of the channel in a point-to-point configuration because only the host at the near end and the host at the remote end are candidates for transmission. Point-to-point connections are very common in computer-to-computer communications, local connections, and remote connections with only one host. For communication among several hosts over a long distance, true point-to-point connections would be quite expensive because each host would require a separate line with a pair of modems. Fortunately, there are hardware components that allow terminals to share a communications channel while logically operating in a point-to-point manner. The methodology for controlling which station is allowed to use the communications link is sometimes referred to as a line discipline. In a point-to-point configuration, data flow can be managed in several ways: contention, pure contention, and supervisory-secondary.

Contention

One mechanism for managing data flow is known as *contention*. In the contention mode the host and the terminal contend for control of the medium. Each is considered to have an equal right to transmit to the other. To transmit, one station issues a bid for the channel, which means it asks the other party for control. If the other is ready to receive data, control is granted to the requestor. Upon completion of the transfer, control is relinquished and the link goes into an idle state, awaiting the next bid for control. A collision can occur when both stations simultaneously bid for the line. If this occurs, either one station is granted the request based on some predetermined priority scheme or each station waits awhile and then reattempts the bid. With the latter approach the time-out intervals must not be the same, because having the same time-out intervals is likely to result in another collision.

> ● In the **contention** mode the host and the terminal contend for control of the medium.

Pure Contention

A second method for point-to-point communication is also a contention method, but without line bids. This technique, known as *pure contention,* allows either of the two devices to transmit data whenever it is ready, the assumption being that the other station is ready to receive the data when it arrives. The sending station is made aware that the data were received correctly via a positive acknowledgement message from the recipient. Without a positive acknowledgement, and after a designated time-out interval, the sending station must assume the data were not received correctly and

> ● **Pure contention** allows either of the two devices to transmit data whenever it is ready.

resend it. Collisions can also occur with this methodology. Pure contention is good in situations where modem turnaround times make line bids costly—dumb terminals often deploy this technique.

Multipoint Connections

In a multipoint connection multiple hosts share the same communications channel. The number of hosts allowed to share the medium is a function of channel capacity and the workload at the hosts themselves, and sometimes of other hardware employed. As the number of hosts on the line increases, the average time each host has access to the link decreases. A contention type line discipline would work in a multipoint configuration. In fact, contention is a popular technique in local area networks consisting of hundreds of stations. However, studies have shown that as the number of stations communicating in this manner increases so does the number of collisions on the channel. When the number of collisions is high, the effective rate of data transfer declines because of the time required resolving the collisions.

Polling

The most common technique of establishing line discipline in multipoint terminal networks is referred to as poll/select, or simply *polling*. In the polled configuration one station is designated as the supervisor or *primary station*. The host computer almost always assumes this role, although pieces of hardware equipment such as controllers or concentrators may be used instead. There is only one primary station per multipoint link; all other stations are referred to as *secondary stations*. In the discussion that follows it is assumed that the host computer is the primary.

The primary station is in complete control of the link. Secondary stations may transmit data only when given permission by the primary station. This process of asking terminals whether they have data to transmit is referred to as polling. Each secondary station is given a unique address, and each terminal must be able to recognize its own address.

Although there are several distinct methods of polling, essentially the process works as follows:

1. The primary is provided a list of addresses for terminals on a particular link. Several multipoint lines may be controlled by one primary, although addresses on a given line are unique.
2. The primary picks an address from the list and sends a poll message across the link using that address. The poll message is very short, consisting of the poll address and a string of characters that has been designated as a poll message.
3. All secondary stations receive the poll message, but only the addressee responds.
4. The poll message is an inquiry to the secondary station as to whether it has any data to transmit to the primary. If it has data to transmit, the secondary responds either with the data or with a positive acknowledgement and then the data.

In the **polling** method, the primary station inquires whether a secondary station has data to transmit.

The **primary station** is in complete control of the link.

The **secondary station** transmits data only when given permission by the primary station.

5. If the secondary station has no data to send, it responds with a negative acknowledgement; the primary sends another station's polling address and repeats the process.

For a terminal to operate in this manner it must have memory in which to buffer data until it is asked to transmit.

Selection

When the primary has data to send to one or more secondary stations, it selects the station in much the same manner as that used for polling. Some terminals have two addresses, one for polling and one for selecting. In the selection process the following steps occur:

1. The primary sends a selection message to the terminal. A selection message consists of the terminal's selection address and an inquiry to determine whether the terminal is ready to accept data.
2. The terminal may respond positively or negatively. For instance, if the terminal's buffer is full it cannot accept additional data and responds negatively.
3. After a positive acknowledgement to the selection message, the primary transmits the data to the terminal.
4. In some multipoint networks the primary can send a message to all stations simultaneously via a broadcast address, which is one address that all terminals recognize as theirs. Other addressing schemes allow terminals to be divided into broadcast groups by using a common prefix for their addresses.

A fast select is a variation of selection. With fast select, the data accompanies the select message. This is effective when the receiving stations are usually able to accept the data being transmitted. If a station is not ready to receive data, the message is lost and must be retransmitted. In this situation the primary could revert to the previously defined selection method for that terminal.

Types of Polling

There are three basic types of polling: roll call, hub polling, and token passing. In *roll call polling* the primary obtains a list of addresses for terminals on the line and then proceeds sequentially down the list, polling each terminal in turn. If one or more stations on the link are of higher priority or are more likely to have data to send, their addresses could be included in the list multiple times so they can be polled more frequently. Roll call polling is illustrated in Figure 14.13.

Hub polling requires the terminals to become involved in the polling process. The primary sends a poll message to one station on the link. If that station has data to transmit, it does so. After transmitting its data or if it has no data to transmit, the secondary terminal passes the poll to an adjacent terminal. This process is repeated

In **roll call polling** the primary obtains a list of addresses for terminals on the line and then proceeds sequentially down the list, polling each terminal in turn.

Hub polling requires the terminals to become involved in the polling process.

FIGURE 14.13

Roll call polling. (Reproduced with permission of Cisco Systems, Inc. Copyright © 2001 Cisco Systems, Inc. All rights reserved.)

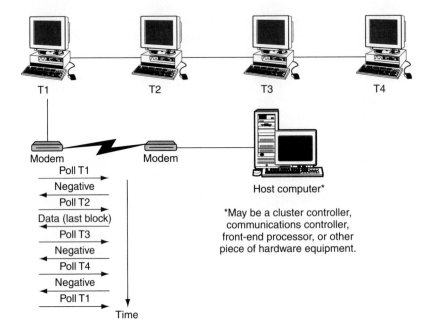

until all terminals have had the opportunity to transmit. The primary then starts the process again. Hub polling is more efficient than roll call polling because less of the line capacity is devoted to polling and acknowledgement messages. It also requires more intelligence at the terminal end because the terminal must recognize the address of its neighbor and pass along the poll. Hub polling is illustrated in Figure 14.14. In the diagram, if T2 were not operational, T3 would pass the poll to T1.

Device polling can be implemented in hardware or software. Software polling can have extensive overhead because it uses a significant number of CPU cycles and reduces the number of CPU cycles for application processing. When software solutions are employed, the function is usually placed in a front-end processor or communications con-

FIGURE 14.14

Hub polling. (Reproduced with permission of Cisco Systems, Inc. Copyright © 2001 Cisco Systems, Inc. All rights reserved.)

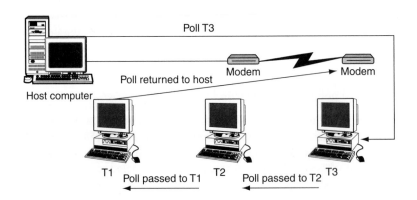

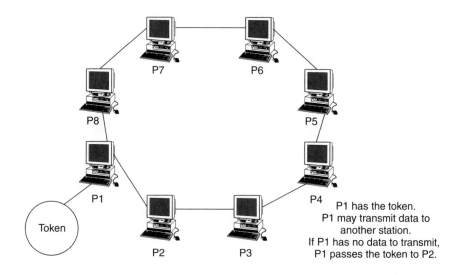

P1 has the token.
P1 may transmit data to
another station.
If P1 has no data to transmit,
P1 passes the token to P2.

troller to keep the main CPU from becoming bogged down with controlling lines. An additional overhead of polling is that a portion of the available line capacity is needed to exchange poll messages, which reduces the data carrying capacity of the medium.

Token passing is more common in local area networks and multipoint configurations of intelligent terminals. This technique could be considered separate from polling because there is no primary station and all stations have equal status. Token passing is similar in some respects to hub polling. A particular bit sequence is designated as the token. Whichever station has the token is the station that is allowed to transmit. The token is passed from station to station, together with the privilege of transmitting information. Because the token eventually arrives at each station, each receives the opportunity to send. Token passing is depicted in Figure 14.15.

In **token passing** there is no primary station and all stations have equal status.

Difficulty of Dealing with Errors

When we consider the differences between the data communications and centralized mainframe environments, it becomes clear that data communications systems are more subject to error because the transmissions take place through a hostile environment. A microwave signal is illustrative of this concept. During transmission it may encounter varying temperatures, fog, rain, or snow, as well as other microwave signals that tend to distort the signal. This is one reason, among others, that optical fiber is replacing some of the microwave routes.

A transmission through a communications systems travels through several components, and each component introduces an added probability of the occurrence of errors. For example, as a signal moves through a network, it must pass through switches, modems, multiplexers, and other instruments. If the interfaces among these components are not established properly, an error can occur. Moreover, the components themselves can introduce errors.

The centralized communications system exercises considerable control over its resources. Events usually do not occur without the permission of the host computer. In the event an error occurs, the operating system interrupts the work in progress, suspends the problem program, stores its registers and buffers, and executes the requisite analysis to uncover the problem.

A data communications system may not allow this type of control. First, it is often impractical to suspend and freeze resources because they may be used by other components. Second, their condition may have changed by the time a network control component receives the error indication. Third, networks often are distributed in their control mechanisms and do not always operate under the tight centralized manner found in the centralized computing system.

Clearly, a distributed networking environment poses formidable challenges to a network administrator. However, a wide array of remedial measures can be taken to mitigate the effects of these problems.

Major Types of Impairments

Transmission impairments can be defined broadly as random or nonrandom events. The random events cannot be predicted. Nonrandom impairments are predictable and, therefore, are subject to preventive maintenance efforts. These errors may be bothersome only for a voice or video transmission. However, they may be quite serious for a data transmission because they can result in distortions of the bit streams that represent user data.

White noise is a signal whose energy is uniformly distributed among all frequencies.

The nonrandom movement of electrons creates electric current on a hardwire that is used for a transmission signal. Along with these signals, all electrical components also experience the vibrations of the random movement of electrons. These vibrations create thermal energy and cause the emission of electromagnetic waves of many frequencies. The phenomenon is called *white noise* because it contains an average of all the spectral frequencies equally, just as white light does. Other names are gaussian noise, random noise, or background noise. Excessive noise can undermine the transmission integrity of the link and can prevent the receiving device from detecting the incoming data stream accurately.

Crosstalk is the interference of signals from another channel.

Most of us who use the public telephone network have experienced the interference of another party's faint voice on our line. This is *crosstalk*—the interference of signals from another channel. One source of crosstalk is in physical circuits that run parallel to each other in building ducts and telephone facilities. The electromagnetic radiation of the signals on the circuits creates an inductance and capacitance effect on the nearby circuits. Crosstalk can also occur with the coupling of a transmitter and receiver at the same location, which is called near-end crosstalk or NEXT. The coupling of a transmitter to an incorrect remote receiver is called far-end crosstalk or FEXT. Near-end and far-end crosstalk travel in different directions. Although crosstalk may be only a minor irritation during a voice call, it can lead to data distortions.

In addition, almost everyone using a telephone has experienced echoes during a conversation. The effect sounds like one is in an echo chamber; the talker's voice is actually echoed back to the telephone handset. Echoes are caused by the changes in impedances in communications circuits. Impedance is the combined effect of inductance, resistance, and capacitance of a signal at a particular frequency. For example, connecting two wires of different gauges could create an impedance mismatch. Circuit junctions that allow a portion of the signal to erroneously find its way into the return side of a four wire circuit also cause echoes. The telephone company must undertake special measures in dealing with echo on data circuits, and if these measures are carried out incorrectly, the link is not adequate for the transmission of data.

The strength of a signal attenuates or decays as it travels through a transmission path. The amount of attenuation depends on the frequency of the signal, the transmission medium, and the length of the circuit. Unfortunately, signal attenuation is not the same for all frequencies. The nonuniform loss across the bandwidth can create amplitude distortion, also referred to as attenuation distortion on a voice channel.

To combat the loss on the line, inductive loading is utilized. The use of inductive loading provides a means of reducing the natural loss in the cable. By placing loading coils at regular intervals in the subscriber loop, the electrical loss throughput yields a specific range of frequencies can be better managed. Because the natural loss of a cable increases rapidly with frequency and distance, the loading systems can be used to create a consistent performance across the bandwidth in the channel. However, loaded cable acts like a low pass filter and severely attenuates frequencies above 3000 Hz. Loaded cables reduce the signal propagation by as much as a factor of 3, which increases transmission delay as well. Loading also introduces significant delay distortion, which requires equalization on longer lines. In addition, repeaters are required to obtain additional range because of the attenuation. These operations must be crafted carefully for data transmissions because data transmission is more sensitive to attenuation than voice transmission.

A signal consists of many frequencies. Because these frequencies do not travel at the same speed they arrive at the receiver at different times. Excessive delays create errors known as delay distortion or envelope delay. The problem is not serious for voice transmissions because a human ear is not very sensitive to phase. However, delay distortion creates problems for data transmissions.

Marking and spacing distortion occurs when the receiving component samples the incoming signal at the wrong interval or threshold, and/or the signal takes too long to build up and decay on the channel. This problem can occur in interfaces such as EIA-RS232-C and ITU-T's V.28, if the standard is not followed. EIA specifies that the capacitance of the cable and the terminator shall be less than 2,500 picofarads. If this specification is violated, the transition from a 1 (mark) to a 0 (space), and vice versa, may exceed the EIA-232-C standard. The standard states that time required to pass through the -3 V to $+3$ V or $+3$ V to -3 V transition region is not to exceed 1 millisecond or 4 percent of a bit time, whichever is smaller.

The net effect of this impairment is that the signal takes too long to complete the transitions from marks and spaces. Spacing distortion results when the receiver produces space bits longer than the mark bits. Marking distortion occurs when the mark bits are elongated. These problems may cause data errors, especially if the sampling clock is inaccurate and noise exists on the line. The spreading of a pulse signal can affect timing and lead to accumulated timing jitter.

Bit and Block Error Rates

Because one of the data link layer's primary functions is to recover from errors that are encountered on the link, the quality of the link from the standpoint of errors encountered determines how much work the data link protocol must do. Retransmissions can lead to reduced throughput and increased delay on the communications link. Therefore, the network manager is quite interested in knowing about the quality of the links in the network.

The number of erroneous bits received during a given period often measures channel or link quality. Dividing the number of bits received in error by the number of bits transmitted derives this bit error rate (BER). A typical error rate on a high quality leased telephone line consisting of just copper wire is as low as $1:10^6$. In most cases, the errors occur in bursts and cannot be predicted precisely. The BER is a useful measure for determining the quality of the channel, calibrating the channel, and pinpointing its problems.

BER should be measured over a finite time interval, and the time measurement should be included in the description of the error rate. The BER can be calculated as follows:

$$BER = B_e / RT_m$$

where R = channel speed in bits per second, B_e = number of bits in error, and T_m = measurement period in seconds. The actual bit sequences are important to BER. Pseudorandom bit sequences have all the appearance of random digital data. These sequences, generated in repeating lengths of $2^n - 1$ bits, will generate all but one possible word combination of bit length n. The most common sequences defined by the ITU-T are 511 and 2,047 bits long, representing n = 9 and n = 11, respectively.

The **block error rate** is a ratio of the number of blocks or frames received that contain at least one erroneous bit to the total blocks received.

The *block error rate (BLER)* is a ratio of the number of blocks or frames received that contain at least one erroneous bit to the total blocks received. Thus, BLER is calculated by dividing the number of blocks received in error by the number of blocks transmitted. The BLER is an effective calculation for determining overall throughput on the channel and often is used by network designers to perform line loading and network topology configuration.

The error rate on the link is an important component in determining the size of a data link frame. Another useful calculation is to determine the percentage of seconds during a stated period in which no errors occur. Error-free seconds (EFSs) are calculated by:

$$\%EFS = [S - S_e \,/\, S] \times 100\%$$

where S = measurement period in seconds and S_e = the number of 1 second intervals during which one bit error occurred. The S parameter is important because like T_m in BER, it is necessary to specify a measuring interval. The period tested may be hours or even days if problems are experienced on the line.

EFS is a valuable measure of performance for channels on which data are transmitted in blocks. BER is not a very good measure of performance of throughput, but it is used widely to evaluate the performance of modems and other DCEs.

Error Detection

Because errors are inevitable in a data communications system, some method of detecting the errors is required. Although the methods may vary, they all have one goal: detecting the corrupted bits.

Parity Checking

Parity checking is one of the oldest forms of error detection. Like most data communication error techniques, parity checking requires the insertion of redundant bits that are used to model the content of the bits in the frame. Although the techniques vary on how parity detection is implemented, generally it consists of inserting one bit to a character or a number of bits to a number of characters at the transmitter and checking these bits at the receiver to determine whether distortion has altered any of the bits in the data stream.

Parity checking is a form of error detection that requires the insertion of redundant bits that are used to model the content of the bits in the frame.

Single Parity

Single parity, also called vertical redundancy checking (VRC), is a simple parity technique. It consists of adding a single bit (a parity bit) to each string of bits that comprise a character. The bit is set to 1 or 0 to give the character bits an odd or even number of bits that are:

- 1s for a 1-parity protocol
- 0s for a 0-parity protocol

This parity bit is inserted at the transmitting station, sent with each character in the message, and checked at the receiver to determine whether each character is the correct parity. If transmission impairment caused a bit flip of 1 to 0 or 0 to 1, the parity check would so indicate. Figure 14.16 shows the single parity check logic. The

FIGURE 14.16

Single parity checking. (Reproduced with permission of Cisco Systems, Inc. Copyright © 2001 Cisco Systems, Inc. All rights reserved.)

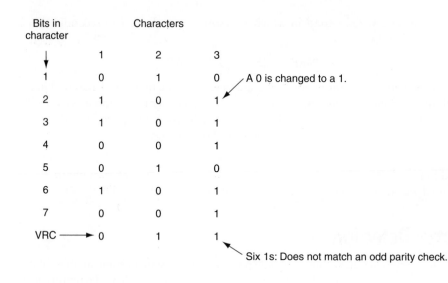

second bit of the third character is distorted due to noise. Because the VRC bit was set to a 1 to give the third character an odd number of 1 bits (in this case, five 1 bits), the detection of an even number of 1 bits serves as an indicator of an error. The protocol may then ask for a retransmission or may ignore the error, if the error is of no consequence.

The single parity check works well enough for a single bit error, but errors occur in clusters as well. Figure 14.17 illustrates that this type of error will not be detected. At

FIGURE 14.17

An undetected error. (Reproduced with permission of Cisco Systems, Inc. Copyright © 2001 Cisco Systems, Inc. All rights reserved.)

Bits in character	Characters		
	1	2	3
1	0	1	0
2	1	0	1
3	1	0	0
4	0	0	1
5	0	1	0
6	1	0	1
7	0	0	1
VRC	0	1	1

A 0 is changed to a 1 and a 1 is changed to a 0.

Five 1s: Error is not detected.

Bits in character	Characters		LRC
	1	2	3
1	0	1	0
2	1	0	(0) Errors
3	1	0	(1) detected
4	0	0	0
5	0	1	0
6	1	0	1
7	0	0	0
VRC →	0	1	1

FIGURE 14.18

Multiple parity checking. (Reproduced with permission of Cisco Systems, Inc. Copyright © 2001 Cisco Systems, Inc. All rights reserved.)

a transmission rate of 1,200 bits/second, the probability of more than one errored bit occurring within 7 bits is almost 50 percent. Fortunately, the multiple bit errors may not occur within one character in the data stream, but may fall between two successive characters. Nonetheless, the problem is serious enough to apply to other methods. A common solution to the single parity check problem is the use of additional parity checks.

Multiple or Block Parity

Multiple or block parity, also called longitudinal redundancy check (LRC), is a refinement of the character parity approach. In addition to a parity bit on each character, LRC places a parity (odd or even) on a block of characters. The block check of a double parity provides a better method to detect for errors across characters. LRC usually is implemented with VRC and is then called a two-dimensional parity check code, as illustrated in Figure 14.18. The VRC-LRC combination provides a substantial improvement over a single method. A typical telephone line with an error rate of $1:10^5$ can be improved to a perceived range of $1:10^7$ to $1:10^9$ with the two-dimensional check.

The parity check is expressed mathematically by the use of the Exclusive OR operation. The Exclusive OR of the two binary digits is 1 if the digits differ. If the digits are the same (both 0s or both 1s), the Exclusive OR result is 0.

The following equation demonstrates the use of parity:

$$P_j = b_{ij} \, b_{2j} \, ... \, b_{nj}$$

where P_j = parity of jth character, b_{ij} = ith bit in the jth character, and n = number of bits in the character.

Checksum

The realization that checking groups of bit streams rather than individual characters would yield better results led to another error checking technique. Although this technique is known by various names in the field, the process usually is called the *checksum*. The field used by the checksum frequently is called a block check character (BCC).

The checksum is a simple sum of the binary numerical values of the characters in the block of data. At the transmitter these characters are summed to produce the block check character. At the receiver an identical checksum is performed and compared with the BCC field that was transmitted with the frame. If the two sums are not equal, it is assumed that an error has occurred during the transmission. Remedial action is then taken.

Checksums are used widely in the industry because of their simplicity and the fact that they can be implemented with a few lines of software code. They are not perfect, however, because they will not detect certain types or error sequences in the bit stream.

The **checksum** is a simple sum of the binary numerical values of the characters in the block of data.

Echoplex

Echoplex is a primitive technique that is used in many asynchronous devices, notably personal computers. Each character is transmitted to the receiver where it is sent back or echoed to the original station. The echoed character is compared with a copy of the transmitted character. If they are the same, a high probability exists that the transmission is correct.

Echoplex requires the use of full duplex facilities. In the event a full duplex configuration is not available, a device usually is switched to local echo, which permits the system to operate as if it were echoplex. However, be aware that local echo does not check for errors across the link.

Echoplex is a primitive technique that requires the use of full duplex facilities.

Error Checking Codes

Any discussion of error control must include two widely used error checking codes: block codes and convolutional codes. The purpose of these codes is to detect errors that have corrupted the signals. With some codes, the errors also are corrected.

Block Codes

A block code consists of an information sequence referred to as a block or blocks. Each block contains k information bits. The block is also referred to as a message. For binary based codes, a total of 2^k possible messages exist. An encoder transforms the bits in the message into a code, and 2^k code words are possible from the output of the encoder. The term block code or [n,k] block codes is associated with the set of 2^k code words of length n.

Another term important to this discussion is the code rate (R), defined as $R = k/n$. It is the ratio of the number of information bits per transmitted symbol. For a different code word to be assigned to each symbol, k < n or R < 1. Redundant bits can be added to a message to form a code word if k < n. The number of redundant bits can be $n - k$. It is these redundant bits that enable a communications system to deal with transmission impairments. The encoder outputs the code based only on the input value of the k-bit message. Therefore, the encoder is said to be memoryless and can be implemented with a combinational logic.

A block code contains an important attribute called the minimum distance. It determines its error detecting and error correcting capabilities and is described as the Hamming weight or Hamming distance. As an example, the Hamming distance between binary $n - $ tuples of $x = (11010001)$ and $y = (01000101)$, written as $d(x,y)$, is 3. Reading left to right they differ in the first, fourth, and sixth positions. It is possible to calculate the Hamming distance between any two code words. The minimum distance is defined as:

$$d_{min} = min \{d(x,y): x,y \ C,x = W\}$$

where C = a block code, and x and y = code vectors.

When a code $v = $ Vector v is transmitted over a channel, a number of L errors will result in the received Vector r, which is different from Vector v in L places. Hamming theory states that for Code C, no error of $d_{min} - 1$ (or fewer) can change one code into another. It follows that any code word received with an error pattern of $d_{min} - 1$ or fewer errors will be detected as an error. In other words, the received code word is not a code word of C. The Hamming code detects all of $d_{min} - 1$ or fewer errors and also can detect a large number of errors with d_{min} or more errors. It is preferable to develop a block code with the Hamming distance to be as large as possible. A large minimum distance increases the likelihood of detecting errors.

Convolutional Codes

The convolutional encoder has a memory. The convolutional code accepts blocks of k-bit information as well and produces an encoded value of n-symbol blocks. However, in contrast to block coding, the code word value depends on the current k-bit message and on m previous message blocks. The encoded value is called a (n,k,m) convolutional code and is implemented with sequential logic.

The sequential aspect of convolutional codes allows us to describe its operation with a state diagram as illustrated in Figure 14.19. Each branch on the state diagram is labeled with the k inputs causing the transition and its corresponding n inputs. Starting at S_0 the path is followed through the state diagram according to the k-bits input

FIGURE 14.19

Convolutional codes. (Reproduced with permission of Cisco Systems, Inc. Copyright © 2001 Cisco Systems, Inc. All rights reserved.)

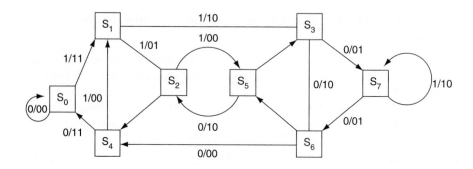

[u(1) .... u9k0]. The output label is the resulting code symbol [v(1) .... v(n)]. After the last nonzero bit, the path is followed back to S_0, with all zeros appended to the code. For example, a bit stream of 11101 produces the code word: 11. 10, 01, 01, 11,10,11,11. The path through the state diagram is:

$$S_0\ S_1\ S_3\ S_7\ S_6\ S_5\ S_2\ S_4\ S_0$$

Convolutional codes are also evaluated in relation to their distance properties. This is achieved by several analyses, but the simplest is in the minimum free distance d_{free}:

$$d_{free} = min\ \{d(v^1, v^{11}): u^1 = u^{11}\}$$

where v^1 and v^{11} = code words corresponding to information $u^1 = u^{11}$, respectively. We have only touched the surface on block and convolutional codes regarding error control coding. However, it provides enough background to understand the cyclic codes using block coding concepts.

Cyclic Redundancy Check (CRC)

The cyclic redundancy check is a technique that is memoryless and can be implemented with combined logic.

Many error detection codes for data transmission use logic based on the *cyclic redundancy check (CRC)*. It is so named because the bits in a message v = (v0, v1 .... Vn −1) are cyclically shifted through a register one place at a time: v(one place) = (Vn−1, V0 .... Vn−2). The CRC method is derived from the concepts of block coding. Thus, the CRC technique is memoryless and can be implemented with combined logic. Moreover, cyclic codes possess algebraic properties that lend themselves to relatively simple error detection implementations.

With this technique, the transmitter generates a bit pattern called a frame check sequence (FCS) based on the contents of the frame. The combined contents of the FCS and the frame (F) are exactly divisible by some predetermined number with no remainder or, as an alternative, are divisible by the number with a known remainder. If the contents of the frame are damaged during transmission, the receiver's division will yield a nonzero or a value other than a known remainder—an indication of an error. CRC detects all of the following errors:

- All single-bit errors
- All double-bit errors, if the divisor is at least three terms
- Any odd number of errors, if the divisor contains a factor $(x + 1)$
- Any error in which the length of the error (an error burst) is less than the length of the FCS
- Most errors with larger bursts

An algebraic notation is used to describe the FCS generation and checking process. The modulo 2 divisor is called a generator polynomial. A polynomial is an algebraic expression consisting of two or more terms. The divisor actually is one bit longer than the FCS. Both high order and low order bits of the divisor must be 1s. For example, a bit sequence of 110010001 is represented by the polynomial: $f(x) = x^7 + x^6 + x^3 + 1$.

The leading bits on the left-hand side correspond to the higher order coefficients of the polynomial.

In general terms, the following rules apply to the CRC operations, although later examples show these rules implemented with Exclusive OR cyclic shift registers.

- The frame contents are appended by a set of zeros equal in number to the length of the FCS.
- This value is divided modulo 2 by the generator polynomial, which contains one more digit than the FCS and must have high and low order bits of 1s. The divisor can be divided into the dividend, if the dividend has as many significant bits as the divisor.
- Each division is carried out in the conventional manner, except the next step subtraction is done modulo 2. Subtraction and addition are identical to Exclusive OR, no borrows or carries:

1	1	0	0	1	1	0	0
-1	-0	-1	-0	$+1$	$+0$	$+1$	$+0$
0	1	1	0	0	1	1	0

- The answer provides a quotient, which is discarded, and a remainder, which becomes the FCS field.
- The FCS is placed at the back of the frame contents and sent to the receiver.
- The receiver performs the same division on the frame contents and the FCS field. The FCS replaces the appended zeros used at the transmitter.
- If the result equals the expected number (a zero) or the predetermined number, the transmission is considered error free.

To illustrate CRC, let us assume the following:

Frame contents:	111011	
Polynomial:	11101	$(x^4 + x^3 + x^2 + 1)$
Frame contents and appended zeros:	1110110000	

FIGURE 14.20

CRC-ITU-T and CRC-16 polynomials. (Reproduced with permission of Cisco Systems, Inc. Copyright © 2001 Cisco Systems, Inc. All rights reserved.)

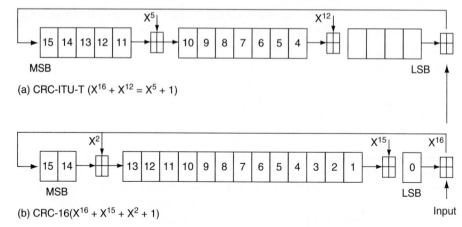

(a) CRC-ITU-T ($X^{16} + X^{12} = X^5 + 1$)

(b) CRC-16($X^{16} + X^{15} + X^2 + 1$)

The transmitter performs the following calculations:

$$\begin{array}{r} \overline{100001} \quad \text{(quotient ignored)} \\ 11101 \overline{)\ 1110110000} \\ \underline{11101} \\ 10000 \\ \underline{11101} \\ 1101 \quad \text{(remainder becomes FCS)} \end{array}$$

At the receiver, the frame contents and the FCS are divided by the same polynomial. As this example shows, if the remainder equals zero, the transmission is accepted as having no errors:

$$\begin{array}{r} 100001 \\ 11101 \overline{)\ 1110111101} \quad \text{(FCS)} \\ \underline{11101} \\ 11101 \\ \underline{11101} \\ 00000 \quad \text{(remainder no errors)} \end{array}$$

The division operation is equivalent to performing an Exclusive OR operation. The CRC calculations actually are performed in a cyclic shift register that uses Exclusive OR gates. Exclusive OR logic outputs 0 for inputs of 1 or 0; it outputs 1 if the inputs differ. The setup of the register depends on the type of generator polynomial.

Figure 14.20 shows the registers for CRC-ITU-T ($X^{16} + X^{12} + X^5 + 1$) and CRC-16 ($X^{16} + X^{15} + X^2 + 1$). The former detects errors of bursts up to 16 bits in length. It detects more than 99 percent of error bursts greater than 12 bits. The CRC-16 detects more than 99 percent of error bursts greater than 16 bits. It is evident from Figure 14.20a and 14.20b that the feedback logic means that the register contents at a given instant are dependent on the past transmission of bits. Consequently, the chances of a multiple bit burst creating an undetected error a zero remainder is very unlikely.

Binary Synchronous Control (BSC)

This section examines character oriented link protocols. The emphasis is on a family of protocols known as binary synchronous control (BSC). The BSC frame formats are illustrated as well as the functions provided by the fields in the frames. BSC line modes, control modes, and timers are also examined.

BSC Characteristics

In the mid-1960s, IBM introduced the first general-purpose data link control to support multipoint or point-to-point configurations. This product, the *Binary Synchronous Control Protocol (BSC),* also known simply as BISYNC, found widespread use throughout the world in the 1970s and 1980s. BSC became one of the most commonly used synchronous line protocols in existence. IBM's product families, originally designated as 3270 and 3780, were based on BSC. Practically every vendor has some version of BSC implemented in a product line, and the standards organizations publish specifications on the protocol. Despite this, it should be understood that link control products that use BSC are gradually being phased out of most organizations' systems. Replacements are either an HDLC type protocol or a LAN protocol.

BSC is a half-duplex protocol in that transmissions are provided two ways, but a station alternates the use of the link with another station. The protocol supports point-to-point and multipoint connections, as well as switched and nonswitched channels. BSC is a code sensitive protocol, and every character transmitted across a channel must be decoded at the receiver to determine whether it is a control character or a data character. Code dependent protocols are also called byte or character protocols.

> The **Binary Synchronous Control Protocol** is a character oriented data link layer protocol for half-duplex applications.

BSC Formats and Control Codes

The BSC frame formats and control codes are illustrated in Figure 14.21. Although the figure does not show all the possibilities for the format of a BSC frame, it offers a sampling of some major implementations of the BSC frame format.

Because BSC is a character oriented protocol it is possible that a code recognized as BSC control could be created by the user application process. For instance, assume a user program creates a bit sequence that is the same as the ETX (end of text) control code.

FIGURE 14.21

BSC control character. (Reproduced with permission of Cisco Systems, Inc. Copyright © 2001 Cisco Systems, Inc. All rights reserved.)

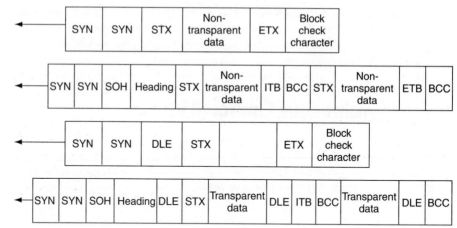

SYN - Synchronous idle keeps channel active
DLE - Data link escape used to achieve code transparency
ENQ - Enquiry used with polls/selects and bids
SOH - Start of heading
STX - Start of text puts line in text mode
ITB - End of intermediate block
ETB - End of transmission block
ETX - End of text
EOT - End of transmission puts line in control mode
BCC - Block check count

The receiving station, upon encountering the ETX inside the user data, would assume that the end of the transmission is signified by the user generated ETX. BSC would then accept the ETX as a protocol control character and would attempt to perform an error check on an incomplete BSC frame. This situation would result in an error.

Control codes must be excluded from the text and header fields, and BSC addresses the problem with the DLE control code. This code is placed in front of the control codes to identify these characters as valid line control characters. The DLE is not placed in front of user generated data. Consequently, if bit patterns resembling any of these control characters are created in the user text and encountered by the receiving station, the receiving station assumes they are valid user data because the DLE does not precede them. The DLE places the line into a transparent text mode, which allows the transmission of any bit pattern.

The DLE presents a special problem if it is generated by the end user application process because it could be recognized as a control code. BSC handles this situation by inserting a DLE next to a data DLE character. The receiver discards the first of two successive DLEs and accepts the second DLE as a valid user data. The headers illustrated in Figure 14.21 are optional. If they are included, the SOH code is placed in front of the header.

Line Modes

The BSC channel or link operates in one of two modes. The control mode is used by a master station to control the operations on the link, such as the transmission of polling and selection frames. The message or text mode is used for the transmittal of an information block or blocks. Upon receiving an invitation to send data a poll, the slave station transmits user data either with an STX or an SOH in front of the data or heading. These control characters place the channel in the message or text mode. Thereafter, data are exchanged under the text mode until an EOT is received, which changes the mode back to control. During the time the channel is in text mode, it is dedicated to the exchange of data between two stations only. All other stations must remain passive. The polls and selects are initiated by a frame with the contents: Address,ENQ where address is the address of the station. The control (master) station is responsible for sending polls and selects.

BSC provides for contention operation on a point-to-point circuit, as well. In this situation one of the stations can become the master by bidding to the other station. The station accepting the bid becomes the slave. A point-to-point line enters the contention mode following the transmission of the EOT.

The ENQ code plays an important role in BSC control modes. To summarize its functions:

- Poll: Control station sends with an address prefix.
- Select: Control station sends with an address prefix.
- Bid: Point-to-point stations send to contend for control station status.

BSC uses several EBCDIC or ASCII values to indicate whether a frame is a poll or a select. For example, one approach for the 3274 control unit is to use EBCDIC hex 7F as a poll to a device and EBCDIC C4 as a select.

Line Control

The transmitting station knows the exact order of frames it transmits, and it expects to receive ACKs to its transmissions. The receiving site transmits the ACKs with sequence numbers. Only two numbers are used. The ACK0 is represented as 1070 in EBCDIC hexadecimal notation and 1030 in ASCII hexadecimal notation. The ACK1 is represented as 1061 in EBCDIC hexadecimal notation and as 1031 in ASCII hexadecimal notation.

This sequencing technique is sufficient because the channel inherently is half duplex. Only one frame can be outstanding at one time. An ACK0 indicates the correct receipt of even-numbered frames; an ACK1 indicates the receipt of odd-numbered frames.

In addition to the frame format control codes in Figure 14.21, BSC uses several other line codes:

- ACK0: Positive acknowledgement to even-sequenced blocks of data or a response to a select or bid.
- ACK1: Positive acknowledgement to odd-sequenced blocks of data.
- WACK: Wait Before Transmit (Positive Acknowledgement) Receiving station temporarily unable to continue to process or receive transmissions. Signifies a line reversal. Also used as a positive acknowledgement of a transmission. Station will continue to send WACK until it is ready to receive.
- RVI: (Reverse Interrupt) Indicates station has data to send at the earliest opportunity. This causes an interrupt of the transmission process.
- TTD: (Temporary Text Delay) Indicates sending DTE cannot send data immediately, but wishes to maintain control of line such as its buffer is being filled or its card hopper is empty.

BSC Timers and Time-Outs

Like other line protocols, BSC uses timers and time-outs to manage errors and unusual conditions. A receiving station usually will time-out after three seconds of no activity from a transmitting station (receive time-out). On a switched line, a station will disconnect if it is inactive for a specified period, usually twenty seconds (disconnect time-out). In the event of a delay, a station must issue a WACK or TTD (continue time-out). The continue time-out value must be less than the value of the receive time-out and usually is set to two seconds. The transmitting station periodically transmits SYN SYN to maintain synchronization between stations. Table 14.4 provides several examples of BSC control traffic flow on the link.

High Level Data Link Control (HDLC) Protocol

High Level Data Link Control (HDLC) Protocol is the most widely used synchronous data link control protocol in existence. The architecture of the protocol from the standpoint of operating line modes and link configuration options will be examined. The many features of HDLC are contrasted and compared. Most vendors have a version of HDLC available, although the protocol often is renamed by the vendor or designated by different initials.

Characteristics of High Level Data Link Control (HDLC) Specifications

HDLC is a bit oriented line protocol specification published by the International Organization for Standardization (ISO). It has achieved widespread use throughout the world. The recommended standard provides for many functions and covers a broad

Table 14.4 Examples of BSC Link Activity

Transmission:	ENQ SYN SYN
Meaning:	A point-to-point station seeks a connection establishment and acquires control of the line. If both stations transmit this signal at the same time, a predesignated primary station is allowed to retry.
Transmission:	*ENQ Station Address SYN SYN EOT SYN SYN*
Meaning:	With a multipoint link, the primary station uses this format to send a poll (station address is uppercase) or a select (station address is lowercase).
Transmission:	*ACK0 SYN SYN*
Meaning:	Station is ready to receive traffic.
Transmission:	*ACK0/1 SYN SYN*
Meaning:	Station ACKs transmission and/or is ready to receive traffic.
Transmission:	*NAK SYN SYN*
Meaning:	Station NAKs transmission and/or is not ready to receive traffic.
Transmission:	*WACK SYN SYN*
Meaning:	Station requests other station try again because it is temporarily not ready. This format is also used to acknowledge previous traffic.
Transmission:	RVI SYN SYN
Meaning:	ACKs previous traffic and requests other station to suspend transmissions because receiving station needs the link itself.
Transmission:	*ENQ STX SYN SYN*
Meaning:	Sending station is experiencing a temporary delay but does not wish to relinquish the line. Also used when a polled station is not ready to transmit but wants the polling station to wait. This format is called a temporary text delay (TTD).
Transmission:	*EOT SYN SYN*
Meaning:	This is a negative response to a poll and is also used when a sending station has completed its transmission.

FIGURE 14.22

The HDLC family. (Reproduced with permission of Cisco Systems, Inc. Copyright © 2001 Cisco Systems, Inc. All rights reserved.)

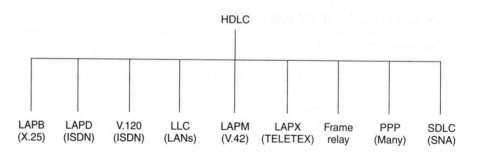

range of applications. Frequently, it serves as a foundation for other protocols that use specific options in the HDLC repertoire. The Advanced Data Communication Control Procedures (ADCCP) are published as ANSI X3.66. With minor variations they are identical to HDLC. However, because HDLC is more widespread only HDLC will be examined.

The HDLC Family

The HDLC protocol has been used as a basis for the development of a number of other widely used link layer protocols. Figure 14.22 provides an illustration of how pervasive HDLC is in the industry. This figure is not all-inclusive, but it represents some of the major implementations of HDLC. These protocols are shown at the node of the tree of the figure. The terms in parentheses indicate the technologies, usually upper layer protocols, and most often a network layer that the data link protocol supports. The exception to this statement is the node labeled Frame Relay. Frame Relay supports no overlying network layer because it is strictly a data link layer protocol.

Although many systems use the HDLC protocol, it does not mean that these systems are compatible. Often they are not compatible. HDLC has a broad range of options that can be implemented in different vendors' products.

The Link Access Procedure Balanced (LAPB) Protocol is a link layer protocol used on X.25 interfaces. LAPB operates within an X.25 three layer stack of protocols at the data link layer and is used to ensure that the X.25 packet is delivered safely between the user device and the packet network.

The link access procedure for the D channel (LAPD) is employed on ISDN interfaces. Its purpose is to deliver ISDN messages safely between user devices and the ISDN node.

The V.120 recommendation published by the ITU-T contains an HDLC protocol. It is used on ISDN terminal adapters for multiplexing operations. It uses many of the concepts of LAPD for addressing and it allows the multiplexing of multiple users across one link.

The Logical Link Control (LLC) Protocol is employed on IEEE 802 and ISO 8802 local area networks (LANs). It is configured in a variety of ways to provide different types of HDLC services. It rests atop any 802 or 8802 LAN.

The Link Access Procedure for Modems (LAPM) Protocol is relatively new and gives modems a powerful HDLC capability. It operates within V.42 modems and, as one might expect, is responsible for the safe delivery of traffic across communications links between two modems.

The Link Access Procedure for Half-Duplex Links (LAPX) is, as its name suggests, a half-duplex link control protocol. It is used in the teletex technology.

The Frame Relay Protocol uses an HDLC procedure for its link operations. Frame relay is so named because its purpose is to relay an HDLC type frame across a network. Frame relay was derived from many of the operations of LAPD and V.120.

The Point-to-Point Protocol (PPP) also is a derivation of HDLC. It is employed on a number of internet point-to-point links. Its primary purpose is to encapsulate network PDUs and to identify the different types of network protocols that may be carried in the I field of the PPP frame.

The Synchronous Data Link Control (SDLC) Protocol is the layer 2 protocol for IBM's Systems Network Architecture (SNA), which is a multiplayer protocol suite.

Now you can see the importance of understanding the features of HDLC and its basic operations. By knowing, one also comprehends the operation of many widely used protocols. If one learns HDLC, one learns about other protocols as well. The bad news is that each of these protocols is a derivation of HDLC. The protocols use various combinations of HDLC features and, in many instances, add their own protocol-specific operations. Although learning HDLC is very valuable, it does not ensure knowledge concerning the specific features of the other protocols. Therefore, to learn about a specific protocol there is no substitute for the actual source document.

HDLC Characteristics

HDLC provides a number of options to satisfy a wide variety of user requirements. It supports both half-duplex and full duplex transmission, point-to-point and multipoint configuration, and switched or nonswitched channels. In reading the following section study Figure 14.23.

An HDLC station is classified as one of three types. The primary station is in control of the data link. This station acts as a master and transmits command frames to the secondary stations on the channel. In turn, it receives response frames from those stations. If the link is multipoint, the primary station is responsible for maintaining a separate session with each station attached to the link. It uses poll commands to solicit data from the secondary stations. HDLC does not need a select command because it establishes windows and buffers during link setup, thereby ensuring that the receiving station has adequate resources to handle the traffic within the negotiated window size.

FIGURE 14.23
HDLC link configurations. (Reproduced with permission of Cisco Systems, Inc. Copyright © 2001 Cisco Systems, Inc. All rights reserved.)

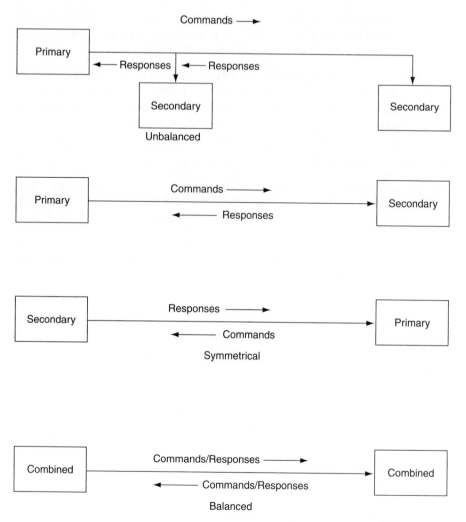

The secondary station acts as a slave to the primary station. It responds to the command frames from the primary station in the form of response frames. It maintains only one session, that being with the primary station, and it has no responsibility for control of the link. On a multipoint link, secondary stations cannot communicate directly with each other; they must first transfer their frames to the primary station.

The combined station transmits both commands and responses and receives both commands and responses from another combined station. It maintains a session with the combined station. The stations are peers on the link.

HDLC provides three methods to configure the channel for primary, secondary, and combined station use. An *unbalanced configuration* provides for one primary station and one or more secondary stations to operate as point-to-point or multipoint, half

An **unbalanced configuration** provides for one primary station and one or more secondary stations to operate as point-to-point or multipoint, half duplex, full duplex, and switched or nonswitched.

duplex, full duplex, and switched or nonswitched. The configuration is called unbalanced because the primary station is responsible for controlling each secondary station and for establishing and maintaining the link.

The symmetrical configuration is used very little today. The configuration provides for two independent, point-to-point unbalanced station configurations. Each station has a primary and secondary status; therefore, each station is considered logically to be two stations—a primary station and a secondary station. The primary station transmits commands to the secondary station at the other end of the channel and vice versa. Even though the stations have both primary and secondary capabilities, the actual commands and responses are multiplexed onto one physical channel.

A *balanced configuration* consists of two combined stations connected point-to-point only, half duplex or full duplex, and switched or nonswitched. The combined stations have equal status on the channel and may send unsolicited frames to each other. Each station has equal responsibility for link control. Typically, a station uses a command in order to solicit a response from the other station. The other station can send its own command as well. The terms unbalanced and balanced have nothing to do with the electrical characteristics of the circuit. In fact, data link controls should not be aware of the physical circuit attributes. The two terms are used in a completely different context at the physical and link levels.

While the stations are transferring data, they communicate in one of three modes of operation. *Normal response mode (NRM)* requires the secondary station to receive explicit permission from the primary station before transmitting. After receiving permission, the secondary station initiates a response transmission that may contain data. The transmission may consist of one or more frames while the secondary station is using the channel. After the last frame transmission, the secondary station must again wait for explicit permission before it can transmit again.

Asynchronous response mode (ARM) allows a secondary station to initiate transmission without receiving explicit permission from the primary station. The transmission may contain data frames or control information reflecting status changes of the secondary station. ARM can decrease overhead because the secondary station does not need a poll sequence in order to send data. A secondary station operating in ARM can transmit only when it detects an idle channel state for a two-way alternate half-duplex data flow, or at any time for a two-way simultaneous duplex data flow. The primary station maintains responsibility for tasks such as error recovery, link setup, and link disconnections.

Asynchronous balanced mode (ABM) uses combined stations. The combined station may initiate transmissions without receiving prior permission from the other combined station.

NRM is used frequently on multipoint lines. The primary station controls the link by issuing polls to the attached stations usually terminals, personal computers, and cluster controllers. ABM is a better choice on point-to-point links because it incurs no overhead and delay in polling. ARM is used very little today.

A **balanced configuration** consists of two combined stations connected point-to-point only, half duplex or full duplex, and switched or nonswitched.

Normal response mode requires the secondary station to receive explicit permission from the primary station before transmitting.

Asynchronous response mode allows a secondary station to initiate transmission without receiving explicit permission from the primary station.

Asynchronous balanced mode uses combined stations.

The term asynchronous has nothing to do with the physical interface of the stations. It is used to indicate that the stations need not receive a preliminary signal from another station before sending traffic. HDLC uses synchronous formats in its frames.

Frame Format

HDLC uses the term frame to indicate the independent entity of the protocol data unit transmitted across the link from one station to another. Figure 14.24 illustrates the frame format. The frame consists of four or five fields:

Flag fields (F)	8 bits
Address Field (A)	8, or multiples of 8 bits
Control Field (C)	8, or multiples of 8 bits
Information Field (I)	variable length; not used in some frames
Frame Check Sequence Field (FCS)	16 or 32 bits

All frames must start and end with the flag (F) sequence fields. The stations attached to the data link are required continuously to monitor the link for the flag sequence. The flag sequence consists of 011111110. Flags are continuously transmitted on the link between frames to keep the link in an active condition.

FIGURE 14.24
Control field formats. (Reproduced with permission of Cisco Systems, Inc. Copyright © 2001 Cisco Systems, Inc. All rights reserved.)

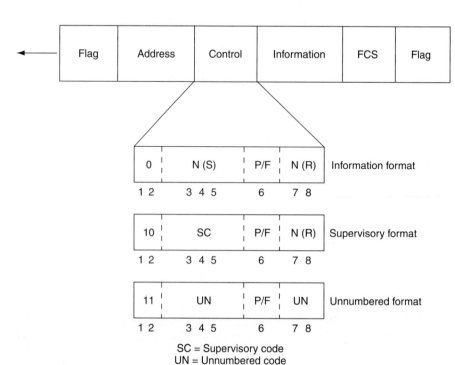

Other bit sequences are also used. At least 7, but fewer than 15 continuous 1s, which is an abort signal that indicates a problem on the link. Fifteen or more 1s keep the channel in an idle condition. One use of the idle state is in support of a half-duplex session. A station can detect the idle pattern and reverse the direction of the transmission. Once the receiving station detects a nonflag sequence, it is aware it has encountered the beginning of the frame, an abort condition, or an idle condition. Upon encountering the next flag sequence, the station recognizes it has found the full frame. In summary, the link recognizes the following bit sequences as follows:

01111110   = Flags
At least 7, but fewer than 15 1s = Abort
15 or more 1s   = Idle

The time between the actual transmissions of the frames on the channel is called interframe time fill. This time fill is accomplished by transmitting continuous flags between the frames. The flags may be 8-bit multiples, and they can combine the ending 0 of the preceding flag with the starting 0 of the next flag.

HDLC is a code-transparent protocol. It does not rely on a specific code for the interpretation of line control. For example, bit position n within an octet has a specific meaning, regardless of the other bits in the octet. On occasion, a flaglike field 01111110 may be inserted into the user data stream (I field) by the application process. More frequently, the bit patterns in the other fields appear flaglike. To prevent phony flags from being inserted into the frame, the transmitter inserts a zero bit after it encounters five continuous 1s anywhere between the opening and closing flags of the frame. Consequently, zero insertion applies to the address, control, information, and FCS fields. This technique is called bit stuffing. As the frame is stuffed, it is transmitted across the link to the receiver.

The framing receiver logic can be summarized as follows: The receiver continuously monitors the bit stream. After it receives a 0 bit with five continuous, succeeding 1 bits, it inspects the next bit. If it is a 0 bit, it pulls this bit out; in other words, it unstuffs the bit. However, if the seventh bit is a 1, the receiver inspects the eighth bit. If it is a 0, it recognizes that a flag sequence of 01111110 has been received. If it is a 1, then it knows an abort or idle signal has been received and counts the number of succeeding 1 bits to take appropriate action.

In this manner, HDLC achieves code and data transparency. The protocol is not concerned about any particular bit code inside the data stream. Its main concern is to keep the flags unique. Figure 14.25 summarizes the framing operations of HDLC at the receiver.

Many systems use bit stuffing and the non-return-to-zero-inverted (NRZI) encoding technique to keep the receiver clock synchronized. With NRZI, binary 1s do not cause a line transition, but binary 0s do cause a change. It might appear that a long sequence of 1s could present synchronization problems because the receiver clock would not receive the line transitions necessary for the clock adjustment. However,

FIGURE 14.25

Framing operations at the receiver. (Reproduced with permission of Cisco Systems, Inc. Copyright © 2001 Cisco Systems, Inc. All rights reserved.)

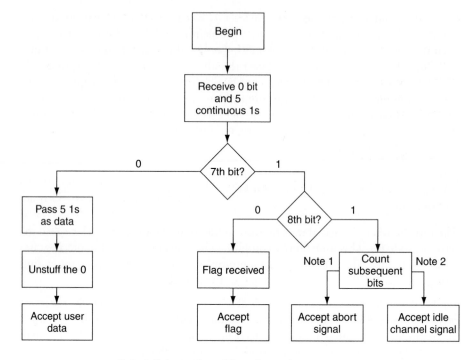

Note 1: Detect at least 7 but <15 continuous 1s.
Note 2: Detect 15 or more continuous 1s.

bit stuffing ensures a 0 bit exists in the data stream at least every 5 bits. The receiver can use them for clock alignment.

The address (A) field identifies the primary or secondary station involved in the frame transmission or reception. A unique address is associated with each station. In an unbalanced configuration, the address field in both commands and responses contains the address of the secondary station. In balanced configurations, a command frame contains the destination station address and the response frame contains the sending station address.

The control (C) field contains the commands, responses, and sequence numbers used to maintain the data flow accountability of the link between the primary stations. The format and the contents of the control field vary, depending on the use of the HDLC frame.

The information (I) field contains the actual user data. The information field resides only in the frame under the Information frame format. It usually is not found in the Supervisory or Unnumbered frame, although one option of HDLC allows the I field to be used with an Unnumbered frame.

The frame check sequence (FCS) field is used to check for transmission errors between the two data link stations. The FCS field is created by a cyclic redundancy check. The transmitting station performs modulo 2 division based on an established polynomial on the A, C, I, and FCS fields. If the remainder equals a predetermined value, the chances are quite good the transmission occurred without any errors. If the comparisons do not match it indicates a probable transmission error, in which case the receiving station sends a negative acknowledgement, requiring a retransmission of the frame.

The Control Field

The control field determines how HDLC controls the communication process previously illustrated in Figure 14.24. The control field defines the function of the frame and therefore invokes the logic to control the movement of the traffic between the receiving and sending stations. The field can be in one of three formats: Unnumbered, Supervisory, and Information transfer.

The Information format frame is used to transmit end user data between the two devices. The Information frame also may acknowledge the receipt of data from a transmitting station. It can perform certain other functions as well, such as a poll command.

The Supervisory format frame performs control functions such as the acknowledgement of frames, the request for the retransmission of frames, and the request for the temporary suspension of the transmission frames. The actual usage of the supervisory frame is dependent on the operational mode of the link (normal response mode, asynchronous balanced mode, asynchronous response mode).

The Unnumbered format is also used for control purposes. The frame is used to perform link initialization, link disconnection, and other link control functions. The frame uses five bit positions, which allow up to 32 commands and 32 responses. The particular type of command and response depends on the HDLC class of procedure.

The actual format of the HDLC determines how the control field is coded and used. The simplest format is the Information transfer format. The N(S) (send sequence) number indicates the sequence number associated with a transmitted frame. The N(R) (receive sequence) number indicates the sequence number that is expected at the receiving site.

Piggybacking, Flow Control, and Accounting for Traffic

HDLC maintains accountability of the traffic and controls the flow of frames by state variables and sequence numbers. The traffic at both the transmitting and receiving sites is controlled by these state variables. The transmitting site maintains a send state variable (V[S]), which is the sequence number of the next frame. The V(S) is incremented with each frame transmitted and placed in the send sequence field in the frame.

Upon receiving the frame, the receiving site checks the send sequence number with its V(R). If the CRC passes and if V(R) = N(S), it increments V(R) by one, places the value in the sequence number field in a frame, and sends it to the original transmitting site to complete the accountability for the transmission. If the V(R) does not match the sending sequence number in the frame or the CRC does not pass, an error has occurred, and a reject or selective reject with a value in V(R) is sent to the original transmitting site. The V(R) value informs the transmitting DTE of the next frame that it is expected to send, that is, the number of the frame to be retransmitted.

The Poll/Final Bit

The **P/F bit** is used by the primary and secondary stations to provide a dialogue with each other.

The fifth bit position in the control field is called the *P/F or poll/final bit.* It is recognized only when set to 1 and is used by the primary and secondary stations to provide a dialogue with each other:

- The primary station uses the P bit = 1 to solicit a status response from a secondary station. The P bit signifies a poll.
- The secondary station responds to a P bit with data or a status frame, and with the F bit = 1. The F bit also can signify end of transmission from the secondary station under NRM.

The P/F bit is called the P bit when used by the primary station and the F bit when used by the secondary station. Most versions of HDLC permit one P bit to be outstanding at any time on the link. Consequently, a P set to 1 can be used as a checkpoint. Checkpoints are quite important in all forms of automation. It is the machine's technique to clear up ambiguity and, perhaps, to discard copies of previously transmitted frames. Under some versions of HDLC, the device may not proceed further until the F-bit frame is received, but other versions of HDLC such as LAPB do not require the F-bit frame to interrupt the full duplex operations.

The fifth bit is a P bit and the frame is a command if the address field contains the address of the receiving station; it is an F bit and the frame is a response if the address is that of the transmitting station. Some implementations of HDLC, such as LAPD, have expanded the address or control field to permit the use of a command/response bit. The use of a command or response frame can be quite important because a station may react very differently to the two types of frames. For example, a command P = 1 usually requires the station to send back specific types of frames.

HDLC Commands and Responses

Table 14.5 shows the HDLC commands and responses. They are summarized briefly here.

The Receive Ready (RR) is used by the primary or secondary station to indicate that it is ready to receive an information frame and/or to acknowledge previously received frames by using the N(R) field. The primary station may also use the Receive Ready command to poll a secondary station by setting the P bit to 1.

Table 14.5 HDLC Control Field Format

Format	*Encoding*								*Commands*	*Responses*
	1	*2*	*3*	*4*	*5*	*6*	*7*	*8*		
Information	0	–	N(S)	–		–	N(R)	–	1	1
Supervisory	1	0	0	0	•	–	N(R)	–	RR	RR
	1	0	0	1	•	–	N(R)	–	REJ	REJ
	1	0	1	0	•	–	N(R)	–	RNR	RNR
	1	0	1	1	•	–	N(R)	–	SREJ	SREJ
Unnumbered	1	1	0	0	•	0	0	0	UI	UI
	1	1	0	0	•	0	0	1	SNRM	
	1	1	0	0	•	0	0	0	DISC	RD
	1	1	0	0	•	1	1	0	UP	
	1	1	0	0	•	1	0	0		UA
	1	1	0	1	•	0	0	0	NR0	NR0
	1	1	0	1	•	0	1	1	NR1	NR1
	1	1	0	1	•	0	1	0	NR2	NR2
	1	1	0	1	•	0	0	1	NR3	NR3
	1	1	1	0	•	0	0	0	SIM	SIM
	1	1	1	0	•	0	0	1		FRMR
	1	1	1	1	•	0	0	0	SARM	DM
	1	1	1	1	•	0	0	1	RSET	
	1	1	1	1	•	0	0	0	SARME	
	1	1	1	1	•	0	0	1	SNRME	
	1	1	1	1	•	0	1	0	SABM	
	1	1	1	1	•	1	0	1	XID	XID
	1	1	1	1	•	1	1	0	SABME	

I	Information	NR0 Non Reserved 0	
RR	Receive Ready	NR1 Non Reserved 1	
REJ	Reject	NR2 Non Reserved 2	
SREJ	Selective Reject	NR3 Non Reserved 3	
UI	Unnumbered Information	SIM Set Initialization Mode	
SNRM	Set Normal Response Mode	RIM Request Initialization Mode	
DISC	Disconnect	FRMR Frame Reject	
RD	Request Disconnect	SARM Set Asynchronous Response Mode	
UP	Unnumbered Poll	SARME Set ARM Extended Mode	
RSET	Reset	SNRME Set Normal Response Mode Extended	
XID	Exchange Identification	SABM Set Asynchronous Balanced Mode	
DM	Disconnect Mode	SABME Set ABM Extended Mode	
•	The P/F bit		

The Receive Not Ready (RNR) frame is used by the station to indicate a busy condition. This informs the transmitting station that the receiving station is unable to accept additional incoming data. The RNR frame may acknowledge previously transmitted frames by using the N(R) field. The busy condition can be cleared by sending the RR frame.

The Selective Reject (SREJ) is used by a station to request the retransmission of a single frame identified in the N(R) field. As with inclusive acknowledgement, all information frames are accepted and held for the retransmitted frame. The SREJ condition is cleared upon receipt of an I frame with an N(S) equal to V(R). An SREJ frame must be transmitted for each errored frame; each frame is treated as a separate error. Only one SREJ frame can be outstanding at a time because the N(R) field in the frame inclusively acknowledges all preceding frames. To send a second SREJ would contradict the first SREJ, because all I frames with N(S) lower than N(R) of the second SREJ would be acknowledged.

The Reject (REJ) is used to request the retransmission of frames starting with the frame numbered in the N(R) field. Frames numbered N(R) − 1 are all acknowledged. The REJ frame can be used to implement the Go-Back-N technique.

The Unnumbered Information (UI) format allows for transmission of user data in an unnumbered or unsequenced frame. The UI frame actually is a form of connectionless-mode link protocol in that the absence of the N(S) and N(R) fields precludes flow-controlling and acknowledging frames. The IEEE 802.2 logical link control (LLC) uses this approach with its LLC Type 1 version of HDLC.

The Request Initialization Mode (RIM) format is a request from a secondary station for initialization to a primary station. Once the secondary station sends RIM, it can monitor frames, but it can respond only to SIM, DISC, TEST, or XID.

The Set Normal Response Mode (SNRM) places the secondary station in the Normal Response Mode (NRM). The NRM precludes the secondary station from sending any unsolicited frames. This means the primary station controls all frame flow on the line.

The Disconnect (DISC) places the secondary station in the disconnected mode. This command is valuable for switched lines; the command provides a similar function to indicate it is in the disconnect mode (not operational).

The Test (TEST) frame is used to solicit testing responses from the secondary station. HDLC does not stipulate how the TEST frames are to be used. An implementation can use the I field for diagnostic purposes.

The Set Asynchronous Response Mode (SARM) allows a secondary station to transmit without a poll from the primary station. It places the secondary station in the information transfer state (IS) of ARM.

The Set Asynchronous Balanced Mode (SABM) sets the mode to ABM, in which stations are peers with each other. No polls are required to transmit because each station is a combined station.

The Set Normal Response Mode Extended (SNRME) sets SNRM with two octets in the control field. This is used for extended sequencing and permits the N(S) and N(R) to be seven bits in length, thus increasing the window to a range of 1–127.

The Set Asynchronous Balanced Mode Extended (SABME) sets SABM with two octets in the control field for extended sequencing.

The Unnumbered Poll (UP) polls a station without regard to sequencing or acknowledgement. Response is optional if the poll bit is set to 0. This provides for one response opportunity.

The Reset (RSET) is used as follows: the transmitting station resets its N(S) and receiving station resets its N(R). The command is used for recovery. Previously unacknowledged frames remain unacknowledged.

Summary

In this chapter you learned about the data link operations and protocols. The transfer of data across the communications link must flow in a controlled, orderly manner. Because communication links experience distortions such as noise, a method must be provided to deal with periodic errors that occur. The data communications system must provide each station on the link with the capability to send data to another station, and the sending station must be assured that the data arrive error free at the receiving station. The sending and receiving stations must maintain complete accountability for all traffic. In the event the data are distorted, the receiver site must be able to direct the originator to resend the erroneous frame or correct the errors.

In the past, error checking codes used single or double parity checking techniques, and they are still quite prevalent today. During the 1970s and 1980s, checksums became widely accepted in the industry and provided better error checking performance than the parity operations. Increasingly, systems have developed cyclic redundancy check (CRC) techniques, which yield vastly superior results in the older technologies. Some organizations choose to perform multiple error checking, with a CRC check at the data link layer and a checksum at a higher layer protocol.

Character oriented link control protocols were the pervasive technology for managing data links in the 1960s and 1970s. BSC was offered practically in all vendors' link layer products. With some exceptions, link control products that use BSC have been replaced by bit oriented protocols because of the relative inefficiency of their half-duplex operations and the awkward nature of code dependent operations.

HDLC serves as a foundation for many widely used data link protocols. Its features of asynchronous balanced mode and normal response mode provide a flexible means of configuring point-to-point, peer-to-peer operations or multipoint master/slave operations. The HDLC family has spread into the inventory of practically every vendor offering a data link control product.

Review Questions

1. What is the purpose of DLCs?
2. List the functions of the DLC.
3. Define asynchronous transmission.
4. Define synchronous transmission.
5. What is the difference between asynchronous and synchronous transmission?
6. What are character oriented protocols?
7. Give an example of a character oriented protocol.
8. What are the acknowledgements that make up the character oriented protocols?
9. What is the purpose of the DLE control code?
10. What is the purpose of the SYN control code?
11. What are the types of protocols that operate at the data link layer?
12. What are the characteristics of a count oriented protocol?
13. What are the characteristics of a bit oriented protocol?
14. Describe the process called bit stuffing.
15. Give two examples of asynchronous systems.
16. What are the processes used by a synchronous system?
17. What is an ACK and a NAK and how do they work?
18. How are the timers used in a synchronous system?
19. List some commonly used timers in a synchronous system.
20. What is the process that makes up the ARQ flow control protocol?

Summary Questions

1. What is an inclusive acknowledgement?
2. What is piggybacking?
3. Describe the sliding window flow control protocol process.
4. Give some examples of negative acknowledgements and retransmissions.
5. What are the forms of data flow in a point-to-point system?
6. What is polling?
7. List the different types of polling and define each type.
8. What are the major types of data transmission impairments?
9. What are the major types of error checking performed at the data link layer?
10. What is line control?

Internetworking Design Basics

Objectives

- Explain basic internetworking concepts.
- Identify and select internetwork capabilities.
- Identify and select internetworking devices.
- Explain LAN switching.
- Examine Virtual LANs (VLANs).
- Discuss the backbone services of path traffic prioritization, load balancing, alternative paths, switched access, and encapsulation.
- Explain backbone bandwidth management, area and service filtering, policy based distribution, gateway service, interprotocol route redistribution, and media translation.
- Explain redundant power systems, fault-tolerant media implementations, and backup hardware.
- Explain designing switched networks to include LAN switching, VLANs, and ATM switching.

Key Terms

Introduction

Designing an internetwork can be a challenging task. An internetwork that consists of only 50 meshed routing nodes can pose complex problems that lead to unpredictable results. Attempting to optimize internetworks that feature thousands of nodes can pose even more complex problems.

Despite improvements in equipment performance and media capabilities, internetwork design is becoming more difficult. The trend is toward increasingly complex environments involving multiple media, multiple protocols, and interconnection to networks outside any single organization's dominion of control. Carefully designing internetworks can reduce the hardships associated with growth as a networking environment evolves.

This chapter provides an overview of planning and design guidelines for internetworks and switched LAN networks as well. Discussions are divided into the following general topics:

- Understanding basic internetworking concepts
- Identifying and selecting internetwork capabilities
- Identifying and selecting internetworking devices
- LAN switching
- Virtual LANs (VLANs)
- ATM switching

Understanding Basic Internetworking Concepts

This section covers the following basic internetworking concepts:

- Overview of internetworking devices
- Switching overview

Overview of Internetworking Devices

Network designers faced with designing an internetwork have four basic types of internetworking devices available to them:

- *Hubs* (concentrators)
- *Bridges*
- *Switches*
- *Routers*

Table 15.1 summarizes these four internetworking devices.

Data communications experts generally agree that network designers are moving away from bridges and concentrators and primarily using switches and routers to build internetworks.

Switching Overview

In today's data communications systems, all switching and routing equipment perform two basic operations:

- Switching data frames: This is generally a store-and-forward operation in which a frame arrives on an input media and is transmitted to an output media.
- Maintenance of switching operations: In this operation, switches build and maintain switching tables and search for loops. Routers build and maintain both routing tables and service tables.

There are two methods of switching data frames: layer 2 and layer 3 switching.

Hubs are used to connect multiple users to a single physical device, which connects to the network.

Bridges are used to logically separate network segments within the same network.

Switches provide a unique network segment on each port, thereby separating collision domains.

Routers separate broadcast domains and are used to connect different network.

Table 15.1 Summary of Internetworking Devices

Device	Description
Hubs (concentrators)	Hubs (concentrators) are used to connect multiple users to a single physical device, which connects to the network. Hubs and concentrators act as repeaters by regenerating the signal as it passes through them.
Bridges	Bridges are used to logically separate network segments within the same network. They operate at the OSI data link layer (layer 2) and are independent of higher layer protocols.
Switches	Switches are similar to bridges but usually have more ports. Switches provide a unique network segment on each port, thereby separating collision domains. Today, network designers are replacing hubs in their wiring closets with switches to increase their network performance and bandwidth while protecting their existing wiring investments.
Routers	Routers separate broadcast domains and are used to connect different networks. Routers direct network traffic based on the destination network layer address (layer 3) rather than the workstation data link layer or MAC address. Routers are protocol dependent.

Layer 2 and Layer 3 Switching

Switching is the process of taking an incoming frame from one interface and delivering it out through another interface. Routers use layer 3 switching to route a packet, and switches use layer 2 switching to forward frames.

Switches use **layer 2 switching** to forward frames, while routers use **layer 3 switching** to route a packet.

The difference between *layer 2* and *layer 3 switching* is the type of information inside the frame that is used to determine the correct output interface. With layer 2 switching, frames are switched based on network layer information. Layer 2 switching does not look inside a packet for network layer information as does layer 3 switching. Layer 2 switching is performed by looking at a destination MAC address within a frame. It looks at the frame's destination address and sends it to the appropriate interface if it knows the destination address location. Layer 2 switching builds and maintains a switching table that keeps track of which MAC addresses belong to each port or interface. If the layer 2 switch does not know where to send the frame, it broadcasts the frame out all its ports to the network to learn the correct destination. When the frame's reply is returned, the switch learns the location of the new address and adds the information to the switching table.

Layer 2 addresses are determined by the manufacturer of the data communications equipment used. They are unique addresses that are derived in two parts: the manufacturing (MFG) code and the unique identifier. The MFG code is assigned to each vendor by the IEEE. The vendor assigns a unique identifier to each board it produces. Except for Systems Network Architecture (SNA) networks, users have little or no control over layer 2 addressing because layer 2 addresses are fixed with a device, whereas layer 3 addresses can be changed. In addition, layer 2 addresses assume a flat address space with universally unique addresses.

Layer 3 switching operates at the network layer. It examines packet information and forwards packets based on their network layer destination addresses. Layer 3 switching also supports router functionality. For the most part, layer 3 addresses are determined by the network administrator who installs a hierarchy on the network. Protocols such as IP, IPX, and AppleTalk use layer 3 addressing. By creating layer 3 addresses, a network administrator creates local areas that act as single addressing units and assigns a number to each local entity. If users move to another building their end stations will obtain new layer 3 addresses, but their layer 2 addresses remain the same.

As routers operate at layer 3 of the OSI model, they can adhere to and formulate a hierarchical addressing structure through TCP/IP subnets or IPX networks for each segment. Traffic flow in a switched (flat) network is therefore inherently different from traffic flow in a routed (hierarchical) network. Hierarchical networks offer more flexible traffic flow than flat networks because they can use the network hierarchy to determine optimal paths and contain broadcast domains.

Implications of Layer 2 and Layer 3 Switching

The increasing power of desktop processors and the requirements of client-server and multimedia applications have driven the need for greater bandwidth in traditional shared-media environments. These requirements are prompting network designers to replace hubs in wiring closets with switches.

Although layer 2 switches use microsegmentation to satisfy the demands for more bandwidth and increased performance, network designers are now faced with increasing demands for intersubnet communication. For example, every time a user accesses servers and other resources, which are located on different subnets, the traffic must go through a layer 3 device. Figure 15.1 illustrates the route of intersubnet traffic with layer 2 switches and layer 3 switches.

As Figure 15.1 illustrates, for Client X to communicate to Server Y, which is on another subnet, it must traverse through the following route: first go through Switch A (a layer 2 switch) and then through Router A (a layer 3 switch) and finally through Switch B (a layer 2 switch). Potentially there is a tremendous bottleneck, which can threaten network performance, because the intersubnet traffic must pass from one network to another.

FIGURE 15.1

Flow of Intersub-net traffic with layer 2 switches and routers. (Reproduced with permission of Cisco Systems, Inc. Copyright © 2001 Cisco Systems, Inc. All rights reserved.)

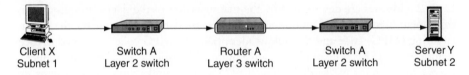

Client X
Subnet 1

Switch A
Layer 2 switch

Router A
Layer 3 switch

Switch A
Layer 2 switch

Server Y
Subnet 2

To relieve this bottleneck network designers can add layer 3 capabilities throughout the network. They are implementing layer 3 switching on edge devices to alleviate the burden on centralized routers. Figure 15.2 illustrates how deploying layer 3 switching throughout the network allows Client X to directly communicate with Server Y without passing through Router A.

FIGURE 15.2

Flow of intersub-net traffic with layer 3 switches. (Reproduced with permission of Cisco Systems, Inc. Copyright © 2001 Cisco Systems, Inc. All rights reserved.)

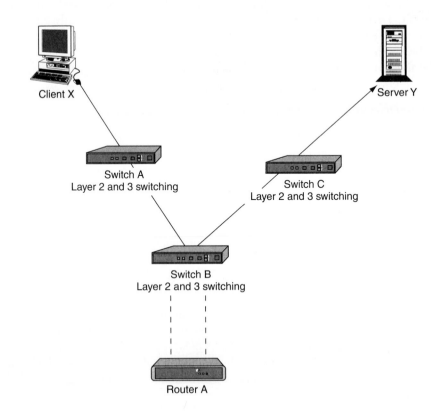

Client X

Server Y

Switch A
Layer 2 and 3 switching

Switch C
Layer 2 and 3 switching

Switch B
Layer 2 and 3 switching

Router A

Identifying and Selecting Internetworking Capabilities

After you understand your internetworking requirements, you must identify and then select the specific capabilities that fit your computing environment. The following discussions provide a starting point for making these decisions:

- Identifying and selecting an internetworking model
- Choosing internetworking reliability options

Identifying and Selecting an Internetworking Model

Hierarchical models for internetwork design allow you to design internetworks in layers. To understand the importance of layering, consider the OSI reference model, which is a layered model for understanding and implementing computer communications. By using layers, the OSI model simplifies the task required for two computers to communicate. Hierarchical models for internetwork design also use layers to simplify the task required for internetworking. Each layer can be focused on specific functions, thereby allowing the networking designer to choose the right systems and features for the layer.

Using a hierarchical design can facilitate changes. Modularity in network design allows you to create design elements that can be replicated as the network grows. As each element in the network design requires change, the cost and complexity of making the upgrade are constrained to a small subset of the overall network. In large flat or meshed network architectures changes tend to impact a large number of systems. Improved fault isolation is also facilitated by modular structuring of the network into small, easy-to-understand elements. Network managers can easily understand the transition points in the network, which helps identify failure points.

Using the Hierarchical Design Model

A *hierarchical network design model* includes the following three layers:

- The *backbone* (core) layer that provides optimal transport between sites.
- The distribution layer that provides policy based connectivity.
- The local access layer that provides workgroup and user access to the network.

Figure 15.3 shows a high level view of the various aspects of a hierarchical network design. A hierarchical network design presents three layers—core, distribution, and access—with each layer providing different functionality.

The **hierarchical network design model** presents three layers: core, distribution, and access.

The **backbone** is the part of the network that carries the heaviest traffic.

FIGURE 15.3

Hierarchical network design model. (Reproduced with permission of Cisco Systems, Inc. Copyright © 2001 Cisco Systems, Inc. All rights reserved.)

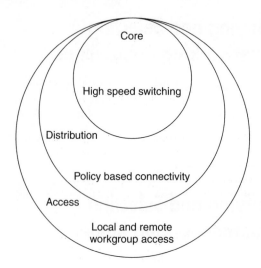

Function of the Core Layer

● The **core layer** is a high speed switching backbone that provides optimal transport between sites

The *core layer* is a high speed switching backbone and should be designed to switch packets as fast as possible. This layer of the network should not perform any packet manipulation, such as access lists and filtering that would slow down the switching of packets. Access lists are security lists that can filter out packets coming from unwanted hosts from foreign networks. Access lists are beyond the scope of this book.

Function of the Distribution Layer

● The **distribution layer** helps to define and differentiate the core. It provides policy based connectivity.

The *distribution layer* of the network is the demarcation point between the access and core layers and helps to define and differentiate the core. The purpose of this layer is to provide boundary definition and is the place at which packet manipulation can take place. In the campus environment, the distribution layer can include several functions such as the following:

- ● Address or area aggregation
- ● Departmental or workgroup access
- ● Broadcast/multicast domain definition
- ● Virtual LAN (VLAN) routing
- ● Any media transitions that need to occur
- ● Security

In the noncampus environment, the distribution layer can be a redistribution point between routing domains or the demarcation between static and dynamic routing protocols. It can also be the point at which remote sites access the corporate network. The distribution layer can be summarized as the layer that provides policy based connectivity.

Function of the Access Layer

The *access layer* is the point at which the local end users are allowed into the network. This layer may also use access lists or filters to further optimize the needs of a particular set of users. In the campus environment, access layer functions can include the following:

- Shared bandwidth
- Switched bandwidth
- MAC layer filtering
- Microsegmentation

In the noncampus environment, the access layer can give remote sites access to the corporate network via some wide area technology, such as Frame Relay, ISDN, or leased lines.

It is sometimes mistakenly thought that the three layers (core, distribution, and access) must exist in clear and distinct physical entities, but this does not have to be the case. The layers are defined to aid successful network design and to represent functionality that must exist in a network. The instantiation of each layer can be in distinct routers or switches, be represented by a physical media, be combined in a single device, or be omitted altogether. The way the layers are implemented depends on the needs of the network being designed. Note, however, that for a network to function optimally, hierarchy must be maintained.

The discussions that follow outline the capabilities and services associated with backbone, distribution, and local access internetworking services.

Evaluating Backbone Services

This section addresses internetworking features that support backbone services. The following topics are examined:

- Path optimization
- Traffic prioritization
- Load balancing
- Alternative paths
- Switched access
- Encapsulation

Path Optimization

One of the primary advantages of a router is its capability to help you implement a logical environment in which optimal paths for traffic are automatically selected. Routers rely on routing protocols that are associated with the various network layer protocols to accomplish this automated *path optimization*.

> The **access layer** provides workgroup and user access to the network.

> In **path optimization** the optimal paths for traffic are selected.

Depending on the network protocols implemented, routers permit you to implement routing environments that suit your specific requirements. For example, in an IP internetwork, routers can support all widely implemented routing protocols, including OSPF, RIP, IGRP, BGP, EGP, and HELLO. Key built-in capabilities that promote path optimization include rapid and controllable route convergence and tunable routing metrics and timers.

Convergence is the process of agreement by all routers on optimal routes.

Convergence is the process of agreement by all routers on optimal routes. When a network event causes routes to either halt operation or become available, routers distribute routing update messages. Routing update messages permeate networks, stimulating recalculation of optimal routes and eventually causing all routers to agree on these routes. Routing algorithms that converge slowly can cause routing loops or network outages.

Many different metrics are used in routing algorithms. Some sophisticated routing algorithms base route selection on a combination of multiple metrics, resulting in the calculation of a single hybrid metric. IGRP uses one of the most sophisticated distance vector routing algorithms. IGRP combines values for bandwidth, load, and delay to create a composite metric value. Link state routing protocols, such as OSPF and IS-IS, employ a metric that represents the cost associated with a given path.

Traffic Prioritization

Traffic prioritization enables policy based routing and ensures that protocols carrying important data take precedence over less important traffic.

Although some network protocols can prioritize internal homogeneous traffic, the router prioritizes the heterogeneous traffic flows. Such *traffic prioritization* enables policy based routing and ensures that protocols carrying mission-critical data take precedence over less important traffic.

Priority Queuing

Priority queuing allows the network administrator to prioritize traffic.

Priority queuing allows the network administrator to prioritize traffic. Traffic can be classified according to various criteria, including protocol and subprotocol type, and then queued on one of four output queues (high, medium, normal, or low priority). For IP traffic additional fine-tuning is possible.

Priority queuing is most useful on low speed serial links. Figure 15.4 shows how priority queuing can be used to segregate traffic by priority level, speeding the transit of certain packets through the network.

You can also use intraprotocol traffic prioritization techniques to enhance internetwork performance. IP's type-of-service (TOS) feature and prioritization of IBM logical units (LUs) are intraprotocol prioritization techniques that can be implemented to improve traffic handling over routers. Figure 15.5 illustrates LU prioritization.

In Figure 15.5, the IBM mainframe is channel-attached to a 3745 communications controller, which is connected to a 3174 cluster controller via remote source route

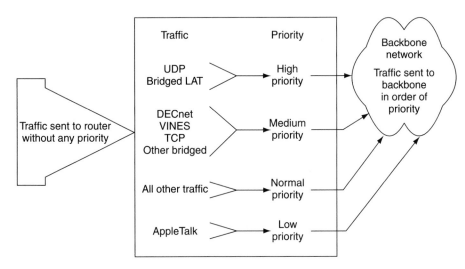

bridging (RSRB). Multiple 3270 terminals and printers, each with a unique local LU address, are attached to the 3174. By applying LU address prioritization, you can assign a priority to each LU associated with a terminal or printer; that is, certain users can have terminals that have better response time than others, and printers can have lowest priority. This function increases application availability for those users running extremely important applications.

Finally, most routed protocols such as AppleTalk, IPX, and DECnet employ a cost based routing protocol to access the relative merit of the different routes to a destination. By tuning associated parameters, you can force particular kinds of traffic to take particular routes, thereby performing a type of manual traffic prioritization.

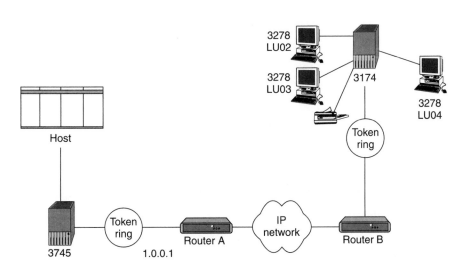

Custom Queuing

Custom queuing reserves bandwidth for a specific protocol, thus allowing mission-critical traffic to receive a minimum amount of bandwidth at any time.

Priority queuing introduces a fairness problem in that all packets classified to lower priority queues might not get serviced in a timely manner, or at all. Custom queuing is designed to address this problem. *Custom queuing* allows more granularity than priority queuing. In fact, this feature is commonly used in the internetworking environment in which multiple higher layer protocols are supported. Custom queuing reserves bandwidth for a specific protocol, thus allowing mission-critical traffic to receive a guaranteed minimum amount of bandwidth at any time.

The intent is to reserve bandwidth for a particular type of traffic. For example, in Figure 15.6 SNA has 40 percent of the bandwidth reserved using custom queuing,

FIGURE 15.6

Custom queuing. (Reproduced with permission of Cisco Systems, Inc. Copyright © 2001 Cisco Systems, Inc. All rights reserved.)

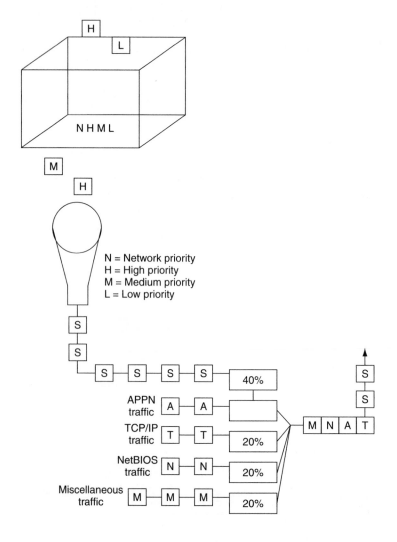

TCP/IP 20 percent, NetBIOS 20 percent, and the remaining protocols 20 percent. The APPN protocol itself has the concept of class of service (COS), which determines the transmission priority for every message. APPN prioritizes the traffic before sending it to the DLC transmission queue.

Weighted Fair Queuing

Weighted fair queuing is a traffic priority management algorithm that uses the time division multiplexing (TDM) model to divide the available bandwidth among clients that share the same interface. In time division multiplexing, each client is allocated a time slice in a round-robin fashion. In weighted fair queuing, the bandwidth is distributed evenly among clients so that each client gets a fair share if every one has the same weighting. You can assign a different set of weights through type-of-service so that more bandwidth is allocated.

If every client is allocated the same bandwidth independent of the arrival rates, the low volume traffic has effective priority over high volume traffic. The use of weighting allows time-delay sensitive traffic to obtain additional bandwidth, thus consistent response time is guaranteed under heavy traffic. There are different types of data stream converging on a wire as illustrated in Figure 15.7. Both C and E are FTP sessions and they are high volume traffic, while A, B, and D are interactive sessions and they are low volume traffic. Every session in this case is termed a conversation. If each conversation is serviced in a cyclic manner and gets a slot regardless of its arrival rate, the FTP sessions do not monopolize the bandwidth. Round-trip delays for the interactive traffic therefore become predictable.

Weighted fair queuing provides an algorithm to identify data streams dynamically using an interface and sorts them into separate logical queues. The algorithm uses various discriminators based on whatever network layer protocol information is available and sorts among them. For example, for IP traffic the discriminators are source and

> **Weighted fair queuing** is a traffic priority management algorithm that uses the time division multiplying model.

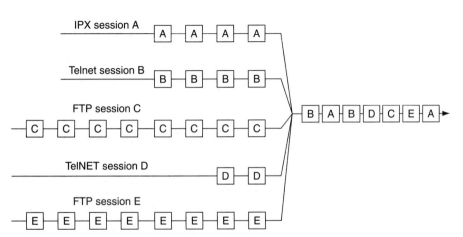

FIGURE 15.7
Weighted fair queuing. (Reproduced with permission of Cisco Systems, Inc. Copyright © 2001 Cisco Systems, Inc. All rights reserved.)

destination addresses, protocol type, socket numbers, and TOS. This is how the two Telnet sessions (sessions B and D) are assigned to different logical queues, as illustrated in Figure 15.7.

Ideally, the algorithm would classify every conversation that is sharing the wire so that each conversation receives its fair share of the bandwidth. Unfortunately, with such protocols as SNA, you cannot distinguish one SNA session from another. For example, in DLSw+, SNA traffic is multiplexed onto a single TCP session. Similarly in APPN, SNA sessions are multiplexed onto a single LLC2 session. The weighted fair queuing algorithm treats these sessions as a single conversation. If you have many TCP sessions, the TCP sessions get the majority of the bandwidth and the SNA traffic gets the minimum. For this reason, this algorithm is not recommended for SNA using DLSw+ TCP/IP encapsulation and APPN.

Weighted fair queuing, however, has many advantages over priority queuing and custom queuing. Priority queuing and custom queuing require the installation of access lists—the bandwidth has to be preallocated and priorities have to be defined. This is clearly a burden. Sometimes network administrators cannot identify and prioritize network traffic in real time. Weighted fair queuing sorts among individual traffic streams without the administrative burden associated with the other two types of queuing.

Load Balancing

● **Load balancing** is the ability of a router to distribute traffic over all its network parts that are the same distance from the destination address.

The easiest way to add bandwidth in a backbone network is to implement additional links. Routers provide built-in *load balancing* for multiple links and paths. You can use up to four paths to a destination network. In some cases, the paths need not be of equal cost.

Within IP, routers provide load balancing on both a per packet and a per destination basis. For per destination load balancing, each router uses its route cache to determine the output interface. If IGRP or Enhanced IGRP routing is used, unequal cost load balancing is possible. The router uses metrics to determine which paths the packets will take—the amount of load balancing can be adjusted by the user.

Load balancing bridged traffic over serial lines is also supported. Serial lines can be assigned to circuit groups. If one of the serial links in the circuit group is in the spanning tree for a network, any of the serial links in the circuit group can be used for load balancing. Data ordering problems are avoided by assigning each destination to a serial link. Reassignment is done dynamically if interfaces go down or come up.

Alternative Paths

Many internetwork backbones carry mission-critical information. Organizations running such backbones are usually interested in protecting the integrity of this information at virtually any cost. Routers must offer sufficient reliability so that they are not the weak link in the internetwork chain. The key is to provide *alternative paths* that can come on line whenever link failures occur along active networks.

End-to-end reliability is not ensured simply by making the backbone fault tolerant. If communication on a local segment within any building is disrupted for any reason that information will not reach the backbone. End-to-end reliability is only possible when redundancy is employed throughout the internetwork. Because this is usually cost prohibitive, most companies prefer to employ redundant paths only on those segments that carry mission-critical information.

What does it take to make the backbone reliable? Routers hold the key to reliable internetworking. Depending on the definition of reliability, this can mean duplicating every major system on each router and possibly every component. However, hardware component duplication is not the entire solution because extra circuitry is necessary to link the duplicate components to allow them to communicate. This solution is usually very expensive, but more importantly, it does not completely address the problem. Even assuming all routers in your network are completely reliable systems, link problems between nodes within a backbone can still defeat a redundant solution.

To really address the problem of network reliability, links must be redundant. Further, it is not enough to simply duplicate all links. Dual links must terminate at multiple routers unless all backbone that are not fault tolerant become single points of failure. The inevitable conclusion is that a completely redundant router is not the most effective solution to the reliability problem because it is expensive and still does not address link reliability. Most network designers do not implement a completely redundant network. Instead, network designers implement partially redundant internetworks.

Switched Access

Switched access provides the capability to enable a WAN link on an as-needed basis via automated router controls. One model for a reliable backbone consists of dual dedicated links and one switched link for idle hot backup. Under normal operational conditions, you can load balance over the dual links, but the switched link is not operational until one of the dedicated links fails.

Traditionally, WAN connections over the *Public Switched Telephone Network (PSTN)* have used dedicated lines. This can be very expensive when an application requires only low volume, periodic connections. To reduce the need for dedicated circuits, a feature called *dial-on-demand routing (DDR)* is available. Figure 15.8 illustrates a DDR connection.

Using DDR, low volume periodic network connections can be made over the PSTN. A router activates the DDR feature when it receives a bridged or routed IP packet destined for a location on the other side of the dial-up line. After the router dials the destination phone number and establishes the connection, packets of any supported protocol can

Alternative paths can come on line whenever link failures occur along active networks.

Switched access provides the capability to enable a WAN link on an as-needed basis via automated router controls.

The **Public Switched Telephone Network** refers to the variety of telephone networks and services in place worldwide.

Dial-on-demand routing is a technique whereby a Cisco router can automatically initiate and close a circuit-switched session as transmitting stations demand.

FIGURE 15.8

The dial-on-demand routing environment. (Reproduced with permission of Cisco Systems, Inc. Copyright © 2001 Cisco Systems, Inc. All rights reserved.)

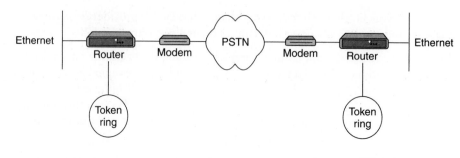

be transmitted. When the transmission is complete, the line is automatically disconnected. By terminating unneeded connections, DDR reduces cost of ownership.

Encapsulation (Tunneling)

Encapsulation, also called tunneling, takes packets or frames from one network system and places them inside frames from another network system.

Encapsulation takes packets or frames from one network system and places them inside frames from another network system. This method is sometimes called *tunneling*. Tunneling provides a means for encapsulating packets inside a routable protocol via virtual interfaces. Synchronous Data Link Control (SDLC) transport is also an encapsulation of packets in a routable protocol. In addition, transport provides enhancements to tunneling such as local data link layer termination, broadcast avoidance, media conversion, and other optimizations.

IP tunneling provides communication between subnetworks that have invalid or discontiguous network addresses. With tunneling, virtual network addresses are assigned to subnetworks, making discontiguous subnetworks reachable. Figure 15.9 illustrates that with tunneling it is possible for the two subnetworks of network 131.108.0.0 to talk to each other even though they are separated by another network.

FIGURE 15.9

Connecting discontiguous networks with tunnels. (Reproduced with permission of Cisco Systems, Inc. Copyright © 2001 Cisco Systems, Inc. All rights reserved.)

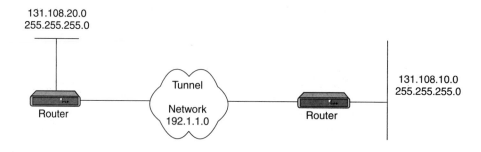

Because encapsulation requires handling of the packets, it is generally faster to route protocols natively than to use tunnels. Tunneled traffic is switched at approximately half the typical process switching rates. This means approximately 1,000 packets per second (pps) aggregate for each router. Tunneling is CPU intensive, and as such should be turned on cautiously. Routing updates, SAP updates, and other administrative traffic may be sent over each tunnel interface. It is easy to saturate a physical link with routing information if several tunnels are configured over it. Performance depends on the passenger protocol, broadcasts, routing updates, and bandwidth of the physical interfaces. It is also difficult to debug the physical link if problems occur. This problem can be mitigated in several ways. In IPX environments, route filters and SAP filters cut down on the size of the updates that travel over tunnels. In AppleTalk networks, keeping zones small and using route filters can limit excess bandwidth requirements.

Tunneling can disguise the nature of a link, making it look slower, faster, or more or less costly than it may actually be in reality. This can cause unexpected or undesirable route selection. Routing protocols that make decisions based only on hop count will usually prefer a tunnel to a real interface. This may not always be the best routing decision because an IP cloud can compromise several different media with very disparate qualities; for example, traffic may be forwarded across both 100-Mbps Ethernet lines and 9.6-Kbps serial lines. When using tunneling, pay attention to the media over which virtual tunnel traffic passes and the metrics used by each protocol.

If a network has sites that use protocol based packet filters as part of a firewall security scheme, be aware that because tunnels encapsulate unchecked passenger protocols, you must establish filtering on the firewall router so that only authorized tunnels are allowed to pass. If tunnels are accepted from unsecured networks, it is a good idea to establish filtering at the tunnel destination or to place the tunnel destination outside the secure area of your network so that the current firewall scheme will remain secure.

When tunneling IP over IP, you must be careful to avoid inadvertently configuring a recursive routing loop. A routing loop occurs when the passenger protocol and the transport protocol are identical. The routing loop occurs because the best path to the tunnel destination is via the tunnel interface. A routing loop can occur when tunneling IP over IP as follows:

1. The packet is placed in the output queue of the tunnel interface.
2. The tunnel interface includes one protocol header and enqueues the packet to the transport protocol (IP) for the destination address of the tunnel interface.
3. IP looks up the route to the tunnel destination address and learns that the path is the tunnel interface.
4. Once again the packet is placed in the output queue of the tunnel interface, as described in step 1, hence the routing loop.

When a router detects a recursive routing loop, it shuts down the tunnel interface for one to two minutes and issues a warning message before it goes into the recursive loop. Another indication that a recursive route loop has been detected is if the tunnel interface is up and the line protocol is down.

To avoid recursive loops, keep passenger and transport routing information in separate locations by implementing the following procedures:

- Use separate routing protocol identifiers.
- Use different routing protocols.
- Assign the tunnel interface a very low bandwidth so that routing protocols, such as IGRP, will recognize a very high metric for the tunnel interface and will, therefore, choose the correct next hop (that is, choose the best physical interface instead of the tunnel).
- Keep the two IP address ranges distinct; that is, use a major address for your tunnel network that is different from your actual IP network. Keeping the address ranges distinct also aids in debugging because it is easy to identify an address as the tunnel network instead of the physical network and vice versa.

Evaluating Distribution Services

This section addresses internetworking features that support distribution services. The following topics are discussed:

- Backbone bandwidth management
- Area and service filtering
- Policy based distribution
- Gateway service
- Interprotocol route redistribution
- Media translation

Backbone Bandwidth Management

To optimize backbone network operations, routers offer several performance tuning features. Examples include priority queuing, routing protocol metrics, and local session termination.

You can adjust the output queue length on priority queues. If a priority queue overflows, excess packets are discarded and quench messages that halt packet flow are sent, if appropriate for that protocol. You can also adjust routing metrics to increase control over the paths that the traffic takes through the internetwork.

A **proxy** is a device that acts on behalf of another device.

Local session termination allows routers to act as proxies for remote systems that represent session endpoints. A *proxy* is a device that acts on behalf of another device. Figure 15.10 illustrates an example of local session termination in an IBM environment.

In Figure 15.10, the routers locally terminate Logical Link Control type 2 (LLC2) data link control sessions. Instead of end-to-end sessions, during which all session

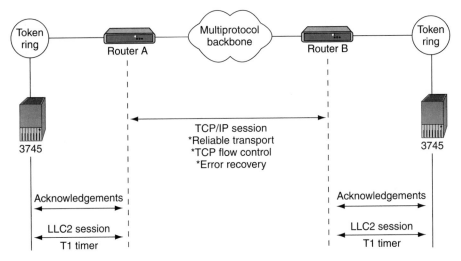

control information is passed over the multiprotocol backbone, the routers take responsibility for acknowledging packets that come from hosts on directly attached LANs. Local acknowledgement saves WAN bandwidth and therefore WAN utilization costs, solves session time-out problems, and provides faster response to users.

Area and Service Filtering

Traffic filters based on area or service type are the primary distribution service tools used to provide policy based access control into backbone services. Both area and service filtering are implemented using access lists. An *access list* is a sequence of statements, each of which either permits or denies certain conditions or addresses. Access lists can be used to permit or deny messages from particular network nodes and messages sent using particular protocols and services.

Area or network access filters are used to enforce the selective transmission of traffic based on network address. You can apply these on incoming or outgoing ports. Service filters use access lists applied to protocols such as IP's UDP, applications such as the Simple Mail Transfer Protocol (SMTP), and specific protocols.

Suppose you have a network connected to the Internet, and you want any host on an Ethernet to be able to form TCP connections to any host on the Internet. However, you do not want Internet hosts to be able to form TCP connections to hosts on the Ethernet except to the SMTP port of a dedicated mail host. SMTP uses TCP port 25 on one end of the connection and a random port number on the other end. The same two port numbers are used throughout the life of the connection. Mail packets

An **access list** is a sequence of statements, each of which either permits or denies certain conditions or addresses.

coming in from the Internet will have a destination port of 25. Outbound packets will have the port numbers reversed. The fact that the secure system behind the router always accepts mail connections on port 25 is what makes it possible to separately control incoming and outgoing services. The access list can be configured on either the outbound or inbound interface.

In the following example, the Ethernet network is a Class B network with the address 128.88.0.0, and the mail host's address is 128.88.1.2. The keyword establish is used only for the TCP protocol to indicate an established connection. A match occurs if the TCP datagram has the ACK or RST bits set, which indicate that the packet belongs to an existing connection.

```
Access-list 102 permit tcp 0.0.0.0 255.255.255.255
128.88.0.0.255.255 establish
Access-list 102 permit tcp 0.0.0.0 255.255.255.255
128.88.1.2 0.0.0.0 eq 25
interface ethernet 0
IP access-group 102
```

Policy Based Distribution

Policy based distribution is based on the premise that different departments within a common organization might have different policies regarding traffic dispersion.

Policy based distribution is based on the premise that different departments within a common organization might have different policies regarding traffic dispersion through the organization-wide internetwork. Policy based distribution aims to meet the differing requirements without compromising performance and information integrity.

A policy within this internetworking context is a rule or set of rules that governs end-to-end distribution of traffic to and subsequently through a backbone network. One department might send traffic representing three different protocols to the backbone, but it might want to expedite one particular protocol's transit through the backbone because it carries mission-critical application information. To minimize already excessive internal traffic, another department might want to exclude all backbone traffic except electronic mail and one key custom application from entering its network segment.

These examples reflect policies specific to a single department. However, policies can reflect overall organizational goals. For example, an organization might want to regulate backbone traffic to a maximum of 10 percent average bandwidth during the workday and 1-minute peaks of 30 percent utilization. Another corporate policy might be to ensure that communication between two remote departments can freely occur, despite differences in technology.

In addition to support for internetworking technologies, there must be a means to not only keep separate but also integrate these technologies as appropriate. The different technologies should be able to coexist or combine intelligently as the situation warrants. Consider the situation depicted in Figure 15.11. Assume that a corporate policy limits unnecessary backbone traffic. One way to do this is to restrict the

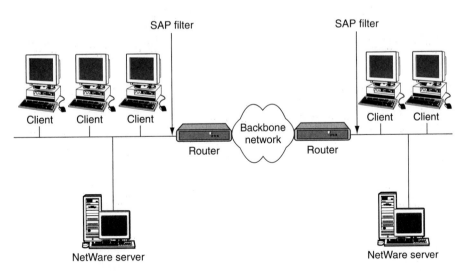

transmission of SAP messages. SAP messages allow NetWare servers to advertise services to clients. The organization might have another policy stating that all NetWare services should be provided locally. If this is the case, there should be no reason to be advertised remotely. SAP filters prevent SAP traffic from leaving a router interface, thereby fulfilling this policy.

Gateway Service

Protocol *gateway* capabilities are part of each router's standard software. For example, DECnet is currently in Phase V. DECnet Phase V addresses are different from DECnet Phase IV addresses. For those networks that require both types of hosts to coexist, two-way Phase IV/Phase V translation conforms to Digital specified guidelines. The routers interoperate with Digital routers, and Digital hosts do not differentiate between the different devices.

> **Gateway** capabilities are part of each router's standard software.

The connection of multiple independent DECnet networks can lead to addressing problems. Nothing precludes two independent DECnet administrators from assigning node address 10 to one of the nodes in their respective networks. When the two networks are connected at some later time, conflicts result. DECnet *address translation gateways (ATGs)* address this problem. The ATG solution provides router based translation between addresses in two different networks connected by a router. Figure 15.12 illustrates an example of this operation.

> **Address translation gateways** allow a router to route multiple independent DECnet networks.

In Network 0, the router is configured at address 19.4 and is a level 1 router. In Network 1, the router is configured at address 50.5 and is an area router. At this point, no routing information is exchanged between the two networks. The router maintains a separate routing table for each network. By establishing a translation map,

FIGURE 15.12

Sample DECnet ATG implementation. (Reproduced with permission of Cisco Systems, Inc. Copyright © 2001 Cisco Systems, Inc. All rights reserved.)

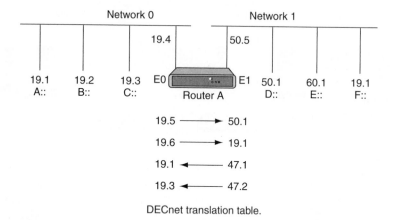

DECnet translation table.

packets in Network 0 sent to address 19.5 will be routed to Network 1 as 19.1; packets sent to address 47.1 in Network 1 will be routed to Network 0 as 19.1; and packets sent to 47.2 in Network 1 will be sent to Network 0 as 19.3.

AppleTalk is another protocol with multiple revisions, each with somewhat different addressing characteristics. AppleTalk Phase 1 addresses are simple local forms, while AppleTalk Phase 2 uses extended multinetwork addressing. Normally a Phase 1 node cannot understand information sent from a Phase 2 node if Phase 2 extended addressing is used. Routers support routing between Phase 1 and Phase 2 nodes on the same cable by using translational routing.

You can accomplish translational routing by attaching two router ports to the same physical cable. Configure one port to support nonextended AppleTalk and the other to support extended AppleTalk. Both ports must have unique network numbers. Packets are translated and sent out the other port as necessary.

Interprotocol Route Redistribution

Route redistribution allows routing information discovered through one routing protocol to be distributed in the update messages of another routing protocol.

The preceding section, "Gateway Service," discussed how routed protocol gateways such as one that translates between AppleTalk Phase 1 and Phase 2 allow two end nodes with different implementations to communicate. Routers can also act as gateways for routing protocols. Information derived from one routing protocol, such as the IGRP, can be passed to, and used by, another routing protocol, such as RIP. This is useful when running multiple protocols in the same internetwork.

Routing information can be exchanged between any supported IP routing protocols. These include RIP, IGRP, OSPF, HELLO, EGP, and BGP. Similarly, *route redistribution* is supported by ISO CLNS for route redistribution between ISO IGRP and IS-IS. Static route information can also be redistributed. Defaults can be assigned so that one routing protocol can use the same metric for all redistributed routes, thereby simplifying the routing redistribution mechanism.

Media Translation

Media translation techniques translate frames from one network system into frames of another. Such translations are rarely 100 percent effective because one system might have attributes with no corollary to the other. For example, Token Ring networks support a built-in priority and reservation system whereas Ethernet networks do not. Translations between Token Ring and Ethernet networks must somehow account for this discrepancy. It is possible for two vendors to make different decisions about how this discrepancy will be handled, which can prevent multivendor interoperation.

For those situations in which communication between end stations on different media is required, routers can translate between Ethernet and Token Ring frames. Source route translational bridging translates between Token Ring and Ethernet frame formats; SRT allows routers to use both SRB and the transparent bridging algorithm used in standard Ethernet bridging. When bridging from the SRB domain to the transparent bridging domain, the SRB fields of the frames are removed. RIFs are cached for use by subsequent return traffic. When bridging in the opposite direction, the router checks the packet to determine whether it has a multicast or broadcast destination or unicast destination. If it has a multicast or broadcast destination, the packet is sent as a spanning tree explorer. If it has unicast destination, the router looks up the path to the destination in the RIF cache. If a path is found, it will be used; otherwise, the router looks up the path to the destination in the RIF cache. If a path is found, it will be used; otherwise, the router will send the packet as a spanning tree explorer. A simple example of this topology is shown in Figure 15.13. Routers support SRT through implementation of both transparent bridging and SRB algorithms on each SRT interface. If an interface notes the presence of an RIF, it uses the SRB algorithm; if not, it uses the transparent bridging algorithm.

Translation between serial links running the SDLC protocol and Token Ring running LLC2 is also available. This is referred to as SDLLC frame translation. SDLLC frame translation allows connections between serial lines and Token Rings. This is useful for consolidating traditionally disparate SNA/SDLC networks into a LAN based, multi-protocol, multimedia backbone network. Using SDLLC, routers terminate SDLC sessions, translate SDLC frames to LLC2 frames, and then forward the LLC2 frames using RSRB over a point-to-point or IP network. Because a router based IP network can use arbitrary media, such as FDDI, Frame Relay, X.25, or leased lines, routers support SDLLC over all such media through IP encapsulation. A complex SDLLC configuration is shown in Figure 15.14.

Evaluating Local Access Services

The following discussion addresses internetworking features that support local access services. Local access service topics outlined here include the following:

- Value added network addressing
- Network segmentation

FIGURE 15.13

Source route translational bridging topology. (Reproduced with permission of Cisco Systems, Inc. Copyright © 2001 Cisco Systems, Inc. All rights reserved.)

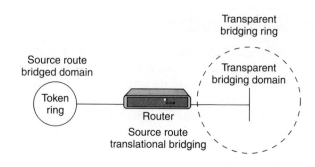

Transparent bridging ring

Source route bridged domain

Transparent bridging domain

Token ring

Router

Source route translational bridging

Frames lose their RIFs in this direction

Frames gain their RIFs in this direction

FIGURE 15.14

Complex SDLLC configuration. (Reproduced with permission of Cisco Systems, Inc. Copyright © 2001 Cisco Systems, Inc. All rights reserved.)

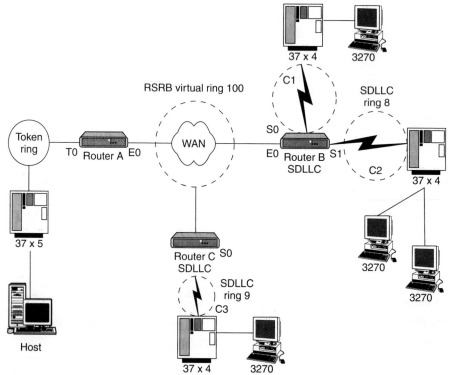

- Broadcast and multicast capabilities
- Naming, proxy, and local cache capabilities
- Media access security
- Router discovery

Value Added Network Addressing

Address schemes for LAN based networks such as NetWare and others do not always adapt perfectly to use over multisegment LANs or WANs. One tool routers implement to ensure operation of such protocols is protocol-specific helper addressing. *Helper addressing* is a mechanism to assist the movement of specific traffic through a network when that traffic might not otherwise transit the network. The use of helper addressing is best illustrated with an example. Consider the use of helper addresses on Novell IPX internetworks. Novell clients send broadcast messages when looking for a server. If the server is not local, broadcast traffic must be sent through routers. Helper addresses and access lists can be used together to allow broadcasts from certain nodes on one network to be directed specifically to certain servers on another network. Multiple helper addresses on each interface are supported so that broadcast packets can be forwarded to multiple hosts. Figure 15.15 illustrates the use of NetWare based helper addressing. NetWare clients on Network AA are allowed to broadcast to any server on Network BB. An applicable access list would specify that broadcasts of type 10 will be permitted from all nodes on Network AA. A configuration specified helper address identifies the addresses on Network BB to which these broadcasts are directed. No other nodes on Network BB receive the broadcasts. No other broadcasts other than type 10 broadcasts are routed.

> **Helper addressing** assists the movement of specific traffic through a network when that traffic might not otherwise transit the network.

Any downstream networks beyond AA, for example, some Network AA1, are not allowed to broadcast to Network BB through Router C1, unless the routers partitioning Networks AA and AA1 are configured to forward broadcasts with a series of configuration entries. These entries must be applied to the input interfaces and be set to forward broadcasts between directly connected networks. In this way, traffic is passed along in a directed manner from network to network.

Network Segmentation

The splitting of networks into more manageable pieces is an essential role played by local access routers. In particular, local access routers implement local policies and limit unnecessary traffic. Examples of capabilities that allow network designers to use local access routers to *segment* networks include IP subnets, DECnet area addressing, and AppleTalk zones.

> **Network segmentation** reduces overall network congestion.

You can use local access routers to implement local policies by placing the routers in strategic locations and by configuring specific segmenting policies. For example, you

FIGURE 15.15

Sample network map illustrating helper address broadcast control. (Reproduced with permission of Cisco Systems, Inc. Copyright © 2001 Cisco Systems, Inc. All rights reserved.)

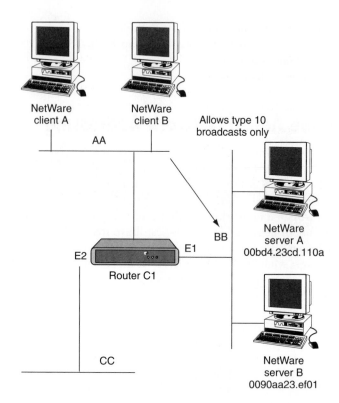

can set up a series of LAN segments with different subnet addresses—routers would be configured with suitable interface addresses and subnet masks. In general, traffic on a given segment is limited to local broadcasts, traffic intended for a specific end station on that segment, or traffic intended for another specific router. By distributing hosts and clients carefully, you can use this simple method of dividing up a network to reduce overall network congestion.

Broadcast and Multicast Capabilities

Broadcasts are messages that are sent out to all network destinations.

Many protocols use broadcast and multicast capabilities. *Broadcasts* are messages that are sent out to all network destinations. Multicasts are messages sent to a specific subset of network destinations. Routers inherently reduce broadcast proliferation by default. However, routers can be configured to relay broadcast traffic if necessary. Under certain circumstances, passing along broadcast information is desirable and possibly necessary. The key is controlling broadcasts and multicasts using routers.

In the IP world, as with many other technologies, broadcast requests are very common. Unless broadcasts are controlled, network bandwidth can be seriously reduced. Routers offer various broadcast limiting functions that reduce network traffic and minimize broadcast storms. For example, directed broadcasting allows broadcasts to a specific network or a series of networks, rather than to the entire internetwork. When flooded broadcasts (broadcasts sent through the entire internetwork) are necessary, routers support a technique by which these broadcasts are sent over a spanning tree of the network. The spanning tree ensures complete coverage without excessive traffic because only one packet is sent over each network segment.

As discussed in the section "Value Added Network Addressing," broadcast assistance is accommodated with the helper address mechanisms. You can allow a router or series of routers to relay broadcasts that would otherwise be blocked by using helper addresses. For example, you can permit retransmission of SAP broadcasts using helper addresses, thereby notifying clients on different network segments of certain NetWare services available from specific remote servers.

The IP multicast feature allows IP traffic to be propagated from one source to any number of destinations. Rather than sending one packet to each destination, one packet is sent to a multicast group identified by a single IP destination group address. IP multicast provides excellent support for such applications as video- and audio-conferencing, resource discovery, and stock market traffic distribution.

For full support of IP multicast, IP hosts must run the *Internet Group Management Protocol (IGMP)*. IGMP is used by IP hosts to report their multicast group memberships to an immediately neighboring multicast router. The membership of a multicast group is dynamic. Multicast routers send IGMP query messages on their attached local networks. Host members of a multicast group respond to a query by sending IGMP reports for multicast groups to which they belong. Reports sent by the first host in a multicast group suppress the sending of identical reports from other hosts of the same group. The multicast router attached to the local network takes responsibility for forwarding multicast datagrams from one multicast group to all other networks that have members in the group. Routers build multicast group distribution trees (routing tables) so that multicast packets have loop-free paths to all multicast group members so that multicast packets are not duplicated. If no reports are received from a multicast group after a set of IGMP queries, the multicast routers assume the group has no local members and stop forwarding multicasts intended for that group.

Internet Group Management Protocol is used by IP hosts to report their multicast group memberships to an immediately neighboring multicast router.

Naming, Proxy, and Local Cache Capabilities

Three key router capabilities help reduce network traffic and promote efficient internetworking operation: name service support, proxy services, and local caching of network information.

Network applications and connection services provided over segmented internetworks require a rational way to resolve names to addresses. Various facilities accommodate this requirement. Any router you select must support the name services implemented for different end system environments. Examples of supported name services include NetBIOS, IP's Domain Name System (DNS) and IEN116, and AppleTalk Name Binding Protocol (NBP).

The router can also act as a proxy for a name server. The router's support of NetBIOS name caching is one example of this kind of capability. NetBIOS name caching allows the router to maintain a cache of NetBIOS names, which avoids the overhead of transmitting all of the broadcasts between client and server NetBIOS PCs (IBM PCs or PS/2s) in an SRB environment. When NetBIOS name caching is enabled, the router does the following:

- Notices when any host sends a series of duplicate query frames and limits retransmission to one frame per period. The time period is a configuration parameter.
- Keeps a cache of mappings between NetBIOS server and client names and their MAC addresses. As a result, broadcast requests sent by clients to find servers and by servers in response to their clients can be sent directly to their destinations, rather than being broadcast across the entire bridged network.

When NetBIOS name caching is enabled and default parameters are set on the router, the NetBIOS name server and the NetBIOS name client, approximately 20 broadcast packets per login, are kept on the local ring where they are generated. In most cases, the NetBIOS name cache is best used when large amounts of NetBIOS broadcast traffic might create bottlenecks on a WAN that connects local internetworks to distant locations.

The router can also save bandwidth or handle nonconforming name resolution protocols by using a variety of other proxy facilities. By using routers to act on behalf of other devices to perform various functions, you can more easily scale networks. Instead of being forced to add bandwidth when a new workgroup is added to a location, you can use a router to manage address resolution and control message services. Examples of this kind of capability include the proxy ARP address resolution for IP internetworks and NBP proxy in AppleTalk internetworks.

Local caches store previously learned information about the network so that new information requests do not need to be issued each time the same piece of information is desired. A router's ARP cache stores physical address and network address mappings so that it does not need to broadcast ARP requests more than once within a given time period for the same address. Address caches are maintained for many other protocols as well, including DECnet, Novell IPX, and SRB, where RIF information is cached.

Media Access Security

If all corporate information is readily available to all employees, security violations and inappropriate file access can occur. To prevent this, routers must do the following:

- Keep local traffic from inappropriately reaching the backbone.
- Keep backbone traffic from exiting the backbone into an inappropriate department or workgroup network.

These two functions require packet filtering. Packet filtering capabilities should be tailored to support a variety of corporate policies. Packet filtering methods can reduce traffic levels on a network, thereby allowing a company to continue using its current technology rather than investing in more network hardware. In addition, packet filters can improve security by keeping unauthorized users from accessing information and can minimize network problems caused by excessive congestion.

Routers support many filtering schemes designed to provide control over network traffic that reaches the backbone. Perhaps the most powerful of these filtering mechanisms is the access list. Each of the following possible local access services can be provided through access lists:

- You have an Ethernet-to-Internet routing network and you want any host on the Ethernet to be able to form TCP connections to any host on the Internet. However, you do not want Internet hosts to be able to form TCP connections into the Ethernet except to the SMTP port of a dedicated mail host.
- You want to advertise only one network through an RIP routing process.
- You want to prevent packets that originated on any Sun workstation from being bridged on a particular Ethernet segment.
- You want to keep a particular protocol based on Novell IPX from establishing a connection between a source network or source port combination and a destination network or destination port combination.

Access lists logically prevent certain packets from traversing a particular router interface, thereby providing a general tool for implementing network security. In addition to this method, several specific security systems already exist to help increase network security. For example, the U.S. government has specified the use of an optional field within the IP packet header to implement a hierarchical packet security system called the *Internet Protocol Security Option (IPSO)*. IPSO support on routers addresses both the basic and extended security options described in a draft of the IPSO circulated by the Defense Communications Agency. This draft document is an early version of Request for Comments (RFC) 1108. IPSO defines security levels, for example, TOP SECRET, SECRET, and others on a per interface basis and accepts or rejects messages based on whether they include adequate authorization. Some security systems are designed to keep remote users from accessing the network unless they have adequate authorization. For example, the *Terminal Access Controller Access Control System (TACACS)* is a means of protecting modem access into a network. The Defense Data Network (DDN) developed TACACS to control access to its TAC terminal servers.

The router's TACACS support is patterned after the DDN application. When a user attempts to start an EXEC command interpreter on a password protected line, TACACS prompts for a password. If the user fails to enter the correct password, access is denied. Router administrators can control various TACACS parameters, such as the number of retries allowed, the time-out interval, and the enabling of TACACS accounting.

The *Challenge Handshake Authentication Protocol (CHAP)* is another way to keep unauthorized remote users from accessing a network. It is also commonly used to

The **Internet Protocol Security Option** is a hierarchical packet security system.

The **Terminal Access Controller Access Control System** is a means of protecting modem access into a network.

The **Challenge Handshake Authentication Protocol** is commonly used to control router-to-router communications.

control router-to-router communications. When CHAP is enabled, a remote device such as a PC, workstation, router, or communication server attempting to connect to a local router is challenged to provide an appropriate response. If the correct response is not provided, network access is denied. CHAP is becoming popular because it does not require a secret password to be sent over the network. CHAP is supported on all router serial lines using PPP encapsulation.

Router Discovery

Hosts must be able to locate routers when they need access to devices external to the local network. When more than one router is attached to a host's local segment, the host must be able to locate the router that represents the optimal path to the destination. This process of finding routers is called *router discovery*.

The following are router discovery protocols:

Router discovery is the process of finding the router that represents the optimal path to the destination.

- End System-to-Intermediate System (ES-IS): This protocol is defined by the ISO OSI protocol suite. It is dedicated to the exchange of information between intermediate systems (routers) and end systems (hosts). ESs send hello messages to all ISs on the local subnetwork. Both types of messages convey the subnetwork and network layer addresses of the systems that generate them. Using this protocol, end systems and intermediate systems can locate one another.
- ICMP Router Discovery Protocol (IRDP): Although the issue is currently under study, at this time there is no single standardized manner for end stations to locate routers in the IP world. In many cases, stations are simply configured manually with the address of a local router. However, RFC 1256 outlines a router discovery protocol using the Internet Control Message Protocol (ICMP). This protocol is commonly referred to as IRDP.
- Proxy Address Resolution Protocol (ARP): ARP uses broadcast messages to determine the MAC layer address that corresponds to a particular internetwork address. ARP is sufficiently generic to allow the use of IP with virtually any type of underlying media access mechanism. A router that has proxy ARP enabled responds to ARP requests for those hosts for which it has a route, which allows hosts to assume that all other hosts are actually on their network.
- RIP: RIP is a routing protocol that is commonly available on IP hosts. Many hosts use RIP to find the address of the routers on a LAN or, when there are multiple routers, to pick the best router to use for a given internetwork address.

Choosing Internetworking Reliability Options

One of the first concerns of most network designers is to determine the required level of application availability. In general, this key consideration is balanced against implementation cost. For most organizations, the cost of making a network completely

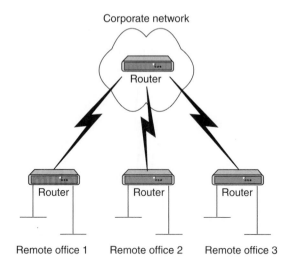

Corporate network

Router

Router
Router
Router

Remote office 1   Remote office 2   Remote office 3

FIGURE 15.16

Typical nonredundant internetwork design. (Reproduced with permission of Cisco Systems, Inc. Copyright © 2001 Cisco Systems, Inc. All rights reserved.)

fault tolerant is prohibitive. Determining the appropriate level of fault tolerance to be included in a network and where redundancy should be used is not trivial. The nonredundant internetwork design illustrated in Figure 15.16 illustrates the considerations involved with increasing levels of internetwork fault tolerance.

The internetwork shown in Figure 15.16 has two levels of hierarchy: a corporate office and remote offices. Assume the corporate office has eight Ethernet segments, to which approximately 400 users an average of 50 per segment are connected. Each Ethernet segment is connected to a router. In the remote offices, two Ethernet segments are connected to the router in the corporate office through a T1 link.

The following section addresses various approaches to creating redundant internetworks, provides some context for each approach, and contrasts their relative merits and drawbacks. The following topics are discussed:

- *Redundant links* versus meshed topologies
- Redundant power systems
- Fault-tolerant media implementation
- Backup hardware

Redundant links are two or more links going to the same destination or path.

Redundant Links Versus Meshed Topologies

Typically, WAN links are the least reliable components in an internetwork, usually because of problems in the local loop. In addition to being relatively unreliable, these links are often an order of magnitude slower than the LANs they connect. However, because they are capable of connecting geographically diverse sites, WAN links often make up the backbone network and are therefore critical to corporate operations. The

FIGURE 15.17

Internetwork with dual links to remote offices. (Reproduced with permission of Cisco Systems, Inc. Copyright © 2001 Cisco Systems, Inc. All rights reserved.)

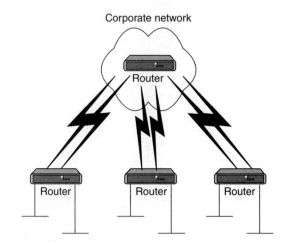

combination of potentially suspect reliability, lack of speed, and high importance makes the WAN link a good candidate for redundancy.

As a first step in making the example internetwork more fault tolerant, you might add a WAN link between each remote office and the corporate office. This results in the topology shown in Figure 15.17.

The new topology has several advantages. First it provides a backup link that can be used if a primary link connecting any remote office and the corporate office fails. Second, if the routers support load balancing, link bandwidth has now been increased, lowering response times for users and increasing application availability. Load balancing in transparent bridging and IGRP environments is another tool for increasing fault tolerance. Routers also support load balancing on either a per packet or a per destination basis in all IP environments. Per packet load balancing is recommended if the WAN links are relatively slow, for example, less than 56 Kbps. If WAN links are faster than 56 Kbps, enabling fast switching on the routers is recommended. When fast switching is enabled, load balancing occurs on a per destination basis.

Routers can automatically compensate for failed WAN links through routing algorithms of protocols, such as IGRP, OSPF, and IS-IS. If one link fails, the routing software recalculates the routing algorithm and begins sending all traffic through another link. This allows applications to proceed in the face of WAN link failure, improving application availability.

The primary disadvantage of duplicating WAN links to each remote office is cost. In the example outlined in Figure 15.17, three new WAN links are required. In large star networks with more remote offices, 10 or 20 new WAN links might be needed, as well as new equipment including new WAN router interfaces. A lower cost alternative that is becoming increasingly popular links the remote offices using a meshed topology as illustrated in Figure 15.18.

Before

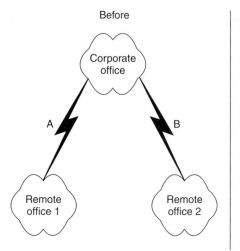

After

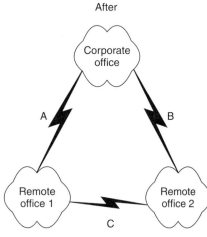

FIGURE 15.18
Evolution from a star to a meshed topology. (Reproduced with permission of Cisco Systems, Inc. Copyright © 2001 Cisco Systems, Inc. All rights reserved.)

In the before portion of Figure 15.18, any failure associated with either Link A or B blocks access to a remote site. The failure might involve the link connection equipment, such as a data service unit (DSU) or a channel service unit (CSU), either the entire router or a single router port, or the link itself. Adding Link C as shown in the after portion of the figure offsets the effect of a failure in any single link.

If Link A or B fails, the affected remote site can still access the corporate office through Link C and the other site's link to the corporate office. Note also that if Link C fails, the two remote sites can communicate through their connections to the corporate office.

A *meshed topology* has three distinct advantages over a redundant star topology:

- A meshed topology is usually slightly less expensive at least by the cost of one WAN link.
- A meshed topology provides more direct and potentially faster communication between remote sites, which translates to greater application availability. This can be useful if direct traffic volumes between remote sites are relatively high.
- A meshed topology promotes distributed operation, preventing bottlenecks on the corporate router and further increasing application availability.

A redundant star is a reasonable solution under the following conditions:

- Relatively little traffic must travel between remote sites.
- Traffic moving between corporate and remote offices is delay sensitive and mission-critical. The delay and potential reliability problems associated with making an extra hop when a link between a remote office and the corporate office fails might not be tolerable.

A **meshed topology** promotes distributed operation, preventing bottlenecks on the corporate router and increasing application availability.

Redundant Power Systems

Power failures are common in large-scale networks. Because they can strike across a very local or a very wide scale, power faults are difficult to preempt. Simple power problems include dislodged power cords, tripped circuit breakers, and local power supply failures. More extensive power problems include large-scale outages caused by natural phenomena such as lightning strikes or brownouts. Each organization must assess its needs and the probability of each type of power outage before determining which preventative actions to take.

You can take many precautions to try to ensure that problems, such as dislodged power cords, do not occur frequently. From the standpoint of internetworking devices, dual power systems can prevent otherwise debilitating failures. Imagine a situation where the so-called backbone-in-a-box configuration is being used. This configuration calls for the connection of many networks to a router being used as a connectivity hub. Benefits include a high speed backbone, essentially the router's backplane and cost efficiency less media. Unfortunately, if the router's power system becomes faulty, each network connected to that router loses its capability to communicate with all other networks connected to that router.

General power outages are usually more common than failures in a router's power system. Consider the effect of a site-wide power failure on redundant star and meshed topologies. If the power fails in the corporate office, the organization might be seriously inconvenienced. Key network applications are likely to be placed at a centralized corporate location. The organization could easily lose revenue for every minute its network is down. The meshed network configuration is superior. In both cases, all other remote offices will still be able to communicate with the corporate office. Generally, power failures in a remote office are more serious when network services are widely distributed.

To protect against local and site-wide power outages, some companies have negotiated an agreement with local power companies to use multiple power grids within their organization. Failure within one power grid will not affect the network if all critical components have access to multiple power grids. Unfortunately, this arrangement is very expensive and should only be considered by companies with substantial resources, extremely mission-critical operations, and a relatively high likelihood of power failures. The effect of highly localized power failures can be minimized with prudent network planning. Wherever possible, redundant components should use power supplied by different circuits. Further, these redundant components should not be physically collocated. For example, if redundant routers are employed for all stations on a given floor, these routers can be physically stationed in wiring closets on different floors. This prevents local wiring closet power problems from affecting the capability of all stations on a given floor to communicate. Figure 15.19 shows such a configuration.

For some organizations, the need for fault tolerance is so great that potential power failures are protected against with a duplicate corporate data center. Organizations

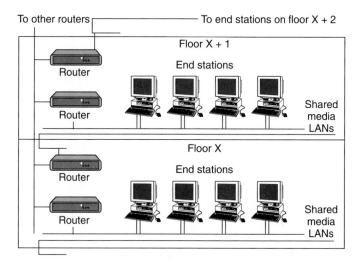

FIGURE 15.19

Redundant components on different floors. (Reproduced with permission of Cisco Systems, Inc. Copyright © 2001 Cisco Systems, Inc. All rights reserved.)

with these requirements often locate a redundant data center in another city, or in part of the same city that is some distance from the primary data center. All backend services are duplicated, and transactions coming in from remote offices are sent to both data centers. This configuration requires duplicate WAN links from all remote offices, duplicate network hardware, duplicate servers and server resources, and leasing another building. Because this approach is so costly, it is typically the last step taken by companies desiring the ultimate in fault tolerance.

Partial duplication of the data center is also a possibility. Several key servers and links to those servers can be duplicated. This is a common compromise to the problem presented by power failures.

Fault-Tolerant Media Implementations

Media failure is another possible network fault. Included in this category are all problems associated with the media and its link to each individual end station. Under this definition, media components include network interface controller failures—lobe or attachment unit interface (AUI) cable failures are caused by operator negligence and cannot easily be eliminated.

One way to reduce the havoc caused by failed media is to divide existing media into smaller segments and support each segment with different hardware. This minimizes the effect of a failure on a particular segment. For example, if you have 100 stations attached to a single switch, move some of them to other switches. This reduces the effect of a hub failure and of certain subnetwork failures. If you place an internetworking device such as a router between segments, you protect against additional problems and cut subnetwork traffic. As illustrated in Figure 15.19, redundancy can be employed to help minimize media failures. Each station in this figure is attached to two different media segments. NICs, hub ports, and interface cables are all redundant. This approach doubles the cost of network connectivity for each end station as well as the port usage on all internetworking devices, and is therefore only recommended in situations where complete redundancy is required. It also assumes that end station software, including both the network and the application subsystems, can handle and effectively use the redundant components. The application software or the networking software or both must be able to detect network failures and initiate use of the other network.

Certain media access protocols have some fault-tolerant features built in. Token Ring multistation access units (MAUs) can detect certain media connection failures and bypass the failure internally. FDDI dual rings can wrap traffic onto the backup ring to avoid portions of the network with problems.

From a router's standpoint, many media failures can detect certain media connection failures and bypass the failure internally. FDDI dual rings can wrap traffic onto the backup ring to avoid portions of the network with problems. If routing updates or routing keepalive messages have not been received from devices that would normally be reached through a particular router port, the router will soon declare that route down and will look for alternative routes. Meshed networks provide these alternative paths allowing the router to compensate for media failures.

Backup Hardware

Like all complex devices, routers, switches, and other internetworking devices develop hardware problems. When serious failures occur, the use of dual devices can effectively reduce the adverse effects of a hardware failure. After a failure, discovery protocols help end stations choose new paths with which to communicate across the network. If each network connected to the failed device has an alternative path out of the local area, complete connectivity will still be possible. For example, when backup routers are used, routing metrics can be set to ensure that the backup routers will not be used unless the primary routers are not functioning.

Switchover is automatic and rapid. For example, consider the situation shown in Figure 15.20. In this network, dual routers are used at all sites with dual WAN links. If Router R1 fails, the routers on FDDI will detect the failure by the absence of messages from Router R1. Using any of several dynamic routing protocols, Router A, Router B, and Router C will designate Router R3 as the new next hop on the way to remote resources accessible via Router R4.

Many networks are designed with multiple routers connecting particular LANs in order to provide redundancy. In the past, the effectiveness of this design was limited by the speed at which the hosts on those LANs detected a topology update and changed

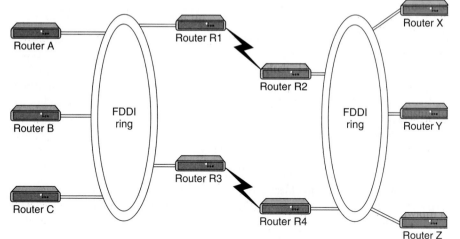

routers. In particular, IP hosts tend to be configured with a default gateway or con-figured to use proxy ARP to find a router on their LAN. Convincing an IP host to change its router usually required manual intervention to clear the ARP cache or to change the default gateway.

The *Hot Standby Router Protocol (HSRP)* is a solution that allows network topol-ogy changes to be transparent to the host. HSRP typically allows hosts to reroute in approximately 10 seconds. HSRP is supported on Ethernet, Token Ring, fast Ethernet, and ATM.

> The **Hot Standby Router Protocol** allows network topology changes to be transparent to the host.

An HSRP group can be defined on each LAN. All members of the group know the standby IP address and the standby MAC address. One member of the group is elected the leader. The lead router services all packets sent to the HSRP group ad-dress. The other routers monitor the leader and act as HSRP routers. If the lead router becomes unavailable, the HSRP router elects a new leader who inherits the HSRP MAC address and IP address.

Designing Switched LAN Internetworks

This section describes three technologies that network designers can use to design switched LAN internetworks:

- LAN switching
- Virtual LANs (VLANs)
- ATM switching

Evolution from Shared to Switched Networks

In the past, network designers had only a limited number of hardware options when purchasing a technology for their campus networks. Hubs were for wiring closets and routers were for the data center or main telecommunications operations. The increasing power of desktop processors and the requirements of client-server and multimedia applications, however, have driven the need for greater bandwidth in traditional shared-media environments. These requirements are prompting network designers to replace hubs in their wiring closets with switches, as illustrated in Figure 15.21. This strategy allows network managers to protect their existing wiring investments and boost network performance with dedicated bandwidth to the desktop for each user.

LAN emulation
is a technology
that allows an
ATM network to
function as a LAN
backbone.

Coinciding with the wiring closet evolution is a similar trend in the network backbone. Here, the role of Asynchronous Transfer Mode (ATM) is increasing as a result of standardizing protocols, such as *LAN emulation (LANE),* that enable ATM devices to coexist with existing LAN technologies. Network designers are collapsing their router backbones with ATM switches, which offer the greater backbone bandwidth required by high throughput data services.

FIGURE 15.21
*Evolution from
shared to switched
internetworks.
(Reproduced with
permission of
Cisco Systems, Inc.
Copyright
© 2001 Cisco
Systems, Inc. All
rights reserved.)*

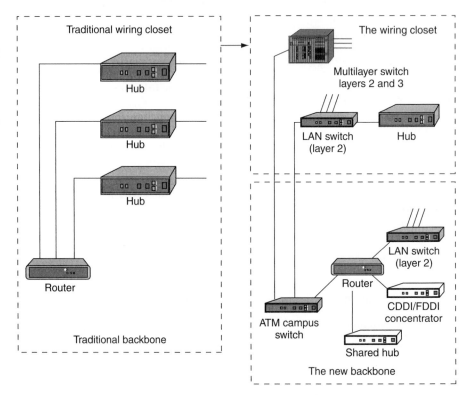

Technologies for Building Switched LAN Internetworks

With the advent of such technologies as layer 3 switching, LAN switching, and VLANs, building campus LANs is becoming more complex than in the past. Today, the following three technologies are required to build successful campus networks:

- LAN switching technologies
  - Ethernet switching: Provides layer 2 switching and offers broadcast domain segmentation using VLANs. This is the base fabric of the network.
  - Token Ring switching: Offers the same functionality as Ethernet switching but uses Token Ring technology. You can use a Token Ring switch as either a transparent bridge or source route bridge.
  - *Copper Data Distributed Interface (CDDI):* Provides a *single attachment station (SAS)* or *dual attachment station (DAS)* to two Category 5 unshielded twisted-pair (UTP), 100 Mbps RJ-45 connectors.
  - Fiber Distributed Data Interface (FDDI): Provides an SAS or DAS connection to the FDDI backbone network using two multimode, *media interface connector (MIC)* fiber optic connections.
- ATM switching technologies
  - ATM switching offers high speed switching technology for voice, video, and data. Its operation is similar to LAN switching technologies for data operations. However, ATM offers superior voice, video, and data integration today.
- Routing technologies
  - Routing is a key technology for connecting LANs in a campus network. It can be either layer 3 switching or more traditional routing with layer 3 switching features and enhanced layer 3 software features. Switched LAN internetworks are also referred to as campus LANs.

Role of LAN Switching Technology in Campus Networks

Most network designers are beginning to integrate switching devices into their existing shared-media networks to achieve the following goals:

- Increase the bandwidth that is available to each user, thereby alleviating congestion in their shared-media networks.
- Employ the manageability of VLANs by organizing network users into logical workgroups that are independent of the physical topology of wiring closet hubs. This, in turn, can reduce the cost of moves, adds, and changes while increasing the flexibility of the network.
- Deploy emerging applications across different switching platforms and technologies, making them available to a variety of users.
- Provide a smooth evolution path to high performance switching solutions, such as fast Ethernet and ATM.

Copper Data Distributed Interface provides a single attachment station or dual attachment station to two Category 5 unshielded twisted-pair connectors.

A **single attachment station** is a device attached only to the primary ring of an FDDI ring.

A **dual attachment station** is a device attached to both the primary and the secondary FDDI rings.

A **media interface connector** is an FDDI de facto standard connector.

LAN switching
technology employs microsegmentation, which segments the LAN to fewer users and ultimately to a single user.

Segmenting shared-media LANs divides the users into two or more separate LAN segments, reducing the number of users contending for bandwidth. *LAN switching* technology, which builds upon this trend, employs microsegmentation, which further segments the LAN to fewer users and ultimately to a single user with a dedicated LAN segment. Each switch port provides a dedicated 10 MB Ethernet segment or dedicated 4/16 Token Ring segment.

Segments are interconnected by internetworking devices that enable communication between LANs while blocking other types of traffic. Switches have the intelligence to monitor traffic and compile address tables, which then allows them to forward packets directly to specific ports in the LAN. Switches also usually provide nonblocking service, which allows multiple conversations traffic between two ports to occur simultaneously.

Switching technology is quickly becoming the preferred solution for improving LAN traffic for the following reasons:

- Unlike hubs and repeaters, switches allow multiple data streams to pass simultaneously.
- Switches have the capability through microsegmentation to support the increased speed and bandwidth requirements of emerging technologies.
- Switches deliver dedicated bandwidth to users through high density group switched and switched 10Base-T or 100Base-T Ethernet, flexible 10/100Base-T Ethernet, fiber based fast Ethernet, fast EtherChannel, Token Ring, CDDI/FDDI, and ATM LAN emulation.

Switched Internetwork Solutions

Network designers are discovering, however, that many products offered as switched internetwork solutions are inadequate. Some offer a limited number of hardware platforms with little or no system integration with the current infrastructure. Others require complete abandonment of all investments in the current network infrastructure. To be successful, a switched internetwork solution must accomplish the following:

- Leverage strategic investments in the existing communications infrastructure while increasing available bandwidth.
- Reduce the costs of managing network operations.
- Offer the option to support multimedia applications and other high demand traffic across a variety of platforms.
- Provide scalability, traffic control, and security that is at least as good or better than that of today's router based internetworks.
- Provide support for embedded remote monitoring (RMON) agent.

The key to achieving these benefits is to understand the role of the internetworking software infrastructure within the switched internetworks. Within today's networks, routers allow for the interconnection of disparate LAN and WAN technologies, while also implementing security filters and logical firewalls. It is these capabilities that have allowed current internetworks to scale globally while remaining stable and robust. As networks evolve toward switched internetworks, similar logical internetworking capabilities are required for stability and scalability. Although LAN and ATM switches provide great performance improvements, they also raise new internetworking challenges. Switched internetworks must integrate with existing LAN and WAN networks. Such services as VLANs, which will be deployed with switched internetworks, also have particular internetworking requirements.

A true-switched internetwork, therefore, is more than a collection of boxes. Rather, it consists of a system of devices integrated and supported by an intelligent internetworking software infrastructure. Presently, this network intelligence is centralized within routers. However, with the advent of switched internetworks, the intelligence will often be dispersed throughout the network, reflecting the decentralized nature of switching systems. The need for an internetworking infrastructure, however, will remain.

Components of the Switched Internetworking Model

A switched internetwork is composed of the following three basic components:

- Physical switching platforms
- A common software infrastructure
- Network management tools and applications

Scalable Switching Platforms

The first component of the switched internetworking model is the physical switching platform. This can be an ATM switch, a LAN switch, or a router.

ATM Switches

Although switched internetworks can be built with a variety of technologies, many network designers will deploy ATM in order to utilize its unique characteristics. ATM provides scalable bandwidth that spans both LANs and WANs. It also promises Quality of Service (QoS) guarantees—bandwidth on demand—that can map into and support higher level protocol infrastructures for emerging multimedia applications and provide a common multiservice network infrastructure.

ATM switches are one of the key components of ATM technology. All ATM switches, however, are not alike. Even though all ATM switches perform cell relay, ATM switches differ markedly in the following capabilities:

- Variety of interfaces and services that are supported
- Redundancy

- Depth of ATM internetworking software
- Sophistication of traffic management mechanism
- Blocking and nonblocking switching fabrics
- SVC and PVC support

Just as there are routers and LAN switches available at various price or performance points with different levels of functionality, ATM switches can be segmented into the following four distinct types that reflect the needs of particular applications and markets:

- Workgroup ATM switches
- Campus ATM switches
- Enterprise ATM switches
- Multiservice access switches

Workgroup and Campus ATM Switches

Workgroup ATM switches are optimized for deploying ATM to the desktop over low cost ATM desktop interfaces, with ATM signaling interoperability for ATM adapters and QoS support for multimedia applications.

Campus ATM switches are generally used for small-scale ATM backbones, for example, to link ATM routers or LAN switches. This use of ATM switches can alleviate current backbone congestion while enabling the deployment of such new services as VLANs. Campus switches need to support a wide variety of both local backbone and WAN types but be price or performance optimized for the local backbone function. In this class of switches, ATM routing capabilities that allow multiple switches to be tied together are very important. Congestion control mechanisms for optimizing backbone performance are also important.

Enterprise ATM Switches

Enterprise ATM switches are sophisticated multiservice devices that are designed to form the core backbones of large enterprise networks. They are intended to complement the role played by today's high end multiprotocol routers. Enterprise ATM switches, much as campus ATM switches, are used to interconnect workgroup ATM switches and other ATM connected devices, such as LAN switches. Enterprise-class switches, however, can not only act as ATM backbones but also serve as the single point of integration for all of the disparate services and technology found in enterprise backbones today. By integrating all of these services onto a common platform and a common ATM transport infrastructure, network designers can gain greater manageability while eliminating the need for multiple overlay networks.

Multiservice Access Switches

Beyond private networks, ATM platforms will also be widely deployed by service providers both as *customer premises equipment (CPE)* and within public networks. Such equipment will be used to support multiple *metropolitan area network (MAN)* and WAN services such as Frame Relay switching, LAN interconnect, or public ATM services on a common ATM infrastructure. Enterprise ATM switches will often be used in these public network applications because of their emphasis on high availability and redundancy, and their support of multiple interfaces.

LAN Switches

A LAN switch is a device that typically consists of many ports that connect LAN segments (Ethernet and Token Ring) and a high speed port such as 100-Mbps Ethernet, FDDI, or 155-Mbps ATM. The high speed port, in turn, connects the LAN switch to other devices in the network.

A LAN switch has dedicated bandwidth per port, and each port represents a different segment. For best performance, network designers often assign just one host to a port, giving that host dedicated bandwidth of 10 Mbps, as shown in Figure 15.22, or 16 Mbps for Token Ring networks.

When a LAN switch first initializes and as the devices that are connected to it request services from other devices, the switch builds a table that associates the MAC addresses of each local device with the port number through which that device is reachable. That way, when Host A on Port 1 needs to transmit to Host B on Port 2, the LAN switch forwards frames from Port 1 to Port 2, thus sparing other hosts on Port 3 from responding to frames destined for Host B, it can do so because the LAN switch can forward frames from Port 3 to Port 4 at the same time it forwards frames from Port 1 to Port 2.

Customer premises equipment
refers to equipment such as terminals, telephones, and modems that are installed at customer sites and connected to the telephone company network.

The metropolitan area network
spans a larger geographic area than a LAN, but a smaller geographic area than a WAN.

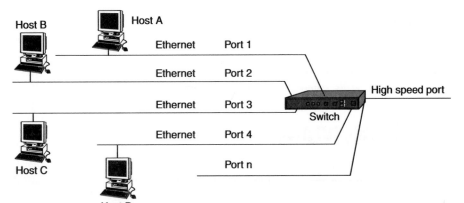

FIGURE 15.22
Sample LAN switch configuration. (Reproduced with permission of Cisco Systems, Inc. Copyright © 2001 Cisco Systems, Inc. All rights reserved.)

Flooding occurs whenever a device connected to the LAN switch sends the packet out all ports except for the port from which the packet originated.

Because they work like traditional transparent bridges, LAN switches dissolve previously well-defined workgroup or department boundaries. A network built and designed only with LAN switches appears as a flat network topology consisting of a single broadcast domain. Consequently, these networks are liable to suffer the problems inherent in flat or bridged networks, which is they do not scale well. LAN switches that support VLANs are more scalable than traditional bridges.

Routing Platforms

In addition to LAN switches and ATM switches, network designers typically use routers as one of the components in a switched internetwork infrastructure. While LAN switches are being added to wiring closets to increase bandwidth and to reduce congestion in existing shared-media hubs, high speed backbone technologies such as ATM switching and ATM routers are being deployed in the backbone. Within a switched internetwork, routing platforms also allow for the interconnection of disparate LAN and WAN technologies while also implementing broadcast filters and logical firewalls. In general, if you need advanced internetworking services, such as broadcast firewalling and communication between dissimilar LANs, routers are necessary.

Common Software Infrastructure

The second level of a switched internetworking model is a common software infrastructure. The function of this software is to unify the variety of the physical switching platforms: LAN switches, ATM switches, and multiprotocol routers. Specifically, the software infrastructure should perform the following tasks:

- Monitor the logical topology of the network
- Logically route traffic
- Manage and control sensitive traffic
- Provide firewalls, gateways, filtering, and protocol translation

A **Virtual LAN** consists of several end systems, all of which are members of a single logical broadcast domain.

Virtual LANs (VLANs)

A *Virtual LAN (VLAN)* consists of several end systems, either hosts or network equipment such as switches and routers, all of which are members of a single logical broadcast domain. A VLAN no longer has physical proximity constraints for the broadcast domain. This VLAN is supported on various pieces of network equipment

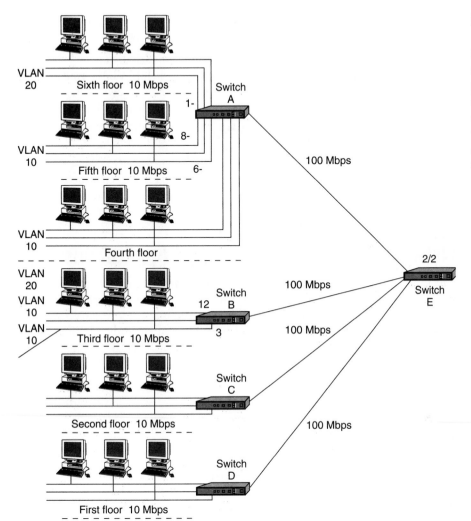

such as LAN switches that support VLAN trunking protocols between them. Each VLAN supports a separate spanning tree (IEEE 802.1d).

First-generation VLANs are based on various OSI layer 2 bridging and multiplexing mechanisms, such as IEEE 802.10, LANE, and inter-switch link (ISL), that allow the formation of multiple, disjointed, overlaid broadcast groups on a single network infrastructure. Figure 15.23 shows an example of a switched LAN network that uses VLANs. Layer 2 of the OSI reference model provides reliable transit of data across a physical link. The data link layer is concerned with physical addressing, network topology, line discipline, error notification, ordered delivery frames, and flow control.

The IEEE has divided this layer into two sublayers: the MAC sublayer and the LLC sublayer, sometimes simply called link layer.

In Figure 15.23, 10-Mbps Ethernet connects the hosts on each floor to switches A, B, C, and D. 100-Mpbs fast Ethernet connects these to Switch E. VLAN 10 consists of those hosts that are on Ports 6 and 8 of Switch A and Port 2 on Switch B. VLAN 20 consists of those hosts that are on Port 1 of Switch A and Ports 1 and 3 of Switch B.

VLANs can be used to group a set of related users, regardless of their physical connectivity. They can be located across a campus environment or even across geographically dispersed locations. The users might be assigned to a VLAN because data flow patterns among them are such that it makes sense to group them together. Note, however, that without a router the hosts in one VLAN cannot communicate with the hosts in another VLAN.

Network Management Tools and Applications

The third and last component of a switched internetworking model consists of network management tools and applications. Because switching is integrated throughout the network, network management becomes crucial at both the workgroup and backbone levels. Managing a switch based network requires a radically different approach from managing traditional hub and router based LANs.

As part of designing a switched internetwork, network designers must ensure that their design takes into account network management applications needed to monitor, configure, plan, and analyze switched internetwork devices and services.

Switched LAN Network Designs

A successful switched internetworking solution must combine the benefits of both routers and switches in every part of the network, as well as offer a flexible evolution path from shared-media networking to switched internetworks.

In general, incorporating switches in campus network designs results in the following benefits:

- High bandwidth
- Quality of Service (QoS)
- Low cost
- Easy configuration

If you need advanced internetworking services, however, routers are necessary. Routers offer the following services:

- Broadcast firewalling
- Hierarchical addressing

- Communication between dissimilar LANs
- Fast convergence
- Policy routing
- QoS routing
- Security
- Redundancy and load balancing
- Traffic flow management
- Multimedia group membership

In the future, some of these router services will be offered by switches. For example, support for multimedia often requires a protocol, such as Internet Group Management Protocol (IGMP), that allows workstations to join a group that receives multimedia multicast packets. One router will still be necessary, but you will not need a router in each department because IGMP switches can communicate with the router to determine whether any of their attached users are part of a multicast group.

Switching and bridging sometimes can result in nonoptimal routing of packets. This is because every packet must go through the root bridge of the spanning tree. When routers are used, the routing of packets can be controlled and designed for optimal paths. When designing switched LAN networks, you should consider the following:

- Comparison of LAN switches and routers
- Benefits of LAN switches (layer 2 services)
- Benefits of routers (layer 3 services)
- Benefits of VLANs
- VLAN implementation
- General network design principles
- Switched LAN network design principles

Comparison of LAN Switches and Routers

The fundamental difference between a LAN switch and a router is that the LAN switch operates at layer 2 of the OSI model and the router operates at layer 3. This difference affects the way that LAN switches and routers respond to network traffic. This section compares LAN switches and routers with regard to the following network design issues:

- Loops
- Convergence
- Broadcasts
- Subnetworking
- Security
- Media dependence

Because routers implement layer 2 functionality and switches are beginning to implement layer 3 functionality, the functions of a LAN switch and a router are merging.

FIGURE 15.24

Switched LAN topology with loops. (Reproduced with permission of Cisco Systems, Inc. Copyright © 2001 Cisco Systems, Inc. All rights reserved.)

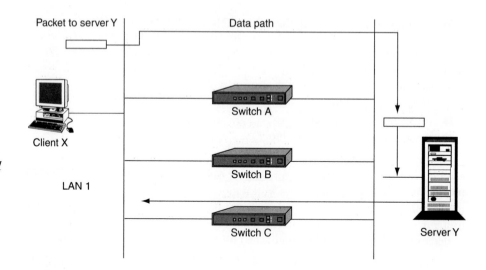

Loops

A **loop** is a route where packets never reach their destination, but simply cycle repeatedly through a constant series of network nodes.

Switched LAN topologies are susceptible to *loops* as illustrated in Figure 15.24.

In Figure 15.24 it is possible for packets from Client X to be switched by Switch A and then for Switch B to put the same packet back on to LAN 1. In this situation, packets loop and undergo multiple replications. To prevent looping and replication, topologies that may contain loops need to run the Spanning Tree Protocol (STP). STP uses the spanning tree algorithm to construct topologies that do not contain any loops. Because the spanning tree algorithm places certain connections in blocking mode, only a subset of the network topology is used for forwarding data. In contrast, routers provide freedom from loops and make use of optimal paths.

Convergence

In transparent switching, neighboring switches make topology decisions locally based on the exchange of BPDUs. This method of making topology decisions means that convergence on an alternative path can take an order of magnitude longer than in a routed network environment.

In a routed environment, sophisticated routing protocols such as OSPF maintain concurrent topological databases of the network and allow the network to converge quickly.

Broadcasts

LAN switches do not filter broadcasts, multicasts, or unknown address frames. The lack of filtering can be a serious problem in modern distributed networks in which broadcast messages are used to resolve addresses and dynamically discover network resources such as file servers. Broadcasts originating from each segment are received by every computer in the switched internetwork. Most devices discard broadcasts because they are irrelevant, which means that large amounts of bandwidth are wasted by the transmission of broadcasts.

In some cases, the circulation of broadcasts can saturate the network so that there is no bandwidth left for application data. In this case, new network connections cannot be established and existing connections may be dropped—a situation known as a broadcast storm. The probability of broadcast storms increases as the switched internetwork grows. Routers do not forward broadcasts and, therefore, are not subject to broadcast storms.

Subnetworking

Transparently switched networks are composed of physically separate segments, but are logically considered to be one large network, for example, one IP subnet. This behavior is inherent to the way the LAN switches work: they operate at OSI layer 2 and have to provide connectivity to hosts as if each host were on the same cable. Layer 2 addressing assumes a flat address space with universally unique addresses.

Routers operate at OSI layer 3, so they can formulate and adhere to a hierarchical addressing structure. Routed networks can associate a logical addressing structure to a physical infrastructure so that each network segment has a TCP/IP subnet or IPX network. Traffic flow on routed networks is inherently different from traffic flow on switched networks. Routed networks have more flexible traffic flow because they can use hierarchy to determine optimal paths depending on dynamic factors such as network congestion.

Security

Information is available to routers and switches that can be used to create more secure networks. LAN switches may use custom filters to provide access control based on destination address, source address, protocol type, packet length, and offset bits within the frame. Routers can filter on logical network addresses and provide control based on options available in layer 3 protocols. For example, routers can permit or deny traffic based on specific TCP/IP socket information for a range of network addresses.

Media Dependence

Two factors need to be considered with regard to mixed media. Table 15.2 lists the maximum frame size for various network media.

First, when LANs of dissimilar media are switched, hosts must use the MTU that is the lowest common denominator of all the switched LANs that make up the internetwork.

Table 15.2 MTUs for Various Network Media

Media	Minimum Valid Frame	Maximum Valid Size
Ethernet	64 bytes	1,518 bytes
Token Ring	32 bytes	16 KB theoretical, 4 KB normal
Fast Ethernet	64 bytes	1,518 bytes
FDDI	32 bytes	4,400 bytes
ATM Lane	64 bytes	1,518 bytes
ATM Classical IP	64 bytes	9,180 bytes
Serial HDLC	14 bytes	No limit, 4.5 KB normal

This requirement limits throughput and can seriously compromise performance over a relatively fast link, such as FDDI or ATM. Most layer 3 protocols can fragment and reassemble packets that are too large for a particular subnetwork; therefore, routed networks can accommodate different MTUs, which maximizes throughput.

Second, because they operate at layer 2, switches must use a translation function to switch between dissimilar media. The translation function can result in serious problems such as noncanonical versus canonical Token Ring-to-Ethernet MAC format conversion. One issue with moving data from a Token Ring to an Ethernet network is layer 2 addressing. Token Ring devices read the layer 2 MAC address as most significant bit starting from left to right. Ethernet devices read the layer 2 MAC address as most significant bit starting from right to left.

By working at layer 3, routers are essentially independent of the properties of any physical media and can use a simple address resolution algorithm or a protocol, such as ARP, to resolve differences between layer 2 and layer 3 addresses.

Benefits of LAN Switches (Layer 2 Services)

An individual layer 2 switch might offer some or all of the following benefits:

Automatic packet recognition and translation technology recognizes and converts a variety of Ethernet protocol formats into industry standard formats.

- Bandwidth: LAN switches provide excellent performance for individual users by allocating dedicated bandwidth to each switch port, for example, each network segment. This technique is known as microsegmenting.
- VLANs: LAN switches can group individual ports into logical switched workgroups called VLANs, thereby restricting the broadcast domain to designated VLAN member ports. VLANs are also known as switched domains and autonomous switching domains. Communication between VLANs requires a router.
- Automated packet recognition and translation: *automatic packet recognition and translation (APaRT)* technology recognizes and converts a variety of Ether-

net protocol formats into industry standard CDDI/FDDI formats. With no changes needed in either client or server end stations the switched solution can provide easy migration to 100-Mbps server access while preserving the user's investment in existing shared 10Base-T LANs.

Benefits of Routers (Layer 3 Services)

Because routers use layer 3 addresses, which typically have structure, routers can use techniques such as address summarization to build networks that maintain performance and responsiveness as they grow in size. By imposing hierarchical structure on a network, routers can effectively use redundant paths and determine optimal routes even in a dynamically changing network. This section describes the router functions that are vital in LAN designs:

● Broadcast and multicast control
● Broadcast segmentation
● Media transition

Broadcast and Multicast Control

Routers control broadcasts and multicasts in the following ways:

● By caching the address of remote hosts: When a host sends a broadcast packet to obtain the address of a remote host that the router already knows about, the router responds on behalf of the remote host and drops the broadcast packet, sparing hosts from having to respond to it.

● By caching advertised network services: When a router learns of a new network service, it caches the necessary information and does not forward broadcasts related to it. When a client of that network service sends a broadcast to locate that service, the router responds on behalf of that service and drops the broadcast packet. For example, Novell clients use broadcasts to find local services. In a network without a router, every server responds to every client broadcast by multicasting its list of services. Routers manage Novell broadcasts by collecting services not local to the switch and sending out periodic updates that describe the services offered on the entire network. Each router sends out one frame for every seven services on the network.

● By providing special protocols, such as IGMP and *Protocol Independent Multicast (PIM)*. These new protocols allow a multicasting application to negotiate with routers, switches, and clients to determine the devices that belong to a multicast group. This negotiation helps limit the scope and impact of the multicast stream on the network as a whole. Successful network designs contain a mix of appropriately scaled switching and routing. Given the effects of broadcast radiation on CPU performance, well-managed switched LAN designs must include routers for broadcast and multicast management.

●
Protocol Independent Multicast allows the addition of IP multicast routing on existing IP networks.

Broadcast Segmentation

In addition to preventing broadcasts from radiating throughout the network, routers are also responsible for generating services to each LAN segment. The following are examples of services that the router provides to the network for a variety of protocols:

- IP: Proxy ARP and Internet Control Message Protocol (ICMP)
- IPX: SAP table updates
- AppleTalk: ZIP table updates
- Network Management: SNMP queries

In a flat virtual network, a single router would be bombarded by myriad requests needing replies, severely taxing its processor. Therefore, the network designer needs to consider the number of routers that can provide reliable services to a given subset of VLANs. Some type of hierarchical design needs to be considered.

Media Translation

In the past, routers have been used to connect networks of different media types, taking care of the OSI layer 3 address translations and fragmentation requirements. Routers continue to perform this function in switched LAN designs. Most switching is done within like media such as Ethernet, Token Ring, and FDDI switches with some capability of connecting to another media type. However, if a requirement for a switched campus network design is to provide high speed connectivity between unlike media, routers play a significant part in the design.

Benefits of VLANs

In a flat bridged network all broadcast packets generated by any node in the network are sent to and received by all other network nodes. The ambient level of broadcasts generated by the higher layer protocols in the network, known as broadcast radiation, will typically restrict the total number of nodes that the network can support. In extreme cases, the effects of broadcast radiation can be so severe that an end station spends all of its CPU power on processing broadcasts.

VLANs have been designed to address the following problems inherent in a flat bridged network:

- Scalability issues of a flat network topology
- Simplification of network management by facilitating network reconfigurations

VLANs solve the scalability problems of large flat networks by breaking a single bridged domain into several smaller bridged domains, each of which is a virtual LAN. Note that each virtual LAN itself is constrained by the scalability issues described in the section on *broadcasts in switched LAN internetworks*. It is insufficient to solve the broadcast problems inherent to a flat-switched network by superimposing VLANs and reducing broadcast domains. VLANs without routers do not scale to the large campus environments. Routing is instrumental in the building of scalable VLANs and is the only way to impose hierarchy on the switched VLAN internetwork.

VLANs offer the following features:

- Broadcast control: Just as switches isolate collision domains for attached hosts and only forward appropriate traffic out a particular port, VLANs refine this concept further and provide complete isolation between VLANs. A VLAN is a bridging domain, and all broadcast and multicast traffic is contained within it.
- Security: VLANs provide security in two ways:
  - High security users can be grouped into a VLAN, possibly on the same physical segment, and no users outside of that VLAN can communicate with them.
  - Because VLANs are logical groups that behave like physical separate entities, inter-VLAN communication is achieved through a router. When inter-VLAN communication occurs through a router, all the security and filtering functionality that routers traditionally provide can be used because routers are able to look at OSI layer 3 information. In the case of nonroutable protocols, there can be no inter-VLAN communication. All communication must occur within the same VLAN.
- Performance: The logical grouping of users allows an engineer making intensive use of a networked CAD/CAM station or testing a multicast application to be assigned to a VLAN that contains just that engineer and the services that they may need. The engineer's work does not affect the rest of the engineering group, which results in improved performance for the engineer by being on a dedicated LAN and improved performance for the rest of the engineering group whose communications are not slowed down by the engineer's use of the network.
- Network management: The logical grouping of users, divorced from their physical or geographic locations, allows easier network management. It is no longer necessary to pull cables to move a user from one network to another. Adds, moves, and changes are achieved by configuring a port into the appropriate VLAN. Expensive time-consuming recabling to extend connectivity in a switched LAN environment is no longer necessary because network management can be used to logically assign a user from one VLAN to another.

VLAN Implementation

This section describes the different methods of creating the logical groupings or broadcast domains that make up various types of VLANs. There are three ways of defining a VLAN:

1. By port: Each port on the switch can support only one VLAN. With port based VLANs, no layer 3 address recognition takes place, so IP, Novell, and AppleTalk networks must share the same VLAN definition. All traffic within the VLAN is

switched, and traffic between VLANs is routed by an external router or by a router within the switch. This type of VLAN is also known as a segment based VLAN.

2. By protocol: VLANs based on network address such as OSI layer 3 addresses can differentiate between different protocols, allowing the definition of VLANs to be made on a per protocol basis. With network address based VLANs it will be possible to have a different virtual topology for each protocol, with each topology having its own set of rules, firewalls, and so forth. Routing between VLANs comes automatically without the need for an external router or card. Network address based VLANs will mean that a single port on a switch can support more than one VLAN. This type of VLAN is also known as a virtual subnet VLAN.

3. By a user defined value: This type of VLAN is typically the most flexible, allowing VLANs to be defined based on the value of any field in a packet. For example, VLANs could be defined on a protocol basis or could be dependent on a particular IPX or NetBIOS service. The simplest form of this type of VLAN is to group users according to their MAC addresses.

The best method of implementing VLANs is by port. To efficiently operate and manage protocols, such as IP, IPX, and AppleTalk, all nodes in a VLAN should be in the same subnet or network.

There are three technologies to implement VLANs:

1. IEEE 802.10
2. Inter-switch link (ISL)
3. LAN emulation

The three technologies are similar in that they are based on OSI layer 2 bridge multiplexing mechanisms. VLANs are differentiated by assigning each VLAN a color or VLAN ID. For example, Engineering might be the blue VLAN and Manufacturing might be the green VLAN.

IEEE 802.10

IEEE 802.10 defines a method for secure bridging of data across a shared MAN backbone. Companies have initially implemented the relevant portions of the standard to allow the coloring of bridged traffic across high speed backbones such as FDDI, fast Ethernet, Token Ring, and serial links. There are two strategies using IEEE 802.10 to implement VLANs, depending on how traffic is handled through the backbone: switched backbone and routed backbone.

Switched Backbone

In the switched backbone topology shown in Figure 15.25, you want to ensure that intra-VLAN traffic goes only between Segment A and Segment D both in VLAN 10 and Segment B and Segment C both in VLAN 20.

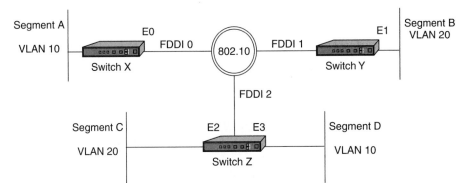

In Figure 15.25, all Ethernet ports on Switches X, Y, and Z are in a VLAN and are to be VLAN interfaces. All FDDI interfaces in Switches X, Y, and Z are called VLAN trunk interfaces.

The coloring of traffic across the FDDI backbone is achieved by inserting a 16-byte header between the source MAC address and the Link Service Access Point (LSAP) of frames leaving a switch. This header contains a 4-byte VLAN ID or color. The receiving switch removes the header and forwards the frame to interfaces that match that VLAN color.

Routed Backbone

In the routed backbone topology illustrated in Figure 15.26, the goal is the same as for the switched topology—that is, to ensure that intra-VLAN traffic goes only between Segment A and Segment D both in VLAN 10 and Segment B and Segment C both in VLAN 20.

It is important that a single VLAN use only one subnet. In Figure 15.26, VLAN 10 (subnet 10) is split and therefore must be glued together by maintaining a bridged path for it through the network. For Switch X and nodes in VLAN 20 (subnet 20), traffic is switched locally if appropriate. If traffic is destined for a node in VLAN 30 (subnet 30) from a node in VLAN 20, Router Y routes it through the backbone to Router Z. If traffic from Segment D on VLAN 10 is destined for a node in VLAN 20, Router Y routes it back out the FDDI interface.

Fast EtherChannel

Fast EtherChannel is a trunking technology based on grouping together multiple full duplex 802.3 fast Ethernets to provide fault-tolerant high speed links between

FIGURE 15.26

*IEEE 802.10
routed backbone
implementation.
(Reproduced with
permission of
Cisco Systems, Inc.
Copyright
© 2001 Cisco
Systems, Inc. All
rights reserved.)*

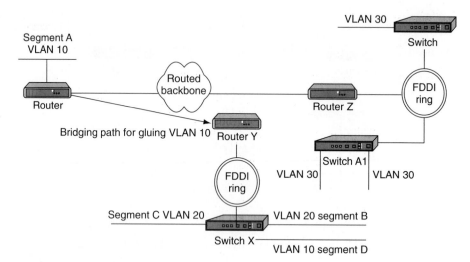

switches, routers, and servers. Fast EtherChannels can be composed of two to four industry standard fast Ethernet links to provide load sharing of traffic with up to 800 Mbps of usable bandwidth. Fast EtherChannels can interconnect LAN switches, routers, servers, and clients. Because its load balancing is integrated inside the LAN switch architectures, there is no performance degradation for adding links to a channel—high throughput and low latencies can be maintained while gaining more total available bandwidth. Fast EtherChannel provides link resiliency within a channel, if links should fail the traffic is immediately directed to the remaining links. Finally, fast EtherChannel is not dependent on any type of media. It can be used with fast Ethernet running on existing UTP wiring or single mode and multimode fiber. Figure 15.27 illustrates a collapsed backbone topology design using the fast EtherChannel modules to provide links of 400 MB between switches in the wiring closets and the data center.

IEEE 802.10 Design Issues

- Routers fast switch IEEE 802.10, which means that the fast-switching throughput of the platform must be considered.
- VLANs must be consistent with the routed model—that is, subnets cannot be split.
  1. If subnets must be split, they must be glued together by a bridged path.
  2. Normal routed behavior needs to be maintained for end nodes to correctly achieve routing between VLANs.
  3. Networks need to be designed carefully when integrating VLANs; the simplest choice is to avoid splitting VLANs across a routed backbone.

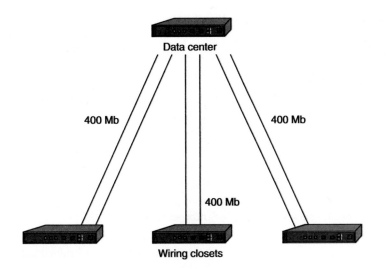

The difference between these two strategies is subtle. Table 15.3 compares the advantages and disadvantages of the two strategies.

A VLAN interface can have only one VLAN ID, and VLAN trunk interfaces support multiple VLANs across them.

Inter-Switch Link

ISL is a vendor proprietary protocol for interconnecting multiple switches and maintaining VLAN information as traffic goes between switches. This technology is similar to IEEE 802.10 in that it is a method of multiplexing bridge groups over a high speed backbone. It is defined only on fast Ethernet.

With ISL, an Ethernet frame is encapsulated with a header that maintains VLAN IDs between switches. A 30-byte header is prepended to the Ethernet frame, and it contains a 2-byte VLAN ID. In Figure 15.28, Switch Y switches VLAN 20 traffic between Segments A and B is appropriate. Otherwise, it encapsulates traffic with an ISL header that identifies it as traffic for VLAN 20 and sends it through the interim switch to Router X. Router X routes the packet to the appropriate interface, which could be through a routed network beyond Router X (as in this case) out the fast Ethernet interface to Switch Z. Switch Z receives the packet, examines the ISL header noting that this packet is destined for VLAN 20, and switches it to all ports in VLAN 20 if the packet is a broadcast or multicast or the appropriate port if the packet is a unicast. Routers fast switch ISL, which means that the fast-switching throughput of the platform must be considered.

Table 15.3 Advantages and Disadvantages of Switched and Routed Backbones

Switched Backbone		*Routed Backbone*	
Advantages	*Disadvantages*	*Advantages*	*Disadvantages*
Propagates color information across entire network	Backbone is running bridging	No bridging in backbone	Color information is not propagated across backbone and must be configured manually
Allows greater scalability by extending bridge domains	Broadcast traffic increases drastically on the backbone	Easy to integrate into existing internetwork	
		Can run native protocols in the backbone	If subnets are split, a bridged path has to be set up between switches

FIGURE 15.28

Inter-switch link design. (Reproduced with permission of Cisco Systems, Inc. Copyright © 2001 Cisco Systems, Inc. All rights reserved.)

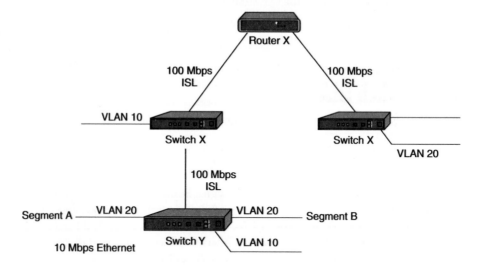

LAN Emulation

LANE is a service that provides interoperability between ATM based workstations and devices connected to existing legacy LAN technology. The ATM Forum has defined a standard for LANE that provides the same capabilities to workstations attached via ATM that they are used to obtaining from legacy LANs.

LANE uses MAC encapsulation (OSI layer 2) because this approach supports the largest number of existing OSI layer 3 protocols. The end result is that all devices attached to an emulated LAN appear to be on one bridged segment. In this way, AppleTalk, IPX, and other protocols should have similar performance characteristics as in a traditional bridged environment. In ATM LANE environments, the ATM switch handles traffic that belongs to the same emulated LAN (ELAN), and routers handle inter-ELAN traffic.

Virtual Multihomed Servers

In traditional networks, there are usually several well-known servers such as e-mail and corporate servers that almost everyone in an enterprise needs to access. If these servers are located in only one VLAN, the benefits of VLANs will be lost because all of the different workgroups will be forced to route to access this common information source.

This problem can be solved with LANE and virtual multihomed servers as illustrated in Figure 15.29. Network interface cards (NICs) allow workstations and servers to join up to eight different VLANs. This means that the server will appear in eight different ELANs, and that to other members of each ELAN the server appears to be like any other member. This capability greatly increases the performance of the network as a whole because common information is available directly through the optimal Data Direct VCC and does not need to be routed. This also means that the server must process all broadcast traffic in each VLAN that it belongs to, which can decrease performance.

To multihome servers in non-ATM environments there are two possible choices:

1. Use servers with multiple NICs with different connections to each VLAN.
2. Use servers with NICs that support the VLAN trunking technology (IEEE 802.10 or ISL) used in the backbone.

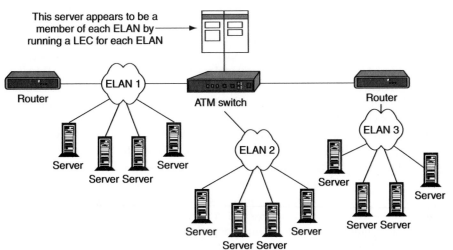

This server appears to be a member of each ELAN by running a LEC for each ELAN

Router

ELAN 1

ATM switch

Router

Server

Server Server

Server

ELAN 2

ELAN 3

Server

Server

Server Server

Server

Server Server

Server

FIGURE 15.29

Multihomed servers in an ATM network. (Reproduced with permission of Cisco Systems, Inc. Copyright © 2001 Cisco Systems, Inc. All rights reserved.)

Virtual Trunk Protocol (VTP)

Virtual Trunk Protocol establishes a virtual terminal connection across a network.

Virtual Trunk Protocol (VTP) is a vendor proprietary protocol and is the industry's first protocol implementation specifically designed for large VLAN deployments. VTP enhances VLAN deployment by providing the following:

- Integration of ISL, 802.10, and ATM LAN based VLANs
- Autointelligence within the switches for configuring VLANs
- Configuring consistency across the network
- An automapping scheme for going across mixed-media backbones
- Accurate tracking and monitoring of VLANs
- Dynamic reporting of added VLANs across the network
- Plug-and-play setup and configuration when adding new VLANs

Summary

In this chapter you learned the aspects of network design for routed and switched networks. In order to design a state-of-the-art network you should follow the design principles listed below.

General Network Design Principles

Good network design is based on many concepts that are summarized by the following key principles:

- Examine single points of failure carefully: There should be redundancy in the network so that a single failure does not isolate any portion of the network. There are two aspects of redundancy that need to be considered: backup and load balancing. In the event of a failure in the network, there should be an alternative or backup path. Load balancing occurs when two or more paths to a destination exist and can be utilized depending on the network load. The level of redundancy required in a particular network varies from network to network.
- Characterize application and protocol traffic: For example, the flow of application data will profile client-server interaction and is crucial for efficient resource allocation, such as the number of clients using a particular server or the number of client workstations on a segment.
- Analyze bandwidth availability: For example, there should not be an order of magnitude difference between the different layers of the hierarchical model. It is important to remember that the hierarchical model refers to conceptual layers that provide functionality. The actual demarcation between layers does not have to be a physical link. It can be the backplane of a particular device.
- Build networks using a hierarchical or modular model: The hierarchy allows autonomous segments to be internetworked together.

The best way to design a network is through practice. Instead of the normal review and summary questions, there is a practical hands-on case study to give experience in network design. There are no right or wrong answers. Take the concepts learned from this chapter and put them to use.

Case Study

Grobet File Corporation of America

The case study involves a manufacturing company, Grobet File Corporation of America.

The Client

Mr. Sidney Witzer of Grobet File Corporation is responsible for updating the network to use new technologies. Although he has a broad understanding of the options available to him, he needs your help to plan a good network design.

Company Background

Grobet is a manufacturer of high speed carbide steel tools and also makes brushes for jeweler and dental instruments. Although Grobet operates as an independent business, Grobet is owned by The Bank of Boston. It has three locations. The manufacturing facility consisting of 600 employees is located in Charlestown, New Hampshire. The main headquarters is located in Hamburg, Germany, and consists of human resources, payroll, marketing, and other key departments totaling 1,000 employees. The third facility employing 350 people is located in Carlstadt, New Jersey. This is the warehouse/distribution center, which is used primarily for filling orders and shipping products to customers. In addition, there are three satellite sales offices located in New York, Chicago, and Los Angeles.

Applications

Administration, production, and support of the company's products and services are accomplished with LAN based applications. Manufacturing products consist of a CRT based ordering system. At each facility is a mainframe computer designed to perform remote job entry. The manufacturing and inventory data are collected and maintained on an IBM AS400 system. The IBM system, TN3270 terminals, and PCs are connected to a single Token Ring network. Grobet has standardized on Microsoft Office applications and Microsoft Exchange for internal e-mail and, therefore, will use the SMTP Connector for SMTP mail to Germany and the Internet. The accounting function is maintained by Quark Express and a customized SQL Server application has been developed in-house for order processing and shipping and receiving functions.

Departmental Servers

Each of the departments (Sales and Marketing, Production, Shipping and Receiving, Accounting, Finance, and Human Resources) will have its own Windows NT file and print server, which means adding three servers because Sales and Marketing and Distribution share one server and the remaining departments share a second server. The IS staff is divided over where to locate the servers and wants a recommendation on the physical location of the servers.

Clients and Terminals

The company now wants to convert the 73 TN3270 terminals to PCs or network computers (NCs). In all, 800 PCs will be on the internetwork spread out between the various locations. Approximately 300 will be dialing in remotely to access the corporate LAN in Germany. Ninety-seven terminals will need access to the AS400 system via Open Connect TN3270 emulation. The warehouse has 10 terminals that will be replaced with PCs or NCs.

Existing Network

The company has leased line connections between all three sites and satellite offices and the warehouse. A T1/E1 point-to-point circuit is maintained between Grobet in Charlestown, New Hampshire, and Hamburg, Germany. This circuit is part of a much larger WAN owned and maintained by The Bank of Boston. SNA is the only traffic on this line. The new PCs and NCs will create a great deal of terminal emulation traffic over the LAN. The Token Ring environment will be completely replaced with Ethernet. Logistics issues require that the mainframe remain up until such time that the entire staff is moved and is operational.

The new routed traffic will be shared with SNA traffic using a load sharing CSU with a built-in multiplexer so that SNA traffic is not routed and is maintained separate from the routed traffic. Only IP will be allowed to be routed over the WAN. The Bank of Boston's Network Operations personnel will be responsible for all WAN functions from Grobet's FEP and router outward.

Goals of the Network

The IS manager has asked for an internetwork design to connect all the systems within the existing sites and The Bank of Boston's high end routers located in Boston. The Bank of Boston's Network Operations personnel will be responsible for the router purchase. The Network Operations group has specified OSPF as the routing protocol used by the corporate router.

Cabling Intermediate Distribution Facilities are located in the center of each floor with a Main Distribution Facility located in the computer room. The computer room is located on the first floor and takes up half of that floor. The IS department is located in the remaining half of the first floor. Good luck in your network design.

ATM

Objectives

- Explain the benefits of ATM.
- Describe what a connection oriented network is.
- Explain the ATM cell size and header format.
- Define the ATM adaption layers.
- Discuss and explain ATM multicasting.
- Define and explain permanent virtual circuits and switched virtual circuits.
- Explain ATM signaling and addressing.
- Define ATM and its relationship to the OSI reference model.
- Explain the ATM routing protocols.
- Explain ATM traffic management and QoS parameters.
- Define LAN emulation (LANE).

Key Terms

asynchronous transfer mode
(ATM), 680
ATM adaption layer (AAL), 685
available bit rate (ABR), 704
available cell rate, 705
Broadband Integrated Services Digital
Networks, Inc. (B-ISDN), 696
cell delay variation tolerance
(CDVT), 708

cell loss priority, 683
connection oriented network service
(CONS), 694
constant bit rate (CBR), 704
Interim Local Management Interface
(ILMI), 693
maximum cell transfer delay
(max CTD), 709
minimum cell rate (MCR), 708

Introduction

Asynchronous transfer mode delivers important advantages over existing LAN and WAN technologies.

Asynchronous transfer mode (ATM) technology will play a central role in the evolution of current workgroup, campus, and enterprise networks in the very near future. ATM delivers important advantages over existing LAN and WAN technologies, including the promise of scalable bandwidths at unprecedented price and performance points and quality of service guarantees that facilitate new classes of applications such as multimedia.

Of course, these benefits come at a price. Contrary to common misconceptions, ATM is a very complex technology—perhaps the most complex technology developed to date by the networking industry. The structures of ATM cells and cell switching do facilitate the development of hardware intensive, high performance ATM switches. Furthermore, the deployment of ATM networks requires the overlay of a highly complex, software intensive, protocol infrastructure. This infrastructure is required to allow individual ATM switches to be linked to a network, and for such networks to inter-network with the vast installed base of existing local and wide area networks.

Benefits of ATM

ATM has several key benefits.

1. ATM brings the benefit of one network. ATM provides a single network for all traffic types including voice, data, and video. ATM allows for the integration of networks and thus improves efficiency and manageability.
2. ATM enables new applications. Due to ATM's high speed and the integration of traffic types, ATM will enable the creation and expansion of new applications such as multimedia to the desktop.
3. ATM has the benefit of compatibility. Because ATM is not based on a specific type of physical transport, ATM is compatible with currently deployed physical networks. ATM can be transported over twisted-pair wires, coaxial cable, and fiber optics.
4. There can be an incremental migration to ATM. Efforts within the standards organizations and the ATM Forum continue to ensure that embedded networks will be able to gain the benefits of ATM incrementally, thereby upgrading portions of the network based on new application requirements and business needs.

5. ATM creates simplified network management. ATM is evolving into a standard technology for local, campus/backbone, and public and private wide area services. This uniformity is intended to simplify network management through the use of the same technology for all levels of the network.

6. ATM will have a long architectural lifetime. The information systems and telecommunications industries are focusing on and standardizing based on ATM. ATM has been designed from the outset to be scalable and flexible in geographic distance, number of users, access, and trunk bandwidths. Today, the speeds range from Mbps to Gbps.

This flexibility and scalability ensures that ATM will be around for a long time. ATM has moved from concept to reality. Products and services are available today. The ATM Forum has sponsored interoperability demonstrations to prove the technology, and it continues to meet to discuss the evolution of ATM. ATM coexists with current LAN/WAN technology. ATM specifications are being written to ensure that ATM smoothly integrates numerous existing network technologies at several levels such as Frame Relay, Ethernet, and TCP/IP. Equipment, services, and applications are available today and are being used in live networks. The telecommunications industry is converging on ATM.

Asynchronous transfer mode is an extremely high speed, low delay, multiplexing and switching technology that can support any type of user traffic including voice, data, and video applications. ATM is ideally suited to applications that cannot tolerate time delay, as well as to transport Frame Relay and IP traffic, which are characterized as bursty. Before ATM, separate networks were required to carry voice, data, and video information. The unique profiles of these traffic types make significantly different demands on network speeds and resources. Data traffic can tolerate delay, but voice and video cannot. With ATM, all of these traffic types can be transmitted or transported across one network because ATM can adapt the transmission of cells to the information generated.

ATM works by breaking information into fixed length 53-byte data cells. The cells are transported over traditional wire or fiber optic networks at extremely high speeds. Because information can be moved in small, standard sized cells, switching can take place in the hardware of a network, which is much faster than software switching. As a result, cells are routed through the network in a predictable manner with very low delay and with determined time intervals between cells. For network users, this predictability means extremely fast, real-time transmission of information on the same network infrastructure that supports non-real-time traffic.

ATM is a connection oriented protocol. This means that ATM must establish a logical connection to a defined endpoint before this connection can transport data. Cells on each port are assigned a path and a channel identifier that indicates the path or channel over which the cell is to be routed. The connections are called virtual paths or virtual channels. These connections can be permanently established or they can be set up and released dynamically depending upon the requirements of the user. A path can be an end-to-end connection in itself, or it can be a logical association or bundle of virtual channels. ATM creates a common way of transmitting any type of digital information from any other intelligent device over a system of networks. Many types of networks can be consolidated using one technology. This gives ATM a substantial advantage over other transport technologies.

ATM was designed for users and network providers who require guaranteed real-time transmission for voice, data, and images while also requiring efficient, high performance transport of bursty packet data. Hospitals are using ATM to share real-time video and images for long-distance consultation during diagnosis and operations. Schools are using ATM to bring students and instructors together, regardless of their locations. Corporations whose employees are in different locations benefit by using ATM to effortlessly share even the largest data files. The Internet runs on high speed ATM backbones. ATM is also proving to be an important technology for managing the demands placed on overburdened networks by the surge of Internet traffic. By adding a rich layer of traffic engineering to the IP, ATM supports the transportation of traffic according to priority level and class of service, neither of which are supported by IP alone. The result is a more effective means of ensuring the transmission of mission-critical traffic, a growing concern of service providers and businesses alike.

ATM Network Operation

A **user network interface** connects ATM end systems such as hosts and routers to an ATM switch.

A **network node interface** connects two ATM switches together.

An ATM network consists of a set of ATM switches interconnected by point-to-point ATM links or interfaces. ATM switches support two kinds of interfaces: *user network interfaces (UNIs)* and *network node interfaces (NNIs),* which are sometimes referred to as network-to-network interfaces. UNIs connect ATM end systems such as hosts and routers to an ATM switch, while an NNI may be imprecisely defined as an interface connecting two ATM switches together. Slightly different cell formats are defined across UNIs and NNIs. In NNI cells, unlike UNI cells, there is no generic flow control (GFC) field, and an expanded (12-bit) VPI field uses the first four bits of the cell. Since GFC is rarely used, there is no functional difference between UNI and NNI cells, other than the fact that NNI cells can support a larger VPI space. ATM does not have redundant physical links provided by FDDI, with its dual attached stations. Any end system requiring a redundant connection to an ATM network will need to support two separate UNIs and either operate one link in a standby mode or perform local connection level load sharing between the links. More precisely, an NNI interface is any physical or logical link across which two ATM switches exchange the NNI protocol. As a result, the connection between a private ATM switch and a public ATM switch is a UNI, which is referred to as a public UNI because these switches do not typically exchange NNI information.

The Cell

The basic idea of ATM is to segment data into small cells and then transfer them by the use of cell switching. Such cells have a uniform layout and a fixed size of 53 bytes, which greatly simplifies switching. Being more complex, packet switching is not nearly fast enough to be of use for isochronous data, for example, real-time video and sound. Cell switching gives maximum utilization of the physical resources.

The basic principles of this technology were first formulated by AT&T and the French Telecompany in the first half of the 1980s. Researchers found that if data were segmented into small, fixed size cells, switching could be done with simple, specially designed ICs. With sufficient intelligence to handle routing information in each cell, such ICs could be used to build very fast switch-matrixes. By connecting several switch-matrixes, highly efficient networks with small and predictable transmission delays were built. The fact that ATM can be used efficiently in both WANs and LANs shows how powerful ATM is.

Cell Size

An ATM cell contains 5 bytes of header (for administrative information) and 48 bytes of payload (data), for a total of 53 bytes. The choice of the cells caused much controversy during CCITT standardization. The U.S. computer branch wanted 64 bytes on payload, because they were considering the bandwidth utilization for data networks, and an efficient memory transfer length of payload should be a power of 2 or at least a multiple of 4. Sixty-four bytes would have fit both requirements.

Representatives from the European and Japanese telephony branch were taking voice applications into consideration and wanted smaller cells. At cell sizes greater than 151 bytes, there is a talker echo problem. Cell sizes between 32 and 152 bytes result in listener echo. Cell sizes less than or equal to 32 overcome both problems, under ideal conditions. European telecommunications companies had no desire to invest money in echo canceling equipment, and thus they advocated 32 bytes of payload. Because of the great distances involved in the United States, U.S. telecommunications companies had already installed echo cancelers across the country and didn't consider this to be a problem. After a lot of discussion, a value of 48 bytes of payload was agreed upon. As far as the header size, 10 percent of payload was perceived as an upper bound on the acceptable overhead, so a value of 5 bytes was chosen. Although the header of 5 bytes is a lot for a 53-byte cell, the flexibility is worth the cost.

Cell Header Format

In the cell header format, GFC: generic flow control; VPI/VCI: 0/0 implies idle cell; 0/n implies signaling; VCI: virtual channel identifier; VPI: virtual path identifier; HEC: header error correction, calculated as $1 + x + x^2 + x^8$; PTI: payload type indicator; and CLP: *cell loss priority*. The CLP indicates whether the cell should have high or low priority in case of overload.

Figure 16.1 illustrates the header consisting of 5 bytes that contain 24 bits of VCI label, 8 bits of control field, and 8 bits of checksum. The 48 bytes of payload may

Cell loss priority indicates whether the cell should have high or low priority in case of overload.

FIGURE 16.1
ATM cell header format.

GFC/VPI		VPI	
VPI		VCI	
VCI			
VCI	PTI		CLP
Payload			

optionally contain a 4-byte ATM adaption layer and 44 bytes of actual data, or all 48 bytes may be data. This depends on a bit in the control field of the header. This enables fragmentation, and reassembly, of cells into larger packets at the source and destination, respectively. The control field may also contain a bit to specify whether there is a flow control cell or an ordinary cell, and an advisory bit to indicate whether a cell is droppable in the face of congestion in the network or not. There is no sync character in the ATM cell. In order to stay in sync with the data stream, the receiving host calculates the checksum for everything received from the wire. When the checksum has been computed correctly for 16 subsequent cells, it is reasonable to assume that the host is in sync.

Virtual Paths and Channels

The **virtual path identifier** is a number between 1 and 255.

The **virtual channel identifier** is a 16-bit field in the header of an ATM cell.

The **quality of service** specifies which requirements are wanted for a connection. It is given when a virtual path is established.

In the cells, there are 5 bytes for addressing and other control information. With only 5 bytes, it is not desirable, because of performance loss, to use the addressing mechanisms found in Ethernet. Instead, the concepts of virtual channel and virtual path are introduced into ATM. Before two endpoints can communicate with each other, a logical or virtual path must be established between them. When the path is established, many applications can communicate independently through it, using virtual channels. Such connections are made between all ATM switches and equipment. The first 8 bits of the VCI label is the *virtual path identifier (VPI)*, which is a number between 1 and 255, and the last 16 bits is the *virtual channel identifier (VCI)*, which means there can be 65,535 active channels.

It is necessary to be able to specify which requirements are wanted for a connection, such as traffic type, peak and average bandwidth requirements, delay and cell loss requirements, and how much money to spend. This is called the *quality of service (QoS)* and it is given when a virtual path is established. All the virtual channels in a path have the same QoS, which simplifies the switching of the cells through the network. It is also possible to establish more than one virtual path between endpoints if there is a need for different levels of quality between the same endpoints.

ATM Switches

An ATM switch is a switching-matrix built by simple binary switching elements. The cells go through the switching-matrix at high speed controlled by a time sharing mechanism that synchronizes and manages the switch. The technology is scalable, in the sense that when needed, more of the simple switching elements can be added. An ATM switch receives traffic from many different users and then multiplexes the cells to the next switch over one or more physical connections. When a switch receives a cell, the switch looks at the destination address to see where to send the cell. A connection has already been set up between the station that sent the cell and the station in the destination field. When several stations want to use the available bandwidth, as when two stations (A and B) want to communicate with another station (C) simultaneously, they have to share the bandwidth. How the priority of the traffic is assigned depends on the configurations of the switches. When too much bandwidth is requested, cells are discarded. Whether cells should be buffered depends on the type of the cells, as it does not make sense to buffer cells that are going to be obsolete when they finally arrive, as in the case of real-time sound.

ATM Adaption Layers

Two watchwords of ATM are configurability and flexibility. One of the primary goals of ATM is that it should support different kinds of services with different traffic characteristics. The different services in ATM have been split into four classes. Classes A and B are for time critical applications such as speech, sound, and video. In these classes, variable delay cannot be tolerated. Class C is for the typical data channel between two endpoints. Timing is not important, but cell loss is unacceptable. Class D simulates connectionless data transmission, which is typical of LANs.

By definition, it is impossible to transmit connectionless data in ATM, but it is easy to simulate, as is done with classical IP over ATM. For services that require correct timing rather than correctness of data, such as Class A voice transmission, it may make sense to compensate for cell loss by using previously received information. For services that require correctness of data, such as Class C computer communication, it makes sense to retransmit data in the case of cell loss. Different services have different timing and delay requirements. For audio and video purposes, it is of paramount importance that cells are dispatched quickly, rather than being buffered in ATM switches.

ATM adaption layers (AALs) map applications onto ATM cell services. These AALs specify how data are encapsulated into cells and how data are reconstructed from cells for higher layer services. A separate AAL may be defined for each type of service provided such as voice, video, and data.

An AAL is generally composed of two primary components: the segmentation and reassembly (SAR) sublayer and the convergence sublayer. The SAR accepts data from a higher layer and segments that data into cells for transmission over the ATM

ATM adaption layers map applications onto ATM cell services.

network. When receiving data, the SAR must demultiplex incoming cells and re-assemble them into their original form. This may be a complex task since cells from many sources can be interleaved. Therefore, the SAR may be simultaneously building several data buffers from multiple source incoming cells.

The convergence sublayer of an AAL is responsible for providing a service interface for higher layer applications or protocols. In traditional data networks, the convergence sublayer will provide ATM driver level services to a protocol stack. The convergence sublayer also acts as the communications vehicle between the SAR sublayer and the higher level application or protocols. The ATM layer provides cell services for ATM switches and end stations. This layer controls the management and transmission of individual cells based upon VCI/VPI and the services provided to the AAL. The ATM layer has cell routing responsibility, requiring it to direct cells to the appropriate media on an end station or switch. Thus, the ATM layer will interact directly with driver level services on connected media. ATM is currently defined to run over DS3 and E3, over multimode and single mode fiber optic, and over STP and UTP using SONET/SDH framing.

Currently, four distinct adaption layers have been defined for ATM networks. These adaption layers are known as AAL1, AAL2, AAL3/4, and AAL5.

AAL1 provides a fixed data rate, low latency, and jitter service. AAL1 will be used by carriers to transport voice data over the public ATM infrastructure as it evolves. AAL1 may also be used for higher bandwidth real-time services such as video teleconferencing. AAL1 provides fixed rate, connection oriented service.

AAL2 is a variable data rate, connection oriented, low latency service. Its intended use is real-time applications such as high definition television (HDTV).

AAL3/4 is generally used to provide ATM transport for computer or network data. The most common implementations using AAL3/4 provide transport of Frame Relay and switched multimegabit data services (SMDS) over ATM. Frame Relay is commonly used today to minimize POP/POI connections in private WANs. Transport of Frame Relay over ATM will minimize the cost of ATM transition for current Frame Relay users. SMDS provides a function similar to Frame Relay, but it is generally associated with a public data internetwork. Subscribers may connect to private networks in unrelated organizations, as with a telephone call, if both parties agree to do so. Frame Relay networks, in contrast, are entirely private and cannot be accessed by organizations that have not subscribed to a particular frame relay network.

A new adaption layer, AAL5, also referred to as the simple and efficient adaptation layer (SEAL), has been adopted as a method of connecting data oriented and LAN oriented systems such as workstations. AAL5 has some advantages over AAL3/4 for computer data transport, with support for large datagrams up to 64 K as opposed to 9 K with SMDS, and 10 percent less overhead than AAL3/4. AAL5 is well suited to the workstation ATM market for which it was designed.

Table 16.1 ATM Adaption Layers

Layer	Description	Bit Rate	Type	Class
AAL1	ATM adaption layer supports connection oriented services that require constant bit rates and have specific timing and delay requirements. Typically this translates to uncompressed digitized voice or video, which consists of a constant stream of data.	Constant	Connection oriented	A
AAL2	ATM adaption layer 2 supports connection oriented services that do not require constant bit rates but have timing and delay requirements. It is targeted on compressed video or sound.	Variable	Connection oriented and connectionless	B
AAL3/4	ATM adaption layer 3/4 is intended for both connectionless and connection oriented services with variable bit rates. AAL3/4 was originally two separate adaption layers that have since been merged into a single AAL.	Variable	Connection oriented and connectionless	C/D
AAL5	ATM adaption layer 5 supports connection oriented services that require variable bit rates. Typically this translates to computer communication, which is characterized by short bursts of data transfer. Delay and timing are not crucial. AAL5 is a simpler AAL than AAL 3/4, at the expense of error correction and automatic retransmission, but pays off with less bandwidth overhead and reduced implementation complexity. AAL5 can be implemented in silicon.	Variable	Connection oriented and connectionless	C/D

Overall, as shown in Table 16.1, ATM adaption layers provide an organized method of supporting multiple services over a common network infrastructure. ATM adaption layers may continue to be defined as unanticipated services become required. The design of ATM includes multiple provisions, such as flexible AALs, to accommodate these new services. Table 16.1 gives a distinctive description of each of the ATM adaption layers.

Multicasting in ATM

ATM networks are fundamentally connection oriented. Connection oriented means that a virtual circuit needs to be set up across the ATM network prior to any data transfer. ATM circuits are of two types: virtual paths, which are identified by a virtual path identifier; and virtual channels, identified by the combination of a VPI and

FIGURE 16.2

ATM switch operation. (Adapted from Stratacom Frame Relay Course on IPX Networks, p. 10, © 1999, by permission of Cisco Systems, Inc.)

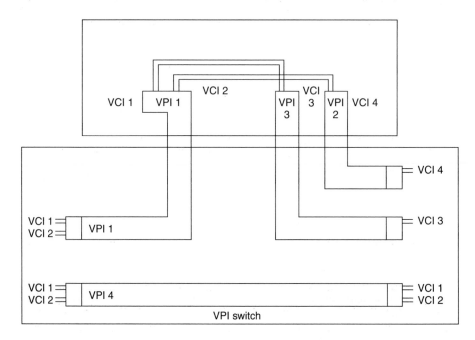

Input		Output	
Port	VPI/VCI	Port	VPI/VCI
1	29	2	45
2	45	1	29
1	64	3	29
3	29	1	64

a virtual channel identifier. A virtual path is a bundle of virtual channels, as illustrated in Figure 16.2, all of which are switched transparently across the ATM network on the basis of common VPI. All VCI and VPI have only local significance across a particular link and are remapped at each switch. In normal operation, switches allocate all UNI connections within VPI = 0.

As illustrated in Figure 16.3, the basic operation of an ATM switch is very simple as it receives a cell across a link on a known VCI or VPI value; looks up the connection value in a local translation table to determine the outgoing port or ports of the connection and the new VPI/VCI value of the connection on that link; and then retransmits the cell on that outgoing link with the appropriate connection identifiers.

FIGURE 16.3

Virtual circuit and virtual path switching. (Adapted from Stratacom Frame Relay Course on IPX Networks, p.104, © 1999, by permission of Cisco Systems, Inc.)

The switch operation is simple because external mechanisms set up the local translation tables prior to the translation of any data. The manner in which these tables are set up determine the two fundamental types of ATM connections.

- A PVC is a connection set up by some external mechanism, typically network management, in which a set of switches between an ATM source and destination ATM system are programmed with the appropriate VPI/VCI values. ATM signaling can facilitate the setup of PVCs, but by definition, PVCs always require some manual configuration.
- An SVC is a connection that is set up automatically through a signaling protocol. SVCs do not require the manual interaction needed to set up PVCs, and thus, SVCs are likely to be much more widely used. Higher layer protocols operating over ATM primarily use SVCs.

An ATM end system that is attempting to set up a connection through an ATM network initiates this through ATM signaling. Signaling packets are sent on a well-known virtual channel, VPI = 0, VCI = 5. This means that this virtual channel is reserved for signaling traffic, and no other types of information may be transmitted across the connection. All switches are also preconfigured to receive any signaling packets sent across this connection and pass them to a signaling process associated with the switch. All VCIs below 32 are reserved within each VPI for control purposes; data connections are allocated through VCIs outside this range.

Figure 16.4 illustrates how ATM signaling is routed through the network, from switch to switch, setting up the connection identifiers as it travels until it reaches the

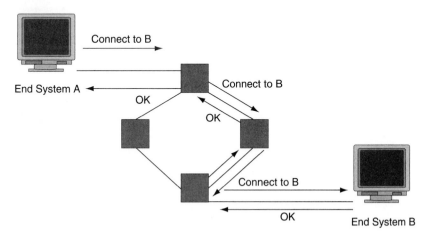

FIGURE 16.4

Connection setup through ATM signaling. (Adapted from Stratacom Frame Relay Course on IPX Networks, p. 21, © 1999, by permission of Cisco Systems, Inc.)

Signaling request
Connection routed—setup path
Connection accepted/rejected
Data flow—along same path
Connection tear down

destination end system. The virtual channel is reserved for signaling traffic, and no other types of information may be transmitted across the connection. All switches are also preconfigured to receive any signaling packets sent across this connection and pass them to a signaling process associated with the switch. The signaling requests are passed between the signaling or call control processes associated with the switches, and it is these that set up the connection through the switches. For the sake of robustness and performance, most vendors will integrate the call control capability into each switch, rather than supporting them on an off-board processor. The connection identifiers, that is, VPI/VCI values for a particular connection, are typically allocated, across any given link, by the node to which the request is sent, as opposed to the requesting node. Connection identifiers with the same VPI/VCI values are always allocated in each direction of a connection, but the traffic parameters in each direction can be different. The end system can either accept and confirm the connection request or reject it, clearing the connection. Because the connection is set up along the path of the connection request, the data flow along this same path.

There are two fundamental types of ATM connections:

- Point-to-point connections, illustrated in Figure 16.5, connect two ATM end systems. Such connections can be unidirectional or bidirectional.
- Point-to-multipoint connections connect a single source end system, which is referred to as the root node, to multiple destination end systems, referred to

FIGURE 16.5

Types of ATM connections. (Adapted from Stratacom Frame Relay Course on IPX Networks, p. 23, © 1999, by permission of Cisco Systems, Inc.)

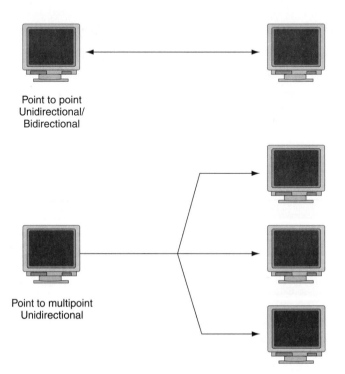

Point to point
Unidirectional/
Bidirectional

Point to multipoint
Unidirectional

as leaves. The ATM does cell replication within the network switches at which the connection splits into two or more branches. Such connections are unidirectional, permitting the root to transmit to the leaves, but not the leaves to transmit to the root or to each other on the same connection.

End systems could also replicate cells and send them to multiple end systems across multiple point-to-point links, but ATM switches can perform replication much more efficiently than end systems.

What is notably missing from these types of ATM connections in Figure 16.5 is an analog to the multicasting or broadcasting capability common in many shared medium LAN technologies such as Ethernet and Token Ring. Broadcasting is a single system transmitting to all other systems. In such technologies, multicasting allows multiple end systems to both receive data from other multiple systems and transmit data to these multiple systems. Such capabilities are easy to implement in shared media technologies such as LANs, where all nodes on a single LAN segment must necessarily process all packets sent on that segment. The analog in ATM to a multicast LAN group would be a bidirectional multipoint-to-multipoint connection. However, this solution cannot be implemented when using AAL5, the most common AAL used to transmit data across ATM networks.

Unlike AAL3/4, with its message identifier (MID) field, AAL5 does not have any provision within its cell format for the interleaving of cells from different AAL5 packets on a single connection. What this means is that all AAL5 packets sent to a particular destination across a particular connection must be received in sequence, with no interleaving between the cells of different packets on the same connection, or the destination reassembly process would not be able to reconstruct the packets. Despite the problems that AAL5 has with multicast support, it is not really feasible to use AAL3/4 for data transport. This is because AAL3/4 is a much more complex protocol than AAL5 and would lead to much more complex and expensive implementations. AAL5, however, was developed to replace AAL3/4. While the MID field of AAL3/4 could preclude cell interleaving problems, allowing for bidirectional, multipoint-to-multipoint connections, it would also require some mechanism for ensuring that all nodes in the connection use a unique MID value. There is no such mechanism currently in existence or development. The number of possible nodes within a given multicast group would also be severely limited due to the small size of the MID space.

This is why AAL5 point-to-multipoint connections can only be unidirectional, for if a leaf node were to transmit an AAL5 packet onto the connection, it would be received by the root node and all other leaf nodes. At these nodes, the packet sent by the leaf could well be interleaved with packets sent by the root, and possibly other leaf nodes, which would preclude the reassembly of any of the interleaved packets, which is not acceptable. Aside from this problem, ATM does require some form of multicast capability since most existing protocols, being developed initially for LAN technologies, rely upon the existence of a low level multicast/broadcast facility. Three methods have been proposed to solve this problem.

FIGURE 16.6

*Multicast server
operation.*

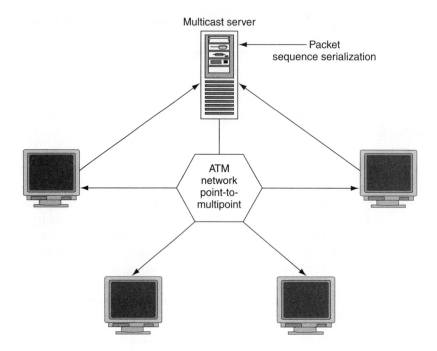

VP Multicasting: In this mechanism, a multipoint-to-multipoint VP links all
nodes in the multicast group, and each node is given a unique VCI value within the
VP. Interleaved packets can be identified by the unique VCI value of the source. This
mechanism requires a protocol to uniquely allocate VCI values to nodes. Such a
mechanism does not currently exist. It is also not clear whether current SAR devices
could easily support such a mode of operation. Furthermore, there is no support for
switched virtual paths in the existing signaling specifications. This capability will be
added to the UNI 4.0 signaling protocols currently under development.

Multicast Server: The multicast server is illustrated in Figure 16.6. The figure
shows that in this mechanism, all nodes wishing to transmit onto a multicast group
set up a point-to-point connection with an external device known as a multicast
server. The multicast server, in turn, is connected to all nodes wishing to receive the
multicast packets through a point-to-multipoint connection. The multicast server
could also connect to each of the destinations using point-to-point connections and
replicate the packets before transmission. In general, ATM networks can perform
replication through point-to-multipoint connections much more efficiently than
through a multicast server.

Overlaid Point-to-Multipoint Connections: In this mechanism, illustrated in
Figure 16.7, all nodes in the multicast group establish a point-to-multipoint connection with every other node in the group, and, in turn, each node becomes a leaf in

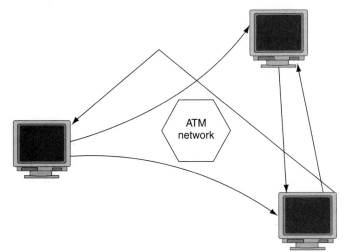

FIGURE 16.7
Multicast through overlaid point-to-multipoint connections.

the equivalent connections of all other nodes. Thus, all nodes can both transmit to and receive from all other nodes. This mechanism requires each node to maintain N connections for each group, where N is the total number of transmitting nodes within the group, while the multicast server mechanism requires only two connections. The overlaid mechanism also requires a registration process for telling nodes that join a group what the other nodes in the group are, so that it can form its own point-to-multipoint connection. The other nodes also need to know about the new node so they can add the new node to their own point-to-multipoint connections. The multicast server mechanism is more scalable in terms of connection resources, but it has the problem of requiring a centralized resequencer, which is both a potential bottleneck and a potential single point of failure.

In short, there is as yet no ideal solution within ATM for multicast. Higher layer protocols within ATM networks use both multicast server operations and multicast through overlaid point-to-multipoint connections for multicast. This problem is just one example of why internetworking under existing protocols with ATM is so complex. Most current protocols, particularly those developed for LANs, implicitly assume a network infrastructure very similar to existing LAN technologies, which is a shared-medium, connectionless technology with implicit broadcast mechanisms. ATM violates all of these assumptions.

ATM does contain an *Interim Local Management Interface (ILMI)* protocol. The ILMI protocol uses SNMP format packets across the UNI and also across NNI links to access an ILMI management information base (MIB) associated with the link, within each node. The ILMI protocol is run across a well-known virtual channel, VPI = 0, VCI = 16. The ILMI protocol allows adjacent nodes to determine various characteristics of other nodes, for example, the size of each other's connection space, the type of signaling used, and hooks for network management autodiscovery. One

The **Interim Local Management Interface** is developed by the ATM Forum for incorporating network management capabilities into the ATM UNI.

of its most useful features, address registration, greatly facilitates the administration of ATM addresses. The ILMI protocol will likely be extended in the future to support other autoconfiguration capabilities, such as group addressing.

Connection Oriented Network Services

Connection oriented network services require the establishment of a connection between the source and the destination before data are transmitted.

Connection oriented network services (CONSs) require the establishment of a connection between the source and the destination before data are transmitted. The connection is established as a single path of one or multiple transmission links through intermediate nodes in a network. Once the path is established, all data travel over the same predetermined path throughout the network. The fact that data arrive at the destination node in the same order as sent by the source node is fundamental to connection oriented services.

If network management or provisioning actions establish the connection and leave the connection operational indefinitely, then the connection is referred to as a permanent virtual circuit (PVC). If control signaling of any type dynamically establishes and tears down the connection, then the connection is referred to as a switched virtual circuit (SVC).

A PVC connection may be established by physical wiring, equipment configuration commands, service provider provisioning procedures, or any combination of these actions. These actions may take several minutes to several weeks, depending upon exactly what is required. Once the PVC is established, then data may be transmitted over the connection. Usually, PVCs are established for long periods of time. Examples of PVCs are analog private lines, DTE-to-DCE connections, and digital private lines. Examples of logical PVCs are the X.25 PVC, the Frame Relay PVC, and the ATM PVC.

In the case of an SVC service, only the access line and address for the source node and each destination endpoint are provisioned ahead of time. The use of a control signaling protocol plays a central role in SVC services. Via the signaling protocol, the source node requests that the network make a connection to a destination. The network determines the physical and logical location of the destination node and attempts to establish the connection through intermediary nodes to the destination. The success or failure of the attempt is indicated to the originator. There may be a progress indication to the source, alerting the destination or other handshaking elements of the signaling protocol as well. Often the destination utilizes signaling to either accept or reject the call. In the case of a failed attempt, the signaling protocol usually informs the source node of the reason that the attempt failed. Once the connection is established, the transfer of data can then take place.

Networks employ SVCs to efficiently share resources by providing dynamic connections and disconnections in response to signaling protocol instructions generated by end users. End users favor the use of SVCs as they can dynamically allocate expensive bandwidth resources without a prior reservation. The simplest way to explain an SVC is to compare it to the traditional telephone call. After ordering the service and receiving address assignments, the communication device picks up the phone and requests a connection to a destination address. The network establishes the call or rejects it with a busy signal. After call establishment, the connected devices send data until one of the parties takes the call down.

ATM Signaling and Addressing

The current and planned ATM signaling protocols and their associated ATM addressing models will be discussed in this section. ATM signaling protocols vary by the type of ATM link. ATM UNI signaling is used between an ATM end system and an ATM switch across an ATM UNI. ATM NNI signaling is used across NNI links. At present, standards exist only for UNI signaling, and work is continuing on NNI signaling. The current standard for ATM UNI signaling is described in the ATM Forum UNI 3.1 specification titled Forum 1, which is a slight modification of the earlier UNI 3.0 specification. The only substantive difference between UNI 3.0 and UNI 3.1 is in the data link service specific convergence protocol (SSCOP) used for the reliable transport of the ATM signaling packets. UNI 3.1 brought the ATM Forum signaling specification into alignment with the ITU-T's Q.2931 signaling protocol stack. There are no functional differences between UNI 3.0 and UNI 3.1, but unfortunately, the two are not interoperable due to the differences in the data link protocol. UNI signaling requests are carried across the UNI in a well-known default connection: VPI = 0, VCI = 5.

The UNI 3.1 specification is based upon Q.2931, a public network signaling protocol developed by the ITU-T, which in turn was based on the Q.931 signaling protocol used with *Narrowband ISDN (N-ISDN)*. The ATM signaling protocols run on top of an SSCOP defined by the ITU-T recommendations Q.2100, Q.2110, and Q.2130; SSCOP is a data link protocol that guarantees delivery through the use of windows and retransmissions.

> **Narrowband ISDN** is a communication standard developed by the ITU-T for baseband networks.

ATM signaling uses the "one-pass" method of connection setup, which is the model used in all common telecommunications networks. The one-pass method is a connection request from the source end system that is propagated through the network, setting up the connection as it travels until it reaches the final destination end system. The routing of the connection request and any subsequent data flow is governed by the ATM routing protocols. Such protocols route the connection request based upon both the destination address and the traffic and QoS parameters requested by the source end system. The destination end system may choose to accept or reject the connection request. But since the call routing is based purely on the parameters in the initial connection request message, the scope of connection parameter negotiation between source and destination that may in turn affect the connection routing is limited.

A number of message types are defined in the UNI 3.0/3.1 specification, together with a number of state machines defining the operation of the protocol, and cause error codes defining reasons for connection failure. Data elements used in the signaling protocol addresses are carried within information elements within the signaling packets.

A source end system wishing to set up a connection will formulate and send into the network, across its UNI, a setup message containing the destination end system address, desired traffic and QoS parameters, and various IEs defining particular desired higher layer protocol bindings. This setup message is sent to the first ingress switch across the UNI, which responds with a local call proceeding acknowledgment. The switch will then invoke an ATM routing protocol to propagate the signaling request across the network to the egress switch, to which is attached the destination end system.

This egress switch will then forward the setup message to the end system across its UNI. The UNI may choose to either accept or reject the connection request. If the UNI accepts the connection request, it returns a connect message back to the network, along the same path, to the source end system. Once the source end system receives and acknowledges the connect message, either node can then start transmitting data on the connection. If the destination end system rejects the connection request, it returns a release message, which is also sent back to the source end system, clearing the connection with any connection identifiers as it proceeds. Release messages are also used by either of the end systems, or by the network, to clear an established connection.

The ATM Forum greatly simplified the Q.2931 protocol but also extended it to add support for point-to-multipoint connection setup. In particular, UNI 3.1 allows for a root node to set up a point-to-multipoint connection, and to subsequently add a leaf node. While a leaf node can autonomously leave such a connection, it cannot add itself.

The ATM Forum in 1995 standardized a new signaling capability, which was released as part of its UNI 4.0 specification Forum 3. UNI 4.0 added support for leaf initiated connections to a multipoint connection. Even though some would like to use UNI 4.0 to allow for a true multipoint-to-multipoint connection, signaling support for such connections does not imply the existence of a suitable mechanism for such connections.

The most important contribution of UNI 3.0/3.1 in terms of internetworking across ATM was its addressing structure. Any signaling protocol requires an addressing scheme to allow the signaling protocol to identify the sources and destination of connections. The ITU-T long ago settled upon the use of telephone number like E. 164 addresses as the addressing structure for public ATM *Broadband Integrated Services Digital Networks, Inc. (B-ISDN)*. Since E. 164 addresses are a public resource and cannot typically be used within private networks, the ATM Forum extended ATM addressing to include private networks. In developing such a private network addressing scheme for UNI 3.0/3.1, the ATM Forum evaluated two fundamentally different models for addressing.

Broadband Integrated Services Digital Networks, Inc. is a communication standard designed to handle high bandwidth applications.

These two models differed in the way in which the ATM protocol layer was viewed in relation to existing protocol layers, in particular, existing network layer protocols such as IP and IPX. These existing protocols all have their own addressing schemes and associated routing protocols. One proposal, known as the peer model, was to also use these same addressing schemes within ATM networks. Thus, existing network layer addresses, such as IP addresses, would identify ATM endpoints. ATM signaling requests would carry such addresses. Existing network layer routing protocols such as IGRP and OSPF would also be used within the ATM network to route the ATM signaling requests, since these requests, using existing network layer addresses, would essentially look like connectionless packets. The reason why this proposal was referred to as the peer model was that it essentially treats the ATM layer as a peer of existing network layers.

An alternate model sought to decouple the ATM layer from any existing protocol, defining for it an entirely new addressing structure. By implication, all existing protocols would operate over the ATM network. For this reason, the model is known as the subnetwork or overlay model. The subnetwork or overlay model is the manner in which such protocols as IP operate over such protocols as X.25 or over dial-up lines. The overlay model requires the definition of both a new addressing structure and an associated routing protocol. All ATM systems would need to be assigned an ATM address in addition to any higher layer protocol addresses it would also support. The ATM addressing space would be logically disjoint from the addressing space of whatever protocol would run over the ATM layer, and typically would not bear any relationship to it. All protocols operating over an ATM subnet would also require some form of ATM address resolution protocol to map higher layer addresses such as IP addresses to their corresponding ATM addresses.

The peer model does not require such address resolution protocols. By using existing routing protocols, the peer model also may have precluded the need for the development of a new ATM routing protocol. Nonetheless, it was the overlay model that was finally chosen by the ATM Forum for use with UNI 3.0/3.1 signaling. Among other reasons, the peer model, while simplifying end system address administration, greatly increases the complexity of ATM switches, since they must essentially act like multiprotocol routers and support address tables for all current protocols, as well as supporting all of their existing routing protocols. Current routing protocols, being originally developed for current LAN and WAN networks, also do not map well into ATM or allow use of ATM's unique QoS properties.

The overlay model, by decoupling ATM from other higher protocol layers, allows each to be developed independently of the other. This is very important from a practical engineering viewpoint. Both ATM and evolving higher layer protocols are extremely complex, and coupling their development would likely have slowed the deployment of ATM considerably. The price to pay for such layering is the need for disjoint address spaces and routing protocols, and possibly suboptimal end-to-end routing. However, the practical benefits arguably greatly exceed the theoretical costs. This may happen in large meshed networks consisting of both packet routers and ATM switches because the higher layer packet routing protocols operate independently of the ATM level routing protocol. Once a path is chosen that crosses the ATM network, a change in the topology or characteristics of the ATM layer would not become known to the higher layer routing protocol, even if that change would result in the choice of a different, more optimal, end-to-end path that bypasses the ATM network. While this is indeed a potential drawback of the overlay model, in practice it is unlikely to be a major problem since it is likely that in any practical network, the ATM network will always remain the preferred path.

Given the choice of the overlay model, the ATM Forum then defined an address format for private networks based on the syntax of an OSI network service access point (NSAP) address. An ATM address is not an NSAP, despite the similar structure.

FIGURE 16.8

Peer model of ATM addressing. (Adapted from Stratacom Frame Relay Course on IPX Networks, p. 29, © 1999, by permission of Cisco Systems, Inc.)

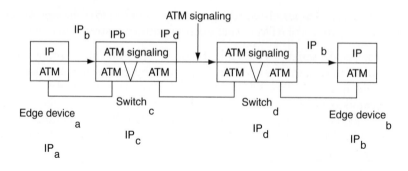

While in common usage such addresses are often referred to as NSAP addresses, they are better described as ATM private network addresses, or ATM endpoint identifiers; and they are often identified not as NSAPs, but as subnetwork points of attachment.

The 20-byte NSAP format illustrated in Figure 16.8 shows that ATM addresses are designed for use within private ATM networks, while public networks typically use E. 164 addresses that are formatted as defined by ITU-T. The Forum did specify an NSAP encoding for E. 164 addresses. This will be used for encoding E. 164 addresses within private networks but may also be used by some private networks. Such networks may base their own NSAP format addressing on the E. 164 address of the public UNI to which they are connected and take the address prefix from the E. 164 number to identify local nodes by the lower order bits.

All NSAP format ATM addresses illustrated in Figure 16.9 consist of three components: an authority and format identifier (AFI), which identifies the type and format of the initial domain identifier (IDI); the IDI, which identifies the address allocation and administration authority; and the domain specific part (DSP), which contains

FIGURE 16.9

Overlay model of ATM addressing. (Adapted from Stratacom Frame Relay Course on IPX Networks, p. 33, © 1999, by permission of Cisco Systems, Inc.)

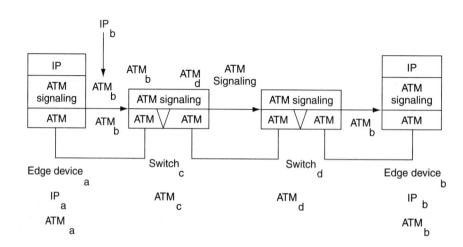

actual routing information. The Q.2931 protocol defines source and destination address fields for signaling requests, and also defines subaddress fields for each.

There are three formats of private ATM addressing that differ by the nature of the AFI and IDI. The first is the NSAP encoded E. 164 format, in which the IDI is an E. 164 number. The second is the DCC format, in which the IDI is a data country code. These identify particular countries, as specified in ISO 3166. The ISO national member body administers such addresses in each country. The third is the ICD format, in which the IDI is an international code designator (ICD). The ISO 6523 registration authority allocates ICDs. ICD codes identify particular international organizations.

The ATM Forum recommends that organizations or private network service providers use either the DCC or ICD formats to form their own numbering plans. Organizations that want to obtain ATM addresses would do so through the same mechanism used to obtain NSAP addresses. For example, in the U.S., this is the ANSI. Once obtained, such addresses can be used for ATM addresses and for NSAP addresses. If CLNP is run over ATM, the same value might well be used to identify a node's NSAP address and its ATM address.

The ESI field illustrated in Figure 16.10 is specified to be a 48-bit MAC address, as administered by the IEEE. This facilitates the support of LAN equipment, which is typically hardwired with such addresses, and of LAN protocols such as IPX, which rely on MAC addresses. The final one octet selector (SEL) field is meant to be used for local multiplexing within end stations and has no network significance.

To facilitate the administration and configuration of ATM addresses into ATM end systems across UNI, the ATM Forum defined an address registration mechanism using the ILMI. This allows an ATM end system to inform an ATM switch across the UNI of its unique MAC address and to receive the remainder of the node's full ATM address in return. This mechanism not only facilitates the autoconfiguration of a node's ATM addressing, but may also be extended, in the future, to allow for the autoconfiguration of other types of information such as higher layer addresses and server addresses.

The addressing formats defined in UNI 3.0/3.1 identify only single endpoints. These can also be used to set up point-to-multipoint connections because in UNI 3.0/3.1, such connections are set up a leaf at a time, using unicast addressing. UNI 4.0 adds support for group addresses and permits point-to-multipoint connections to be set up to multiple leaves in one request.

The notion of anycast addressing is supported in UNI 4.0. Similar to multicasting, in anycast, the destination is a group of addresses but the packet is delivered to one address instead of many addresses. A well-known anycast address that may be shared by multiple end systems is used to route a request to a node providing a particular node per se. A call made to an anycast address is routed to the nearest end system that registered itself with the network to provide the associated service. Anycast is a powerful mechanism for autoconfiguration and operation of networks since it precludes the need for manual configuration or service location protocols. While a few details

FIGURE 16.10
IEEE ESI field.
(Adapted from
IEEE Standards
document, p. 28,
© 1999, by per-
mission of IEEE.)

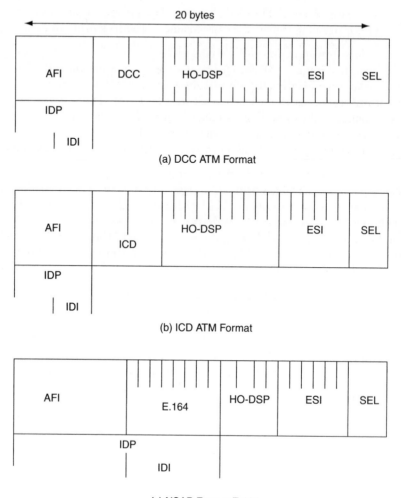

(a) DCC ATM Format

(b) ICD ATM Format

(c) NSAP Format E.164

IDI, initial domain identifier
ICD, international code designator
DSP, domain specific part
DCC, data country code

IDP, initial domain part
ESI, end system identifier
AFI, authority and format identifier

of ATM group addressing have been determined, the ATM Forum has decided that anycast will be addressed as a special case of group addressing.

Specifically, nodes will use an extension of the ILMI address registration mechanism to inform the network that they support a particular group address. As part of this registration, the node also informs the network of the desired scope of registration, which is the extent of the network to which the existence of the multicast node should be advertised. This scope is administrative within a single building, within the local site, or within the enterprise network. The network must map this information through administrative policy to the ATM routing protocol's own hierarchy. Once a node has registered its membership within a multicast group, other nodes may set up connections to these nodes.

If the requesting node initiates a point-to-multipoint connection to the group address, the network will connect all nodes that are registered with that particular ATM address. Conversely, if the requesting node specifies a point-to-point connection, the network will set up a connection to the nearest registered node. In this way, anycast can be supported as a special case of group addressing, and a new format is not required.

ATM and the OSI Model

An issue that often causes great confusion is which layer in the OSI seven layer model ATM corresponds to. The adoption of the overlay model by the ATM Forum sometimes causes some to describe ATM as a layer 2 protocol or a data link protocol, akin to a MAC protocol such as Ethernet or Token Ring. Yet others who note that ATM possesses most, if not all, of the characteristics of the network layer protocol such as IP or IPX often contest this description. These characteristics include a hierarchical address space and a complex routing protocol.

Much of the controversy arises from the limitations of the OSI model and from an incomplete understanding of the complexities of practical network operation. The basic OSI model did not incorporate the concept of overlay networks, where one network layer must overlie another, though such concepts were later added as addenda to the model. Such a model is often used when one type of network protocol must be carried transparently across another. For example, layer 3 protocols as IP and IPX are often carried transparently across other network layer protocols such as X.25 or the telephone network, since this is generally much simpler than attempting to interoperate the protocols through a protocol gateway. The ATM overlay model was chosen so as to separate and facilitate the engineering efforts involved in both. In this way, the ATM layer protocols were completed to operate with ATM. The overlay model also simplifies switch operation, at the arguable cost of redundancy in protocol functions and suboptimality in routing. The overlay model also leverages the existing installed application base and facilitates future application portability, since it builds upon and extends today's ubiquitous network layer protocol infrastructure. Such trade-offs were felt by the forum to be defensible but in no way to detract from the fact that ATM is a full fledged network layer protocol, which is at least as complex as any that exists today.

What makes ATM a network layer protocol is indeed the very complexity of its addressing and routing protocols, and this is independent of the fact that other network layer protocols are run over ATM. The LAN emulation (LANE) protocols actually operate a MAC layer protocol over ATM, but this does not make ATM a physical layer.

A related issue that also causes confusion is the notion of flat addressing and whether or not ATM can be used to build a simpler network than today's network layer protocol based routed internetworks. This issue, along with the layering issue, draws a correspondence between ATM and layer 2 MAC protocols. Layer 2 MAC protocols have a flat address space, which is 48-bit MAC addresses, and MAC layer internetworking devices such as MAC bridges do offer plug-and-play capabilities and do not require the complex configuration of layer 3 internetworking devices such as routers.

Since MAC addresses are indeed flat, meaning they have no logical hierarchy, packets must be flooded throughout the network using bridging protocols. While this requires no network configuration, it also greatly reduces the scalability and stability of such bridged networks. A hierarchical address space together with address assignment policies that minimize flat host routes permit the use of address aggregation, where reachability for entire sets of end systems can be summarized by a single address prefix or, equivalently, by subnet masks. Combined with a routing protocol that disseminates such address prefixes, hierarchical addressing precludes the need for flooding and greatly reduces the amount of reachability information that must be exchanged.

Protocols with hierarchical, aggregatable address spaces do indeed generally require more configuration for address and subnet assignment, but by the same token this very hierarchy permits the operation of routing protocols and thus the deployment of much more scalable and stable networks. Flat addressing, by definition, precludes routing and requires bridging, with a consequent lack of scalability.

Very few networks, outside of bridged LANs, actually have a truly flat address space. The telephone network, for instance, which is often thought of as a flat network, actually incorporates a very structured hierarchy within its address space to include country code, area code, and so on, and it is only this rigid hierarchy that has permitted the telephone network to scale globally as it has. ATM networks certainly do not have a flat address space. The ATM address space has scope for an unprecedented level of hierarchical structure, and this structure is exploited in the ATM routing protocols to support greater degrees of scalability within ATM networks than is possible within any other network.

Much of the discussion about flat addressing and ATM actually revolve around the perception that ATM networks can be made easier to administer than existing layer 3 networks. Ease of administration argues not for flat addressing, but for a systematic focus on supporting autoconfiguration within protocols, as is now being done for the IPv6 protocol. This has been a prime focus for the ATM Forum from its inception, and by building on such mechanisms as the ILMI, most of the protocols developed for ATM do not incorporate such support.

ATM Routing Protocols

Network node interface protocols used within ATM networks have the function of routing ATM signaling between ATM switches. Since ATM is connection oriented, a connection request needs to be routed through the ATM network and to the destination node, much as packets are routed within a packet-switched network. The NNI protocols are to ATM networks as routing protocols such as OSPF and IGRP are to current routed networks.

The ATM Forum is making an ongoing effort to define a private NNI (P-NNI) protocol. The goal is to define NNI protocols for use within private ATM networks or, more specifically, within networks that use NSAP format ATM addresses. Public networks that use E. 164 numbers for addressing will be interconnected using a different NNI protocol stack based upon the ITU-T B-ISUP signaling protocol and the ITU-T MTP level 3 routing protocol. This work is being carried out by the broadband inter-carrier interface (B-ICI) subworking group of the ATM Forum (Forum 4) and other international standards bodies.

The P-NNI protocol consists of two components. The first is a P-NNI signaling protocol used to relay ATM connection requests within the networks, between the source and destination UNI. The UNI signaling request is mapped into NNI signaling at the source ingress switch. The NNI signaling is remapped back into UNI signaling at the destination (egress) switch. The ingress switch is known as the DTL originator, and the final egress switch as the DTL terminator, since these nodes respectively insert and remove the DTLs used to route the connection request through the network.

The P-NNI protocol operates between ATM switching systems that can represent either physical switches or entire networks operating as a single P-NNI entity that are connected by P-NNI links. A private ATM network might use proprietary NNI protocols internally and use P-NNI protocol for external connectivity and interoperability. P-NNI links can be physical links or virtual multi-hop links. A typical example of a virtual link is a virtual path that connects two nodes together. Since all virtual channels, including the connection carrying the P-NNI signaling, would be carried transparently through any intermediate switches between these nodes on this virtual path, the nodes are logically adjacent in relation to the P-NNI protocols.

The ILMI protocol, first defined for use across UNI links, will also be used across both physical and virtual NNI links. Enhancements to the ILMI MIBs allow for automatic recognition of NNI versus UNI links, and of private versus public UNI.

The P-NNI signaling protocol illustrated in Figure 16.11 is an extension of UNI signaling and incorporates additional information elements for such NNI-related parameters as designated transit lists (DTLs). P-NNI signaling is carried across NNI links on the same virtual channel, VCI = 5, that is used for signaling across the UNI. The VPI value depends on whether the NNI link is physical or virtual.

The second component of the P-NNI protocol is a virtual circuit routing protocol. This is used to route the signaling request through the ATM network. This is also the route on which the ATM connection is set up, and along which the data will flow. The operation of routing a signaling request through an ATM network is similar to that of routing connectionless packets within existing network layer protocols such as IP. This is because prior to connection setup, there is no connection for the signaling request to follow.

A VC routing protocol can use some of the concepts underlying many of the connectionless routing protocols that have been developed over the last few years. The P-NNI protocol is much more complex than any existing routing protocol. This

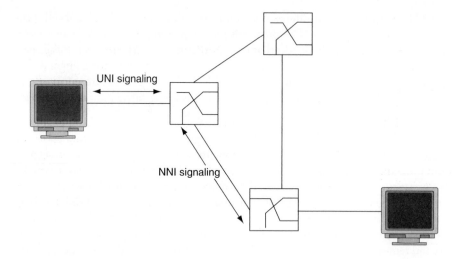

FIGURE 16.11
UNI and NNI signaling. (Adapted from Stratacom Frame Relay Course on IPX Networks, p. 36, © 1999, by permission of Cisco Systems, Inc.)

UNI signaling

NNI signaling

● **Constant bit rate** is used for connections that depend on precise clocking to ensure undistorted delivery.

● **Variable bit rate** is used for connections in which there is a fixed timing relationship between samples.

● **Available bit rate** is used for connections that do not require timing relationships between source and destination.

● **Unspecified bit rate** allows any amount of data up to a specified maximum to be sent across the network.

complexity arises from two goals of the protocol: to allow for much greater scalability than is possible with any existing protocol, and to support true QoS based routing.

P-NNI Phase 1 QoS Support

One of the great advantages of ATM is its support for guaranteed QoS connections. A node requesting a connection setup can request a certain QoS from the network and can be assured that the network will deliver that QoS for the life of the connection. Such connections are categorized into various ATM QoS types: *constant bit rate (CBR), variable bit rate (VBR), available bit rate (ABR),* and *unspecified bit rate (UBR),* depending upon the nature of the QoS guarantee desired and the characteristics of the expected traffic types. Depending upon the type of ATM service requested, the network is expected to deliver guarantees on the particular mix of QoS elements that are specified at the connection setup, such as cell loss ratio, cell delay, and cell delay variation.

To deliver such QoS guarantees, ATM switches implement a function known as connection admission control (CAC). Whenever the switch receives a connection request, the switch performs the CAC function. The connection request is based upon the traffic parameters and requested QoS of the connection. The switch determines whether setting up the connection violates the QoS guarantees of established connections such as excessive contention for switch buffering. The switch accepts the connection only if violations of current guarantees are not reported. CAC is a local switch function and makes decisions based on the strictness of QoS guarantees.

The *virtual channel (VC)* routing protocol must ensure that a connection request is routed along a path that leads to the destination and has a high probability of meeting the QoS requested in the connection setup of traversing switches whose local CAC will not reject the call.

To accomplish the criteria just described, the protocol uses a topology state routing protocol in which nodes flood QoS and reachability information so that all nodes obtain knowledge about reachability within the network and the available traffic resources within the network. Such information is passed within P-NNI topology state elements (PTSEs). This is similar to current link state routing protocols such as OSPF. Unlike these, however, which only have rudimentary support for QoS, the P-NNI protocol supports a large number of link and node state parameters that are transmitted by nodes to indicate their current state at regular intervals, or when triggered by particular events.

There are two types of link parameters: nonadditive link attributes, which are used to determine whether a given network link or node can meet a requested QoS; and additive link metrics, which are used to determine whether a given path consisting of a set of concatenated links and nodes with summed link metrics can meet the requested QoS.

The current set of link metrics is as follows:

- Maximum cell transfer delay (MCTD) per traffic class. Nodes can ensure adequate levels of separation between the different types of traffic passing through the node so that one traffic class does not consume the resources reserved for another traffic class.
- Maximum cell delay variation (MCDV) per traffic class.
- Maximum cell loss ratio (MCLR) for CLP = 0 cells, for the CBR and VBR traffic classes.
- Administrative weight, which is a value set by the network administrator and is used to indicate the desirability or undesirability of a network link.

The current set of link attributes is:

- *Available cell rate (ACR),* a measure of the available bandwidth in cells per second, per traffic class.
- Cell rate margin (CRM), a measure of the difference between the effective bandwidth allocation per traffic class and the allocation for sustainable cell rate. This is a measure of the safety margin allocated above the aggregate sustained rate.
- Variance factor, a relative measure of CRM normalized by the variance of the aggregate cell rate of the link.

All network nodes can obtain an estimate of the current state of the entire network through flooded PTSPs that contain such information as was just described. Unlike most current link state protocols, the P-NNI protocol advertises not only link metrics, but also nodal information. Typically, PTSPs include bidirectional information about the transit behavior of particular nodes based upon entry and exit ports and the current internal state. This is particularly important in cases where the node represents an aggregated network. In such cases, the node metrics must attempt to approximate the state of the entire aggregated network. This internal state is often at least as important as that of the connecting links for QoS routing purposes.

The need to aggregate network elements and their associated metric information also has important consequences for the accuracy of such information.

Two approaches are possible for routing a connection through the network: hop-by-hop routing and source routing. Hop-by-hop routing is used by most current network layer protocols such as IP and IPX, where a packet is routed at any given node only to another node, the next hop closer to the final destination. In source routing, the initial node in the path determines the entire route to the final destination.

Hop-by-hop routing is a good match for current connectionless protocols because it imposes little packet processing at each intermediate node. The P-NNI protocol uses source routing for a number of reasons. It is very difficult to do true QoS based routing with a hop-by-hop protocol since each node needs to perform local CAC and evaluate the QoS across the entire network to determine the next hop. Hop-by-hop routing also requires a standard route determination algorithm at each hop to preclude the danger of looping.

In a source routed protocol, only the first node would ideally need to determine a path across the network, based upon the requested QoS and its knowledge of the network state, which is gained from the PTSPs. It could then insert a full source routed path into the signaling request that would route it to the final destination. Ideally, intermediate nodes would only need to perform local CAC before forwarding the request. Because it is easy to preclude loops when calculating a source route, a particular route determination algorithm does not need to be standardized, leaving this as another area for vendor differentiation.

In practice, what actually happens is that the source routed path that is determined by a node can only be a best guess. This is because in any practical network, any node can have only an imperfect approximation of the true network state because of the necessary latencies and periodicity in PTSP flooding. The need for hierarchical summarization of reachability information means that link parameters must also be aggregated. Aggregated link parameters lead to inaccuracies. CAC is a local matter, which means that the CAC algorithm performed by any given node is both system dependent and open to vendor differentiation.

The P-NNI protocol tackles these problems by defining a generic connection admission control (GCAC) algorithm. This is a standard function that any node can use to calculate the expected CAC behavior of another node, given that node's advertised additive link metrics and the requested QoS of the new connection request. The GCAC is an algorithm that was chosen to provide a good prediction of a typical node-specific CAC algorithm, while requiring a minimum number of link state metrics. Individual nodes can control the degree of stringency of the GCAC calculation involving the particular node by controlling the degree of laxity or conservativeness in the metrics advertised by the node.

The GCAC actually uses the additive metrics to support the GCAC algorithm chosen for the P-NNI protocol. Individual nodes, both physical and logical, will need to determine and then advertise the values of these parameters for themselves based upon their internal structure and loading. The P-NNI phase 1 GCAC algorithm is primarily designed for CBR and VBR connections; variants of the GCAC are used depending upon the type of QoS guarantees requested and the types of link metrics available, yielding greater or lesser degrees of accuracy.

The only form of GCAC done for UBR connections is used to determine whether a node can support such connections. For ABR connections, a check is made to determine whether the link or node is authorized to carry any additional ABR connections and to ensure that the ACR for the ABR traffic class for the node is greater than the minimum cell rate (MCR) specified by the connection.

ATM Traffic Management

One of the primary benefits of ATM networks is that they can provide users with a guaranteed QoS. To do this, the user must inform the network upon connection setup of both the expected nature of the traffic that will be sent along the connection and the type of quality of service that the connection requires. The quality of service is described by a set of traffic parameters, while the nature of traffic is specified by a set of desired QoS parameters. The source node must inform the network of the traffic parameters and desired QoS for each direction of the requested connection upon initial setup. These parameters may be different in each direction of the connection.

ATM networks offer a specific set of service classes; and at connection setup, the user must request a specific service class from the network for that connection. Service classes are used by ATM networks to differentiate among specific types of connections, each with a particular mix of traffic and QoS parameters, since such traffic may need to be differentiated within the network by using priorities to allow for the requested behavior. The current set of QoS classes is as follows:

CBR: End systems use CBR connection types to carry constant bit rate traffic with a fixed timing relationship between data samples for circuit emulation.

Real Time VBR (VBR [RT]):   The VBR (RT) service class is used for connections that carry variable bit rate traffic in which there is a fixed timing relationship between samples for such applications as variable bit rate video compression.

Non-Real-Time VBR (VBR [NRT]):   The VBR (NRT) service class is used for connections that carry variable bit rate traffic in which there is no timing relationship between data samples, but a guarantee of QoS on bandwidth or latency is still required. Such a service class might be used for Frame Relay internetworking, in which the committed information rate (CIR) of the Frame Relay connection is mapped into a bandwidth guarantee within the ATM network.

ABR: ABR supports variable rate data transmissions and does not preserve timing relationships between source and destination. Unlike the VBR (NRT) service, the ABR service does not provide any bandwidth guarantee to the user. The network provides a best-effort service in which feedback (i.e., a flow control mechanism) is used to increase the bandwidth available to the user—the allocated ACR—if the network is not congested and to reduce the bandwidth when there is congestion. Through such flow control mechanisms, the network can control the amount of traffic that it allows into the network and minimize cell loss within the network due to congestion.

ABR supports a rate based mechanism for congestion control where resource management (RM) cells or the explicit forward congestion indication (EFCI) bit within ATM cells is used to indicate the presence of congestion within the network to the source system. A specified traffic pacing algorithm, controlling the ACR, is used at the source to control the traffic rate into the network, based either upon the number of RM cells received with a congestion indication or on an explicit rate indication from the network.

ABR is designed to map existing LAN protocols that opportunistically use as much bandwidth as is available from the network but can either back off or be buffered in the presence of congestion. ABR is ideal for carrying LAN traffic across ATM networks.

A **minimum cell rate** specifies the minimum value for the ACR.

The ABR service can optionally provide a guaranteed *minimum cell rate (MCR)* for an ABR connection.

UBR: The UBR service does not offer any service guarantees. The user is free to send any amount of data up to a specified maximum while the network makes no guarantees at all on the cell loss rate, delay, or delay variation that might be experienced.

Because UBR does not have any flow control mechanisms, to control or limit congestion the ATM switches have to implement pre-standard congestion control mechanisms, or support adequate buffering to minimize the probability of cell loss when multiple large data bursts are received concurrently at a switch, as might be expected. Such a case might be a typical client-server environment.

QoS Parameters

There is no explicit priority field associated with ATM connection types, but such priorities are required within ATM switches. The only indication of relative priority within an ATM cell is the CLP bit that is carried within the cell header; setting this bit to 1 (CLP = 1) indicates that the cell may be dropped in preference to cells with CLP = 0. This bit can be set by the end systems but is more likely to be set by the network.

The traffic parameter fields that affect priority within ATM are as follows:

- Peak cell rate (PCR)

- *Cell delay variation tolerance (CDVT)*
- Sustainable cell rate (SCR)
- Burst tolerance (BT)
- MCR for ABR only

These parameters define an envelope around a traffic stream, but not all parameters are valid for all service classes. For CBR connections, only the PCR, which determines how often data samples are sent, and the CDVT, which determines how much jitter is tolerable for such samples, are relevant. For VBR connections, the SCR and BT together determine the long-term average cell rate and the size of the maximum burst of contiguous cells that can be transmitted. In the case of ABR service, the PCR determines the maximum value of the ACR, which is dynamically controlled by the network, through congestion control mechanisms, to vary between the MCR and the PCR.

When setting up a connection, the requesting node informs the network of the type of service required, the traffic parameters of the data flows in each direction of the connection, and the QoS requested for each direction. Together these form the traffic descriptors for the connection. The current set of QoS parameters consists of three delay parameters and one dependability parameter. The three delay parameters are as follows:

- Peak-to-peak cell delay variation (CDV)
- *Maximum cell transfer delay (max CTD)*
- Mean cell transfer delay (mean CTD)

The dependability parameter is the cell loss ratio (CLR).

The CDV, max CTD, and mean CDV are treated as dynamic, additive metrics, and their expected values through the network will be cumulated in signaling requests, while the CLR is considered to be a configured link and node attribute, which local CAC algorithms will strive to meet. Particular combinations of the CDV, max CTD, mean CTD, and CLR (for CLP = 0 streams only) parameters will be negotiable, depending upon the service class, between the end system and the network. As with the traffic parameters, not all QoS parameters will apply to all service classes.

Table 16.2 illustrates the various services and applicable parameters incorporated in ATM.

An ATM connection that is set up with specified traffic descriptors constitutes a traffic contract between the user and the network. The network offers the type of guarantee appropriate to the service class, as long as the user keeps the traffic on the connection within the envelope defined by the traffic parameters. The network can enforce the traffic contract by a mechanism known as usage parameter control (UPC), better referred to as traffic policing. UPC is a set of algorithms performed by

Table 16.2 ATM Service Classes and Applicable Parameters

Attribute	CBR	VBR RT	VBR NRT	ABR	UBR	Parameter
CLR	Specified	Specified	Specified	Specified	Unspecified	QoS
CTD and CDV	DCV and max CTD	DCV and max CTD	Mean CTD only	Unspecified	Unspecified	QoS
PCR and CDVT	Specified	Specified	Specified	Specified	Specified	Traffic
SCR and BT	N/A	Specified	Specified	N/A	N/A	Traffic
MCR	N/A	N/A	N/A	Specified	N/A	Traffic
Congestion control	No	No	No	Yes	No	

an ATM switch on the receipt of cells within a connection that determine whether or not the cell stream is compliant with the traffic contract. The UNI 3.1 specification called for a dual leaky bucket algorithm for traffic policing.

The dual leaky bucket mechanism can best be thought of as a means of pacing the transmission of cells along a link so that the traffic stream meets the specified PCR and CDVT and, optionally, the SCR and BT for the connection for various combinations of CLP = 0, CLP = 1, and CLP = 0 + 1 cell streams. The UNI 3.1 UPC mechanism measures cell arrivals as if they were generated by such a leaky bucket based generic cell rate algorithm (GCRA). Any type of traffic shaping can be used, as long as the traffic envelope fits within the traffic contract parameters. In practice, traffic sent across ATM links that are controlled by UPC are sometimes actually shaped by using a leaky bucket algorithm and the requested traffic parameters, which ensures that cells will not be inadvertently marked as nonconformant. Traffic shaping can also help control and reduce congestion within a network by limiting the peak rate of a connection to that of the slowest link along the path.

Upon the detection of a nonconformant cell, a switch can choose either to selectively discard the cell or, if local resources and policies permit, to tag the cell as nonconformant by setting its CLP to 1. The cell would then be more likely to be discarded further within the ATM network if more congestion is experienced. UPC is primarily designed to be used across UNI, since passage through ATM switches will change the shape of the traffic stream due to buffering delays. UPC is likely to be used across public UNI since public ATM networks will likely base their tariffs on traffic usage. This may require ATM switches that are connected to public UNI to reshape the traffic sent across public UNI.

The ATM routing protocols performed by ATM switches use the traffic descriptors associated with a signaling request both to route the connection appropriately to meet the traffic guarantees, and to control connection admission, which ensures that establishing a new connection will not adversely affect established connections. To support multiple traffic classes, ATM switches internally must implement a mechanism for isolating traffic flows of particular connection types from each other. The switch may allocate different priority levels to the different service classes, so that the cells of some connection types gain preferential access to scarce resources. Typically CBR connections receive high priority to minimize the amount of latency and jitter experienced by the cells on such connections.

LAN Emulation

In the discussion of LAN emulation (LANE), the internetworking of existing protocols across ATM networks also will be discussed. Given the vast installed base of LANs and WANs today and the network and link layer protocols operating on these networks, a key to ATM success will be the ability for interoperability between these technologies and ATM. Few users will tolerate islands of ATM without connectivity to the remainder of the enterprise network. The key to such connectivity is the use of the same network layer protocols, such as IP and IPX, on both existing networks and on ATM, since the function of the network layer is to provide a uniform network view to higher level protocols and applications. There are, however, two fundamentally different ways of running network layer protocols across an overlay mode ATM network. In one method, known as native mode operation, address resolution mechanisms are used to map network layer addresses directly into ATM addresses, and the network layer packets are then carried across the ATM network. The alternate method of carrying network layer packets across an ATM network is LANE. The function of the LANE protocol is to emulate a local area network on top of an ATM network. The LANE protocol defines mechanisms for emulating either IEEE 802.3 Ethernet or a Token Ring 802.5 LAN. The current LANE protocol does not define a separate encapsulation for FDDI. FDDI packets must be mapped into either Ethernet or Token Ring emulated LANs, using existing translational bridging techniques. The two most prominent new LAN standards under consideration, fast Ethernet (100Base-T) and 802.12 (100VG-AnyLAN) can both be mapped unchanged into either Ethernet or Token Ring LANE formats and procedures, as appropriate, since they use the same packet formats.

What LAN emulation, as illustrated in Figure 16.12, means is that the LANE protocol defines a service interface for higher layer (network layer) protocols that is identical to that of existing LANs, and that data sent across the ATM network are encapsulated in the appropriate LAN MAC packet format. It does not mean that any attempt is made to emulate the actual media access control protocol of the specific LAN concerned, which is CSMA/CD for Ethernet or token passing 802.5. The LANE protocol supports a range of minimum packet sizes (MPDU) corresponding to maximum size Ethernet, to 4 Mbps and 16 Mbps Token Ring packets, and to the

FIGURE 16.12
ATM internet-working.

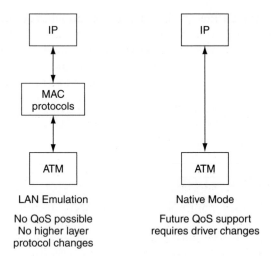

value of the default MPDU for IP over ATM. Typically the appropriate MPDU is used, depending upon what type of LAN is being emulated, and is supported on the LAN switches bridged to the ELAN. An ELAN with only native ATM hosts, however, may optionally use any of the available MPDU sizes, even if this does not correspond to the actual MPDU in a real LAN of the type being emulated. All LECs within a given ELAN must use the same MPDU size.

In other words, the LANE protocols make an ATM network look and behave like an Ethernet or Token Ring LAN—albeit one operating much faster than these real networks.

The rationale for using emulation is that it requires no modifications to higher layer protocols to enable their position over an ATM network. Because the LANE service presents the same service interface of existing MAC protocols to network layer drivers, no changes are required in those drivers. The intention behind implementing emulation is to accelerate the deployment of ATM because considerable work remains to be done in defining native mode operation for the plethora of existing network layer protocols.

It is envisaged that the LANE protocol will be deployed in two types of ATM attached equipment.

ATM Network Interface Cards: ATM NICs will implement the LANE protocol and interface to the ATM network, but they will present the current LAN service interface to the higher level protocol drivers within the attached end system. The network layer protocols on the end system will continue to communicate as if they were on a known LAN, using known procedures. They will be able to use the vastly greater bandwidth of ATM networks.

Internetworking and LAN Switching Equipment: The second class of network gear that will implement LANE will be ATM attached LAN switches and routers.

These devices, together with directly attached ATM hosts equipped with ATM NICs, will be used to provide a virtual LAN service, where ports on the LAN switches will be assigned to particular virtual LANs, independent of physical location. LAN emulation is a particularly good fit for the first generation of LAN switches that effectively act as fast multiport bridges because LANE is essentially a protocol for bridging across ATM.

The LANE protocol does not directly impact ATM switches. LANE, as with most of the other ATM internetworking protocols, builds upon the overlay model. As such, the LANE protocols operate transparently over and through ATM switches, using only standard ATM signaling procedures. ATM switches may well be used as convenient platforms upon which to implement some of the LANE server components, but this is independent of the cell relay operation of the ATM switches themselves. This logical decoupling is one of the great advantages of the overlay model, since it allows ATM switch designs to proceed independently of the operation of overlying internetworking protocols.

The basic function of the LANE protocol is to resolve MAC addresses into ATM addresses. By doing so, it actually implements a protocol for MAC bridging on ATM, hence the close fit with current LAN switches. The goal of LANE is to perform such address mapping so that LANE end systems can set up direct connections between themselves and forward data. The element that adds significant complexity to LANE is supporting LAN switches, which is to say, LAN bridges. The function of the LAN bridge as defined by ISO and IEEE is to shield LAN segments from each other. Although bridges learn about MAC addresses on the LAN segments to which they are connected, such information is not propagated.

LANE Components and Connection Types

The LANE protocol illustrated in Figure 16.13 defines the operation of a single emulated LAN (ELAN). Multiple ELANs may coexist simultaneously on a single ATM network because ATM connections do not collide. A single ELAN emulates either Ethernet or Token Ring and consists of a LANE client (LEC), a LANE server (LES), a broadcast and unknown server (BUS), and a LANE configuration server (LECS).

LEC: An LEC is the entity in an end system that performs data forwarding, address resolution, and other control functions for a single end system within a single ELAN. An LEC also provides a standard LAN service interface to any higher layer entity that interfaces with the LEC. An ATM NIC or LAN switch interfacing with an ELAN supports a single LEC for each ELAN to which they are connected. An end system that connects to multiple ELANs will have one LEC per ELAN.

Each LEC is identified by a unique ATM address and is associated with one or more MAC addresses reachable through that ATM address. In the case of an ATM NIC, for instance, the LEC may be associated with all the MAC addresses reachable through the ports of that LAN switch which are assigned to the particular ELAN. This set of

FIGURE 16.13

LANE protocol architecture. (Adapted from Stratacom Frame Relay Course on IPX Networks, p. 44, © 1999, by permission of Cisco Systems, Inc.)

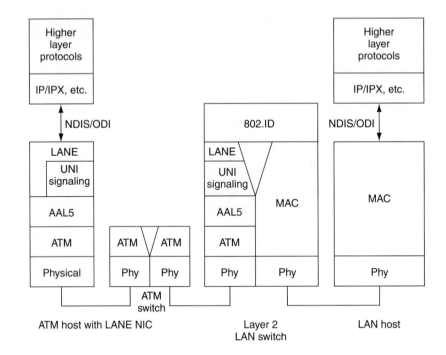

addresses may change, both as MAC nodes come up and down, and as particular paths are reconfigured by logical or physical changes in the LAN network topology.

Although the current LANE specification defines two types of emulated LANs, one for Ethernet and one for Token Ring, it does not permit direct connectivity between a LEC that implements an Ethernet ELAN and one that implements a Token Ring ELAN. In other words, LANE does not attempt to solve the mixed media bridging problem, which is intractable for Ethernet-to-Token Ring interconnection. Two such ELANs can only be interconnected through an ATM router that acts as a client on each ELAN.

LES: The LES implements the control function for a particular ELAN. There is only one logical LES per ELAN, and to belong to a particular ELAN means to have a control relationship with that ELAN's particular LES. Each LES is identified by a unique ATM address.

BUS: The BUS is a multicast server that is used to flood unknown destination address traffic and forward multicast and broadcast traffic to clients within a particular ELAN. Each LEC is associated with only a single BUS per ELAN, but there may be multiple BUSs within a particular ELAN that communicate and coordinate in some vendor-specific manner. The BUS to which an LEC connects is identified by a unique ATM address. In the LES, this is associated with the broadcast MAC address, which is all 1s, and this mapping is normally configured into the LES.

LECS:　The LECS is an entity that assigns individual LANE clients to particular ELANs by directing them to the LESs that correspond to the ELAN. There is logically one LECS per administrative domain, and this serves all ELANs within that domain.

The LANE protocol does not specify where any of the server components described here should be located. Any device or devices with ATM would suffice. For the purposes of reliability and performance, it is likely that most vendors will implement these server components on networking equipment, such as ATM switches or routers, rather than on a workstation or host. This also applies to all other ATM server components.

The LANE protocol specifies only the operation of the LANE user-to-network interface (LUNI) between an LEC and the network providing the LANE service. This may be contrasted with the LANE NNI (LNNI) interface, which operates between the server components within a single ELAN system. These components represent single points of failure and potential bottlenecks. The interactions between each of the server components in the LANE Phase 1 protocol are currently left in a proprietary manner by vendors.

The ATM Forum has just finished work on Phase 2 LANE protocol, which will specify LNNI protocols so as to allow redundant LESs and replicated BUSs in order to address concerns about these limitations. The LNNI protocols will specify open interfaces between the various LANE server entities—LES/LES, LES/LECS, and BUS/BUS—and will allow hierarchies of BUSs for greater scalability within ELANs. The fundamental limit to the scalability of an ELAN is not the number of BUSs, but the fact that all broadcast and flood traffic must be sent to all LECs. In the case where the LEC is within a LAN switch, this limits the amount of such traffic to be much less than the speed of the associated LAN, such as 10 Mbps in the case of an Ethernet ELAN.

Summary

ATM is an ITU-T and ANSI standard for worldwide services that is experiencing rapid growth in both the telecommunications and local networking marketplaces. ATM offers benefits in all areas of networking, including architectural stability through scalability, manageability, and design simplicity as well as support for multimedia and other real-time applications. Low complexity and cost-effective carrier support for telecommunications users is another significant benefit.

Review Questions

1. Discuss how a corporation can benefit from using ATM to integrate its corporate network.
2. Define the basic ATM cell.
3. What is CONS and how does it work in the ATM world?
4. What is the ATM basic header format?

5. What are virtual paths and virtual channels?
6. What are the four distinct ATM adaption layers?
7. What is a permanent virtual circuit?
8. What is a switched virtual circuit?
9. How does ATM multicasting work?
10. What is an ATM multicast server?
11. What are the two ATM models originally proposed by the ATM forum?
12. Describe how ATM signaling and addressing work.
13. How does ATM integrate with the OSI model?
14. What is P-NNI and what is its significance in the ATM world?
15. What is the GCAC algorithm?
16. What are the various ATM traffic management classes and how do they work?
17. What are the various ATM QoS parameters?
18. How does the carrier provision an ATM circuit?
19. What is ATM traffic policing?
20. Define LANE.
21. What are the two types of LANE that will be deployed in ATM?
22. What are the LANE component and connection types?

Summary Questions

1. What is meant by a connection oriented protocol?
2. Explain the ATM network operation.
3. How was the standard size of 53 bytes derived for ATM?
4. What is the purpose of the ATM adaption layers?
5. What are the two fundamental types of ATM connections?
6. What are the various link metrics defined in ATM?

T1 Digital Communications

Objectives

- Explain what T1 is.
- Explain why T1 is in widespread use today in the telecommunications market.
- Explain what a T1 does.
- Discuss the evolution of T1 since the 1960s.
- Explain the switching centers of the telephone network prior to divestiture.
- Explain the relationship between OSI and T1.
- Explain the DS1 signal level and D4 protocol.
- Explain DS1 framing and line coding.
- Explain the North American digital hierarchy.
- Explain pulse code modulation.
- Explain the voice channel bank.
- Explain differential PCM and adaptive DPCM.
- Explain CVSD continuous variable slope delta modulation.
- Explain vector quantizing code.
- Explain ESF extended superframe.

Key Terms

adaptive differential pulse code modulation (ADPCM), 735

alternate mark inversion (AMI), 723

Introduction

What Is T1?

T1 is a digital transmission facility operating at 1.544 Mbps in a full duplex, time-division-multiplexing (TDM) mode. T1 is the first level or primary rate within the T carrier system, a digital transmission and switching hierarchy defined by AT&T during the late 1950s and early 1960s.

T1 networks are the most widely installed wide area networks for integrated voice, data, and video image transmissions. The digital signals used within T1 permit the concurrent transmission of voice, data, and image (video, facsimile, graphics, etc.) information in a digitally encoded form. The full duplex feature of T1 permits two-way simultaneous transmission, the transmit path and the receive path both operating independently at 1.544 Mbps. The TDM feature of T1 permits the derivation of logical channels from the serial by bit Mbps stream. Whether via static or dynamic bandwidth allocation, the T1 bit stream can be viewed as 24 64 Kbps (thousand bits per second) channels as: 4 384 Kbps, as 1 1.536-Mbps channel, a fractional T1 at 768 Mbps, or any variation of a fraction of T from 56 Kbps to 128 to 256 to 384 Kbps. Each of these derived logical channels can be independently transmitted and switched, and may independently transport different types of information such as voice, data, or image.

Why T1?

There are three basic factors that account for the increasingly widespread implementation of T1 networks.

The first factor is the potential for decreased telecommunications costs. With the TDM feature, T1 can integrate several currently self-standing links into a single link. This integrated link approach reduces the cost of in-house wiring and connections, the cost of in-house data circuit terminating equipment (DCE), and the recurring monthly cost of switched and leased lines.

The second factor is the potential for increased network availability. T1 provides for a derived dedicated channel for network management. Functions such as testing, maintenance, diagnostics, and performance measurement are supported on this channel. These functions are available concurrently with other logical channels for user information transport, thus overlapping network management functions with user production time. In theory, network downtime for testing, diagnostics, and fault isolation should approach zero, leaving defective component replacement time as the major portion of downtime. Further, with proactive switching via the network management channel, deteriorating paths can be bypassed before failure. Again, in theory, using this scenario, total network downtime should approach zero, thus increasing availability.

The third factor is the potential for improvement in the quality of the information being transmitted. The relatively high bandwidth provides for higher throughput rates and more timely information. The error detection capability of T1 improves the integrity of the network as compared to the lack of error detection within the traditional analog circuits. T1 error detection and associated performance measurement reporting provide the network user with effective tools and techniques to improve the accuracy of all transmitted information, including voice. Thus, the quality of transmitted information in terms of timeliness and accuracy can be improved with T1 transmission facilities.

What Does T1 Do?

The T1 facility provides the mechanical, electrical, functional, and procedural capability to activate, maintain, and deactivate a physical transmission link to transmit bits of information between two nodes. The T1 resides at the physical layer of the OSI reference model of the ISO.

How Does T1 Accomplish Its Goals?

T1 transmission facilities and functions are implemented within a set of T1 protocols. A protocol is the set of rules by which transmission in this case takes place. This set of protocols covers the T1 hierarchy consisting of the physical digital signal (DS1), the T1 framing conventions (D1, D2, D3, D4, and SF [super frame] and ESF [extended super frame] formats), the channelized and nonchannelized formats, and various T1 operating procedures. AT&T developed these protocols in the very beginning, when it developed the T1 exclusively. Currently, standards organizations are addressing T1 protocols.

Evolution of T1

The initial implementation of T1 was as an interoffice trunk between two telco (telephone company) central offices. The four elements that made up the network in a metropolitan area beginning at the subscriber location or terminal site were as follows:

1. Terminal equipment
   Telephone
   Voice jack
   Data jack
2. Local loop
3. Central office
4. Interoffice trunk

Figure 17.1 reflects a typical configuration of the earlier means of interconnecting two metropolitan areas. The local loop served as the link between the telephone equipment at the user's site to a switch at the central office (CO). This local loop was characterized by analog, half-duplex (two-way alternating), 4 kHz transmission. At the CO, the analog signals were converted to digital signals, multiplexed, and transmitted across the CO-CO links in accordance with T1 protocols. At the terminating end of the call, the CO would convert the T1 digital signals back to analog format and transmit across the local loop in that analog format to the called user.

Thus, in this manner T1 links were directly available only to the COs as an interoffice trunk. Local loops employed analog transmission, while interoffice trunks employed digital transmission. This was how the first generation of T1s conducted transmission proceedings.

FIGURE 17.1

Connecting metropolitan areas.

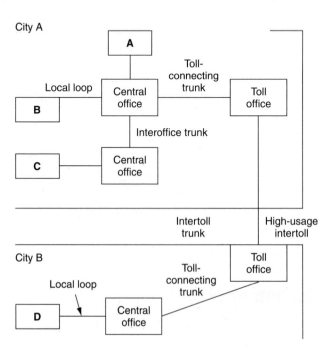

The metropolitan areas were connected through expanded digital networks employing digital toll collecting trunks, digital intertoll trunks, and digital high usage intertoll trunks. The digital toll-collecting trunks ran at the T1 rate of 1.544 Mbps. The digital intertoll and high usage intertoll trunks ran at the T3 rate of 44.736 Mbps: T1 on copper wires, and T3 on microwave.

The digital trunks terminated in five classes of digital *switching centers* as illustrated in Table 17.1 and Figure 17.2. These centers operated in sequential and hierarchical

Switching centers operated in sequential and hierarchical nodes.

Table 17.1 Switching Centers Before the Divestiture of AT&T

Switching Center	*Type*
Class 5	Central Office
Class 4	Toll Center
Class 3	Primary Center
Class 2	Sectional Center
Class 1	Regional Center

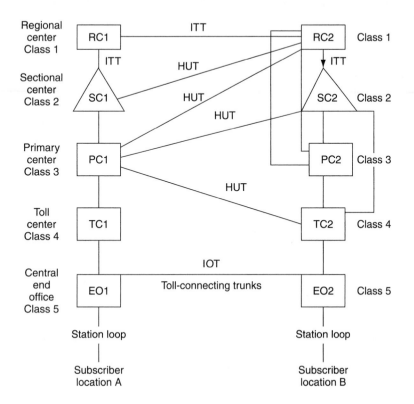

FIGURE 17.2
Switching centers.

nodes. At one point in time, there were over 60,000 class 5 central offices, over 1,500 class 4 toll centers, and 10 regional centers. A fixed hierarchical routing scheme was used with the numbers indicating the order of choice of route at the center for calls originating at end office 1.

The 1970s brought about two additional changes in the evolution of the T1 world, digital loops and integrated voice/data T1 links. The initial T1 facility offered as a local loop was a 64-Kbps digital link with a net 56-Kbps throughput rate from the customer's perspective. AT&T tariffed this offering as dataphone digital services (DDS). The DDS links were accessed via a digital service unit (DSU) that resided on the user side of the link and provided the proper signaling to the telco digital network. These DDS links were fitted into the T1 interoffice trunks as a logical channel, multiplexed along with other voice and data channels. The DDD (direct distance dialing) and leased networks would employ T1 facilities as interoffice trunks and analog facilities for non-DDS local loops and digital facilities for DDS local loops.

The T1 evolution of the 1980s, referred to as third generation, brought forth a 1.544-Mbps digital local loop, and the user now drives the T1 link directly.

T1 equipment at the customer premises (CPE) accesses the T1 local loop. The CPE is typically composed of a T1 network processor such as a channel bank, a multiplexer, or PBX (private branch exchange), a DSU that performs a signal conversion, and a CSU that formats the signals in accordance with telco specifications. The CSU can be combined with the DSU. The provision of the T1 equipment at the customer's premises requires the customer to be in compliance with T1 protocols. The customer must acquire a hardware/software complex that performs in accordance with the T1 protocols.

These user controlled networks may be implemented in one of three major network categories: private, public, or hybrid. A private network employs T1 facilities that are owned and used exclusively by the customer or that are owned by a telco but controlled via routing or switching by the customer. A public network employs telco provided T1 links and facilities. A hybrid network employs a mix of private and public T1 links and facilities.

OSI and T1

The OSI reference model identifies four basic components of any telecommunications network: the application process, the system, the physical medium for transmission, and the protocols.

With the OSI reference model, the system is a set of one or more computers, associated software, peripherals, terminals, human operators, physical processes, information transfer means, and so forth, which form an autonomous whole capable of performing information processing and/or information transfer.

An application process is an element within a system that performs the information processing for a particular application. The application process may be a manual process, a computerized process, or a physical process.

The physical medium provides the means for the transfer of information among the systems.

Because the model is concerned with the exchange of information among systems and not the internal functioning of each system, the objective of OSI is to define a set of standards allowing the various systems to communicate in an open manner.

These OSI standards evolved from a seven layer hierarchical model. The application layer (layer 7) supports the application process, the physical layer (layer 1) supports transmission, and the intermediate layers (layers 2–6) provide support for other network functions.

In the past, T1/T3 protocols were physical layer functions such as connector definitions, cabling, voltage levels, timing requirements, and so forth.

Currently, T1/T3 protocols are being defined to incorporate layer 2 functions of data link control primarily in the areas of network fault and performance management.

DS1/D4 Protocol

The prevailing protocol in today's T1 networks is the *DS1/D4 Protocol*. This protocol uses the DS1 signal and the D4 framed format on T1 transmission facilities.

The DS1 signal generates a 1.544 Mbps bit stream employing a bipolar signal format. The primary line code used with the DS1 signal is *alternate mark inversion (AMI)* or *bipolar with eight zeros substitution (B8ZS)*.

D4 framing involves synchronizing the 1.544-Mbps bit stream into a series of 193-bit groups, designated as D4 frames. The frames are further grouped into a 12-frame superframe by a 12-bit pattern. These groupings are required to synchronize the transmitting and receiving equipment on the T1 link to permit signaling and channelization.

The DS1/D4 framed formats can be broadly categorized as channelized and nonchannelized. In the channelized format, the transmitting and receiving equipment derive 24 64-Kbps channels (referred to as DS0s) from the 1.544-Mbps DS1 bit stream. In the nonchannelized framed format, the transmitting and receiving equipment may allocate the 1.544 Mbps into 64 Kbps channels on a dynamic basis.

There are two modes of signaling defined for the DS1/D4 protocol, referred to as in-band and out-of-band signaling. Signaling is required to set up, maintain, and tear down connections in a public T1 network. The in-band technique uses designated bit slots within the user channels (in-band) to convey signaling information. The out-of-band technique uses designated bit slots outside of the user channels (out-of-band).

Within the public common carrier T1 networks, a wide range of services exist based on the DS1/D4 protocol. These services include the M24, M44, CCR/DACS, and fractional T1 services.

The **DS1/D4 Protocol** uses the DS1 signal and the D4 framed format on T1 transmission facilities.

Alternate mark inversion or bipolar with eight zeros substitution is the primary line code used with the DS1 signal.

FIGURE 17.3

DS1/D4 protocols.

DS1/D4 Protocols

The DS1 signal

D4 framing

Channelized and nonchannelized formats

In-band signaling

Out-of-service maintenance

The DS1/D4 protocol illustrated in Figure 17.3 is based upon two platforms. AT&T's publication 62411 has served as the de facto standard for DS1/D4. In 1989, ANSI published its standard T1.403-1989, which is currently emerging as the industry standard.

The DS1 Signal

The DS1 signal employs digital transmission. Binary digital signals are transmitted in a discrete fashion, meaning a defined signal for transmitting values of 0 and a defined signal for transmitting values of 1. Any input information that can be represented digitally may be transported. Consequently, digital data, voice, image, and other information encoded digitally can be transmitted via the DS1 signal.

The DS1 signal employs TDM. The signal is a serial bit by bit transmission scheme with assigned time slots. Each user is assigned a time slot. Consequently, the digital TDM DS1 signal can concurrently support a range of data, voice, and other digitized input streams.

The DS1 signal is employed within a full duplex transmission facility. Either physically or logically, there is a path for transmitting a DS1 signal toward the network and a path for receiving a DS1 signal from the network. Therefore, the digital, time-division-multiplexed, full duplex DS1 signal can concurrently support a set of data, voice, and other digitized input on a two-way simultaneous transmission basis.

The signal is referred to as a 1.544-Mbps bipolar 50 percent pulse train. There are 1,544,000 time slots per second, and each time slot carries a binary digit or bit. If a pulse is present within a time slot, a 1 bit has been transmitted. If there is no pulse within a time slot, a 0 bit has been transmitted. When the pulse is present, it occupies 50 percent of the time slot. On the time-division T1 links, time is divided into 648 ns time slots. Each time slot may or may not carry a pulse with an amplitude of 3 V. When the pulse is present, it occupies 50 percent of the time slot and has a polarity opposite that of the last transmitted pulse.

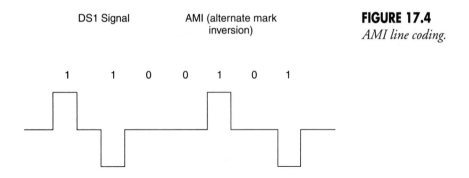

DS1 Signal   AMI (alternate mark
             inversion)

FIGURE 17.4

AMI line coding.

In order to maintain synchronization within a T1 network, common carriers require a minimum number of 1 bits within a given amount of time or within a given number of bits transmitted. Figure 17.4 shows an example of AMI signaling. In the AMI signaling format, a square wave or pulse represents the binary value of 1, and a straight line represents the binary value of 0. Each pulse alternates between positive and negative polarity, making the signal bipolar in format. The primary advantage of the bipolar format is that it allows the DS1 signal to travel twice as far on a pair of copper wires. Another advantage of the bipolar format is its ability to offer a built-in method of error detection. When consecutive pulses of the same polarity are detected, this constitutes a bipolar violation (BPV). BPVs indicate that signal input has been disrupted due to defective equipment or poor environmental conditions. One technique for ensuring compliance with the particular 1's density rule is coding. The DS1 line codes are listed in Table 17.2. The DS1 format is summarized in Figure 17.5.

CSUs/DSUs are data circuit-terminating equipment located at the customer's premises that provide overall compliance with the T1 physical signal specifications. The DSU component of the CSU/DSU equipment primarily provides the unipolar to bipolar signal conversions, and the CSU component of the CSU/DSU primarily addresses the output pulse requirements and 1's density rules.

To ensure network synchronization, T1 networks must impose certain pulse density requirements such that there are a minimum number of 1s within a given time frame

Table 17.2 DS1 Line Codes

DS1 Line Codes	Meaning
AMI	Alternate mark inversion
B8ZS	Bipolar with eight zeros substitution
ZBTSI	Zero byte time slot interchange
ZCS	Zero code suppression

FIGURE 17.5

DS1 format.

DS1 line rate = 1.544 Mbps
DS1 bit time slot = 648 n
DS1 pulse width = 324 n
DS1 pulse amplitude = 3 Volts
DS1 line codes = AMI, B8ZS

or number of bits transmitted. The dominant requirement is that there must not be more than 15 0s in a row, and in each and every time frame of $x(n + 1)$ bits, where n can equal from 1 to 23, there should be at least n 1s. To simplify matters, there can be no more than 15 0s in a row and there must be at least 3 1s in every 24 bits.

For certain networks, there are even more stringent requirements. For selected Acunet T1.5 services, there must be at least one 1 in every 8 bits transmitted. According to AT&T, the purpose of this requirement is to preclude the production of long strings of 0s and to ensure adequate timing energy.

A recommended technique to meet the minimum pulse density requirements in all cases is B8ZS line coding, illustrated in Figure 17.6. With B8ZS, each block of eight consecutive 0s is removed and the B8ZS code is inserted in their place. If the preceding pulse is positive, the inserted code is $000 + - 0 - +$. If the preceding code is negative, the inserted code is $000 - + 0 + -$. In both cases, bipolar violations appear in the fourth and seventh positions that line equipment can detect and process accordingly. As a result, the T1 link will exhibit bit sequence independent of the minimum pulse requirements.

Pulse code modulation is the transmission of analog information in digital form through sampling and encoding the samples with a fixed number of bits.

DS1/D4 Framing

The serial 1.544 Mbps stream of T1 must be interpreted in a formatted fashion in order to derive the logical DS/64-Kbps channels. Since the inception of T1 and the introduction of the *pulse code modulation (PCM)* voice digitization scheme, the basic format of

FIGURE 17.6

Sample of B8ZS line coding.

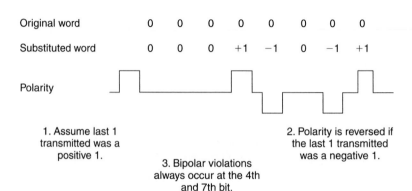

| Original word | 0 | 0 | 0 | 0 | 0 | 0 | 0 | 0 |
| Substituted word | 0 | 0 | 0 | +1 | −1 | 0 | −1 | +1 |

Polarity

1. Assume last 1 transmitted was a positive 1.

2. Polarity is reversed if the last 1 transmitted was a negative 1.

3. Bipolar violations always occur at the 4th and 7th bit.

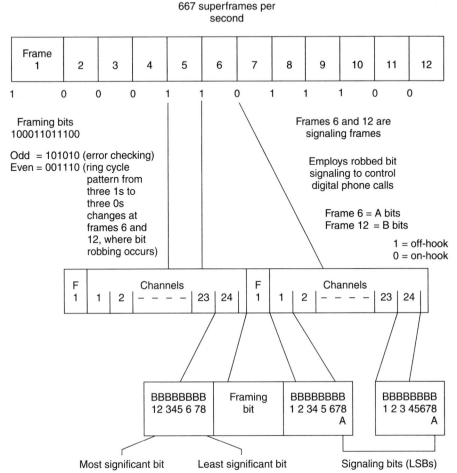

FIGURE 17.7
*Superframe for-
mat D4/SF.*

the T1 bit was a framing bit, used to denote the beginning of a frame. The remaining 192 bits represented 8 bits from each of the logical 24 channels. The D1 format from the D1 channel bank equipment used a framing pattern of alternate 1s and 0s for the framing bits. The D2 format introduced the *superframe* concept, where within each group of 12 frames (a superframe), a finite 12-bit pattern would be employed for framing. The D3 format introduced a sequenced arrangement of the 192 bits within a frame, meaning the first 8 bits from channel 1, the second 8 bits from channel 2, and so forth.

Currently there are two major frame format standards for T1. One is mature and widely used, and the other is evolving and promises higher order link level features. They are the D4/SF format illustrated in Figure 17.7 and the extended superframe (ESF) format, which will be discussed later. Both retain the basic concept of a 193-bit frame with one bit for framing and 192 for user information and signals.

A **superframe** consists of 12 frames.

FIGURE 17.8

Summary of D4 protocol.

8,000 frames/s
125 µs/frame
24 channels/frame
8 bits per channel
1 framing bit
193 bits per frame

Figure 17.7 depicts the basic structure of the T1 format, which is 193 bits per frame, one bit assigned as a framing bit, and 192 bits of control and user information. The 192 bits may be channelized into 24 8-bit channels (DSs) or nonchannelized in the sense that the CPE derives time slots from the 192 bits but the network transports the 192 bits as a composite.

A summary of D4 protocol is given in Figure 17.8. Table 17.3 gives information about the DS0 through DS4 levels of T1. The DS1/D4 framing protocol uses a superframe concept. Each superframe consists of 12 frames. There is a specified framing bit pattern, which denotes the frames and the superframe format. This 12-bit pattern is used for frame synchronization and to mark frames 6 and 12, which contain the signaling bits. This 12-bit sequence is divided into two sequences, one for framing and the other for signaling. The framing bit alternates between 1 and 0 for every other frame. The signaling bits denote frames 6 and 12 in the following manner: frame 6 occurs when the signal bit 1 is preceded by three 0 bits, and frame 12 occurs when three 1 bits precede the signal bit 0. The 12-bit sequence is:

```
1 0 0 0 1 1 0 1 1 1 0 0
```

with bit 1 denoting framing, bits 2 through 5 denoting framing, bit 6 denoting framing and signaling, bits 7 through 11 denoting framing, and bit 12 denoting framing

Table 17.3 North American T1 Digital Hierarchy

Level	Bit Rate	Voice Channels	Media
DS0	64 Kbps	1	No interface defined
DS1	1.544 Mbps	24	2 copper paths
DS1C	3.152 Mbps	48	2 copper paths
DS2	6.312 Mbps	96	2 pairs "LoCap"
DS3	44.736 Mbps	672	Fiber, coax, radio
DS4	274.176 Mbps	4,032	Coax (T4M)

and signaling. Frames 6 and 12 carry voice-signaling information when voice transmission is used, and those signals are distinct from the signal bits for framing.

Once the DS1 bit stream is framed, the next step is to derive logical channels from the physical serial bit stream. There are two broad categories of framed formats that facilitate this derivation of logical channels: the channelized and the nonchannelized formats.

The channelized formats are based on the DS0 concept of 64 Kbps time slots. The nonchannelized formats are based on variable time slots. Table 17.4 shows a summary of the two formats. The two nonchannelized formats are from Stratacom, now owned by Cisco Systems, which is a T1/T3 equipment manufacturer, and its fast-packet switch is considered a proprietary format.

Figures 17.9 and 17.10 illustrate the frame structure of a channelized T1. Following the framing bit, the first eight bits comprise channel 1, the next eight bits channel 2, and so on, with the last eight bits comprising channel 24. There are eight bits per channel, 24 channels per frame, the channels are sequential within a frame, and the bits are weighted high order to low order within the channel. Figure 17.11 shows the standard T1 D4 frame format.

Table 17.4 Channelized and Nonchannelized Formats

Channelized Formats	*Nonchannelized Formats*
24–64 Kbps DS0s	Vendor proprietary
$n \times 64$ Kpbs	Subrate voice (32 K, 16 K, 8 K, 4.8 Kpbs)
Subrate data (2.4 K, 4.8 K, 9.6 K, 19.2 Kbps)	Fast packet
	Frame Relay
	Cell relay

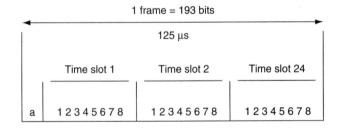

FIGURE 17.9

Frame structure of channelized 1.544 Mbps interface.

FIGURE 17.10

Fast packet frame format voice packet (a) and fast packet frame format data packet (b).

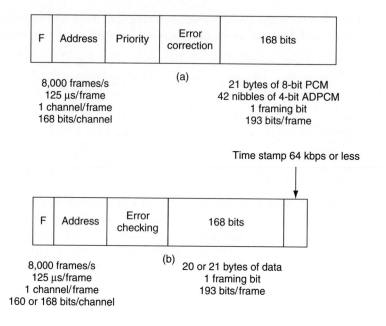

| F | Address | Priority | Error correction | 168 bits |

(a)

8,000 frames/s
125 µs/frame
1 channel/frame
168 bits/channel

21 bytes of 8-bit PCM
42 nibbles of 4-bit ADPCM
1 framing bit
193 bits/frame

Time stamp 64 kbps or less

| F | Address | Error checking | 168 bits | |

(b)

8,000 frames/s
125 µs/frame
1 channel/frame
160 or 168 bits/channel

20 or 21 bytes of data
1 framing bit
193 bits/frame

FIGURE 17.11

Standard T1 D4 frame format

| F | 1 | 2 | 3 | | 22 | 23 | 24 |

8,000 frames/s
125 µs/frame
24 channels/frame
8 bits/channel
1 framing bit
193 bits/frame

DS1/D4 Signaling

In the world of telephony, signaling refers to the process of establishing, maintaining, terminating, and accounting for a circuit-switched connection between two endpoints. Currently, the most common signaling schemes are dial pulse (DP) and dual tone multifrequency (DTMF). Both schemes employ an in-band signaling approach in which the signaling information occupies the same channel as the user's information. In employing this concept of signaling, an extra overhead burden exists, consequently robbing the user of the limited available bandwidth.

An alternative approach to these methods of signaling is common channel signaling. With common channel signaling, all information is carried out-of-band on a special signaling channel, which is a low bandwidth channel as compared with the DP and DTMF overhead. Common channel signaling also allows for quicker establishment and routing. This is because the routing is performed over a separate high speed digital channel. In addition, the out-of-band signaling channel is shared by a number of different network users, eliminating the excessive overhead burden.

The DS1/D4 signaling protocol offers two signaling options that are referred to as robbed-bit signaling, which is an in-band approach, and a transparent approach that can be used for common channel signaling. Robbed-bit signaling occupies the least significant bit of every time slot in the 6th and 12th frames of a superframe. The state of the bit, 0 or 1, tells whether the device at the other end is on-hook or off-hook.

Network signaling is used to:

1. Identify called and calling stations.
2. Set up connections in the network.
3. Identify the status of lines.
4. Provide billing information.
5. Release connections.

The signals are generally categorized as either supervisory, address, or audible tone signals. Supervisory signals cover line status and off-hook/on-hook. Address signaling pertains to call destination and dial pulse/touch tone signaling. Audible tone signals pertain to dial tones, busy tones, or ringing tones.

In the DS1/D4 superframe format (SF) the F-bits, as illustrated in Table 17.5, are divided into two groups, which are Ft (terminal framing) bits that are used to identify frame boundaries, and Fs (signaling framing) bits that are used to identify the robbed-bit signaling frames and associated signaling channels A and B.

DS1/D4 Maintenance

DS1/D4 maintenance is characterized by an out-of-service in-band approach. ANSI T1.403 specifies framed codes that may be used within the network to support out-of-service maintenance and protection switch operations. These codes (11000, 11100, and 10100) are used in repetitive pulse patterns of more than 5 s.

T1.403 also defines three in-band alarm signals, which are yellow alarm, alarm indication signal, and line loopback (LLB).

The yellow alarm signal is transmitted in the outgoing direction when a DS1 terminal determines that it has effectively lost the incoming signal. For the duration of the alarm condition, but for at least 1 s, the second bit in every 8-bit time slot using a channelized framing format will be a 0.

An AIS should be transmitted to the network interface when the loss of the originating signal occurs, or when any action is taken that would cause a signal disruption.

Table 17.5 Superframe Format

Frame No.	F-Bits			Bit Use in Each Time Slot		Signal Bit Use Options	
	Bit No.	Terminal Frame F1	Signaling Frame F1	Traffic	Signal	T	Signal Channel
1	0	1	—	1–8	—		
2	193	—	0	1–8	—		
3	386	0	—	1–8	—		
4	579	—	0	1–8	—		
5	772	1	—	1–8	—		
6	965	—	1	1–7	8	—	A
7	1,158	0	—	1–8	—		
8	1,351	—	1	1–8	—		
9	1,544	1	—	1–8	—		
10	1,737	—	1	1–8	—		
11	1,930	0	—	1–8	—		
12	2,123	—	0	1–7	8	—	B

The AIS will be removed when the condition triggering the AIS has been terminated. The AIS signal will be an all-1s signal.

Carriers and users use loopbacks as a maintenance tool to aid in problem resolution. To activate a loopback, a framed DS1 signal consisting of repetitions of four 0s followed by a 1, lasting for at least 5 s, with the framing bits overwriting the pattern, is sent across the network. To deactivate, a framed DS1 signal of two 0s followed by a 1 is sent, for at least 5 s, with the framing bits overwriting the previous signal.

Pulse Code Modulation

The present standard for transmitting digital voice across wire is pulse code modulation, also referred to as G.711. The PCM conversion process of taking an analog signal and modulating the signal onto a digital signal is done using a single integrated circuit chip called a CODEC, which stands for COder/DECoder. PCM is derived from the old Nyquist Theorem developed by a Bell Labs engineer named Harry Nyquist. The incoming analog signal represents a variation in voice tone and quality and is sampled 8,000 times per s. The modulator uses the sample to send a square wave pulse whose voltage level is the same as the original analog signal at that point. A coder then takes the height of the pulse and converts it to a digital value. A coder is an analog to digital converter. The output from the coder is an 8-bit code word, which represents the voltage of the pulse and the result of the analog input at the time of the sample.

$$8,000 \text{ samples per s} \times 8 \text{ bits/word} = 64,000 \text{ bits per s}$$

The value of 64,000 bits per s is equivalent to one DS0 channel on a T1.

The Nyquist Theorem demonstrated that reconstruction of an analog signal from digital data could contain all the information of the original analog signal provided the sampling rate is faster than twice the highest frequency in the original signal. Nyquist determined that the highest frequency level in speech was 4,000 Hz. Thus,

$$4,000 \text{ Hz} \times 2 = 8,000 \text{ Hz}$$

If the sampling rate is not rapid enough, the result can be a distorted analog signal creating overlapping signal waves instead of one solid wave. The sound then becomes unintelligible or difficult to understand. The size of the sample (8 bits) was decided as a result of optimizing a trade-off between bit rate and voice quality. The digital representation of a signal can only take on a small number of discrete values. An analog signal, on the other hand, has almost infinite variability. Consequently, at the precise time of sampling, the analog signal is seldom exactly the same as a digital output of the same signal.

Voice inputs are low-pass filtered to block any amount of signal at a frequency above 4,000 Hz. The filter affects adjacent frequencies. The usable upper limit for most bandwidth frequency ranges is 3,300 Hz. Filtering out the low end of the bandwidth spectrum blocks out unwanted noise around 60 Hz, limiting the practical lower limit of the bandwidth spectrum to around 300 Hz.

The Voice Channel Bank

Voice channels are digitized and multiplexed in PCM voice *channel banks*. Channel banks perform pulse amplitude modulation at the individual analog input ports. The pulse streams from 24 channels are interleaved on a single *pulse amplitude modulation (PAM)* bus. Each of the 24 signals in turn appears on the bus. Digital coding is done by one CODEC between the analog bus and a high speed T1 line. Synchronization from a control circuit inside the channel bank ensures that each modulator sends its pulses at the correct times.

The CODEC located inside the channel bank on the remote side will convert the T1 bit stream to pulses interleaved on the analog bus. Individual ports obtain their own pulses through monitoring a specific repetitive time slot. The PAM signal is then filtered at the assigned port to re-create the original analog signal. A device designed for voice that also must support pulse amplitude modulation must be modified to handle digital signals. Channel banks today have many of the built-in features and functions of a T1 multiplexer due to new hardware and advanced semiconductor technology.

Channel banks perform pulse amplitude modulation at the individual analog input ports.

Pulse amplitude modulation is a modulation scheme where the modulating wave is caused to modulate the amplitude of a pulse stream.

Table 17.6 Comparing Different Voice Encoding Methods from Guide to T1 Networking

Encoding Method	Bits per Sample	Sampling Frequency	Total Bit Rate	Interpretation of the Digital Signal
PCM	8	8,000	64,000	Value of analog signal at this time (this sample)
ADPCM	4	8,000	32,000	Change in analog signal level since last sample
CVSD	1	9.6 K–32 K	9.6 K–32 K	Does slope of the signal curve increase or decrease?
VQL	3.5	6,667	32,000	Loudness of this sample compared with the loudness of recent sample
VQC	varies	8,000	8 K, 16 K	Which vectors or fixed patterns of 1s and 0s does PCM signal look like most recently?
HVC	1.2	8,000	8 K, 16 K	Similar to VQC with more waveform coding and speech modeling

Adapted from Flannigan, W., *Guide to T1 Networking*, Flatiron Publishing, 1990, p. 46.

The Nature of Voice Signals

There are several forms of PCM, compared in Table 17.6, that can make voice quality better and allow service providers the ability to offer new services using voice. PCM, although it has been a mainstay for years for modulating an analog signal to a digital carrier wave, does have its drawbacks. A PCM signal does contain redundant information. By retaining redundant information, PCM uses more bandwidth than is really necessary. PCM, because of its very nature, sends a complete description of the sample each time a sample is taken. As a direct result of carrying a complete description, which allows PCM to handle analog waves that jump from one extreme to another, PCM introduces a minimum amount of quantizing noise. However, this ability is used only when a loud passage occurs at a high frequency. Such a situation can occur with a high speed modem, but the human voice does not change that rapidly. The average human voice will be a repetitive monotone pattern unless we experience deep emotions, which cause us to raise or lower the pitch of our speech pattern. Consequently, to save bandwidth and improve overall voice quality, several variations of PCM have been developed since the original PCM.

Differential Pulse Code Modulation

By nature, voice signals vary very slowly. The difference between successive samples in regard to change in voice pattern and quality is less than the full volume range. For sample accuracy this means that fewer bits are required to compensate for the change from one sample to the next. To take advantage of this, *differential pulse code modulation (DPCM)* only transmits the change since the last sample was taken. The beauty of the differential pulse code modulation technique is that DPCM can work with any number of bits. However, DPCM generally uses a 4-bit sample. A problem with using DPCM is that when the input signal really does change drastically between samples, the DPCM technique is not able to follow the input precisely. Sharp changes in high frequency and full volume may exceed capacity. As a result, the discrepancy between samples creates a large amount of quantizing noise and distortion. Consequently, it is very difficult for a 4-bit DPCM modulation to pass a modem signal of 4,800 bps.

● **Differential pulse code modulation** works with any number of bits. However, it generally uses a 4-bit sample.

Adaptive DPCM

DPCM can be improved without sacrificing an increase in bandwidth. There is a sophisticated algorithm that can assign different meanings to the 4-bit digital words to suit different conditions. The range of volume represented by the 4-bit word can be increased or decreased. The DPCM algorithm can be made to adapt to the situation of loud volume by increasing the range represented by 4 bits. It is possible in the extreme to have 4 bits cover a range typically covered by 8 bits.

While this method will reduce quantizing noise for large signals, for normal signals the noise level will increase. When the volume drops, *adaptive differential pulse code modulation (ADPCM)* reduces the volume range covered by the 4-bit signal. Thus, the difference between samples depends on the recent history of the level of the analog input signal.

● **Adaptive differential pulse code modulation** reduces the volume range covered by the 4-bit signal.

The standard ADPCM implementation sends 4 bits per sample but still samples at 8,000 times per s. Thus, the digital bandwidth needed would be 32,000 Kbps. This is how voice is compressed over a 64,000 Kbps circuit. ADPCM sends only the change since the last sample.

Continuously Variable Slope Delta Modulation

Now we will consider a form of DPCM in which the length of the digital word per sample is one bit. By using a small sample, many samples can be taken and sent in the same digital bandwidth. The only problem is that when using a one-bit word, it becomes difficult to measure an increase in volume or the loudness of a word. The one-bit that is being used is either a 1 or a 0. *Continuously variable slope delta modulation (CVSD)* will look at each individual bit and then adjust how fast the voltage is changing. Instead of sending the height of the analog signal curve or the change in the height of the curve, CVSD sends information about changes in the slope of the curve. CVSD controls the rate of change in the output. CVSD measures the units in V/s, and the ranges of values encountered are a few volts and 1/32,000 s (32,000 bps).

● **Continuously variable slope delta modulation** will look at each bit and then adjust how fast the voltage is changing.

At the transmitting end, CVSD compares the input analog voltage with an internal reference voltage. If the input signal is greater than the reference, a 1 is sent, and the slope of the reference curve is shaped upward. If the input is less than the reference, a 0 is sent and the slope of the reference reduced curve is shaped downward. CVSD tries to bring the reference to equal the input signal by steering the reference curve to follow the shape of the inputted analog signal.

On the opposite side, at the receiving end, the receiver will reconstruct the transmitter's reference voltage. The operation is exactly the same as at the transmitting end. The receiver reconstructs an increase in the slope when a 1 is received and a decrease in the slope when a 0 is received. As a result, there is an unfiltered reconstruction that is almost identical to the reference voltage, and that approximates the inputted analog signal. Filtering smooths this signal to a replica of the input. When the reconstruction is plotted on a graph, the result is a changing slope. The steeper the changes in the slope, whether up or down, the larger the output change between samples. Consequently, each time a change occurs in the same direction, the CVSD algorithm increases the size of the step taken between samples. Thus, with a good implementation of CVSD, voices are recognizable at 16,000 bps and still understandable at 9,600 bps.

Vector Quantizing Code

Vector quantizing code transmits a series of samples that in digital form is a string of 1s and 0s.

Vector quantizing code (VQC) starts with PCM. But instead of transmitting individual samples, VQC transmits a series of samples that in digital form is a string of 1s and 0s. A string of numbers (even 1s and 0s) is a vector. The vector that results from encoding the voice signal is compared to a set of vectors stored in the system in a lookup table. Once the closest match is identified, then a shorter identification (a code) of that vector is sent to the far end. Again, as in CVSD, the receiver performs the same function in reverse. The receiver has its own lookup table and the receiver code indicates which full-length vector is wanted. That vector is used exactly as if it were PCM information, meaning that it is converted to analog audio. As in all speech encoding systems, the VQC approach requires that both transmitting and receiving systems are the same. VQC transmits voice at 16 Kbps.

High Capacity Voice

If we want to have better voice quality at 8,000 bps, we need a considerable amount of processing under a powerful algorithm. High capacity voice (HCV) expands upon VQC by adding additional forms of waveform coding to model the vocal process. What this means is that this processing intensity leads to an implementation on digital signal processor (DSP) chips. There is a processing delay of 40 µs. This processing delay is enough to require echo cancellation, which is included in part of the algorithm. HCV does include echo cancellation. Integral echo cancellation compensates for satellite delay as well. As a result of tight bandwidth limitations, transition signaling is preferred. By sending a signal only when such state changes as someone hanging up the phone occur, HCV avoids robbing bits that would add to a significant portion of the 8,000 bps channel.

Extended Superframe

While the DS1/D4 protocols are prevailing in today's T1 networks, the *extended superframe (ESF)* format protocol, illustrated in Figure 17.12, will prevail in tomorrow's T1 networks. ESF adds substantial enhancements to T1. These enhancements are particularly seen in the areas of network management and telephony signaling.

Presently, the industry is in the midst of a transition from DS1/D4 to ESF. This transition will necessitate the upgrading of network equipment such as channel service units, and digital termination equipment (DTE) for full implementation of the ESF features. Through the upgrading of equipment to accommodate ESF, the T1 network will yield lower long-haul transmission costs, increased network availability, and increased network data integrity.

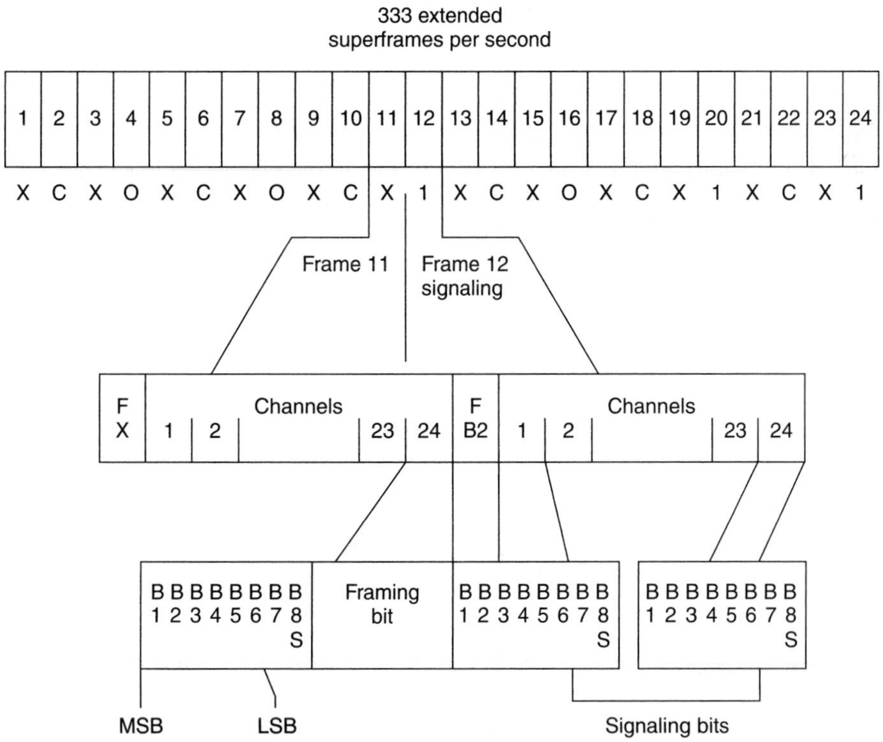

The **extended superframe** format protocol is a framing type used on T1 circuits.

FIGURE 17.12
ESF format.

ESF employs the same DS1 signal as DS1/D4 and the same 193-bit frame as DS1/D4. However, ESF specifies a new line code (ZBTSI) and will use an extended superframe of 24 frames as opposed to the 12-frame superframe of DS1/D4.

The two key areas of difference between ESF and SF are in signaling and maintenance. ESF offers a total of four signaling options to provide for in-band and out-of-band signaling. ESF offers a derived channel of 4 Kbps for in-service, nonintrusive diagnostics, testing, and performance measurement.

At present, three references exist for the ESF protocols. The initial reference for ESF was the AT&T publication 54016 in 1986 that was later revised in 1989. The standard reference for ESF is the ANSI T1.403-1989 standard. A future reference for ESF will be the ITU-T I.431 recommendation as T1 evolves to the ISDN primary rate interface.

The DS1/ESF protocols include the DS1 signal, ESF framing, channelized and non-channelized formats, out-of-band signaling, and in-service maintenance. The ESF maintenance facilities are implemented primarily within the CSU, while the ESF signaling facilities are implemented within the DTE.

The DS1 Signal

The same DS1 signal employed with DS1/D4 is employed with DS1/ESF. There is a standard line code, zero byte time slot interchange (ZBTSI) that is defined within ANSI T1.403 for ESF. ZBTSI is not specified for D4 and is recommended as an interim line code for ESF. The standard line code for ESF is in the same format as D4, which is B8ZS. The DS1 signal is transmitted on the T1 link in a binary format (1s and 0s). The ability to recognize the proper format of the DS1 signal is why regenerative repeaters can distinguish valid input from line noise.

DS1/ESF Framing

The framed DS1 signal illustrated in Table 17.7 consists of one framing bit and 24 8-bit channels. The extended superframe format consists of 24 frames. The S-bit of each frame forms an 8-Kbps channel that is subdivided into three subchannels. The three subchannels are designated FE (forward error), DL (data link), and BC (block control).

A 2-Kbps subchannel is derived from the FE bit positions of the S-bits. This subchannel is used for framing and signaling synchronization. The FE bit position of every fourth frame forms a repetitive pattern of 001011. This pattern facilitates the location of the 24 8-bit channels within a frame and the location of the frames carrying signaling information.

A 4-Kbps subchannel is derived from the m-bit position of the S-bits. The application of the m-bits varies by ESF standards that are provisioned. The m-bit position appears in every odd-numbered frame. The m-bits as a group are referred to as the data link channel.

Table 17.7 Assignment of S-Bits in a DS1 Signal

Frame Number	S-Bits		
	FE	DL	BC
1	—	E	—
2	—	—	C1
3	—	E	—
4	0	—	—
5	—	E	—
6	—	—	C2
7	—	E	—
8	0	—	—
9	—	E	—
10	—	—	C3
11	—	E	—
12	1	—	—
13	—	E	—
14	—	—	C4
15	—	E	—
16	0	—	—
17	—	E	—
18	—	—	C5
19	—	E	—
20	1	—	—
21	—	E	—
22	—	—	C6
23	—	E	—
24	1	—	—

A 2-Kbps subchannel is derived from the C1 through C6 positions of the S-bits. These bits are referred to as the block control channel. The BC channel employs a cyclic redundancy check scheme for error detection called CRC6. At the sender, a CRC6 sequence is generated for each outgoing ESF and transmitted immediately following the ESF frame. At the receiver, a CRC6 sequence is also checked against the incoming CRC6 for error detection.

ZBTSI Line Coding

There is another ESF variant when the zero byte time slot interchange (see Table 17.7) line code is used on a T1 facility. ZBTSI is an alternative to the B8ZS line code; both techniques provide a clear channel capability to the user of T1 facilities. The ZBTSI processing indicator flag bit (Z-bit) is associated with each 4-frame set. When ZBTSI is provisioned with a DS1/ESF link, m-bits from frames 1, 5, 9, 13, 17, and 21 are reassigned as Z-bits. In other words, 2K of the 4K data link channel is determined using the following procedure. The 4-frame data are scrambled to provide random distribution. The scrambled data are then scanned on a byte boundary to detect pulse density violations. If a pulse density violation is found within the 4-frame set, an address chain is set up to link the all-0 bytes. The addresses are inserted into the zero-byte slots. The last byte of the 4-frame set (96 bytes) is displaced by the first address of the chain of 0 bytes yielding a constant search start point. The original value of this last byte is stored in the location of the last 0 byte. If the value is a 0, then the address is used as the value. The Z-bit for that 4-frame set is encoded as a 0. If no pulse density violation is found after scrambling, the Z-bit is encoded as a 1. The framing bit definitions for ZBTSI line coding can be found in Table 17.8.

In general, the channelized and nonchannelized formats that are available with DS1/D4 are also available with ESF. One key exception involves a special format when implementing a common channel signaling facility within ESF.

The signaling bits within framed DS1s are used to identify called and calling situations, set up connections in the network, identify the status of lines, provide billing information, release connections, and provide other supervisory functions.

The DS1/ESF protocols offer four signaling options (see Table 17.9). Three of the four options use the in-band signaling structure and processes of the DS1/D4 protocols. Option 2 robs the low order, least significant bit (LSB) of each user channel (DS0) in frames 6, 12, 18, and 24 of the ESF and uses that bit position as an A signaling bit to reflect on-hook/off-hook status. Option 4 robs the same LSB but uses frame 6 LSBs as A-bits, frame 12 LSBs as B-bits, frame 18 LSBs as C-bits, and frame 24 LSBs as D-bits. Option 16 robs the same LSB but uses frame 6 LSBs as A-bits, frame 12 LSBs as B-bits, frame 18 LSBs as C-bits, and frame 24 LSBs as D-bits. The actual information content of these LSBs is a function of the DS0/trunk type and the carrier's network services.

Table 17.8 ZBTSI Line Coding

FRAMING BIT DEFINITIONS				
Frame No.	Bit No.	FPS	FDL	ZBTSI	CRC

Table 17.9 Extended Framing Format

Frame Number	S-Bits			Bit Use in Each Channel Time Slot		Signaling Bit Use Options			
	FE	DL	BC	Traffic	Signaling	2	4	16	T
1	—	m	—	Bits 1–8					
2	—	—	C1	Bits 1–8					
3	—	m	—	Bits 1–8					
4	0	—	—	Bits 1–8					
5	—	m	—	Bits 1–8					
6	—	—	C2	Bits 1–7	Bit 8	A	A	A	—
7	—	m	—	Bits 1–8					
8	0	—	—	Bits 1–8					
9	—	m	—	Bits 1–8					
10	—	—	C3	Bits 1–8					
11	—	m	—	Bits 1–8					
12	1	—	—	Bits 1–7	Bit 8	A	B	B	—
13	—	m	—	Bits 1–8					
14	—	—	C4	Bits 1–8					
15	—	m	—	Bits 1–8					
16	0	—	—	Bits 1–8					
17	—	m	—	Bits 1–8					
18	—	—	C5	Bits 1–7	Bit 8	A	A	C	—
19	—	m	—	Bits 1–8					
20	1	—	—	Bits 1–8					
21	—	m	—	Bits 1–8					
22	—	—	C6	Bits 1–8					
23	—	m	—	Bits 1–8					
24	1	—	—	Bits 1–7	Bit 8	A	B	D	—

The fourth signaling option of DS1/ESF is option T. Option T does not rob the LSBs of each DS0 in frames 6, 12, 18, and 24. With option T, DS1/ESF can offer a clear channel capability to the end user by using either a bit oriented signaling approach (BOS) or a message oriented signaling approach (MOS). With common channel signaling and option T, the 24th channel can be used as the signaling channel. The protocol for the 24th channel can be a bit oriented signal or a message oriented signal. I.431 calls for a message oriented signaling approach and designates this 24th slot as a D-channel.

DS1/ESF Maintenance: The Data Link Channel

The primary application of the DS1/ESF data link channel is in the general area of network management, primarily in fault detection and performance reporting. T1/ESF equipment will detect faults such as out of frame (OOF), bipolar violations (BPV), line code violation, and frame slips and report these conditions to the far end via the data link channel. T1/ESF equipment will monitor the performance of the T1 links in terms of parameters such as error-free seconds (EFS), errored seconds (ES), and severely errored seconds (SES) and report these conditions to the far end via the data link channel.

The data link channel (the m-bits) is used in two basic modes: a bit oriented and a message oriented mode.

In the bit oriented mode, referred to as bit oriented signaling, the m-bits will be grouped into a fixed sized word and the bit sequence within that fixed sized word denotes a signal, such as an alarm or alert signal. The bit pattern 0000000011111111 is widely used as a yellow alarm signal. This signal denotes that the far end node has detected an OOF condition and is informing the transmitting near end node of that condition.

In the message oriented mode, referred to as message oriented signaling, the m-bits will be grouped to form a multielement record or message. The message would convey error conditions, performance parameters, performance reports, and other network management information.

Internal standards groups such as the ITU-T and national standards groups such as ANSI are defining the protocols for the data link channel. Vendors and carriers have defined earlier protocols, but these are viewed as proprietary approaches.

An ESF device such as a channel service unit or multiplexer internal unit extracts the m-bits off the layer 1 bit stream of a T1 link and groups the m-bits into a layer 2 framed format in accordance with the format defined by ITU-T layer 2 protocol, high level data link control (HDLC).

Causes of T1 Transmission Impairments

There are four main causes of T1 impairments: faulty equipment, improper connections, environmental problems, and data specific problems.

Faulty Equipment: T1 equipment can cause errors when components fail or when they operate outside of manufacturer specifications. Errors that may suggest faulty equipment include bipolar violations, bit errors, frame errors, jitter, slips, and excess 0s. BPVs can occur due to faulty clock recovery circuitry. These errors occur as equipment becomes older and begins to drift out of specification.

Improper Connections: Improper connections or configurations create transmission errors. Intermittent errors can occur when component or cable connections are loose, and timing errors can occur when improper or conflicting timing sources are connected together. The circuit may not work at all due to mislabeled pins on terminating cable blocks and to crossed wires, for example, transmit-to-transmit instead of transmit-to-receive. These errors are typically discovered upon circuit installation and possibly during circuit acceptance when end-to-end tests are performed.

Environmental Problems: Electrical storms, power lines, electrical noise, interference, and crosstalk between transmission links can cause BPVs as well as bit, frame, and CRC errors. Typically, these conditions cause intermittent, bursty errors, which are among the most difficult to locate. Although the cause of these impairments is obvious, the one that is difficult to pinpoint is crosstalk. Improperly separated cable pairs sometimes cause crosstalk. Transmit and receive pairs should be between 25 and 100 pairs apart.

Data Specific Problems: Data characteristics, such as repetitive patterns, can force equipment to create pattern-dependent jitter and code errors. These errors may not exist when testing the transmission path with standard pseudorandom patterns. To facilitate timing recovery, there must be a sufficient number of transitions—no more than 15 consecutive 0s—to allow the equipment time to recover.

Testing

There are basically two methods to perform out-of-service testing on T1s. The first method is loopback testing and the second method is end-to-end testing. The difference between loopback and end-to-end testing is the equipment needed and the establishment of a loopback.

End-to-end testing, illustrated in Figure 17.13, is performed with two pieces of test equipment simultaneously in both directions. End-to-end testing does have an advantage over loopback testing because the direction of errors can be more quickly isolated, so that you can determine whether the transmit or receive leg is faulty.

Loopback testing is performed with one test set. If CSU loopbacks are established to perform the test, it is important to realize that the far end CSU in loopback affects the results. Most CSUs by design remove received BPVs, frame errors, and other errors before transmitting the data. This affects analysis interpretation because the near end technician will not be aware of BPVs on the far end's metallic loop and may have inconclusive results. For further information on T1 testing, please see Appendix A.

FIGURE 17.13
End-to-end testing.

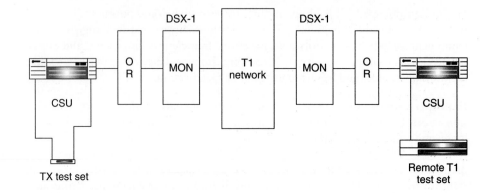

Review Questions

1. Briefly explain what a T1 is.
2. What are the three factors that account for the increasingly widespread implementation of T1 networks?
3. What are the four elements that make up a network in a metropolitan area?
4. What were the five classes of switching centers prior to the divestiture of AT&T?
5. What two platforms are T1 DS1/D4 protocols based on?
6. What are the two line coding formats used for D4 framing?
7. What are the two common channel signaling schemes?
8. What does network signaling do?
9. Describe how pulse code modulation works.
10. What is a channel bank?
11. Describe how differential pulse code modulation works.
12. Describe how adaptive pulse code modulation works.
13. What is CVSD?
14. What is VQC?
15. Describe the differences between superframe format and extended superframe format.
16. Describe how alternate line coding and B8ZS line coding work.
17. What are the four signaling options for ESF?
18. Describe the difference between bit oriented and message oriented signaling approaches.
19. What are the four main causes of T1 impairments?
20. What is the difference between loopback testing and end-to-end testing?

Summary Questions

1. How does T1 accomplish its goals?
2. What type of multiplexing does the DS1 signal employ?
3. What is the job of a CSU?
4. What is signaling?

For Further Reading

Flannigan, W. *Guide to T1 Networking*. New York: Telecom Library, 1990.

OSPF Design Guide

Introduction

Open Shortest Path First (OSPF) protocol was developed because of a need in the Internet community to introduce a high functionality, nonproprietary Internal Gateway Protocol (IGP) for the TCP/IP protocol family. The discussion of creating a common interoperable IGP for the Internet started in 1988 and did not get formalized until 1991. At that time the OSPF Working Group requested that OSPF be considered for advancement to Draft Internet Standard.

The OSPF protocol is based on link state technology, which is a departure from the Bellman-Ford vector based algorithms used in traditional Internet routing protocols such as RIP. OSPF has introduced new concepts such as authentication of routing updates, Variable Length Subnet Masks (VLSMs), route summarization, and so on.

The following will discuss the OSPF terminology, algorithm, and the pros and cons of the protocol in designing the large and complicated networks of today.

OSPF Versus RIP

The rapid growth and expansion of today's networks has pushed RIP to its limits. RIP has certain limitations that could cause problems in large networks:

- RIP has a limit of 15 hops. A RIP network that spans more than 15 hops (15 routers) is considered unreachable.
- RIP cannot handle VLSM. Given the shortage of IP addresses and the flexibility VLSM gives in the efficient assignment of IP addresses, this is considered a major flaw.

- Periodic broadcasts of the full routing table will consume a large amount of bandwidth. This is a major problem with large networks especially on slow links and WAN clouds.
- RIP converges slower than OSPF. In large networks, convergence tends to be in the order of minutes. RIP routers will go through a hold-down period and garbage collection and will slowly time-out information that has not been received recently. This is inappropriate in large environments and could cause routing inconsistencies.
- RIP has no concept of network delays and link costs. Routing decisions are based on hop counts. The path with the lowest hop count to the destination is always preferred even if the longer path has a better aggregate link bandwidth and slower delays.
- RIP networks are flat networks. There is no concept of areas or boundaries. With the introduction of classless routing and the intelligent use of aggregation and summarization, RIP networks seem to have fallen behind.

Some enhancements were introduced in a new version of RIP called RIP2. RIP2 addresses the issues of VLSM, authentication, and multicast routing updates. RIP2 is not a big improvement over RIP because it still has the limitations of hop counts and slow convergence, which are essential in today's large networks.

OSPF, on the other hand, addresses most of the preceding issues:

- With OSPF there is no limitation on the hop count.
- The intelligent use of VLSM is very useful in IP address allocation.
- OSPF uses IP multicast to send link state updates. This ensures less processing on routers that are not listening to OSPF packets. Also, updates are only sent in case routing changes occur, instead of periodically. This ensures better utilization of bandwidth.
- OSPF has better convergence than RIP. This is because routing changes are propagated instantaneously and not periodically.
- OSPF allows better load balancing.
- OSPF allows a logical definition of networks where routers can be divided into areas. This will limit the explosion of link state updates over the whole network. This also provides a mechanism for aggregating routes and cutting down on the unnecessary propagation of subnet information.
- OSPF allows routing authentication by using different methods of password authentication.
- OSPF allows the transfer and tagging of external routes injected into an Autonomous System. This keeps track of external routes injected by protocols such as BGP.

This of course would lead to more complexity in configuring and troubleshooting OSPF networks. Administrators that are used to the simplicity of RIP will be challenged with the amount of new information they have to learn to keep up with OSPF networks. Also, this will introduce more overhead in memory allocation and CPU utilization. Some of the routers running RIP might have to be upgraded in order to handle the overhead caused by OSPF.

What Do We Mean by Link States?

OSPF is a link state protocol. We could think of a link as being an interface on the router. The state of the link is a description of that interface and of its relationship to its neighboring routers. A description of the interface would include, for example, the IP address of the interface, the mask, the type of network it is connected to, the routers connected to that network, and so on. The collection of all these link states forms a link state database.

Link State Algorithm

OSPF uses a link state algorithm to build and calculate the shortest path to all known destinations. The algorithm by itself is quite complicated. The following is a very high level, simplified way of examining the various steps of the algorithm:

Routing Hierarchy: Unlike RIP, OSPF can operate within a hierarchy. The largest entity within the hierarchy is the autonomous system (AS), which is a collection of networks under a common administration that share a common routing strategy. OSPF is an intra-AS (interior gateway) routing protocol, although it is capable of receiving routes from and sending routes to other ASs.

An AS can be divided into a number of areas, which are groups of contiguous networks and attached hosts. Routers with multiple interfaces can participate in multiple areas. These routers, which are called area border routers, maintain separate topological databases for each area.

A topological database is essentially an overall picture of networks in relationship to routers. The topological database contains the collection of LSAs received from all routers in the same area. Because routers within the same area share the same information, they have identical topological databases.

The term domain is used to describe a portion of the network in which all routers have identical topological databases. Domain is frequently used interchangeably with AS.

An area's topology is invisible to entities outside the area. By keeping area topologies separate, OSPF passes less routing traffic than it would if the AS were not partitioned. Area partitioning creates two different types of OSPF routing, depending on whether the source and destination are in the same area—interarea routing occurs when they are in different areas.

An OSPF backbone is responsible for distributing routing information between areas. It consists of all area border routers, networks not wholly contained in any area, and their attached routers. Figure A.1 shows an example of an internetwork with

FIGURE A.1

An OSPF AS consists of multiple areas linked by routers. (Reproduced with permission of Cisco Systems, Inc. Copyright © 2001 Cisco Systems, Inc. All rights reserved.)

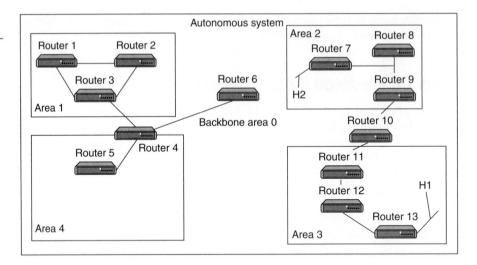

several areas. In Figure A.1, Routers 4,5,6,10,11, and 12 make up the backbone. If Host H1 in Area 3 wants to send a packet to Host 2 in Area 2, the packet is sent to router 13, which forwards the packet to router 12, which sends the packet to router 11. Router 11 then forwards the packet along the backbone to area border router 10, which sends the packet through two intra-area routers, router 9 and router 7, to be forwarded to Host H2.

The backbone itself is an OSPF area, so all backbone routers use the same procedures and algorithms to maintain routing information within the backbone that any area router would. The backbone topology is invisible to all intra-area routers, as are individual area topologies to the backbone.

Area border routers running OSPF learn about exterior routes through exterior gateway protocols such as Exterior Gateway Protocol (EGP) or Border Gateway Protocol (BGP), or through configuration information.

SPF Algorithm

The shortest path first (SPF) routing algorithm is the basis for OSPF operations. When an SPF router is powered up, it initializes its routing protocol data structures and then waits for indications from lower layer protocols that its interfaces are functional. After a router is assured that its interfaces are functioning, it uses the OSPF Hello protocol to acquire neighbors, which are routers with interfaces to a common network. The router sends Hello packets to its neighbors, which are routers with interfaces to a common network. The router sends Hello packets to its neighbors and receives their Hello packets. In addition to helping acquire neighbors, Hello packets also act as keepalives to let routers know that other routers are still functional.

On multiaccess networks, which are networks supporting more than two routers, the Hello protocol elects a designated router and a backup designated router. Among other things, the designated router is responsible for generating LSAs for the entire multiaccess network. Designated routers allow a reduction in network traffic and in size of the topological database.

When the link state databases of two neighboring routers are synchronized, the routers are said to be adjacent. On multiaccess networks, the designated router determines which routers should become adjacent. Topological databases are synchronized between pairs of adjacent routers. Adjacencies control the distribution of routing protocol packets, which are sent and received only on adjacencies.

Each router periodically sends an LSA to provide information on a router's adjacencies or to inform others when a router's state changes. By comparing established adjacencies to link states, failed routers can be detected quickly and the network's topology altered appropriately. From the topological database generated from LSAs, each router calculates a shortest path tree, with itself as root. The shortest path tree, in turn, yields a routing table.

Designing and Implementing an OSPF Network

There are a variety of RFCs that deal with OSPF. You should be familiar with the many different features available within the OSPF protocol. But which RFCs are commonly referred to in terms of OSPF design?

- RFC 1253: Open Shortest Path First (OSPF) MIB: This RFC contains information that provides management information relating to OSPF.
- RFC 1583: OSPF version 2: This RFC contains information regarding stub areas, route redistribution, authentication, tunable interface parameters, and virtual links.
- RFC 1587: Not-So-Stubby Areas (NSSA): This RFC explains the concept of NSSA.

Network Design Goals

It is not necessary to get into the reasons behind a decision to build an OSPF network or any of the previously covered definitions of what a network is. However, the five basic goals that you should keep in mind while designing an OSPF network, or any network for that matter, should be adhered to:

- Functionality
- Scalability
- Adaptability
- Manageability
- Cost effectiveness

FIGURE A.2

*Route summa-
rization affects
network scalabil-
ity. (Reproduced
with permission
of Cisco Systems,
Inc. Copyright
© 2001 Cisco
Systems, Inc. All
rights reserved.)*

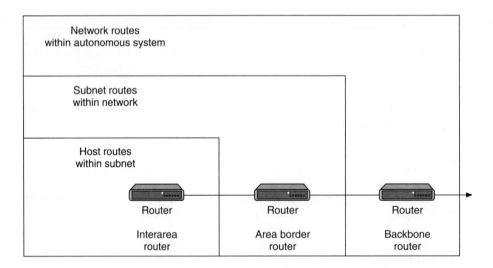

Functionality

The bottom line is that the network must work. Because networks are an integral part of enabling users to do their jobs, this is essential. It is here that the use of Service Level Agreements (SLAs) is essential. You must know what is expected of the network in order to design it properly.

Scalability

As your organization grows, the network must be able to keep pace. Your network and its initial design must be able to expand accordingly. A network that cannot keep pace with the organization's needs is not much use.

Routing summarization is a major factor in the success of designing your network. If you want to ensure that your network can scale properly, the summarization is the biggest factor of your success. Without summarization, you will have a flat address design with specific route information for every host being transmitted across the network, a very bad thing in large networks. To briefly discuss summarization, remember that routers summarize at several levels as illustrated in Figure A.2.

For example, hosts are grouped into subnetworks, subnetworks are then grouped into major networks, and these are then consolidated into areas. The network can then be grouped into an autonomous system. There are many smaller networks that desire to use a standard routing protocol such as OSPF. These networks can, for example, have 100 or less routers with a relatively small IP space. In these situations, summarization may not be possible and might not gain much if it were implemented.

Adaptability

Adaptability refers to your network's capability to respond to changes. In most cases, adaptability refers to your network's ability to embrace new technologies in a timely and efficient manner. This becomes extremely important as the network ages because change within networking is racing forward at breakneck speeds. Although it is not necessary to always be on the leading or bleeding edge, there is a lot to be said for letting others find the bugs!

Manageability

To provide true proactive management is the goal here. The network must have the proper tools and design to ensure that you are always aware of its operation and current status.

Cost Effectiveness

In this case, the true bottom line of network design is saved for last. The reality of life is that budgets and resources are limited, and building or expanding the network while staying within the predetermined budget is always a benefit to your career and proper network design.

Although there are five basic goals of network design that can be followed in any situation, there also should be a certain mind-set during the process. This mind-set is in regard to the actual technology that will be used. It is very important to use state-of-the-art technology. The reasoning behind this is that by spending a little extra money up front you are investing with an eye toward the future, knowing that the network that is being built will be able to grow longer than otherwise possible, at least from a technological standpoint.

Network Design Issues

Up until this point, the various network design goals and the methodology needed to make the goals become a reality have been discussed. There are also certain design issues that must be considered when working through the network design process:

- Reliability: When designing networks, reliability is usually the most important goal because the WAN is often the backbone of any network.
- Latency: Another big concern with users occurs when network access requests take a long time to be granted. Users should be notified about a latency problem in the network.
- Cost of WAN resources: WAN resources are expensive and, as such, frequently involve a trade-off between cost efficiency and full network redundancy. Usually cost efficiency wins.

● Amount of traffic: This is a very straightforward consideration. You must be able to accurately determine the amount of traffic that will be on the network in order to properly size the various components that will make up the network. As the network is implemented, a baseline should be developed that can be used to project future growth.

● Allowing multiple protocols on the WAN: The simplicity of IP is of great benefit to any network. For example, by allowing only IP based protocols on the network you will avoid the unique addressing and configuration issues relating to other protocols.

● Compatibility with standards or legacy systems. Compatibility is always going to be an issue within your network throughout its life. As a network designer this must always be kept in mind as you proceed forward.

● Simplicity and easy configuration: This feature is a very important one, even though you might be involved only in the design and implementation of the network and not the management. In that case the knowledge developed will need to be passed onto those who manage the network. Ensure that the ideas of simplicity and ease of configuration are kept in mind while the design documents are developed for the network.

● Support for remote offices and telecommuters: In today's telecommunications environment, remote satellite offices are becoming commonplace and require network connectivity, so you must plan accordingly. The estimates say that companies have increased the number of telecommuters almost every day. This must be kept in mind as the placement of network components is determined to ensure that they can handle this requirement when it becomes a priority for your organization.

Network Design Methodology

There are six common steps that can be used to design your OSPF network, or any network for that matter. These steps are not set in stone and will not guarantee the perfect network, but they will provide you with realistic steps and considerations that if taken into account will make for a well-designed network. These steps will also help you avoid getting caught up in all the bells and whistles available in the new-enhanced-ultra-secret-turbo-series-network equipment, which is the answer to all of your networking needs.

The steps to designing a network have been proven not only over time but also through countless networks that have been designed and implemented based on this standard.

1. Analyze the requirements.
2. Develop the network topology.
3. Determine addressing and naming conventions.
4. Provision the hardware.
5. Deploy protocol and IOS features.
6. Implement, monitor, and maintain the network.

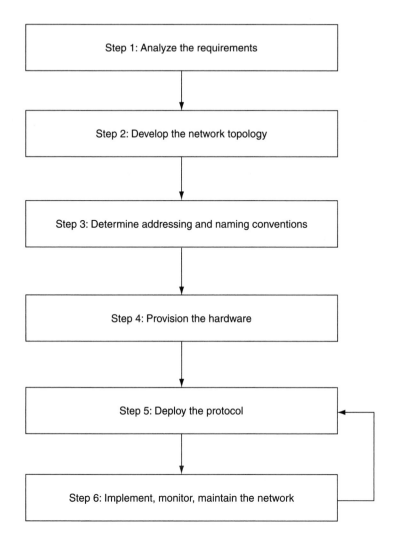

FIGURE A.3

Network design methodology. (Reproduced with permission of Cisco Systems, Inc. Copyright © 2001 Cisco Systems, Inc. All rights reserved.)

Although the network might not have the technology du jour, it might not really need it if you objectively determine the needs of a network by following the design methodology illustrated in Figure A.3.

Step 1: Analyze the Requirements

This step will detail the process of determining expectations and then converting those into a real network or explaining why everyone cannot have videoconferencing at the desktop.

Note: What do you know? Going into Step 1, you know that an OSPF network is required but not what it will need to accomplish for your users or how you will physically design the network.

Granted, the needs of users are always changing, and sometimes they do not know what they need. However, it is true that they know what they want and when they want it, which is always now or yesterday. Nevertheless, from a network design prospective, they do not always know what they need or why they need it.

Nevertheless, as the network engineer involved in the design of the network, you must still objectively listen and determine user needs. In the end, they are going to be the customers of the network, and the customer is always right. You must also take into consideration what the future might hold for them. Therefore, you should ask the users what needs they see themselves having in the future. This question should be directed toward their jobs because it is your responsibility to take their response and turn that into the requirements of the network.

A corporate vision is always important. For example, do the long-range corporate plans include having a Web site? If so, what will it be doing? How about running voice over the network? What about videoconferencing; is that going to be a corporate need? Additional data you might want to consider gathering are the current organizational structure, locations, and flow of information within the organization and any internal or external resources available to you. Armed with this information, your networks needs analysis, you should then begin determining the cost and benefit analysis. Of course in many cases you will not be able to get all the equipment or bandwidth you think is necessary. Therefore, it is also advisable to create a risk assessment detailing the potential problems or areas of concern regarding the network design.

OSPF Deployment: As you go through the process of determining the network requirements, keep in mind some important questions regarding the requirements of OSPF. The answers to these questions will help you further define the requirements of your OSPF network.

- How should the OSPF autonomous system be delineated? How many areas should it have and what should the boundaries be?
- Does the network and its data need to have built-in security?
- What information from other autonomous systems should be imported into your network?
- Which sites will have links that should be preferred (lower costs)?
- Which sites will have links that should be avoided (higher costs)?

Load Balancing with OSPF: While determining the network requirements, keep in mind the load balancing feature of OSPF. Certain vendors in their implementation of OSPF feel that any router can support up to four equal-cost routes to a destination. When a failure to the destination is recognized, OSPF immediately switches to the remaining paths. OSPF will automatically perform load balancing, allowing equal-cost paths. The cost associated is determined (default) by the interface bandwidth statement unless otherwise configured to maximize multiple path routing. In other words, the interface bandwidth statement defines a cost for that particular link. The default cost can be calculated by dividing 1,000,000,000 by the default bandwidth of the interface.

OSPF Convergence: OSPF convergence is extremely fast when compared with other protocols—this was one of the main features included within its initial design. To keep this desirable feature fully functional in your network, you need to consider the three components that determine how long it takes for OSPF to converge:

- The length of time it takes OSPF to detect a link or interface failure.
- The length of time it takes the routers to exchange routing information via LSAs, rerun the SPF algorithm, and build a new routing table.
- A built-in SPF delay time of 5 seconds (default value).

Thus, the average time for OSPF to propagate LSAs and rerun the SPF algorithm is approximately 1 second. Then the SPF delay timer of 5 seconds must elapse. Therefore, OSPF convergence can be anything from 6 to 46 seconds depending on the type of failure, SPF timer settings, size of the network, and size of the LSA database. The worst case scenario involves a link failure in which the destination is still reachable via an alternate route. The 40-second default dead timer will need to expire before the SPF is rerun.

Step 2: Develop the Network Topology

This step will cover the process of determining the network's physical layout. There are generally only two common design topologies: meshed or hierarchical.

Note: What do you know? Coming into Step 2, you have developed a list of requirements associated with this OSPF network. You have also begun to lay out the financial costs associated with the network based on this information. These costs could include equipment, memory, and associated media.

Meshed Topology: In a meshed structure, the topology is flat and all routers perform essentially the same function, so there is no clear definition of where specific functions are performed. Network expansion tends to proceed in a haphazard, arbitrary manner. This type of topology is not acceptable to the operation of OSPF. It will not correctly support the use of areas or designated routers.

Hierarchical Topology: In a hierarchical topology, the network is organized in layers that will have clearly defined functions. In this type of network there are three layers:

- Core Layer: This would make an excellent place for OSPF backbone routers that are connected through area 0. All of these routers would be interconnected, and there should not be any host connections. This is because its primary purpose is to provide connectivity between other areas.
- Distribution Layer: It is here that you would locate other OSPF areas connected through Area Border Routers (ABRs) back to the core layer (area 0).

FIGURE A.4

OSPF hierarchical topology.
(Reproduced with permission of Cisco Systems, Inc. Copyright © 2001 Cisco Systems, Inc. All rights reserved.)

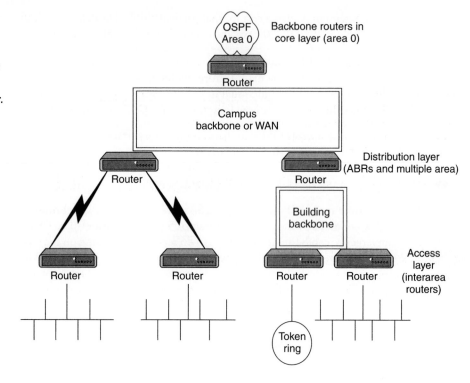

This is also a good location to begin implementing various network policies such as security, DNS, and so forth.

- Access Layer: This is where the interarea routers that provide connections to the users would be located. This layer ID is where the majority of the hosts and servers should be connecting to the network.

By using this type of layered network design, you will gain some benefits that will help you design the network as illustrated in Figure A.4.

The benefits of the OSPF hierarchical topology as implemented in Figure A.4 are as follows:

- Scalable: Networks can grow easily because functionality is localized so that additional sites can be added easily and quickly.
- Ease of implementation: Because functionality is localized, it is easier to recognize problem locations and isolate them.
- Predictability: Because of the layered approach, the functionality of each layer is much more predictable. This makes capacity planning and modeling that much easier.
- Protocol support: Because an underlying physical architecture is already in place if you want to incorporate additional protocols, such as BGP, or if your

organization acquires a network running a different protocol, such as Token Ring, you will be able to add it easily.

- Manageability: The physical layout of the network lends itself toward logical areas that make network management much easier.

From using the three layered hierarchical design approach, you can see that this model fits perfectly into OSPF's logical design, and it is this model on which you will be basing your network design. Before discussing how to implement and design this type of model, you need some basic OSPF design suggestions.

OSPF Backbone Design in the Hierarchical Model

The process of designing the backbone area should be as simple as possible by avoiding a complex mesh. Consider using a LAN solution for the backbone. The transit across the backbone is always one hop, latency is minimized, and it is a simple design that converges quickly. Figure A.5 illustrates a simple OSPF backbone design. You know that users should be kept off the backbone because it is only a transit area, but that is not all. The backbone should be secured physically. As a network-critical shared resource, the routers need to be physically secured. If you use the previously

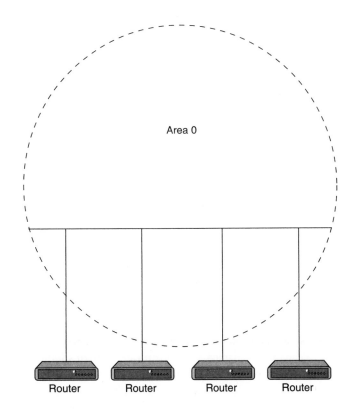

FIGURE A.5

Simple OSPF backbone design. (Reproduced with permission of Cisco Systems, Inc. Copyright © 2001 Cisco Systems, Inc. All rights reserved.)

FIGURE A.6

Isolate the backbone and secure it both physically and logically. (Reproduced with permission of Cisco Systems, Inc. Copyright © 2001 Cisco Systems, Inc. All rights reserved.)

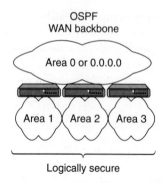

OSPF
WAN backbone

Area 0 or 0.0.0.0

Area 1   Area 2   Area 3

Logically secure

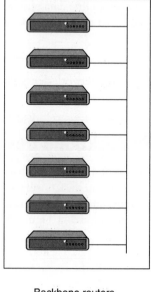

LAN backbone in secure closet

Backbone routers

Physically secure

mentioned LAN backbone solution, then securing the network can be relatively easy; just place it in a secure closet or rack as illustrated in Figure A.6.

Areas: Stub, Totally Stubby, or Not-So-Stubby

You will have to design your OSPF network with areas to make the network scalable and efficient. Areas should be kept simple and stubby, with less than 100 routers—optimally between 40 to 50 routers—and have maximum summarization for ease of routing. The network illustrated in Figure A.7 demonstrates these suggestions.

Even though these design suggestions are helpful, what are you really going to gain in your network by adding stub areas? Simply, they will summarize all external LSAs as one single default LSA that applies only to the external links from outside the autonomous system.

The stub area border router sees all the LSAs for the entire network and floods them to other stub area routers. They keep the LSA database for the stub area with this additional information and the external route. Figure A.8 illustrates the operations in a stub area.

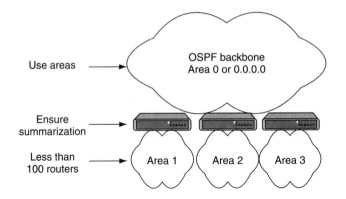

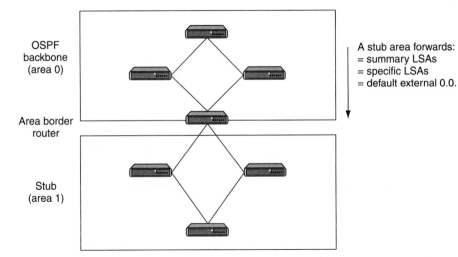

There are also totally stubby areas that you can design within your network. Totally stubby areas are a vendor-specific feature available within their implementation of the OSPF standard. If an area is configured as totally stubby, only the default summary link is propagated into the area by the ABR. It is important to note that an ASBR cannot be part of a totally stubby area, nor can redistribution of routes from other protocols take place in this area. Figure A.9 shows the operations in a totally stubby area.

The main difference between a stub area and a not-so-stubby area (NSSA) is that the NSSA imports a limited number of external routes. The number of routes is limited to only those required to provide connectivity between the backbone areas. You may configure areas that redistribute routing information from another protocol to the OSPF backbone as an NSSA.

FIGURE A.9
Totally stubby area operation. (Reproduced with permission of Cisco Systems, Inc. Copyright © 2001 Cisco Systems, Inc. All rights reserved.)

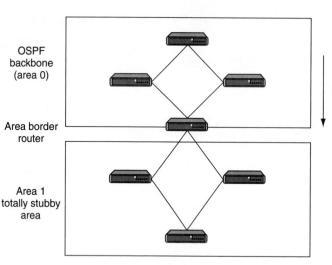

OSPF backbone (area 0)

Area border router

Area 1 totally stubby area

A totally stubby area forward = default link 0.0.0.0 blocks LSA types 3, 4, 5 and 6

Case Study

An Example of an OSPF Network with a Hierarchical Structure

To design this type of network model, you should gather a list of the different locations requiring network connectivity within your organization. For purposes of this example and ease of understanding, let us consider your organization as an international corporation, and you have determined that you have the following divisions each with various business units within it with the units grouped by location and then by function:

I. Headquarters: Washington, DC (all in the same building)
   A. United States Executives
   B. Legal Department

II. Human Resources (located at headquarters but in different building)
   A. Payroll
   B. Benefits
   C. Corporate Recruiting

III. Sales and Marketing
   A. Northern Division (6 offices)
   B. Southern Division (6 offices)
   C. Eastern Division (5 offices)
   D. Western Division (7 offices)

IV. Manufacturing
   A. Engineering located in western United States
   B. Widgets division located in northern United States (4 factories)

C. Gidgets division located in southern United States (3 factories)

D. Tomgets division located in western United States (3 factories)

The listed units will become the basis of OSPF areas. Contained within the areas will be OSPF interarea routers that connect to the various hosts.

Of these groupings, you should select essential locations at which to locate the backbone routers. For our example, you know that the Headquarters will have a backbone router that will be connected to area 0. You have been given several requirements based on traffic flow and corporate requirements:

- All divisions must be within the same area, regardless of geographic location
- All divisions must be able to connect to headquarters
- In our company, area 0 links all major continental locations throughout the globe
- All region clusters must have alternate routes
- Internet connectivity for entire company
- If backbone router fails, network operation with U.S. division must continue
- Engineering and Manufacturing must communicate quickly and easily

Begin separating the sites into areas and picking one location within each

area at which will reside the ABR. This will result in a proposed set of OSPF routers deployed as follows:

- Backbone router (area 0): Connects to global area 0
- ABR (area 1): Executives and Legal Department
- ABR (area 2): Human Resources
- ABR (area 3): Sales
- ABR (area 4): Manufacturing and Engineering
- ASBR: Internet connectivity

The remaining sites will each be assigned an interarea router to connect them to the network. One main site within each geographical area will be the hub site for that geographic area, thereby reducing bandwidth costs.

At this point, you should have your organization separated into areas or layers and an overall topology map laid out. Figure A.10 illustrates the described example network. Let us remember the requirements before tearing up this design layout: (1) All divisions must be within the same area, regardless of geographic location. (2) There are many ways of designing a network and this is just one way and one person's opinion. (3) There is no substitute for actual network design experience, because everyone makes mistakes. (4) Now that you think you have a solid network design, have someone else look at it and consider modeling it in a software package.

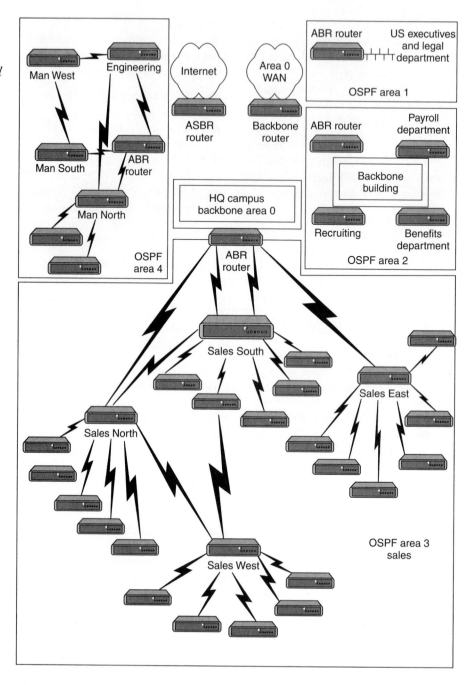

FIGURE A.10

Proposed network design: topology map. (Reproduced with permission of Cisco Systems, Inc. Copyright © 2001 Cisco Systems, Inc. All rights reserved.)

Step 3: Determine Addressing and Naming Conventions

Step 3 covers the actual process of assigning the overall network addressing scheme. By assigning blocks of addresses to portions of the network, you are able to simplify addressing, administration, and routing, and increase scalability.

Because OSPF supports VLSM, you can really develop a true hierarchical addressing scheme. This hierarchical addressing results in very efficient summarization of routes throughout the network.

Note: What do you know? Coming into Step 3 you have determined your network's requirements and developed a physical network topology. While planning, you have continued to keep track of the one-time costs and recurring costs. In this step, you will determine the addressing and naming conventions that you plan on using.

Public or Private Address Space: A good rule of thumb to remember when determining whether to use public or private address space is that your address scheme must be able to scale enough to support a larger network because your network will most likely continue to grow.

You must determine what range of IP addresses you are going to deploy within your network. The first question you need to answer is: Do I have public address space assigned to me by the InterNIC or am I going to be using private address space as specified in RFC 1918 and 1597? Either choice will have its implications on the design of your network. By choosing to use private address space and connecting to the Internet, you will be faced with having to include the capability to do address translation as part of your network design.

To further complicate the issue, you might also have to deal with a preexisting addressing scheme and/or the need to support the use of Dynamic Host Configuration Protocol (DHCP) or Domain Naming System (DNS). DHCP is a broadcast technique used to obtain an IP address for an end station. DNS is used for translating the names of network nodes into IP addresses. Figure A.11 shows a good example of how to lay out the IP addresses and network names for the example network.

Plan Now for OSPF Summarization: You should realize the importance of proper route summarization on your network. The OSPF network in Figure A.12 does not have summarization turned on. Notice that by not using summarization, every specific link LSA will be propagated into the OSPF backbone and beyond, causing unnecessary network traffic and router overhead. Whenever an LSA is sent, all affected OSPF routers will have to recompute their LSA database and routes using the SPF algorithm.

OSPF will provide some added benefits if you design the network with summarization. For example, summary link LSAs will propagate into the backbone (area 0). This is very important because it prevents every router from having to rerun the SPF algorithm, increases the network's stability, and reduces unnecessary traffic. Figure A.13 demonstrates this principle.

FIGURE A.11
Address and naming conventions. (Reproduced with permission of Cisco Systems, Inc. Copyright © 2001 Cisco Systems, Inc. All rights reserved.)

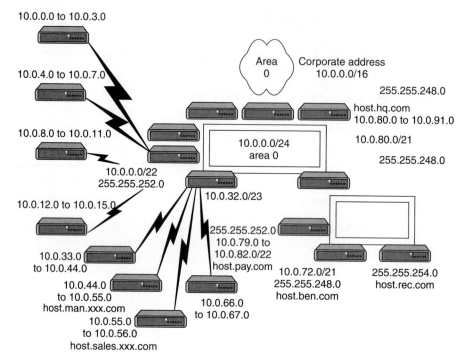

FIGURE A.12
No route summarization will cause network problems. (Reproduced with permission of Cisco Systems, Inc. Copyright © 2001 Cisco Systems, Inc. All rights reserved.)

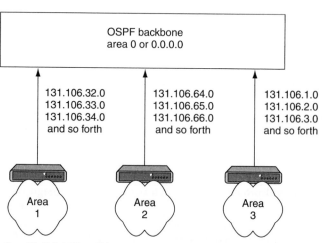

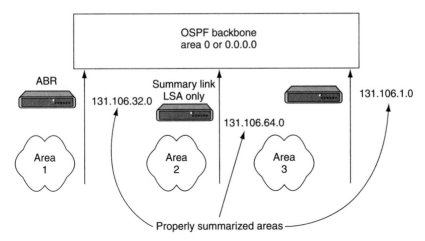

Specific link LSAs are not propagated to the backbone so flopping links are hidden inside an area.

IP addresses in an OSPF network should be grouped by area, and you can expect to see areas with some or all of the following characteristics:

- Major network numbers
- Fixed subnet mask(s)
- Random combination of networks, subnets, and host addresses

It is important that hosts, subnets, and networks be allocated in a controlled manner during the design and implementation of your OSPF network. The allocation should be in the form of contiguous blocks that are adjacent so that OSPF LSAs can easily represent the address space. Figure A.14 shows an example of this.

Note: Allocation of IP addresses should be done in powers of two so that these blocks can be represented by a single summary link advertisement. Large blocks of contiguous blocks of addresses can be summarized. To minimize the number of blocks you should make them as large as possible.

Bit Splitting: Bit splitting is a very useful technique if you have to split a large network number across more than one OSPF area. Simply, bit splitting borrows some subnet bits for designated areas such as:

- To differentiate two areas, split 1 bit.
- To differentiate 16 areas, split 4 bits.

Figure A.15 demonstrates this bit splitting technique. The example uses 4 bits for the area and uses 32-bit numbers to represent 4 of the 16 possible areas. The area numbers correspond to the summary advertisement that represents the area.

FIGURE A.14

Configure OSPF for summariza-tion. (Reproduced with permission of Cisco Systems, Inc. Copyright © 2001 Cisco Systems, Inc. All rights reserved.)

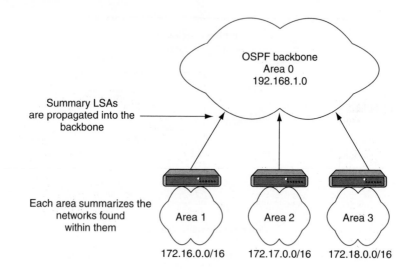

FIGURE A.15

Bit splitting ad-dress space. (Reproduced with permission of Cisco Systems, Inc. Copyright © 2001 Cisco Systems, Inc. All rights reserved.)

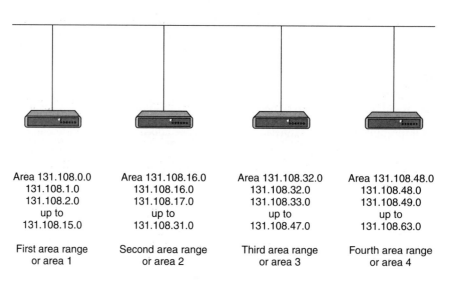

MAP OSPF Address for VLSM: The reasons behind VLSM are similar to bit split-ting. Remember to keep small subnets in a contiguous block and increase the num-ber of subnets for a serial meshed network. Figure A.16 provides a good example of VLSM OSPF mappings.

Discontiguous Subnets: Subnets become discontiguous when they are separated by one or more segments represented by a different major network number. Discontiguous subnets are supported by OSPF because subnet masks are part of the link state database.

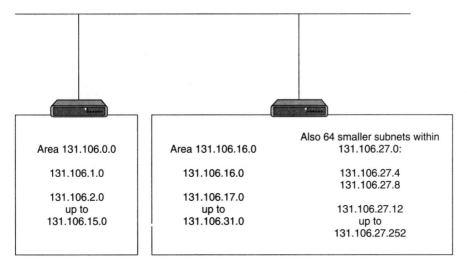

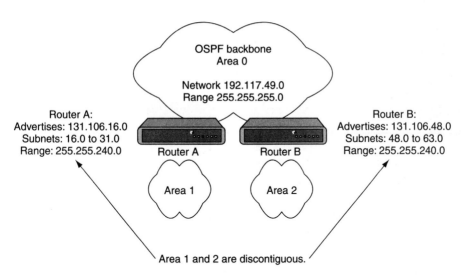

Consider the following example: The OSPF backbone area 0 could be a class C address while all the other areas could consist of address ranges from a class B major network illustrated in Figure A.17.

Note: OSPF supports discontiguous subnets regardless of whether summarization is configured within the network. However, everything within your network will route better and have a more stable design if summarization is configured.

Naming Schema: The naming scheme used in your network should also be designed in a systematic way. By using common prefixes for names within an organization, you will make the network much easier to manage and more scalable.

It is also important to carry a naming convention into your routers as well. This will assist everyone in dealing with your network because the router names actually hold some meaning, instead of an abstract like order number.

Step 4: Provision the Hardware

In Step 4, you must use vendor documentation, salespeople, and system engineers to determine the hardware necessary for your network. This is for both LAN and WAN components. For LANs you must select and provision router models, switch models, cabling systems, and backbone connections. For WANs you must select and provision router models, modems, CSUs/DSUs, and remote servers.

Note: What do you know? Coming into Step 4 you have determined your network requirements, developed a physical network topology, and laid out your addressing and naming scheme for the network. In this step, you will begin selecting and provisioning the necessary network equipment to implement the design.

When selecting and provisioning routing and switching hardware, consider the following areas:

- Expected CPU usage
- Minimum RAM
- BUS budget
- Forwarding budget
- Required interface types and density

Step 5: Deploy Protocol and IOS Features

In Step 5, you will need to deploy the more specific OSPF and IOS features. It is not necessary to have a network with every single option turned, nor is it something you are likely to see. Some of the features you should consider implementing are covered next.

Note: What do you know? Coming into Step 5 you have determined your network requirements, developed a physical network topology, laid out your addressing and naming scheme, and begun the provisioning of the network equipment. In this step, you will begin deploying the OSPF and IOS features that you will be using within the network.

OSPF Features: This area covers some of the features of OSPF (authentication and route redistribution between protocols) that you should consider deploying within your network. There can be only one choice concerning which feature should be considered first.

Protecting corporate resources, security, policing the network, ensuring correct usage of the network, and authentication are different labels for a similar need within every network: network security. Network security should be built into the network from day one, not added as an afterthought. Mistakes have already happened in the networking environment you know today. How could they not with the most required Internet presence and WWW logo seen on almost every business card? The open, unsecure protocols such as Simple Mail Transfer Protocol (SMTP) or Simple Network Management Protocol (SNMP) are essential for business and network management, although they are also vulnerable for exploitation. Hopefully the respective working groups will move toward solving this problem. However, OSPF does come with built-in authentication.

OSPF's built-in authentication set is extremely useful and flexible. In the OSPF specification, MD5 is the only cryptographic algorithm that has been completely specified. The overall implementation of security within OSPF is rather straightforward. For example, you assign a key to OSPF. This key can either be the same throughout your network or different on each router's interface or a combination of the two. The bottom line is that each router directly connected to each other must have the same key for communication to take place.

Route redistribution is another useful feature. Redistribution is the exchange of routing information between two different routing processes (protocols). This feature should be enabled in your routers if you have separate routing domains within your autonomous system and you need to exchange routes between them. For example, the engineering department might be running OSPF and the accounting department might be running IGRP, as illustrated in Figure A.18.

Figure A.18 depicts one router connecting the two separate touring processes (protocols), which need to share routing information. This sharing process is called redistribution. The router shown in Figure A.18 is configured to run both IGRP and

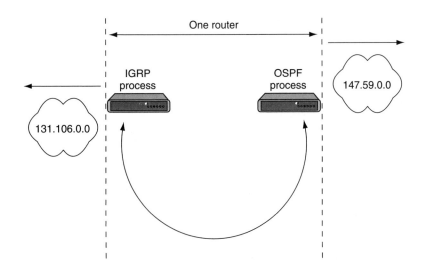

FIGURE A.18

Redistributing routing information between protocols. (Reproduced with permission of Cisco Systems, Inc. Copyright © 2001 Cisco Systems, Inc. All rights reserved.)

OSPF. When routes are redistributed between major networks, no subnet information is required.

IOS Features: Some of the features of the IOS that you should consider deploying in your network are as follows:

- Access lists
- Queuing
- Route maps
- Limit certain routes from being propagated

Step 6: Implement, Monitor, and Manage the Network

The last step is also the first step to continually managing the growth of your network. In the context of this step you should consider the following actions:

- Using network management tools for monitoring
- Performing proactive data gathering
- Knowing when to scale the network to meet new demands (new hardware, upgrade circuit speeds, support new applications)

Note: What do you know? Coming into Step 6 you have determined your network requirements, developed a physical network topology, laid out your addressing and naming scheme, provisioned your network equipment, and deployed the necessary OSPF and IOS features. In this step, you will begin to answer the question of how to actually make your routers do all this, which is vendor-specific configuration. Thus, we have discussed the entire process of designing and building an OSPF network.

BGP Attributes and Policy Routing

Controlling BGP Routes

Chapter 12 focused on the existence of policy engines that provide attribute manipulation and route filtering. Appendix B focuses on attribute manipulation and route filtering, the keys to controlling routing information. Each BGP attribute is examined to determine what it does and how to use it.

Traffic inside and outside an AS always flows according to the road map laid out by routes. Altering the routes translates to changes in traffic behavior. Among the questions that organizations and service providers ask about controlling routes are: How do I prevent my private networks from being advertised? How do I filter routing updates coming from a particular neighbor? How do I make sure that I use this link or this provider rather than another one? BGP provides the necessary hooks and attributes to address all these questions.

BGP Attributes

The BGP attributes are a set of parameters that describe the characteristics of a prefix (route). The BGP decision process uses these attributes to select its best routes. Attributes are part of each BGP UPDATE packet. The next few sections cover these attributes and how they can be manipulated to affect the routing behavior.

The Next_Hop Attribute

The Next_Hop Attribute is a well-known mandatory (type code 3). In IGP, the next hop to reach a route is the IP address of the connected interface that has announced the route. The Next_Hop concept with BGP is slightly more elaborate and takes one of the following three forms:

1. For EBGP sessions: the next hop is the IP address of the neighbor that announced the route.
2. For IBGP sessions: for routes originated inside the AS, the next hop is the IP address of the neighbor that announced the route. For routes injected into the AS via EBGP, the next hop learned from EBGP is carried unaltered into IBGP. The next hop is the IP address of the EBGP neighbor from which the route was learned.
3. When the route is advertised on a multiaccess media, the next hop is usually the IP address of the interface of the router connected to that media, which originated the route.

Figure B.1 illustrates the BGP Next_Hop attribute environment. The SF router is running an EBGP session with the LA router and an IBGP session with the SJ router. The SF router is learning route 128.213.1.0/24 from the LA router. In turn, the SF router is injecting the local route 192.212.1.0/24 into BGP.

The SJ router learns route 192.212.1.0/24 via 2.2.2.2, the IP address of the IBGP peer announcing the route. Thus, 2.2.2.2 is the next hop, according to the definition, for SJ to reach 192.212.1.0/24. Similarly, the SF router sees 128.213.1.0/24 coming from the LA router via next hop 1.1.1.1. When it passes this route update to the SJ router via IBGP, SF includes the next hop information unaltered. Thus, the SJ router

FIGURE B.1

BGP Next_Hop example. (Reproduced with permission of Cisco Systems, Inc. Copyright © 2001 Cisco Systems, Inc. All rights reserved.)

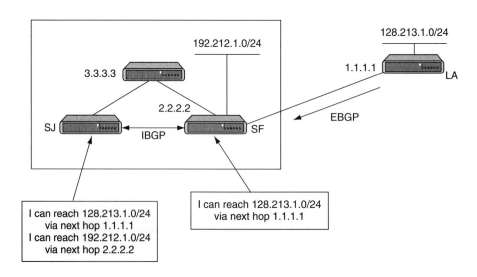

would receive the BGP update about 128.213.1.0/24 with the next hop 1.1.1.1. This is an example of the EBGP next hop being carried into IBGP.

As you can see from the preceding example, the next hop is not necessarily reachable via a direct connection. SJ's next hop for 128.213.1.0/24, for example, is 1.1.1.1, but reaching it requires a pathway through 3.3.3.3. Thus, the next hop behavior mandates a recursive IP lookup for a router to know where to send the packet. To reach the next hop 1.1.1.1, the SJ router will recursively look into its IGP routing table to see if and how 1.1.1.1 is reachable. This recursive search continues until the router associates destination 1.1.1.1 with an outgoing interface. The same recursive behavior is performed to reach next hop 2.2.2.2. If a hop is not reachable, BGP would consider the route as being inaccessible.

The following is a sample of how IP recursive lookup is used to direct the traffic toward the final destination. Tables B.1 and B.2 list the BGP and IP routing tables for the SJ router illustrated in Figure B.1.

Table B.2 indicates that 128.213.1.0/24 is reachable via next hop 1.1.1.1. Looking into the IP routing table, network 1.1.1.0/24 is reachable via next hop 3.3.3.3. Another recursive lookup in the IP routing table indicates that network 3.3.3.0/24 is directly connected via Serial 0. This would indicate that traffic toward next hop 1.1.1.1 should go via Serial 0. The same reasoning applies to deliver traffic toward next hop 2.2.2.2. Care must be taken to make sure that reachability of the next hop is advertised via some IGP or static routing. In case the BGP next hop cannot be reached, the BGP route would be considered inaccessible.

Table B.1 BGP Table of SJ Router

Destination	Next Hop
192.212.1.0/24	2.2.2.2
128.213.1.0/24	1.1.1.1

Table B.2 IP Routing Table of SJ Router

Destination	Next Hop
192.212.1.0/24	2.2.2.2
2.2.2.0/24	3.3.3.3
3.3.3.0/24	Connected, Serial 0
128.213.1.0/24	1.1.1.1
1.1.1.0/24	3.3.3.3

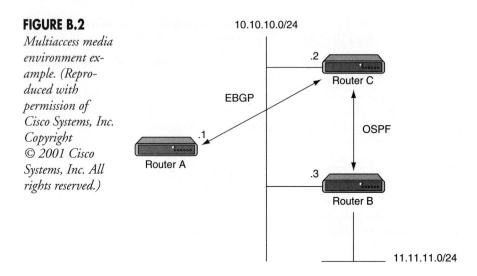

FIGURE B.2

Multiaccess media environment example. (Reproduced with permission of Cisco Systems, Inc. Copyright © 2001 Cisco Systems, Inc. All rights reserved.)

Next_Hop Behavior on Multiaccess Media

A media is considered multiaccess (MA) if routers connected to that media have the capability to exchange data in a many-to-many relationship. Routers on an MA media share the same IP subnet and can physically access all other routers on the media in one hop (directly connected). Ethernet, FDDI, Token Ring, Frame Relay, and ATM are examples of multiaccess media. IP has a rule on MA media that states that a router should always advertise the actual source of the route in case the source is on the same MA media as the router. In other words, if RTA (router A) is advertising a route learned from RTB, and RTA and RTB share a common media, when RTA advertises the route it should indicate RTB as being the source of the route. If not, routers on the same media would have to make an unnecessary hop via RTA to get to a router that is sitting in the same segment.

In Figure B.2, RTA, RTB, and RTC share a common multiaccess media. RTA and RTC are running EBGP, while RTC and RTB are running OSPF. RTC has learned network 11.11.11.0/24 from RTB via OSPF and is advertising it to RTA via EBGP. Because RTA and RTB are running different protocols, you might think that RTA would consider RTC 10.10.10.2 as its next hop to reach 11.11.11.0/24, but this is incorrect. The correct behavior is for RTA to consider RTB 10.10.10.3 as the next hop because RTB shares the same media with RTC.

In situations where the media is broadcast, such as Ethernet and FDDI, physical connectivity is a given and the next hop behavior is no problem. On the contrary, in situations where the media is nonbroadcast, such as Frame Relay and ATM, special care should be taken.

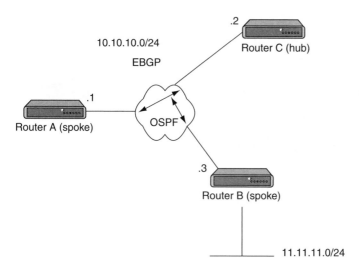

FIGURE B.3

Nonbroadcast multiaccess Next_Hop example. (Reproduced with permission of Cisco Systems, Inc. Copyright © 2001 Cisco Systems, Inc. All rights reserved.)

Next_Hop Behavior over Nonbroadcast Multiaccess Media (NBMA)

Media such as Frame Relay and ATM are nonbroadcast multiaccess. The many-to-many direct interaction between routers is not guaranteed unless virtual circuits are configured from each router to all other routers. This is called a fully meshed topology, and it is not always implemented for a number of reasons. In practice, the access carrier at a certain dollar amount per circuit provides Frame Relay or ATM virtual circuits, and additional circuits translate into extra money. In addition to this cost disincentive, most organizations use a hub and spoke approach, where multiple remote sites have virtual circuits built to one or more concentration routers at a central site (the hub site) where information resides. Figure B.3 illustrates an example of next hop behavior in a nonbroadcast multiaccess environment.

The only difference between the environments illustrated in Figure B.3 and Figure B.2 is that the media in Figure B.3 is a Frame Relay cloud that is NBMA. RTC is the hub router; RTA and RTB are the spokes. Notice how the virtual circuits are laid out between RTC and RTA, and between RTC and RTB, but not between RTA and RTB. This is called a partially meshed topology. RTA gets a BGP routing update about 11.11.11.0/24 from RTC and would try to use RTB 10.10.10.3 as the next hop—the same behavior as on MA media. Routing will fail because no virtual circuit exists between RTA and RTB.

The next-hop-self parameter when configured as part of the BGP neighbor connection forces the router in this case, RTC, to advertise 11.11.11.0/24 with itself as the next hop 10.10.10.2. RTA would then direct its traffic to RTC to reach destination 11.11.11.0/24.

FIGURE B.4

Next-hop-self parameter. (Reproduced with permission of Cisco Systems, Inc. Copyright © 2001 Cisco Systems, Inc. All rights reserved.)

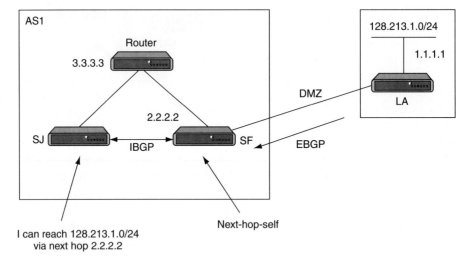

I can reach 128.213.1.0/24
via next hop 2.2.2.2

Use of Next-Hop-Self Versus Advertising DMZ

The demilitarized zone (DMZ) defines the shared network resources between ASs. The IP subnet used for the DMZ link might be part of any of the networked ASs or might not belong to any of them. The next hop address learned from the EBGP peer is carried inside IBGP. It is important for the IGP to be able to reach the next hop. One way of doing so is for the DMZ subnet to be part of the IGP and have the subnet advertised in the AS. The other way is to override the next hop address by forcing the next hop to be the IP address of the border IBGP neighbor.

In Figure B.4, the SJ router is receiving updates about 128.213.1.0/24 with next hop 1.1.1.1 part of the DMZ. For the SJ router to be able to reach this next hop, one option is for network 1.1.1.0/24 to be advertised inside the AS by the SF border router. The other option is to have the SF router set the next-hop-self parameter as part of the IBGP neighbor connection to the SJ router. This will set the next hop address of all EBGP routes to 2.2.2.2, that is already part of the IGP. The SJ router can now reach the next hop with no problem.

Choosing one method over the other depends on whether you want to reach the DMZ. An example could be an operator trying to do a ping from inside the AS to a router interface that belongs to the DMZ. For the ping to succeed, the DMZ must be injected in the IGP. In other cases, the DMZ might be reachable via some suboptimal route external to the AS. Instead of reaching the DMZ from inside the AS, the router might attempt to use another EBGP link to reach the DMZ. In this case, using next-hop-self ensures that the next hop is reachable from within the AS. In all other cases, both methods are similar as far as the BGP routing functionality.

The AS_Path Attribute

An AS_Path attribute is a well-known mandatory attribute (type code 2). It is a sequence of autonomous system numbers a route has traversed to reach a destination. The AS that originates the route adds its own AS number when sending the route to its external BGP peers. Therefore, each AS that receives the route and passes it on to other BGP peers will prepend its own AS number to the list. Prepending is the act of adding the AS number to the beginning of the list. The final list represents all the AS numbers that a route has traversed with the AS number of the AS that originated the route all the way at the end of the list. This type of AS_Path list is called an AS_Sequence, because all the AS numbers are ordered sequentially.

BGP uses the AS_Path attribute as part of the routing updates (UPDATE packet) to ensure a loop-free topology on the Internet. Each route that gets passed between BGP peers will carry a list of all AS numbers that the route has already been through. If the route is advertised to the AS that originated it, that AS will see itself as part of the AS_Path attribute list and will not accept the route. BGP speakers prepend their AS numbers when advertising routing updates to other ASs (external peers). When the route is passed to a BGP speaker within the same AS, the AS_Path information is left intact. Figure B.5 illustrates the AS_Path attribute at each instance of the route 172.16.10/24, originating in AS1 and passed to AS2 then AS3 and AS4 and back to AS1.

Note how each AS that passes the route to other external peers adds its own AS number to the beginning of the list. When the route gets back to AS1, the BGP border router will realize that this route has already been through its AS (AS number 1 appears in the list) and would not accept the route.

AS_Path information is one of the attributes BGP looks at to determine the best route to take to get to a destination. In comparing two or more different routes, given that all other attributes are identical, a shorter path is always preferred. In case of a tie, other attributes are used to make the decision.

Using Private ASs

To conserve AS numbers, InterNIC generally does not assign a legal AS number to customers whose routing policies are an extension of the policies of their provider. Thus, in the situation where a customer is single-homed or multihomed to the same provider, the provider generally requests that the customer use an AS number taken from the private pool of ASs (64512-65535). As such, all BGP updates the provider receives from its customer contain private AS numbers.

Private AS numbers cannot be leaked to the Internet because they are not unique. For this reason certain vendors have implemented a feature in their IOS software to strip private AS numbers out of the AS_Path list before the routes get propagated to the Internet. This is illustrated in Figure B.6.

FIGURE B.5

Example loop con-dition addressed by AS_Path at-tribute. (Repro-duced with permission of Cisco Systems, Inc. Copyright © 2001 Cisco Systems, Inc. All rights reserved.)

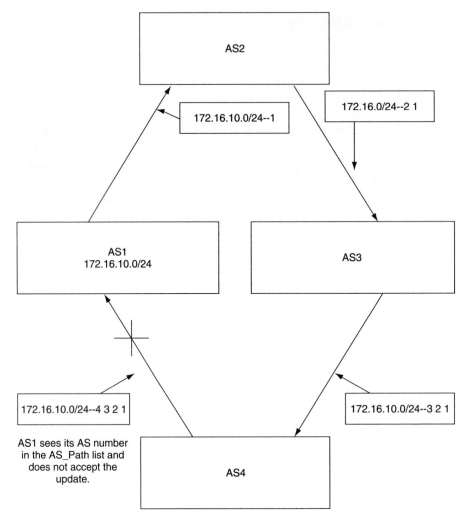

In Figure B.6, AS1 is providing Internet connectivity to its customer AS65001. Because the customer has only this provider and no plans for having an additional provider in the near future, the customer has been allocated a private AS number. If the customer needs to connect to another provider later, a legal AS number should be assigned. Prefixes originating from AS65001 have an AS_Path of 65001. Note prefix 172.16.220.0/24 in Figure B.6 as it leaves AS65001. For AS1 to propagate the prefix to the Internet, it would have to strip the private AS number. When the prefix reaches the Internet, it would look like it has originated from the provider's AS. Note how prefix 172.16.220.0/24 has reached the NAP with AS_Path 1.

BGP will strip private ASs only when propagating updates to the external peers. This means that the AS stripping would be configured on RTC as part of its neighbor con-

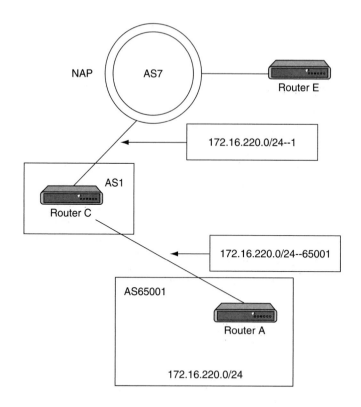

nection to RTE. Private ASs should only be connected to a single provider. If the AS_Path contains a mixture of private and legal AS numbers, BGP will view this as an illegal design and will not strip the private AS numbers from the list, and the update will be treated as usual. Only AS_Path lists that contain private AS numbers in the range 64512 to 65535 are stripped.

AS_Path and Route Aggregation Issues

Route aggregation involves summarizing ranges of routes into one or more aggregates or CIDR blocks to minimize the number of routes in the global routing tables. A drawback of route aggregation is the loss of granularity that existed in the specific routes that form the aggregate. For example, the AS_Path information that exists in multiple routes will be lost when these routes get summarized into one single advertisement. This would lead to potential routing loops because a route that has passed through an AS might be accepted by the same AS as a new route.

BGP defines another type of AS_Path list called an AS-SET where the ASs are listed in an unordered set. The set includes all the ASs a route has traversed. Aggregates carrying the AS-SET information would have a collective set of the attributes that form the individual routes they summarize.

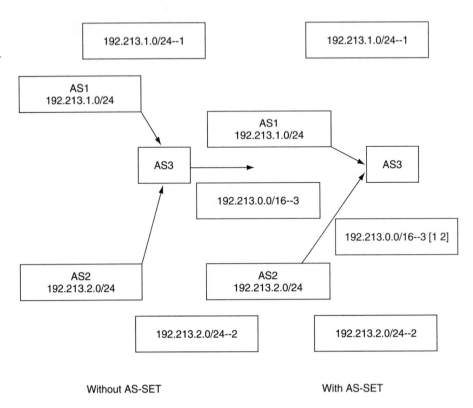

Without AS-SET   With AS-SET

In Figure B.7, AS1 is advertising 192.213.1.0/24, and AS2 is advertising 192.213.2.0/24. AS3 is aggregating both routes into 192.213.0.0/16.

An AS that advertises an aggregate considers itself the originator of that route, irrespective of where that route came from. When AS3 advertises the aggregate 192.213.0.0/16, the AS_Path information would be just 3. This would cause a loss of information because the originators of the route AS1 and AS2 are no longer listed in the AS_Path. In a situation where the aggregate is somehow advertised back to AS1 and AS2 by some other AS, AS1 and AS2 would accept the route that would potentially lead to routing loops. With the notion of AS-SET, it is possible to have AS3 advertise the aggregate 192.213.0.0/16 while keeping information about the components of the aggregate. The set {1 2} indicates that the aggregate has come from both of these ASs in no particular order. The AS_Path information of the aggregate with the AS-SET option would be 3 {1 2}.

AS_Path Manipulation

AS_Path information is manipulated to affect interdomain routing behavior. Because BGP prefers a shorter path over a longer one, system operators are tempted to change the path information by including dummy AS path numbers that would increase the path length and influence the traffic trajectory one way or the other. The adminis-

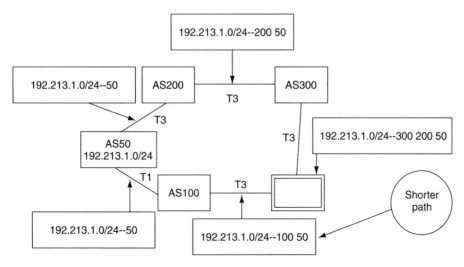

trator has the capability to insert AS numbers at the beginning of an AS_Path to make the path length longer. Figures B.8 and B.9 show how this can be implemented.

In Figure B.8, AS50 is connected to two providers, AS200 and AS100. AS100 is directly connected to the NAP, whereas AS200 has to go through an extra hop via AS300 to reach the NAP. Figure B.8 shows instances of prefix 192.213.1.0/24 as it traverses the ASs in its way to the NAP. When the 192.213.1.0/24 prefix reaches the NAP via AS300, it would have an AS_Path of 300 200 50. If the same prefix reaches the NAP via AS300, it would have an AS_Path of 100 50, which is shorter. ASs upstream from the NAP would prefer the shorter AS_Path length and would direct their traffic toward AS100 at all times for destination 192.213.1.0/24.

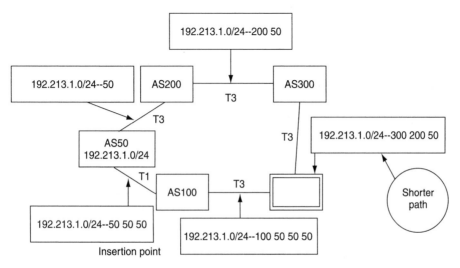

AS50 is not happy with this behavior because it prefers the traffic to come via its higher bandwidth T3 link to AS200. AS50 will manipulate the AS_Path information by inserting dummy AS numbers when sending routing updates to AS100. One common practice is for AS50 to repeat its AS number as many times as necessary to tip the balance and make the path via AS200 become shorter.

In Figure B.9, AS50 will insert two AS numbers 50 50 at the beginning of the AS_Path of prefix 192.213.1.0/24. When the prefix 192.213.1.0/24 reaches the NAP via AS100, it would have the AS_Path 100 50 50 50, which is longer than the AS_Path 300 200 50 via AS300. AS upstream of the NAP would prefer the shortest path and would direct the traffic toward AS300 for destination 192.213.1.0/24.

The bogus number should always be a duplicate of the AS announcing the route or the neighbor the route is learned from in case an AS is increasing the path length for incoming updates. Adding any other number is misleading and could potentially lead to routing points. Note the insertion point in Figure B.9.

The Local Preference Attribute

The local preference is a well-known discretionary attribute (type code 5). The local preference attribute is a degree of preference given to a route to compare it with other routes for the same destination. A higher local preference value is an indication that the route is more preferred. Local preference, as indicated by the name, is local to the autonomous system and gets exchanged between IBGP peers only and is not passed to EBGP peers.

An AS connected via BGP to multiple other ASs will get routing updates about the same destination from different ASs. Local preference is usually used to set the exit point of an AS to reach a certain destination. Because this attribute is communicated within all BGP routers inside the AS, all BGP routers will have a common view on how to exit the AS.

Consider the environment in Figure B.10. Suppose that company TMI has purchased Internet connections via two service providers MCI and AT&T. TMI is connected to AT&T via a primary T3 link and to MCI via a backup T1 link. It is important for TMI to decide what path its outbound traffic is going to take. Of course, TMI prefers to use the T3 link via AT&T in normal operation because it is a high speed link.

This is where local preference comes into play: The LA router will give the routes coming from AT&T a local preference of 300. The SJ router will give the routes coming from MCI a lower value, say 200. Because both the SJ and LA routers are exchanging routing updates via IBGP, they both agree that the exit point of the AS is going to be via AT&T because of the higher local preference. In Figure B.10, TMI learns route 128.213.0.0/16 via MCI and AT&T. The SJ and LA routers will agree on using AT&T as the exit point for destination 128.213.0.0/16 because of the higher local preference value of 300. The local preference manipulation discussed in

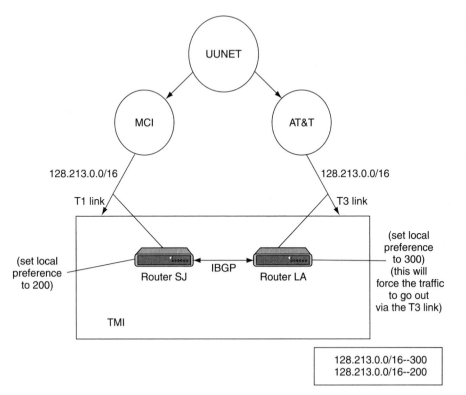

FIGURE B.10
Local preference attribute example. (Reproduced with permission of Cisco Systems, Inc. Copyright © 2001 Cisco Systems, Inc. All rights reserved.)

this example affects the traffic going out of the AS and not the traffic coming into the AS. Inbound traffic will still come in via the T1 link.

The Multi_Exit_DISC (MED) Attribute

The Multi_Exit_DISC (MED) attribute is an optional nontransitive attribute (type code 4). It is a hint to external neighbors about the preferred path into an AS that has multiple entry points. The MED is also known as the external metric of a route. A lower MED value is preferred over a higher MED value.

Unlike local preference, the MED attribute is exchanged between ASs, but a MED attribute that comes into an AS does not leave the AS. When an update enters the AS with a certain MED value, that value is used for decision making within the AS. When BGP passes on the routing update to another AS, the MED is reset to zero unless the outgoing MED is set to a specific value.

When the route is originated by the AS itself, the MED value follows the internal IGP metric of the same route. This becomes useful when a customer has multiple

FIGURE B.11

Effects of the MED attribute. (Reproduced with permission of Cisco Systems, Inc. Copyright © 2001 Cisco Systems, Inc. All rights reserved.)

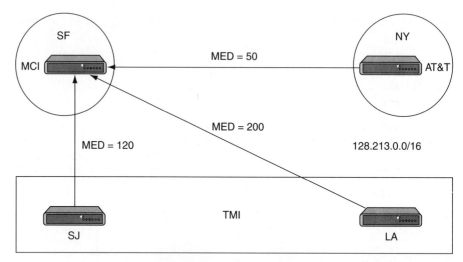

connections to the same provider. The IGP metric reflects how close or how far a network is to a certain exit point. A network that is closer to point A than to exit point B will have a lower IGP metric in the border router connected to A. When the IGP metric translates to MED, traffic coming into the AS can enter from the link closer to the destination because a lower MED is preferred for the same destination. This can be used by providers and customers to balance the traffic over multiple links between two ASs.

In the local preference example illustrated in Figure B.10, an AS was shown determining how to influence its own outbound decision. In the example illustrated in Figure B.11, the MED shows how an AS can influence the outbound decision of another AS.

In the example illustrated in Figure B.11, TMI and AT&T try to influence MCI's outbound traffic by sending it different metric values. MCI is receiving routing updates about 128.213.0.0/16 from three different sources: SJ (metric 120), LA (metric 200), and NY (metric 50). SF will compare the two metric values coming from TMI and will prefer the SJ router because it is advertising a lower metric (120). The SF router will compare the metric 120 with metric 50 coming from NY and will prefer NY to reach 128.213.0.0/16. Note that SF could have influenced its decision by using local preference inside MCI to override the metrics coming from outside ASs. Nevertheless, MED is still useful in case MCI prefers to base its BGP decisions on outside factors to simplify router configuration on its end. Customers that connect to the same provider in multiple locations could exchange metrics with their providers to influence each other's outbound traffic, which leads to better load balancing.

The Community Attribute

In the context of BGP, a community is a group of destinations that share some common property. A community is not restricted to one network or one autonomous system—it has no physical boundaries. An example is a group of networks that belong to the educational or government communities. These networks can belong to any autonomous system. Communities are used to simplify routing policies by identifying routes based on a logical property rather than an IP prefix or an AS number. A BGP speaker can use this attribute in conjunction with other attributes to control which routes to accept, prefer, and pass on to other BGP neighbors.

The community attribute (type code 8) is an optional transitive attribute that is of variable length and consists of a set of 4-byte values. Communities in the range 0×00000000 through 0×0000FFFF and 0×FFFFF0000 through 0×FFFFFFFF are reserved. These communities are well-known; that is, they have a global meaning. An example of well-known communities are:

- NO_EXPORT (0×FFFFFF01): A route carrying this community value should not be advertised to peers outside a confederation or the AS if it is the only AS in the confederation.
- NO_ADVERTISE (0×FFFFFF02): A route carrying this community value, when received, should not be advertised to any BGP peer.

Besides well-known community attributes, private community attributes can be defined for special uses. A common practice is to use the first two bytes of the community attribute for the AS number and the last two bytes to define a value in relation to that AS. A provider (AS256), for example, who wants to define a private community called my-peer-routers could use the community 256:1 represented in decimal notation. The 256 indicates that this particular provider has defined the community, and the 1 has special meaning to the provider—in this case it is my-peer-routers.

A route can have more than one community attribute. A BGP speaker that sees multiple community attributes in a route can act based on one, some, or all the attributes. A router has the option of adding or modifying community attributes before passing routes on to other peers.

Figure B.12 shows a simple use of the community attribute. MCI is sending toward AT&T routes X and Y with a NO_EXPORT community attribute, and route Z with no modification. The BGP router in AT&T will propagate route Z only toward UUNET. Routes X and Y will not be propagated because of the NO_EXPORT community attribute.

The ATOMIC_AGGREGATE Attribute

Route aggregation causes a loss of information because the aggregate is coming from different sources that have different attributes. The ATOMIC_AGGREGATE is a well-known discretionary attribute (type code 6) that gets set as an indication of information loss. If a system propagates an aggregate that causes loss of information, it is required to attach the ATOMIC_AGGREGATE attribute to the route.

FIGURE B.12

Simple application of community attribute. (Reproduced with permission of Cisco Systems, Inc. Copyright © 2001 Cisco Systems, Inc. All rights reserved.)

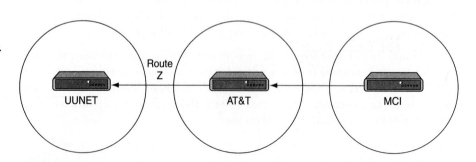

The ATOMIC_AGGREGATE should not be set when the aggregate carries some extra information that gives an indication of where the aggregated information came from. An example is an aggregate with the AS-SET parameter. An aggregate that carries the set of ASs that form the aggregate is not required to attach the ATOMIC_AGGREGATE attribute.

The AGGREGATOR Attribute

The AGGREGATOR attribute is an optional transitive attribute (type code 7). It specifies the autonomous system and the router that has generated an aggregate. A BGP speaker that performs route aggregation might add the AGGREGATOR attribute, which contains the speaker's AS number and IP address. The IP address is actually the Router ID (RID), which is the highest IP address on the router or loopback address if it exists. Figure B.13 illustrates the AGGREGATOR attribute.

FIGURE B.13

AGGREGATOR implementation example. (Reproduced with permission of Cisco Systems, Inc. Copyright © 2001 Cisco Systems, Inc. All rights reserved.)

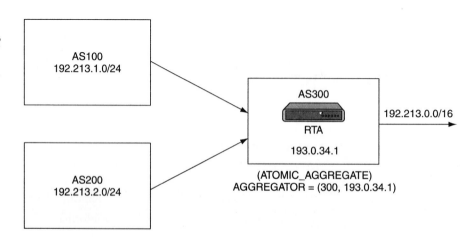

AS300 is receiving routes 192.213.1.0/24 and 192.213.2.0/24 from AS100 and AS200, respectively. When RTA generates aggregate 192.213.0.0/16, it has the option of including the AGGREGATOR attribute, which consists of the AS number 300 and the RID 193.0.34.1 of the router (RTA) that originated the aggregate.

The ORIGIN Attribute

The ORIGIN attribute is a well-known mandatory attribute (type code 1). It indicates the origin of the routing update (NLRI, which indicates prefix and mask) with respect to the autonomous system that originated it. BGP considers three types of origins:

- IGP—The Network Layer Reachability Information (NLRI) is internal to the originating AS.
- EGP—The Network Layer Reachability Information is learned via the EGP.
- INCOMPLETE—The Network Layer Reachability Information is learned by some other means.

BGP considers the ORIGIN attribute in its decision-making process to establish a preference ranking among multiple routes. Specifically, BGP prefers the path with the lowest origin type, where IGP is lower than EGP and EGP is lower than INCOMPLETE.

BGP Decision Process Summary

BGP bases its decision process on the attribute values. When faced with multiple routes to the same destination, BGP chooses the best route for routing traffic toward the destination. The following process summarizes how BGP chooses the best route.

1. If the next hop is inaccessible, the route is ignored (this is why it is important to have an IGP route to the next hop).
2. Prefer the path with the largest weight (weight is a vendor proprietary parameter).
3. If the weights are the same, prefer the route with the largest local preference.
4. If the routes have the same local preference, prefer the route that was locally originated (originated by this router).
5. If the local preference is the same, prefer the route with the shortest AS_Path.
6. If the AS_Path length is the same, prefer the route with the lowest origin type (where IGP is lower than EGP and EGP is lower than INCOMPLETE).
7. If the origin type is the same, prefer the route with the lowest MED.
8. If the routes have the same MED, prefer the route in the following manner: External (EBGP) is better than Confederation External, which is better than Internal (IBGP).
9. If all the preceding scenarios are identical, prefer the route that can be reached via the closest IGP neighbor—that is, take the shortest internal path inside the AS to reach the destination, then follow the shortest path to the BGP Next_Hop.

10. If the internal path is the same, the BGP router ID will be a tiebreaker. Prefer the route coming from the BGP router with the lowest router ID. The router ID is usually the highest IP address on the router or the loopback (virtual) address. The router ID could be implementation specific.

Route Filtering and Attribute Manipulation

The concept of route filtering is straightforward. A BGP speaker can choose what routes to send and what routes to receive from any of its BGP peers. Route filtering is essential in defining routing policies. An autonomous system can identify the inbound traffic it is willing to accept from other neighbors by specifying the list of routes it advertises to its neighbors. Conversely, an AS can control what routes its outbound traffic uses by specifying the routes it accepts from its neighbors.

Filtering is also used on the protocol level to limit routing updates flowing from one protocol to another. Filtering is essential in specifying exactly what goes into the BGP and IGP and vice versa. Routes permitted through a filter can have their attributes manipulated. Manipulating the attributes affects the BGP decision process of identifying best routes.

Inbound and Outbound Filtering

Both inbound and outbound filtering concepts can be applied to the peer and to the protocol level. Figure B.14 illustrates this behavior.

At the peer level, inbound filtering indicates that the BGP speaker is filtering routing updates coming from other peers, whereas outbound filtering limits the routing updates advertised from the BGP speaker to other peers. Filtering behavior is the same whether the BGP peers are external (EBGP) or internal (IBGP). At the protocol level, inbound filtering limits the routing updates being injected into a protocol. Outbound filtering limits the routing updates being injected from this protocol. With respect to BGP, for example, inbound filtering limits the updates being redistributed from other protocols such as IGP and static into BGP. Outbound filtering limits the updates being redistributed from BGP into IGP.

Policy Routing

Policy routing is a means of controlling routes, which relies on the source, or source and destination, of traffic rather than destination alone. Policy routing can be used to control traffic inside an AS as well as between ASs. Policy routing is a glorified form of static routing. It is used when you want to force a routing behavior different from what the dynamic routing protocols dictate.

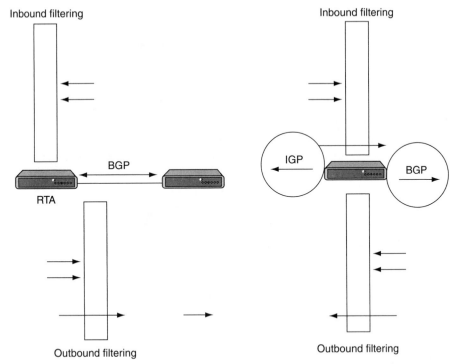

Inbound filtering

Inbound filtering

RTA

BGP

IGP

BGP

Outbound filtering

Outbound filtering

At the peer level with respect to RTA

At the protocol level with respect to BGP

FIGURE B.14

Inbound and outbound filtering example. (Reproduced with permission of Cisco Systems, Inc. Copyright © 2001 Cisco Systems, Inc. All rights reserved.)

Static routing enables you to direct traffic based on the traffic destination. Traffic toward destination 1 can go via point A whereas traffic toward destination 2 can go via point B. Consider the example illustrated in Figure B.15.

Assume that AS1 was assigned network numbers from two different providers. The 10.10.10.0/24 range was taken from AS3, and the 11.11.11.0/24 range was taken from AS4. AS1 wants to have any traffic originated from its 10.10.10.0/24 networks to be directed toward AS3 and traffic from its 11.11.11.0/24 networks directed to AS4, irrespective of destination of the traffic. AS1 could use policy routing to achieve this requirement by forcing all traffic with a source IP address belonging to 10.10.10.0/24 to have a next hop of 1.1.1.1 and traffic with source IP belonging to 11.11.11.0/24 to have a next hop of 2.2.2.2.

Policy routing can also be based on a source and destination combination. This is illustrated in Figure B.16.

Assume RTA wants to use the SF link for any traffic originating from network 10.10.10.0/24 and reaching network 12.12.12.0/24 in NY. Also RTA wants to use the SJ link for any traffic originating from network 10.10.10.0/24 and reaching

FIGURE B.15

Policy routing scenario based on source. (Reproduced with permission of Cisco Systems, Inc. Copyright © 2001 Cisco Systems, Inc. All rights reserved.)

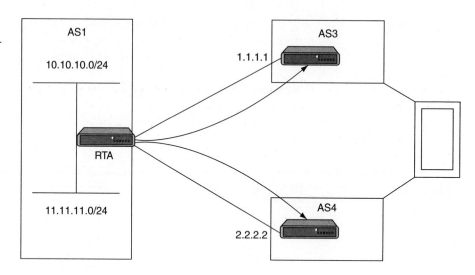

FIGURE B.16

Policy routing based on source and destination. (Reproduced with permission of Cisco Systems, Inc. Copyright © 2001 Cisco Systems, Inc. All rights reserved.)

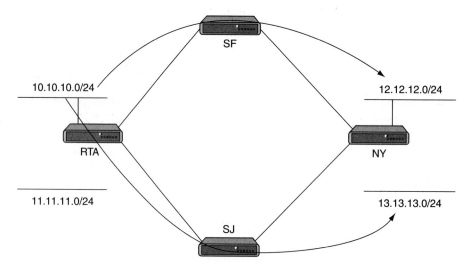

network 13.13.13.0/24 in NY. Policy routing can be used to set the next hop for the traffic combination (Source = 10.10.10.0/24, Destination = 13.13.13.0/24) will be set with next hop 2.2.2.2.

Policy routing should not replace dynamic routing, but instead should complement it. Policy routing has the following drawbacks:

1. Extra configuration is needed to identify sources of traffic or a combination of source and destination. Care should be taken not to disrupt other traffic and to specify other alternatives for traffic in case of backup situations.

2. Policy routing is CPU intensive because it is based on the source IP addresses, unlike dynamic and static routing, which are based on the destination IP address. Sophisticated caching and switching techniques have been implemented all along based on the destination of the traffic. Most implementations have not yet optimized routing and caching techniques based on the source of the IP packet. As such, policy routing takes additional CPU cycles to detect source addresses. This behavior should change as implementations move toward better understanding of IP traffic flows that enable caches to keep track of source and destination information. This new caching methodology would alleviate routers from disruptive processing on matching sources of IP traffic and make policy routing much more effective and practical.

Signaling Interfaces

Introduction

There are many standards and recommended practices used to define data communications interfaces and signaling. In fact, the entire problem of data communications can be looked at as the task of passing information through a series of interfaces and transmission channels without the loss of meaning.

The interface that is of the greatest interest is the one between the equipment that originates and/or receives the data, the Data Terminal Equipment (DTE), and the equipment that handles the problem of transmitting them from place to place, the Data Communications Equipment (DCE). Computers are called Data Terminal Equipment and modems are called Data Communications Equipment, also called data circuit-terminating equipment because their function is to communicate the data rather than consider, compute, or change them in any way.

Terms such as RS-232, V.24, RS-422, RS-423, RS-449, X.21, X.25, current loop, and several others are the designations of the various standards and recommendations designed to make the task of connecting computers, terminals, modems, and networks easier. Manufacturers have in many cases used the standardized interfaces for functions that were never intended. This method helps by not increasing the number of interfaces, but it adds to the confusion of the user.

In today's computer and communications world, the Electronic Industries Association (EIA) RS-232 is the standard. RS-232 will be described and used as a way to explain interfaces in general. Some of the limitations of RS-232 will be addressed as well as some of the newer developments that eventually might replace RS-232.

The RS-232 and V.24 Interfaces

The proper name of RS-232 is Interface Between Data Terminal Equipment and Data Communication Equipment Employing Serial Binary Data Interchange. Currently, the most popular version of RS-232 is revision C, which is formally referred to as RS-232-C. In the late 1980s, revision D was introduced, which was followed by revision E in the early 1990s. The latest version will eventually supersede RS-232-C and RS-232-D. All three versions of the RS-232 standard have a large core of common functions and operating features that will collectively be referred to as RS-232. An appropriate revision will be designated prior to describing specific functions associated with each version of RS-232. What is written here also applies to ITU-T Recommendation V.24, which is almost identical to RS-232, but the electrical signal characteristics are specified separately in ITU-T Recommendation V.28.

In addition to explanatory notes and a short glossary, the RS-232 standard covers four areas:

- The mechanical characteristics of the interface
- The electrical signals across the interface
- The function of each signal
- Subsets of signals for certain applications

The RS-232 standard covers the mechanical and signal interface between data terminal equipment and data communications equipment employing serial binary data interchange.

Let us look at some sample configurations:

1. Transmit Only
2. Transmit Only With RTS
3. Receive Only
4. Half Duplex
5. Full Duplex
6. Full Duplex With RTS
7. Special

RS-232-C Standard Configuration

RS-232-C interchange circuit	(1)	(2)	(3)	(4)	(5)	(6)	(7)
1 Protective Ground	–	–	–	–	–	–	–
7 Signal Ground	X	X	X	X	X	X	X
2 Transmitted Data	X	X		X	X	X	
3 Received Data			X	X	X	X	
4 Request to Send		X		X		X	
5 Clear to Send	X	X		X	X	X	

```
 6 Data Set Ready                     X   X   X   X   X   X
20 Data Terminal Ready                S   S   S   S   S   S
22 Ring Indicator                     S   S   S   S   S   S
 8 Receive Line Signal Detector       X   X   X   X
```

```
X = required for any configuration
S = required for using PSDN (public switched telephone network)
o = specified by cable designer
```

Notice that only *one* circuit is an *absolute requirement* for any such cable, this is the Signal Ground on pin 7. That means that as long as Signal Ground is included on pin 7 this cable is "RS-232-C compliant."

Although most microcomputer systems have one-way devices such as transmit-only joysticks, or receive-only printers, the most common situation is a full-duplex, two-way communication. The classical application for the above configuration is a send and receive terminal where characters are transmitted from the keyboard to a microcomputer and echoed back to a display screen. The data are traveling in both directions from the DTE (keyboard and screen) to the DCE (computer serial I/O port). There are two configurations for full duplex, one with and one without the Request to Send line implemented. A "safe" strategy is to always include it. Figure C.1 shows a schematic full-duplex configuration.

RS-232-C Standard Full-Duplex Cable

The signals present in a standard full-duplex RS-232-C cable are:

- Signal Ground
- Transmitted Data
- Received Data
- Request to Send
- Clear to Send

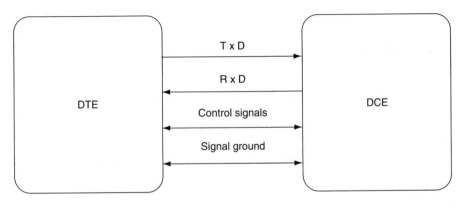

FIGURE C.1
RS-232 full-duplex configuration.

- Data Set Ready
- Received Line Signal Detector

For a modem over a switched telephone network we must add two more signals:

- Data Terminal Ready
- Ring Indicator

To understand the function of each signal let us look at the events taking place to eventually exchange data. Each event will cause a transition from state to state—from the idle state through the data exchange and communicating state and finally back to idle state.

The events and transitions are grouped in phases:

- Alerting
- Equipment Readiness
- Circuit Assurance
- Channel Readiness

Alerting

In the Alerting phase, the call originating station dials the phone number of the call answering station. The remote telephone begins to ring, and the Ring Indicator Signal of the answering DCE makes an OFF to ON transition. This ends this phase.

Equipment Readiness

The Equipment Readiness phase involves events associated with turning Data Set Ready and Data Terminal Ready to ON, and thus completing this phase.

Circuit Assurance

The Circuit Assurance phase consists of events associated with turning ON the Received Line Signal Detector signals at both communicating stations.

Channel Readiness

The Channel Readiness phase uses Request to Send and Clear to Send handshaking to arrive at the target state of active data exchange. All these events result in the equipment being disconnected from the telephone network and getting back to original idle state. The diagram in Figure C.2 illustrates the process.

Other (Nonstandard) Common Configurations

Three-Wire Economy Model

This model involves only a minimum number of circuits for full-duplex communications. The circuits present are Transmitted Data on pin 2, Received Data on pin 3,

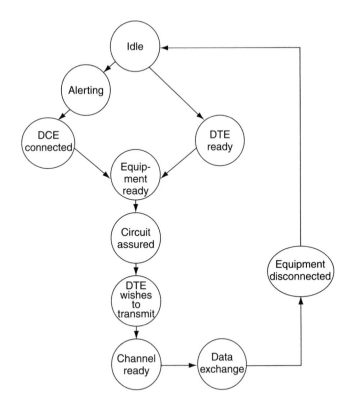

FIGURE C.2

Process of RS-232 signals.

and Signal Ground on pin 7. There are configurations for which this cable is entirely adequate, but many common microsystems components use the RTS and CTS circuits. This equipment will not transmit until it receives an asserted CTS signal. For this we use the next model to trick the USART based I/O ports into transmission. The diagram in Figure C.3 illustrates the three-wire economy model.

Three-Wire Model with Luxury LoopBack

This cable has the following loopback jumpers:

- Request To Send–> (jumpered to) –> Clear to Send
- Request To Send–> (jumpered to) –> Received Line Signal Detector
- Data Terminal Ready–> (jumpered to) –> Data Set Ready

By jumpering Data Set Ready, the Equipment Readiness phase is completed as soon as the DTE asserts its Data Terminal Ready line. This is achieved when the DTE is powered-up. Also, when the DTE is powered-up the Request To Send is asserted and the Circuit Assurance phase is completed because the Request To Send is jumpered to the Received Line Signal Detector. Because Request To Send is jumpered to Clear To Send, it also implies the completion of the Channel Readiness phase. The bottom

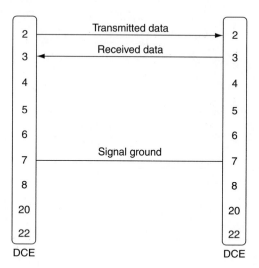

FIGURE C.3

The three-wire economy model.

line is that Data Terminal Ready and Request To Send are the only two events required to achieve the target data exchange state.

Notice that this implementation omits some features from the full implementation. Most of them concern the prevention of overrun errors. The diagram in Figure C.4 illustrates the three-wire luxury loopback model.

Null Modem with Luxury LoopBack and the Null Modem with Double-Cross

This model is designed to answer the requirement to trick two DTEs into communicating over a strictly *local* RS-232-C interface, with no modems or DCEs. This concept is identified as the "crossover technique." The word *luxury* is used because there are some modest models in which some loopbacks are omitted. When none of the loopbacks are present it is simply the three-wire economy model. Notice that this model is exactly the three-wire with luxury loopback model with null modem.

The double-cross variant involves a crossover between the following two pairs of control signals:

- Request To Send<–> Received Line Signal Detector
- Data Terminal Ready<–> Ring Indicator

The following loopbacks are included:

- Request To Send–> (jumpered to) –> Clear To Send
- Ring Indicator –> (jumpered to) –> Data Set Ready

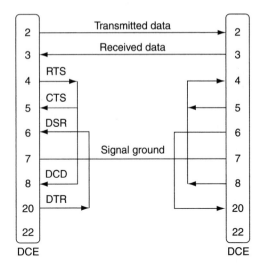

FIGURE C.4
Three-wire luxury loopback model.

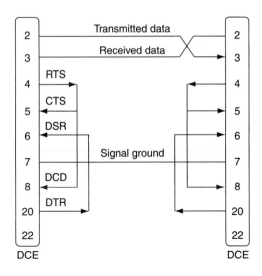

FIGURE C.5
Null modem with luxury loopback.

And finally, a crossover for null modem function:

● Transmitted Data<–> Received Data

Figure C.5 illustrates the null modem with luxury loopback.

RS-232 Characteristics

● RS 232 Signals
  ● Signal Travel Direction
  ● Electrical Signal Characteristics

- Voltage Levels Defined in the Standard
- The Noise Margin Issue
- Interface Mechanical Issues
- Pin Designation for the DB Connector
- Diagram of the DB Connector

RS-232 Signals

The number preceding each signal name corresponds to the pin number defined in the standard.

1. Protective Ground
2. Transmitted Data
3. Received Data
4. Request To Send
5. Clear To Send
6. Data Set Ready
7. Signal Ground
8. Received Line Signal Detect (Carrier Detect)
9. +P For Testing Only
10. −P For Testing Only
11. (unassigned)
12. Secondary Received Line Signal Detect
13. Secondary Clear To Send
14. Secondary Transmitted Data
15. Transmission Signal Element Timing
16. Secondary Received Data
17. Receiver Signal Element Timing
18. (unassigned)
19. Secondary Request To Send
20. Data Terminal Ready
21. Signal Quality Detector
22. Ring Indicator
23. Data Signal Rate Selector
24. Transmitter Signal Element Timing
25. (unassigned)

RS-232 Signals Functional Description

The first letter of the EIA signal name categorizes the signal into one of five groups, each representing a different "circuit":

- A—Ground
- B—Data

- C—Control
- D—Timing
- S—Secondary Channel
- 1 Protective Ground
  - Name: AA
  - Direction: -
  - CCITT: 101

This pin is usually connected to the frame of one of the devices, either the DCE or the DTE, which is properly grounded. The sole purpose of this connection is to protect against accidental electric shock, and usually this pin should not be tied to Signal Ground. This pin should connect the chassis (shields) of the two devices, but this connection is made only when the connection of chassis grounds is safe (see ground loops following). It is considered optional.

Ground loops are low impedance closed electric loops composed from ground conductors. When two grounded devices are connected together, say by an RS-232 cable, the alternating current on the lines in the cable induces an electric potential across the ends of the grounding line (either Protective Ground or Signal Ground), and an electric current will flow across this line and through the ground.

Because the impedance of the loops is low, this current can be quite high and easily burn out electric components. Electrical storms could also cause a burst of destructive current across such a loop. Therefore, connection of the Protective Ground pin is potentially hazardous. Furthermore, not all signal grounds are necessarily isolated from the chassis ground, and using an RS-232 interface, especially across a long distance, is unreliable and could be hazardous. The maximum distance at which the grounding signals can be connected safely is 30 meters.

- 2 Transmitted Data
  - Name: BA
  - Direction: DTE –> DCE
  - CCITT: 103

Serial data (primary) are sent on this line from the DTE to the DCE. The DTE holds this line at logic 1 when no data are being transmitted. An "On" (logic 0) condition must be present on the following signals, where implemented, before data can be transmitted on this line: CA, CB, CC, and CD (Request To Send, Clear To Send, Data Set Ready, Data Terminal Ready).

- 3 Received Data
  - Name: BB
  - Direction: DTE <– DCE
  - CCITT: 104

Serial data (primary) are sent on this line from the DCE to the DTE. This pin is held at logic 1 (Mark) when no data are being transmitted, and is held

"Off" for a brief interval after an "On" to "Off" transition on the Request To Send line to allow the transmission to complete.

- 4 Request To Send
  - Name: CA
  - Direction: DTE –> DCE
  - CCITT: 105

This signal enables transmission circuits. The DTE uses this signal when it wants to transmit to the DCE. This signal, in combination with the Clear To Send signal, coordinates data transmission between the DTE and the DCE.

A logic 0 on this line keeps the DCE in transmit mode. The DCE will receive data from the DTE and transmit it to the communication link.

The Request To Send and Clear To Send signals relate to a half-duplex telephone line. A half-duplex line is capable of carrying signals on both directions but only one at a time. When the DTE has data to send, it raises Request To Send, and then waits until the DCE changes from receive to transmit mode. This "On" to "Off" transition instructs the DCE to move to "transmit" mode, and when a transmission is possible the DCE sets Clear To Send and transmission can begin.

On a full-duplex line, like a hard wired connection where transmission and reception can occur simultaneously, the Clear To Send and Request To Send signals are held to a constant "On" level.

An "On" to "Off" transition on this line instructs the DCE to complete the transmission of data that is in progress, and to move to a "receive" (or "no transmission") mode.

- 5 Clear To Send
  - Name: CB
  - Direction: DTE <– DCE
  - CCITT: 106

This is an answer signal to the DTE. When this signal is active, it tells the DTE that it can now start transmitting (on Transmitted Data line). When this signal is "On" and the Request To Send, Data Set Ready, and Data Terminal Ready are all "On," the DTE is assured that its data will be sent to the communications link. When this signal is "Off," it is an indication to the DTE that the DCE is not ready, and therefore data should not be sent.

When the Data Set Ready and Data Terminal Ready signals are not implemented in a local connection, which does not involve the telephone network, the Clear To Send and Request To Send signals are sufficient to control data transmission.

- 6 Data Set Ready
  - Name: CC
  - Direction: DTE <– DCE
  - CCITT: 107

On this line the DCE tells the DTE that the communication channel is available (i.e., in an automatic calling system the DCE [modem] is not in the dial, test, or talk modes and therefore is available for transmission and reception). It reflects the status of the local data set and does not indicate that an actual link has been established with any remote data equipment.

- 7 Signal Ground
    - Name: AB
    - Direction: -
    - CCITT: 102

This pin is the reference ground for the other signals, data, and control.

- 8 Received Line Signal Detect or Data Carrier Detect
    - Name: CF
    - Direction: DTE <– DCE
    - CCITT: 109

The DCE uses this line to signal the DTE that a good signal is being received (a "good signal" means a good analog carrier that can ensure demodulation of received data).

- 9 +P

This pin is held at +12 volts DC for test purposes.

- 10 −P

This pin is held at −12 volts DC for test purposes.

- 12 Secondary Received Line Signal Detect
    - Name: SCF
    - Direction: DTE <– DCE
    - CCITT: 122

This signal is active when the secondary communication channel is receiving a good analog carrier (same function as the Received Line Signal Detect).

- 13 Secondary Clear To Send
    - Name: SCB
    - Direction: DTE <– DCE
    - CCITT: 121

An answer signal to the DTE. When this signal is active, it tells the DTE that it can now start transmitting on the secondary channel (on the Secondary Transmitted Signal Line).

- 14 Secondary Transmitted Data
    - Name: SBA
    - Direction: DTE –> DCE
    - CCITT: 118

Serial data (secondary channel) are sent on this line from the DTE to the DCE. This signal is equivalent to the Transmitted Data line except that it is used to transmit data on the secondary channel.

- 15 Transmission Signal Element Timing
  - Name: DB
  - Direction: DTE <– DCE
  - CCITT: 114

The DCE sends the DTE a clock signal on this line. This enables the DTE to clock its output circuitry, which transmits serial data on the Transmitted Data line.

  The clock signal frequency is the same as the bit rate of the Transmitted Data line. An "On" to "Off" transition should mark the center of each signal element (bit) on the Transmitted Data line.

- 16 Secondary Received Data
  - Name: SBB
  - Direction: DTE <– DCE
  - CCITT: 119

Serial data (secondary channel) are received on this line from the DCE to the DTE. When the secondary channel is being used only for diagnostic purposes or to interrupt the flow of data in the primary channel, this signal is normally not provided.

- 17 Receiver Signal Element Timing
  - Name: DD
  - Direction: DTE <– DCE
  - CCITT: 115

The DCE sends the DTE a clock signal on this line. This clocks the reception circuitry of the DTE, which receives serial data on the Received Data line.

  The clock signal frequency is the same as the bit rate of the Received Data line (BB). The "On" to "Off" transition should indicate the center of each signal element (bit) on the Received Data line.

- 19 Secondary Request To Send
  - Name: SCA
  - Direction: DTE –> DCE
  - CCITT: 120

The DTE uses this signal to request transmission from the DCE on the secondary channel. It is equivalent to the Request To Send signal.

  When the secondary channel is only used for diagnostic purposes or to interrupt the flow of data in the primary channel, this signal should turn "On" the secondary channel unmodulated carrier.

- 20 Data Terminal Ready
  - Name: CD
  - Direction: DTE –> DCE
  - CCITT: 108.2

When "On," it tells the DCE that the DTE is available for receiving. This signal must be "On" before the DCE can turn Data Set Ready "On," thereby indicating that it is connected to the communications link.

The Data Terminal Ready and Data Set Ready signals deal with the readiness of the equipment, as opposed to the Clear To Send and Request To Send signals that deal with the readiness of the communication channel.

When "Off," it causes the DCE to finish any transmission in progress and to be removed from the communication channel.

- 21 Signal Quality Detector
  - Name: CG
  - Direction: DTE <– DCE
  - CCITT: 110

This line is used by the DCE to indicate whether or not there is a high probability of an error in the received data. When there is a high probability of an error, it is set to "Off," and is "On" at all other times.

- 22 Ring Indicator
  - Name: CE
  - Direction: DTE <– DCE
  - CCITT: 125

On this line the DCE signals the DTE that there is an incoming call. This signal is maintained "Off" at all times except when the DCE receives a ringing signal.

- 23 Data Signal Rate Selector
  - Name: CH/CI
  - Direction: DTE –> DCE
  - CCITT: 111/112

The DTE uses this line to select the transmission bit rate of the DCE. The selection is between two rates in the case of a dual rate synchronous connection, or between two ranges of data rates in the case of an asynchronous connection.

Typically, when this signal is "On," it tells the DCE (modem) that the receive speed is greater than 600 baud.

- 24 Transmitter Signal Element Timing
  - Name: DA
  - Direction: DTE –> DCE
  - CCITT: 113

The DTE sends the DCE a transmit clock on this line. This is only when the master clock is in the DTE.

An "On" to "Off" transition should indicate the center of each signal element (bit) on the Transmitted Data line.

● A Note on Signal Travel Direction

The pin names are the same for the DCE and DTE. The Transmit Data (pin number 2) is a transmit line on the DTE and a receive line on the DCE, Data Set Ready (pin number 6) is a receive line on the DTE and a transmit line on the DCE, and so forth.

Electrical Signal Characteristics

- Voltage levels defined in the standard
- Data signals   "0","Space"   "1","Mark"
- Driver (Required)   5 - 15   -5 - -15 Volts
- Terminator (Expected)   3 - 25   -3 - -25 Volts
- Control signals   "Off"   "On"
- Driver (Required)   -5 - -15   5 - 15 Volts
- Terminator (expected)   -3 - -25   3 - 25 Volts

The Noise Margin Issue

Note that terminator (receiving end) voltages are not the same as driver required voltages. This voltage level definition compensates for voltage losses across the cable.

Signals traveling along the cable are attenuated and distorted as they pass. Attenuation increases as the length of the cable increases. This effect is largely a result of the electrical capacitance of the cable.

The maximum load capacitance is specified as 2,500 pf (picofarad) by the standard. The capacitance of one meter of cable is typically around 130 pf, thus the maximum cable length is limited to around 17 meters. However, this is a nominal length defined by the standard, and it is possible to use longer cables up to 30 meters with low capacitance cables or with slow data rates and a proper error correction mechanism.

Interface Mechanical Characteristics

The connection of the DCE and the DTE is done with a pluggable connector. The female connector should be associated with the DCE. The following table lists the pin assignments defined by the standard. The type of connector to be used is not mentioned in the standard, but the DB-25 (or on IBM-ATs, a minimal DB-9) connectors are almost always used.

Pin Designation for the 25-Pin and 9-Pin DB Connector Includes Equivalent CCITT V.24 Identification and Signal Direction

DB-25 Pin #	DB-9 Pin #	Common Name	EIA Name	CCITT	DTE-DCE	Formal Name
1		FG	AA	101	–	Frame Ground
2	3	TD	BA	103	→	Trans. Data, TxD
3	2	RD	BB	104	←	Received Data, RxD
4	7	RTS	CA	105	→	Request To Send
5	8	CTS	CB	106	←	Clear To Send
6	6	DSR	CC	107	←	Data Set Ready
7	5	SG	AB	102	——	Signal Ground, GND
8	1	DCD	CF	109	←	Data Carrier Detect
9		—	—	-	–	+P
10		—	—	-	–	−P
11		—	—	-	–	unassigned
12		SDCD	SCF	122	←	Secondary Data Carrier Detect
13		SCTS	SCB	121	←	Secondary Clear To Send
14		STD	SBA	118	→	Secondary Transmitted Data
15		TC	DB	114	←	Transmission Signal Element Timing
16		SRD	SBB	119	←	Secondary Received Data
17		RC	DD	115	→	Receiver Signal Element Timing
18		—	—	-	–	unassigned
19		SRTS	SCA	120	→	Secondary Request To Send
20	4	DTR	CD	108.2	→	Data Terminal Ready
21		SQ	CG	110	←	Signal Quality Detector
22	9	RI	CE	125	←	Ring Indicator
23		—	CH/CI	111/112	→	Data Signal Rate Selector
24		—	DA	113	←	Transmitter Signal Element Timing
25		—	—	-	–	unassigned

Diagram of the DB-25 and DB-9 connectors.

Male connectors, front view

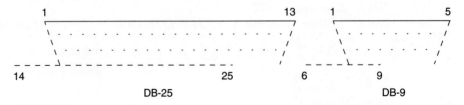

An Introduction to IGRP

Goals for IGRP

IGRP is a protocol that allows a number of gateways to coordinate their routing. Its goals are:

- Stable routing even in very large or complex networks. No routing loops should occur, even as transients.
- Fast response to changes in network topology.
- Low overhead. That is, IGRP itself should not use more bandwidth than what is actually needed for its task.
- Splitting traffic among several parallel routes when they are of roughly equal desirability.
- Taking into account error rates and level of traffic on different paths.
- The ability to handle multiple "types of service" with a single set of information.

The current implementation of IGRP handles routing for TCP/IP. However, the basic design is intended to handle a variety of protocols.

During the last few years, routing has suddenly become a more difficult problem than it used to be. A few years ago, protocols such as RIP were sufficient to handle most real networks. However, growth in the Internet and decentralization of control of its structure have now resulted in a system of networks that is nearly beyond our capabilities to manage. Similar situations are occurring in large corporate networks as well. IGRP is one tool intended to help attack this problem.

No one tool is going to solve all routing problems. Conventionally, the routing problem is broken into several pieces. Protocols such as IGRP are called internal gateway protocols (IGPs). They are intended for use within a single set of networks, either under a single management or closely coordinated managements. Such sets of networks are connected by external gateway protocols (EGPs). An IGP is designed to keep track of a good deal of detail about network topology. Priority in designing an IGP is placed on producing optimal routes and responding quickly to changes. An EGP is intended to protect one system of networks against errors or intentional misrepresentation by other systems. Priority in designing an EGP is on stability and administrative controls. Often it is sufficient for an EGP to produce a reasonable route, rather than the optimal route. In fact, there are features that allow IGRP to be used as an EGP in some circumstances. However, the emphasis is on its use as an IGP.

IGRP has some similarities to older protocols such as Xerox's Routing Information Protocol, Berkeley's RIP, and Dave Mills' Hello. It differs from these protocols primarily because it is designed for larger and more complex networks.

Like these older protocols, IGRP is a distance vector protocol. In such a protocol, gateways exchange routing information only with adjacent gateways. This routing information contains a summary of information about the rest of the network. It can be shown mathematically that all of the gateways taken together are solving an optimization problem by what amounts to a distributed algorithm. Each gateway only needs to solve part of the problem, and it only has to receive a portion of the total data.

The major alternative is a class of algorithms referred to as shortest path first (SPF). These are based on a flooding technique in which every gateway is kept up to date about the status of every interface on every other gateway. Each gateway independently solves the optimization problem from its point of view using data for the entire network. There are advantages to each approach. In some circumstances SPF may be able to respond to changes more quickly. To prevent routing loops, IGRP has to ignore new data for a few minutes after certain kinds of changes. Because SPF has information directly from each gateway, it is able to avoid these routing loops. Thus, it can act on new information immediately. However, SPF has to deal with substantially more data than IGRP, both in internal data structures and in messages between gateways. Thus, SPF implementations can be expected to have higher overhead than IGRP implementations.

The Routing Problem

IGRP is intended for use in gateways connecting several networks. We assume that the networks use packet based technology. In effect the gateways act as packet switches. When a system connected to one network wants to send a packet to a system on a different network, it addresses the packet to a gateway. If the destination is on one of the networks connected to the gateway, the gateway will forward the packet to the destination. If the destination is more distant, the gateway will forward the packet to another gateway that is closer to the destination. Gateways use routing tables to help them decide what to do with packets. Here is a simple example routing table. Note that the basic routing problem is similar for other protocols as well, but this description will assume that IGRP is being used for routing IP. (Actual IGRP routing tables have additional information for each gateway.)

Network	Gateway	Interface
128.6.4	None	Ethernet 0
128.6.5	None	Ethernet 1
128.6.21	128.6.4.1	Ethernet 0
128.121	128.6.5.4	Ethernet 1
10	128.6.5.4	Ethernet 1

This gateway is connected to two Ethernets called 0 and 1. They have been given IP network numbers (actually subnet numbers) 128.6.4 and 128.6.5. Thus, packets addressed for these specific networks can be sent directly to the destination simply by using the appropriate Ethernet interface. There are two nearby gateways: 128.6.4.1 and 128.6.5.4. Packets for networks other than 128.6.4 and 128.6.5 will be forwarded to one or the other of those gateways. The routing table indicates which gateway should be used for which network. For example, packets addressed to a host on network 10 should be forwarded to gateway 128.6.5.4. One hopes that this gateway is closer to network 10 (i.e., that the best path to network 10 goes through this gateway). The primary purpose of IGRP is to allow the gateways to build and maintain routing tables like this.

Summary of IGRP

As mentioned earlier, IGRP is a protocol that allows gateways to build up their routing table by exchanging information with other gateways. A gateway starts out with entries for all of the networks that are directly connected to it. It gets information about other networks by exchanging routing updates with adjacent gateways. In the simplest case, the gateway will find one path that represents the best way to get to each network. A path is characterized by the next gateway to which packets should be sent, the network interface that should be used, and metric information. Metric information is a set of numbers that characterize how good the path is. This allows the gateway to compare paths that it has heard from various gateways and decide which one to use. There are often cases where it makes sense to split traffic between two or more paths. IGRP will do this whenever two or more paths are equally good. The user can also configure it to split traffic when paths are almost equally good. In this case, more traffic will be sent along the path with the better metric. The intent is that traffic can be split between a 9,600-bps line and a 19,200-bps line, and the 19,200 line will get roughly twice as much traffic as the 9,600-bps line.

The metric used by IGRP includes:

- The topological delay time
- The bandwidth of the narrowest bandwidth segment of the path

- The channel occupancy of the path
- The reliability of the path.

Topological delay time is the amount of time it would take to get to the destination along that path, assuming an unloaded network. Of course, there is additional delay when the network is loaded. However, load is accounted for by using the channel occupancy figure, not by attempting to measure actual delays. The path bandwidth is simply the bandwidth in bits per second of the slowest link in the path. Channel occupancy indicates how much of that bandwidth is currently in use. It is measured and will change with load. Reliability indicates the current error rate. It is the fraction of packets that arrive at the destination undamaged.

Although they are not used as part of the metric, two additional pieces of information are passed with it: hop count and MTU. The hop count is simply the number of gateways that a packet will have to go through to get to the destination. MTU is the maximum packet size that can be sent along the entire path without fragmentation. (That is, it is the minimum of the MTUs of all the networks involved in the path.)

Based on the metric information, a single "composite metric" is calculated for the path. The composite metric combines the effect of the various metric components into a single number representing the "goodness" of that path. It is the composite metric that is actually used to decide on the best path.

Periodically, each gateway broadcasts its entire routing table (with some censoring because of the split horizon rule) to all adjacent gateways. When a gateway gets this broadcast from another gateway, it compares the table with its existing table. Any new destinations and paths are added to the gateway's routing table. Paths in the broadcast are compared with existing paths. If a new path is better it may replace the existing one. Information in the broadcast is also used to update channel occupancy and other information about existing paths. This general procedure is similar to that used by all distance vector protocols. It is referred to in the mathematical literature as the Bellman-Ford algorithm. For a detailed development of the basic procedure see RFC 1058, which describes RIP—an older distance vector protocol.

In IGRP, the general Bellman-Ford algorithm is modified in three critical aspects. First, instead of a simple metric, a vector of metrics is used to characterize paths. Second, instead of picking a single path with the smallest metric, traffic is split among several paths whose metrics fall into a specified range. Third, several features are introduced to provide stability in situations where the topology is changing. The best path is selected based on a composite metric:

$$[(K1 / Be) + (K2 * Dc)]\ r$$

where:

K1, K2 = constants
Be = unloaded path bandwidth $\times$ (1 − channel occupancy)
Dc = topological delay
r = reliability

The path having the smallest composite metric will be the best path. Where there are multiple paths to the same destination, the gateway can route the packets over more than one path. This is done in accordance with the composite metric for each data path. For instance, if one path has a composite metric of 1 and another path has a composite metric of 3, three times as many packets will be sent over the data path having the composite metric of 1. However, only paths whose composite metrics are with a certain range of the smallest composite metric will be used. K1 and K2 indicate the weight to be assigned to bandwidth and delay. These will depend on the type of service. For example, interactive traffic would normally place a higher weight on delay and file transfer on bandwidth.

There are two advantages to using a vector of metric information. The first is that it provides the ability to support multiple types of service from the same set of data. The second advantage is improved accuracy. When a single metric is used, it is normally treated as if it were a delay. Each link in the path is added to the total metric. If there is a link with a low bandwidth, it is normally represented by a large delay. However, bandwidth limitations do not really cumulate the way delays do. By treating bandwidth as a separate component, it can be handled correctly. Similarly, the load can be handled by a separate channel occupancy number.

IGRP provides a system for interconnecting computer networks, which can stably handle a general graph topology including loops. The system maintains full path metric information (i.e., it knows the path parameters to all other networks to which any gateway is connected). Traffic can be distributed over parallel paths and multiple path parameters can be simultaneously computed over the entire network.

Comparison with RIP

This section compares IGRP with RIP. This comparison is useful because RIP is used widely for purposes similar to IGRP. However, doing this is not entirely fair. RIP was not intended to meet all of the same goals as IGRP. RIP was intended for use in small networks with reasonably uniform technology. In such applications it is generally adequate.

The most basic difference between IGRP and RIP is the structure of their metrics. Unfortunately, this is not a change that can simply be retrofitted to RIP. It requires the new algorithms and data structures present in IGRP.

RIP uses a simple "hop count" metric to describe the network. Unlike IGRP, where every path is described by a delay, bandwidth, and so on, in RIP it is described by a number from 1 to 15. Normally this number is used to represent how many gateways the path goes through before getting to the destination. This means that no distinction is made between a slow serial line and an Ethernet. In some implementations of RIP, it is possible for the system administrator to specify that a given hop should be counted more than once. Slow networks can be represented by a large hop count. But since the maximum is 15, this cannot be done very often (e.g., if an Ethernet is represented by 1 and a 56-Kb line by 3, there can be at most 5 56-Kb lines in a path or the maximum of 15 is exceeded). In order to represent the full range of available network speeds and allow for a large network, studies suggest that a 24-bit metric is needed. If the maximum metric is too small, the system administrator is presented with an unpleasant choice: either he cannot distinguish between fast and slow routes, or he cannot fit his whole network into the limit. In fact, a number of national networks are now large enough that RIP cannot handle them even if every hop is counted only once. RIP simply cannot be used for such networks.

The obvious response would be to modify RIP to allow a larger metric. Unfortunately, this will not work. Like all distance vector protocols, RIP has the problem of "counting to infinity." This is described in more detail in RFC 1058. When topology changes, spurious routes will be introduced. The metrics associated with these spurious routes slowly increase until they reach 15, at which point the routes are removed. Because 15 is a small enough maximum this process will converge fairly quickly, assuming that triggered updates are used. If RIP were modified to allow a 24-bit metric, loops would persist long enough for the metric to be counted up to $2^{**}24$. This is not tolerable. IGRP has features designed to prevent spurious routes from being introduced. It is not practical to handle complex networks without introducing such features or changing to a protocol such as SPF.

IGRP does a bit more than simply increase the range of allowable metrics. It restructures the metric to describe delay, bandwidth, reliability, and load. It is possible to represent such considerations in a single metric such as RIP's. However, the approach taken by IGRP is potentially more accurate. For example, with a single metric, several successive fast links will appear to be equivalent to a single slow one. This may be the case for interactive traffic, where delay is the primary concern. However, for bulk data transfer the primary concern is bandwidth, and adding metrics together is not the right approach. IGRP handles delay and bandwidth separately, cumulating delays but taking the minimum of the bandwidths. It is not easy to see how to incorporate the effects of reliability and load into a single-component metric.

One of the big advantages of IGRP is ease of configuration. It can directly represent quantities that have physical meaning. This means that it can be set up automatically, based on interface type, line speed, and so on. With a single-component metric, the metric is more likely to have to be "cooked" to incorporate effects of several different things.

Other innovations are more a matter of algorithms and data structures than of the routing protocol. For example, IGRP specifies algorithms and data structures that support splitting traffic among several routes. It is certainly possible to design an implementation of RIP that does this. However, once routing is being re-implemented, there is no reason to stick with RIP.

So far a generic IGRP has been described—a technology that could support routing for any network protocol. However, in this section it is worth mentioning more about the specific TCP/IP implementation. That is, the implementation that is going to be compared with RIP.

RIP update messages simply contain snapshots of the routing table. That is, they have a number of destinations and metric values, and little else. The IP implementation of IGRP has additional structure. First, the update message is identified by an autonomous system number. This terminology comes out of the ARPANET tradition and has specific meaning there. However, for most networks what it means is that you can run several different routing systems on the same network. This is useful for places where networks from several organizations converge. Each organization can maintain its own routing. Because each update is labeled, gateways can be configured to pay attention only to the right one. Certain gateways are configured to receive updates from several autonomous systems. They pass information between the systems in a controlled manner. Note that this is not a complete solution to problems of routing security. Any gateway can be configured to listen to updates from any autonomous system. However, it is still a very useful tool in implementing routing policies where there is a reasonable degree of trust between the network administrators.

The second structural feature about IGRP update messages affects the way default routes are handled by IGRP. Most routing protocols have a concept of default route. It is often not practical for routing updates to list every network in the world. Typically, a set of gateways need detailed routing information for networks within their organization. All traffic for destinations outside their organization can be sent to one of a few boundary gateways. Those boundary gateways may have more complete information. The route to the best boundary gateway is a default route. It is a default in the sense that it is used to get to any destination that is not listed specifically in the internal routing updates. RIP and other routing protocols circulate information about the default route as if it were a real network. IGRP takes a different approach. Rather than a single fake entry for the default route, IGRP allows real networks to be flagged as candidates for being a default. This is implemented by placing information about those networks in a special exterior section of the update message. However, it might as well be thought of as turning on a bit associated with those networks. Periodically IGRP scans all the candidate default routes and chooses the one with the lowest metric as the actual default route.

Potentially this approach to defaults is somewhat more flexible than the approach taken by most RIP implementations. Most typically, RIP gateways can be set to generate a default route with a certain specified metric. The intention is that this would be done at boundary gateways. However, there are situations where this is not really good enough. Suppose that a company is directly connected to two regional networks, say NYSERnet at 56 Kbps and JvNCnet at T1. Normally the link to JvNCnet would be the best default. Using RIP we would configure our gateway to JvNCnet to advertise a default route with a lower metric than our gateway to NYSERnet. However, suppose something happens within JvNCnet so that its access to the national networks is either cut off or becomes much worse (e.g., it might end up passing traffic through a backup link to some other regional). In that case it might be better to use NYSERnet. The typical RIP implementations would not adapt to that change because the boundary gateway is generating a default with a fixed metric. With IGRP we would configure our gateways so that the NSFnet national backbone was flagged as a candidate default route. In that case, we would choose the default route at each gateway based on metric information that takes into account the entire path back to the national backbone. So the default would change to take into account events within our regional networks, assuming that they pass full IGRP metric information to us.

Detailed Description

Overall Description

When a gateway is first turned on, its routing table is initialized. This may be done by an operator from a console terminal or by reading information from configuration files. A description of each network connected to the gateway is provided, including the topological delay along the link (i.e., how long it takes a single bit to transverse the link) and the bandwidth of the link.

For instance, in Figure D.1, gateway S would be told that it is connected to networks 2 and 3 via the corresponding interfaces. Thus, initially, gateway 2 only knows that it can reach any destination computer in networks 2 and 3. All the gateways are programmed to periodically transmit to their neighboring gateways the information that they have been initialized with, as well as information gathered from other gateways. Thus, gateway S would receive updates from gateways R and T and learn that it can reach computers in network 1 through gateway R and computers in network 4 through gateway T. Because gateway S sends its entire routing table, in the next cycle gateway T will learn that it can get to network 1 through gateway S. It is easy to see that information about every network in the system will eventually reach every gateway in the system, providing only that the network is fully connected.

FIGURE D.1

A simple example network. (Reproduced with permission of Cisco Systems, Inc. Copyright © 2001 Cisco Systems, Inc. All rights reserved.)

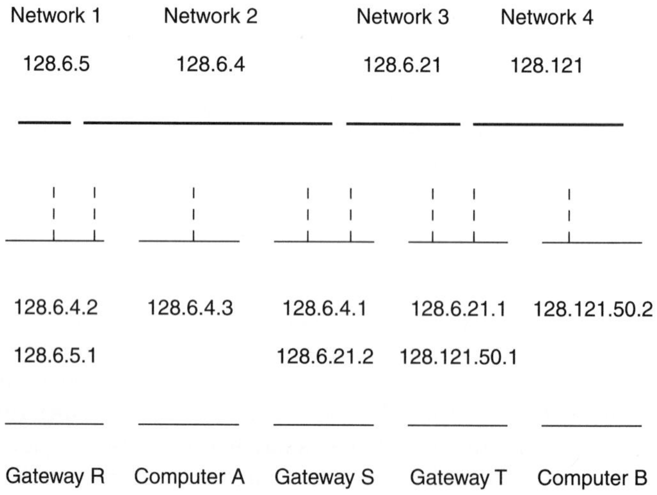

Network 1	Network 2	Network 3	Network 4
128.6.5	128.6.4	128.6.21	128.121

128.6.4.2	128.6.4.3	128.6.4.1	128.6.21.1	128.121.50.2
128.6.5.1		128.6.21.2	128.121.50.1	

| Gateway R | Computer A | Gateway S | Gateway T | Computer B |

Each gateway computes a composite metric to determine the desirability of the data paths to destination computers. For instance, in Figure D.2, for a destination in network 6, gateway A would compute metric functions for two paths via gateways B and C. Note that paths are defined simply by the next hop. There are actually three possible routes from A to network 6:

- Direct to B
- To C and then to B
- To C and then to D

However, gateway A need not choose between the two routes involving C. The routing table in A has a single entry representing the path to C. Its metric represents the best way of getting from C to the final destination. If A sends a packet to C, it is up to C to decide whether to use B or D. The composite metric function computed for each data path is as follows:

$$[(K1 / Be) + (K2 * Dc)]\, r \qquad\qquad \text{(Eq. 1)}$$

where:

r = fractional reliability (% of transmissions that are successfully received at the next hop)

Dc = composite delay

Be = effective bandwidth: unloaded bandwidth $\times$ (1 − channel occupancy)

$K1, K2$ = constants

In principle, the composite delay, Dc, could be determined as follows:

$$Dc = Ds + Dc_{ir} + DT \qquad\qquad \text{(Eq. 2)}$$

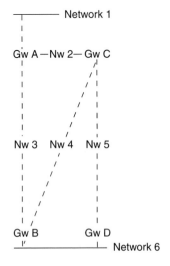

FIGURE D.2

Example of alternate paths. (Reproduced with permission of Cisco Systems, Inc. Copyright © 2001 Cisco Systems, Inc. All rights reserved.)

where:

Ds = switching delay
Dc_{ir} = circuit delay (propagation delay of 1 bit)
DT = transmission delay (no load delay for a 1,500 bit message)

However, in practice a standard delay figure is used for each type of network technology. For instance, there will be a standard delay figure for Ethernet and for serial lines at any particular bit rate.

Here is an example of how gateway A's routing table might look in the case of Figure D.2. (Note that individual components of the metric vector are not shown, for simplicity.)

The basic process of building up a routing table by exchanging information with neighbors is described by the Bellman-Ford algorithm. The algorithm has been used in earlier protocols such as RIP (RFC 1058). In order to deal with more complex networks, IGRP adds four features to the basic Bellman-Ford algorithm:

- Instead of a simple metric, a vector of metrics is used to characterize paths. A single composite metric can be computed from this vector according to Eq. 1. Use of a vector allows the gateway to accommodate different types of service by using several different coefficients in Eq. 1. It also allows a more accurate representation of the characteristics of the network than a single metric.
- Instead of picking a single path with the smallest metric, traffic is split among several paths with metrics falling into a specified range. This allows several routes to be used in parallel, providing a greater effective bandwidth than any single route. A variance V is specified by the network administrator. All paths with minimal composite metric M are kept. In addition, all paths whose met-

FIGURE D.3

An example routing table. (Reproduced with permission of Cisco Systems, Inc. Copyright © 2001 Cisco Systems, Inc. All rights reserved.)

	Interface	Next gateway	Metric
Network 1	NW 1	None	Directly connected
Network 2	NW 2	None	Directly connected
Network 3	NW 3	None	Directly connected
Network 4	NW 2	C	1270
	NW 3	B	1180
Network 5	NW 2	C	1270
	NW 3	B	2130
Network 6	NW 2	C	2040
	NW 3	B	1180

ric is less than V × M are kept. Traffic is distributed among multiple paths in inverse proportion to the composite metrics.

- There are some problems with this concept of variance. It is difficult to come up with strategies that make use of variance values greater than 1 and do not also lead to packets looping. The effect of this is to set the variance permanently to 1.
- Several features are introduced to provide stability in situations where the topology is changing. These features are intended to prevent routing loops and "counting to infinity," which have characterized previous attempts to use Bellman-Ford-type algorithms for this type of application. The primary stability features are "holddowns," "triggered updates," "split horizon," and "poisoning." These will be discussed in more detail.

Traffic splitting (point 2) raises a rather subtle danger. The variance V is designed to allow gateways to use parallel paths of different speeds. For example, there might be a 9,600-bps line running in parallel with a 19,200-bps line, for redundancy. If the variance V is 1, only the best path will be used. So the 9,600-bps line will not be used if the 19,200-bps line has a reasonable reliability. (However, if several paths are the same, the load will be shared among them.) By raising the variance, we can allow traffic to be split between the best route and other routes that are nearly as good. With a large enough variance, traffic will be split between the two lines. The danger is that with a large enough variance, paths become allowed that are not just slower but actually "in the wrong direction." Thus, there should be an additional rule to prevent traffic from being sent "upstream": No traffic is sent along paths whose remote composite metric (the composite metric calculated at the next hop) is greater than the composite metric calculated at the gateway. In a general system, administrators are encouraged not to set the variance above 1 except in specific situations where parallel paths need to be used. In this case, the variance is carefully set to provide the right results.

IGRP is intended to handle multiple types of service and multiple protocols. Type of service is a specification in a data packet that modifies the way paths are to be evaluated. For example, the TCP/IP protocol allows the packet to specify the relative importance of high bandwidth, low delay, or high reliability. Generally, interactive applications will specify low delay, whereas bulk transfer applications will specify high bandwidth. These requirements determine the relative values of K1 and K2 that are appropriate for use in Eq. 1. Each combination of specifications in the packet that is to be supported is referred to as a type of service. For each type of service, a set of parameters K1 and K2 must be chosen. A routing table is kept for each type of service. This is done because paths are selected and ordered according to the composite metric defined by Eq. 1. This is different for each type of service. Information from all of these routing tables is combined to produce the routing update messages exchanged by the gateways, as described.

Stability Features

This section describes holddowns, triggered updates, split horizon, and poisoning. These features are designed to prevent gateways from picking up erroneous routes. As described in RFC 1058, this can happen when a route becomes unusable due to failure of a gateway or a network. In principle, the adjacent gateways detect failures. They then send routing updates that show the old route as unusable. However, it is possible for updates not to reach some parts of the network at all, or to be delayed in reaching certain gateways. A gateway that still believes the old route is good can continue spreading that information, thus reentering the failed route into the system. Eventually this information will propagate through the network and come back to the gateway that reinjected it. The result is a circular route.

In fact, there is some redundancy among the countermeasures. In principle, holddowns and triggered updates should be sufficient to prevent erroneous routes in the first place. However, in practice, communications failures of various kinds can cause them to be insufficient. Split horizon and route poisoning are intended to prevent routing loops in any case.

Normally, new routing tables are sent to neighboring gateways on a regular basis (every 90 seconds by default, although this can be adjusted by the system administrator). A triggered update is a new routing table that is sent immediately, in response to some change. The most important change is removal of a route. This can happen because a time-out has expired (probably a neighboring gateway or line has gone down) or because an update message from the next gateway in the path shows that the path is no longer usable. When a gateway G detects that a route is no longer usable, it triggers an update immediately. This update will show that route as unusable. Consider what happens when this update reaches the neighboring gateways. If the neighbor's route pointed back to G, the neighbor must remove the route. This causes the neighbor to trigger an update, and so on. Thus a failure will trigger a wave of update messages. This wave will propagate throughout that portion of the network in which routes went through the failed gateway or network.

Triggered updates would be sufficient if we could guarantee that the wave of updates reached every appropriate gateway immediately. However, there are two problems. First, packets containing the update message can be dropped or corrupted by some link in the network. Second, the triggered updates do not happen instantaneously. It is possible that a gateway that has not yet gotten the triggered update will issue a regular update at just the wrong time, causing the bad route to be reinserted in a neighbor that had already gotten the triggered update. Holddowns are designed to get around these problems. The holddown rule says that when a route is removed, no new route will be accepted for the same destination for some period of time. This gives the triggered updates time to get to all other gateways so that we can be sure any new routes we get are not just some gateway reinserting the old one. The holddown period must be long enough to allow for the wave of triggered updates to go throughout the network. In addition, it should include a couple of regular broadcast cycles to handle dropped packets. Consider what happens if one of the triggered updates is dropped or corrupted. The gateway that issued that update will issue another update at the next regular update. This will restart the wave of triggered updates at neighbors that missed the initial wave.

The combination of triggered updates and holddowns should be sufficient to get rid of expired routes and prevent them from being reinserted. However, some additional precautions are worth doing anyway. They allow for very lossy networks, and networks that have become partitioned. The additional precautions called for by IGRP are split horizon and route poisoning. Split horizon arises from the observation that it never makes sense to send a route back in the direction from which it came. Consider the following situation:

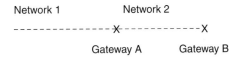

Gateway A will tell B that it has a route to network 1. When B sends updates to A, there is never any reason for it to mention network 1. Because A is closer to 1, there is no reason for it to consider going via B. The split horizon rule says a separate update message should be generated for each neighbor (actually each neighboring network). The update for a given neighbor should omit routes that point to that neighbor. This rule prevents loops between adjacent gateways. For example, suppose A's interface to network 1 fails. Without the split horizon rule, B would be telling A that it can get to 1. Because it no longer has a real route, A might pick up that route. In this case, A and B would both have routes to 1. But A would point to B and B would point to A. Of course triggered updates and holddowns should prevent this from happening. But since there is no reason to send information back to the place it came from, split horizon is worth doing anyway. In addition to its role in preventing loops, split horizon keeps down the size of update messages.

Split horizon should prevent loops between adjacent gateways. Route poisoning is intended to break larger loops. The rule is that when an update shows that the metric for an existing route has increased sufficiently, there is a loop. The route should be removed and put into holddown. Currently, the rule is that a route is removed if the composite metric increases more than a factor of 1.1. It is not safe for just any increase in composite metric to trigger removal of the route because small metric changes can occur due to changes in channel occupancy or reliability. Therefore, the factor of 1.1 is just a heuristic. The exact value is not critical. We expect this rule to be needed only to break very large loops because small ones will be prevented by triggered updates and holddowns.

Disabling Holddowns

The disadvantage of holddowns is that they delay adoption of a new route when an old route fails. With default parameters, it can take several minutes before a router adopts a new route after a change. However, for the reasons explained earlier, it is not safe simply to remove holddowns. The result would be count to infinity, as described in RFC 1058. It is conjecture, but it cannot be proven that with a stronger version of route poisoning, holddowns are no longer needed to stop count to infinity. Thus, disabling holddowns enables this stronger form of route poisoning. Note that split horizon and triggered updates are still in effect.

The stronger form of route poisoning is based on a hop count. If the hop count for a path increases, the route is removed. This will obviously remove routes that are still valid. If something elsewhere in the network changes so that the path now goes through one more gateway, the hop count will increase. In this case, the route is still valid. However, there is no completely safe way to distinguish this case from routing loops (count to infinity). Thus, the safest approach is to remove the route whenever the hop count increases. If the route is still legitimate it will be reinstalled by the next update, and that will cause a triggered update that will reinstall the route elsewhere in the system.

In general, distance vector algorithms adopt new routes easily. The problem is completely purging old ones from the system. Thus, a rule that is overly aggressive about removing suspicious routes should be safe.

The primary difference from protocol to protocol will be the format of the routing update packet, which must be designed to be compatible with a specific protocol. Note that the definition of a destination may vary from protocol to protocol. The method described here can be used for routing to individual hosts, to networks, or for more complex hierarchical address schemes. Which type of routing is used will depend on the addressing structure of the protocol. The current TCP/IP implementation supports only routing to IP networks. Thus, destination really means IP network or subnet number. Subnet information is only kept for connected networks.

At the start of the program, acceptable protocols and parameters describing each interface are entered. The gateway will only handle certain protocols, which are listed. Any communication from a system using a protocol not on the list will be ignored. The data inputs are:

- The networks to which the gateway is connected.
- The unloaded bandwidth of each network.
- Topological delay of each network.
- Reliability of each network.
- Channel occupancy of each network.
- MTU of each network.

The metric function for each data path is then computed according to Eq. 1. Note that the first three items are reasonably permanent. They are a function of the underlying network technology and do not depend on load. They could be set from a configuration file or by direct operator input. Note that IGRP does not use measured delay. Both theory and experience suggest that it is very difficult for protocols that use measured delay to maintain stable routing. There are two measured parameters: reliability and channel occupancy. Reliability is based on error rates reported by the network interface hardware or firmware.

In addition to these inputs, the routing algorithm requires a value for several routing parameters. This includes timer values, variance, and whether holddowns are enabled. This would normally be specified by a configuration file or operator input.

Once initial information is entered, operations in the gateway are triggered by events—either the arrival of a data packet at one of the network interfaces or expiration of a timer. There are four critical time constants that control route propagation and expiration. These time constants may be set by the system administrator. However, there are default values. These time constants are:

- Broadcast time—updates are broadcast by all gateways on all connected interfaces for this amount of time. The default is once every 90 seconds.
- Invalid time—if no update has been received for a given path within this amount of time, it is considered to have timed out. It should be several times the broadcast time to allow for the possibility that packets containing an update could be dropped by the network. The default is 3 times the broadcast time.
- Hold time—when a destination has become unreachable (or the metric has increased enough to cause poisoning), the destination goes into holddown. During this state, no new path will be accepted for the same destination for this amount of time. The hold time indicates how long this state should last. It should be several times the broadcast time. The default value is 3 times the broadcast time plus 10 seconds.
- Flush time—if no update has been received for a given destination within this amount of time, the entry for it is removed from the routing table. Note the difference between invalid time and flush time: After the invalid time a path is timed out and removed. If there are no remaining paths to a destination, the destination is now unreachable. However, the database entry for the destination remains. It has to remain in order to enforce the holddown. After the flush time the database entry is removed from the table. It should be somewhat longer than the invalid time plus the holddown time. The default is 7 times the broadcast time.

These figures presuppose the following major data structures. A separate set of these data structures is kept for each protocol supported by the gateway. Within each protocol, a separate set of data structures is kept for each type of service to be supported.

- For each destination known to the system there is a (possibly null) list of paths to the destination, a holddown expiration time, and a last update time. The last update time indicates the last time any path for this destination was included in an update from another gateway. Note that there are also update times kept for each path. When the last path to a destination is removed, the destination is put into holddown, unless holddowns are disabled. The holddown expiration time indicates the time at which the holddown expires. The fact that it is nonzero indicates that the destination is in holddown. To save calculation time, it is also a good idea to keep a best metric for each destination. This is simply the minimum of the composite metrics for all the paths to the destination.

- For each path to a destination there is the address of the next hop in the path, the interface to be used, and a vector of metrics characterizing the path including topological delay, bandwidth, reliability, and channel occupancy. Other information is also associated with each path including hop count, MTU, source of information, the remote composite metric, and a composite metric calculated from these numbers according to Eq. 1. There is also a last update time. The source of information indicates where the most recent update for that path came from. In practice this is the same as the address of the next hop. The last update time is simply the time at which the most recent update arrived for this path. It is used to expire timed-out paths.

Note that an IGRP update message has three portions: interior, system (meaning "this autonomous system" but not interior), and exterior. The interior section is for routes to subnets. Not all subnet information is included. Only subnets of one network are included. This is the network associated with the address to which the update is being sent. Normally updates are broadcast on each interface, so this is simply the network on which the broadcast is being sent. (Other cases arise for responses to an IGRP request and point-to-point IGRP.) Major networks (i.e., non-subnets) are put into the system portion of the update message unless they are specifically flagged as exterior.

A network will be flagged as exterior if it was learned from another gateway and the information arrived in the exterior portion of the update message. The system administrator can declare specific networks as exterior. Exterior routes are also referred to as candidate default. They are routes that go to or through gateways that are considered to be appropriate as defaults, to be used when there is no explicit route to a destination. For example, some companies configure the gateway that connects to its regional network so that it flags the route to the NSFnet backbone as exterior. You can manipulate the router to choose a default route by picking that exterior route with the smallest metric.

The specifications of a protocol include a procedure for determining the destination of a packet, a procedure for comparing the destination with the gateway's own addresses to determine whether the gateway itself is the destination, a procedure for determining whether a packet is a broadcast, and a procedure for determining whether the destination is part of a specified network. A path is usable for the purposes of its remote composite metric is less than its composite metric. A path whose remote composite metric is greater than its composite metric is a path whose next hop is farther away from the destination, as measured by the metric. This is referred to as an upstream path. Normally, one would expect that the use of metrics would prevent upstream paths from being chosen. It is easy to see that an upstream path can never be the best one. However, if a large variance is allowed, paths other than the best one can be used. Some of those could be upstream.

Paths whose remote composite metric is not less than their composite metrics are not considered. If more than one path is acceptable, such paths are used in a weighted form of round-robin alternation. The frequency with which a path is used is inversely proportional to its composite metric.

Reception of Routing Updates

Such updates consist of a list of entries, each of which gives information for a single destination. More than one entry for the same destination can occur in a single routing update to accommodate multiple types of service. Each of these entries is processed individually. If an entry is in the exterior section of the update, the exterior flag will be set for the destination if it is added as a result of this process.

The entire process must be repeated once for each type of service supported by the gateway, using the set of destination/path information associated with that type of service. The entire routing update must be processed once for each type of service. (Note that the current implementation of IGRP does not support multiple types of service. Therefore, the outermost loop is not actually implemented.)

Basic acceptability tests are done on the path. This should include reasonableness tests for the destination. Impossible ("Martian") network numbers should be rejected. (See RFC 1009 and RFC 1122 for details.) Updates are also rejected if the destination they refer to is in holddown (i.e., the holddown expiration time is nonzero and later than the current time).

The routing table is searched to see whether this entry describes a path that is already known. A path in the routing table is defined by the destination with which it is associated, the next hop listed as part of the path, the output interface to be used for the path, and the information source (the address from which the update came—in practice normally the same as the next hop). The entry from the update packet describes a path whose destination is listed in the entry, whose output interface is the interface that the update came in, and whose next hop and information source are the address of the gateway that sent the update (the "source" S).

The update process will actually run after the entire process has been finished. That is, the update process will only happen once, even if it is triggered several times during the processing. Furthermore, precautions must be taken to keep updates from being issued too often, if the network is changing rapidly.

The router compares the new composite metric computed from data in the update packet with the best composite metric for the destination. Note that the best composite metric is not recomputed at this time. So if the path being considered is already in the routing table, this test may compare new and old metrics for the same path.

If the path is worse than the existing best composite metric, this includes both new paths that are worse than existing ones and existing paths whose composite metric has increased. Tests are conducted to see whether the new path is acceptable. Note that this test implements both the test for whether a new path is good enough to keep and route poisoning. In order to be acceptable, the delay value must not be the special value that indicates an unreachable destination (for the current IP implementation, all ones in a 24-bit field), and the composite metric (calculated as specified) must be acceptable. To determine whether the composite metric is acceptable, compare it with the composite metrics of all other paths to the destination. Let M be the minimum of these. The new path is acceptable if it is < V × M, where V is the variance set when the gateway was initialized. If V = 1, then a metric any worse than the existing one is not acceptable. There is one exception to this: If the path already exists and is the only path to the destination, the path will be retained if the metric has not increased by more than 10 percent (or where holddowns are disabled if the hop count has not increased).

The new information for a path indicates that the composite metric will be decreased. The composite metrics of all paths to a particular destination D are compared. In this comparison, the new composite metric for P is used, rather than the one appearing in the routing table. The minimum composite metric M is calculated. Then all paths to D are examined again. If the composite metric for any path > M × V, that path is removed. V is the variance, entered when the gateway was initialized.

Periodic Processing

The process is triggered once a second. It examines various timers in the routing table to see if any has expired. These timers have been described earlier.

The composite metrics stored in the routing table depend on the channel occupancy, which changes over time, based on measurements. Periodically the channel occupancy is recalculated, using a moving average of measured traffic through the interface. If the newly calculated value differs from the existing one, all composite metrics involving that interface must be adjusted. Every path shown in the routing table is examined. Any path whose next hop uses interface "I" has its composite metric recalculated. This is done in accordance with Eq. 1, using as the channel occupancy the maximum of the value stored in the routing table as part of the path's metric and the newly calculated channel occupancy of the interface.

Generating Update Messages

A separate message is generated for each network interface attached to the gateway. That message is then sent to all other gateways that are reachable through the interface. Generally, this is done by sending the message as a broadcast. However, if the network technology or protocol does not allow broadcasts, it may be necessary to send the message individually to each gateway.

In general, the message is built up by adding an entry for each destination in the routing table. Note that the destination/path data associated with each type of service must be used. In the worst case, a new entry is added to the update for each destination for each type of service. However, before adding an entry to the update message the entries already added are scanned. If the new entry is already present in the update message, it is not added again. A new entry duplicates an existing one when the destinations and next hop gateways are the same.

For the sake of simplicity, the pseudocode omits one thing: IGRP update messages have three parts: interior, system, and exterior. Thus, there are actually three loops over destinations. The first includes only subnets of the network to which the update is being sent. The second includes all major networks (i.e., non-subnets) that are not flagged as exterior. The third includes all major networks that are flagged as exterior.

In the normal case, the split horizon test fails for routes whose best path goes out the same interface in which the update is being sent. However, if the update is being sent to a specific destination (e.g., in response to an IGRP request from another gateway or as part of point-to-point IGRP), split horizon fails only if the best path originally came from that destination (its "information source" is the same as the destination) and its output interface is the same as the one the request came in from.

Computing Metric Information

The metric information is processed from update messages received by the gateway, and it is generated for update messages being sent by the gateway. Note that the entry is based on one particular path to the destination. If there is more than one path to the destination, a path whose composite metric is minimum is chosen. If more than one path has the minimum composite metric, an arbitrary tie-breaking rule is used. (For most protocols this will be based on the address of the next hop gateway.)

```
Processing incoming packets
   Data packet arrives using interface I
A Determine protocol used by packet
   If protocol is not supported
     then discard packet

B If destination address matches any of gateway's addresses
   or the broadcast address
       then process packet in protocol-specific way

C If destination is on a directly connected network
       then send packet direct to the destination, using
         the encapsulation appropriate to the protocol and link type

D If there are no paths to the destination in the routing
     table, or all paths are upstream
        then send protocol-specific error message and discard the packet

E Choose the next path to use. If there are more than
     one, alternate round-robin with frequency proportional
     to inverse of composite metric.
```

Get next hop from path chosen in previous step.
Send packet to next hop, using encapsulation appropriate
to protocol and data link type.

Processing incoming routing updates

Routing update arrives from source S. For each type of service supported by the gateway:

   Use routing data associated with this type of service
   For each destination D shown in update

A If D is unacceptable or in holddown
   then ignore this entry and continue loop with next destination D

B Compute metrics for path P to D via S
   If destination D is not already in the routing table
   then Begin
      Add path P to the routing table, setting last
      update times for P and D to current time.

H Trigger an update
   Set composite metric for D and P to new composite
   metric computed in step B.
   End
   Else begin (dest.. D is already in routing table)

K Compare the new composite metric for P with best
   existing metric for D.
   New > old:

L If D is shown as unreachable in the update,
   or holddowns are enabled and
    the new composite metric >
       (the existing metric for D) * V
       [use 1.1 instead of V if V = 1,]

O or holddowns are disabled and
   P has a new hop count > old hop count
   then Begin
   Remove P from routing table if present
   If P was the last route to D
      then Unless holddowns are disabled
          Set holddown time for D to
            current time + holddown time

```
T        and Trigger an update
   End
   else Begin
   Compute new best composite metric for D
   Put the new metric information into the
   entry for P in the routing table
   Add path P to the routing table if it
   was not present.
   Set last update times for P and D to
   current time.
   End
   New <= OLD:

V  Set composite metric for D and P to new
   composite metric computed in step B.
   If any other paths to D are now outside the
   variance, remove them.
   Put the new metric information into the
   entry for P in the routing table
   Set last update times for P and D to
   current time.
   End
   End of for
   End of for
```

Periodic processing

```
   Process is activated by regular clock, e.g., once per second
   For each path P in the routing table (except directly
   connected interfaces)
      If current time < P'S LAST UPDATE TIME + INVALID TIME
         THEN CONTINUE WITH THE NEXT PATH P
      Remove P from routing table
      If P was the last route to D
         then Set metric for D to inaccessible
            Unless holddowns are disabled,
               Start holddown timer for D and
            Trigger an update
         else Recompute the best metric for D
      End of for
```

```
     For each destination D in the routing table
        If D's metric is inaccessible
           then Begin
           Clear all paths to D
           If current time >= D's last update time + flush time
              then Remove entry for D
           End
        End of for
     For each network interface I attached to the gateway
R  Recompute channel occupancy and error rate
S  If channel occupancy or error rate has changed,
        then recompute metrics
        End of for
     At intervals of broadcast time
U  Trigger update
```

Generate update

```
     Process is caused by "trigger update"
     For each network interface I attached to the gateway
        Create empty update message
        For each type of service S supported
           Use path/destination data for S
           For each destination D
E  If any paths to D have a next hop reached through I
        then continue with the next destination
     If any paths to D with minimal composite metric are
     already in the update message
        then continue with the next destination
G  Create an entry for D in the update message, using
   metric information from a path with minimal
   composite metric
   End of for
   End of for
J  If there are any entries in the update message
        then send it out interface I
   End of for
```

Details of Metric Computations

This section describes the procedure for computing metrics and hop counts from an arriving routing update.

The input to this function is the entry for a specific destination in a routing update packet. The output is a vector of metrics, which can be used to compute the composite metric, and a hop count. If this path is added to the routing table, the entire vector of metrics is entered in the table. The interface parameters used in the following definitions are those set when the gateway was initialized for the interface on which the routing update arrived, except that the channel occupancy and reliability are based on a moving average of measured traffic through the interface.

> delay = delay from packet + interface topological delay
> bandwidth = max (bandwidth from packet, interface bandwidth)
> reliability = min (reliability from packet, interface reliability)
> channel occupancy = max (channel occupancy from packet, interface channel occupancy)

(Max is used for bandwidth because the bandwidth metric is stored in inverse form. Conceptually, we want the minimum bandwidth.) Note that the original channel occupancy from the packet must be saved because it will be needed to recompute the effective channel occupancy whenever the interface channel occupancy changes.

The following are not part of the metric vector, but are also kept in the routing table as characteristics of the path:

> hop count = hop count from packet
> MTU = min (MTU from packet, interface MTU)
> remote composite metric—calculated from Eq. 1 using the metric values from the packet. That is, the metric components are those from the packet and are not updated as shown above. Obviously, this must be calculated before the adjustments shown above are done.
> composite metric—calculated from Eq. 1 using the metric values calculated as described in this section

This section describes the procedure for computing metrics and hop count for routing updates to be sent. This function determines the metric information and hop count to be put into an outgoing update packet. It is based on a specific path to a destination, if there are any usable paths. If there are no paths, or the paths are all upstream, the destination is called "inaccessible."

```
If destination is inaccessible, this is indicated by using a specific
    value in the delay field. This value is chosen to be larger
    than the largest valid delay. For the IP implementation this is
    all ones in a 24-bit field.
```

```
If destination is directly reachable through one of the interfaces,
use
     the delay, bandwidth, reliability, and channel occupancy of the
     interface. Set hop count to 0.
Otherwise, use the vector of metrics associated with the path in the
     routing table. Add one to the hop count from the path in the
     routing table.
```

Details of the IP Implementation

This section gives a brief description of the packet formats used by IGRP. IGRP is sent using IP datagrams with IP protocol 9 (IGP). The packet begins with a header. It starts immediately after the IP header.

```
unsigned version: 4;    /* protocol version number */
unsigned opcode: 4;     /* opcode */
uchar edition;          /* edition number */
ushort asystem;         /* autonomous system number */
ushort ninterior;       /* number of subnets in local net */
ushort nsystem;         /* number of networks in AS */
ushort nexterior;       /* number of networks outside AS */
ushort checksum;        /* checksum of IGRP header and data */
```

For update messages, routing information follows immediately after the header.

The version number is currently 1. Packets having other version numbers are ignored.

The opcode is either:

1 update
2 request

This indicates the type of message. The format of the two message types will be discussed later.

Edition is a serial number that is incremented whenever there is a change in the routing table. (This is done in those conditions in which the pseudocode above says to trigger a routing update.) The edition number allows gateways to avoid processing updates containing information that they have already seen. (This is not currently implemented. That is, the edition number is generated correctly, but it is ignored on input. Because it is possible for packets to be dropped, it is not clear that the edition number is sufficient to avoid duplicate processing. It would be necessary to make sure that all of the packets associated with the edition had been processed.)

Asystem is the autonomous system number. In any vendor's implementation of IGRP, a gateway can participate in more than one autonomous system. Each such system runs its own IGRP protocol. Conceptually, there are completely separate routing tables for each autonomous system. Routes that arrive via IGRP from one autonomous system are sent only in updates for that AS. This field allows the gateway to select which set of routing tables to use for processing this message. If the gateway receives an IGRP message for an AS that it is not configured for, it is ignored. In fact some vendor IOS implementation allows information to be leaked from one AS to another. However, this may be regarded as an administrative tool and not part of the protocol.

Ninterior, nsystem, and nexterior indicate the number of entries in each of the three sections of update messages. These sections have been described earlier. There is no other demarcation between the sections. The first ninterior entries are taken to be interior, the next nsystem entries as being system, and the final nexterior as exterior.

Checksum is an IP checksum, computed using the same checksum algorithm as a UDP checksum. The checksum is computed on the IGRP header and any routing information that follows it. The checksum field is set to zero when computing the checksum. The checksum does not include the IP header, nor is there any virtual header as in UDP and TCP.

Requests

An IGRP request asks the recipient to send its routing table. The request message has only a header. Only the version, opcode, and asystem fields are used. All other fields are zero. The recipient is expected to send a normal IGRP update message to the requester.

Updates

An IGRP update message contains a header, followed immediately by routing entries. As many routing entries are included as will fit into a 1,500-byte datagram (including IP header). With current structure declarations, this allows up to 104 entries. If more entries are needed, several update messages are sent. Because update messages are simply processed entry by entry, there is no advantage to using a single fragmented message rather than several independent ones.

Here is the structure of a routing entry:

```
uchar number[3];      /* 3 significant octets of IP address */
uchar delay[3];       /* delay, in tens of microseconds */
uchar bandwidth[3];   /* bandwidth, in units of 1 Kbit/sec */
uchar mtu[2];         /* MTU, in octets */
uchar reliability;    /* percent packets successfully tx/rx */
uchar load;           /* percent of channel occupied */
uchar hopcount;       /* hop count */
```

The fields defined uchar[2] and uchar[3] are simply 16- and 24-bit binary integers in normal IP network order.

Number defines the destination being described. It is an IP address. To save space only the first 3 bytes of the IP address are given, except in the interior section. In the interior section, the last 3 bytes are given. For system and exterior routes no subnets are possible, so the low order byte is always zero. Interior routes are always subnets of a known network, so the first byte of that network number is supplied.

Delay is in units of 10 microseconds. This gives a range of 10 microseconds to 168 seconds, which seems sufficient. A delay of all ones indicates that the network is unreachable.

Bandwidth is inverse bandwidth in bits per sec scaled by a factor of 1.0e10. The range is from a 1,200-bps line to 10 Gbps. (That is, if the bandwidth is N Kbps, the number used is 10000000 / N.)

MTU is in bytes.

Reliability is given as a fraction of 255. That is, 255 is 100 percent.

Load is given as a fraction of 255.

Hopcount is a simple count.

Because of the somewhat weird units used for bandwidth and delay, some examples seem in order. These are the default values used for several common media.

	Delay		Bandwidth	
Satellite	200,000	(2 sec)	20	(500 Mbit)
Ethernet	100	(1 ms)	1,000	
1.544 Mbit	2000	(20 ms)	6,476	
64 Kbit	2000		156,250	
56 Kbit	2000		178,571	
10 Kbit	2000		1,000,000	
1 Kbit	2000		10,000,000	

Metric Computations

Here is a description of the way the composite metric is actually computed.

```
metric = [K1*bandwidth + (K2*bandwidth)/(256 - load) + K3*delay] *
[K5/(reliability + K4)]
If K5 == 0, the reliability term is not included.
The default version of IGRP has K1 == K3 == 1, K2 == K4 == K5 == 0
```

Enhanced Interior Gateway Routing Protocol

Background

The Enhanced Interior Gateway Routing Protocol (EIGRP) represents an evolution from its predecessor IGRP. This evolution resulted from changes in networking and the demands of diverse, large-scale internetworks. Enhanced IGRP integrates the capabilities of link state protocols into distance vector protocols. It incorporates the *Diffusing Update Algorithm* (DUAL) developed at SRI International by Dr. J. J. Garcia-Luna-Aceves.

Enhanced IGRP provides compatibility and seamless interoperation with IGRP routers. An automatic redistribution mechanism allows IGRP routes to be imported into Enhanced IGRP, and vice versa, so it is possible to add Enhanced IGRP gradually into an existing IGRP network. Because the metrics for both protocols are directly translatable, they are as easily comparable as if they were routes that originated in their own autonomous systems (ASs). In addition, Enhanced IGRP treats IGRP routes as external routes and provides a way for the network administrator to customize them.

This appendix provides an overview of the basic operations and protocol characteristics of Enhanced IGRP.

Enhanced IGRP Capabilities and Attributes

Key capabilities that distinguish Enhanced IGRP from other routing protocols include fast convergence, support for variable length subnet masks, support for partial updates, and support for multiple network layer protocols.

A router running Enhanced IGRP stores all its neighbors' routing tables so that it can quickly adapt to alternate routes. If no appropriate route exists, Enhanced IGRP queries its neighbors to discover an alternate route. These queries propagate until an alternate route is found.

Its support for variable length subnet masks permits routes to be automatically summarized on a network number boundary. In addition, Enhanced IGRP can be configured to summarize on any bit boundary at any interface.

Enhanced IGRP does not make periodic updates. Instead, it sends partial updates only when the metric for a route changes. Propagation of partial updates is automatically bounded so that only those routers that need the information are updated. As a result of these two capabilities, Enhanced IGRP consumes significantly less bandwidth than IGRP.

Enhanced IGRP includes support for AppleTalk, IP, and Novell NetWare. The AppleTalk implementation redistributes routes learned from the Routing Table Maintenance Protocol (RTMP). The IP implementation redistributes routes learned from OSPF, Routing Information Protocol (RIP), IS-IS, Exterior Gateway Protocol (EGP), or Border Gateway Protocol (BGP). The Novell implementation redistributes routes learned from Novell RIP or Service Advertisement Protocol (SAP).

Underlying Processes and Technologies

To provide superior routing performance, Enhanced IGRP employs four key technologies that combine to differentiate it from other routing technologies: neighbor discovery/recovery, Reliable Transport Protocol (RTP), DUAL finite-state machine, and protocol-dependent modules.

Neighbor discovery/recovery is used by routers to dynamically learn about other routers on their directly attached networks. Routers also must discover when their neighbors become unreachable or inoperative. This process is achieved with low overhead by periodically sending small Hello packets. As long as a router receives Hello packets from a neighboring router, it assumes that the neighbor is functioning, and the two can exchange routing information.

Reliable Transport Protocol (RTP) is responsible for guaranteed, ordered delivery of Enhanced IGRP packets to all neighbors. It supports intermixed transmission of multicast or unicast packets. For efficiency, only certain Enhanced IGRP packets are transmitted reliably. On a multiaccess network that has multicast capabilities, such as Ethernet, it is not necessary to send Hello packets reliably to all neighbors individually. For that reason, Enhanced IGRP sends a single multicast Hello packet containing an indicator that informs the receivers that the packet need not be acknowledged. Other types of packets, such as updates, indicate in the packet that acknowledgment is required. RTP contains a provision for sending multicast packets quickly when unacknowledged packets are pending, which helps ensure that convergence time remains low in the presence of varying speed links.

DUAL finite-state machine embodies the decision process for all route computations by tracking all routes advertised by all neighbors. DUAL uses distance information to select efficient, loop-free paths and selects routes for insertion in a routing table based on feasible successors. A *feasible successor* is a neighboring router used for packet forwarding that is a least-cost path to a destination that is guaranteed not to be part of a routing loop. When a neighbor changes a metric, or when a topology change occurs, DUAL tests for feasible successors. If one is found, DUAL uses it to avoid recomputing the route unnecessarily. When no feasible successors exist but neighbors still advertise the destination, a recomputation (also known as a diffusing computation) must occur to determine a new successor. Although recomputation is not processor intensive, it does affect convergence time, so it is advantageous to avoid unnecessary recomputations.

Protocol-dependent modules are responsible for network layer protocol-specific requirements. The IP-Enhanced IGRP module, for example, is responsible for sending and receiving Enhanced IGRP packets that are encapsulated in IP. Likewise, IP-Enhanced IGRP is also responsible for parsing Enhanced IGRP packets and informing DUAL of the new information that has been received. IP-Enhanced IGRP asks DUAL to make routing decisions, the results of which are stored in the IP routing table. IP-Enhanced IGRP is responsible for redistributing routes learned by other IP routing protocols.

Routing Concepts

Enhanced IGRP relies on four fundamental concepts: neighbor tables, topology tables, route states, and route tagging. Each of these is summarized in the discussions that follow.

Neighbor Tables

When a router discovers a new neighbor, it records the neighbor's address and interface as an entry in the *neighbor table*. One neighbor table exists for each protocol-dependent module. When a neighbor sends a Hello packet, it advertises a hold time, which is the amount of time a router treats a neighbor as reachable and operational. If a Hello packet is not received within the hold time, the hold time expires and DUAL is informed of the topology change.

The neighbor table entry also includes information required by RTP. Sequence numbers are employed to match acknowledgments with data packets, and the last sequence number received from the neighbor is recorded so that out-of-order packets can be detected. A transmission list is used to queue packets for possible retransmission on a per neighbor basis. Round-trip timers are kept in the neighbor-table entry to estimate an optimal retransmission interval.

Topology Tables

The *topology table* contains all destinations advertised by neighboring routers. The protocol-dependent modules populate the table, and the table is acted on by the DUAL finite-state machine. Each entry in the topology table includes the destination address and a list of neighbors that have advertised the destination. For each neighbor, the entry records the advertised metric, which the neighbor stores in its routing table. An important rule that distance vector protocols must follow is that if the neighbor advertises this destination, it must use the route to forward packets.

The metric that the router uses to reach the destination is also associated with the destination. The metric that the router uses in the routing table, and to advertise to other routers, is the sum of the best advertised metric from all neighbors plus the link cost to the best neighbor.

Route States

A topology table entry for a destination can exist in one of two states: active or passive. A destination is in the passive state when the router is not performing a recomputation or in the active state when the router is performing a recomputation. If feasible successors are always available, a destination never has to go into the active state, thereby avoiding a recomputation.

A recomputation occurs when a destination has no feasible successors. The router initiates the recomputation by sending a query packet to each of its neighboring routers. The neighboring router can send a reply packet, indicating it has a feasible successor for the destination, or it can send a query packet, indicating that it is participating in the recomputation. While a destination is in the active state, a router cannot change the destination's routing table information. After the router has received a reply from each neighboring router, the topology table entry for the destination returns to the passive state, and the router can select a successor.

Route Tagging

Enhanced IGRP supports internal and external routes. Internal routes originate within an Enhanced IGRP AS. Therefore, a directly attached network that is configured to run Enhanced IGRP is considered an internal route and is propagated with this information throughout the Enhanced IGRP AS. External routes are learned by another routing protocol or reside in the routing table as static routes. These routes are tagged individually with the identity of their origin.

External routes are tagged with the following information:

- Router ID of the Enhanced IGRP router that redistributed the route
- AS number of the destination

- Configurable administrator tag
- ID of the external protocol
- Metric from the external protocol
- Bit flags for default routing

Route tagging allows the network administrator to customize routing and maintain flexible policy controls. Route tagging is particularly useful in transit ASs, where Enhanced IGRP typically interacts with an interdomain routing protocol that implements more global policies, resulting in a very scalable policy based routing.

Enhanced IGRP Packet Types

Enhanced IGRP uses the following packet types: Hello and acknowledgment, update, and query and reply.

Hello packets are multicast for neighbor discovery/recovery and do not require acknowledgment. An *acknowledgment* packet is a hello packet that has no data. Acknowledgment packets contain a nonzero acknowledgment number and are always sent by using a unicast address.

Update packets are used to convey reachability of destinations. When a new neighbor is discovered, unicast update packets are sent so that the neighbor can build up its topology table. In other cases, such as a link cost change, updates are multicast. Updates always are transmitted reliably.

Query and reply packets are sent when a destination has no feasible successors. *Query* packets are always multicast. Reply packets are sent in response to query packets to instruct the originator not to recompute the route because feasible successors exist. *Reply* packets are unicast to the originator of the query. Both query and reply packets are transmitted reliably.

EIGRP Theory of Operation

Enhanced Interior Gateway Routing Protocol (EIGRP) is an interior gateway protocol suited for many different topologies and media. In a well-designed network, EIGRP scales well and provides extremely quick convergence times with minimal network traffic. Some of the many advantages of EIGRP are:

- Very low usage of network resources during normal operation; only hello packets are transmitted on a stable network.
- When a change occurs, only routing table changes are propagated, not the entire routing table; this reduces the load the routing protocol itself places on the network.
- Rapid convergence times for changes in the network topology (in some situations convergence can be almost instantaneous).

EIGRP is an enhanced distance vector protocol, relying on the Diffused Update Algorithm (DUAL) to calculate the shortest path to a destination within a network.

Major Revisions of the Protocol

There are two major revisions of EIGRP: versions 0 and 1. IOS versions earlier than run the earlier version of EIGRP; some explanations in this appendix may not apply to that earlier version. It is highly recommended to use the later version of EIGRP because it includes many performance and stability enhancements.

Basic Theory

A typical distance vector protocol saves the following information when computing the best path to a destination: the distance (total metric or distance, such as hop count) and the vector (the next hop). For instance, all the routers in the network in Figure E.1 are running Routing Information Protocol (RIP). Router Two chooses the path to Network A by examining the hop count through each available path. Because the path through Router Three is three hops, and the path through Router One is two hops, Router Two chooses the path through One and discards the information it learned through Three. If the path between Router One and Network A goes down, Router Two loses all connectivity with this destination until it times out the route of its routing table (three update periods, or 90 seconds), and Router Three readvertises the route (which occurs every 30 seconds in RIP). Not including any hold-down time, it will take between 90 and 120 seconds for Router Two to switch the path from Router One to Router Three.

EIGRP, instead of counting on full periodic updates to reconverge, builds a topology table from each of its neighbor's advertisements (rather than discarding the data), and

FIGURE E.1

Basic EIGRP diagram. (Reproduced with permission of Cisco Systems, Inc. Copyright © 2001 Cisco Systems, Inc. All rights reserved.)

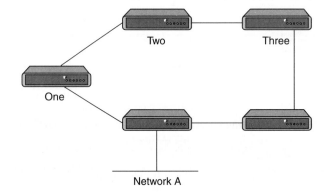

converges by either looking for a likely loop-free route in the topology table or, if it knows of no other route, querying its neighbors. Router Two saves the information it received from both Routers One and Three. It chooses the path through One as its best path (the successor) and the path through Three as a loop-free path (a feasible successor). When the path through Router One becomes unavailable, Router Two examines its topology table and, finding a feasible successor, begins using the path through Three immediately.

From this brief explanation, it is apparent that EIGRP must provide:

- A system where it sends only the updates needed at a given time—this is accomplished through neighbor discovery and maintenance
- A way of determining which paths a router has learned are loop-free
- A process to clear bad routes from the topology tables of all routers on the network
- A process for querying neighbors to find paths to lost destinations

We will cover each of these requirements in turn.

Neighbor Discovery and Maintenance

To distribute routing information throughout a network, EIGRP uses nonperiodic incremental routing updates. That is, EIGRP only sends routing updates about paths that have changed when those paths change.

The basic problem with sending only routing updates is that you may not know when a path through a neighboring router is no longer available. You cannot time out routes, expecting to receive a new routing table from your neighbors. EIGRP relies on neighbor relationships to reliably propagate routing table changes throughout the network—two routers become neighbors when they see each other's Hello packets on a common network.

EIGRP sends Hello packets every 5 seconds on high bandwidth links and every 60 seconds on low bandwidth multipoint links.

- 5-second hello:
    - broadcast media, such as Ethernet, Token Ring, and FDDI
    - point-to-point serial links, such as PPP or HDLC leased circuits, Frame Relay point-to-point subinterfaces, and ATM
    - point-to-point subinterfaces
    - high bandwidth (greater than T1) multipoint circuits, such as ISDN PRI and Frame Relay

- 60-second hello:
    - multipoint circuits T1 bandwidth or slower, such as Frame Relay multipoint interfaces, ATM multipoint interfaces
    - switched virtual circuits and ISDN BRIs

The rate at which EIGRP sends Hello packets is called the hello interval, and you can adjust it. The hold time is the amount of time that a router will consider a neighbor alive without receiving a Hello packet. The hold time is typically three times the hello interval, by default, 15 seconds and 180 seconds. You can adjust the hold time.

Note that if you change the hello interval, the hold time is not automatically adjusted to account for this change—you must manually adjust the hold time to reflect the configured hello interval.

It is possible for two routers to become EIGRP neighbors even though the hello and hold timers do not match. The hold time is included in the Hello packets so each neighbor should stay alive even though the hello interval and hold timers do not match.

While there is no direct way of determining what the hello interval is on a router, you can infer it from the output on the neighboring router.

The value in the Hold column of the command output should never exceed the hold time and should never be less than the hold time minus the hello interval (unless, of course, you are losing Hello packets). If the Hold column usually ranges between 10 and 15 seconds, the hello interval is 5 seconds and the hold time is 15 seconds. If the Hold column usually has a wider range—between 120 and 180 seconds—the hello interval is 60 seconds and the hold time is 180 seconds. If the numbers do not seem to fit one of the default timer settings, check the interface in question on the neighboring router—the hello and hold timers may have been configured manually.

Note: EIGRP does not build peer relationships over secondary addresses. All EIGRP traffic is sourced from the primary address of the interface.

Building the Topology Table

Now that these routers are talking to each other, what are they talking about? Their topology tables, of course! EIGRP, unlike RIP and IGRP, does not rely on the routing (or forwarding) table in the router to hold all of the information it needs to operate. Instead, it builds a second table, the topology table, from which it installs routes in the routing table.

Note: RIP maintains its own database from which it installs routes into the routing table.

The topology table contains the information needed to build a set of distances and vectors to each reachable network including:

- Lowest bandwidth on the path to this destination as reported by the upstream neighbor
- Total delay
- Path reliability
- Path loading
- Minimum path maximum transmission unit (MTU)
- Feasible distance
- Reported distance
- Route source (external routes are marked)

EIGRP Metrics

EIGRP uses the minimum bandwidth on the path to a destination network and the total delay to compute routing metrics. Although you can configure other metrics, we do not recommend it because it can cause routing loops in your network. The bandwidth and delay metrics are determined from values configured on the interfaces of routers in the path to the destination network.

For instance, in Figure E.2, Router One is computing the best path to Network A. It starts with the two advertisements for this network: one through Router Four, with a minimum bandwidth of 56 and a total delay of 2,200; and the other through Router Three, with a minimum bandwidth of 128 and a delay of 1,200. Router One chooses the path with the lowest metric.

Let us compute the metrics. EIGRP calculates the total metric by scaling the bandwidth and delay metrics. EIGRP uses the following formula to scale the bandwidth:

```
bandwidth = (10000000/bandwidth) * 256
```

EIGRP uses the following formula to scale the delay:

```
delay = delay * 256
```

EIGRP uses these scaled values to determine the total metric to the network:

```
metric = [K1 * bandwidth + (K2 * bandwidth) / (256 -
load) + K3 * delay] * [K5 / (reliability + K4)]
```

Note: These **K** values should be used after careful planning. Mismatched **K** values prevent a neighbor relationship from being built, which can cause your network to fail to converge.

The default values for **K** are:

K1 = 1
K2 = 0
K3 = 1
K4 = 0
K5 = 0

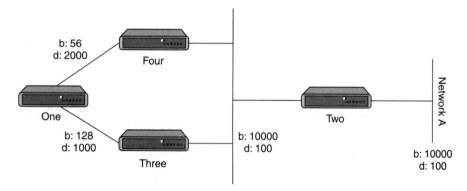

FIGURE E.2

EIGRP metrics example. (Reproduced with permission of Cisco Systems, Inc. Copyright © 2001 Cisco Systems, Inc. All rights reserved.)

For default behavior, you can simplify the formula as follows:

```
metric = bandwidth + delay
```

Combine this formula with the scaling factors and you have:

$$\left[\left(\frac{10^7}{\text{min bandwidth}}\right) + \text{sum of delays}\right] \times 256$$

These formulas assume you are using the delay as configured on the interface, which is in tens of microseconds. The delay is shown in microseconds, so you must divide by 10 before you use it in this formula. Throughout this appendix use delay as it is configured and shown on the interface.

Routers do not perform floating-point math, so at each stage in the calculation, you need to round down to the nearest integer to properly calculate the metrics.

In this example, the total cost through Router Four is:

```
minimum bandwidth = 56k
total delay = 100 + 100 + 2000 = 2200
[(10000000/56) + 2200] X 256 = (178571 + 2200) X 256
= 180771 X 256 = 46277376
```

And the total cost through Router Three is:

```
minimum bandwidth = 128k
total delay = 100 + 100 + 1000 = 1200
[(10000000/128) + 1200] X 256 = (78125 + 1200) X 256
= 79325 X 256 = 20307200
```

So to reach Network A, Router One chooses the route through Router Three.

Note that the bandwidth and delay values we used are those configured on the interface through which the router reaches its next hop to the destination network. For example, Router Two advertised Network A with the delay configured on its Ethernet interface, Router Four added the delay configured on its Ethernet, and Router One added the delay configured on its serial.

Feasible Distance, Reported Distance, and Feasible Successor

Feasible distance is the best metric along a path to a destination network, including the metric to the neighbor advertising that path. Reported distance is the total metric along a path to a destination network as advertised by an upstream neighbor. A feasible successor is a path whose reported distance is less than the feasible distance. Figure E.3 illustrates this process.

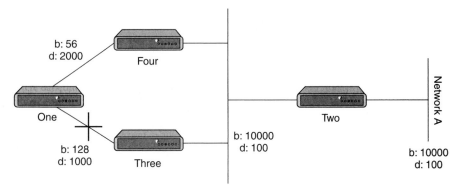

Router One sees that it has two routes to Network A: one through Router Three and another through Router Four.

- The route through Router Four has a cost of 46021376 and a reported distance of 307200.
- The route through Router Three has a cost of 20307200 and a reported distance of 307200.

Note that in each case EIGRP calculates the reported distance from the router advertising the route to the network. In other words, the reported distance from Router Four is the metric to get to Network A from Router Four, and the reported distance from Router Three is the metric to get to Network A from Router Three. EIGRP chooses the route through Router Three as the best path, and uses the metric through Router Three as the feasible distance. Because the reported distance to this network through Router Four is less than the feasible distance, Router One considers the path through Router Four a feasible successor.

When the link between Routers One and Three goes down, Router One examines each path it knows to Network A and finds that it has a feasible successor through Router Four. Router One uses this route, using the metric through Router Four as the new feasible distance. The network converges instantly, and updates to downstream neighbors are the only traffic from the routing protocol.

Let us look at a more complex scenario, shown in Figure E.4.

There are two routes to Network A from Router One: one through Router Two with a metric of 46789376 and another through Router Four with a metric of 20307200. Router One chooses the lower of these two metrics as its route to Network A, and this metric becomes the feasible distance. Next, let us look at the path through Router Two to see if it qualifies as a feasible successor. The reported distance from Router Two is 46277376, which is higher than the feasible distance—so this path is not a feasible successor. If you were to look in the topology table of Router One at this point, you would only see one entry for Network A—through Router Four.

FIGURE E.4

*Complex scenario.
(Reproduced with
permission of
Cisco Systems, Inc.
Copyright
© 2001 Cisco
Systems, Inc. All
rights reserved.)*

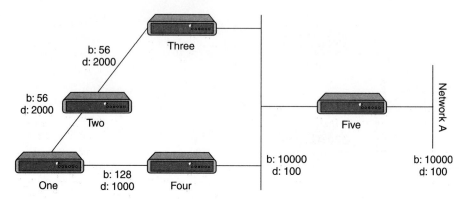

Let us suppose that the link between Router One and Router Four goes down. Router One sees that it has lost its only route to Network A and queries each of its neighbors (in this case, only Router Two) to see if they have a route to Network A. Because Router Two does have a route to Network A, it responds to the query. Because Router One no longer has the better route through Router Four, it accepts this route through Router Two to Network A.

Deciding If a Path Is Loop-Free

How does EIGRP use the concepts of feasible distance, reported distance, and feasible successor to determine if a path is valid and not a loop? In Figure E.5, Router Three examines routes to Network A. Because split horizon is disabled (for example, if these are multipoint Frame Relay interfaces), Router Three shows three routes to Network A: through Router Four, through Router Two (path is two, one, three, four), and through Router One (path is one, two, three, four).

If Router Three accepts all of these routes, it results in a routing loop. Router Three thinks it can get to Network A through Router Two, but the path through Router Two passes through Router Three to get to Network A. If the connection between Router Four and Router Three goes down, Router Three believes it can get to Network A through one of the other paths, but because of the rules for determining feasible successors, it will never use these paths as alternates. Let us look at the metrics to see why:

● total metric to Network A through Router Four: 20281600
● total metric to Network A through Router Two: 47019776
● total metric to Network A through Router One: 47019776

Because the path through Router Four has the best metric, Router Three installs this route in the forwarding table and uses 20281600 as its feasible distance to Network A.

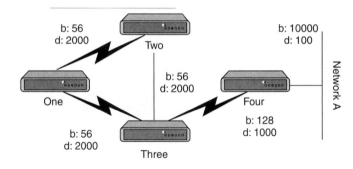

Router Three then computes the reported distance to Network A through Routers Two and One: 47019776 for the path through Router Two and 47019776 for the path through Router One. Because both of these metrics are greater than the feasible distance, Router Three does not install either route as a feasible successor for Network A.

Suppose that the link between Routers Three and Four goes down. Router Three queries each of its neighbors for an alternative route to Network A. Router Two receives the query and, because the query is from its successor, searches each of the other entries in its topology table to see if there is a feasible successor. The only other entry in the topology table is from Router One, with a reported distance equal to the last known best metric through Router Three. Because the reported distance through Router One is not less than the last known feasible distance, Router Two marks the route as unreachable and queries each of its neighbors—in this case, only Router One—for a path to Network A.

Router Three also sends a query for Network A to Router One. Router One examines its topology table and finds that the only other path to Network A is through Router Two with a reported distance equal to the last known feasible distance through Router Three. Once again, because the reported distance through Router Two is not less than the last known feasible distance, this route is not a feasible successor. Router One marks the route as unreachable and queries its only other neighbor, Router Two, for a path to Network A.

This is the first level of queries. Router Three has queried each of its neighbors in an attempt to find a route to Network A. In turn, Routers One and Two have marked the route unreachable and queried each of their remaining neighbors in an attempt to find a path to Network A. When Router Two receives Router One's query, it examines its topology table and notes that the destination is marked as unreachable. Router Two replies to Router One that Network A is unreachable. When Router One receives Router Two's query, it also sends back a reply that Network A is unreachable. Now Routers One and Two have both concluded that Network A is unreachable, and they reply to Router Three's original query. The network has converged, and all routes return to the passive state.

Basic Terminology for EIGRP

Successor
A neighbor that has been selected as the next hop for a destination based on the Feasibility Condition.

Feasible Successor (FS)
A neighbor that has satisfied the Feasibility Condition and has a path to the destination.

Feasibility Condition (FC)
A condition that is met when the lowest of all the neighbors' costs plus the link cost to that neighbor is found, and the neighbor's advertised cost is less than the current successor's cost.

Active State
A router's state for a destination when it has lost its successor to that destination and has no other FS available. The router is forced to compute a route to the destination.

Passive State
A router's state after losing its successor when it has an FS to the destination available in its topology table.

Hello/ACKs
Periodic Hello packets are exchanged between EIGRP neighbors. ACKs are sent to acknowledge that packets were reliably received.

Update
Sent in the following instances:

- When a neighbor first comes up.
- When a router transitions from Active to Passive for a destination.
- When there is a metric change for a certain destination.

Query
Sent to all neighbors when a router goes into Active for a destination and is asking for information on that destination. Unless it receives replies back from *all* its neighbors, the router will remain in Active state and not start the computation for a new successor.

Replies
Sent by every EIGRP neighbor that receives a query. If the neighbor does not have the information, it queries its neighbors.

Neighbor Table
Maintained with a Hold Timer (3 × Hello Timer) for each of the neighbors, based on the Hellos received from adjacent EIGRP routers.

Topology Table
Stores routing information that neighbors exchange after the first Hello exchange. DUAL acts on the topology table to determine Successors and FSs.

Troubleshooting Overview

Internetworks come in a variety of topologies and levels of complexity—from single-protocol, point-to-point links connecting crosstown campuses, to highly meshed, large-scale wide area networks (WANs) traversing multiple time zones and international boundaries. The industry trend is toward increasingly complex environments involving multiple media types, multiple protocols, and often interconnection to "unknown" networks. Unknown networks may be defined as a transit network belonging to an Internet service provider (ISP) or a telco that interconnects your private networks. In these unknown networks, you do not have control of such factors as delay, media types, or vendor hardware.

More complex network environments mean that the potential for connectivity and performance problems in internetworks is high, and the source of problems is often elusive. The keys to maintaining a problem-free network environment, as well as maintaining the ability to isolate and fix a network fault quickly, are documentation, planning, and communication. This requires a framework of procedures and personnel to be in place long before any network changes take place. The goal of this book is to help you isolate and resolve the most common connectivity and performance problems in your network environment.

Symptoms, Problems, and Solutions

Failures in internetworks are characterized by certain symptoms. These symptoms might be general (such as clients being unable to access specific servers) or more specific (routes not in routing table). Each symptom can be traced to one or more problems or causes by using specific troubleshooting tools and techniques. Once identified, each problem can be remedied by implementing a solution consisting of a series of actions.

This book describes how to define symptoms, identify problems, and implement solutions in generic environments. You should always apply the specific context in which you are troubleshooting to determine how to detect symptoms and diagnose problems for your specific environment.

General Problem-Solving Model

When you are troubleshooting a network environment, a systematic approach works best. Define the specific symptoms, identify all potential problems that could be causing the symptoms, and then systematically eliminate each potential problem (from most likely to least likely) until the symptoms disappear.

Figure F.1 illustrates the process flow for the general problem-solving model. This process flow is not a rigid outline for troubleshooting an internetwork; it is a foundation from which you can build a problem-solving process to suit your particular environment.

The following steps detail the problem-solving process outlined in Figure F.1:

> **Step 1.** When analyzing a network problem, make a clear problem statement. You should define the problem in terms of a set of symptoms and potential causes.

FIGURE F.1

General problem-solving model. (Reproduced with permission of Cisco Systems, Inc. Copyright © 2001 Cisco Systems, Inc. All rights reserved.)

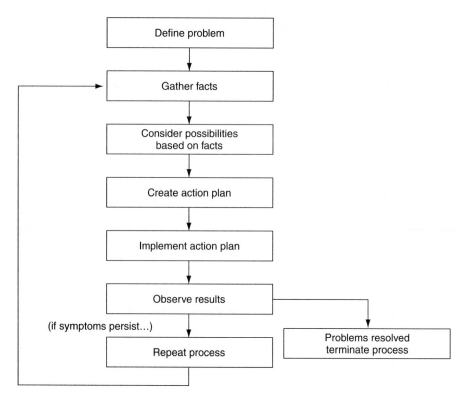

To properly analyze the problem, identify the general symptoms and then ascertain what kinds of problems (causes) could result in these symptoms. For example, hosts might not be responding to service requests from clients (a symptom). Possible causes might include a misconfigured host, bad interface cards, or missing router configuration commands.

Step 2. Gather the facts you need to help isolate possible causes.

Ask questions of affected users, network administrators, managers, and other key people. Collect information from sources such as network management systems, protocol analyzer traces, output from router diagnostic commands, or software release notes.

Step 3. Consider possible problems based on the facts you gathered. Using the facts you gathered, you can eliminate some of the potential problems from your list.

Depending on the data, you might, for example, be able to eliminate hardware as a problem so that you can focus on software problems. At every opportunity, try to narrow the number of potential problems so that you can create an efficient plan of action.

Step 4. Create an action plan based on the remaining potential problems. Begin with the most likely problem and devise a plan in which only *one* variable is manipulated.

Changing only one variable at a time allows you to reproduce a given solution to a specific problem. If you alter more than one variable simultaneously, you might solve the problem, but identifying the specific change that eliminated the symptom becomes far more difficult and will not help you solve the same problem if it occurs in the future.

Step 5. Implement the action plan, performing each step carefully while testing to see whether the symptom disappears.

Step 6. Whenever you change a variable, be sure to gather results. Generally, you should use the same method of gathering facts that you used in Step 2 (that is, working with the key people affected in conjunction with utilizing your diagnostic tools).

Step 7. Analyze the results to determine whether the problem has been resolved. If it has, then the process is complete.

Step 8. If the problem has not been resolved, you must create an action plan based on the next most likely problem in your list. Return to Step 4, change one variable at a time, and reiterate the process until the problem is solved.

Preparing for Network Failure

It is always easier to recover from a network failure if you are prepared ahead of time. Possibly the most important requirement in any network environment is to have current and accurate information about that network available to the network support personnel at all times. Only with complete information can intelligent decisions be made about network change, and only with complete information can troubleshooting be done as quickly and easily as possible. During the process of troubleshooting the network it is most critical to ensure that this documentation is kept up-to-date.

To determine whether you are prepared for a network failure, answer the following questions:

- Do you have an accurate physical and logical map of your internetwork?
    Does your organization or department have an up-to-date internetwork map that outlines the physical location of all the devices on the network and how they are connected, as well as a logical map of network addresses, network numbers, subnetworks, and so forth?
- Do you have a list of all network protocols implemented in your network?
    For each of the protocols implemented, do you have a list of the network numbers, subnetworks, zones, areas, and so on that are associated with them?
- Do you know which protocols are being routed?
    For each routed protocol, do you have correct, up-to-date router configuration?
- Do you know which protocols are being bridged?
    Are there any filters configured in any bridges, and do you have a copy of these configurations?
- Do you know all the points of contact to external networks, including any connections to the Internet?
    For each external network connection, do you know what routing protocol is being used?
- Do you have an established baseline for your network?
    Has your organization documented normal network behavior and performance at different times of the day so that you can compare the current problems with a baseline?

If you can answer yes to all questions, you will be able to recover from a failure more quickly and more easily than if you are not prepared.

RJ 45 Wiring Pinouts and Hints

How to wire a 10Base-T or 100Base-T connector with Category 5 cable and RJ45 connectors using USOC 568B wiring standards:

EIA/TIA 568B Wiring Standard

PIN	Wire Color
1	White w/**Orange** Stripe
2	**Orange** w/White Stripe
3	White w/**Green** Stripe
4	**Blue** w/White Stripe
5	White w/**Blue** Stripe
6	**Green** w/White Stripe
7	White w/**Brown** Stripe
8	**Brown** w/White Stripe

(EIA/TIA 658A) Crossover Cable Wiring Wire ONE End Using 568B and One End as Follows (Swap Orange and Green Pairs):

PIN	Wire Color
1	White w/**Green** Stripe
2	**Green** w/White Stripe
3	White w/**Orange** Stripe
4	**Blue** w/White Stripe
5	White w/**Blue** Stripe
6	**Orange** w/White Stripe
7	White w/**Brown** Stripe
8	**Brown** w/White Stripe

How to wire a 10Base-T or 100Base-T connector with Category 5 cable and RJ45 connectors using USOC 568B wiring standards:

EIA/TIA 568B Wiring Standard

PIN	Wire Color
1	White w/**Orange** Stripe
2	**Orange** w/White Stripe
3	White w/**Green** Stripe
4	**Blue** w/White Stripe
5	White w/**Blue** Stripe
6	**Green** w/White Stripe
7	White w/**Brown** Stripe
8	**Brown** w/White Stripe

(EIA/TIA 658A)
Crossover Cable Wiring
Wire ONE End Using 568B and One End as Follows (Swap Orange and Green Pairs):

PIN	Wire Color
1	White w/**Green** Stripe
2	**Green** w/White Stripe
3	White w/**Orange** Stripe
4	**Blue** w/White Stripe
5	White w/**Blue** Stripe
6	**Orange** w/White Stripe
7	White w/**Brown** Stripe
8	**Brown** w/White Stripe

Step 1. Cut the outer jacket of the wire about 1.5" to 2" from the end. This will give you room to work with the wire pairs. Separate the pairs and align them in the order shown following. Begin flattening the wires into a "ribbon" as shown so that it will easily slip into the connector and into the individual channeled areas.

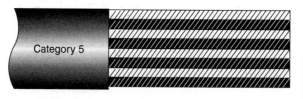 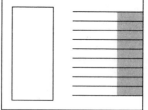

Step 2. Once you have all the wires aligned and ready to insert, you must trim them to approximately 1/2" in order to have as little "untwisted" wire in the connection as possible. Category 5 specifications require a certain number of twists per inch and even the connector counts!

Key Words:

10Base-T
Unshielded Twisted Pair (UTP)
RJ45
110 Blocks
Category 5
Level 5
EIA 568B
Straight Through
Reversed
Crossover
MDI
MDI-X Hub Ports

Wiring Tutorial for 10Base-T Unshielded Twisted Pair (UTP)

One of the most common and most puzzling problems a network engineer/technician may face is what is the proper way to make up a 10Base-T cable. Usually, to confound the learning process, someone introduces the need for a *reversed* or *crossover* cable at the same time. What these are and how to make them is the subject of this online tutorial.

Selection of Cabling Category

Because the overwhelming bulk of network cabling done today uses Unshielded Twisted Pair (UTP) wiring, that is what we will discuss. The process begins with the selection of the proper wiring level or category. Today it is basically inexcusable to use or install anything at less than Level V or Category 5.

While technically *Category 5* and *Level V* are not the same, they are identical in practice. Both support up to 100 megabit per second data transmission, and their physical cable assembly requirements are the same. Throughout this tutorial we will refer to them both as CAT5.

When you order CAT5 unshielded twisted pair (UTP) cable you will receive a cable containing 4 twisted pairs of wires, a total of 8 wires. The strands that constitute each wire will be either a single strand or multiple strands, usually referred to as *solid* or *flex*. Typically the solid is used to run through walls and ceilings and the flex is used to make drop cables (the cable from the wall plate to the desktop computer) and patch cables (the cable from the patch panel to the hub). Whether the exterior portion of the cable that contains the 4 twisted pairs, the jacket, is *Plenum grade* or *Non-plenum grade* is very important because it refers to the Fire Codes, but is outside the scope of this tutorial.

Ordering Pairs

The pairs of wires in UTP cable are colored so that you can identify the same wire at each end. Furthermore, they are usually color coded by pair so that the pairs can also be identified from end to end. Typical CAT5 UTP cables contain 4 pairs made up of a solid color and the same solid color striped onto a white background. The most common color scheme is the one that corresponds to the Electronic Industry Association/ Telecommunications Industry Association's Standard 568B.

The following table demonstrates the proper color scheme.

Wire pair #1:	White/Blue Blue
Wire pair #2:	White/Orange Orange
Wire pair #3:	White/Green Green
Wire pair #4:	White/Brown Brown

Connectors

The cable connectors and jacks that are most commonly used with CAT5 UTP cables are RJ45. The RJ simply means *Registered Jack* and the *45* designation specifies the pin numbering scheme. The connector is attached to the cable and the jack is the device that the connector plugs into, whether it is in the wall, the network interface card in the computer, or the hub.

Now that we are ready to insert the cable into the RJ45 plug, the wire number and color sequence become more complicated.

The IEEE Specification for Ethernet 10Base-T requires that two twisted pairs be used and that one pair is connected to pins 1 and 2, and that the second pair is connected to pins 3 and 6. Yes, that is right—pins 4 and 5 are skipped and are connected to one of the remaining twisted pairs.

According to the EIA/TIA-568B RJ-45 Wiring Scheme, it gets even more odd because wire Pair#2 (white/orange, orange) and Pair#3 (white/green, green) are the only two pairs used for 10Base-T data.

Pair #2 is connected to pins 1 and 2 like this:

Pin 1 wire color: white/orange

Pin 2 wire color: orange

Pair #3 is connected to pins 3 and 6 like this:

Pin 3 wire color: white/green

Pin 6 wire color: green

The remaining two twisted pairs are connected as such:

Pair #1

Pin 4 wire color: blue

Pin 5 wire color: white/blue

Pair #4

Pin 7 wire color: white/brown

Pin 8 wire color: brown

This is illustrated in the following diagram:

Now the wires forming the pairs must be gathered together and trimmed so that they can be inserted into the RJ45 plug. This is illustrated in the following diagram:

Then when the pairs are inserted into the RJ45 plug they should look like this:

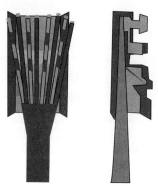

Crossover Cables

To make what is commonly referred to as a "crossover" cable one must change the pinout connections on *one* end of the cable. If you do it on both ends of the cable you have crossed over the crossover and now have a straight through cable, albeit a very nonstandard one. In this case two negatives do make a positive.

You need to make a cable where pins 1 and 2 from one end are connected to pins 3 and 6 on the other end, and pins 3 and 6 from the first end are connected to pins 1 and 2 on the other end. Pins 4 and 5 and 7 and 8 are unchanged.

The two ends look like this:

Standard End   Crossover End
  Pin 1 White/Orange   Pin 1 White/Green
  Pin 2 Orange   Pin 2 Green
  Pin 3 White/Green   Pin 3 White/Orange
  Pin 4 Blue   Pin 4 Blue
  Pin 5 White/Blue   Pin 5 White/Blue
  Pin 6 Green   Pin 6 Orange
  Pin 7 White/Brown   Pin 7 White/Brown
  Pin 8 Brown   Pin 8 Brown

The following is the proper pinout and cable pair/color order for the crossover end.

Pair #2 is connected to pins 1 and 2 like this:

Pin 1 wire color: white/green

Pin 2 wire color: green

Pair #3 is connected to pins 3 and 6 like this:

Pin 3 wire color: white/orange

Pin 6 wire color: orange

The crossover pairs are illustrated in the following diagram:

Cross over
cable order

Then when the pairs are inserted into the RJ45 plug they should look like this:

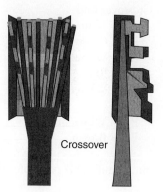

Crossover

Note: Even though we are only interested in attaching the connectors to the cable in this tutorial, we must take into account the wiring of the jacks as well so that we connect the proper wires from the cable to the proper pins in the connectors. And that is determined by the wiring in the jack the connectors will be plugged into.

Wireless LAN

What Is a Wireless LAN?

In the last few years, a new type of local area network has appeared. This new type of LAN, which is the wireless LAN, provides an alternative to the traditional LANs based on twisted pair, coaxial cable, and optical fiber. The wireless LAN serves the same purpose as that of a wired or optical LAN: to convey information among the devices attached to the LAN. But with the lack of physical cabling to tie down the location of a node on a network, the network can be much more flexible, as moving a wireless node is easy. This is in stark contrast to the large amount of labor required to add or move the cabling in any other type of network. In addition, going wireless may be a better alternative in cases where it is difficult or impossible to run wire in a building.

Wireless networks can be ideal for portable computers and laptops. Using wireless connections allows portable computers to still be portable without sacrificing the advantages of being connected to a network. Wireless portable clients may be installed virtually anywhere within a building.

Wireless networks can be used in combination with cabled LANs. Those machines that require relative mobility can be connected wirelessly, while the more permanent stations can be connected through cable.

A wireless LAN is a flexible data communications system implemented as an extension to or an alternative to a wired LAN. Using radio frequency (RF) technology, wireless LANs transmit and receive data over the air, minimizing the need for wired connections. Thus, wireless LANs combine data connectivity with user mobility.

Wireless LANs have gained strong popularity in a number of vertical markets, including the health care, retail, manufacturing, and warehousing industries, and in academia. These groups have profited from the productivity gains of using handheld terminals and notebook computers to transmit real-time information to centralized hosts for processing. Today, wireless LANs are becoming more widely recognized as a general-purpose connectivity alternative for a broad range of business customers. A 6-fold expansion of the worldwide wireless LAN market occurred by the year 2000, reaching more than $2 billion in revenues.

Network communication links (NCLs) is the terminology used in reference to wireless information connectivity for LANs and MANs. These wireless links are generally in unlicensed bands of the radio spectrum throughout the world. NCLs augment rather than replace other access technologies (e.g., dial-up, ISDN, T1/E1, ATM, XSDL, fiber optic, satellite, and cable).

The types of wireless technologies can be grouped into three categories: infrared, microwave, and radio.

Infrared (IR) systems are simple in design and therefore inexpensive. They use the same signal frequencies used on fiber optic links. IR systems detect only the amplitude of the signal and so interference is greatly reduced. These systems are not bandwidth limited and thus can achieve transmission speeds greater than the other systems. Infrared transmission operates in the light spectrum and does not require a license from the FCC to broadcast.

There are two conventional ways to set up an IR LAN. The first option is to aim the infrared transmissions. This provides a range of 2 km and can be used outdoors. It also offers the highest bandwidth and throughput. The other option is to transmit omnidirectionally and bounce the signals off everything in every direction. This reduces coverage to 10–20 mi, but it is area coverage. IR technology can deliver high data rates at a relatively cheap price. The drawbacks to IR systems are that the transmission spectrum is shared with the sun outdoors and fluorescent lights indoors. Enough interference from other sources can render the LAN useless. IR systems require an unobstructed line of sight (LOS). IR signals cannot penetrate opaque objects. This means that walls, dividers, curtains, and even fog can obstruct the signal. InfraLAN is an example of a wireless LAN using infrared technology.

Microwave (MW) systems typically operate at less than 500 mW of power. They use narrowband transmission with single frequency modulation and are set up mostly in the 5.8 GHz and 23 GHz band. The big advantage to MW systems is the higher throughput achieved by avoiding the overhead involved with spread spectrum systems. MW systems require an unobstructed LOS and Fresnel zone clearance. The Fresnel zone is the zone in front of the host wireless antenna that provides LOS to the remote antenna. The Fresnel zone is also thought of as the first 60 ft in front of the antenna.

Radio frequency systems for NCLs are generally divided into two categories: narrowband technology and wideband technology. A narrowband transceiver transmits and receives on the same frequency. The bandwidth of the communication channel is kept narrow to minimize the spectrum while maximizing the number of users. To avoid interference with adjacent channels, channel allocation is very important when deploying these types of systems.

Broadband technology for the wireless information connection uses spread spectrum technology. Spread spectrum technology is commercially available using one of two techniques: direct sequence spread spectrum (DSSS) and frequency hopping spread spectrum (FHSS). Requirements for this application are very different from desktop systems operating inside buildings. Many desktop wireless systems are unsuitable for outdoor applications that are exposed to greater distances. Systems must be individually designed to fit the requirements for NCL applications.

Why Wireless?

The widespread reliance on networking in business and the meteoric growth of the Internet and online services are strong testimonies to the benefits of shared data and shared resources. With wireless LANs, users can access shared information without looking for a place to plug in, and network managers can set up or augment networks without installing or moving wires. Wireless LANs offer the following productivity, convenience, and cost advantages over traditional networks:

Mobility.   Wireless LAN systems can provide LAN users with access to real-time information anywhere in their organization. This mobility supports productivity and service opportunities not possible with wired networks.

Installation Speed and Simplicity.   Installing a wireless LAN system can be fast and easy and can eliminate the need to pull cable through walls and ceilings.

Installation Flexibility.   Wireless technology allows the network to go where wire cannot go.

Reduced Cost-of-Ownership.   While the initial investment required for wireless LAN hardware can be higher than the cost of wired LAN hardware, overall installation expenses and life cycle costs can be significantly lower. Long-term cost benefits are greatest in dynamic environments requiring frequent moves and changes.

Scalability.   Wireless LAN systems can be configured in a variety of topologies to meet the needs of specific applications and installations. Configurations are easily changed and range from peer-to-peer networks suitable for a small number of users to full infrastructure networks of thousands of users that enable roaming over a broad area.

How Wireless LANs Are Used in the Real World

Wireless LANs frequently augment rather than replace wired LAN networks, often providing the final few meters of connectivity between a wired network and the mobile user. The following list describes some of the many applications made possible through the power and flexibility of wireless LANs:

- Doctors and nurses in hospitals are more productive because handheld or notebook computers with wireless LAN capability deliver patient information instantly.
- Consulting or accounting audit teams or small workgroups increase their productivity with quick network setup.
- Students holding a class on a campus green would access the Internet to consult the catalog of the Library of Congress.
- Training groups at corporations and students at universities use wireless connectivity to ease access to information, information exchanges, and learning.
- Network managers in dynamic environments minimize the overhead caused by moves, extensions to networks, and other changes with wireless LANs.
- Network managers installing networked computers in older buildings find that wireless LANs are a cost-effective network infrastructure solution.
- Trade show and branch office workers minimize setup requirements by installing preconfigured wireless LANs to provide backup for mission-critical applications running on wired networks.
- Senior executives in meetings make quicker decisions because they have real-time information at their fingertips.

Wireless LAN Technology

Manufactures of wireless LANs have a range of technologies to choose from when designing a wireless LAN solution. Each technology comes with its own set of advantages and limitations. The technologies covered in this section will be narrowband, spread spectrum, microwave, and infrared.

Narrowband Technology

UHF wireless data communication systems have been available since the early 1980s. These systems normally transmit in the 430 to 470 MHz frequency range, with rare systems using segments of the 800 MHz range. The lower portion of this band, 430–450 MHz, is often referred to as unprotected (unlicensed) and 450–470 MHz is referred to as the protected (licensed) band. In the unprotected band, radio frequency (RF) licenses are not granted for specific frequencies and anyone is allowed to use any frequency in the band. In the protected band, RF licenses are granted for specific frequencies, giving customers some assurance that they will have complete use of that frequency. Other terms for UHF include narrowband and 400 MHz RF.

Because independent narrowband RF systems cannot coexist on the same frequency, government agencies allocate specific radio frequencies to users through RF site licenses. A limited amount of an unlicensed spectrum is also available in some countries. In order to have many frequencies that can be allocated to users, the bandwidth given to a specific user is very small.

The term narrowband is used to describe this technology because the RF signal is sent in a very narrow bandwidth, typically 12.5 kHz or 25 kHz. Power levels range from 1 to 2 W for narrowband RF data systems. This narrow bandwidth combined with high power results in larger transmission distances than are available from 900 MHz or 2.4 GHz spread spectrum systems, which have lower power levels and wider bandwidths.

UHF radio shipments accounted for 23 percent of the wireless revenue and 28 percent of the wireless device units in 1995. From 1994 to 1995 the growth rate for the UHF radio devices was 9.1 percent, compared with 19.5 percent for spread spectrum devices during the same period.

Many modern UHF systems use synthesized radio technology, which refers to the way channel frequencies are generated in the radio. The crystal controlled products in legacy UHF products require factory installation of unique crystals for each possible channel frequency. Synthesized technology uses a single, standard crystal frequency and derives the required channel frequency by dividing the crystal frequency down to a small value, then multiplying it up to the desired channel frequency. The division and multiplication factors are programmed into digital memory in the radio at the time of manufacture.

Synthesized UHF based solutions provide the ability to install equipment without the complexity of hardware crystals. Common equipment can be purchased for the specific location requirements. Additionally, synthesized UHF radios do not exhibit the frequency drift problem experienced in crystal controlled UHF radios, a feature that eliminates tuning problems after installations have been running for a period of use.

The advantages of UHF radio are that it has the longest range and is a low cost solution for large sites with low to medium data throughput requirements. The disadvantages include low throughput, a lack of multivendor interoperability, and the potential for interference. In addition, an RF site license is required for protected bands, and large radios and antennas increase wireless client size.

A narrowband radio system transmits and receives user information on a specific radio frequency. Narrowband radio keeps the radio signal frequency as narrow as possible to pass information. Undesirable crosstalk between communication channels is avoided by carefully coordinating different users on different frequency channels.

A private telephone line is much like a radio frequency. When each home in a neighborhood has its own private telephone line, people in one home cannot listen to calls made to other homes. In a radio system, privacy and noninterference are accomplished by the use of separate radio frequencies. The radio receiver filters out all radio signals except the ones on its designated frequency.

From a customer standpoint, one drawback of narrowband technology is that the end user must obtain an FCC license for each desired frequency.

Spread Spectrum

Spread spectrum is currently the most widely used transmission technique for wireless LANs. Spread spectrum was initially developed by the military to avoid jamming signals and eavesdropping. It is very reliable, is secure, and can handle mission critical applications. Spread spectrum is designed to trade off bandwidth efficiency for reliability, integrity, and security. Spread spectrum is a modulation scheme whereby the signal is spread over a very wide bandwidth. This results in a dilution of the signal energy such that the power density (W per Hz) is lowered by the same amount that the spectrum is widened. Beyond a certain distance from the transmitter, the spread signal can be below the noise level yet still be recovered with the proper spread spectrum receiver. Only the intended receiver can recover the signal and know when it was started in time. This technique is accomplished by spreading the signal over a range of frequencies that consist of the industrial, scientific, and medical (ISM) bands of the electromagnetic spectrum. The ISM bands include the frequency ranges at 902 MHz to 928 MHz and at 2.4 GHz to 2.48 GHz, and they do not require an FCC license.

Spreading the signal over a range of frequencies is commonly referred to as a frequency hopping spectrum. A frequency hopping spectrum broadcasts the signal over a random series of radio frequencies. The carrier frequency is hopping according to a unique sequence over the spectrum bandwidth. Both the transmitter and receiver visit the same frequencies at the same time and must stay in exact synchronization. Each holds the same list of pseudorandom ordered frequencies, and the transmitter and receiver start hopping together using the same starting point on the list. A receiver hopping between frequencies in synchronization with the transmitter receives the message. The message can only be fully received if the series of frequencies is known. Because only the intended receiver knows the transmitter's hopping sequence, only that receiver can successfully receive all of the data. This technique splits the band into many small subchannels (1 MHz). The signal then hops from subchannel to subchannel transmitting short bursts of data on each channel for a set period of time-dwell time. The band is split into 75 subchannels and the dwell time is no longer than 400 ms. Frequency hopping is less susceptible to interference because the frequency is constantly shifting. This makes frequency hopping systems extremely difficult to intercept. This feature gives hopper systems a high degree of security. But security is not foolproof; many FHSS LANs can be colocated if an orthogonal hopping sequence is used.

Even though no license is required, the FCC has made some rules for frequency hopping spread spectrum technologies. The FCC dictates that the transmitters must not spend more than 0.4 s on any one channel every 20 s in the 902 MHz band and 75 channels in the 2.4 GHz band. A channel consists of a frequency width that is determined by the FCC. The IEEE 802.11 committee has drafted a standard that limits frequency hopping spread spectrum to the 2.4 GHz band.

The other type of spread spectrum communication is called direct sequence spread spectrum (DSSS), or pseudonoise. The DSSS method seems to be the one that most wireless spread spectrum LANs use. In DSSS, a high speed pseudorandom binary data stream is used to shift the carrier phase between 0 and 180 degrees. The phase shifting is normally done in a balanced mixer, and the information being transmitted is normally added to the high-speed code sequence. With DSSS, the transmission signal is spread over an allowed band—25 MHz. A random binary string is used to modulate the transmitted signal. Direct sequence transmitters spread their transmissions by adding redundant data bits, called chips. Direct sequence spread spectrum adds at least 10 chips to each data bit. Like a frequency hopping receiver, a direct sequence receiver must know a transmitter's spreading code to decipher the data. This spreading code is what allows multiple direct sequence transmitters to operate in the same area without interference. The data bits are mapped into a pattern of chips and mapped back into the bits at the destination. The number of chips represents a spreading ratio. Once the receiver has the entire data signal, it uses a correlator to remove the chips and collapse the signal to its original length.

As with FHSS, the FCC has also set rules for direct sequence transmitters. Each signal must have 10 or more chips. This rule limits the practical raw data rate of 8 Mbps in the 2–4 GHz band. Unfortunately, the number of chips is directly related to a signal's immunity to interference. The higher the spreading ratio, the more the signal is resistant to interference. The lower the spreading ratio, the more bandwidth is available to the user. In an area with lots of radio interference, throughput is compromised to avoid interference. The IEEE 802.11 committee has drafted a standard of 11 chips for DSSS. The transmitter and the receiver must be synchronized with the same spreading code. If orthogonal spreading codes are used, then more than one LAN can share the same band. However, because DSSS systems use wide subchannels, the number of colocated LANs is limited by the size of those subchannels.

Frequency hopping radios currently use less power than direct sequence radios and generally cost less. While direct sequence radios have a practical raw data rate of 8 Mbps, frequency hopping radios have a practical limit of 2 Mbps. If high performance is key and interference is not a problem, it is advisable to go with direct sequencing. If a small inexpensive portable wireless adapter for a notebook or PDA is needed, the frequency hopping method should be sufficient. With either method of spread spectrum, the end result is a system that is extremely difficult to detect, does not interfere with other devices, and still carries a large bandwidth of data.

Direct Sequence Versus Frequency Hopping Spread Spectrum

Because DSSS and FHSS are inherently different, the properties of each must be considered when choosing which technique to use. Table H.1 illustrates some of the advantages and disadvantages of each method.

Table H.1 Comparing DSSS and FHSS

Direct Sequence	*Frequency Hopping*
ADVANTAGES	**ADVANTAGES**
• Good range resolution.	• Reduced RF power levels.
• Noiselike spectrum, low power spectral density, signal can be hidden in noise.	• Low susceptibility to typical RF problems such as jamming and interception.
• Can operate below ambient noise.	• Operates in unlicensed bands.
• Can operate without error correction codes.	• Multiple access within frequency band is facilitated.
• More robust than FHSS.	• Signals can be difficult to detect by unauthorized listeners.
• Supports more equal-power users per unit of bandwidth.	• Higher permitted transmitting power by FCC.
• Low interference to coexisting discrete frequency or spread spectrum systems within near-far constraints.	• Relatively large unused bandwidth: 85 MHz.
• Less susceptible to local TX WB noise including other DS units.	• Best near-far performance. Can co-site multiple transceivers especially if T&R channel sets are separated and bandpassed.
• Initial synchronization more difficult but faster. For even faster synchronization, use a short synchronization code and a long spreading code. Care is needed to avoid false locks. Need tracking loop to maintain synchronization.	• Operates well with nonlinearity in signal path.
• Good voice quality with analog or digital modulation.	• Easier to synchronize.
	• Higher speeds than DS.
• More multipath resistant when delays exceed chip time. Inherent frequency diversity resists narrowband flat fading.	• Fewer users per unit of bandwidth.
	• Narrow instantaneous bandwidth; therefore signal presence easily detected.
• High resolution in ranging.	• Discrete frequencies or sub-bands can be deleted to circumvent interference.
• Signal paths must be reasonably linear. The receiver circuits before the correlator should limit at least P.G. above largest desired signal.	• Can withstand several narrowband in-band interferes much stronger than the desired signal. However, it is susceptible to local TX WB noise.

(continued)

Table H.1 Comparing DSSS and FHSS *(continued)*

Direct Sequence	*Frequency Hopping*
ADVANTAGES	**ADVANTAGES**
• More amenable to general purpose IC implementation than frequency hopping.	• Less critical code design. CDMA possible with codes of moderate orthogonality.
	• Code balance is not as important. The codes are typically slower and shorter according to number of channels and output in parallel.
	• Initial synchronization is easier but takes longer. Uses a short preamble for acquisition; time hops from there. Can track for longer intervals by means of local timer than DS.
	• Secure against casual eavesdroppers, especially if data modulate.
	• Delay spreads may impact very fast data. Inherent frequency and diversity resist fading but whole hops may be lost in narrowband flat fades.
	• Accommodates signal path nonlinearities well at least up to compression level.
DISADVANTAGES	**DISADVANTAGES**
• Near-far performance worse than frequency hopping. Cannot co-site multiple transceivers unless T&R channels are separated and bandpassed. Power control can help. TDMA reduces near-far and cross-correlation problems of CDMA, but could introduce synchronization problems.	• Relatively complex system as compared to system using standard modulation schemes—leading to increased costs.
• Requires linear signal path.	• SS chip sets (standard components) not widely available yet.
• Synchronization more difficult.	• Low data rate bandwidth limited— even with multiple frequencies, total bandwidth available to spread spectrum systems is inadequate to support more than 1 Mbps, only a tenth of the minimum 10 Mbps
• Harder to implement than FHSS.	

(continued)

Table H.1 Comparing DSSS and FHSS *(continued)*

Direct Sequence	*Frequency Hopping*
DISADVANTAGES	**DISADVANTAGES**
• Employs contiguous spectrum and therefore must accept all signals in this spectral window.	required to be compatible with current wired LAN data rates. Needs fast-switching synchronization, and data buffers for continuous data.
• Susceptible to single in-band above jamming margin. This is a major weakness but can be offset by using FDMA or TDMA.	• Boosting the data rate tends to reduce the range. Adding base stations solves problem but adds cost and can cause interference problems with cellular phones, FM radio, and upper end of the UHF television band.
• Code design more critical. Needs good index of discrimination for reliable synchronization detection and good CDMA performance as well to aid near-far performance. Code must be balanced to avoid discrete spectra. The codes are typically fast and long and are output serially.	• Higher power consumption.
	• Requires error correction codes.
	• Poor range resolution.
	• Narrow instantaneous bandwidth.
	• Not as robust as DS.
• Less secure against casual eavesdropping, especially if analog modulated.	• Must have a positive signal-to-noise ratio.
	• Will interfere pathologically with co-sited DS systems sharing common spectrum.
	• Susceptible to local TX WB noise including DS transmitters.
	• Voice quality is poor unless digitized.
	• Low resolution in ranging.

To say that one method is better than the other would be meaningless without considering the intended application. In practice, the decision to use direct sequence or frequency hopping modulation is dependent on the following:

- What features the product must possess.
- What physical and electromagnetic interference is anticipated, such as what other systems occupy the same frequency band.
- The technical expertise of the designer (frequency hopping is considered the more difficult of the two).
- The price of the product.

Microwave

Microwave technology is not really a LAN technology. The main use of MW is to interconnect LANs between buildings. This requires MW dishes on both ends of the link. The dishes must be in LOS to transmit and collect the MW signals. MW is used to bypass the telephone company when connecting LANs between buildings.

One major drawback to the use of MW technology is that the frequency band cannot be licensed to anyone else, for any purpose, within a 17.5-m radius.

Infrared

Infrared is an invisible band of radiation that exists at the lower end of the visible electromagnetic spectrum. This type of transmission is most effective when a clear LOS exists between the transmitter and the receiver.

Two types of infrared WLAN solutions are available: diffused-beam and direct-beam, or LOS. Currently, direct-beam WLANs offer a faster data rate than diffused-beam networks, but direct-beam is more directional. Since diffused-beam technology uses reflected rays to transmit/receive a data signal, it achieves lower data rates in the 1–2 Mbps range.

Infrared optical signals are often used in remote control device applications. Users who continuously set up temporary offices, such as auditors, salespeople, consultants, and managers, connect to the local wired network via an infrared device for retrieving information or using fax and print functions on a server.

A group of users may also set up a peer-to-peer infrared network while on location to share printer, fax, or other server facilities within their own LAN environment. The education and medical industries commonly use this configuration to easily move networks.

Infrared is a range technology. When used indoors, solid objects such as walls, doors, merchandise, and racks can limit infrared. In addition, the lighting environment can affect signal quality.

For example, loss of communications may occur as a result of the large amount of sunlight or background light in an environment. Fluorescent lights also may contain large amounts of infrared radiation. Using high signal power and an optical bandwidth filter, which lessens the infrared signals coming from outside sources, may solve this problem. In an outdoor environment, snow, ice, and fog may affect the operation of an infrared based system.

Infrared's advantages are that there are no government regulations controlling its use, and that infrared is immune to (RF) and electromagnetic interference (EMI). Its disadvantages are that infrared is generally a short-range technology as its range is from a 30–50 ft radius under ideal conditions; signals cannot penetrate solid objects; signals are affected by light, snow, and ice; and even dirt can interfere with infrared.

Infrared LANs use infrared signals to transmit data. This is the same technology used in products such as remote controls for televisions and VCRs. These LANs can be set up using either a point-to-point configuration or a sun-and-moon configuration, whereby the signals are diffused by being reflected off some type of surface.

Because of its many limitations, infrared is not a very popular technology for WLANs. According to a recent survey, infrared has less than 14 percent of the in-building WLAN market, and this market is expected to drop in the near future.

Line-of-Sight/Fresnel Zone

In general, LOS is necessary for communication in the 2.4 GHz frequency region. Strictly speaking, this is not true. The fact of the matter is that at any frequency, it is possible for communication to occur without LOS conditions. Why? Diffraction loss is a phenomenon that may be predicted and observed for all electromagnetic waves.

Diffraction loss causes some hindrance to systems design and to operation at lower frequencies. Diffraction loss is less destructive to such properties as signal attenuation at higher frequencies. This means that relative to normal system design goals, communications coverage may be allowed in shadowed areas for signals above approximately 1 GHz. Actually, for relatively short distances (< 100 m), providing the transmitter is of sufficiently high frequency, communication is possible through walls and around corners.

Even so, for all practical purposes, for point-to-point or point-to-multipoint links, we must assume that LOS is a mandatory requirement.

How is LOS determined? The first step in establishing that optical LOS exists is to determine whether one proposed antenna location is visible from the other. This can be easily confirmed using a high-power flashlight or, if the weather conditions permit, bright sunlight and a mirror. If conditions are foggy or hazy, a strobe light may suffice.

With optical LOS established, the next step is to qualify the path in terms of Fresnel zone clearance. In other words, does radio LOS exist?

```
Radio LOS = Optical LOS + First Fresnel Zone Clearance
```

Fresnel zones can be viewed as a series of concentric ellipsoids surrounding the entire microwave path. A cross-sectional view of the microwave path would be a series of concentric circles. From a propagation point of view, the first Fresnel zone is defined as the surface containing every point for which the sum of the distances from that point to the two ends of the path is exactly one-half wavelength longer than the direct end-to-end path. The nth Fresnel zone is defined in the same manner, except that the difference is in n half-wavelengths.

The Fresnel zones are successive regions where secondary waves (waves other than the direct or straight-line wave) have a path length from transmitter to receiver that is greater than the total path length of a line-of-sight path. Successive Fresnel zones have the effect of alternatively providing constructive and destructive interference to the total received signal.

Thus, the Fresnel zones exist as coaxial ellipsoids connecting the two antennae. The maximum diameter of each ellipsoid is located at the center point of the ellipsoid's axis and increases as the distance between the antennae increases. This zone is antenna independent and is based on distance between the two endpoints.

Typically, for optimally designed microwave paths, at least 60 percent of the first Fresnel zone must be free of obstructions. Any further clearance will yield negligible improvement in the received signal level. For maximum link reliability, the first Fresnel zone should be free of obstructions. A 3-m value is added to buffer this height.

It is worth noting that the radius of the first Fresnel zone will be the greatest at the midpath point. For a straight line from the center of one antenna to the center of the other, following the 60 percent clearance requirement (0.6F1), midpath clearance for this path is slightly more than 10.5 m (34.7 ft) from the centerline between the antennae.

A graphical representation of this area is defined by the illustration in Figure H.1.

For the purpose of simplicity, imagine the area of signal transmission between the two antennae to be a cylinder. Thus, if any obstacle appears within the first Fresnel zone, even though visual LOS is established, the radio link may not be viable. The reason is that part of the radiated energy is absorbed or diverted by the obstacle. If the link fade margin is not large enough to accommodate this signal attenuation, operation is not possible.

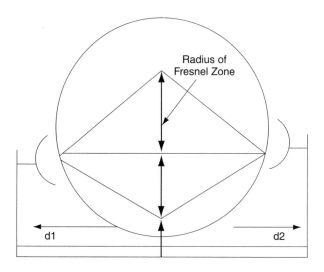

FIGURE H.1

Fresnel zone clearance.

Radio Path Analysis

Proper operation of any microwave radio communications system is dependent upon a LOS path between the microwave antennae at each end of the radio link. In general, if the path between the two sites is unobstructed and within an allowable distance, the microwave system will provide reliable service. However, further investigation is recommended to ensure that other path characteristics will not affect propagation of the microwave signal.

Assuming an appropriate LOS path from radio site to radio site can be established, both the feasibility and viability of a point-to-point microwave radio link will be dependent upon the gains, losses, and receiver sensitivity corresponding to the system. Gains are associated with the transmitter power output of the radio and the gains of both the transmitting and receiving antennae. Losses are associated with the cabling path between the radios and their respective antennae, and with the path between the antennae. Other losses can also occur if the path is partially obstructed or if path reflections cancel a portion of the normal receive signal.

One of the first items to consider for any microwave path is the actual distance from antenna to antenna. The farther a microwave signal must travel, the greater the signal loss. This form of attenuation is termed free space loss (FSL). Assuming an unobstructed path, only two variables need to be considered in FSL calculations: (1) the frequency of the microwave signal and (2) the actual path distance.

Wireless LAN Topologies

Wireless LANs can be built with either of two topologies: peer-to-peer or access point based. In a peer-to-peer wireless LAN topology, client devices within the wireless cell communicate directly to each other. Figure H.2 illustrates the peer-to-peer configuration.

An access point is a bridge that connects a wireless client device to the wired network. An access point based topology uses access points to bridge traffic onto a wired Ethernet or Token Ring, or a wireless backbone, as illustrated in Figure H.3.

The access point enables a wireless client device to communicate with any other wired or wireless device on the network. The access point topology is more commonly used. This demonstrates that WLANs do not replace wired LANs; they extend connectivity to mobile devices.

Another popular wireless network topology is the point-to-multipoint bridge. A bridge is defined as a node or pair of nodes with a transceiver client device that connects two networks using similar protocols. Wireless bridges connect a LAN in one building to a LAN in another, even if the buildings are many miles apart. These connections require a clear LOS. The LOS varies based on the type of wireless bridge and antennae used as well as environmental conditions.

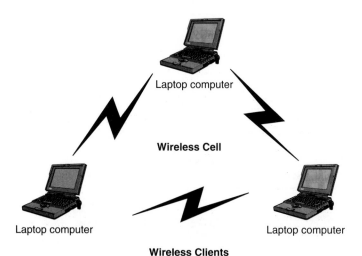

FIGURE H.2
Peer-to-peer configuration.

Laptop computer

Wireless Cell

Laptop computer

Laptop computer

Wireless Clients

WLAN clients are available in a number of ways for use in any of these network topologies. PCs can connect to a WLAN using ISA and PC adapter cards. Wireless modems can attach to parallel ports, RS232, 10BASE-T, IRDA, or other popular physical interfaces on a PC or other device. In this configuration, the client device communicates via the physical interface such as an ISA or PC adapter (RS232), to the radio device, which in turn provides the physical interface to the WLAN. For portable applications, the most common configurations are integrated LAN modules for application-specific, handheld terminals.

Systems Planning, Analysis, and Implementation

Professional systems planning is fundamental to the successful installation, operation, and performance of any communications system. The installation of every radio system can vary widely. Compared with visible light, the spread spectrum RF bands are not appreciably affected by fog, clouds, rain, snow, hail, smoke, smog, and so on. This is because the wavelength of the components that make up these items (water droplets, smoke particles, etc.) are virtually transparent to the radio signal.

When RF energy is transmitted from a parabolic antenna, the energy spreads outward, much like a beam of light. This microwave beam can be influenced by the terrain between the antennae, as well as by objects along the path.

Some level of signal loss will occur due to diffraction when the centerline of a beam from one antenna to another just grazes an obstacle along the path. To minimize the possibility of signal loss due to refraction, diffraction, and reflection, as well as other

FIGURE H.3
*Access point
configuration.*

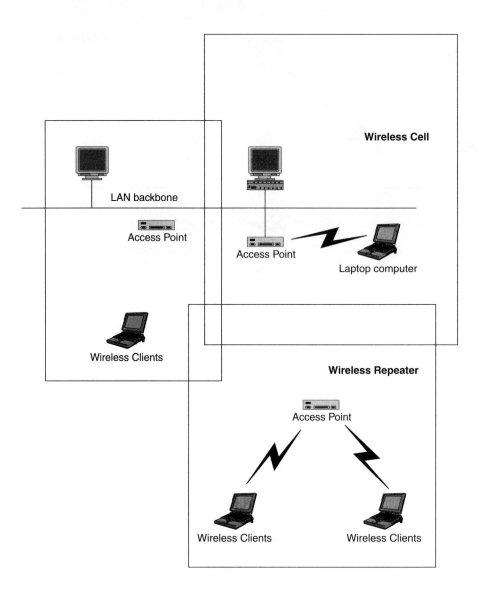

effects of obstructions and terrain, microwave paths must be properly engineered to address all of these issues.

For relatively short 2.4 or 5.7 GHz microwave paths, only reflection points and obstructions are of real concern. The effects of atmosphere, Earth curvature, and so on, will not usually come into play, so the engineering of these paths is quite straightforward.

Wireless Application Protocol

The wireless application protocol (WAP) is a hot topic that has been hyped within the mobile industry and outside of it. WAP is simply a protocol—a standardized way that a mobile phone talks to a server installed in the mobile phone network. The growth of WAP has been amazing. In the six months from August 1999 to January 2000, it became imperative for all IT companies in Nordic countries to have a WAP division. Many advertising agencies and dot-coms have announced WAP services. WAP is hot for several reasons:

- It provides a standardized way of linking the Internet to mobile phones, thereby linking two of the hottest commodities at any two locations.
- WAP's founding members include the major wireless vendors of Nokia, Ericsson, and Motorola, plus a newcomer, Phone.com.
- The WAP forum has over 200 member companies.
- Mobile information services, a key application for WAP, have not been as successful as many network operators expected. WAP is seen as a way to rectify this situation.

WAP also has its detractors and controversies:

- It is very difficult to configure WAP phones for new WAP services, with 20 or so different parameters needing to be entered to gain access to a WAP service. This is described in the details for the Nokia 7110 and Motorola L series in the new edition of Data on WAP.
- There are few mobile phones that currently support WAP. Widespread WAP support in handsets is unlikely to occur for a long time.
- WAP is a protocol that runs on top of an underlying bearer. None of the existing global system for mobiles (GSM) bearers for WAP—the short message service (SMS), unstructured supplementary services data (USSD), and circuit switched data (CSD)—are optimized for WAP.
- The WAP standard is incomplete, with key elements such as push technology (proactive sending of information to mobile devices) and wireless telephony updating address reports not yet standardized.
- There are many WAP gateway vendors competing against each other with largely the same standardized product. This has led to consolidation in the industry such as the pending acquisition of APON by Phone.com.
- Other protocols such as SIM application tool kit and mobile station application execution environment (MexE) are either already widely supported or designed to supersede WAP.
- WAP services are expected to be expensive to use. The expected trend is toward long online calls utilizing CSD and end user uses features such as interactivity and selection of more information. Without specific tariff initiatives, the mobile phone bills are likely to be surprisingly large for first time WAP users.

Motorola, Nokia, Ericsson, and the U.S. software company Phone.com (formerly Unwired Planet) were the initial partners that teamed up in mid-1997 to develop and deploy the WAP. WAP is an attempt to define the standard for how content from the Internet is filtered for mobile communications. Content is now readily available on the Internet; and WAP was designed to make this content easily available on mobile terminals.

The WAP forum was formed after a U.S. network operator, Omnipoint, requested a supply of mobile information services in early 1997. It received several responses from different suppliers using proprietary techniques for delivering the information, such as smart messaging from Nokia and HDML (handheld device markup language) from Phone.com. Omnipoint informed the vender responders that it would not accept a proprietary approach and recommended that the various vendors get together to explore defining a common standard. After all, there was not a great deal of difference between the different approaches and they could be combined and extended to form a powerful standard. These events were the impetus behind the development of WAP, with Ericsson and Motorola joining Nokia and Unwired Planet as the founding members of the WAP forum.

WAP takes a client server approach. WAP incorporates a relatively simple microbrowser into a phone, requiring only limited resources on the mobile phone. This makes WAP suitable for thin clients and early smart phones. WAP puts the intelligence in the WAP gateways while adding just a microbrowser to the mobile phones themselves. Microbrowser-based services and applications reside temporarily on servers, not permanently in phones. WAP is aimed at turning a mass-market mobile phone into a network based smart phone. The philosophy behind the WAP approach is to utilize as few resources as possible on the handheld device and compensate for the constraints of the device by enriching the functionality of the network.

WAP is envisaged as a comprehensive and scalable protocol designed for use with

- Any mobile phone, from those with a one-line display to a smart phone.
- Any existing or planned wireless service, such as SMS, CSD, USSD, or general packet radio service (GPRS). Indeed, the importance of WAP is that it provides an evolutionary path for application developers and network operators to offer their services on different network types, bearers, and terminal capabilities. The design of the WAP standard separates the application elements from the bearer being used. This helps in the migration of some applications from SMS or CSD to GPRS.
- Any mobile network standard, such as code division multiple access (CDMA), GSM, or universal telephone mobile system (UTMS). WAP has been designed to work with all cellular standards and is supported by major worldwide wireless leaders such as AT&T and NTT DoCoMo.
- Multiple input terminals such as keypads, keyboards, touch-screens, and styluses.

WAP embraces and extends the previously conceived and developed wireless data protocols. Phone.com created a version of the standard HTML (hypertext markup language) Internet protocols designed specifically for effective and cost-effective information transfer across mobile networks. Wireless terminals incorporated an HDML microbrowser, and Phone.com's handheld device transport protocol (HDTP) then linked the terminal to the UP.Link Server Suite, which connected to the Internet or intranet, where the information being requested resided. The Internet site content was tagged with HDML.

This technology was incorporated into WAP and renamed using some of the many WAP related acronyms such as WMLS, WTP, and WSP. Anyone with a WAP compliant phone can use the built-in microbrowser to perform the following sequence of tasks.

1. Make a request in WML (Wireless Markup Language), a language derived from HTML especially for wireless network characteristics.
2. This request is passed to a WAP gateway that then retrieves the information from an Internet server either in standard HTML format or, preferably, directly prepared for wireless terminals using WML. If the content being retrieved is in HTML format, a filter in the WAP gateway may translate it into WML. A WML scripting language is available to format data such as calendar entries and electronic business cards for direct incorporation into the client service.
3. The requested information is then sent from the WAP gateway to the WAP client, using whatever mobile network bearer service is available and most appropriate.

WAP Protocol Stack

WAP has a layered architecture, as shown in Figure H.4. Following is a description of each layer in the WAP protocol stack:

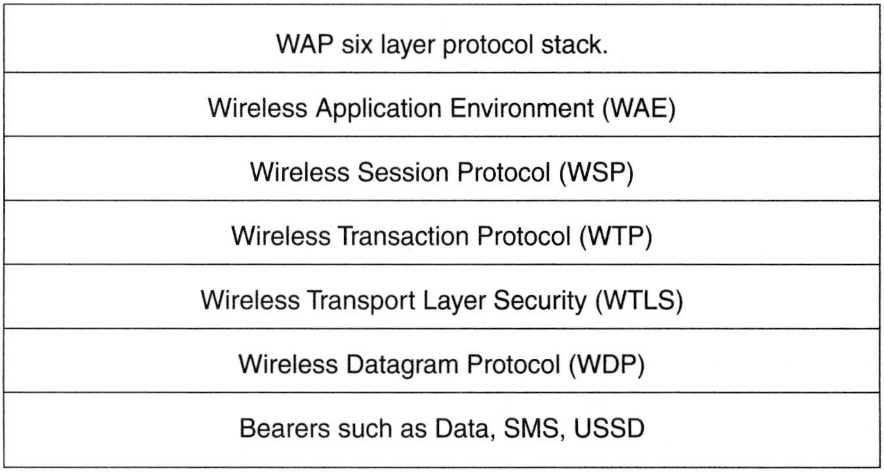

WAP six layer protocol stack.
Wireless Application Environment (WAE)
Wireless Session Protocol (WSP)
Wireless Transaction Protocol (WTP)
Wireless Transport Layer Security (WTLS)
Wireless Datagram Protocol (WDP)
Bearers such as Data, SMS, USSD

FIGURE H.4
WAP six layer protocol stack.

Wireless Application Environment.   The WAE defines the user interface on the phone. The application development facilitates the development of services that support multiple bearers. To achieve this, the WAE contains the WML; WMLScript, a scripting microlanguage similar to JavaScript; and the wireless telephony application (WTA). These are the tools that allow WAP based applications to be developed.

Wireless Session Protocol.   The WSP is a sandwich layer that links the WAE to two session services: a connection oriented service operating above the WTP and a connectionless service operating above the WDP.

Wireless Transaction Protocol.   The WTP runs on top of a datagram service such as UDP. The WTP is part of the standard suite of TCP/IP protocols, used to provide a simplified protocol suitable for low bandwidth mobile stations. WTP offers three classes of transaction service: unreliable one way request, reliable one way request, and reliable two way request respond. Interestingly, WTP supports protocol data unit concatenation and delayed acknowledgment to help reduce the number of messages sent. This protocol therefore tries to optimize the user experience by providing the information that is needed when it is needed—it can be confusing to receive confirmation of delivery messages when the information itself is expected. By stringing several messages together, the end user may well be able to get a better feel more quickly for what information is being communicated.

Wireless Transport Layer Security.   WTLS incorporates security features based upon the established transport layer security (TLS) protocol standard, including data integrity checks, privacy on the WAP gateway to the client leg, and authentication.

Wireless Datagram Protocol.   WDP allows WAP to be bearer independent by adapting the transport layer of the underlying bearer. WDP presents a consistent data format to the higher layers of the WAP protocol stack, thereby conferring the advantage of bearer independence to application developers.

Optimal WAP Bearers

There are several options for WAP bearers. This section covers SMS, CSD, USSD, and GPRS.

SMS.   Given its limited length of 160 characters per short message, SMS may not be an adequate bearer for WAP because of the weight of the protocol. The overhead of the WAP protocol that would be required to transmit an SMS message would mean that even for the simplest of transactions, several SMS messages might in fact have to be sent. This means that using SMS as a bearer can be a time-consuming and expensive exercise. Only one network operator, SBC of the United States, is known to be developing WAP services based on SMS.

CSD.   Most of the trial WAP based services use CSD as the underlying bearer. Since CSD has relatively few users currently, WAP could kick start usage of and traffic generated by this bearer. However, CSD lacks immediacy—a dial-up connection taking about 10 s is required to connect the WAP client to the WAP gateway, and this is the best case scenario when there is a complete end-to-end digital call. In the case where there is a need for analog modem handshaking because the WAP phone does not support V.110, the digital protocol, or when the WAP gateway does not have a digital direct connection, such as ISDN, into the mobile network, the connect time can be increased to about 30 s.

USSD. USSD is a means of transmitting information or instructions over a GSM network. USSD has some similarities to SMS since both use the GSM network's signaling path. Unlike SMS, USSD is not a store and forward service and is session-oriented such that when a user accesses a USSD service, a session is established and the radio connection stays open until the user, application, or time-out releases it. This has more in common with CSD than SMS. USSD text messages can be up to 182 characters in length.

USSD has some advantages and disadvantages as a tool for deploying services on mobile networks:

- Turnaround response times for interactive applications are shorter for USSD than SMS because of the session-based feature of USSD, and because it is not a store and forward service. According to Nokia, USSD can be up to seven times faster than SMS to carry out the same two way transaction.
- Users do not need to access any particular phone menu to access services with USSD—they can enter the USSD command directly from the initial mobile phone screen.
- Because USSD commands are routed back to the home mobile network's home location register (HLR), services based on USSD do not vary when users are roaming.
- USSD works well on all existing GSM mobile phones.
- Both SIM Application Tool Kit and WAP support USSD.
- USSD stage 2 (USSD2) has been incorporated into the GSM standard. Wireless USSD was previously a one way bearer useful for administrative purposes such as service access. Stage 2 is more advanced and interactive. By sending in a USSD2 command, the user can receive an information services menu. As such, USSD2 provides WAP-like features on existing phones.
- USSD strings are typically complicated for the user to remember, involving the use of characters to denote the start and finish of the USSD string. However, USSD strings for regularly used services can be stored in the phonebook, reducing the need to remember and render them.

Thus, USSD could be an ideal bearer for WAP on GSM networks.

GPRS. The GPRS is a new packet based bearer that is currently being introduced on many GSM and TDMA mobile networks. It is an exciting new bearer because it is immediate. There is no dial-up connection, and GPRS is relatively fast, up to 177.2 kbps in a theoretical extreme. GPRS supports virtual connectivity, allowing relevant information to be sent from the network as and when it is generated.

As of this writing, there has been no confirmation from any handset vendor that the initial GPRS terminals will support mobile terminated GPRS traffic such as direct receipt of GPRS packets on mobile phones. Whether GPRS MT (mobile terminate) is available or not is a central question with a critical impact on GPRS business scenarios, such as application migration from other nonvoice bearers.

There are two efficient means of delivering proactively sent pushing content to a mobile phone: by the SMS, which is one of the WAP bearers, or by the user maintaining more or less a permanent GPRS mobile originated session with the content server. Mobile terminated IP traffic might allow unsolicited information to reach the terminal. Internet sources originating such unsolicited content may not be chargeable and so the mobile user would have to pay. A possible worst case scenario would be that mobile users would have to pay for receiving unsolicited junk content. This is a potential reason for a mobile vendor not to support GPRS MT in its GPRS terminals. However, by originating the session themselves from their handset, users confirm their agreement to pay for the delivery of content from that service. Users could make their requests via a WAP session, which would not therefore need to be blocked. As such, a WAP session initiated from the WAP microbrowser could well be the only way that GPRS users can receive information at their mobile terminals.

Because all but the early WAP enabled phones will also support the GPRS, WAP and GPRS could well be synergistic and be used widely together. For the kinds of interactive, menu based information exchanges that WAP anticipates, CSD is not immediate enough because of the need to set up a call. Early prototypes of WAP services based on CSD were therefore close to unusable. SMS, on the other hand, is immediate but is always store and forward, such that even when a subscriber has just requested information from his or her microbrowser, the SMS center resources are used in the information transfer. Thus, GPRS and WAP are ideal bearers for each other.

Additionally, WAP incorporates two different connection modes—WSP connection mode or WSP connectionless protocol. This is very similar to the two GPRS point-to-point services—connection oriented and connectionless.

The predominant bearer for WAP based services will depend on delays in availability of WAP handsets and delays in the availability of GPRS terminals. If WAP terminals are delayed, most WAP terminals will support GPRS as well. If the first WAP terminals support SMS and CSD, but not GPRS, then SMS could become the predominant initial WAP bearer.

WAP certainly will be important for the development of GPRS based applications. Because the bearer level is separated from the application layer in the WAP protocol stack, WAP provides the ideal, defined, and standardized means to port the same application to different bearers. Thus, many application developers will use WAP to facilitate the migration of their applications across bearers once GPRS based WAP protocols are supported.

Considerations for Selecting a Wireless LAN Solution

The choice of radio technology may be less important than the wireless networking software, which can also have a substantial impact on system performance and throughout. In addition to these technical factors, business factors also should be considered when selecting a WLAN vendor, as illustrated in Table H.2.

Not all vendors in the industry today support all of the capabilities described in Table H.2. To properly evaluate a WLAN vendor, it is important to make a weighted list of requirements, including current and anticipated needs. Matching these requirements against the offerings of each WLAN vendor will allow a business to shorten the list of prospective WLAN solutions considerably. It is possible that no single vendor will match all of the current and future product plans. A business must work with vendors to determine which vendor offers the development and migration plans closest to its needs.

Business Considerations

Cost.   Calculating the total cost of a WLAN system involves several elements, including client devices, access points, and ongoing maintenance. The cost of access points can vary depending on the range and throughput of the selected technology. Support costs include equipment service, problem diagnosis, and software upgrades. If a customer has installed WLAN systems in multiple locations, the ability to remotely access those systems, upgrade the access point, and upgrade client devices can offer substantial savings for support costs.

Table H.2 Business Factors for Selecting a WLAN Vendor

Factor	Key Considerations
Wireless experience	Total system cost: equipment, support, future upgrades
Full provider	Vendor provides full system, installation, and support solution
Business stability	Company strength and commitment to wireless market
Wireless clients	Wide selection of client devices
System integration	Software and support to integrate devices into a complete system
Service and support	On-site service options
Global presence	Country approvals and local support programs

Another cost consideration may be whether a technology migration path will reuse the customer's existing equipment. As the system vendor offers new capabilities, such as faster data rates, they should be backward compatible with older technologies. Other cost factors include product quality, reliability, ease-of-use, and availability. All of these factors must be compared to determine the true cost of competitive product offerings.

Wireless Experience.   The installation of WLAN systems is sometimes more of an art than a science. Walls, ceilings, racks, and merchandise—even the composition of the building's construction—affect the coverage patterns of WLAN radios inside buildings. Installation decisions must take into account the multitude of antenna options including omnidirectional, flat panel, directional, and yagi antennas. Splitters, couplers, and lightning arrestors are also commonly used in WLAN installations.

Working with a vendor experienced in installing WLAN systems in comparable facilities can provide a less costly, trouble-free installation. An experienced WLAN solution provider can complete an RF site survey, provide a detailed installation plan, and complete the installation to provide full wireless coverage where it is needed.

Full System and Solution Provider.   Wireless LAN systems can become very complex as they grow with organizations. The system infrastructure includes access points, PC cards, ISA cards, wireless client devices, and WLAN enabled software. While it is possible to purchase these products from many companies, few are truly full system and solution vendors. By working with a full system and solution provider, customers can select and purchase most system components from a single vendor, including installation and wireless client devices. In the future, a single vendor will protect mission-critical applications by providing a single point of contact for resolving system problems.

Business Stability of the LAN Vendor

Selection of a WLAN vendor should take into account the vendor's strength for supporting future expansion and new capabilities in the system. The risk of selecting a vendor that may not be around in a few years is significant. Many WLAN vendors are small, start-up companies without the critical customer base that will sustain the continued investment required to survive. The industry has already seen a number of pioneers retrench and become very specialized niche players, or go out of business entirely.

Even large companies with substantial financial and technical resources may exit the business if the WLAN segment fails to reach the company's internal expectations. Motorola, which was one of the industry pioneers but withdrew from the business in 1995, provides a good example.

Selecting a vendor that may not be around in the future makes it likely that long-term system costs will increase. The best way to evaluate the stability of a vendor, large or small, is to assess the company's success in the WLAN market. A company with a substantial WLAN market share and a long roster of major customers is most likely to thrive and be a reliable supplier. Another criterion is whether WLAN products are a key business segment for the vendor.

Wide Selection of Wireless Clients.   No single client device meets the requirements of all WLAN applications in use today. While PC ISA adapter cards can be installed to stand alone in computing devices, most customers demand client devices with integrated radios. Integration provides secure mounting of the radio for everyday use, tuned antennae for the wireless client, battery and power management, and software driver support.

A vendor that provides a wide selection of wireless client devices enables customers to select the most appropriate device for each application. WLAN client devices include handheld, stationary, vehicle mount, and pen based terminals, as well as wireless modems and PC/ISA adapter cards.

System Integration.   Wireless LAN systems include much more than applications and client devices. A complete system includes software and services that integrate client devices with the host and supporting software devices that enable successful installation of a WLAN application. In turn, this application gives users a trouble-free solution that provides immediate benefits from the use of wireless technologies. The installation should include the hardware and software integration necessary to get the application running quickly.

Service and Support.   Once a WLAN system is installed and operational, service and support become critical factors for the continued success of the system. A vendor that provides on-site service and support best handles changes in system requirements, the physical environment, applications, and system components.

Other Important Considerations

Global Presence.   Check with the vendor to ensure that equipment is approved for use in all countries where a WLAN solution will be required today or in the future. In addition, verify that the vendor's support and service programs in each country are adequate for the needs of the local installations.

Technical Considerations.   The technical considerations for selecting a WLAN product should not be based on a theoretical argument about one technology and another. Instead, the decision should be based on how the product meets business needs for capabilities, features, and performance, as illustrated in Table H.3. Because each customer's needs will differ, the factors discussed in the following paragraphs should be weighted in making the individual decision.

Seamless Integration into the Corporate Network   A critical aspect of a WLAN system is seamless integration into the corporate network. WLANs are not a replacement for, but instead an extension of, the wired LAN. The WLAN should seamlessly integrate into the wired network whether based on Ethernet or Token

Table H.3 Technical Factors for Selecting a WLAN Solution

Factor	Key Considerations
Seamless integration	Integration with wired LANs; roaming support
Cost and coverage	Device range for total system cost
Performance	Data rates in the target environment
Response time	Delay management techniques in the WLAN software
Size and convenience	Use and placement of access points and client devices
Interfaces	Availability of hardware and software options
Interoperability	Communication with other WLAN systems and client devices
Power management	Minimized battery changes for client devices
Security	Encryption of data transmission

Ring. As the network topology changes, the WLAN system should support new networking requirements. The WLAN should also support a variety of topologies from wired access points to wireless bridging to offer flexibility for implementation regardless of the physical requirements of the facility.

A WLAN system should support roaming across the enterprise regardless of where a client user roams. As the client moves from an access point attached to one subnet of the network to another subnet, the WLAN system should allow seamless communications for the client device. The system should not produce a lost connection or, worse, a lost connection plus reinitialization of the router or intelligent bridge. By selecting a WLAN system that has built-in support for roaming across subnets, wireless applications work seamlessly whenever users roam in the facility.

Cost and Coverage. A number of factors will affect the range of a WLAN device, including receiver sensitivity, transmit power output, multipath immunity, and antenna system performance, including the proper use of antenna diversity. The greater the range of WLAN devices, the fewer the number of access points that will be required to cover a given building or installation. Device range may ultimately become the key factor in determining total system cost.

Performance (Throughput). WLAN products typically specify the over-the-air data rates provided, such as 1 or 2 Mbps. What matters most to the user is the actual throughput of the system in a specific application and environment. Performance tests can determine actual throughput rates under differing conditions. Among the variables to consider in a throughput test are the range of the device from the access point, the system load number, the data traffic of clients using an access point, the typical packet size on the network, and the network operating system (NOS). Another performance factor is the networking software used in each implementation, including how that system supports roaming, the use of dynamic load balancing, and a number of other factors. Each WLAN product will provide substantially different results that may have little correlation with the over-the-air data rates specified by the product vendor.

Response Time and Delay.   Another consideration for determining system performance is response time of the wireless client transaction. Response time includes host and network delay when delivering individual packets for a given system.

Again, the wireless networking software will have the most substantial impact on this performance attribute. Software factors that determine this impact include the technique used to support roaming, the use of dynamic load balancing, and the access point forwarding of buffered packets during roaming. Also important is the reliability of the WLAN device transmission at various ranges, which determines the number of packet errors and retransmissions required. Different WLAN products may have wide variations in their delay performance.

Size and Convenience.   The attributes of size and convenience are related for determining user satisfaction with a wireless system. Small access points that can be mounted in or on ceilings, attached to walls, and placed on desks or cubicle walls are much more convenient to implement than access points without the form factor of a desktop PC.

For client devices, a single-piece, PCMCIA Type II card form factor will usually be more convenient than a PCMCIA interface card with a separate cable-attached radio. Additionally, for specialized handheld terminals or personal digital assistants (PDAs), a built-in WLAN device will usually be better than using a PCMCIA slot with an external wireless device.

With a true PCMCIA form factor, the potential availability of specialized handheld and other wireless devices will be greatly increased. It is likely that once all WLAN client devices reach a true PCMXCIA Type II form factor and all access points are small, consideration of equipment size will become less important for choosing a system vendor.

Availability of Hardware and Software Interfaces.   A number of different hardware or software options may be supported by a WLAN system. The importance of these options depends on a user's short- and long-term requirements. Tables H.4 and H-5 describe the most important options.

Interoperability.   Until a few years ago, interoperability among WLAN systems was not possible. A particular WLAN system could communicate only with the client devices offered by the same vendor. With the introduction of the WLI forum 2.4 GHz open air specification in 1995, customers gained the ability to purchase WLAN systems and client devices from a variety of vendors. In the future, the IEEE 802.11, Bluetooth, and HomeRF standards and specifications will further interoperability.

WLAN systems that support interoperability give customers the freedom to choose equipment from a variety of vendors. Interoperability increases competition, a

Table H.4 Access Point Options

Factor	Key Considerations
Topologies supported	Access point, wireless access point (repeater), wireless bridge
Networks supported	Ethernet, Token Ring
Connections	10BASE-T, 10BASE-2, AUI
Wireless technologies supported	900 MHz, 2.4 GHz Open Air, IEEE 802.11 frequency hopping, IEEE 802.11 direct sequence future high speed wireless technologies

result that historically has yielded lower costs, increased features, and new product selections. Customers may assume that when a standard is in place and interoperability is supported, products from different vendors will work and perform identically. However, even with standards-compliant products, substantial performance differences may exist between products offered by different vendors. The reason for such variations is typically that although the communications protocol is specified by the standard, each vendor determines the functionality of access points and client devices.

Power Management.   Most WLAN client devices are battery-operated, with a limited battery life. The use of radio communications can significantly affect battery life. Most users would not be satisfied if they had to change batteries frequently. Wireless client devices should offer advanced power management support to maximize battery life and minimize battery changes, optimally once per work shift.

Security Considerations.   Security has always been a concern in wireless communications, as evidenced by the issues around interception of cellular telephone transmissions. Radios utilizing a broadcast-mode transmission scheme allow the possibility of interception by unintended receivers.

Much of today's technology for spread spectrum radio was developed with exactly this problem in mind. The unique spreading patterns in DSSS transmissions were meant to be difficult to decode for a receiver that didn't know the specific pattern. Similarly, the pseudorandom hopping patterns of the frequency hopping technique were designed to avoid casual decoding. Spread spectrum radio technologies offer a great deal of protection because of these techniques. However, because interoperable WLAN solutions use the same spread spectrum patterns, this security protection is diminished.

If a customer requires more security, the IEEE 802.11 wireless LAN standard specifies use of WEP encryption. This specification utilizes the RSA Data Security, Inc. RC4 encryption algorithm to encrypt over-the-air data transmissions. Encryption operates on top of the security provided by spread spectrum techniques. With WEP

Table H.5 Access Point Capabilities

Factor	Key Considerations
Basic	Basic support for required IEEE 802.11 protocols; access point filtering of packets (multicast, broadcast, etc.)
	Access point filtering of traffic to wireless devices associated with an access point
Advanced	Roaming across subnet boundaries
	Forwarding of buffered packets between access points when roaming
	Dynamic load balancing of client devices between access points; fault-tolerant setup possible
	Capability for self-configuration and firmware upgrades via BOOTP does not support firmware upgrades by itself
	Remote diagnostics, configuration, software upgrades and management via Telnet, FTP, and/or SNMP
	WEP (wired equivalent privacy) encryption support
	Support for the optional IEEE 802.11 specified PCT (point coordination Function), which enables real-time multimedia
Environmental considerations	Heating, cooling, sealing
Client Device Connections	PCMCIA card and socket services, ISA cards, RS232, 10BASE-T, 10BASE-2, AUI, RS422, parallel port
Client Device/ capabilities	Drivers, NDIS2, NDIS3, NDIS4, ODI; operating systems: DOS, Windows, Windows 95, Windows NT, OS 2, MAC, Unix
	Advanced: Advanced packet power management, serial communications support, and software upgradable via flash memory

encryption, a user's wireless transmission is meant to be as secure as an encrypted transmission over a wired LAN.

An important consideration is that the IEEE 802.11 standard secures only over-the-air transmissions. An access point will send information over the Ethernet or Token Ring network without encryption. For higher-level security requirements, customers can use an end-to-end encryption technique such as that specified by the IEEE 802.10 standard. With end-to-end encryption layered on top of the security measures in the wireless system, user data should be totally secure. The choice of security should be based on individual application requirements and consideration of the trade-off of cost, performance, and complexity.

Case Studies

Case Study

BGP4

Introduction

The Border Gateway Protocol (BGP), defined in RFC 1771, allows you to create loop-free interdomain routing between autonomous systems (AS). An AS is a set of routers under a single technical administration. Routers in an AS can use multiple interior gateway protocols to exchange routing information inside the AS and an exterior gateway protocol to route packets outside the AS.

How Does BGP Work?

BGP uses TCP as its transport protocol (port 179). Two BGP routers form a TCP connection between one another (peer routers) and exchange messages to open and confirm the connection parameters.

BGP routers exchange network reachability information. This information is mainly an indication of the full paths (BGP AS numbers) that a route should take in order to reach the destination network. This information helps in constructing a graph of ASs that are loop-free and where routing policies can be applied in order to enforce some restrictions on the routing behavior.

Any two routers that have formed a TCP connection in order to exchange BGP routing information are called peers, or neighbors. BGP peers initially exchange their full BGP routing tables. After this exchange, incremental updates are sent as the routing table changes. BGP keeps a version number of the BGP table, which should be the same for all of its BGP peers. The version number changes whenever BGP updates the table due to routing information changes. Keepalive packets are sent to ensure that the connection is alive between the BGP peers and notification packets are sent in response to errors or special conditions.

EBGP and IBGP

If an AS has multiple BGP speakers, it could be used as a transit service for other ASs. As you can see, AS200 is a transit AS for AS100 and AS300.

It is necessary to ensure reachability for networks within an AS before sending the information to external ASs. This is done by a combination of internal BGP (IBGP) peering between routers inside an AS and by redistributing BGP information to Internal Gateway Protocols (IGPs) running in the AS.

As far as this case study is concerned, when BGP is running between routers belonging to two different ASs, we call this exterior BGP (EBGP). When BGP is running between routers in the same AS, we call this IBGP.

Enabling BGP Routing

Use these steps to enable and configure BGP.

Let us assume you want to have two routers, RTA and RTB, talk BGP. In the first example RTA and RTB are in different ASs and in the second example both routers belong to the same AS.

We start by defining the router process and the AS number to which the routers belong. Use this command to enable BGP on a router:

```
router bgp autonomous-system
RTA#
router bgp 100

RTB#
router bgp 200
```

The above statements indicate that RTA is running BGP and it belongs to AS100 and RTB is running BGP and it belongs to AS200.

The next step in the configuration process is to define BGP neighbors, which indicates the routers that are trying to talk BGP.

Forming BGP Neighbors

Two BGP routers become neighbors once they establish a TCP connection between each other. The TCP connection is essential in order for the two peer routers to start exchanging routing updates.

Once the TCP connection is up, the routers send open messages in order to exchange values such as the AS number, the BGP version they're running, the BGP router ID and the keepalive hold time. After these values are confirmed and accepted the neighbor connection is established. Any state other than "established" is an indication that the two routers did not become neighbors, and BGP updates will not be exchanged.

Use this neighbor command to establish a TCP connection:

```
neighbor ip-address remote-as number
```

The remote-as *number* is the AS number of the router we are trying to connect to using BGP. The *ip-address* is the next hop directly connected address for eBGP and any IP address on the other router for iBGP.

It is essential that the two IP addresses used in the **neighbor** command of the peer routers be able to reach one another. One sure way to verify reachability is an extended ping between the two IP addresses. The extended ping forces the pinging router to use as source the IP address specified in the **neighbor** command rather than the IP address of the interface the packet is going out from.

It is important to reset the neighbor connection in case any BGP configuration changes are made in order for the new parameters to take effect.

```
clear ip bgp address (where address is the neighbor address)
clear ip bgp * (clear all neighbor connections)
```

By default, BGP sessions begin using BGP version 4 and negotiating downward to earlier versions if necessary. To prevent negotiations and force the BGP version used to communicate with a neighbor, perform the following task in router configuration mode:

```
neighbor {ip address|peer-group-name} version value
```

An example of the **neighbor** command configuration follows:

```
RTA#
router bgp 100
neighbor 129.213.1.1 remote-as 200

RTB#
router bgp 200
neighbor 129.213.1.2 remote-as 100
neighbor 175.220.1.2 remote-as 200

RTC#
router bgp 200
neighbor 175.220.212.1 remote-as 200
```

In the above example RTA and RTB are running EBGP. RTB and RTC are running IBGP. The difference between EBGP and IBGP is manifested by having the remote-as number pointing to either an external or an internal AS.

Also, the eBGP peers are directly connected while the iBGP peers are not. iBGP routers do not have to be directly connected, as long as there is some IGP running that allows the two neighbors to reach one another.

The following is an example of the information that the **show ip bgp neighbors** command displays. Pay special attention to the BGP state, since anything other than state "established" indicates the peers are not up. You should also note the BGP version is 4, the remote router ID (highest IP address on the router or the highest loopback interface in case it exists) and the table version (this is the state of the table, any time new information comes in, the table increases the version and a version that keeps incrementing indicates that some route is flapping causing routes to continuously be updated).

```
#show ip bgp neighbors
BGP neighbor is 129.213.1.1, remote AS 200, external link
BGP version 4, remote router ID 175.220.12.1
BGP state = Established, table version = 3, up for 0:10:59
Last read 0:00:29, hold time is 180, keepalive interval is 60
 seconds
Minimum time between advertisement runs is 30 seconds
Received 2828 messages, 0 notifications, 0 in queue
Sent 2826 messages, 0 notifications, 0 in queue
Connections established 11; dropped 10
```

BGP and Loopback Interfaces

Using a loopback interface to define neighbors is common with iBGP, but not with eBGP. Normally the loopback interface is used to make sure the IP address of the neighbor stays up and is independent of hardware functioning properly. In the case of eBGP, peer routers are frequently directly connected and loopback doesn't apply.

If you use the IP address of a loopback interface in the **neighbor** command, you need some extra configuration on the neighbor router. The neighbor router needs to tell BGP it's using a loopback interface rather than a physical interface to initiate the BGP neighbor TCP connection. The command used to indicate a loopback interface is:

neighbor *ip-address* **update-source** *interface*

The following example illustrates the use of this command.

```
RTA#
router bgp 100
neighbor 190.225.11.1 remote-as 100
neighbor 190.225.11.1 update-source loopback 1

RTB#
router bgp 100
neighbor 150.212.1.1 remote-as 100
```

In the above example, RTA and RTB are running iBGP inside AS 100. RTB is using in its **neighbor** command the loopback interface of RTA (150.212.1.1); in this case RTA has to force BGP to use the loopback IP address as the source in the TCP neighbor connection. RTA does this by adding the update-source int loopback configuration (neighbor 190.225.11.1 update-source loopback 1) and this statement forces BGP to use the IP address of its loopback interface when talking to neighbor 190.225.11.1.

Note that RTA has used the physical interface IP address (190.225.11.1) of RTB as a neighbor, which is why RTB doesn't need any special configuration.

eBGP Multihop

In some cases, a Cisco router can run eBGP with a third party router that doesn't allow the two external peers to be directly connected. To achieve this, you can use eBGP multihop, which allows the neighbor connection to be established between two non-directly-connected external peers. The multihop is used only for eBGP and not for iBGP. The following example illustrates of eBGP multihop.

```
RTA#
router bgp 100
neighbor 180.225.11.1 remote-as 300
neighbor 180.225.11.1 ebgp-multihop

RTB#
router bgp 300
neighbor 129.213.1.2 remote-as 100
```

RTA is indicating an external neighbor that is not directly connected. RTA needs to indicate that it is using **ebgp-multihop.** On the other hand, RTB is indicating a neighbor that is directly connected (129.213.1.2), which is why it does not need the **ebgp-multihop** command. You should also configure an IGP or static routing to allow the non-connected neighbors to reach each other.

The following example shows how to achieve load balancing with BGP in a particular case where we have eBGP over parallel lines.

eBGP Multihop (Load Balancing)

```
RTA#
int loopback 0
ip address 150.10.1.1 255.255.255.0
router bgp 100
neighbor 160.10.1.1 remote-as 200
neighbor 160.10.1.1 ebgp-multihop
neighbor 160.10.1.1 update-source loopback 0
network 150.10.0.0
ip route 160.10.0.0 255.255.0.0 1.1.1.2
ip route 160.10.0.0 255.255.0.0 2.2.2.2

RTB#
int loopback 0
ip address 160.10.1.1 255.255.255.0
router bgp 200
neighbor 150.10.1.1 remote-as 100
neighbor 150.10.1.1 update-source loopback 0
neighbor 150.10.1.1 ebgp-multihop
network 160.10.0.0
ip route 150.10.0.0 255.255.0.0 1.1.1.1
ip route 150.10.0.0 255.255.0.0 2.2.2.1
```

The above example illustrates the use of loopback interfaces, update-source and ebgp-multihop. This is a workaround in order to achieve load balancing between two eBGP speakers over parallel serial lines. In normal situations, BGP picks one of the lines to send packets on, and load balancing would not happen. By introducing loopback interfaces, the next hop for eBGP is the loopback interface. We use static routes (we could also use an IGP) to introduce two equal cost paths to reach the destination. RTA has two choices to reach next hop 160.10.1.1: one via 1.1.1.2 and the other one via 2.2.2.2, and the same for RTB.

Route-Maps

At this point let us introduce route maps because they will be used heavily with BGP. In the BGP context, route-map is a method used to control and modify routing information. This is done by defining conditions for redistributing routes from one routing protocol to another or controlling routing information when injected in and out of BGP. The format of the route-map follows:

route-map *map-tag* [[**permit** | **deny**] | [*sequence-number*]]

The map-tag is just a name you give to the route-map. Multiple instances of the same route-map (same name-tag) can be defined. The sequence number is just an indication of the position a new route-map is to have in the list of route-maps already configured with the same name.

For example, let us define two instances of the route-map, and let us call it MYMAP, the first instance will have a sequence-number of 10, and the second will have a sequence-number of 20.

```
route-map MYMAP permit 10
    (first set of conditions goes here.)
route-map MYMAP permit 20
    (second set of conditions goes here.)
```

When applying route-map MYMAP to incoming or outgoing routes, the first set of conditions will be applied via instance 10. If the first set of conditions is not met then we proceed to a higher instance of the route-map **match** and **set** configuration commands. Each route-map will consist of a list of match and set configurations. The match will specify a **match** criteria and set specifies a **set** action if the criteria enforced by the match command are met.

For example, you could define a route-map that checks outgoing updates and if there is a match for IP address 1.1.1.1 then the metric for that update will be set to 5. The above explanation can be illustrated by the following commands:

```
match ip address 1.1.1.1
set metric 5
```

Now, if the match criteria are met and we have a **permit** then the routes will be redistributed or controlled as specified by the set action and we break out of the list.

If the match criteria are met and we have a **deny** then the route will not be redistributed or controlled and we break out of the list.

If the match criteria are not met and we have a **permit or deny** then the next instance of the route-map (instance 20 for example) will be checked, and so on until we either break out or finish all the instances of the route-map. If we finish the list without a match then the route we are looking at will **not be accepted nor forwarded.**

When you use route-maps for filtering BGP updates, rather redistributing between protocols, you *cannot* filter on the inbound when using a **match** command on the IP address. Filtering on the outbound is acceptable.

The related commands for **match** are:

```
match as-path
match community
match clns
match interface
match ip address
match ip next-hop
match ip route-source
match metric
match route-type
match tag
```

The related commands for **set** are:

```
set as-path
set clns
```

set automatic-tag
set community
set interface
set default interface
set ip default next-hop
set level
set local-preference
set metric
set metric-type
set next-hop
set origin
set tag
set weight

Let us look at some route-map examples:

Example 1:

Assume RTA and RTB are running RIP and RTA and RTC are running BGP. RTA is getting up-dates via BGP and redistributing them to RIP. If RTA wants to redistribute to RTB routes about 170.10.0.0 with a metric of 2 and all other routes with a metric of 5 then we might use the follow-ing configuration:

```
RTA#
router rip
network 3.0.0.0
network 2.0.0.0
network 150.10.0.0
passive-interface Serial0
redistribute bgp 100 route-map SETMETRIC
router bgp 100
neighbor 2.2.2.3 remote-as 300
network 150.10.0.0
route-map SETMETRIC permit 10
match ip-address 1
set metric 2
route-map SETMETRIC permit 20
set metric 5
access-list 1 permit 170.10.0.0 0.0.255.255
```

In the above example if a route matches the IP address 170.10.0.0 it will have a metric of 2 and then we break out of the route-map list. If there is no match then we go down the route map list which says, set everything else to metric 5. It is always very important to ask the question, what will happen to routes that do not match any of the match statements because they will be dropped by default.

Example 2:

Suppose in the above example we did not want AS100 to accept updates about 170.10.0.0. Because route maps cannot be applied on the inbound when matching based on an IP address, we have to use an outbound route-map on RTC:

```
RTC#
router bgp 300
network 170.10.0.0
neighbor 2.2.2.2 remote-as 100
neighbor 2.2.2.2 route-map STOPUPDATES out
route-map STOPUPDATES permit 10
match ip address 1
access-list 1 deny 170.10.0.0 0.0.255.255
access-list 1 permit 0.0.0.0 255.255.255.255
```

Now that you feel more comfortable with how to start BGP and how to define a neighbor, let us look at how to start exchanging network information.

There are multiple ways to send network information using BGP. I will go through these methods one by one.

Network Command

The format of the network command follows:

network *network-number* [**mask** *network-mask*]

The network command controls what networks are originated by this box. This is a different concept from what you are used to configuring with IGRP and RIP. With this command we are not trying to run BGP on a certain interface, rather we are trying to indicate to BGP what networks it should originate from this box. The mask portion is used because BGP4 can handle subnetting and supernetting. A maximum of 200 entries of the network command are accepted.

The network command will work if the network you are trying to advertise is known to the router, whether connected, static or learned dynamically.

An example of the network command follows:

```
RTA#
router bgp 1
network 192.213.0.0 mask 255.255.0.0
ip route 192.213.0.0 255.255.0.0 null 0
```

The above example indicates that router A, will generate a network entry for 192.213.0.0/16. The /16 indicates that we are using a supernet of the class C address and we are advertizing the first two octets (the first 16 bits).

Note that we need the static route to get the router to generate 192.213.0.0 because the static route will put a matching entry in the routing table.

Redistribution

The network command is one way to advertise your networks via BGP. Another way is to redistribute your IGP (IGRP, OSPF, RIP, EIGRP, etc.) into BGP. This sounds scary because now you are dumping all of your internal routes into BGP, some of these routes might have been learned via BGP and you do not need to send them out again. Careful filtering should be applied to make sure you are sending to the internet only routes that you want to advertise and not everything you have. Let us look at the following example.

RTA is announcing 129.213.1.0 and RTC is announcing 175.220.0.0. Look at RTC's configuration:

If you use a network command you will have:

```
RTC#
router eigrp 10
network 175.220.0.0
redistribute bgp 200
default-metric 1000 100 250 100 1500
router bgp 200
neighbor 1.1.1.1 remote-as 300
network 175.220.0.0 mask 255.255.0.0 (this will limit the net-
  works
originated by your AS to 175.220.0.0)
```

If you use redistribution instead you will have:

```
RTC#
router eigrp 10
network 175.220.0.0
redistribute bgp 200
default-metric 1000 100 250 100 1500
router bgp 200
neighbor 1.1.1.1 remote-as 300
redistribute eigrp 10 (eigrp will inject 129.213.1.0 again
  into BGP)
```

This will cause 129.213.1.0 to be originated by your AS. This is misleading because you are not the source of 129.213.1.0 but AS100 is. So you would have to use filters to prevent that network from being sourced out by your AS. The correct configuration would be:

```
RTC#
router eigrp 10
network 175.220.0.0
redistribute bgp 200
default-metric 1000 100 250 100 1500
router bgp 200
neighbor 1.1.1.1 remote-as 300
neighbor 1.1.1.1 distribute-list 1 out
```

```
redistribute eigrp 10
access-list 1 permit 175.220.0.0 0.0.255.255
```

The access-list is used to control what networks are to be originated from AS200.

Static Routes and Redistribution

You could always use static routes to originate a network or a subnet. The only difference is that BGP will consider these routes as having an origin of incomplete (unknown). In the above example the same could have been accomplished by doing:

```
RTC#
router eigrp 10
network 175.220.0.0
redistribute bgp 200
default-metric 1000 100 250 100 1500
router bgp 200
neighbor 1.1.1.1 remote-as 300
redistribute static
 . . .
ip route 175.220.0.0 255.255.255.0 null0
 . . .
```

The null 0 interface means disregard the packet. So if I get the packet and there is a more specific match than 175.220.0.0 (which exists of course) the router will send it to the specific match otherwise it will disregard it. This is a nice way to advertise a supernet.

We have discussed how we can use different methods to originate routes out of our autonomous system. Please remember that these routes are generated in addition to other BGP routes that BGP has learned via neighbors (internal or external). BGP passes on information that it learns from one peer to other peers. The difference is that routes generated by the network command, or redistribution or static, will indicate your AS as the origin for these networks.

Injecting BGP into IGP is always done by redistribution.

Example:

```
RTA#
router bgp 100
neighbor 150.10.20.2 remote-as 300
network 150.10.0.0

RTB#
router bgp 200
neighbor 160.10.20.2 remote-as 300
network 160.10.0.0

RTC#
router bgp 300
```

```
neighbor 150.10.20.1 remote-as 100
neighbor 160.10.20.1 remote-as 200
network 170.10.00
```

Note that you do not need network 150.10.0.0 or network 160.10.0.0 in RTC unless you want RTC to also generate these networks on top of passing them on as they come in from AS100 and AS200. Again the difference is that the network command will add an extra advertisement for these same networks indicating that AS300 is also an origin for these routes.

An important point to remember is that BGP will not accept updates that have originated from its own AS. This is to ensure a loop-free interdomain topology.

For example, assume AS200 above had a direct BGP connection into AS100. RTA will generate a route 150.10.0.0 and will send it to AS300 then RTC will pass this route to AS200 with the origin kept as AS100, RTB will pass 150.10.0.0 to AS100 with origin still AS100. RTA will notice that the update has originated from its own AS and will ignore it.

iBGP

iBGP is used if an AS wants to act as a transit system to other ASs. You might ask, why can we not do the same thing by learning via eBGP redistributing into IGP and then redistributing again into another AS? We can, but iBGP offers more flexibility and more efficient ways to exchange information within an AS; for example iBGP provides us with ways to control what is the best exit point out of the AS by using local preference.

```
RTA#
router bgp 100
neighbor 190.10.50.1 remote-as 100
neighbor 170.10.20.2 remote-as 300
network 150.10.0.0

RTB#
router bgp 100
neighbor 150.10.30.1 remote-as 100
neighbor 175.10.40.1 remote-as 400
network 190.10.50.0

RTC#
router bgp 400
neighbor 175.10.40.2 remote-as 100
network 175.10.0.0
```

Note: An important point to remember, is that when a BGP speaker receives an update from other BGP speakers in its own AS (IBGP), the receiving BGP speaker will not redistribute that information to other BGP speakers in its own AS. The receiving BGP speaker will redistribute that information to other BGP speakers outside of its AS. That is why it is important to sustain a full mesh between the IBGP speakers within an AS.

In the above example, RTA and RTB are running iBGP and RTA and RTD are running iBGP also. The BGP updates coming from RTB to RTA will be sent to RTE (outside of the AS) but not to RTD (inside of the AS). This is why an iBGP peering should be made between RTB and RTD in order not to break the flow of the updates.

The BGP Decision Algorithm

After BGP receives updates about different destinations from different autonomous systems, the protocol will have to decide which paths to choose in order to reach a specific destination. BGP will choose only a single path to reach a specific destination.

The decision process is based on different **attributes**, such as next hop, administrative weights, local preference, the route origin, path length, origin code, metric and so on.

BGP will always propagate the best path to its neighbors.

Case Study II

BGP4
CIDR and Aggregate Addresses

One of the main enhancements of BGP4 over BGP3 is Classless Interdomain Routing (CIDR). CIDR or supernetting is a new way of looking at IP addresses. There is no notion of classes anymore (class A, B, or C). For example, network 192.213.0.0 which used to be an illegal class C network is now a legal supernet represented by 192.213.0.0/16 where the 16 is the number of bits in the subnet mask counting from the far left of the IP address. This is similar to 192.213.0.0 255.255.0.0.

Aggregates are used to minimize the size of routing tables. Aggregation is the process of combining the characteristics of several different routes in such a way that a single route can be advertised. In the following example, RTB is generating network 160.10.0.0. We will configure RTC to propagate a supernet of that route 160.0.0.0 to RTA.

```
RTB#
router bgp 200
neighbor 3.3.3.1 remote-as 300
network 160.10.0.0

#RTC
router bgp 300
neighbor 3.3.3.3 remote-as 200
neighbor 2.2.2.2 remote-as 100
network 170.10.0.0
aggregate-address 160.0.0.0 255.0.0.0
```

RTC will propagate the aggregate address 160.0.0.0 to RTA.

Aggregate Commands

There is a wide range of aggregate commands. It is important to understand how each one works in order to have the desired aggregation behavior.

The first command is the one used in the previous example:

aggregate-address *address mask*

This will advertise the prefix route and all of the more specific routes. The command **aggregate-address** 160.0.0.0 will propagate an additional network 160.0.0.0 but will not prevent 160.10.0.0 from being also propagated to RTA. The outcome of this is that both networks 160.0.0.0 and 160.10.0.0 have been propagated to RTA. This is what we mean by advertising the prefix and the more specific route.

Please note that you cannot aggregate an address if you do not have a more specific route of that address in the BGP routing table.

For example, RTB cannot generate an aggregate for 160.0.0.0 if it does not have a more specific entry of 160.0.0.0 in its BGP table. The more specific route could have been injected into the BGP table via incoming updates from other ASs, from redistributing an IGP or static into BGP or via the network command (network 160.10.0.0).

In case we would like RTC to propagate network 160.0.0.0 only and *not* the more specific route then we would have to use the following:

```
aggregate-address address mask summary-only
```

This will a advertise the prefix only; all the more specific routes are suppressed.

The command **aggregate 160.0.0.0 255.0.0.0 summary-only** will propagate network 160.0.0.0 and will suppress the more specific route 160.10.0.0.

Please note that if we are aggregating a network that is injected into our BGP via the network statement (ex: network 160.10.0.0 on RTB) then the network entry is always injected into BGP updates even though we are using "the aggregate summary-only" command. The upcoming CIDR example discusses this situation.

```
aggregate-address address mask as-set
```

This advertises the prefix and the more specific routes but it includes as-set information in the path information of the routing updates.

```
aggregate 129.0.0.0 255.0.0.0 as-set
```

This command will be discussed in an example by itself in the following sections.

In case we would like to suppress more specific routes when doing the aggregation we can define a route map and apply it to the aggregates. This will allow us to be selective about which more specific routes to suppress.

```
aggregate-address address-mask suppress-map map-name
```

This advertises the prefix and the more specific routes but it suppresses advertisement according to a route-map. In the previous diagram, if we would like to aggregate 160.0.0.0 and suppress the more specific route 160.20.0.0 and allow 160.10.0.0 to be propagated, we can use the following route map:

```
route-map CHECK permit 10
match ip address 1
access-list 1 permit 160.20.0.0 0.0.255.255
access-list 1 deny 0.0.0.0 255.255.255.255
```

By definition of the suppress-map, any packets permitted by the access list would be suppressed from the updates.

Then we apply the route-map to the aggregate statement.

```
RTC#
router bgp 300
neighbor 3.3.3.3 remote-as 200
neighbor 2.2.2.2 remote-as 100
neighbor 2.2.2.2 remote-as 100
```

```
network 170.10.0.0
aggregate-address 160.0.0.0 255.0.0.0 suppress-map CHECK
```

Another variation is the:

aggregate-address *address mask* **attribute-map** *map-name*

This allows us to set the attributes (such as metric) when aggregates are sent out. The following route map when applied to the **aggregate attribute-map** command will set the origin of the aggregates to IGP.

```
route-map SETMETRIC
set origin igp
aggregate-address 160.0.0.0 255.0.0.0 attribute-map
SETORIGIN
```

CIDR Example 1

Request: Allow RTB to advertise the prefix 160.0.0.0 and suppress all the more specific routes. The problem here is that network 160.10.0.0 is local to AS200, meaning AS200 is the originator of 160.10.0.0. You cannot have RTB generate a prefix for 160.0.0.0 without generating an entry for 160.10.0.0 even if you use the **aggregate summary-only** command because RTB is the originator of 160.10.0.0. There are two solutions to this problem.

The first solution is to use a static route and redistribute it into BGP. The outcome is that RTB will advertise the aggregate with an origin of incomplete (?).

```
RTB#
router bgp 200
neighbor 3.3.3.1 remote-as 300
redistribute static
!— This generates an update for 160.0.0.0
!— with the origin path as *incomplete*
ip route 160.0.0.0 255.0.0.0 null0
```

In the second solution, in addition to the static route we add an entry for the **network** command. This has the same effect except that the origin of the update will be set to IGP.

```
RTB#
router bgp 200
network 160.0.0.0 mask 255.0.0.0
!— This marks the update with origin IGP
neighbor 3.3.3.1 remote-as 300
redistribute static
ip route 160.0.0.0 255.0.0.0 null0
```

CIDR Example 2 (as-set)

AS-SETS are used in aggregation to reduce the size of the path information by listing the AS number only once, regardless of how many times it may have appeared in multiple paths that were aggregated. The **as-set** aggregate command is used in situations where aggregation of information causes loss of information regarding the path attribute. In the following example RTC is getting updates about 160.20.0.0 from RTA and updates about 160.10.0.0 from RTB. Suppose RTC wants to aggregate network 160.0.0.0/8 and send it to RTD. RTD would not know what the origin of that route is. By adding the aggregate **as-set** statement we force RTC to generate path information in the form of a set { }. All the path information is included in that set irrespective of which path came first.

```
RTB#
router bgp 200
network 160.10.0.0
neighbor 3.3.3.1 remote-as 300

RTA#
router bgp 100
network 160.20.0.0
neighbor 2.2.2.1 remote-as 300
```

1. RTC does not have an **as-set** statement. RTC will send an update 160.0.0.0/8 to RTD with path information (300) as if the route has originated from AS300.

```
RTC#
router bgp 300
neighbor 3.3.3.3 remote-as 200
neighbor 2.2.2.2 remote-as 100
neighbor 4.4.4.4 remote-as 400
aggregate 160.0.0.0 255.0.0.0 summary-only
```

2.
```
RTC#
router bgp 300
neighbor 3.3.3.3 remote-as 200
neighbor 2.2.2.2 remote-as 100
neighbor 4.4.4.4 remote-as 400
aggregate 160.0.0.0 255.0.0.0 summary-only
aggregate 160.0.0.0 255.0.0.0 as-set
```

The next two subjects, confederation and route reflectors, are designed for ISPs who would like to further control the explosion of iBGP peering inside their autonomous systems.

BGP Confederation

BGP confederation is implemented in order to reduce the IBGP mesh inside an AS. The trick is to divide an AS into multiple ASs and assign the whole group to a single confederation. Each AS by itself will have IBGP fully meshed and has connections to other ASs inside the confederation. Even though these ASs will have EBGP peers to ASs within the confederation, they exchange routing as if they were using IBGP; next hop, metric and local preference information are preserved. To the outside world, the confederation (the group of ASs) will look like a single AS.

To configure a BGP confederation use the following:

> **bgp confederation identifier** *autonomous-system*

The confederation identifier will be the AS number of the confederation group. The group of ASs will look to the outside world as one AS with the AS number being the confederation identifier.

Peering within the confederation between multiple ASs is done via the following command:

> **bgp confederation peers *autonomous-system*** [*autonomous-system*]

The following is an example of confederation:

Let us assume that you have an autonomous system 500 consisting of nine BGP speakers (other non-BGP speakers exist also, but we are only interested in the BGP speakers that have EBGP connections to other ASs). If you want to make a full IBGP mesh inside AS500 then you would need nine peer connections for each router, 8 IBGP peers and one EBGP peer to external ASs.

By using confederation we can divide AS500 into multiple ASs: AS50, AS60 and AS70. We give the AS a confederation identifier of 500. The outside world will see only one AS500. For each AS50, AS60 and AS70 we define a full mesh of IBGP peers and we define the list of confederation peers using the bgp confederation peers command.

Let's look at a sample configuration of routers RTC, RTD and RTA. Note that RTA has no knowledge of ASs 50, 60 or 70. RTA has only knowledge of AS500.

```
RTC#
router bgp 50
bgp confederation identifier 500
bgp confederation peers 60 70
neighbor 128.213.10.1 remote-as 50 (IBGP connection within AS50)
neighbor 128.213.20.1 remote-as 50 (IBGP connection within AS50)
neighbor 129.210.11.1 remote-as 60 (BGP connection with confed-
  eration peer 60)
neighbor 135.212.14.1 remote-as 70 (BGP connection with confed-
  eration peer 70)
neighbor 5.5.5.5 remote-as 100 (EBGP connection to external AS100)

RTD#
router bgp 60
bgp confederation identifier 500
bgp confederation peers 50 70
neighbor 129.210.30.2 remote-as 60 (IBGP connection within AS60)
neighbor 128.213.30.1 remote-as 50(BGP connection with confed-
  eration peer 50)
neighbor 135.212.14.1 remote-as 70 (BGP connection with confed-
  eration peer 70)
neighbor 6.6.6.6 remote-as 600 (EBGP connection to external
  AS600)
```

```
RTA#
router bgp 100
neighbor 5.5.5.4 remote-as 500 (EBGP connection to confedera-
  tion 500)
```

Route Reflectors

Another solution for the explosion of IBGP peering within an autonomous system is Route Reflectors (RR). As demonstrated in the Internal BGP section, a BGP speaker will not advertise a route learned via another IBGP speaker to a third IBGP speaker. By relaxing this restriction a bit and by providing additional control, we can allow a router to advertise (reflect) IBGP learned routes to other IBGP speakers. This will reduce the number of IBGP peers within an AS.

In normal cases, a full IBGP mesh should be maintained between RTA, RTB, and RTC within AS100. By utilizing the route reflector concept, RTC could be elected as a RR and have a partial IBGP peering with RTA and RTB. Peering between RTA and RTB is not needed because RTC will be a route reflector for the updates coming from RTA and RTB.

```
neighbor route-reflector-client
```

The router with the above command would be the RR and the neighbors pointed at would be the clients of that RR. In our example, RTC would be configured with the **neighbor route-reflector-client** command pointing at RTA and RTB's IP addresses. The combination of the RR and its clients is called a cluster. RTA, RTB and RTC above would form a cluster with a single RR within AS100.

Other IBGP peers of the RR that are not clients are called non-clients.

An autonomous system can have more than one route reflector; a RR would treat other RRs just like any other IBGP speaker. Other RRs could belong to the same cluster (client group) or to other clusters. In a simple configuration, the AS could be divided into multiple clusters, each RR will be configured with other RRs as non-client peers in a fully meshed topology. Clients should not peer with IBGP speakers outside their cluster.

Consider the above diagram. RTA, RTB, and RTC form a single cluster with RTC being the RR. According to RTC, RTA, and RTB are clients and anything else is a non-client. Remember that clients of an RR are pointed at using the **neighbor route-reflector-client** command. The same RTD is the RR for its clients RTE and RTF; RTG is a RR in a third cluster. Note that RTD, RTC, and RTG are fully meshed but routers within a cluster are not. When a route is received by a RR, it will do the following depending on the peer type:

1. Route from a non-client peer: reflect to all the clients within the cluster.
2. Route from a client peer: reflect to all the non-client peers and also to the client peers.
3. Route from an EBGP peer: send the update to all client and non-client peers.

The following is the relative BGP configuration of routers RTC, RTD and RTB:

```
RTC#
router bgp 100
```

```
neighbor 2.2.2.2 remote-as 100
neighbor 2.2.2.2 route-reflector-client
neighbor 1.1.1.1 remote-as 100
neighbor 1.1.1.1 route-reflector-client
neighbor 7.7.7.7 remote-as 100
neighbor 4.4.4.4 remote-as 100
neighbor 8.8.8.8 remote-as 200

RTB#
router bgp 100
neighbor 3.3.3.3 remote-as 100
neighbor 12.12.12.12 remote-as 300

RTD#
router bgp 100
neighbor 6.6.6.6 remote-as 100
neighbor 6.6.6.6 route-reflector-client
neighbor 5.5.5.5 remote-as 100
neighbor 5.5.5.5 route-reflector-client
neighbor 7.7.7.7 remote-as 100
neighbor 3.3.3.3 remote-as 100
```

As the IBGP learned routes are reflected, it is possible to have the routing information loop. The Route Reflector scheme has a few methods to avoid this:

1. Originator-id: this is an optional, nontransitive BGP attribute that is 4 bytes long and is created by an RR. This attribute will carry the router-id (RID) of the originator of the route in the local AS. Thus, due to poor configuration, if the routing information comes back to the originator, it will be ignored.
2. Cluster-list: this will be discussed in the next section.

Multiple RRs Within a Cluster

Usually, a cluster of clients will have a single RR. In this case, the cluster will be identified by the router ID of the RR. In order to increase redundancy and avoid single points of failure, a cluster might have more than one RR. All RRs in the same cluster need to be configured with a 4 byte cluster-id so that a RR can recognize updates from RRs in the same cluster.

A cluster-list is a sequence of cluster-ids that the route has passed. When a RR reflects a route from its clients to non-clients outside of the cluster, it will append the local cluster-id to the cluster-list. If this update has an empty cluster-list the RR will create one. Using this attribute, a RR can identify if the routing information is looped back to the same cluster due to poor configuration. If the local cluster-id is found in the cluster-list, the advertisement will be ignored.

In the above diagram, RTD, RTE, RTF, and RTH belong to one cluster with both RTD and RTH being RRs for the same cluster. Note the redundancy in that RTH has a fully meshed peering with all the RRs. In case RTD goes down, RTH will take its place. The following are the configuration of RTH, RTD, RTF, and RTC:

```
RTH#
router bgp 100
neighbor 4.4.4.4 remote-as 100
neighbor 5.5.5.5 remote-as 100
neighbor 5.5.5.5 route-reflector-client
neighbor 6.6.6.6 remote-as 100
neighbor 6.6.6.6 route-reflector-client
neighbor 7.7.7.7 remote-as 100
neighbor 3.3.3.3 remote-as 100
neighbor 9.9.9.9 remote-as 300
bgp cluster-id 10

RTD#
router bgp 100
neighbor 10.10.10.10 remote-as 100
neighbor 5.5.5.5 remote-as 100
neighbor 5.5.5.5 route-reflector-client
neighbor 6.6.6.6 remote-as 100
neighbor 6.6.6.6 route-reflector-client
neighbor 7.7.7.7 remote-as 100
neighbor 3.3.3.3 remote-as 100
neighbor 11.11.11.11 remote-as 400
bgp cluster-id 10

RTF#
router bgp 100
neighbor 10.10.10.10 remote-as 100
neighbor 4.4.4.4 remote-as 100
neighbor 13.13.13.13 remote-as 500

RTC#
router bgp 100
neighbor 1.1.1.1 remote-as 100
neighbor 1.1.1.1 route-reflector-client
neighbor 2.2.2.2 remote-as 100
neighbor 2.2.2.2 route-reflector-client
neighbor 4.4.4.4 remote-as 100
neighbor 7.7.7.7 remote-as 100
neighbor 10.10.10.10 remote-as 100
neighbor 8.8.8.8 remote-as 200
```

Note that we did not need the cluster command for RTC because only one RR exists in that cluster.

An important thing to note is that peer groups were not used in the above configuration. If the clients inside a cluster do not have direct IBGP peers among one another and they exchange updates through the RR, peer goups should not be used. If peer groups were to be configured, then a potential withdrawal to the source of a route on the RR would be sent to all clients inside the cluster and could cause problems.

The router subcommand bgp client-to-client reflection is enabled by default on the RR. If BGP client-to-client reflection were turned off on the RR and redundant BGP peering was made between the clients, then using peer groups would be all right.

RR and Conventional BGP Speakers

It is normal in an AS to have BGP speakers that do not understand the concept of route reflectors. We will call these routers conventional BGP speakers. The route reflector scheme will allow such conventional BGP speakers to coexist. These routers could be either members of a client group or a non-client group. This would allow easy and gradual migration from the current IBGP model to the route reflector model. One could start creating clusters by configuring a single router as RR and making other RRs and their clients normal IBGP peers. Then more clusters could be created gradually.

In the above example, RTD, RTE, and RTF have the concept of route reflection. RTC, RTA, and RTB are what we call conventional routers and cannot be configured as RRs. Normal IBGP mesh could be done between these routers and RTD. Later on, when we are ready to upgrade, RTC could be made a RR with clients RTA and RTB. Clients do not have to understand the route reflection scheme; it is only the RRs that would have to be upgraded.

The following is the configuration of RTD and RTC:

```
RTD#
router bgp 100
neighbor 6.6.6.6 remote-as 100
neighbor 6.6.6.6 route-reflector-client
neighbor 5.5.5.5 remote-as 100
neighbor 5.5.5.5 route-reflector-client
neighbor 3.3.3.3 remote-as 100
neighbor 2.2.2.2 remote-as 100
neighbor 1.1.1.1 remote-as 100
neighbor 13.13.13.13 remote-as 300

RTC#
router bgp 100
neighbor 4.4.4.4 remote-as 100
neighbor 2.2.2.2 remote-as 100
neighbor 1.1.1.1 remote-as 100
neighbor 14.14.14.14 remote-as 400
```

When we are ready to upgrade RTC and make it a RR, we would remove the IBGP full mesh and have RTA and RTB become clients of RTC.

Avoiding Looping of Routing Information

We have mentioned so far two attributes that are used to prevent potential information looping: **originator-id** and **cluster-list.**

Another means of controlling loops is to put more restrictions on the set clause of out-bound route-maps. The set clause for out-bound route-maps does not affect routes reflected to IBGP peers.

More restrictions are also put on **nexthop-self**, which is a per neighbor configuration option. When used on RRs the nexthop-self will only affect the nexthop of EBGP learned routes because the nexthop of reflected routes should not be changed.

Route Flap Dampening

Route dampening is a mechanism to minimize the instability caused by route flapping and oscillation over the network. To accomplish this, criteria are defined to identify poorly behaved routes. A route that is flapping gets a penalty for each flap (1000). As soon as the cumulative penalty reaches a predefined "suppress-limit," the advertisement of the route will be suppressed. The penalty will be exponentially decayed based on a preconfigured "half-time." Once the penalty decreases below a predefined "reuse-limit," the route advertisement will be unsuppressed.

Routes, external to an AS, learned via IBGP will not be dampened. This is to avoid the IBGP peers having higher penalty for routes external to the AS.

The penalty will be decayed at a granularity of 5 seconds and the routes will be unsuppressed at a granularity of 10 seconds. The dampening information is kept until the penalty becomes less than half of "reuse-limit," at that point the information is purged from the router.

Initially, dampening will be off by default. This might change if there is a need to have this feature enabled by default. The following are the commands used to control route dampening:

- **bgp dampening** (will turn on dampening)
- **no bgp dampening** (will turn off dampening)
- **bgp dampening** *<half-life-time>* (will change the half-life-time)

A command that sets all parameters at the same time is:

- **bgp dampening** *<half-life-time> <reuse> <suppress> <maximum-suppress-time>*
- *<half-life-time>* (range is 1–45 min, current default is 15 min)
- *<reuse-value>* (range is 1–20,000, default is 750)
- *<suppress-value>* (range is 1–20,000, default is 2000)
- *<max-suppress-time>* (maximum duration a route can be suppressed, range is 1-255, default is 4 times half-life-time)

```
RTB#
hostname RTB
interface Serial0
ip address 203.250.15.2 255.255.255.252
interface Serial1
```

```
ip address 192.208.10.6 255.255.255.252
router bgp 100
bgp dampening
network 203.250.15.0
neighbor 192.208.10.5 remote-as 300

RTD#
hostname RTD
interface Loopback0
ip address 192.208.10.174 255.255.255.192
interface Serial0/0
ip address 192.208.10.5 255.255.255.252
router bgp 300
network 192.208.10.0
neighbor 192.208.10.6 remote-as 100
```

RTB is configured for route dampening with default parameters. Assuming the EBGP link to RTD is stable, RTB's BGP table would look like this:

```
RTB#show ip bgp
BGP table version is 24, local router ID is 203.250.15.2
 Status codes: s
suppressed, d damped, h history, * valid, > best, i - internal
Origin

codes: i - IGP, e - EGP, ? - incomplete
Network            Next Hop          Metric LocPrf  Weight
Path
*> 192.208.10.0    192.208.10.5      0                    0
300 i
*> 203.250.15.0    0.0.0.0           0               32768
i
```

In order to simulate a route flap, use **clear ip bgp 192.208.10.6** on RTD. RTB's BGP table will look like this:

```
RTB#show ip bgp
BGP table version is 24, local router ID is 203.250.15.2
Status codes: s
suppressed, d damped, h history, * valid, > best, i - internal
 Origin
codes: i - IGP, e - EGP, ? - incomplete
Network            Next Hop          Metric LocPrf  Weight
Path
h 192.208.10.0     192.208.10.5      0                    0
300 i
```

```
*> 203.250.15.0      0.0.0.0              0                32768
i
```

The BGP entry for 192.208.10.0 has been put in a "history" state. Which means that we do not have a best path to the route but information about the route flapping still exists.

```
RTB#show ip bgp 192.208.10.0
BGP routing table entry for 192.208.10.0 255.255.255.0, version
 25
Paths: (1 available, no best path)
300 (history entry)
192.208.10.5 from 192.208.10.5 (192.208.10.174)
Origin IGP, metric 0, external
Dampinfo: penalty 910, flapped 1 times in 0:02:03
```

The route has been given a penalty for flapping but the penalty is still below the "suppress limit" (default is 2000). The route is not yet suppressed. If the route flaps few more times we will see the following:

```
RTB#show ip bgp
BGP table version is 32, local router ID is 203.250.15.2 Status
 codes:
s suppressed, d damped, h history, * valid, > best, i - inter-
 nal Origin codes:
i - IGP, e - EGP, ? - incomplete
Network          Next Hop          Metric LocPrf  Weight
Path
*d 192.208.10.0    192.208.10.5      0                   0
300 i
*> 203.250.15.0    0.0.0.0           0                32768 i
RTB#show ip bgp 192.208.10.0
BGP routing table entry for 192.208.10.0 255.255.255.0, version
 32
Paths: (1 available, no best path)
300, (suppressed due to dampening)
192.208.10.5 from 192.208.10.5 (192.208.10.174)
Origin IGP, metric 0, valid, external
Dampinfo: penalty 2615, flapped 3 times in 0:05:18, reuse in
 0:27:00
```

The route has been dampened (suppressed). The route will be reused when the penalty reaches the "reuse value," in our case 750 (default). The dampening information will be purged when the penalty becomes less than half of the reuse-limit, in our case (750/2= 375). The following are the commands used to show and clear flap statistics information:

- **show ip bgp flap-statistics**
      (displays flap statistics for all the paths)

- **show ip bgp-flap-statistics regexp** *<regexp>*
  (displays flap statistics for all paths that match the regexp)
- **show ip bgp flap-statistics filter-list** *<list>*
  (displays flap statistics for all paths that pass the filter)
- **show ip bgp flap-statistics** A.B.C.D m.m.m.m
  (displays flap statistics for a single entry)
- **show ip bgp flap-statistics** A.B.C.D m.m.m.m **longer-prefixes**
  (displays flap statistics for more specific entries)
- **show ip bgp neighbor [dampened-routes] | [flap-statistics]**
  (displays flap statistics for all paths from a neighbor)
- **clear ip bgp flap-statistics**
  (clears flap statistics for all routes)
- **clear ip bgp flap-statistics regexp** *<regexp>*
  (clears flap statistics for all the paths that match the regexp)
- **clear ip bgp flap-statistics filter-list** *<list>*
  (clears flap statistics for all the paths that pass the filter)
- **clear ip bgp flap-statistics** A.B.C.D m.m.m.m
  (clears flap statistics for a single entry)
- **clear ip bgp A.B.C.D flap-statistics**
  (clears flap statistics for all paths from a neighbor)

Case Study

RIP and OSPF Redistribution

This case study addresses the issue of integrating Routing Information Protocol (RIP) networks with Open Shortest Path First (OSPF) networks. Most OSPF networks also use RIP to communicate with hosts or to communicate with portions of the internetwork that do not use OSPF. Cisco supports both the RIP and OSPF protocols and provides a way to exchange routing information between RIP and OSPF networks. This case study provides examples of how to complete the following phases in redistributing information between RIP and OSPF networks.

Configuring an RIP Network

Figure I.1 illustrates an RIP network. Three sites are connected with serial lines. The RIP network uses a Class B address and an 8-bit subnet mask. Each site has a contiguous set of network numbers.

Table I.1 lists the network address assignments for the RIP network, including the network number, subnet range, and subnet masks. All interfaces indicate network 130.10.0.0; however, the specific

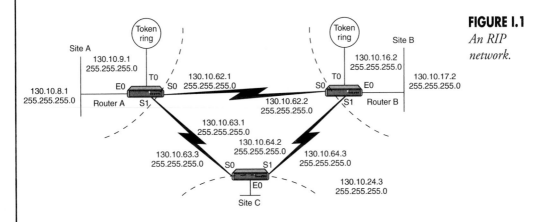

FIGURE I.1
An RIP network.

Table I.1 RIP Network Address Assignments

Network Number	Subnets	Subnet Masks
130.10.0.0	Site A: 8 through 15	255.255.255.0
130.10.0.0	Site B: 16 through 23	255.255.255.0
130.10.0.0	Site C: 24 through 31	255.255.255.0
130.10.0.0	Serial Backbone: 62 through 64	255.255.255.0

address includes the subnet and subnet mask. For example, serial interface 0 on Router C has an IP address of 130.10.63.3 with a subnet mask of 255.255.255.0.

Configuration File Examples

The following commands in the configuration file for Router A determine the IP address for each interface and enable RIP on those interfaces:

```
interface serial 0
ip address 130.10.62.1 255.255.255.0
interface serial 1
ip address 130.10.63.1 255.255.255.0
interface ethernet 0
ip address 130.10.8.1 255.255.255.0
interface tokenring 0
ip address 130.10.9.1 255.255.255.0
router rip
network 130.10.0.0
```

The following commands in the configuration file for Router B determine the IP address for each interface and enable RIP on those interfaces:

```
interface serial 0
ip address 130.10.62.2 255.255.255.0
interface serial 1
ip address 130.10.64.2 255.255.255.0
interface ethernet 0
ip address 130.10.17.2 255.255.255.0
interface tokenring 0
ip address 130.10.16.2 255.255.255.0
router rip
network 130.10.0.0
```

The following commands in the configuration file for Router C determine the IP address for each interface and enable RIP on those interfaces:

```
interface serial 0
ip address 130.10.63.3 255.255.255.0
interface serial 1
ip address 130.10.64.3 255.255.255.0
interface ethernet 0
ip address 130.10.24.3 255.255.255.0
router rip
network 130.10.0.0
```

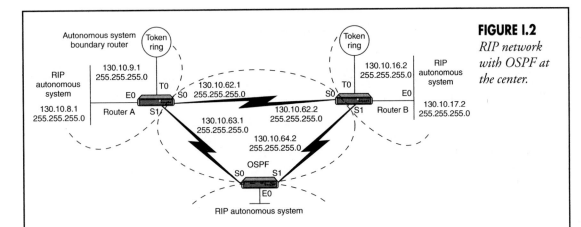

FIGURE I.2
RIP network with OSPF at the center.

Adding OSPF to the Center of an RIP Network

A common first step in converting an RIP network to OSPF is to add backbone routers that run both RIP and OSPF, while the remaining network devices run RIP. These backbone routers are OSPF autonomous system boundary routers. Each autonomous system boundary router controls the flow of routing information between OSPF and RIP. In Figure I.2, Router A is configured as the autonomous system boundary router.

RIP does not need to run between the backbone routers; therefore, RIP is suppressed on Router A with the following commands:

```
router rip
passive-interface serial 0
passive-interface serial 1
```

The RIP routes are redistributed into OSPF by all three routers with the following commands:

```
router ospf 109
redistribute rip subnets
```

The subnets keyword tells OSPF to redistribute all subnet routes. Without the subnets keyword, only networks that are not subnetted will be redistributed by OSPF. Redistributed routes appear as external type 2 routes in OSPF. Each RIP domain receives information about networks in other RIP domains and in the OSPF backbone area from the following commands that redistribute OSPF routes into RIP:

```
router rip
redistribute ospf 109 match internal external 1 external 2
default-metric 10
```

The redistribute command uses the ospf keyword to specify that OSPF routes are to be redistributed into RIP. The keyword internal indicates the OSPF intra-area and interarea routes: External 1 is the external route type 1, and external 2 is the external route type 2. Because the command in the example uses the default behavior, these keywords may not appear when you use the write terminal or show configuration commands.

Because metrics for different protocols cannot be directly compared, you must specify the default metric in order to designate the cost of the redistributed route used in RIP updates. All routes that are redistributed will use the default metric.

In Figure I.2, there are no paths directly connecting the RIP clouds. However, in typical networks, these paths, or "back doors," frequently exist, allowing the potential for feedback loops. You can use access lists to determine the routes that are advertised and accepted by each router. For example, access list 11 in the configuration file for Router A allows OSPF to redistribute information learned from RIP only for networks 130.10.8.0 through 130.10.15.0:

```
router ospf 109
redistribute rip subnet
distribute-list 11 out rip
access-list 11 permit 130.10.8.0 0.0.7.255
access-list 11 deny 0.0.0.0 255.255.255.255
```

These commands prevent Router A from advertising networks in other RIP domains onto the OSPF backbone, thereby preventing other boundary routers from using false information and forming a loop.

Configuration File Examples

The full configuration for Router A follows:

```
interface serial 0
ip address 130.10.62.1 255.255.255.0
interface serial 1
ip address 130.10.63.1 255.255.255.0
interface ethernet 0
ip address 130.10.8.1 255.255.255.0
interface tokenring 0
ip address 130.10.9.1 255.255.255.0
!
router rip
default-metric 10
network 130.10.0.0
passive-interface serial 0
passive-interface serial 1
redistribute ospf 109 match internal external 1 external 2
!
router ospf 109
network 130.10.62.0 0.0.0.255 area 0
network 130.10.63.0 0.0.0.255 area 0
```

```
    redistribute rip subnets
    distribute-list 11 out rip
    !
    access-list 11 permit 130.10.8.0 0.0.7.255
    access-list 11 deny 0.0.0.0 255.255.255.255
```

The full configuration for Router B follows:

```
    interface serial 0
    ip address 130.10.62.2 255.255.255.0
    interface serial 1
    ip address 130.10.64.2 255.255.255.0
    interface ethernet 0
    ip address 130.10.17.2 255.255.255.0
    interface tokenring 0
    ip address 130.10.16.2 255.255.255.0
    !
    router rip
    default-metric 10
    network 130.10.0.0
    passive-interface serial 0
    passive-interface serial 1
    redistribute ospf 109 match internal external 1 external 2
    !
    router ospf 109
    network 130.10.62.0 0.0.0.255 area 0
    network 130.10.64.0 0.0.0.255 area 0
    redistribute rip subnets
    distribute-list 11 out rip
    access-list 11 permit 130.10.16.0 0.0.7.255
    access-list 11 deny 0.0.0.0 255.255.255.255
```

The full configuration for Router C follows:

```
    interface serial 0
    ip address 130.10.63.3 255.255.255.0
    interface serial 1
    ip address 130.10.64.3 255.255.255.0
    interface ethernet 0
    ip address 130.10.24.3 255.255.255.0
    !
    router rip
    default-metric 10
    !
```

```
network 130.10.0.0
passive-interface serial 0
passive-interface serial 1
redistribute ospf 109 match internal external 1 external 2
!
router ospf 109
network 130.10.63.0 0.0.0.255 area 0
network 130.10.64.0 0.0.0.255 area 0
redistribute rip subnets
distribute-list 11 out rip
access-list 11 permit 130.10.24.0 0.0.7.255
access-list 11 deny 0.0.0.0 255.255.255.255
```

Adding OSPF Areas

Figure I.3 illustrates how each of the RIP clouds can be converted into an OSPF area. All three routers are area border routers. Area border routers control network information distribution between OSPF areas and the OSPF backbone. Each router keeps a detailed record of the topology of its area and receives summarized information from the other area border routers on their respective areas.

Figure I.3 also illustrates variable length subnet masks (VLSMs). VLSMs use different size network masks in different parts of the network for the same network number. VLSM conserves address space by using a longer mask in portions of the network that have fewer hosts. Table I.2 lists the network address assignments for the network, including the network number, subnet range, and subnet masks. All interfaces indicate network 130.10.0.0.

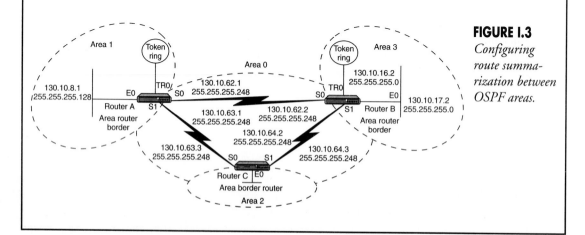

FIGURE I.3

Configuring route summarization between OSPF areas.

Table I.2 OSPF Address Assignments

Network Number	Subnets	Subnet Masks
130.10.0.0	Area 0: 62 through 64	255.255.255.248
130.10.0.0	Area 1: 8 through 15	255.255.255.0
130.10.0.0	Area 2: 16 through 23	255.255.255.0
130.10.0.0	Area 3: 24 through 31	255.255.255.0

To conserve address space, a mask of 255.255.255.248 is used for all the serial lines in area 0. If an area contains a contiguous range of network numbers, an area border router uses the range keyword with the area command to summarize the routes that are injected into the backbone:

```
router ospf 109
network 130.10.8.0 0.0.7.255 area 1
area 1 range 130.10.8.0 255.255.248.0
```

These commands allow Router A to advertise one route, 130.10.8.0 255.255.248.0, which covers all subnets in Area 1 into Area 0. Without the range keyword in the area command, Router A would advertise each subnet individually; for example, one route for 130.10.8.0 255.255.255.0, one route for 130.10.9.0 255.255.255.0, and so forth.

Because Router A no longer needs to redistribute RIP routes, the router rip command can now be removed from the configuration file; however, it is common in some environments for hosts to use RIP to discover routers. When RIP is removed from the routers, the hosts must use an alternative technique to find the routers. Cisco routers support the following alternatives to RIP:

- ICMP Router Discovery Protocol (IRDP)—This technique is illustrated in the example at the end of this section. IRDP is the recommended method for discovering routers. The ip irdp command enables IRDP on the router. Hosts must also run IRDP.
- Proxy Address Resolution Protocol (ARP)—If the router receives an ARP request for a host that is not on the same network as the ARP request sender, and if the router has the best route to that host, the router sends an ARP reply packet giving the router's own local data link address. The host that sent the ARP request then sends its packets to the router, which forwards them to the intended host. Proxy ARP is enabled on routers by default. Proxy ARP is transparent to hosts.

Configuration File Examples

The full configuration for Router A follows:

```
interface serial 0
ip address 130.10.62.1 255.255.255.248
interface serial 1
ip address 130.10.63.1 255.255.255.248
```

```
interface ethernet 0
ip address 130.10.8.1 255.255.255.0
ip irdp
interface tokenring 0
ip address 130.10.9.1 255.255.255.0
ip irdp
router ospf 109
network 130.10.62.0 0.0.0.255 area 0
network 130.10.63.0 0.0.0.255 area 0
network 130.10.8.0 0.0.7.255 area 1
area 1 range 130.10.8.0 255.255.248.0
```

The full configuration for Router B follows:

```
interface serial 0
ip address 130.10.62.2 255.255.255.248
interface serial 1
ip address 130.10.64.2 255.255.255.248
interface ethernet 0
ip address 130.10.17.2 255.255.255.0
ip irdp
interface tokenring 0
ip address 130.10.16.2 255.255.255.0
ip irdp
router ospf 109
network 130.10.62.0 0.0.0.255 area 0
network 130.10.64.0 0.0.0.255 area 0
network 130.10.16.0 0.0.7.255 area 2
area 2 range 130.10.16.0 255.255.248.0
```

The full configuration for Router C follows:

```
interface serial 0
ip address 130.10.63.2 255.255.255.248
interface serial 1
ip address 130.10.64.2 255.255.255.248
interface ethernet 0
ip address 130.10.24.3 255.255.255.0
ip irdp
router ospf 109
network 130.10.63.0 0.0.0.255 area 0
network 130.10.64.0 0.0.0.255 area 0
network 130.10.24.0 0.0.0.255 area 3
area 3 range 130.10.24.0 255.255.248.0
```

FIGURE I.4

*Mutual redistrib-
ution between
RIP and OSPF
networks.*

Setting Up Mutual Redistribution

It is sometimes necessary to accommodate more complex network topologies such as independent RIP and OSPF clouds that must perform mutual redistribution. In this scenario, it is critically important to prevent potential routing loops by filtering routes. The router in Figure I.4 is running both OSPF and RIP.

With the following commands, OSPF routes will be redistributed into RIP. You must specify the default metric to designate the cost of the redistributed route in RIP updates. All routes redistributed into RIP will have this default metric.

```
! passive interface subcommand from previous example is left out
  for clarity!
router rip
default-metric 10
network 130.10.0.0
redistribute ospf 109
```

It is a good practice to strictly control which routes are advertised when redistribution is configured. In the following example, a distribute-list out command causes RIP to ignore routes coming from the OSPF that originated from the RIP domain.

```
router rip
distribute-list 10 out ospf 109
!
access-list 10 deny 130.10.8.0 0.0.7.255
access-list 10 permit 0.0.0.0 255.255.255.255
```

Router A

The full configuration for the router follows:

```
interface serial 0
ip add 130.10.62.1 255.255.255.0
!
interface serial 1
ip add 130.10.63.1 255.255.255.0
!
interface ethernet 0
```

```
ip add 130.10.8.1 255.255.255.0
!
interface tokenring 0
ip add 130.10.9.1 255.255.255.0z
!
router rip
default-metric 10
network 130.10.0.0
passive-interface serial 0
passive-interface serial 1
redistribute ospf 109
distribute-list 10 out ospf 109
!
router ospf 109
network 130.10.62.0 0.0.0.255 area 0
network 130.10.63.0 0.0.0.255 area 0
redistribute rip subnets
distribute-list 11 out rip
!
access-list 10 deny 130.10.8.0 0.0.7.255
access-list 10 permit 0.0.0.0 255.255.255.255
access-list 11 permit 130.10.8.0 0.0.7.255
access-list 11 deny 0.0.0.0 255.255.255.255
```

Summary

Because it is common for OSPF and RIP to be used together, it is important to use the practices described here in order to provide functionality for both protocols on an internetwork. You can configure autonomous system boundary routers that run both RIP and OSPF and redistribute RIP routes into the OSPF, and vice versa. You can also create OSPF areas using area border routers that provide route summarizations. Use VLSM to conserve address space.

GLOSSARY

A

A&B bit signaling Procedure used in T1 transmission facilities in which each of the 24 T1 subchannels devotes one bit of every sixth frame to the carrying of supervisory signaling information. Also called *24th channel signaling*.

AAL (ATM adaptation layer) Service-dependent sublayer of the data link layer. The AAL accepts data from different applications and presents it to the ATM layer in the form of 48-byte ATM payload segments. AALs consist of two sublayers: CS and SAR. AALs differ on the basis of the source destination timing used, whether they use CBR or VBR, and whether they are used for connection oriented or connectionless mode data transfer. At present, the four types of AAL recommended by the ITU-T are AAL1, AAL2, AAL3/4, and AAL5.

AAL1 (ATM adaptation layer 1) One of four AALs recommended by the ITU-T. AAL1 is used for connection oriented, delay-sensitive services requiring constant bit rates, such as uncompressed video and other isochronous traffic.

AAL2 (ATM adaptation layer 2) One of four AALs recommended by the ITU-T. AAL2 is used for connection oriented services that support a variable bit rate, such as some isochronous video and voice traffic.

AAL3/4 (ATM adaptation layer 3/4) One of four AALs (merged from two initially distinct adaptation layers) recommended by the ITU-T. AAL3/4 supports both connectionless and connection oriented links, but is primarily used for the transmission of SMDS packets over ATM networks.

AAL5 (ATM adaptation layer 5) One of four AALs recommended by the ITU-T. AAL5 supports connection oriented VBR services, and is used predominantly for the transfer of classical IP over ATM and LANE traffic. AAL5 uses SEAL and is the least complex of the current AAL recommendations. It offers low bandwidth overhead and simpler processing requirements in exchange for reduced bandwidth capacity and error-recovery capability.

AARP (AppleTalk Address Resolution Protocol) Protocol in the AppleTalk protocol stack that maps a data link address to a network address.

AARP probe packets Packets transmitted by AARP that determine if a randomly selected node ID is being used by another node in a nonextended AppleTalk network. If the node ID is not being used, the sending node uses that node ID. If the node ID is being used, the sending node chooses a different ID and sends more AARP probe packets.

ABM (Asynchronous Balanced Mode) An HDLC (and derivative protocol) communication mode supporting peer oriented, point-to-point communications between two stations, where either station can initiate transmission.

ABR (1) available bit rate. QOS class defined by the ATM Forum for ATM networks. ABR is used for connections that do not require timing relationships between source and destination. ABR provides no guarantees in terms of cell loss or delay, providing only best-effort service. Traffic sources adjust their transmission rate in response to information they receive describing the status of the network and its capability to successfully deliver data. (2) area border router. Router located on the border of one or more OSPF areas that connects those areas to the backbone network. ABRs are considered members of both the OSPF backbone and the attached areas. They therefore maintain routing tables describing both the backbone topology and the topology of the other areas.

access list List kept by routers to control access to or from the router for a number of services (for example, to prevent packets with a certain IP address from leaving a particular interface on the router).

access method (1) Generally, the way in which network devices access the network medium. (2) Software within an SNA processor that controls the flow of information through a network.

access server Communications processor that connects asynchronous devices to a LAN or WAN through network and terminal emulation software. Performs both synchronous and asynchronous routing of supported protocols. Sometimes called a *network access server*.

accounting management   One of five categories of network management defined by ISO for management of OSI networks. Accounting management subsystems are responsible for collecting network data relating to resource usage.

ACF   (Advanced Communications Function) A group of SNA products that provides distributed processing and resource sharing.

ACF/NCP   (Advanced Communications Function/ Network Control Program) The primary SNA NCP. ACF/NCP resides in the communications controller and interfaces with the SNA access method in the host processor to control network communications.

Acknowledgment   Notification sent from one network device to another to acknowledge that some event (for example, receipt of a message) has occurred. Sometimes abbreviated *ACK*.

ACR   (allowed cell rate) Parameter defined by the ATM Forum for ATM traffic management. ACR varies between the MCR and the PCR, and is dynamically controlled using congestion control mechanisms.

ACSE   (association control service element) An OSI convention used to establish, maintain, or terminate a connection between two applications.

active hub   Multiported device that amplifies LAN transmission signals.

active monitor   Device responsible for managing a Token Ring. A network node is selected to be the active monitor if it has the highest MAC address on the ring. The active monitor is responsible for such management tasks as ensuring that tokens are not lost, or that frames do not circulate indefinitely.

ADCCP   (Advanced Data Communications Control Protocol) An ANSI standard bit oriented data link control protocol.

address   Data structure or logical convention used to identify a unique entity, such as a particular process or network device.

addressed call mode   Mode that permits control signals and commands to establish and terminate calls in V.25 bis.

address mapping   Technique that allows different protocols to interoperate by translating addresses from one format to another. For example, when routing IP over X.25, the IP addresses must be mapped to the X.25 addresses so that the IP packets can be transmitted by the X.25 network.

address mask   Bit combination used to describe which portion of an address refers to the network or subnet and which part refers to the host. Sometimes referred to simply as *mask*.

address resolution   Generally, a method for resolving differences between computer addressing schemes. Address resolution usually specifies a method for mapping network layer (layer 3) addresses to data link layer (layer 2) addresses.

adjacency   Relationship formed between selected neighboring routers and end nodes for the purpose of exchanging routing information. Adjacency is based on the use of a common media segment.

adjacent nodes   (1) In SNA, nodes that are connected to a given node with no intervening nodes. (2) In DECnet and OSI, nodes that share a common network segment (in Ethernet, FDDI, or Token Ring networks).

administrative distance   A rating of the trustworthiness of a routing information source. In Cisco routers, administrative distance is expressed as a numerical value between 0 and 255. The higher the value, the lower the trustworthiness rating.

ADPCM   (adaptive differential pulse code modulation) Process by which analog voice samples are encoded into high quality digital signals.

ADSU   (ATM DSU) Terminal adapter used to access an ATM network via an HSSI compatible device.

advertising   Router process in which routing or service updates are sent at specified intervals so that other routers on the network can maintain lists of usable routes.

AEP   (AppleTalk Echo Protocol) Used to test connectivity between two AppleTalk nodes. One node sends a packet to another node and receives a duplicate, or echo, of that packet.

agent   (1) Generally, software that processes queries and returns replies on behalf of an application. (2) In NMSs, process that resides in all managed devices and reports the values of specified variables to management stations. (3) In router hardware architecture, an individual processor card that provides one or more media interfaces.

AGS+   Multiprotocol, high end router optimized for large corporate internetworks. The AGS+ runs the IOS software and features a modular approach that provides for easy and efficient scalability.

AIP   (ATM Interface Processor) ATM network interface for routers are designed to minimize performance bottlenecks at the UNI. The AIP supports AAL3/4 and AAL5.

AIS (alarm indication signal) In a T1 transmission, an all-ones signal transmitted in lieu of the normal signal to maintain transmission continuity and to indicate to the receiving terminal that there is a transmission fault that is located either at, or upstream from, the transmitting terminal.

alarm Message notifying an operator or administrator of a network problem.

a-law The ITU-T companding standard used in the conversion between analog and digital signals in PCM systems. A-law is used primarily in European telephone networks and is similar to the North American mu-law standard.

algorithm Well-defined rule or process for arriving at a solution to a problem. In networking, algorithms are commonly used to determine the best route for traffic from a particular source to a particular destination.

alignment error In IEEE 802.3 networks, an error that occurs when the total number of bits of a received frame is not divisible by eight. Alignment errors are usually caused by frame damage due to collisions.

all routes explorer packet Explorer packet that traverses an entire SRB network, following all possible paths to a specific destination. Sometimes called *all rings explorer packet.*

AM (amplitude modulation) Modulation technique whereby information is conveyed through the amplitude of the carrier signal.

AMI (alternate mark inversion) Line-code type used on T1 and E1 circuits. In AMI, zeros are represented by 01 during each bit cell, and ones are represented by 11 or 00, alternately, during each bit cell. AMI requires that the sending device maintain ones density. Ones density is not maintained independent of the data stream. Sometimes called *binary coded alternate mark inversion.*

amplitude Maximum value of an analog or a digital waveform.

analog transmission Signal transmission over wires or through the air in which information is conveyed through variation of some combination of signal amplitude, frequency, and phase.

ANSI (American National Standards Institute) Voluntary organization comprised of corporate, government, and other members that coordinates standards related activities, approves U.S. national standards, and develops positions for the United States in international standards organizations. ANSI helps develop international and U.S. standards relating to, among other things, communications and networking. ANSI is a member of the IEC and the ISO.

APaRT (automated packet recognition and translation) Technology that allows a server to be attached to CDDI or FDDI without requiring the reconfiguration of applications or network protocols. APaRT recognizes specific data link layer encapsulation packet types and, when these packet types are transferred from one medium to another, translates them into the native format of the destination device.

API (application programming interface) Specification of function-call conventions that defines an interface to a service.

Apollo Domain Proprietary network protocol suite developed by Apollo Computer for communication on proprietary Apollo networks.

APPC (Advanced Program-to-Program Communications) IBM SNA system software that allows high speed communication between programs on different computers in a distributed computing environment. APPC establishes and tears down connections between communicating programs. It consists of two interfaces: a programming interface and a data-exchange interface. The former replies to requests from programs requiring communication; the latter establishes sessions between programs. APPC runs on LU 6.2 devices.

AppleTalk Series of communications protocols designed by Apple Computer. Two phases currently exist. Phase 1, the earlier version, supports a single physical network that can have only one network number and be in one zone. Phase 2, the more recent version, supports multiple logical networks on a single physical network and allows networks to be in more than one zone.

application layer Layer 7 of the OSI reference model. This layer provides services to application processes (such as electronic mail, file transfer, and terminal emulation) that are outside of the OSI model. The application layer identifies and establishes the availability of intended communication partners (and the resources required to connect with them), synchronizes cooperating applications, and establishes agreement on procedures for error recovery and control of data integrity. Corresponds roughly with the transaction services layer in the SNA model.

APPN (Advanced Peer-to-Peer Networking) Enhancement to the original IBM SNA. APPN handles session establishment between peer nodes, dynamic transparent route calculation, and traffic prioritization for APPC traffic.

APPN+ Next-generation APPN that replaces the label swapping routing algorithm with source routing. Also called *high performance routing.*

ARA (AppleTalk Remote Access) Protocol that provides Macintosh users direct access to information and resources at a remote AppleTalk site.

ARCnet (Attached Resource Computer Network) A 2.5-Mbps token bus LAN developed in the late 1970s and early 1980s by Datapoint Corporation.

area Logical set of network segments (either CLNS, DECnet, or OSPF based) and their attached devices. Areas are usually connected to other areas via routers, making up a single autonomous system.

ARM (asynchronous response mode) HDLC communication mode involving one primary station and at least one secondary station, where either the primary or one of the secondary stations can initiate transmissions.

ARP (Address Resolution Protocol) Internet protocol used to map an IP address to a MAC address. Defined in RFC 826.

ARPA (Advanced Research Projects Agency) Research and development organization that is part of DoD. ARPA is responsible for numerous technological advances in communications and networking. ARPA evolved into DARPA, and then back into ARPA again (in 1994).

ARPANET (Advanced Research Projects Agency Network) Landmark packet-switching network established in 1969. ARPANET was developed in the 1970s by BBN and funded by ARPA (and later DARPA). It eventually evolved into the Internet. The term ARPANET was officially retired in 1990.

ARQ (automatic repeat request) Communication technique in which the receiving device detects errors and requests retransmissions.

ASBR (autonomous system boundary router) ABR located between an OSPF autonomous system and a non-OSPF network. ASBRs run both OSPF and another routing protocol, such as RIP. ASBRs must reside in a non-stub OSPF area.

ASCII (American Standard Code for Information Interchange) 8-bit code for character representation (7 bits plus parity).

ASN.1 (Abstract Syntax Notation One) OSI language for describing data types independent of particular computer structures and representation techniques. Described by ISO International Standard 8824.

associative memory Memory that is accessed based on its contents, not on its memory address. Sometimes called *content addressable memory (CAM).*

AST (automatic spanning tree) Function that supports the automatic resolution of spanning trees in SRB networks, providing a single path for spanning explorer frames to traverse from a given node in the network to another. AST is based on the IEEE 802.1 standard.

ASTA (Advanced Software Technology and Algorithms) Component of the HPCC program intended to develop software and algorithms for implementation on high-performance computer and communications systems.

asynchronous transmission Term describing digital signals that are transmitted without precise clocking. Such signals generally have different frequencies and phase relationships. Asynchronous transmissions usually encapsulate individual characters in control bits (called start and stop bits) that designate the beginning and end of each character.

ATDM (asynchronous time-division multiplexing) Method of sending information that resembles normal TDM, except that time slots are allocated as needed rather than preassigned to specific transmitters.

ATG (address translation gateway) DECnet routing software function that allows a router to route multiple independent DECnet networks and to establish a user specified address translation for selected nodes between networks.

ATM (Asynchronous Transfer Mode) International standard for cell relay in which multiple service types (such as voice, video, or data) are conveyed in fixed length (53-byte) cells. Fixed length cells allow cell processing to occur in hardware, thereby reducing transit delays. ATM is designed to take advantage of high speed transmission media such as E3, SONET, and T3.

ATM Forum International organization jointly founded in 1991 by Cisco Systems, NET/ADAPTIVE, Northern Telecom, and Sprint that develops and promotes standards based implementation agreements for ATM technology. The ATM Forum expands on official standards developed by ANSI and ITU-T, and develops implementation agreements in advance of official standards.

ATM layer Service-independent sublayer of the data link layer in an ATM network. The ATM layer receives the 48-byte payload segments from the AAL and attaches a 5-byte header to each, producing standard 53-byte ATM cells. These cells are passed to the physical layer for transmission across the physical medium.

ATMM (ATM management) Process that runs on an ATM switch that controls VCI translation and rate enforcement.

ATM user-user connection Connection created by the ATM layer to provide communication between two or more ATM service users, such as ATMM processes. Such communication can be unidirectional, using one VCC, or bidirectional, using two VCCs.

ATP (AppleTalk Transaction Protocol) Transport level protocol that allows reliable request-response exchanges between two socket clients.

attenuation Loss of communication signal energy.

attribute Configuration data that defines the characteristics of database objects such as the chassis, cards, ports, or virtual circuits of a particular device. Attributes might be preset or user configurable.

AUI (attachment unit interface) IEEE 802.3 interface between an MAU and an NIC (network interface card). The term AUI can also refer to the rear panel port to which an AUI cable might attach. Ethernet access card. Also called *transceiver cable.*

AURP (AppleTalk Update Based Routing Protocol) Method of encapsulating AppleTalk traffic in the header of a foreign protocol, allowing the connection of two or more discontiguous AppleTalk internetworks through a foreign network (such as TCP/IP) to form an AppleTalk WAN. This connection is called an AURP tunnel. In addition to its encapsulation function, AURP maintains routing tables for the entire AppleTalk WAN by exchanging routing information between exterior routers.

AURP tunnel Connection created in an AURP WAN that functions as a single, virtual data link between AppleTalk internetworks physically separated by a foreign network (a TCP/IP network, for example).

authority zone Associated with DNS, an authority zone is a section of the domain name tree for which one name server is the authority.

automatic call reconnect Feature permitting automatic call rerouting away from a failed trunk line.

autonomous confederation Group of autonomous systems that rely on their own network reachability and routing information more than they rely on that received from other autonomous systems or confederations.

autonomous switching Feature on routers that provides faster packet processing by allowing the bus to switch packets independently without interrupting the system processor.

autonomous system Collection of networks under a common administration sharing a common routing strategy. Autonomous systems are subdivided by areas. An autonomous system must be assigned a unique 16-bit number by the IANA. Sometimes abbreviated *AS.*

autoreconfiguration Process performed by nodes within the failure domain of a Token Ring network. Nodes automatically perform diagnostics in an attempt to reconfigure the network around the failed areas.

average rate The average rate, in kilobits per second (Kbps), at which a given virtual circuit will transmit.

B

B8ZS (binary 8 zero substitution) Line-code type used on T1 and E1 circuits in which a special code is substituted whenever 8 consecutive zeros are sent through the link. This code is then interpreted at the remote end of the connection. This technique guarantees ones density independent of the data stream. Sometimes called *bipolar 8 zero substitution.*

backbone The part of a network that acts as the primary path for traffic that is most often sourced from, and destined for, other networks.

back end Node or software program that provides services to a front end.

backoff The (usually random) retransmission delay enforced by contentious MAC protocols after a network node with data to transmit determines that the physical medium is already in use.

backplane Physical connection between an interface processor or card and the data buses and power distribution buses inside a router chassis.

back pressure Propagation of network congestion information upstream through an internetwork.

backward learning Algorithmic process used for routing traffic that surmises information by assuming symmetrical network conditions. For example, if node A receives a packet from node B through intermediate node C, the backward-learning routing algorithm will assume that A can optimally reach B through C.

balanced configuration In HDLC, a point-to-point network configuration with two combined stations.

balun (balanced, unbalanced) Device used for matching impedance between a balanced and an unbalanced line, usually twisted-pair and coaxial cable.

bandwidth The difference between the highest and lowest frequencies available for network signals. The term is also used to describe the rated throughput capacity of a given network medium or protocol.

bandwidth reservation Process of assigning bandwidth to users and applications served by a network. Involves assigning priority to different flows of traffic based on how critical and delay-sensitive they are. This makes the best use of available bandwidth, and if the network becomes congested, lower priority traffic can be dropped. Sometimes called *bandwidth allocation.*

BARRNet   (Bay Area Regional Research Network) Regional network serving the San Francisco Bay Area. The BARRNet backbone is composed of four University of California campuses (Berkeley, Davis, Santa Cruz, and San Francisco), Stanford University, Lawrence Livermore National Laboratory, and NASA Ames Research Center. BARRNet is now part of BBN Planet.

baseband   Characteristic of a network technology where only one carrier frequency is used. Ethernet is an example of a baseband network. Also called *narrowband.*

bash   (Bourne-again shell) Interactive UNIX shell based on the traditional Bourne shell, but with increased functionality. The LynxOS bash shell is presented when you log in to an ATM switch as root (bash#) or fldsup (bash$).

basic configuration   The minimal configuration information entered when a new router, switch, or other configurable network device is installed on a network. The basic configuration for an ATM switch, for example, includes IP addresses, the date, and parameters for at least one trunk line. The basic configuration enables the device to receive a full configuration from the NMS.

baud   Unit of signaling speed equal to the number of discrete signal elements transmitted per second. Baud is synonymous with bits per second (bps), if each signal element represents exactly 1 bit.

BBN   (Bolt, Beranek, and Newman, Inc.) High-technology company located in Massachusetts that developed and maintained the ARPANET (and later, the Internet) core gateway system.

BBN Planet   Subsidiary company of BBN that operates a nationwide Internet access network composed in part by the former regional networks BARRNet, NEARNET, and SURAnet.

Bc   (committed burst) Negotiated tariff metric in Frame Relay internetworks. The maximum amount of data (in bits) that a Frame Relay internetwork is committed to accept and transmit at the CIR.

B channel   (bearer channel) In ISDN, a full-duplex, 64-Kbps channel used to send user data.

Be   (excess burst) Negotiated tariff metric in Frame Relay internetworks. The number of bits that a Frame Relay internetwork will attempt to transmit after Bc is accommodated. Be data are, in general, delivered with a lower probability than Bc data because Be data can be marked as DE by the network.

beacon   Frame from a Token Ring or FDDI device indicating a serious problem with the ring, such as a broken cable. A beacon frame contains the address of the station assumed to be down.

BECN   (backward explicit congestion notification) Bit set by a Frame Relay network in frames traveling in the opposite direction of frames encountering a congested path. DTE receiving frames with the BECN bit set can request that higher level protocols take flow control action as appropriate.

Bellcore   (Bell Communications Research) Organization that performs research and development on behalf of the RBOCs.

BER   (1) bit error rate. The ratio of received bits that contain errors. (2) basic encoding rules. Rules for encoding data units described in the ISO ASN.1 standard.

BERT   (bit error rate tester) Device that determines the BER on a given communications channel.

best-effort delivery   Describes a network system that does not use a sophisticated acknowledgement system to guarantee reliable delivery of information.

BGP   (Border Gateway Protocol) Interdomain routing protocol that replaces EGP. BGP exchanges reachability information with other BGP systems. It is defined by RFC 1163.

BGP4   (BGP Version 4) Version 4 of the predominant interdomain routing protocol used on the Internet. BGP4 supports CIDR and uses route aggregation mechanisms to reduce the size of routing tables.

BIGA   (Bus Interface Gate Array) Technology that allows the Catalyst 5000 to receive and transmit frames from its packet-switching memory to its MAC local buffer memory without the intervention of the host processor.

big-endian   Method of storing or transmitting data in which the most significant bit or byte is presented first.

binary   A numbering system characterized by ones and zeros (1 = on, 0 = off).

biphase coding   Bipolar coding scheme originally developed for use in Ethernet. Clocking information is embedded into and recovered from the synchronous data stream without the need for separate clocking leads. The biphase signal contains no direct current energy.

bipolar   Electrical characteristic denoting a circuit with both negative and positive polarity.

B-ISDN   (Broadband ISDN) ITU-T communication standards designed to handle high bandwidth applications such as video. B-ISDN currently uses ATM technology over SONET based transmission circuits to provide data rates from 155 to 622 Mbps and beyond.

bit Binary digit used in the binary numbering system. Can be 0 or 1.

BITNET ("Because It's Time" Networking Services) Low cost, low speed academic network consisting primarily of IBM mainframes and 9,600-bps leased lines. BIT-NET is now part of CREN.

BITNET III Dial-up service providing connectivity for members of CREN.

bit oriented protocol Class of data link layer communication protocols that can transmit frames regardless of frame content. Compared with byte oriented protocols, bit oriented protocols provide full-duplex operation and are more efficient and reliable.

bit rate Speed at which bits are transmitted, usually expressed in bits per second (bps).

bits per second Abbreviated *bps.*

black hole Routing term for an area of the internetwork where packets enter, but do not emerge, due to adverse conditions or poor system configuration within a portion of the network.

blocking In a switching system, a condition in which no paths are available to complete a circuit. The term is also used to describe a situation in which one activity cannot begin until another has been completed.

block multiplexer channel IBM-style channel that implements the FIPS-60 channel, a U.S. channel standard. This channel is also referred to as *OEMI channel* and *370 block mux channel.*

BNC connector Standard connector used to connect IEEE 802.3 10Base-2 coaxial cable to an MAU.

BNN (boundary network node) In SNA terminology, a subarea node that provides boundary function support for adjacent peripheral nodes. This support includes sequencing, pacing, and address translation. Also called *boundary node.*

BOC Bell operating company.

BOOTP Protocol used by a network node to determine the IP address of its Ethernet interfaces in order to affect network booting.

boot PROM (boot programmable read only memory) Chip mounted on a printed circuit board used to provide executable boot instructions to a computer device.

border gateway Router that communicates with routers in other autonomous systems.

boundary function Capability of SNA subarea nodes to provide protocol support for attached peripheral nodes. Typically found in IBM 3745 devices.

BPDU (bridge protocol data unit) Spanning Tree Protocol Hello packet that is sent out at configurable intervals to exchange information among bridges in the network.

bps bits per second.

BRHR (Basic Research and Human Resources) Component of the HPCC program designed to support research, training, and education in computer science, computer engineering, and computational science.

BRI (Basic Rate Interface) ISDN interface composed of two B channels and one D channel for circuit-switched communication of voice, video, and data.

bridge Device that connects and passes packets between two network segments that use the same communications protocol. Bridges operate at the data link layer (layer 2) of the OSI reference model. In general, a bridge will filter, forward, or flood an incoming frame based on the MAC address of that frame.

bridge forwarding Process that uses entries in a filtering database to determine whether frames with a given MAC destination address can be forwarded to a given port or ports. Described in the IEEE 802.1 standard.

bridge group Cisco bridging feature that assigns network interfaces to a particular spanning-tree group. Bridge groups can be compatible with the IEEE 802.1 or the DEC specification.

bridge number Number that identifies each bridge in an SRB LAN. Parallel bridges must have different bridge numbers.

bridge static filtering Process in which a bridge maintains a filtering database consisting of static entries. Each static entry equates a MAC destination address with a port that can receive frames with this MAC destination address and a set of ports on which the frames can be transmitted. Defined in the IEEE 802.1 standard.

broadband Transmission system that multiplexes multiple independent signals onto one cable. In telecommunications terminology, any channel having a bandwidth greater than a voice-grade channel (4 kHz). In LAN terminology, a coaxial cable on which analog signaling is used. Also called *wideband.*

broadcast Data packet that will be sent to all nodes on a network. Broadcasts are identified by a broadcast address.

broadcast address   Special address reserved for sending a message to all stations. Generally, a broadcast address is a MAC destination address of all ones.

broadcast domain   The set of all devices that will receive broadcast frames originating from any device within the set. Broadcast domains are typically bounded by routers because routers do not forward broadcast frames.

broadcast search   Propagation of a search request to all network nodes if the location of a resource is unknown to the requester.

broadcast storm   Undesirable network event in which many broadcasts are sent simultaneously across all network segments. A broadcast storm uses substantial network bandwidth and, typically, causes network time-outs.

BSC   (binary synchronous communication) Character oriented data link layer protocol for half-duplex applications. Often referred to simply as *bisync.*

BSD   (Berkeley Standard Distribution) Term used to describe any of a variety of UNIX type operating systems based on the UC Berkeley BSD operating system.

BT   (burst tolerance) Parameter defined by the ATM Forum for ATM traffic management. For VBR connections, BT determines the size of the maximum burst of contiguous cells that can be transmitted.

buffer   Storage area used for handling data in transit. Buffers are used in internetworking to compensate for differences in processing speed between network devices. Bursts of data can be stored in buffers until they can be handled by slower processing devices. Sometimes referred to as a *packet buffer.*

BUS   (broadcast and unknown server) Multicast server used in ELANs that is used to flood traffic addressed to an unknown destination, and to forward multicast and broadcast traffic to the appropriate clients.

bus   Common physical signal path composed of wires or other media across which signals can be sent from one part of a computer to another. Sometimes called *highway.*

bus and tag channel   IBM channel developed in the 1960s that incorporated copper multiwire technology. Replaced by the ESCON channel.

bus topology   Linear LAN architecture in which transmissions from network stations propagate the length of the medium and are received by all other stations.

bypass mode   Operating mode on FDDI and Token Ring networks in which an interface has been removed from the ring.

bypass relay   Allows a particular Token Ring interface to be shut down and thus effectively removed from the ring.

byte   Term used to refer to a series of consecutive binary digits that are operated upon as a unit (for example, an 8-bit byte).

byte oriented protocol   Class of data link communications protocols that use a specific character from the user character set to delimit frames. These protocols have largely been replaced by bit oriented protocols.

byte reversal   Process of storing numeric data with the least-significant byte first. Used for integers and addresses on devices with Intel microprocessors.

C

cable   Transmission medium of copper wire or optical fiber wrapped in a protective cover.

cable range   Range of network numbers that is valid for use by nodes on an extended AppleTalk network. The cable range value can be a single network number or a contiguous sequence of several network numbers. Node addresses are assigned based on the cable range value.

call admission control   Traffic management mechanism used in ATM networks that determines whether the network can offer a path with sufficient bandwidth for a requested VCC.

call priority   Priority assigned to each origination port in circuit-switched systems. This priority defines the order in which calls are reconnected. Call priority also defines which calls can or cannot be placed during a bandwidth reservation.

call setup time   The time required to establish a switched call between DTE devices.

CAM   content addressable memory.

carrier   Electromagnetic wave or alternating current of a single frequency, suitable for modulation by another data-bearing signal.

Category 1 cabling   One of five grades of UTP cabling described in the EIA/TIA-586 standard. Category 1 cabling is used for telephone communications and is not suitable for transmitting data.

Category 2 cabling   One of five grades of UTP cabling described in the EIA/TIA-586 standard. Category 2 cabling is capable of transmitting data at speeds up to 4 Mbps.

Category 3 cabling One of five grades of UTP cabling described in the EIA/TIA-586 standard. Category 3 cabling is used in 10Base-T networks and can transmit data at speeds up to 10 Mbps.

Category 4 cabling One of five grades of UTP cabling described in the EIA/TIA-586 standard. Category 4 cabling is used in Token Ring networks and can transmit data at speeds up to 16 Mbps.

Category 5 cabling One of five grades of UTP cabling described in the EIA/TIA-586 standard. Category 5 cabling is used for running CDDI and can transmit data at speeds up to 100 Mbps.

catenet Network in which hosts are connected to diverse networks, which themselves are connected with routers. The Internet is a prominent example of a catenet.

CATV (cable television) Communication system where multiple channels of programming material are transmitted to homes using broadband coaxial cable. Formerly called *Community Antenna Television.*

CBDS (Connectionless Broadband Data Service) European high speed, packet-switched, datagram based WAN networking technology. Similar to SMDS.

CBR (constant bit rate) QoS class defined by the ATM Forum for ATM networks. CBR is used for connections that depend on precise clocking to ensure undistorted delivery.

CCITT (Consultative Committee for International Telegraph and Telephone) International organization responsible for the development of communications standards. Now called the ITU-T.

CCS (common channel signaling) Signaling system used in telephone networks that separates signaling information from user data. A specified channel is exclusively designated to carry signaling information for all other channels in the system.

CD (Carrier Detect) Signal that indicates whether an interface is active. Also, a signal generated by a modem indicating that a call has been connected.

CDDI (Copper Distributed Data Interface) Implementation of FDDI protocols over STP and UTP cabling. CDDI transmits over relatively short distances (about 100 meters), providing data rates of 100 Mbps using a dual-ring architecture to provide redundancy. Based on the ANSI Twisted-Pair Physical Medium Dependent (TPPMD) standard.

CDPD (Cellular Digital Packet Data) Open standard for two-way wireless data communication over high-frequency cellular telephone channels. Allows data transmissions between a remote cellular link and a NAP. Operates at 19.2 Kbps.

CDVT (cell delay variation tolerance) Parameter defined by the ATM Forum for ATM traffic management. In CBR transmissions, determines the level of jitter that is tolerable for the data samples taken by the PCR.

cell The basic unit for ATM switching and multiplexing. Cells contain identifiers that specify the data stream to which they belong. Each cell consists of a 5-byte header and 48 bytes of payload.

cell loss priority Field in the ATM cell header that determines the probability of a cell being dropped if the network becomes congested. Cells with CLP = 0 are insured traffic, which is unlikely to be dropped. Cells with CLP = 1 are best-effort traffic, which might be dropped in congested conditions in order to free up resources to handle insured traffic.

cell relay Network technology based on the use of small, fixed size packets, or cells. Because cells are fixed length, they can be processed and switched in hardware at high speeds. Cell relay is the basis for many high speed network protocols including ATM, IEEE 802.6, and SMDS.

cellular radio Technology that uses radio transmissions to access telephone company networks. Service is provided in a particular area by a low power transmitter.

Centrex AT&T PBX that provides direct inward dialing and automatic number identification of the calling PBX.

CEPT (Conference Européenne des Postes et des Télécommunications) Association of the 26 European PTTs that recommends communication specifications to the ITU-T.

CERFnet (California Education and Research Federation Network) TCP/IP network, based in Southern California, that connects hundreds of higher-education centers internationally while also providing Internet access to subscribers. CERFnet was founded in 1988 by the San Diego Supercomputer Center and General Atomics and is funded by the NSF.

chaining SNA concept in which RUs are grouped together for the purpose of error recovery.

channel (1) A communication path. Multiple channels can be multiplexed over a single cable in certain environments. (2) In IBM, the specific path between large computers (such as mainframes) and attached peripheral devices.

channel-attached Pertaining to attachment of devices directly by data channels (input/output channels) to a computer.

channelized E1 Access link operating at 2.048 Mbps that is subdivided into 30 B channels and 1 D channel. Supports DDR, Frame Relay, and X.25.

channelized T1 Access link operating at 1.544 Mbps that is subdivided into 24 channels (23 B channels and 1 D channel) of 64 Kbps each. The individual channels or groups of channels connect to different destinations. Supports DDR, Frame Relay, and X.25. Also referred to as *fractional T1*.

CHAP (Challenge Handshake Authentication Protocol) Security feature supported on lines using PPP encapsulation that prevents unauthorized access. CHAP does not itself prevent unauthorized access, it merely identifies the remote end. The router or access server then determines whether that user is allowed access.

Cheapernet Industry term used to refer to the IEEE 802.3 10Base-2 standard or the cable specified in that standard.

checksum Method for checking the integrity of transmitted data. A checksum is an integer value computed from a sequence of octets taken through a series of arithmetic operations. The value is recomputed at the receiving end and compared for verification.

choke packet Packet sent to a transmitter to tell it that congestion exists and that it should reduce its sending rate.

CICNet Regional network that connects academic, research, nonprofit, and commercial organizations in the midwestern United States. Founded in 1988, CICNet was a part of the NSFNET and was funded by the NSF until the NSFNET dissolved in 1995.

CICS (Customer Information Control System) IBM application subsystem allowing transactions entered at remote terminals to be processed concurrently by user applications.

CIDR (classless interdomain routing) Technique supported by BGP4 and based on route aggregation. CIDR allows routers to group routes together to cut down on the quantity of routing information carried by the core routers. With CIDR, several IP networks appear to networks outside the group as a single, larger entity.

CIP (Channel Interface Processor) Channel attachment interface for Cisco 7000 series routers. The CIP is used to connect a host mainframe to a control unit, eliminating the need for an FEP for channel attachment.

CIR (committed information rate) The rate at which a Frame Relay network agrees to transfer information under normal conditions, averaged over a minimum increment of time. CIR, measured in bits per second, is one of the key negotiated tariff metrics.

circuit Communications path between two or more points.

circuit group Grouping of associated serial lines that link two bridges. If one of the serial links in a circuit group is in the spanning tree for a network, any of the serial links in the circuit group can be used for load balancing. This load balancing strategy avoids data ordering problems by assigning each destination address to a particular serial link.

circuit switching Switching system in which a dedicated physical circuit path must exist between sender and receiver for the duration of the "call." Used heavily in the telephone company network. Circuit switching can be contrasted with contention and token passing as a channel-access method, and with message switching and packet switching as a switching technique.

classical IP over ATM Specification for running IP over ATM in a manner that takes full advantage of the features of ATM. Defined in RFC 1577. Sometimes called *CIA*.

CLAW (Common Link Access for Workstations) Data link layer protocol used by channel-attached RISC System/6000 series systems and by IBM 3172 devices running TCP/IP off-load. CLAW improves efficiency of channel use and allows the CIP to provide the functionality of a 3172 in TCP/IP environments and support direct channel attachment. The output from TCP/IP mainframe processing is a series of IP datagrams that the router can switch without modifications.

client Node or software program (front-end device) that requests services from a server.

client-server computing Term used to describe distributed computing (processing) network systems in which transaction responsibilities are divided into two parts: client (front end) and server (back end). Both terms (client and server) can be applied to software programs or actual computing devices. Also called *distributed computing (processing)*.

CLNP (Connectionless Network Protocol) OSI network layer protocol that does not require a circuit to be established before data are transmitted.

CLNS (Connectionless Network Service) OSI network layer service that does not require a circuit to be established before data are transmitted. CLNS routes messages to their destinations independently of any other messages.

cluster controller (1) Generally, an intelligent device that provides the connections for a cluster of terminals to a data link. (2) In SNA, a programmable device that controls the input/output operations of attached devices. Typically, an IBM 3174 or 3274 device.

CMI (coded mark inversion) ITU-T line coding technique specified for STS-3c transmissions. Also used in DS-1 systems.

CMIP (Common Management Information Protocol) OSI network management protocol created and standardized by ISO for the monitoring and control of heterogeneous networks.

CMIS (Common Management Information Services) OSI network management service interface created and standardized by ISO for the monitoring and control of heterogeneous networks.

CMNS (Connection Mode Network Service) Extends local X.25 switching to a variety of media (Ethernet, FDDI, Token Ring).

CMT (connection management) FDDI process that handles the transition of the ring through its various states (off, active, connect, and so on), as defined by the ANSI X3T9.5 specification.

CO (central office) Local telephone company office to which all local loops in a given area connect and in which circuit switching of subscriber lines occurs.

coaxial cable Cable consisting of a hollow outer cylindrical conductor that surrounds a single inner wire conductor. Two types of coaxial cable are currently used in LANs: 50-ohm cable, which is used for digital signaling, and 75-ohm cable, which is used for analog signaling and high speed digital signaling.

CODEC (coder-decoder) Device that typically uses PCM to transform analog signals into a digital bit stream and digital signals back into analog.

coding Electrical techniques used to convey binary signals.

collapsed backbone Nondistributed backbone in which all network segments are interconnected by way of an internetworking device. A collapsed backbone might be a virtual network segment existing in a device such as a hub, a router, or a switch.

collision In Ethernet, the result of two nodes transmitting simultaneously. The frames from each device impact and are damaged when they meet on the physical media.

collision domain In Ethernet, the network area within which frames that have collided are propagated. Repeaters and hubs propagate collisions; LAN switches, bridges, and routers do not.

common carrier Licensed, private utility company that supplies communication services to the public at regulated prices.

communication Transmission of information.

communication controller In SNA, a subarea node (such as an IBM 3745 device) that contains an NCP.

communication server Communications processor that connects asynchronous devices to a LAN or WAN through network and terminal emulation software. Performs only asynchronous routing of IP and IPX.

communications line The physical link (such as wire or a telephone circuit) that connects one or more devices to one or more other devices.

community In SNMP, a logical group of managed devices and NMSs in the same administrative domain.

Community Antenna Television Now known as CATV.

community string Text string that acts as a password and is used to authenticate messages sent between a management station and a router containing an SNMP agent. The community string is sent in every packet between the manager and the agent.

companding Contraction derived from the opposite processes of compression and expansion. Part of the PCM process whereby analog signal values are logically rounded to discrete scale-step values on a nonlinear scale. The decimal step number is then coded in its binary equivalent prior to transmission. The process is reversed at the receiving terminal using the same nonlinear scale.

compression The running of a data set through an algorithm that reduces the space required to store or the bandwidth required to transmit the data set.

configuration management One of five categories of network management defined by ISO for management of OSI networks. Configuration management subsystems are responsible for detecting and determining the state of a network.

configuration register In routers, a 16-bit, user-configurable value that determines how the router functions during initialization. The configuration register can be stored in hardware or software. In hardware, the bit position is set using a jumper. In software, the bit position is set by specifying a hexadecimal value using configuration commands.

congestion Traffic in excess of network capacity.

congestion avoidance The mechanism by which an ATM network controls traffic entering the network to minimize delays. To use resources most efficiently, lower priority traffic is discarded at the edge of the network if conditions indicate that it cannot be delivered. Sometimes abbreviated *CA*.

connectionless Term used to describe data transfer without the existence of a virtual circuit.

connection oriented Term used to describe data transfer that requires the establishment of a virtual circuit.

CONP (Connection Oriented Network Protocol) OSI protocol providing connection oriented operation to upper layer protocols.

console DTE through which commands are entered into a host.

contention Access method in which network devices compete for permission to access the physical medium. Contrast with circuit switching and token passing.

convergence The speed and ability of a group of internetworking devices running a specific routing protocol to agree on the topology of an internetwork after a change in that topology.

conversation In SNA, an LU 6.2 session between two transaction programs.

core gateway The primary routers in the Internet.

core router In a packet-switched star topology, a router that is part of the backbone and that serves as the single pipe through which all traffic from peripheral networks must pass on its way to other peripheral networks.

COS (1) class of service. Indication of how an upper layer protocol requires that a lower layer protocol treat its messages. In SNA subarea routing, COS definitions are used by subarea nodes to determine the optimal route to establish a given session. A COS definition comprises a virtual route number and a transmission priority field. Also called *TOS (type of service)*. (2) Corporation for Open Systems. Organization that promulgates the use of OSI protocols through conformance testing, certification, and related activities.

COSINE (Cooperation for Open Systems Interconnection Networking in Europe) European project financed by the European Community (EC) to build a communication network between scientific and industrial entities in Europe. The project ended in 1994.

cost Arbitrary value typically based on hop count, media bandwidth, or other measures, which is assigned by a network administrator and used to compare various paths through an internetwork environment. Cost values are used by routing protocols to determine the most favorable path to a particular destination: the lower the cost, the better the path. Sometimes called *path cost*.

count to infinity Problem that can occur in routing algorithms that are slow to converge, in which routers continually increment the hop count to particular networks. Typically, an arbitrary hop-count limit is imposed to prevent this problem.

CP (control point) In SNA networks, element that identifies the APPN networking components of a PU 2.1 node, manages device resources, and can provide services to other devices. In APPN, CPs are able to communicate with logically adjacent CPs by way of CP-to-CP sessions.

CPCS (common part convergence sublayer) One of the two sublayers of any AAL. The CPCS is service independent and is further divided into the CS and the SAR sublayers. The CPCS is responsible for preparing data for transport across the ATM network, including the creation of the 48-byte payload cells that are passed to the ATM layer.

CPE (customer premises equipment) Terminating equipment such as terminals, telephones, and modems that is supplied by the telephone company, installed at customer sites, and connected to the telephone company network.

CPI-C (Common Programming Interface for Communications) Platform independent API developed by IBM and used to provide portability in APPC applications.

cps cells per second.

CRC (cyclic redundancy check) Error checking technique in which the frame recipient calculates a remainder by dividing frame contents by a prime binary divisor and compares the calculated remainder to a value stored in the frame by the sending node.

CREN (Corporation for Research and Educational Networking) The result of a merger of BITNET and CSNET. CREN is devoted to providing Internet connectivity to its members, which include the alumni, students, faculty, and other affiliates of participating educational and research institutions, via BITNET III.

crosstalk Interfering energy transferred from one circuit to another.

CS (convergence sublayer) One of the two sublayers of the AAL CPCS, responsible for padding and error checking. PDUs passed from the SSCS are appended with an 8-byte trailer (for error checking and other control information) and padded, if necessary, so that the length of the resulting PDU is divisible by 48. These PDUs are then passed to the SAR sublayer of the CPCS for further processing.

CSA (Canadian Standards Association) Agency within Canada that certifies products that conform to Canadian national safety standards.

CSLIP (Compressed Serial Link Internet Protocol) Extension of SLIP that when appropriate allows just header information to be sent across a SLIP connection, reducing overhead and increasing packet throughput on SLIP lines.

CSMA/CD (carrier sense multiple access collision detect) Media access mechanism wherein devices ready to transmit data first check the channel for a carrier. If no carrier is sensed for a specific period of time, a device can transmit. If two devices transmit at once, a collision occurs and is detected by all colliding devices. This collision subsequently delays retransmissions from those devices for some random length of time. CSMA/CD access is used by Ethernet and IEEE 802.3.

CSNET (Computer Science Network) Large internetwork consisting primarily of universities, research institutions, and commercial concerns. CSNET merged with BITNET to form CREN.

CSNP (complete sequence number PDU) PDU sent by the designated router in an OSPF network to maintain database synchronization.

CSU (channel service unit) Digital interface device that connects end user equipment to the local digital telephone loop. Often referred to together with DSU as *CSU/DSU*.

CTS (1) Clear To Send. Circuit in the EIA/TIA-232 specification that is activated when DCE is ready to accept data from DTE. (2) common transport semantic. Cornerstone of the IBM strategy to reduce the number of protocols on networks. CTS provides a single API for developers of network software and enables applications to run over APPN, OSI, or TCP/IP.

cut-through packet switching Packet-switching approach that streams data through a switch so that the leading edge of a packet exits the switch at the output port before the packet finishes entering the input port. A device using cut-through packet switching reads, processes, and forwards packets as soon as the destination address is looked up and the outgoing port determined. Also known as *on-the-fly packet switching*. Contrast with store-and-forward packet switching.

D

DAC (dual-attached concentrator) FDDI or CDDI concentrator capable of attaching to both rings of an FDDI or CDDI network. It can also be dual-homed from the master ports of other FDDI or CDDI concentrators.

DARPA (Defense Advanced Research Projects Agency) U.S. government agency that funded research for and experimentation with the Internet. Evolved from ARPA, and then, in 1994, back to ARPA.

DAS (dual attachment station) Device attached to both the primary and the secondary FDDI rings. Dual attachment provides redundancy for the FDDI ring: If the primary ring fails, the station can wrap the primary ring to the secondary ring, isolating the failure and retaining ring integrity. Also known as a *Class A station*.

database object In general, a piece of information that is stored in a database.

data flow control layer Layer 5 of the SNA architectural model. This layer determines and manages interactions between session partners, particularly data flow. Corresponds to the session layer of the OSI model.

datagram Logical grouping of information sent as a network layer unit over a transmission medium without prior establishment of a virtual circuit. IP datagrams are the primary information units in the Internet. The terms frame, message, packet, and segment are also used to describe logical information groupings at various layers of the OSI reference model and in various technology circles.

data link control layer Layer 2 in the SNA architectural model. Responsible for the transmission of data over a particular physical link. Corresponds roughly to the data link layer of the OSI model.

data link layer Layer 2 of the OSI reference model. This layer provides reliable transit of data across a physical link. The data link layer is concerned with physical addressing, network topology, line discipline, error notification, ordered delivery of frames, and flow control. The IEEE has divided this layer into two sublayers: the MAC sublayer and the LLC sublayer. Sometimes simply called *link layer*. Roughly corresponds to the data link control layer of the SNA model.

data sink Network equipment that accepts data transmissions.

data stream All data transmitted through a communications line in a single read or write operation.

dB decibels.

DB connector (data bus connector) Type of connector used to connect serial and parallel cables to a data bus. DB connector names are of the format DB-*x*, where *x* represents the number of (wires) within the connector. Each line is connected to a pin on the connector, but in many cases not all pins are assigned a function. DB connectors are defined by various EIA/TIA standards.

DCA   (Defense Communications Agency) U.S. government organization responsible for DDN networks such as MILNET. Now called DISA.

DCC   (Data Country Code) One of two ATM address formats developed by the ATM Forum for use by private networks. Adapted from the subnetwork model of addressing in which the ATM layer is responsible for mapping network layer addresses to ATM addresses.

DCE   (data communications equipment [EIA expansion] or data circuit-terminating equipment [ITU-T expansion]) The devices and connections of a communications network that comprise the network end of the user-to-network interface. The DCE provides a physical connection to the network, forwards traffic, and provides a clocking signal used to synchronize data transmission between DCE and DTE devices. Modems and interface cards are examples of DCE.

D channel   (1) data channel. Full-duplex, 16-Kbps (BRI) or 64-Kbps (PRI) ISDN channel. (2) In SNA, a device that connects a processor and main storage with peripherals.

DDM   (Distributed Data Management) Software in an IBM SNA environment that provides peer-to-peer communication and file sharing. One of three SNA transaction services.

DDN   (Defense Data Network) U.S. military network composed of an unclassified network (MILNET) and various secret and top-secret networks. DDN is operated and maintained by *DISA*.

DDP   (Datagram Delivery Protocol) Apple Computer network layer protocol that is responsible for the socket-to-socket delivery of datagrams over an AppleTalk internetwork.

DDR   (dial-on-demand routing) Technique whereby a Cisco router can automatically initiate and close a circuit-switched session as transmitting stations demand. The router spoofs keepalives so that end stations treat the session as active. DDR permits routing over ISDN or telephone lines using an external ISDN terminal adaptor or modem.

DE   discard eligible.

deadlock   (1) Unresolved contention for the use of a resource. (2) In APPN, when two elements of a process each wait for action by or a response from the other before they resume the process.

DECnet   Group of communications products (including a protocol suite) developed and supported by Digital Equipment Corporation. DECnet/OSI (also called DECnet Phase V) is the most recent iteration and supports both OSI protocols and proprietary Digital protocols. Phase IV Prime supports inherent MAC addresses that allow DECnet nodes to coexist with systems running other protocols that have MAC address restrictions.

DECnet routing   Proprietary routing scheme introduced by Digital Equipment Corporation in DECnet Phase III. In DECnet Phase V, DECnet completed its transition to OSI routing protocols (ES-IS and IS-IS).

decryption   The reverse application of an encryption algorithm to encrypted data, thereby restoring that data to its original, unencrypted state.

dedicated LAN   Network segment allocated to a single device. Used in LAN switched network topologies.

dedicated line   Communications line that is indefinitely reserved for transmissions, rather than switched as transmission is required.

de facto standard   Standard that exists by nature of its widespread use.

default route   Routing table entry that is used to direct frames for which a next hop is not explicitly listed in the routing table.

de jure standard   Standard that exists because of its approval by an official standards body.

delay   The time between the initiation of a transaction by a sender and the first response received by the sender. Also, the time required to move a packet from source to destination over a given path.

demand priority   Media access method used in 100VG-AnyLAN that uses a hub that can handle multiple transmission requests and can process traffic according to priority, making it useful for servicing time-sensitive traffic such as multimedia and video. Demand priority eliminates the overhead of packet collisions, collision recovery, and broadcast traffic typical in Ethernet networks.

demarc   Demarcation point between carrier equipment and CPE.

demodulation   Process of returning a modulated signal to its original form. Modems perform demodulation by taking an analog signal and returning it to its original (digital) form.

demultiplexing   The separating of multiple input streams that have been multiplexed into a common physical signal back into multiple output streams.

DES (Data Encryption Standard) Standard cryptographic algorithm developed by the U.S. NBS.

designated bridge The bridge that incurs the lowest path cost when forwarding a frame from a segment to the route bridge.

designated router OSPF router that generates LSAs for a multiaccess network and has other special responsibilities in running OSPF. Each multiaccess OSPF network that has at least two attached routers has a designated router that is elected by the OSPF Hello protocol. The designated router enables a reduction in the number of adjacencies required on a multiaccess network, which in turn reduces the amount of routing protocol traffic and the size of the topological database.

destination address Address of a network device that is receiving data.

deterministic load distribution Technique for distributing traffic between two bridges across a circuit group. Guarantees packet ordering between source-destination pairs and always forwards traffic for a source-destination pair on the same segment in a circuit group for a given circuit-group configuration.

DIA (Document Interchange Architecture) Defines the protocols and data formats needed for the transparent interchange of documents in an SNA network. One of three SNA transaction services.

dial backup Feature supported by Cisco routers that provides protection against WAN downtime by allowing the network administrator to configure a backup serial line through a circuit-switched connection.

dial-up line Communications circuit that is established by a switched-circuit connection using the telephone company network.

differential encoding Digital encoding technique whereby a binary value is denoted by a signal change rather than a particular signal level.

differential Manchester encoding Digital coding scheme where a mid-bit-time transition is used for clocking, and a transition at the beginning of each bit time denotes a zero. The coding scheme used by IEEE 802.5 and Token Ring networks.

digital signal level 1 Framing specification used in transmitting digital signals at 1.544 Mbps on a T1 facility (in the United States) or at 2.108 Mbps on an E1 facility (in Europe).

DIN (Deutsche Industrie Norm) German national standards organization.

DIN connector (Deutsche Industrie Norm connector) Multipin connector used in some Macintosh and IBM PC-compatible computers and on some network processor panels.

directed search Search request sent to a specific node known to contain a resource. A directed search is used to determine the continued existence of the resource and to obtain routing information specific to the node.

directory services Services that help network devices locate service providers.

DISA (Defense Information Systems Agency) U.S. military organization responsible for implementing and operating military information systems, including the DDN.

discovery architecture APPN software that enables a machine configured as an APPN EN to automatically find primary and backup NNs when the machine is brought onto an APPN network.

discovery mode Method by which an AppleTalk interface acquires information about an attached network from an operational node and then uses this information to configure itself. Also called *dynamic configuration*.

distance vector routing algorithm Class of routing algorithms that iterate on the number of hops in a route to find a shortest path spanning tree. Distance vector routing algorithms call for each router to send its entire routing table in each update, but only to its neighbors. Distance vector routing algorithms can be prone to routing loops, but are computationally simpler than link state routing algorithms. Also called *Bellman-Ford routing algorithm*.

distortion delay Problem with a communication signal resulting from nonuniform transmission speeds of the components of a signal through a transmission medium. Also called *group delay*.

DLCI (data link connection identifier) Value that specifies a PVC or SVC in a Frame Relay network. In the basic Frame Relay specification, DLCIs are locally significant (connected devices might use different values to specify the same connection). In the LMI extended specification, DLCIs are globally significant (DLCIs specify individual end devices).

DLSw (data link switching) Interoperability standard, described in RFC 1434, that provides a method for forwarding SNA and NetBIOS traffic over TCP/IP networks using data link layer switching and encapsulation. DLSw uses SSP (Switch-to-Switch Protocol) instead of SRB, eliminating the major limitations of SRB, including hop-count limits, broadcast and unnecessary traffic, time-outs, lack of flow control, and lack of prioritization schemes.

DLSw+ (Data Link Switching Plus) Cisco implementation of the DLSw standard for SNA and NetBIOS traffic forwarding. DLSw+ goes beyond the standard to include the advanced features of the current Cisco RSRB implementation, and provides additional functionality to increase the overall scalability of data link switching.

DLU (Dependent LU) An LU that depends on the SSCP to provide services for establishing sessions with other LUs.

DLUR (Dependent LU Requester) The client half of the Dependent LU Requestor/Server enhancement to APPN. The DLUR component resides in APPN ENs and NNs that support adjacent DLUs by securing services from the DLUS.

DLUR node In APPN networks, an EN or NN that implements the DLUR component.

DLUS (Dependent LU Server) The server half of the Dependent LU Requester/Server enhancement to APPN. The DLUS component provides SSCP services to DLUR nodes over an APPN network.

DLUS node In APPN networks, a NN that implements the DLUS component.

DMA (direct memory access) The transfer of data from a peripheral device, such as a hard disk drive, into memory without that data passing through the microprocessor. DMA transfers data into memory at high speeds with no processor overhead.

DMAC (destination MAC) The MAC address specified in the Destination Address field of a packet.

DNA (Digital Network Architecture) Network architecture developed by Digital Equipment Corporation. The products that embody DNA (including communications protocols) are collectively referred to as DECnet.

DNIC (Data Network Identification Code) Part of an X.121 address. DNICs are divided into two parts: the first specifying the country in which the addressed PSN is located and the second specifying the PSN itself.

DNS (domain naming system) System used in the Internet for translating names of network nodes into addresses.

DNSIX (Department of Defense Intelligence Information System Network Security for Information Exchange) Collection of security requirements for networking defined by the U.S. Defense Intelligence Agency.

DoD (Department of Defense) U.S. government organization that is responsible for national defense. The

DoD has frequently funded communication protocol development.

domain (1) In the Internet, a portion of the naming hierarchy tree that refers to general groupings of networks based on organization type or geography. (2) In SNA, an SSCP and the resources it controls. (3) In IS-IS, a logical set of networks.

Domain Networking system developed by Apollo Computer (now part of Hewlett-Packard) for use in its engineering workstations.

dot address Refers to the common notation for IP addresses in the form <*n.n.n.n*> where each number *n* represents, in decimal, 1 byte of the 4-byte IP address. Also called *dotted notation* or *four-part dotted notation*.

DQDB (Distributed Queue Dual Bus) Data link layer communication protocol specified in the IEEE 802.6 standard and designed for use in MANs. DQDB, which permits multiple systems to interconnect using two unidirectional logical buses, is an open standard that is designed for compatibility with carrier transmission standards and is aligned with emerging standards for B-ISDN. SIP (SMDS Interface Protocol) is based on DQDB.

DRAM (dynamic random access memory) RAM that stores information in capacitors that must be periodically refreshed. Delays can occur because DRAMs are inaccessible to the processor when refreshing their contents. However, DRAMs are less complex and have greater capacity than SRAMs.

drop Point on a multipoint channel where a connection to a networked device is made.

drop cable Generally, a cable that connects a network device (such as a computer) to a physical medium. A type of AUI.

DS-0 (digital signal level 0) Framing specification used in transmitting digital signals over a single channel at 64 Kbps on a T1 facility.

DS-1/DTI (DS-1 domestic trunk interface) Interface circuit used for DS-1 applications with 24 trunks.

DS-3 (digital signal level 3) Framing specification used for transmitting digital signals at 44.736 Mbps on a T3 facility.

DSAP (destination service access point) The SAP of the network node designated in the Destination field of a packet.

DSP (domain specific part) The part of a CLNS address that contains an area identifier, a station identifier, and a selector byte.

DSPU (1) downstream physical unit. In SNA, a PU that is located downstream from the host. (2) Cisco IOS software feature that enables a router to function as a PU concentrator for SNA PU 2 nodes. PU concentration at the router simplifies the task of PU definition at the upstream host while providing additional flexibility and mobility for downstream PU devices. This feature is sometimes referred to as *DSPU concentration*.

DSR (data set ready) EIA/TIA-232 interface circuit that is activated when DCE is powered up and ready for use.

DSU (data service unit) Device used in digital transmission that adapts the physical interface on a DTE device to a transmission facility such as T1 or E1. The DSU is also responsible for such functions as signal timing. Often referred to together with CSU as *CSU/DSU*.

DSX-1 Cross-connection point for DS-1 signals.

DTE (data terminal equipment) Device at the user end of a user-network interface that serves as a data source, destination, or both. DTE connects to a data network through a DCE device (for example, a modem) and typically uses clocking signals generated by the DCE. DTE includes such devices as computers, protocol translators, and multiplexers.

DTMF (dual tone multifrequency) Use of two simultaneous voice-band tones for dialing (such as touch tone).

DTR (data terminal ready) EIA/TIA-232 circuit that is activated to let the DCE know when the DTE is ready to send and receive data.

DUAL (Diffusing Update Algorithm) Convergence algorithm used in Enhanced IGRP that provides loop-free operation at every instant throughout a route computation. Allows routers involved in a topology change to synchronize at the same time, while not involving routers that are unaffected by the change.

dual counter-rotating rings Network topology in which two signal paths, whose directions are opposite one another, exist in a token-passing network. FDDI and CDDI are based on this concept.

dual-homed station Device attached to multiple FDDI rings to provide redundancy.

dual homing Network topology in which a device is connected to the network by way of two independent access points (points of attachment). One access point is the primary connection, and the other is a standby connection that is activated in the event of a failure of the primary connection.

DVMRP (Distance Vector Multicast Routing Protocol) Internetwork gateway protocol, largely based on RIP, that implements a typical dense mode IP multicast scheme. DVMRP uses IGMP to exchange routing datagrams with its neighbors.

DXI (Data Exchange Interface) ATM Forum specification described in RFC 1483 that defines how a network device such as a bridge, router, or hub can effectively act as an FEP to an ATM network by interfacing with a special DSU that performs packet segmentation and reassembly.

dynamic address resolution Use of an address resolution protocol to determine and store address information on demand.

dynamic routing Routing that adjusts automatically to network topology or traffic changes. Also called *adaptive routing*.

E

E1 Wide area digital transmission scheme used predominantly in Europe that carries data at a rate of 2.048 Mbps. E1 lines can be leased for private use from common carriers.

E.164 ITU-T recommendation for international telecommunication numbering, especially in ISDN, B-ISDN, and SMDS. An evolution of standard telephone numbers.

E3 Wide area digital transmission scheme used predominantly in Europe that carries data at a rate of 34.368 Mbps. E3 lines can be leased for private use from common carriers.

early token release Technique used in Token Ring networks that allows a station to release a new token onto the ring immediately after transmitting, instead of waiting for the first frame to return. This feature can increase the total bandwidth on the ring.

EARN (European Academic Research Network) European network connecting universities and research institutes. EARN merged with RARE to form TERENA.

EBCDIC (extended binary coded decimal interchange code) Any of a number of coded character sets developed by IBM consisting of 8-bit coded characters. This character code is used by older IBM systems and telex machines.

E channel (echo channel) A 64-Kbps ISDN circuit-switching control channel. The E channel was defined in the 1984 ITU-T ISDN specification but was dropped in the 1988 specification.

echoplex Mode in which keyboard characters are echoed on a terminal screen upon return of a signal from the other end of the line indicating that the characters were received correctly.

ECMA (European Computer Manufacturers Association) Group of European computer vendors who have done substantial OSI standardization work.

EDI (electronic data interchange) The electronic communication of operational data such as orders and invoices between organizations.

EDIFACT (Electronic Data Interchange for Administration, Commerce, and Transport) Data exchange standard administered by the United Nations to be a multi-industry EDI standard.

EEPROM (electrically erasable programmable read only memory) EPROM that can be erased using electrical signals applied to specific pins.

EGP (Exterior Gateway Protocol) Internet protocol for exchanging routing information between autonomous systems. Documented in RFC 904. Not to be confused with the general term *exterior gateway protocol*. EGP is an obsolete protocol that has been replaced by BGP.

EIA (Electronic Industries Association) Group that specifies electrical transmission standards. The EIA and TIA have developed numerous well-known communications standards, including EIA/TIA-232 and EIA/TIA-449.

EIA-530 Refers to two electrical implementations of EIA/TIA-449: RS-422 (for balanced transmission) and RS-423 (for unbalanced transmission).

EIA/TIA-232 Common physical layer interface standard developed by EIA and TIA that supports unbalanced circuits at signal speeds of up to 64 kbps. Closely resembles the V.24 specification. Formerly known as *RS-232*.

EIA/TIA-449 Popular physical layer interface developed by EIA and TIA. Essentially, a faster (up to 2 Mbps) version of EIA/TIA-232 capable of longer cable runs. Formerly called *RS-449*.

EIA/TIA-586 Standard that describes the characteristics and applications for various grades of UTP cabling.

EIP (Ethernet Interface Processor) Interface processor card on the Cisco 7000 series routers. The EIP provides high speed (10-Mbps) AUI ports that support Ethernet Version 1 and Ethernet Version 2 or IEEE 802.3 interfaces and a high speed data path to other interface processors.

EISA (Extended Industry Standard Architecture) A 32-bit bus interface used in PCs, PC-based servers, and some UNIX workstations and servers.

ELAN (emulated LAN) ATM network in which an Ethernet or Token Ring LAN is emulated using a client-server model. ELANs are composed of an LEC, an LES, a BUS, and an LECS. Multiple ELANs can exist simultaneously on a single ATM network. ELANs are defined by the LANE specification.

electronic mail Widely used network application in which mail messages are transmitted electronically between end users over various types of networks using various network protocols. Often called *e-mail*.

EMA (1) Enterprise Management Architecture. Digital Equipment Corporation network management architecture based on the OSI network management model. (2) Electronic Messaging Association. Forum devoted to standards and policy work, education, and development of electronic messaging systems such as electronic mail, voice mail, and facsimile.

EMI (electromagnetic interference) Interference by electromagnetic signals that can cause reduced data integrity and increased error rates on transmission channels.

EMIF (ESCON Multiple Image Facility) Mainframe I/O software function that allows one ESCON channel to be shared among multiple logical partitions on the same mainframe.

EMP (electromagnetic pulse) Caused by lightning and other high energy phenomena. Capable of coupling enough energy into unshielded conductors to destroy electronic devices.

emulation mode Function of an NCP that enables it to perform activities equivalent to those performed by a transmission control unit.

EN (end node) APPN end system that implements the PU 2.1, provides end user services, and supports sessions between local and remote CPs. ENs are not capable of routing traffic and rely on an adjacent NN for APPN services.

encapsulation The wrapping of data in a particular protocol header. For example, Ethernet data is wrapped in a specific Ethernet header before network transit. Also, when bridging dissimilar networks, the entire frame from one network is simply placed in the header used by the data link layer protocol of the other network.

encapsulation bridging Carries Ethernet frames from one router to another across disparate media, such as serial and FDDI lines.

encoder Device that modifies information into the required transmission format.

encryption The application of a specific algorithm to data so as to alter the appearance of the data making it incomprehensible to those who are not authorized to see the information.

end point Device at which a virtual circuit or virtual path begins or ends.

Enhanced IGRP (Enhanced Interior Gateway Routing Protocol) Advanced version of IGRP developed by Cisco. Provides superior convergence properties and operating efficiency, and combines the advantages of link state protocols with those of distance vector protocols.

Enhanced Monitoring Services Set of analysis tools on the Catalyst 5000 switch, consisting of an integrated RMON agent and the SPAN. These tools provide traffic monitoring, and network segment analysis and management.

enterprise network Large and diverse network connecting most major points in a company or other organization. Differs from a WAN in that it is privately owned and maintained. Generally, an individual, manageable network device. Sometimes called an *alias*.

EOT (end of transmission) Generally, a character that signifies the end of a logical group of characters or bits.

EPROM (erasable programmable read only memory) Nonvolatile memory chips that are programmed after they are manufactured, and, if necessary, can be erased by some means and reprogrammed.

equalization Technique used to compensate for communications channel distortions.

error control Technique for detecting and correcting errors in data transmissions.

error-correcting code Code having sufficient intelligence and incorporating sufficient signaling information to enable the detection and correction of many errors at the receiver.

error-detecting code Code that can detect transmission errors through analysis of received data based on the adherence of the data to appropriate structural guidelines.

ES (1) end system. Generally, an end user device on a network. (2) end system. Nonrouting host or node in an OSI network.

ESCON (Enterprise System Connection) IBM channel architecture that specifies a pair of fiber optic cables, with either LEDs or lasers as transmitters and a signaling rate of 200 Mbps.

ESCON channel IBM channel for attaching mainframes to peripherals such as storage devices, backup units, and network interfaces. This channel incorporates fiber channel technology. The ESCON channel replaces the bus and tag channel.

ESD (electrostatic discharge) Discharge of stored static electricity that can damage electronic equipment and impair electrical circuitry, resulting in complete or intermittent failures.

ESF (Extended Superframe Format) Framing type used on T1 circuits that consists of 24 frames of 192 bits each, with the 193rd bit providing timing and other functions. ESF is an enhanced version of SF.

ES-IS (End System-to-Intermediate System) OSI protocol that defines how end systems (hosts) announce themselves to intermediate systems (routers).

ESnet (Energy Sciences Network) Data communications network managed and funded by the U.S. Department of Energy Office of Energy Research (DOE/OER). Interconnects the DOE to educational institutions and other research facilities.

Ethernet Baseband LAN specification invented by Xerox Corporation and developed jointly by Xerox, Intel, and Digital Equipment Corporation. Ethernet networks use CSMA/CD and run over a variety of cable types at 10 Mbps. Ethernet is similar to the IEEE 802.3 series of standards.

EtherTalk AppleTalk protocols running on Ethernet.

ETSI (European Telecommunication Standards Institute) Organization created by the European PTTs and the European Community (EC) to propose telecommunications standards for Europe.

EUnet (European Internet) European commercial Internet service provider. EUnet is designed to provide electronic mail, news, and other Internet services to European markets.

event Network message indicating operational irregularities in physical elements of a network or a response to the occurrence of a significant task, typically the completion of a request for information.

excess rate Traffic in excess of the insured rate for a given connection. Specifically, the excess rate equals the maximum rate minus the insured rate. Excess traffic is delivered only if network resources are available and can be discarded during periods of congestion.

EXEC The interactive command processor of the Cisco IOS software.

expansion The process of running a compressed data set through an algorithm that restores the data set to its original size.

expedited delivery Option set by a specific protocol layer telling other protocol layers (or the same protocol layer in another network device) to handle specific data more rapidly.

explicit route   In SNA, a route from a source subarea to a destination subarea as specified by a list of subarea nodes and transmission groups that connect the two.

explorer frame   Frame sent out by a networked device in a SRB environment to determine the optimal route to another networked device.

explorer packet   Generated by an end station trying to find its way through an SRB network. Gathers a hop-by-hop description of a path through the network by being marked (updated) by each bridge that it traverses, thereby creating a complete topological map.

exterior gateway protocol   Any internetwork protocol used to exchange routing information between autonomous systems. Not to be confused with *Exterior Gateway Protocol (EGP),* which is a particular instance of an exterior gateway protocol.

exterior router   Router connected to an AURP tunnel, responsible for the encapsulation and deencapsulation of AppleTalk packets in a foreign protocol header (for example, IP).

F

failure domain   Area in which a failure has occurred in a Token Ring, defined by the information contained in a beacon. When a station detects a serious problem with the network (such as a cable break), it sends a beacon frame that includes the station reporting the failure, its NAUN, and everything in between. Beaconing in turn initiates a process called autoreconfiguration.

fan-out unit   Device that allows multiple devices on a network to communicate using a single network attachment.

fast Ethernet   Any of a number of 100-Mbps Ethernet specifications. Fast Ethernet offers a speed increase ten times that of the 10Base-T Ethernet specification while preserving such qualities as frame format, MAC mechanisms, and MTU. Such similarities allow the use of existing 10Base-T applications and network management tools on fast Ethernet networks. Based on an extension to the IEEE 802.3 specification.

fast switching   Switch feature whereby a route cache is used to expedite packet switching through a router.

fault management   One of five categories of network management defined by IOS for management of OSI net-

works. Fault management attempts to ensure that network faults are detected and controlled.

FCC   (Federal Communications Commission) U.S. government agency that supervises, licenses, and controls electronic and electromagnetic transmission standards.

FCS   (frame check sequence) Refers to the extra characters added to a frame for error control purposes. Used in HDLC, Frame Relay, and other data link layer protocols.

FDDI   (Fiber Distributed Data Interface) LAN standard defined by ANSI X3T9.5, specifying a 100-Mbps token passing network using fiber optic cable with transmission distances of up to 2 km. FDDI uses a dual-ring architecture to provide redundancy.

FDDI II   ANSI standard that enhances FDDI. FDDI II provides isochronous transmission for connectionless data circuits and connection oriented voice and video circuits.

FDM   (frequency-division multiplexing) Technique whereby information from multiple channels can be allocated bandwidth on a single wire based on frequency.

FECN   (forward explicit congestion notification) Bit set by a Frame Relay network to inform DTE receiving the frame that congestion was experienced in the path from source to destination. DTE receiving frames with the FECN bit set can request that higher level protocols take flow-control action as appropriate.

FEIP   (Fast Ethernet Interface Processor) Interface processor on the Cisco 7000 series routers. The FEIP supports up to two 100-Mbps 100Base-T ports.

FEP   (front end processor) Device or board that provides network interface capabilities for a networked device. In SNA, typically an IBM 3745 device.

fiber optic cable   Physical medium capable of conducting modulated light transmission. Compared with other transmission media, fiber optic cable is more expensive but is not susceptible to electromagnetic interference, and is capable of higher data rates. Sometimes called *optical fiber.*

FID1   (format indicator 1) One of several formats that an SNA TH can use. An FID1 TH encapsulates messages between two subarea nodes that do not support virtual and explicit routes.

FID2   (format indicator 2) One of several formats that an SNA TH can use. An FID2 TH is used for transferring messages between a subarea node and a PU 2, using local addresses.

FID3   (format indicator 3) One of several formats that an SNA TH can use. An FID3 TH is used for transferring messages between a subarea node and a PU 1, using local addresses.

FID4 (format indicator 4) One of several formats that an SNA TH can use. An FID4 TH encapsulates messages between two subarea nodes that are capable of supporting virtual and explicit routes.

file transfer Popular network application that allows files to be moved from one network device to another.

filter Generally, a process or device that screens network traffic for certain characteristics, such as source address, destination address, or protocol, and determines whether to forward or discard that traffic based on the established criteria.

FIP (FDDI Interface Processor) Interface processor on the Cisco 7000 series routers. The FIP supports SASs, DASs, dual homing, and optical bypass, and contains a 16-mips processor for high speed (100-Mbps) interface rates. The FIP complies with ANSI and ISO FDDI standards.

firewall Router or access server or several routers or access servers designated as a buffer between any connected public networks and a private network. A firewall router uses access lists and other methods to ensure the security of the private network.

firmware Software instructions set permanently or semipermanently in ROM.

flapping Routing problem where an advertised route between two nodes alternates (flaps) back and forth between two paths due to a network problem that causes intermittent interface failures.

Flash memory Technology developed by Intel and licensed to other semiconductor companies. Flash memory is nonvolatile storage that can be electrically erased and reprogrammed. Allows software images to be stored, booted, and rewritten as necessary.

flash update Routing update sent asynchronously in response to a change in the network topology.

flooding Traffic passing technique used by switches and bridges in which traffic received on an interface is sent out all of the interfaces of that device except the interface on which the information was originally received.

flow Stream of data traveling between two endpoints across a network (for example, from one LAN station to another). Multiple flows can be transmitted on a single circuit.

flow control Technique for ensuring that a transmitting entity, such as a modem, does not overwhelm a receiving entity with data. When the buffers on the receiving device are full, a message is sent to the sending device to suspend the transmission until the data in the buffers has been processed. In IBM networks this technique is called *pacing*.

FM (frequency modulation) Modulation technique in which signals of different frequencies represent different data values.

FNC (Federal Networking Council) Group responsible for assessing and coordinating U.S. federal agency networking policies and needs.

FOIRL (fiber optic interrepeater link) Fiber optic signaling methodology based on the IEEE 802.3 fiber optic specification. FOIRL is a precursor of the 10Base-FL specification, which is designed to replace it.

forward channel Communications path carrying information from the call initiator to the called party.

forward delay interval Amount of time an interface spends listening for topology change information after that interface has been activated for bridging and before forwarding actually begins.

forwarding Process of sending a frame toward its ultimate destination by way of an internetworking device.

Fourier transform Technique used to evaluate the importance of various frequency cycles in a time series pattern.

FRAD (Frame Relay access device) Any network device that provides a connection between a LAN and a Frame Relay WAN.

fragment Piece of a larger packet that has been broken down to smaller units.

fragmentation Process of breaking a packet into smaller units when transmitting over a network medium that cannot support the original size of the packet.

frame Logical grouping of information sent as a data link layer unit over a transmission medium. Often refers to the header and trailer—used for synchronization and error control—that surround the user data contained in the unit. The terms datagram, message, packet, and segment are also used to describe logical information groupings at various layers of the OSI reference model and in various technology circles.

Frame Relay Industry standard, switched data link layer protocol that handles multiple virtual circuits using HDLC encapsulation between connected devices. Frame Relay is more efficient than X.25, the protocol for which it is generally considered a replacement.

Frame Relay bridging Bridging technique described in RFC 1490 that uses the same spanning tree algorithm as other bridging functions, but allows packets to be encapsulated for transmission across a Frame Relay network.

FRAS   (Frame Relay Access Support) A router IOS software feature that allows SDLC, Token Ring, Ethernet, and Frame Relay attached IBM devices to connect to other IBM devices across a Frame Relay network.

free-trade zone   Part of an AppleTalk internetwork that is accessible by two other parts of the internetwork that are unable to directly access one another.

frequency   Number of cycles, measured in hertz, of an alternating current signal per unit time.

front end   Node or software program that requests services of a back end.

FST   (Fast Sequenced Transport) Connectionless, sequenced transport protocol that runs on top of the IP protocol. SRB traffic is encapsulated inside of IP datagrams and is passed over an FST connection between two network devices (such as routers). Speeds up data delivery, reduces overhead, and improves the response time of SRB traffic.

FTAM   (File Transfer, Access, and Management) In OSI, an application layer protocol developed for network file exchange and management between diverse types of computers.

FTP   (File Transfer Protocol) Application protocol, part of the TCP/IP protocol stack, used for transferring files between network nodes. FTP is defined in RFC 959.

full duplex   Capability for simultaneous data transmission between a sending station and a receiving station.

full mesh   Term describing a network in which devices are organized in a mesh topology, with each network node having either a physical circuit or a virtual circuit connecting it to every other network node. A full mesh provides a great deal of redundancy, but because it can be prohibitively expensive to implement, it is usually reserved for network backbones.

Fuzzball   Digital Equipment Corporation LSI-11 computer system running IP gateway software. The NSFnet used these systems as backbone packet switches.

G

G.703/G.704   ITU-T electrical and mechanical specifications for connections between telephone company equipment and DTE using BNC connectors and operating at E1 data rates.

G.804   ITU-T framing standard that defines the mapping of ATM cells into the physical medium.

gateway   In the IP community, an older term referring to a routing device. Today, the term *router* is used to de-scribe nodes that perform this function, and *gateway* refers to a special-purpose device that performs an application layer conversion of information from one protocol stack to another.

gateway host   In SNA, a host node that contains a gateway SSCP.

gateway NCP   NCP that connects two or more SNA networks and performs address translation to allow cross-network session traffic.

GDP   (Gateway Discovery Protocol) Cisco protocol that allows hosts to dynamically detect the arrival of new routers as well as determine when a router goes down. Based on UDP.

GGP   (Gateway-to-Gateway Protocol) MILNET protocol specifying how core routers (gateways) should exchange reachability and routing information. GGP uses a distributed shortest path algorithm.

GNS   (Get Nearest Server) Request packet sent by a client on an IPX network to locate the nearest active server of a particular type. An IPX network client issues a GNS request to solicit either a direct response from a connected server or a response from a router that tells it where on the internetwork the service can be located. GNS is part of the IPX SAP.

GOSIP   (Government OSI Profile) U.S. government procurement specification for OSI protocols. Through GOSIP, the government has mandated that all federal agencies standardize on OSI and implement OSI based systems as they become commercially available.

grade of service   Measure of telephone service quality based on the probability that a call will encounter a busy signal during the busiest hours of the day.

GRE   (generic routing encapsulation) Tunneling protocol developed by Cisco that can encapsulate a wide variety of protocol packet types inside IP tunnels, creating a virtual point-to-point link to Cisco routers at remote points over an IP internetwork. By connecting multiprotocol subnetworks in a single-protocol backbone environment, IP tunneling using GRE allows network expansion across a single-protocol backbone environment.

ground station   Collection of communications equipment designed to receive signals from (and usually transmit signals to) satellites. Also called a *downlink station.*

guard band   Unused frequency band between two communications channels that provides separation of the channels to prevent mutual interference.

GUI   (graphical user interface) User environment that uses pictorial as well as textual representations of the input and output of applications and the hierarchical or other data structure in which information is stored. Conventions such as buttons, icons, and windows are typical, and many actions are performed using a pointing device (such as a mouse). Microsoft Windows and Apple Macintosh are prominent examples of platforms utilizing a GUI.

H

half duplex Capability for data transmission in only one direction at a time between a sending station and a receiving station.

handshake Sequence of messages exchanged between two or more network devices to ensure transmission synchronization.

HBD3 Line code type used on E1 circuits.

H channel (high speed channel) Full-duplex ISDN primary rate channel operating at 384 Kbps.

HDLC (High Level Data Link Control) Bit oriented synchronous data link layer protocol developed by ISO. Derived from SDLC, HDLC specifies a data encapsulation method on synchronous serial links using frame characters and checksums.

headend The end point of a broadband network. All stations transmit toward the headend; the headend then transmits toward the destination stations.

header Control information placed before data when encapsulating that data for network transmission.

HELLO Interior routing protocol used principally by NSFnet nodes. HELLO allows particular packet switches to discover minimal delay routes. Not to be confused with the *Hello protocol*.

Hello packet Multicast packet that is used by routers for neighbor discovery and recovery. Hello packets also indicate that a client is still operating and network-ready.

Hello protocol Protocol used by OSPF systems for establishing and maintaining neighbor relationships. Not to be confused with *HELLO*.

helper address Address configured on an interface to which broadcasts received on that interface will be sent.

hertz Measure of frequency, abbreviated *Hz*. Synonymous with *cycles per second*.

heterogeneous network Network consisting of dissimilar devices that run dissimilar protocols and in many cases support dissimilar functions or applications.

hierarchical routing Routing based on a hierarchical addressing system. For example, IP routing algorithms use

IP addresses, which contain network numbers, subnet numbers, and host numbers.

HIP (HSSI Interface Processor) Interface processor on the Cisco 7000 series routers. The HIP provides one HSSI port that supports connections to ATM, SMDS, Frame Relay, or private lines at speeds up to T3 or E3.

HIPPI (High Performance Parallel Interface) High performance interface standard defined by ANSI. HIPPI is typically used to connect supercomputers to peripherals and other devices.

holddown State into which a route is placed so that routers will neither advertise the route nor accept advertisements about the route for a specific length of time (the holddown period). Holddown is used to flush bad information about a route from all routers in the network. A route is typically placed in holddown when a link in that route fails.

homologation Conformity of a product or specification to international standards, such as ITU-T, CSA, TUV, UL, or VCCI. Enables portability across company and international boundaries.

hop Term describing the passage of a data packet between two network nodes (for example, between two routers).

hop count Routing metric used to measure the distance between a source and a destination. RIP uses hop count as its sole metric.

host Computer system on a network. Similar to the term *node* except that *host* usually implies a computer system, whereas node generally applies to any networked system, including access servers and routers.

host node SNA subarea node that contains an SSCP.

host number Part of an IP address that designates which node on the subnetwork is being addressed. Also called a *host address*.

HPR (high performance routing) Second-generation routing algorithm for APPN. HPR provides a connectionless layer with nondisruptive routing of sessions around link failures, and a connection oriented layer with end-to-end flow control, error control, and sequencing.

HSCI (High Speed Communications Interface) Single-port interface developed by Cisco, providing full-duplex synchronous serial communications capability at speeds up to 52 Mbps.

HSRP (Hot Standby Router Protocol) Provides high network availability and transparent network topology changes. HSRP creates a Hot Standby router group with a lead router that services all packets sent to the Hot Standby address. The lead router is monitored by other routers in the group, and if it fails, one of these standby routers inherits the lead position and the Hot Standby group address.

HSSI   (High Speed Serial Interface) Network standard for high speed (up to 52 Mbps) serial connections over WAN links.

HTML   (hypertext markup language) Simple hypertext document formatting language that uses tags to indicate how a given part of a document should be interpreted by a viewing application, such as a WWW browser.

hub   (1) Generally, a term used to describe a device that serves as the center of a star-topology network. (2) Hardware or software device that contains multiple independent but connected modules of network and internetwork equipment. Hubs can be active (where they repeat signals sent through them) or passive (where they do not repeat, but merely split signals sent through them). (3) In Ethernet and IEEE 802.3, an Ethernet multiport repeater, sometimes referred to as a *concentrator*.

hybrid network   Internetwork made up of more than one type of network technology, including LANs and WANs.

hypertext   Electronically stored text that allows direct access to other texts by way of encoded links. Hypertext documents can be created using HTML, and often integrate images, sound, and other media that are commonly viewed using a WWW browser.

I

IAB   (Internet Architecture Board) Board of internetwork researchers who discuss issues pertinent to Internet architecture. Responsible for appointing a variety of Internet related groups such as the IANA, IESG, and IRSG. The IAB is appointed by the trustees of the ISOC.

IANA   (Internet Assigned Numbers Authority) Organization operated under the auspices of the ISOC as part of the IAB. IANA delegates authority for IP address space allocation and domain name assignment to the NIC and other organizations. IANA also maintains a database of assigned protocol identifiers used in the TCP/IP stack, including autonomous system numbers.

ICD   (International Code Designator) One of two ATM address formats developed by the ATM Forum for use by private networks. Adapted from the subnetwork model of addressing in which the ATM layer is responsible for mapping network layer addresses to ATM addresses.

ICMP   (Internet Control Message Protocol) Network layer Internet protocol that reports errors and provides other information relevant to IP packet processing. Documented in RFC 792.

IDI   (initial domain identifier) In OSI, the portion of the NSAP that specifies the domain.

IDP   (initial domain part) The part of a CLNS address that contains an authority and format identifier and a domain identifier.

IDPR   (Interdomain Policy Routing) Interdomain routing protocol that dynamically exchanges policies between autonomous systems. IDPR encapsulates interautonomous system traffic and routes it according to the policies of each autonomous system along the path. IDPR is currently an IETF proposal.

IDRP   (IS-IS Interdomain Routing Protocol) OSI protocol that specifies how routers communicate with routers in different domains.

IEC   (International Electrotechnical Commission) Industry group that writes and distributes standards for electrical products and components.

IEEE   (Institute of Electrical and Electronics Engineers) Professional organization whose activities include the development of communications and network standards. IEEE LAN standards are the predominant LAN standards today.

IEEE 802.1   IEEE specification that describes an algorithm that prevents bridging loops by creating a spanning tree. The algorithm was invented by Digital Equipment Corporation. The Digital algorithm and the IEEE 802.1 algorithm are not exactly the same nor are they compatible.

IEEE 802.12   IEEE LAN standard that specifies the physical layer and the MAC sublayer of the data link layer. IEEE 802.12 uses the demand priority media access scheme at 100 Mbps over a variety of physical media.

IEEE 802.2   IEEE LAN protocol that specifies an implementation of the LLC sublayer of the data link layer. IEEE 802.2 handles errors, framing, flow control, and the network layer (layer 3) service interface. Used in IEEE 802.3 and IEEE 802.5 LANs.

IEEE 802.3   IEEE LAN protocol that specifies an implementation of the physical layer and the MAC sublayer of the data link layer. IEEE 802.3 uses CSMA/CD access at a variety of speeds over a variety of physical media. Extensions to the IEEE 802.3 standard specify implementations for fast Ethernet. Physical variations of the original IEEE 802.3 specification include 10Base-2, 10Base-5, 10Base-F, 10Base-T, and 10Broad-36. Physical variations for fast Ethernet include 100Base-T, 100Base-T4, and 100Base-X.

IEEE 802.4   IEEE LAN protocol that specifies an implementation of the physical layer and the MAC sublayer of the data link layer. IEEE 802.4 uses token passing access over a bus topology and is based on the token bus LAN architecture.

IEEE 802.5 IEEE LAN protocol that specifies an implementation of the physical layer and MAC sublayer of the data link layer. IEEE 802.5 uses token passing access at 4 or 16 Mbps over STP cabling and is similar to IBM Token Ring.

IEEE 802.6 IEEE MAN specification based on DQDB technology. IEEE 802.6 supports data rates of 1.5 to 155 Mbps.

IESG (Internet Engineering Steering Group) Organization appointed by the IAB that manages the operation of the IETF.

IETF (Internet Engineering Task Force) Task force consisting of over 80 working groups responsible for developing Internet standards. The IETF operates under the auspices of ISOC.

IFIP (International Federation for Information Processing) Research organization that performs OSI pre-standardization work. Among other accomplishments, IFIP formalized the original MHS model.

IGMP (Internet Group Management Protocol) Used by IP hosts to report their multicast group memberships to an adjacent multicast router.

IGP (Interior Gateway Protocol) Internet protocol used to exchange routing information within an autonomous system. Examples of common Internet IGPs include IGRP, OSPF, and RIP.

IGRP (Interior Gateway Routing Protocol) IGP developed by Cisco to address the problems associated with routing in large, heterogeneous networks.

IIH (IS-IS Hello) Message sent by all IS-IS systems to maintain adjacencies.

IITA (Information Infrastructure Technology and Applications) Component of the HPCC program intended to ensure U.S. leadership in the development of advanced information technologies.

ILMI (Interim Local Management Interface) Specification developed by the ATM Forum for incorporating network management capabilities into the ATM UNI.

IMP (interface message processor) Old name for ARPANET packet switches. An IMP is now referred to as a PSN (packet-switch node).

in-band signaling Transmission within a frequency range normally used for information transmission.

infrared Electromagnetic waves whose frequency range is above that of microwaves but below that of the visible spectrum. LAN systems based on this technology represent an emerging technology.

INOC (Internet Network Operations Center) BBN group that in the early days of the Internet monitored and controlled the Internet core gateways (routers). INOC no longer exists in this form.

insured burst The largest burst of data above the insured rate that will be temporarily allowed on a PVC and not tagged by the traffic policing function for dropping in the case of network congestion. The insured burst is specified in bytes or cells.

insured rate The long-term data throughput, in bits or cells per second, that an ATM network commits to support under normal network conditions. The insured rate is 100 percent allocated; the entire amount is deducted from the total trunk bandwidth along the path of the circuit.

insured traffic Traffic within the insured rate specified for the PVC. This traffic should not be dropped by the network under normal network conditions.

Integrated IS-IS Routing protocol based on the OSI routing protocol IS-IS, but with support for IP and other protocols. Integrated IS-IS implementations send only one set of routing updates, making it more efficient than two separate implementations. Formerly referred to as *Dual IS-IS.*

interarea routing Term used to describe routing between two or more logical areas.

interface (1) Connection between two systems or devices. (2) In routing terminology, a network connection. (3) In telephony, a shared boundary defined by common physical interconnection characteristics, signal characteristics, and meanings of interchanged signals. (4) The boundary between adjacent layers of the OSI model.

interference Unwanted communication channel noise.

International Standards Organization Erroneous expansion of the acronym ISO.

Internet Term used to refer to the largest global internetwork, connecting tens of thousands of networks worldwide and having a "culture" that focuses on research and standardization based on real-life use. Many leading-edge network technologies come from the Internet community. The Internet evolved in part from ARPANET. At one time called the *DARPA Internet.* Not to be confused with the general term *internet.*

internet Short for internetwork. Not to be confused with the *Internet.*

Internet protocol Any protocol that is part of the TCP/IP protocol stack.

internetwork Collection of networks interconnected by routers and other devices that functions (generally) as a single network. Sometimes called an *internet,* which is not to be confused with the *Internet.*

internetworking General term used to refer to the industry that has arisen around the problem of connecting networks together. The term can refer to products, procedures, and technologies.

interoperability Ability of computing equipment manufactured by different vendors to communicate with one another successfully over a network.

Inverse ARP (Inverse Address Resolution Protocol) Method of building dynamic routes in a network. Allows an access server to discover the network address of a device associated with a virtual circuit.

I/O input/output.

IP (Internet Protocol) Network layer protocol in the TCP/IP stack offering a connectionless internetwork service. IP provides features for addressing, type-of-service specification, fragmentation and reassembly, and security. Documented in RFC 791.

IP address A 32-bit address assigned to hosts using TCP/IP. An IP address belongs to one of five classes (A, B, C, D, or E) and is written as 4 octets separated with periods (dotted decimal format). Each address consists of a network number, an optional subnetwork number, and a host number. The network and subnetwork numbers together are used for routing, while the host number is used to address an individual host within the network or subnetwork. A subnet mask is used to extract network and subnetwork information from the IP address. Also called an *Internet address.*

IP multicast Routing technique that allows IP traffic to be propagated from one source to a number of destinations or from many sources to many destinations. Rather than sending one packet to each destination, one packet is sent to a multicast group identified by a single IP destination group address.

IPSO (IP Security Option) U.S. government specification that defines an optional field in the IP packet header that defines hierarchical packet security levels on a per interface basis.

IPX (Internet Packet Exchange) NetWare network layer (layer 3) protocol used for transferring data from servers to workstations. IPX is similar to IP and XNS.

IPXWAN Protocol that negotiates end-to-end options for new links. When a link comes up the first IPX packets sent across are IPXWAN packets negotiating the options for the link. When the IPXWAN options have been successfully determined, normal IPX transmission begins. Defined by RFC 1362.

IRDP (ICMP Router Discovery Protocol) Enables a host to determine the address of a router that it can use as a default gateway. Similar to ESIS, but used with IP.

IRN (intermediate routing node) In SNA, a subarea node with intermediate routing capability.

IRSG (Internet Research Steering Group) Group that is part of the IAB and oversees the activities of the IRTF.

IRTF (Internet Research Task Force) Community of network experts that consider Internet related research topics. The IRTF is governed by the IRSG and is considered a subsidiary of the IAB.

IS (intermediate system) Routing node in an OSI network.

ISA (Industry Standard Architecture) A 16-bit bus used for Intel based personal computers.

isarithmic flow control Flow control technique in which permits travel through the network. Possession of these permits grants the right to transmit. Isarithmic flow control is not commonly implemented.

ISDN (Integrated Services Digital Network) Communication protocol offered by telephone companies that permits telephone networks to carry data, voice, and other source traffic.

IS-IS (Intermediate System-to-Intermediate System) OSI link state hierarchical routing protocol based on DECnet Phase V routing whereby ISs (routers) exchange routing information based on a single metric to determine network topology.

ISO International Organization for Standardization. International organization that is responsible for a wide range of standards, including those relevant to networking. ISO developed the OSI reference model, a popular networking reference model.

ISO 3309 HDLC procedures developed by ISO. ISO 3309:1979 specifies the HDLC frame structure for use in synchronous environments. ISO 3309:1984 specifies proposed modifications to allow the use of HDLC in asynchronous environments as well.

ISO 9000 Set of international quality management standards defined by ISO. The standards, which are not specific to any country, industry, or product, allow companies to demonstrate that they have specific processes in place to maintain an efficient quality system.

ISOC (Internet Society) International nonprofit organization founded in 1992 that coordinates the evolution and use of the Internet. In addition, ISOC delegates authority to other groups related to the Internet, such as the IAB. ISOC is headquartered in Reston, Virginia, U.S.A.

isochronous transmission Asynchronous transmission over a synchronous data link. Isochronous signals require a constant bit rate for reliable transport.

ISODE (ISO development environment) Large set of libraries and utilities used to develop upper layer OSI protocols and applications.

ISR (Intermediate Session Routing) Initial routing algorithm used in APPN. ISR provides node-to-node connection oriented routing. Network outages cause sessions to fail because ISR cannot provide nondisruptive rerouting around a failure. ISR has been replaced by HPR.

ISSI (InterSwitching System Interface) Standard interface between SMDS switches.

ITU-T (International Telecommunication Union Telecommunication Standardization Sector) International body that develops worldwide standards for telecommunications technologies. The ITU-T carries out the functions of the former CCITT.

J

jabber (1) Error condition in which a network device continually transmits random, meaningless data onto the network. (2) In IEEE 802.3, a data packet whose length exceeds that prescribed in the standard.

jitter Analog communication line distortion caused by the variation of a signal from its reference timing positions. Jitter can cause data loss, particularly at high speeds.

jumper Electrical switch consisting of a number of pins and a connector that can be attached to the pins in a variety of different ways. Different circuits are created by attaching the connector to different pins.

JvNCnet (John von Neumann Computer Network) Regional network owned and operated by Global Enterprise Services, Inc., composed of T1 and slower serial links providing midlevel networking services to sites in the northeastern United States.

K

Karn's algorithm Algorithm that improves round-trip time estimations by helping transport layer protocols distinguish between good and bad round-trip time samples.

keepalive interval Period of time between each keep-alive message sent by a network device.

keepalive message Message sent by one network device to inform another network device that the virtual circuit between the two is still active.

Kermit Popular file-transfer and terminal-emulation program.

L

label swapping Routing algorithm used by APPN in which each router that a message passes through on its way to its destination independently determines the best path to the next router.

LAN (local area network) High speed, low error data network covering a relatively small geographic area (up to a few thousand meters). LANs connect workstations, peripherals, terminals, and other devices in a single building or other geographically limited area. LAN standards specify cabling and signaling at the physical and data link layers of the OSI model. Ethernet, FDDI, and Token Ring are widely used LAN technologies.

LANE (LAN emulation) Technology that allows an ATM network to function as a LAN backbone. The ATM network must provide multicast and broadcast support, address mapping (MAC-to-ATM), SVC management, and a usable packet format. LANE also defines Ethernet and Token Ring ELANs.

LAN Manager Distributed NOS developed by Microsoft that supports a variety of protocols and platforms.

LAN Server Server based NOS developed by IBM and derived from LNM.

LAN switch High speed switch that forwards packets between data link segments. Most LAN switches forward traffic based on MAC addresses. This variety of LAN switch is sometimes called a *frame switch*. LAN switches are often categorized according to the method they use to forward traffic: cut-through packet switching or store-and-forward packet switching. Multilayer switches are an intelligent subset of LAN switches.

LAPB (Link Access Procedure, Balanced) Data link layer protocol in the X.25 protocol stack. LAPB is a bit oriented protocol derived from HDLC.

LAPD (Link Access Procedure on the D channel) ISDN data link layer protocol for the D channel. LAPD was derived from the LAPB protocol and is designed primarily to satisfy the signaling requirements of ISDN basic access. Defined by ITU-T Recommendations Q.920 and Q.921.

LAPM (Link Access Procedure for Modems) ARQ used by modems implementing the V.42 protocol for error correction.

laser light amplification by stimulated emission of radiation. Analog transmission device in which a suitable active material is excited by an external stimulus to produce a narrow beam of coherent light that can be modulated into pulses to carry data. Networks based on laser technology are sometimes run over SONET.

LAT (local area transport) A network virtual terminal protocol developed by Digital Equipment Corporation.

LATA (local access and transport area) Geographic telephone dialing area serviced by a single local telephone company. Calls within LATAs are called "local calls." There are well over 100 LATAs in the United States.

latency (1) Delay between the time a device requests access to a network and the time it is granted permission to transmit. (2) Delay between the time when a device receives a frame and the time that frame is forwarded out the destination port.

leaf internetwork In a star topology, an internetwork whose sole access to other internetworks in the star is through a core router.

learning bridge Bridge that performs MAC address learning to reduce traffic on the network. Learning bridges manage a database of MAC addresses and the interfaces associated with each address.

leased line Transmission line reserved by a communications carrier for the private use of a customer. A leased line is a type of dedicated line.

LEC (1) LAN Emulation Client. Entity in an end system that performs data forwarding, address resolution, and other control functions for a single ES within a single ELAN. An LEC also provides a standard LAN service interface to any higher layer entity that interfaces to the LEC. Each LEC is identified by a unique ATM address and is associated with one or more MAC addresses reachable through that ATM address. (2) local exchange carrier. Local or regional telephone company that owns and oper-

ates a telephone network and the customer lines that connect to it.

LECS (LAN Emulation Configuration Server) Entity that assigns individual LANE clients to particular ELANs by directing them to the LES that corresponds to the ELAN. There is logically one LECS per administrative domain and this serves all ELANs within that domain.

LED (light emitting diode) Semiconductor device that emits light produced by converting electrical energy. Status lights on hardware devices are typically LEDs.

LEN node (low entry networking node) In SNA, a PU 2.1 that supports LU protocols but whose CP cannot communicate with other nodes. Because there is no CP-to-CP session between a LEN node and its NN, the LEN node must have a statically defined image of the APPN network.

LES (LAN Emulation Server) Entity that implements the control function for a particular ELAN. There is only one logical LES per ELAN and it is identified by a unique ATM address.

Level 1 router Device that routes traffic within a single DECnet or OSI area.

Level 2 router Device that routes traffic between DECnet or OSI areas. All Level 2 routers must form a contiguous network.

line code type One of a number of coding schemes used on serial lines to maintain data integrity and reliability. The line code type used is determined by the carrier service provider.

line conditioning Use of equipment on leased voice-grade channels to improve analog characteristics, thereby allowing higher transmission rates.

line driver Inexpensive amplifier and signal converter that conditions digital signals to ensure reliable transmissions over extended distances.

line of sight Characteristic of certain transmission systems such as laser, microwave, and infrared systems in which no obstructions in a direct path between transmitter and receiver can exist.

line turnaround Time required to change data transmission direction on a telephone line.

link Network communications channel consisting of a circuit or transmission path and all related equipment between a sender and a receiver. Most often used to refer to a WAN connection. Sometimes referred to as a *line* or a *transmission link*.

link state routing algorithm Routing algorithm in which each router broadcasts or multicasts information regarding the cost of reaching each of its neighbors to all nodes in the internetwork. Link state algorithms create a consistent view of the network and are therefore not prone to routing loops, but they achieve this at the cost of relatively greater computational difficulty and more widespread traffic (compared with distance vector routing algorithms).

LLC (Logical Link Control) Higher of the two data link layer sublayers defined by the IEEE. The LLC sublayer handles error control, flow control, framing, and MAC sublayer addressing. The most prevalent LLC protocol is IEEE 802.2, which includes both connectionless and connection oriented variants.

LLC2 (Logical Link Control, type 2) Connection oriented OSI LLC sublayer protocol.

LMI (Local Management Interface) Set of enhancements to the basic Frame Relay specification. LMI includes support for a keepalive mechanism, which verifies that data is flowing; a multicast mechanism, which provides the network server with its local DLCI and the multicast DLCI; global addressing, which gives DLCIs global rather than local significance in Frame Relay networks; and a status mechanism, which provides an ongoing status report on the DLCIs known to the switch. Known as *LMT* in ANSI terminology.

LM/X (LAN Manager for UNIX) Monitors LAN devices in UNIX environments.

LNM (LAN Network Manager) SRB and Token Ring management package provided by IBM. Typically running on a PC, it monitors SRB and Token Ring devices and can pass alerts up to NetView.

load balancing In routing, the ability of a router to distribute traffic over all its network ports that are the same distance from the destination address. Good load balancing algorithms use both line speed and reliability information. Load balancing increases the utilization of network segments, thus increasing effective network bandwidth.

local acknowledgment Method whereby an intermediate network node, such as a router, responds to acknowledgments for a remote end host. Use of local acknowledgments reduces network overhead and, therefore, the risk of time-outs. Also known as *local termination*.

local bridge Bridge that directly interconnects networks in the same geographic area.

local explorer packet Generated by an end system in an SRB network to find a host connected to the local ring. If the local explorer packet fails to find a local host, the end system produces either a spanning explorer packet or an all routes explorer packet.

local loop Line from the premises of a telephone subscriber to the telephone company CO.

LocalTalk Apple proprietary baseband protocol that operates at the data link and physical layers of the OSI reference model. LocalTalk uses CSMA/CD media access scheme and supports transmissions at speeds of 230 Kbps.

local traffic filtering Process by which a bridge filters out (drops) frames whose source and destination MAC addresses are located on the same interface on the bridge, thus preventing unnecessary traffic from being forwarded across the bridge. Defined in the IEEE 802.1 standard.

logical channel Nondedicated, packet-switched communications path between two or more network nodes. Packet switching allows many logical channels to exist simultaneously on a single physical channel.

loop Route where packets never reach their destination, but simply cycle repeatedly through a constant series of network nodes.

loopback test Test in which signals are sent and then directed back toward their source from some point along the communications path. Loopback tests are often used to test network interface usability.

lossy Characteristic of a network that is prone to lose packets when it becomes highly loaded.

LPD (line printer daemon) Protocol used to send print jobs between UNIX systems.

LSA (link state advertisement) Broadcast packet used by link state protocols that contains information about neighbors and path costs. LSAs are used by the receiving routers to maintain their routing tables. Sometimes called a *link state packet (LSP)*.

LU (logical unit) Primary component of SNA, an LU is an NAU that enables end users to communicate with each other and gain access to SNA network resources.

LU 6.2 (Logical Unit 6.2) IN SNA, an LU that provides peer-to-peer communication between programs in a distributed computing environment. APPC runs on LU 6.2 devices.

M

MAC (Media Access Control) Lower of the two sublayers of the data link layer defined by the IEEE. The MAC sublayer handles access to shared media, such as whether token passing or contention will be used.

MAC address Standardized data link layer address that is required for every port or device that connects to a LAN.

Other devices in the network use these addresses to locate specific ports in the network and to create and update routing tables and data structures. MAC addresses are 6 bytes long and are controlled by the IEEE. Also known as a *hardware address,* a *MAC layer address,* or a *physical address.*

MAC address learning   Service that characterizes a learning bridge in which the source MAC address of each received packet is stored so that future packets destined for that address can be forwarded only to the bridge interface on which that address is located. Packets destined for unrecognized addresses are forwarded out every bridge interface. This scheme helps minimize traffic on the attached LANs. MAC address learning is defined in the IEEE 802.1 standard.

MacIP   Network layer protocol that encapsulates IP packets in DDS or transmission over AppleTalk. MacIP also provides proxy ARP services.

MAN   (metropolitan area network) Network that spans a metropolitan area. Generally, a MAN spans a larger geographic area than a LAN, but a smaller geographic area than a WAN.

managed object   In network management, a network device that can be managed by a network management protocol.

management services   SNA functions distributed among network components to manage and control an SNA network.

Manchester encoding   Digital coding scheme used by IEEE 802.3 and Ethernet in which a mid-bit-time transition is used for clocking, and a 1 is denoted by a high level during the first half of the bit time.

MAP   (Manufacturing Automation Protocol) Network architecture created by General Motors to satisfy the specific needs of the factory floor. MAP specifies a token passing LAN similar to IEEE 802.4.

MAU   (media attachment unit) Device used in Ethernet and IEEE 802.3 networks that provides the interface between the AUI port of a station and the common medium of the Ethernet. The MAU, which can be built into a station or can be a separate device, performs physical layer functions including the conversion of digital data from the Ethernet interface, collision detection, and injection of bits onto the network. Sometimes referred to as a *media access unit,* also abbreviated *MAU,* or as a *transceiver.* In Token Ring, a MAU is known as a *multistation access unit* and is usually abbreviated *MSAU* to avoid confusion.

maximum burst   Specifies the largest burst of data above the insured rate that will be allowed temporarily on an ATM PVC, but will not be dropped at the edge by the traffic policing function even if it exceeds the maximum rate. This amount of traffic will be allowed only temporarily; on average, the traffic source needs to be within the maximum rate. Specified in bytes or cells.

maximum rate   Maximum total data throughput allowed on a given virtual circuit, equal to the sum of the insured and uninsured traffic from the traffic source. The uninsured data might be dropped if the network becomes congested. The maximum rate, which cannot exceed the media rate, represents the highest data throughput the virtual circuit will ever deliver—measured in bits or cells per second.

MBONE   (multicast backbone) The multicast backbone of the Internet. MBONE is a virtual multicast network composed of multicast LANs and the point-to-point tunnels that interconnect them.

MCA   (micro channel architecture) Bus interface commonly used in PCs and some UNIX workstations and servers.

MCI   (Multiport Communications Interface) Card on the AGS+ that provides two Ethernet interfaces and up to two synchronous serial interfaces. The MCI processes packets rapidly, without the interframe delays typical of other Ethernet interfaces.

MCR   (minimum cell rate) Parameter defined by the ATM Forum for ATM traffic management. MCR is defined only for ABR transmissions and specifies the minimum value for the ACR.

MD5   (Message Digest 5) Algorithm used for message authentication in SNMP v.2. MD5 verifies the integrity of the communication, authenticates the origin, and checks for timeliness.

media   Plural of *medium.* The various physical environments through which transmission signals pass. Common network media include twisted-pair, coaxial and fiber optic cable, and the atmosphere (through which microwave, laser, and infrared transmission occurs). Sometimes called *physical media.*

media rate   Maximum traffic throughput for a particular media type.

mesh   Network topology in which devices are organized in a manageable, segmented manner with many, often redundant, interconnections strategically placed between network nodes.

message   Application layer (layer 7) logical grouping of information, often composed of a number of lower layer logical groupings such as packets. The terms datagram, frame, packet, and segment are also used to describe logical information groupings at various layers of the OSI reference model and in various technology circles.

message switching Switching technique involving transmission of messages from node to node through a network. The message is stored at each node until such time as a forwarding path is available.

message unit Unit of data processed by any network layer.

metasignaling Process running at the ATM layer that manages signaling types and virtual circuits.

MHS (message handling system) ITU-T X.400 recommendations that provide message handling services for communications between distributed applications. NetWare MHS is a different (though similar) entity that also provides message handling services.

MIB (Management Information Base) Database of network management information that is used and maintained by a network management protocol such as SNMP or CMIP. The value of an MIB object can be changed or retrieved using SNMP or CMIP commands. MIB objects are organized in a tree structure that includes public (standard) and private (proprietary) branches.

MIC (media interface connector) FDDI de facto standard connector.

microcode Translation layer between machine instructions and the elementary operations of a computer. Microcode is stored in ROM and allows the addition of new machine instructions without requiring that they be designed into electronic circuits when new instructions are needed.

microsegmentation Division of a network into smaller segments, usually with the intention of increasing aggregate bandwidth to network devices.

microwave Electromagnetic waves in the range 1 to 30 GHz. Microwave based networks are an evolving technology gaining favor due to high bandwidth and relatively low cost.

midsplit Broadband cable system in which the available frequencies are split into two groups: one for transmission and one for reception.

MILNET Military Network. Unclassified portion of the DDN. Operated and maintained by the DISA.

MIP (MultiChannel Interface Processor) Interface processor on the Cisco 7000 series routers that provides up to two channelized T1 or E1 connections via serial cables to a CSU. The two controllers on the MIP can each pro-

vide up to 24 T1 or 30 E1 channel-groups, with each channel-group presented to the system as a serial interface that can be configured individually.

mips (millions of instructions per second) Number of instructions executed by a processor per second.

modem (modulator-demodulator) Device that converts digital and analog signals. At the source, a modem converts digital signals to a form suitable for transmission over analog communication facilities. At the destination, the analog signals are returned to their digital form. Modems allow data to be transmitted over voice-grade telephone lines.

modem eliminator Device allowing connection of two DTE devices without modems.

modulation Process by which the characteristics of electrical signals are transformed to represent information. Types of modulation include AM, FM, and PAM.

MOP (Maintenance Operation Protocol) Digital Equipment Corporation protocol—a subset of which is supported by Cisco—that provides a way to perform primitive maintenance operations on DECnet systems. For example, MOP can be used to download a system image to a diskless station.

MOSPF (Multicast OSPF) Intradomain multicast routing protocol used in OSPF networks. Extensions are applied to the base OSPF unicast protocol to support IP multicast routing.

MSAU (multistation access unit) Wiring concentrator to which all end stations in a Token Ring network connect. The MSAU provides an interface between these devices and the Token Ring interface of, for example, a Cisco 7000 TRIP. Sometimes abbreviated *MAU*.

MTU (maximum transmission unit) Maximum packet size, in bytes, that a particular interface can handle.

mu-law North American companding standard used in conversion between analog and digital signals in PCM systems. Similar to the European a-law.

multiaccess network Network that allows multiple devices to connect and communicate simultaneously.

multicast Single packets copied by the network and sent to a specific subset of network addresses. These addresses are specified in the destination address field.

multicast address Single address that refers to multiple network devices. Synonymous with *group address*.

multicast group Dynamically determined group of IP hosts identified by a single IP multicast address.

multicast router   Router used to send IGMP query messages on their attached local networks. Host members of a multicast group respond to a query by sending IGMP reports noting the multicast groups to which they belong. The multicast router takes responsibility for forwarding multicast datagrams from one multicast group to all other networks that have members in the group.

multicast server   Establishes a one-to-many connection to each device in a VLAN, thus establishing a broadcast domain for each VLAN segment. The multicast server forwards incoming broadcasts only to the multicast address that maps to the broadcast address.

multidrop line   Communications line having multiple cable access points. Sometimes called a *multipoint line.*

multihomed host   Host attached to multiple physical network segments in an OSI CLNS network.

multihoming   Addressing scheme in IS-IS routing that supports assignment of multiple area addresses.

multilayer switch   Switch that filters and forwards packets based on MAC addresses and network addresses. A subset of LAN switch.

multimode fiber   Optical fiber supporting propagation of multiple frequencies of light.

multiple domain network   SNA network with multiple SSCPs.

multiplexing   Scheme that allows multiple logical signals to be transmitted simultaneously across a single physical channel.

multivendor network   Network using equipment from more than one vendor. Multivendor networks pose many more compatibility problems than single-vendor networks.

N

Nagle's algorithm   Actually two separate congestion control algorithms that can be used in TCP based networks. One algorithm reduces the sending window; the other limits small datagrams.

NAK   (negative acknowledgment) Response sent from a receiving device to a sending device indicating that the information received contained errors.

name caching   Method by which remotely discovered host names are stored by a router for use in future packet forwarding decisions to allow quick access.

name resolution   Generally, the process of associating a name with a network location.

name server   Server connected to a network that resolves network names into network addresses.

NAP   (network access point) Location for interconnection of Internet service providers in the United States for the exchange of packets.

NAU   (network addressable unit) SNA term for an addressable entity. Examples include LUs, PUs, and SSCPs. NAUs generally provide upper level network services.

NAUN   (nearest active upstream neighbor) In Token Ring or IEEE 802.5 networks, the closest upstream network device from any given device that is still active.

NBMA   (nonbroadcast multiaccess) Term describing a multiaccess network that either does not support broadcasting (such as X.25) or in which broadcasting is not feasible (for example, an SMDS broadcast group or an extended Ethernet that is too large).

NBP   (Name Binding Protocol) AppleTalk transport level protocol that translates a character string name into an internetwork address.

NCIA   (native client interface architecture) SNA applications access architecture developed by Cisco that combines the full functionality of native SNA interfaces at both the host and client with the flexibility of leveraging TCP/IP backbones. NCIA encapsulates SNA traffic on a client PC or workstation, thereby providing direct TCP/IP access while preserving the native SNA interface at the end user level. In many networks, this capability obviates the need for a stand-alone gateway and can provide flexible TCP/IP access while preserving the native SNA interface to the host.

NCP   (Network Control Program) In SNA, a program that routes and controls the flow of data between a communications controller (in which it resides) and other network resources.

neighboring routers   In OSPF, two routers that have interfaces to a common network. On multiaccess networks, neighbors are dynamically discovered by the OSPF Hello protocol.

NET   (network entity title) Network addresses defined by the ISO network architecture and used in CLNS based networks.

NetBIOS   (Network Basic Input/Output System) API used by applications on an IBM LAN to request services from lower level network processes. These services might include session establishment and termination, and information transfer.

NetWare   Popular distributed NOS developed by Novell. Provides transparent remote file access and numerous other distributed network services.

network Collection of computers, printers, routers, switches, and other devices that are able to communicate with each other over some transmission medium.

network address Network layer address referring to a logical, rather than a physical, network device. Also called a *protocol address.*

network administrator Person responsible for the operation, maintenance, and management of a network.

network analyzer Hardware or software device offering various network troubleshooting features including protocol-specific packet decodes, specific preprogrammed troubleshooting tests, packet filtering, and packet transmission.

Network Node Server SNA NN that provides resource location and route selection services for ENs, LEN nodes, and LUs that are in its domain.

network number Part of an IP address that specifies the network to which the host belongs.

network operator Person who routinely monitors and controls a network, performing such tasks as reviewing and responding to traps, monitoring throughput, configuring new circuits, and resolving problems.

NFS (Network File System) As commonly used, a distributed file system protocol suite developed by Sun Microsystems that allows remote file access across a network. In actuality, NFS is simply one protocol in the suite. NFS protocols include NFS, RPC, XDR (External Data Representation), and others. These protocols are part of a larger architecture that Sun refers to as ONC.

NHRP (Next Hop Resolution Protocol) Protocol used by routers to dynamically discover the MAC address of other routers and hosts connected to a NBMA network. These systems can then directly communicate without requiring traffic to use an intermediate hop, increasing performance in ATM, Frame Relay, SMDS, and X.25 environments.

NIC (1) network interface card. Board that provides network communication capabilities to and from a computer system. Also called an *adapter.* (2) Network Information Center. Organization that serves the Internet community by supplying user assistance, documentation, training, and other services.

NIS (Network Information Service) Protocol developed by Sun Microsystems for the administration of network-wide databases. The service essentially uses two programs: one for finding an NIS server and one for accessing the NIS databases.

N-ISDN (Narrowband ISDN) Communication standards developed by the ITU-T for baseband networks.

Based on 64-Kbps B channels and 16- or 64-Kbps D channels.

NIST (National Institute of Standards and Technology) Formerly the NBS, this U.S. government organization supports and catalogs a variety of standards.

NLM (NetWare Loadable Module) Individual program that can be loaded into memory and function as part of the NetWare NOS.

NLSP (NetWare Link Services Protocol) Link state routing protocol based on IS-IS. The Cisco implementation of NLSP also includes MIB variables and tools to redistribute routing and SAP information between NLSP and other IPX routing protocols.

NMP (Network Management Processor) Processor module on the Catalyst 5000 switch used to control and monitor the switch.

NMS (network management system) System responsible for managing at least part of a network. An NMS is generally a reasonably powerful and well-equipped computer such as an engineering workstation. NMSs communicate with agents to help keep track of network statistics and resources.

NMVT (network management vector transport) SNA message consisting of a series of vectors conveying network management specific information.

NN (network node) SNA intermediate node that provides connectivity, directory services, route selection, intermediate session routing, data transport, and network management services to LEN nodes and ENs. The NN contains a CP that manages the resources of both the NN itself and those of the ENs and LEN nodes in its domain. NNs provide intermediate routing services by implementing the APPN PU 2.1 extensions.

NNI (Network-to-Network Interface) ATM Forum standard that defines the interface between two ATM switches that are both located in a private network or are both located in a public network. The interface between a public switch and private one is defined by the UNI standard. Also, the standard interface between two Frame Relay switches meeting the same criteria.

NOC (Network Operations Center) Organization responsible for maintaining a network.

node (1) Endpoint of a network connection or a junction common to two or more lines in a network. Nodes can be processors, controllers, or workstations. Nodes, which vary in routing and other functional capabilities, can be interconnected by links, and serve as control points in the network. Node is sometimes used generically to refer to any entity that can access a network, and is frequently used interchangeably with *device*. (2) In SNA, the basic component of a network and the point at which one or more functional units connect channels or data circuits.

noise   Undesirable communications channel signals.

nonseed router   In AppleTalk, a router that must first obtain and then verify its configuration with a seed router before it can begin operation.

nonstub area   Resource-intensive OSPF area that carries a default route, static routes, intra-area routes, interarea routes, and external routes. Nonstub areas are the only OSPF areas that can have virtual links configured across them, and are the only areas that can contain an ASBR.

Northwest Net   NSF-funded regional network serving the northwestern United States, Alaska, Montana, and North Dakota. Northwest Net connects all major universities in the region as well as many leading industrial concerns.

NOS   (network operating system) Generic term used to refer to what are really distributed file systems. Examples of NOSs include LAN Manager, NetWare, NFS, and VINES.

NREN   National Research and Education Network. Component of the HPCC program designed to ensure U.S. technical leadership in computer communications through research and development efforts in state-of-the-art telecommunications and networking technologies.

NRM   (normal response mode) HDLC mode for use on links with one primary station and one or more secondary stations. In this mode, secondary stations can transmit only if they first receive a poll from the primary station.

NRZ   (nonreturn to zero) NRZ signals maintain constant voltage levels with no signal transitions (no return to a zero-voltage level) during a bit interval.

NRZI   (nonreturn to zero inverted) NRZI signals maintain constant voltage levels with no signal transitions (no return to a zero-voltage level), but interpret the presence of data at the beginning of a bit interval as a signal transition and the absence of data as no transition.

NSAP   (network service access point) Network addresses as specified by ISO. An NSAP is the point at which OSI Network Service is made available to a transport layer (layer 4) entity.

NSF   (National Science Foundation) U.S. government agency that funds scientific research in the United States. The now-defunct NSFNET was funded by the NSF.

NSFNET   (National Science Foundation Network) Large network that was controlled by the NSF and provided networking services in support of education and research in the United States from 1986 to 1995. NSFNET is no longer in service.

NTRI   (NCP/Token Ring Interconnection) Function used by ACF/NCP to support Token Ring attached SNA devices. NTRI also provides translation from Token Ring attached SNA devices (PUs) to switched (dial-up) devices.

null modem   Small box or cable used to join computing devices directly, rather than over a network.

NVRAM   (nonvolatile RAM) RAM that retains its contents when a unit is powered off.

NYSERNet   Network in New York (United States) with a T1 backbone connecting NSF with many universities and several commercial concerns.

OAM cell   (Operation, Administration, and Maintenance cell) ATM Forum specification for cells used to monitor virtual circuits. OAM cells provide a virtual circuit level loopback in which a router responds to the cells, demonstrating that the circuit is up and the router is operational.

OARnet   (Ohio Academic Resources Network) Internet service provider that connects a number of U.S. sites, including the Ohio supercomputer center in Columbus, Ohio.

object instance   Network management term referring to an instance of an object type that has been bound to a value.

OC   (Optical Carrier) Series of physical protocols (OC-1, OC-2, OC-3, and so on), defined for SONET optical signal transmissions. OC signal levels put STS frames onto multimode fiber optic line at a variety of speeds. The base rate is 51.84 Mbps (OC-1); each signal level thereafter operates at a speed divisible by that number (thus, OC-3 runs at 155.52 Mbps).

ODA   (Open Document Architecture) ISO standard that specifies how documents are represented and transmitted electronically. Formally called *Office Document Architecture*.

ODI   (Open Data Link Interface) Novell specification providing a standardized interface for NICs (network interface cards) that allows multiple protocols to use a single NIC.

OIM   (OSI Internet Management) Group tasked with specifying ways in which OSI network management protocols can be used to manage TCP/IP networks.

OIR (online insertion and removal) Feature that permits the addition, replacement, or removal of interface processors in a Cisco router without interrupting the system power, entering console commands, or causing other software or interfaces to shut down. Sometimes called *hot swapping.*

ONC (Open Network Computing) Distributed applications architecture designed by Sun Microsystems, currently controlled by a consortium led by Sun. The NFS protocols are part of ONC.

ones density Scheme that allows a CSU/DSU to recover the data clock reliably. The CSU/DSU derives the data clock from the data that passes through it. In order to recover the clock, the CSU/DSU hardware must receive at least one 1 bit value for every 8 bits of data that pass through it. Also called *pulse density.*

open architecture Architecture with which third-party developers can legally develop products and for which public domain specifications exist.

open circuit Broken path along a transmission medium. Open circuits will usually prevent network communication.

OSI (Open Systems Interconnection) International standardization program created by ISO and ITU-T to develop standards for data networking that facilitate multivendor equipment interoperability.

OSI reference model (Open Systems Interconnection reference model) Network architectural model developed by ISO and ITU-T. The model consists of seven layers, each of which specifies particular network functions such as addressing, flow control, error control, encapsulation, and reliable message transfer. The highest layer (the application layer) is closest to the user; the lowest layer (the physical layer) is closest to the media technology. The lower two layers are implemented in hardware and software, while the upper five layers are implemented only in software. The OSI reference model is used universally as a method for teaching and understanding network functionality.

OSPF (Open Shortest Path First) Link state hierarchical IGP routing algorithm proposed as a successor to RIP in the Internet community. OSPF features include least-cost routing, multipath routing, and load balancing. OSPF was derived from an early version of the IS-IS protocol.

OUI (Organizational Unique Identifier) The 3 octets assigned by the IEEE in a block of 48-bit LAN addresses.

outframe Maximum number of outstanding frames allowed in an SNA PU 2 server at any time.

out-of-band signaling Transmission using frequencies or channels outside the frequencies or channels normally used for information transfer. Out-of-band signaling is often used for error reporting in situations in which in-band signaling can be affected by whatever problems the network might be experiencing.

P

packet Logical grouping of information that includes a header containing control information and (usually) user data. Packets are most often used to refer to network layer units of data. The terms datagram, frame, message, and segment are also used to describe logical information groupings at various layers of the OSI reference model and in various technology circles.

packet switch WAN device that routes packets along the most efficient path and allows a communications channel to be shared by multiple connections. Sometimes referred to as a *packet switch node (PSN),* and formerly called an *IMP.*

packet switching Networking method in which nodes share bandwidth with each other by sending packets.

PAD (packet assembler/disassembler) Device used to connect simple devices (like character mode terminals) that do not support the full functionality of a particular protocol to a network. PADs buffer data and assemble and disassemble packets sent to such end devices.

PAM (pulse amplitude modulation) Modulation scheme where the modulating wave is caused to modulate the amplitude of a pulse stream.

PAP (Password Authentication Protocol) Authentication protocol that allows PPP peers to authenticate one another. The remote router attempting to connect to the local router is required to send an authentication request. Unlike CHAP, PAP passes the password and host name or username in the clear (unencrypted). PAP does not itself prevent unauthorized access, but merely identifies the remote end. The router or access server then determines if that user is allowed access. PAP is supported only on PPP lines.

parallel channel Channel that uses bus and tag cables as a transmission medium.

parallelism Indicates that multiple paths exist between two points in a network. These paths might be of equal or unequal cost. Parallelism is often a network design goal: if one path fails, there is redundancy in the network to ensure that an alternate path to the same point exists.

parallel transmission Method of data transmission in which the bits of a data character are transmitted simultaneously over a number of channels.

PARC (Palo Alto Research Center) Research and development center operated by XEROX. A number of widely used technologies were originally conceived at PARC, including the first personal computers and LANs.

parity check Process for checking the integrity of a character. A parity check involves appending a bit that makes the total number of binary 1 digits in a character or word (excluding the parity bit) either odd (for *odd parity*) or even (for *even parity*).

partial mesh Term describing a network in which devices are organized in a mesh topology with some network nodes organized in a full mesh, but others that are only connected to one or two other nodes in the network. A partial mesh does not provide the level of redundancy of a full mesh topology, but is less expensive to implement. Partial mesh topologies are generally used in the peripheral networks that connect to a fully meshed backbone.

path control layer Layer 3 in the SNA architectural model. This layer performs sequencing services related to proper data reassembly. The path control layer is also responsible for routing. Corresponds roughly with the network layer of the OSI model.

path control network SNA concept that consists of lower level components that control the routing and data flow through an SNA network and handle physical data transmission between SNA nodes.

path name Full name of a UNIX, DOS, or LynxOS file or directory, including all directory and subdirectory names. Consecutive names in a path name are typically separated by a forward slash (/) or a backslash (\), as in /usr/app/base/config.

payload Portion of a frame that contains upper layer information (data).

PBX (private branch exchange) Digital or analog telephone switchboard located on the subscriber premises and used to connect private and public telephone networks.

PCI (protocol control information) Control information added to user data to comprise an OSI packet. The OSI equivalent of the term header.

PCM (pulse code modulation) Transmission of analog information in digital form through sampling and encoding the samples with a fixed number of bits.

PCR (peak cell rate) Parameter defined by the ATM Forum for ATM traffic management. In CBR transmissions, PCR determines how often data samples are sent. In ABR transmissions, PCR determines the maximum value of the ACR.

PDN (public data network) Network operated either by a government (as in Europe) or by a private concern to provide computer communications to the public, usually for a fee. PDNs enable small organizations to create a WAN without all the equipment costs of long-distance circuits.

PDU (protocol data unit) OSI term for packet.

peak rate Maximum rate, in kilobits per second, at which a virtual circuit can transmit.

peer-to-peer computing Peer-to-peer computing calls for each network device to run both client and server portions of an application. Also describes communication between implementations of the same OSI reference model layer in two different network devices.

performance management One of five categories of network management defined by ISO for management of OSI networks. Performance management subsystems are responsible for analyzing and controlling network performance including network throughput and error rates.

peripheral node In SNA, a node that uses local addresses and is therefore not affected by changes to network addresses. Peripheral nodes require boundary function assistance from an adjacent subarea node.

P/F (poll/final bit) A bit in bit-synchronous data link layer protocols that indicates the function of a frame. If the frame is a command, a 1 in this bit indicates a poll. If the frame is a response, a 1 in this bit indicates that the current frame is the last frame in the response.

PGP (Pretty Good Privacy) Public-key encryption application that allows secure file and message exchanges. There is some controversy over the development and use of this application, in part due to U.S. national security concerns.

phase Location of a position on an alternating wave form.

phase shift Situation in which the relative position in time between the clock and data signals of a transmission becomes unsynchronized. In systems using long cables at higher transmission speeds, slight variances in cable construction, temperature, and other factors can cause a phase shift, resulting in high error rates.

physical control layer Layer 1 in the SNA architectural model. This layer is responsible for the physical specifications for the physical links between end systems. Corresponds to the physical layer of the OSI model.

physical layer Layer 1 of the OSI Reference model. The physical layer defines the electrical, mechanical, procedural and functional specifications for activating, maintaining, and deactivating the physical link between end systems. Corresponds with the physical control layer in the SNA model.

PHYSNET (Physics Network) Group of many DECnet based physics research networks including HEPnet.

piggybacking Process of carrying acknowledgements within a data packet to save network bandwidth.

PIM (Protocol Independent Multicast) Multicast routing architecture that allows the addition of IP multicast routing on existing IP networks. PIM is unicast routing protocol independent and can be operated in two modes: dense mode and sparse mode.

PIM dense mode One of the two PIM operational modes. PIM dense mode is data-driven and resembles typical multicast routing protocols. Packets are forwarded on all outgoing interfaces until pruning and truncation occurs. In dense mode, receivers are densely populated, and it is assumed that the downstream networks want to receive and will probably use the datagrams that are forwarded to them. The cost of using dense mode is its default flooding behavior. Sometimes called *dense mode PIM* or *PIM DM*.

PIM sparse mode One of the two PIM operational modes. PIM sparse mode tries to constrain data distribution so that a minimal number of routers in the network receive it. Packets are sent only if they are explicitly requested at the RP (rendezvous point). In sparse mode, receivers are widely distributed, and the assumption is that downstream networks will not necessarily use the datagrams that are sent to them. The cost of using sparse mode is its reliance on the periodic refreshing of explicit join messages and its need for RPs. Sometimes called *sparse mode PIM* or *PIM SM*.

PING (packet internet groper) ICMP echo message and its reply. Often used to test the reachability of a network device.

ping-ponging Phrase used to describe the actions of a packet in a two-node routing loop.

PLCP (physical layer convergence procedure) Specification that maps ATM cells into physical media, such as T3 or E3, and defines certain management information.

plesiochronous transmission Term describing digital signals that are sourced from different clocks of comparable accuracy and stability.

PLP (packet level protocol) Network layer protocol in the X.25 protocol stack. Sometimes called *X.25 Level 3* or *X.25 Protocol*.

PLU (Primary LU) The LU that is initiating a session with another LU.

PMD (physical medium dependent) Sublayer of the FDDI physical layer that interfaces directly with the physical medium and performs the most basic bit transmission functions of the network.

PNNI (Private Network-Network Interface) ATM Forum specification that describes an ATM virtual circuit routing protocol, as well as a signaling protocol between ATM switches. Used to allow ATM switches within a private network to interconnect. Sometimes called *Private Network Node Interface*.

poison reverse updates Routing updates that explicitly indicate that a network or subnet is unreachable, rather than implying that a network is unreachable by not including it in updates. Poison reverse updates are sent to defeat large routing loops.

policy routing Routing scheme that forwards packets to specific interfaces based on user-configured policies. Such policies might specify that traffic sent from a particular network should be forwarded out one interface, while all other traffic should be forwarded out another interface.

polling Access method in which a primary network device inquires, in an orderly fashion, whether secondaries have data to transmit. The inquiry occurs in the form of a message to each secondary that gives the secondary the right to transmit.

POP (point of presence) Physical access point to a long-distance carrier interchange.

port (1) Interface on an internetworking device (such as a router). (2) In IP terminology, an upper layer process that is receiving information from lower layers. (3) To rewrite software or microcode so that it will run on a different hardware platform or in a different software environment than that for which it was originally designed.

PPP (Point-to-Point Protocol) A successor to SLIP, PPP provides router-to-router and host-to-network connections over synchronous and asynchronous circuits.

presentation layer Layer 6 of the OSI reference model. This layer ensures that information sent by the application layer of one system will be readable by the application layer of another. The presentation layer is also concerned with the data structures used by programs and therefore negotiates data transfer syntax for the application layer. Corresponds roughly with the presentation services layer of the SNA model.

presentation services layer Layer 6 of the SNA architectural model. This layer provides network resource management, session presentation services, and some application management. Corresponds roughly with the presentation layer of the OSI model.

PRI   (Primary Rate Interface) ISDN interface to primary rate access. Primary rate access consists of a single 64-Kbps D channel plus 23 (T1) or 30 (E1) B channels for voice or data.

primary ring   One of the two rings that make up an FDDI or CDDI ring. The primary ring is the default path for data transmissions.

primary station   In bit-synchronous data link layer protocols such as HDLC and SDLC, a station that controls the transmission activity of secondary stations and performs other management functions such as error control through polling or other means. Primary stations send commands to secondary stations and receive responses.

print server   Networked computer system that fields, manages, and executes (or sends for execution) print requests from other network devices.

priority queuing   Routing feature in which frames in an interface output queue are prioritized based on various characteristics such as packet size and interface type.

process switching   Operation that provides full route evaluation and per packet load balancing across parallel WAN links. Involves the transmission of entire frames to the router CPU where they are repackaged for delivery to or from a WAN interface, with the router making a route selection for each packet. Process switching is the most resource-intensive switching operation that the CPU can perform.

PROM   (programmable read only memory) ROM that can be programmed using special equipment. PROMs can be programmed only once.

propagation delay   Time required for data to travel over a network from its source to its ultimate destination.

protocol   Formal description of a set of rules and conventions that govern how devices on a network exchange information.

protocol converter   Enables equipment with different data formats to communicate by translating the data transmission code of one device to the data transmission code of another device.

protocol stack   Set of related communications protocols that operate together and, as a group, address communication at some or all of the seven layers of the OSI reference model. Not every protocol stack covers each layer of the model, and often a single protocol in the stack will ad-

dress a number of layers at once. TCP/IP is a typical protocol stack.

protocol translator   Network device or software that converts one protocol into another, similar protocol.

proxy   Entity that in the interest of efficiency, essentially stands in for another entity.

proxy ARP   (proxy Address Resolution Protocol) Variation of the ARP protocol in which an intermediate device (for example, a router) sends an ARP response on behalf of an end node to the requesting host. Proxy ARP can lessen bandwidth use on slow speed WAN links.

proxy explorer   Technique that minimizes exploding explorer packet traffic propagating through an SRB network by creating an explorer packet reply cache, the entries of which are reused when subsequent explorer packets need to find the same host.

proxy polling   Technique that alleviates the load across an SDLC network by allowing routers to act as proxies for primary and secondary nodes, thus keeping polling traffic off of the shared links. Proxy polling has been replaced by SDLC Transport.

PSDN   packet-switched data network.

PSE   (packet switch exchange) Essentially, a switch. The term PSE is generally used in reference to a switch in an X.25 PSN.

PSN   (1) packet-switched network. Network that utilizes packet-switching technology for data transfer. Sometimes called a *packet-switched data network (PSDN)*. (2) packet-switching node. Network node capable of performing packet switching functions.

PSTN   (Public Switched Telephone Network) General term referring to the variety of telephone networks and services in place worldwide. Sometimes called *plain old telephone service (POTS)*.

PTT   (Post, Telephone, and Telegraph) Government agency that provides telephone services. PTTs exist in most areas outside North America and provide both local and long-distance telephone services.

PU   (physical unit) SNA component that manages and monitors the resources of a node as requested by an SSCP. There is one PU per node.

PU 2   (Physical Unit 2) SNA peripheral node that can support only DLUs that require services from a VTAM host and that are only capable of performing the secondary LU role in SNA sessions.

PU 2.1   (Physical Unit type 2.1) SNA network node used for connecting peer nodes in a peer oriented network.

PU 2.1 sessions do not require that one node reside on VTAM. APPN is based on PU 2.1 nodes, which can also be connected to a traditional hierarchical SNA network.

PU 4 (Physical Unit 4) Component of an IBM FEP capable of full-duplex data transfer. Each such SNA device employs a separate data and control path into the transmit and receive buffers of the control program.

PU 5 (Physical Unit 5) Component of an IBM mainframe or host computer that manages an SNA network. PU 5 nodes are involved in routing within the SNA path control layer.

PUP (PARC Universal Protocol) Protocol similar to IP developed at PARC.

PVC (permanent virtual circuit) Virtual circuit that is permanently established. PVCs save bandwidth associated with circuit establishment and tear down in situations where certain virtual circuits must exist all the time. Called a *permanent virtual connection* in ATM terminology.

PVP (permanent virtual path) Virtual path that consists of PVCs.

Q

Q.920/Q.921 ITU-T specifications for the ISDN UNI data link layer.

Q.922A ITU-T specification for Frame Relay encapsulation.

Q.931 ITU-T specification for signaling to establish, maintain, and clear ISDN network connections.

Q.93B ITU-T specification signaling to establish, maintain, and clear B-ISDN network connections. An evolution of ITU-T recommendation Q.931.

QLLC (Qualified Logical Link Control) Data link layer protocol defined by IBM that allows SNA data to be transported across X.25 networks.

QoS (quality of service) Measure of performance for a transmission system that reflects its transmission quality and service availability.

QoS parameters (quality of service parameters) Parameters that control the amount of traffic the source router in an ATM network sends over an SVC. If any switch along the path cannot accommodate the requested QoS parameters, the request is rejected and a rejection message is forwarded back to the originator of the request.

quartet signaling Signaling technique used in 100VG-AnyLAN networks that allows data transmission at 100 Mbps

over four pairs of UTP cabling at the same frequencies used in 10Base-T networks.

query Message used to inquire about the value of some variable or set of variables.

queue (1) Generally, an ordered list of elements waiting to be processed. (2) In routing, a backlog of packets waiting to be forwarded over a router interface.

queuing delay Amount of time that data must wait before it can be transmitted onto a statistically multiplexed physical circuit.

queuing theory Scientific principles governing the formation or lack of formation of congestion on a network or at an interface.

R

RACE (Research on Advanced Communications in Europe) Project sponsored by the European Community (EC) for the development of broadband networking capabilities.

RAM (random access memory) Volatile memory that can be read and written by a microprocessor.

RARP (Reverse Address Resolution Protocol) Protocol in the TCP/IP stack that provides a method for finding IP addresses based on MAC addresses.

rate queue Value that is associated with one or more virtual circuits, and that defines the speed at which an individual virtual circuit will transmit data to the remote end. Each rate queue represents a portion of the overall bandwidth available on an ATM link. The combined bandwidth of all configured rate queues should not exceed the total bandwidth available.

RBHC (Regional Bell Holding Company) One of seven telephone companies created by the AT&T divestiture in 1984.

RBOC (Regional Bell Operating Company) Local or regional telephone company that owns and operates telephone lines and switches in one of seven U.S. regions. The RBOCs were created by the divestiture of AT&T. Also called *Bell Operating Company (BOC)*.

rcp (remote copy protocol) Protocol that allows users to copy files to and from a file system residing on a remote host or server on the network. The rcp protocol uses TCP to ensure the reliable delivery of data.

rcp server Router or other device that acts as a server for rcp.

reassembly   The putting back together of an IP datagram at the destination after it has been fragmented either at the source or at an intermediate node.

redirect   Part of the ICMP and ES-IS protocols that allows a router to tell a host that using another router would be more effective.

redirector   Software that intercepts requests for resources within a computer and analyzes them for remote access requirements. If remote access is required to satisfy the request, the redirector forms an RPC and sends the RPC to lower layer protocol software for transmission through the network to the node that can satisfy the request.

redistribution   Allowing routing information discovered through one routing protocol to be distributed in the update messages of another routing protocol. Sometimes called *route redistribution*.

redundancy   (1) In internetworking, the duplication of devices, services, or connections so that in the event of a failure the redundant devices, services, or connections can perform the work of those that failed. (2) In telephony, the portion of the total information contained in a message that can be eliminated without loss of essential information or meaning.

relay   OSI terminology for a device that connects two or more networks or network systems. A data link layer (layer 2) relay is a bridge; a network layer (layer 3) relay is a router.

reliability   Ratio of expected to received keepalives from a link. If the ratio is high, the line is reliable. Used as a routing metric.

reload   The event of a Cisco router rebooting, or the command that causes the router to reboot.

remote bridge   Bridge that connects physically disparate network segments via WAN links.

repeater   Device that regenerates and propagates electrical signals between two network segments.

RF   (radio frequency) Generic term referring to frequencies that correspond to radio transmissions. Cable TV and broadband networks use RF technology.

RFC   (Request For Comments) Document series used as the primary means for communicating information about the Internet. Some RFCs are designated by the IAB as Internet standards. Most RFCs document protocol specifications such as Telnet and FTP, but some are humorous or historical. RFCs are available online from numerous sources.

RFI   (radio frequency interference) Radio frequencies that create noise that interferes with information being transmitted across unshielded copper cabling.

RIF   (Routing Information Field) Field in the IEEE 802.5 header that is used by a source-route bridge to determine through which Token Ring network segments a packet must transit. A RIF is made up of ring and bridge numbers as well as other information.

RII   (Routing Information Indicator) Bit used by SRT bridges to distinguish between frames that should be transparently bridged and frames that should be passed to the SRB module for handling.

ring   Connection of two or more stations in a logically circular topology. Information is passed sequentially between active stations. Token Ring, FDDI, and CDDI are based on this topology.

ring group   Collection of Token Ring interfaces on one or more Cisco routers that is part of a one-bridge Token Ring network.

ring latency   Time required for a signal to propagate once around a ring in a token ring or IEEE 802.5 network.

ring monitor   Centralized management tool for Token Ring networks based on the IEEE 802.5 specification.

ring topology   Network topology that consists of a series of repeaters connected to one another by unidirectional transmission links to form a single closed loop. Each station on the network connects to the network at a repeater. While logically a ring, ring topologies are most often organized in a closed-loop star.

RIP   (Routing Information Protocol) IGP supplied with UNIX BSD systems. The most common IGP in the Internet. RIP uses hop count as a routing metric.

RJ connector   (registered jack connector) Standard connectors originally used to connect telephone lines. RJ connectors are now used for telephone connections and for 10Base-T and other types of network connections. RJ-11, RJ-12, and RJ-45 are popular types of RJ connectors.

RJE   (remote job entry) Application that is batch oriented as opposed to interactive. In RJE environments, jobs are submitted to a computing facility and output is received later.

rlogin   (remote login) Terminal emulation program similar to Telnet that is offered in most UNIX implementations.

RMON   (Remote Monitoring) MIB agent specification described in RFC 1271 that defines functions for the remote monitoring of networked devices. The RMON specification provides numerous monitoring, problem detection, and reporting capabilities.

ROM   (read only memory) Nonvolatile memory that can be read, but not written, by the microprocessor.

root bridge Exchanges topology information with designated bridges in a spanning tree implementation in order to notify all other bridges in the network when topology changes are required. This prevents loops and provides a measure of defense against link failure.

ROSE (Remote Operations Service Element) OSI RPC mechanism used by various OSI network application protocols.

route Path through an internetwork.

routed protocol Protocol that can be routed by a router. A router must be able to interpret the logical internetwork as specified by that routed protocol. Examples of routed protocols include AppleTalk, DECnet, and IP.

route extension In SNA, a path from the destination subarea node through peripheral equipment to a NAU.

route map Method of controlling the redistribution of routes between routing domains.

route summarization Consolidation of advertised addresses in OSPF and IS-IS. In OSPF, this causes a single summary route to be advertised to other areas by an area border router.

router Network layer device that uses one or more metrics to determine the optimal path along which network traffic should be forwarded. Routers forward packets from one network to another based on network layer information. Occasionally called a *gateway* (although this definition of gateway is becoming increasingly outdated).

routing Process of finding a path to a destination host. Routing is very complex in large networks because of the many potential intermediate destinations a packet might traverse before reaching its destination host.

routing domain Group of end systems and intermediate systems operating under the same set of administrative rules. Within each routing domain is one or more areas, each uniquely identified by an area address.

routing metric Method by which a routing algorithm determines that one route is better than another. This information is stored in routing tables. Metrics include bandwidth, communication cost, delay, hop count, load, MTU, path cost, and reliability. Sometimes referred to simply as a *metric*.

routing protocol Protocol that accomplishes routing through the implementation of a specific routing algorithm. Examples of routing protocols include IGRP, OSPF, and RIP.

routing table Table stored in a router or some other internetworking device that keeps track of routes to particular network destinations and, in some cases, metrics associated with those routes.

routing update Message sent from a router to indicate network reachability and associated cost information. Routing updates are typically sent at regular intervals and after a change in network topology.

RP (1) Route Processor. Processor module on the Cisco 7000 series routers that contains the CPU, system software, and most of the memory components that are used in the router. Sometimes called a *supervisory processor*. (2) rendezvous point. Router specified in PIM sparse mode implementations to track membership in multicast groups and to forward messages to known multicast group addresses.

RPC (remote procedure call) Technological foundation of client-server computing. RPCs are procedure calls that are built or specified by clients and executed on servers, with the results returned over the network to the clients.

RPM (Reverse Path Multicasting) Multicasting technique in which a multicast datagram is forwarded out of all but the receiving interface if the receiving interface is one used to forward unicast datagrams to the source of the multicast datagram.

RS-232 Popular physical layer interface. Now known as *EIA/TIA-232*.

RS-422 Balanced electrical implementation of EIA/TIA-449 for high speed data transmission. Now referred to collectively with RS-423 as EIA-530.

RS-423 Unbalanced electrical implementation of EIA/TIA-449 for EIA/TIA-232 compatibility. Now referred to collectively with RS-422 as EIA-530.

RS-449 Popular physical layer interface. Now known as *EIA/TIA-449*.

rsh (remote shell protocol) Protocol that allows a user to execute commands on a remote system without having to log in to the system. For example, rsh can be used to remotely examine the status of a number of access servers without connecting to each communication server, executing the command, and then disconnecting from the communication server.

RSP (Route/Switch Processor) Processor module used in the Cisco 7500 series routers that integrates the functions of the RP and the SP.

RSRB (remote source route bridging) SRB over WAN links.

RSUP (Reliable SAP Update Protocol) Bandwidth-saving protocol developed by Cisco for propagating services information. RSUP allows routers to reliably send standard Novell SAP packets only when the routers detect a change in advertised services. RSUP can transport network information either in conjunction with or independently of the Enhanced IGRP routing function for IPX.

RTMP (Routing Table Maintenance Protocol) Apple Computer proprietary routing protocol. RTMP was derived from RIP.

RTP (1) Routing Table Protocol. VINES routing protocol based on RIP. Distributes network topology information and aids VINES servers in finding neighboring clients, servers, and routers. Uses delay as a routing metric. (2) Rapid Transport Protocol. Provides pacing and error recovery for APPN data as it crosses the APPN network. With RTP, error recovery and flow control are done end-to-end rather than at every node. RTP prevents congestion rather than reacts to it.

RTS (Request To Send) EIA/TIA-232 control signal that requests a data transmission on a communications line.

RTT (round-trip time) Time required for a network communication to travel from the source to the destination and back. RTT includes the time required for the destination to process the message from the source and generate a reply. RTT is used by some routing algorithms to aid in calculating optimal routes.

RU (request/response unit) Request and response messages exchanged between NAUs in an SNA network.

S

SAC (single-attached concentrator) FDDI or CDDI concentrator that connects to the network by being cascaded from the master port of another FDDI or CDDI concentrator.

sampling rate Rate at which samples of a particular waveform amplitude are taken.

SAP (1) service access point. Field defined by the IEEE 802.2 specification that is part of an address specification. Thus, the destination plus the DSAP define the recipient of a packet. The same applies to the SSAP. (2) Service Advertising Protocol. IPX protocol that provides a means of informing network clients via routers and servers of available network resources and services.

SAR (segmentation and reassembly) One of the two sublayers of the AAL CPCS, responsible for dividing (at the source) and reassembling (at the destination) the PDUs passed from the CS. The SAR sublayer takes the PDUs processed by the CS and, after dividing them into 48-byte pieces of payload data, passes them to the ATM layer for further processing.

SAS (single attachment station) Device attached only to the primary ring of an FDDI ring. Also known as a *Class B station.*

satellite communication Use of orbiting satellites to relay data between multiple earth based stations. Satellite communications offer high bandwidth and a cost that is not related to distance between earth stations, long propagation delays, or broadcast capability.

SBus Bus technology used in Sun SPARC based workstations and servers. The SBus specification has been adopted by the IEEE as a new bus standard.

SCR (sustainable cell rate) Parameter defined by the ATM Forum for ATM traffic management. For VBR connections, SCR determines the long-term average cell rate that can be transmitted.

SCTE (serial clock transmit external) Timing signal that DTE echoes to DCE to maintain clocking. SCTE is designed to compensate for clock phase shift on long cables. When the DCE device uses SCTE instead of its internal clock to sample data from the DTE, it is better able to sample the data without error even if there is a phase shift in the cable.

SDH (Synchronous Digital Hierarchy) European standard that defines a set of rate and format standards that are transmitted using optical signals over fiber. SDH is similar to SONET, with a basic SDH rate of 155.52 Mbps, designated at STM-1.

SDLC (Synchronous Data Link Control) SNA data link layer communications protocol. SDLC is a bit oriented, full-duplex serial protocol that has spawned numerous similar protocols, including HDLC and LAPB.

SDLC broadcast Feature that allows a Cisco router that receives an all-stations broadcast on a virtual multidrop line to propagate the broadcast to each SDLC line that is a member of the virtual multidrop line.

SDLLC Feature that performs translation between SDLC and IEEE 802.2 type 2.

SDSU (SMDS DSU) DSU for access to SMDS via HSSIs and other serial interfaces.

SDU (service data unit) Unit of information from an upper layer protocol that defines a service request to a lower layer protocol.

SEAL (simple and efficient AAL) Scheme used by AAL5 in which the SAR sublayer segments CS PDUs without adding additional fields.

secondary ring One of the two rings making up an FDDI or CDDI ring. The secondary ring is usually reserved for use in the event of a failure of the primary ring.

secondary station In bit-synchronous data link layer protocols such as HDLC, a station that responds to commands from a primary station. Sometimes referred to simply as a *secondary*.

security management One of five categories of network management defined by ISO for management of OSI networks. Security management subsystems are responsible for controlling access to network resources.

seed router Responds to configuration queries from nonseed routers on its connected AppleTalk network, allowing those routers to confirm or modify their configurations accordingly.

segment (1) Section of a network that is bounded by bridges, routers, or switches. (2) In a LAN using a bus topology, a segment is a continuous electrical circuit that is often connected to other such segments with repeaters. (3) Term used in the TCP specification to describe a single transport layer unit of information. The terms datagram, frame, message, and packet are also used to describe logical information groupings at various layers of the OSI reference model and in various technology circles.

serial transmission Method of data transmission in which the bits of a data character are transmitted sequentially over a single channel.

server Node or software program that provides services to clients.

service point Interface between non-SNA devices and NetView that sends alerts from equipment unknown to the SNA environment.

session (1) Related set of communications transactions between two or more network devices. (2) In SNA, a logical connection enabling two NAUs to communicate.

session layer Layer 5 of the OSI reference model. This layer establishes, manages, and terminates sessions between applications and manages data exchange between presentation layer entities. Corresponds to the data flow control layer of the SNA model.

SF (Super Frame) Common framing type used on T1 circuits. SF consists of 12 frames of 192 bits each, with the 193rd bit providing error checking and other functions. SF has been superseded by ESF, but is still widely used. Also called *D4 framing.*

SGMP (Simple Gateway Monitoring Protocol) Network management protocol that was considered for Inter-net standardization and later evolved into SNMP. Documented in RFC 1028.

shielded cable Cable that has a layer of shielded insulation to reduce EMI.

shortest path routing Routing that minimizes distance or path cost through application of an algorithm.

signaling Process of sending a transmission signal over a physical medium for purposes of communication.

signaling packet Generated by an ATM connected device that wants to establish a connection with another such device. The signaling packet contains the ATM NSAP address of the desired ATM endpoint, as well as any QoS parameters required for the connection. If the endpoint can support the desired QoS, it responds with an accept message, and the connection is opened.

silicon switching Switching based on the SSE, which allows the processing of packets independent of the SSP (Silicon Switch Processor) system processor. Silicon switching provides high speed, dedicated packet switching.

simplex Capability for data transmission in only one direction between a sending station and a receiving station.

single-mode fiber Fiber optic cabling with a narrow core that allows light to enter only at a single angle. Such cabling has higher bandwidth than multimode fiber, but requires a light source with a narrow spectral width (for example, a laser). Also called *monomode fiber.*

single-vendor network Network using equipment from only one vendor. Single-vendor networks rarely suffer compatibility problems.

SIP (SMDS Interface Protocol) Used in communications between CPE and SMDS network equipment. Allows the CPE to use SMDS service for high speed WAN internetworking. Based on the IEEE 802.6 DQDB standard.

sliding window flow control Method of flow control in which a receiver gives transmitter permission to transmit data until a window is full. When the window is full, the transmitter must stop transmitting until the receiver advertises a larger window. TCP, other transport protocols, and several data link layer protocols use this method of flow control.

SLIP (Serial Line Internet Protocol) Standard protocol for point-to-point serial connections using a variation of TCP/IP. Predecessor of PPP.

slotted ring LAN architecture based on a ring topology in which the ring is divided into slots that circulate continuously. Slots can be either empty or full, and transmissions must start at the beginning of a slot.

slow switching   Packet processing performed at process level speeds, without the use of a route cache.

SMAC   (source MAC) MAC address specified in the Source Address field of a packet.

SMB   (Server Message Block) File-system protocol used in LAN Manager and similar NOSs to package data and exchange information with other systems.

SMDS   (Switched Multimegabit Data Service) High speed, packet-switched, datagram based WAN networking technology offered by the telephone companies.

SMI   (Structure of Management Information) Document (RFC 1155) specifying rules used to define managed objects in the MIB.

SMRP   (Simple Multicast Routing Protocol) Specialized multicast network protocol for routing multimedia data streams on enterprise networks. SMRP works in conjunction with multicast extensions to the AppleTalk protocol.

SMT   (Station Management) ANSI FDDI specification that defines how ring stations are managed.

SMTP   (Simple Mail Transfer Protocol) Internet protocol providing electronic mail services.

SNA   (Systems Network Architecture) Large, complex, feature-rich network architecture developed in the 1970s by IBM. Similar in some respects to the OSI reference model, but with a number of differences. SNA is essentially composed of seven layers.

SNADS   (SNA Distribution Services) Consists of a set of SNA transaction programs that interconnect and co-operate to provide asynchronous distribution of information between end users. One of three SNA transaction services.

SNAP   (Subnetwork Access Protocol) Internet protocol that operates between a network entity in the subnetwork and a network entity in the end system. SNAP specifies a standard method of encapsulating IP datagrams and ARP messages on IEEE networks. The SNAP entity in the end system makes use of the services of the subnetwork and performs three key functions: data transfer, connection management, and QoS selection.

SNI   (1) Subscriber Network Interface. Interface for SMDS based networks that connects CPE and an SMDS switch. (2) SNA Network Interconnection. IBM gateway connecting multiple SNA networks.

SNMP   (Simple Network Management Protocol) Network management protocol used almost exclusively in TCP/IP networks. SNMP provides a means to monitor and control network devices, and to manage configurations, statistics collection, performance, and security.

SNMPV2   (SNMP Version 2) Version 2 of the popular network management protocol. SNMP2 supports centralized as well as distributed network management strategies and includes improvements in the SMI, protocol operations, management architecture, and security.

SNPA   (subnetwork point of attachment) A data link layer address (such as an Ethernet address, X.25 address, or Frame Relay DLCI address). SNPA addresses are used to configure a CLNS route for an interface.

socket   Software structure operating as a communications endpoint within a network device.

SONET   (Synchronous Optical Network) High speed (up to 2.5 Gbps) synchronous network specification developed by Bellcore and designed to run on optical fiber. STS-1 is the basic building block of SONET. Approved as an international standard in 1988.

source address   Address of a network device that is sending data.

SPAN   (Switched Port Analyzer) Feature of the Catalyst 5000 switch that extends the monitoring abilities of existing network analyzers into a switched Ethernet environment. SPAN mirrors the traffic at one switched segment onto a predefined SPAN port. A network analyzer attached to the SPAN port can monitor traffic from any of the other Catalyst switched ports.

span   Full-duplex digital transmission line between two digital facilities.

spanning explorer packet   Follows a statically configured spanning tree when looking for paths in an SRB network. Also known as a *limited-route explorer packet* or a *single-route explorer packet.*

spanning tree   Loop-free subset of a network topology.

spanning tree algorithm   Algorithm used by the Spanning Tree Protocol to create a spanning tree. Sometimes abbreviated *STA.*

Spanning Tree Protocol   Bridge protocol that utilizes the spanning tree algorithm, enabling a learning bridge to dynamically work around loops in a network topology by creating a spanning tree. Bridges exchange BPDU messages with other bridges to detect loops and then remove the loops by shutting down selected bridge interfaces. Refers to both the IEEE 802.1 Spanning Tree Protocol standard and the earlier Digital Equipment Corporation Spanning Tree Protocol upon which it is based. The IEEE version supports bridge domains and allows the bridge to construct a loop-free topology across an extended LAN. The IEEE version is generally preferred over the Digital version. Sometimes abbreviated *STP.*

speed matching Feature that provides sufficient buffering capability in a destination device to allow a high speed source to transmit data at its maximum rate, even if the destination device is a lower speed device.

SPF (shortest path first algorithm) Routing algorithm that iterates on length of path to determine a shortest path spanning tree. Commonly used in link state routing algorithms. Sometimes called *Dijkstra's algorithm.*

SPID (Service Profile Identifier) Number that some service providers use to define the services to which an ISDN device subscribes. The ISDN device uses the SPID when accessing the switch that initializes the connection to a service provider.

split-horizon updates Routing technique in which information about routes is prevented from exiting the router interface through which that information was received. Split-horizon updates are useful in preventing routing loops.

spoofing (1) Scheme used by Cisco routers to cause a host to treat an interface as if it were up and supporting a session. The router spoofs replies to keepalive messages from the host in order to convince that host that the session still exists. Spoofing is useful in routing environments such as DDR, in which a circuit-switched link is taken down when there is no traffic to be sent across it in order to save toll charges. (2) The act of a packet illegally claiming to be from an address from which it was not actually sent. Spoofing is designed to foil network security mechanisms such as filters and access lists.

spooler Application that manages requests or jobs submitted to it for execution. Spoolers process the submitted requests in an orderly fashion from a queue. A print spooler is a common example of a spooler.

SPP (Sequenced Packet Protocol) Provides reliable, connection based, flow controlled packet transmission on behalf of client processes. Part of the XNS protocol suite.

SPX (Sequenced Packet Exchange) Reliable, connection oriented protocol that supplements the datagram service provided by network layer (layer 3) protocols. Novell derived this commonly used NetWare transport protocol from the SPP of the XNS protocol suite.

SQE (signal quality error) Transmission sent by a transceiver back to the controller to let the controller know whether the collision circuitry is functional. Also called *heartbeat.*

SRAM Type of RAM that retains its contents for as long as power is supplied. SRAM does not require constant refreshing like DRAM.

SRB (source route bridging) Method of bridging originated by IBM and popular in Token Ring networks. In an SRB network, the entire route to a destination is predetermined, in real time, prior to the sending of data to the destination.

SRT (source route transparent bridging) IBM bridging scheme that merges the two most prevalent bridging strategies: SRB and transparent bridging. SRT employs both technologies in one device to satisfy the needs of all ENs. No translation between bridging protocols is necessary.

SR/TLB (source route translational bridging) Method of bridging where source route stations can communicate with transparent bridge stations with the help of an intermediate bridge that translates between the two bridge protocols.

SRTP (Sequenced Routing Update Protocol) Protocol that assists VINES servers in finding neighboring clients, servers, and routers

SS7 (Signaling System number 7) Standard CCS system used with B-ISDN and ISDN. Developed by Bellcore.

SSAP (source service access point) The SAP of the network node designated in the Source field of a packet.

SSCP (System Services Control Point) Focal points within an SNA network for managing network configuration, coordinating network operator and problem determination requests, and providing directory services and other session services for network end users.

SSCP-PU session Session used by SNA to allow an SSCP to manage the resources of a node through the PU. SSCPs can send requests to, and receive replies from, individual nodes in order to control the network configuration.

SSCS (service specific convergence sublayer) One of the two sublayers of any AAL. SSCS, which is service dependent, offers assured data transmission. The SSCS can be null as well, in classical IP over ATM or LAN emulation implementations.

SSE (silicon switching engine) Routing and switching mechanism that compares the data link or network layer header of an incoming packet to a silicon-switching cache, determines the appropriate action (routing or bridging), and forwards the packet to the proper interface. The SSE is directly encoded in the hardware of the SSP (Silicon Switch Processor) of a Cisco 7000 series router. It can therefore perform switching independently of the system processor, making the execution of routing decisions much quicker than if they were encoded in software.

SSP (1) Silicon Switch Processor. High performance silicon switch for Cisco 7000 series routers that provides distributed processing and control for interface processors. The SSP leverages the high speed switching and routing capabilities of the SSE to dramatically increase aggregate router performance, minimizing performance bottlenecks at the interface points between the router and a high speed backbone. (2) Switch-to-Switch Protocol. Protocol specified in the DLSw standard that routers use to establish DLSw connections, locate resources, forward data, and handle flow control and error recovery.

standard Set of rules or procedures that are either widely used or officially specified.

standby monitor Device placed in standby mode on a Token Ring network in case an active monitor fails.

StarLAN CSMA/CD LAN based on IEEE 802.3 and developed by AT&T.

star topology LAN topology in which end points on a network are connected to a common central switch by point-to-point links. A ring topology that is organized as a star implements a unidirectional closed-loop star instead of point-to-point links.

static route Route that is explicitly configured and entered into the routing table. Static routes take precedence over routes chosen by dynamic routing protocols.

statistical multiplexing Technique whereby information from multiple logical channels can be transmitted across a single physical channel. Statistical multiplexing dynamically allocates bandwidth only to active input channels, making better use of available bandwidth and allowing more devices to be connected than with other multiplexing techniques. Also referred to as *statistical time-division multiplexing* or *stat mux.*

STM-1 (Synchronous Transport Module level 1) One of a number of SDH formats that specifies the frame structure for the 155.52-Mbps lines used to carry ATM cells.

store-and-forward packet switching Packet-switching technique in which frames are completely processed before being forwarded out the appropriate port. This processing includes calculating the CRC and checking the destination address. In addition, frames must be temporarily stored until network resources (such as an unused link) are available to forward the message.

STP (shielded twisted-pair) Two-pair wiring medium used in a variety of network implementations. STP cabling has a layer of shielded insulation to reduce EMI.

STS-1 (Synchronous Transport Signal level 1) Basic building block signal of SONET, operating at 51.84 Mbps. Faster SONET rates are defined as STS-*n,* where *n* is a multiple of 51.84 Mbps.

STS-3c (Synchronous Transport Signal level 3, concatenated) SONET format that specifies the frame structure for the 155.52-Mbps lines used to carry ATM cells.

stub area OSPF area that carries a default route, intra-area routes, and interarea routes, but does not carry external routes. Virtual links cannot be configured across a stub area, and they cannot contain an ASBR.

stub network Network that has only a single connection to a router.

STUN (serial tunnel) Router feature allowing two SDLC- or HDLC-compliant devices to connect to one another through an arbitrary multiprotocol topology (using Cisco routers) rather than through a direct serial link.

subarea Portion of an SNA network that consists of a subarea node and any attached links and peripheral nodes.

subarea node SNA communication controller or host that handles complete network addresses.

subchannel In broadband terminology, a frequency based subdivision creating a separate communications channel.

subinterface One of a number of virtual interfaces on a single physical interface.

subnet address Portion of an IP address that is specified as the subnetwork by the subnet mask.

subnet mask A 32-bit address mask used in IP to indicate the bits of an IP address that are being used for the subnet address. Sometimes referred to simply as *mask.*

subnetwork (1) In IP networks, a network sharing a particular subnet address. Subnetworks are networks arbitrarily segmented by a network administrator in order to provide a multilevel, hierarchical routing structure while shielding the subnetwork from the addressing complexity of attached networks. Sometimes called a *subnet.* (2) In OSI networks, a collection of ESs and ISs under the control of a single administrative domain and using a single network access protocol.

subvector A data segment of a vector in an SNA message. A subvector consists of a length field, a key that describes the subvector type, and subvector specific data.

SURAnet (Southeastern Universities Research Association Network) Network connecting universities and other organizations in the southeastern United States.

SURAnet, originally funded by the NSF and a part of the NSFNET, is now part of BBN Planet.

SVC (switched virtual circuit) Virtual circuit that is dynamically established on demand and is torn down when transmission is complete. SVCs are used in situations where data transmission is sporadic. Called a *switched virtual connection* in ATM terminology.

switch (1) Network device that filters, forwards, and floods frames based on the destination address of each frame. The switch operates at the data link layer of the OSI model. (2) General term applied to an electronic or mechanical device that allows a connection to be established as necessary and terminated when there is no longer a session to support.

synchronization Establishment of common timing between sender and receiver.

synchronous transmission Term describing digital signals that are transmitted with precise clocking. Such signals have the same frequency, with individual characters encapsulated in control bits (called *start bits* and *stop bits*) that designate the beginning and end of each character.

T

T1 Digital WAN carrier facility. T1 transmits DS-1-formatted data at 1.544 Mbps through the telephone-switching network, using AMI or B8ZS coding.

T3 Digital WAN carrier facility. T3 transmits DS-3-formatted data at 44.736 Mbps through the telephone-switching network.

TAC (Terminal Access Controller) Internet host that accepts terminal connections from dial-up lines.

TACACS (Terminal Access Controller Access Control System) Authentication protocol developed by the DDN community that provides remote access authentication and related services, such as event logging. User passwords are administered in a central database rather than in individual routers, providing an easily scalable network security solution.

tagged traffic ATM cells that have their CLP bit set to 1. If the network is congested, tagged traffic can be dropped to ensure delivery of higher-priority traffic. Sometimes called *DE (discard eligible)* traffic.

TAXI 4B/5B (Transparent Asynchronous Transmitter/Receiver Interface 4-byte/5-byte) Encoding scheme used for FDDI LANs as well as for ATM. Supports speeds of up to 100 Mbps over multimode fiber. TAXI is the chipset that generates 4B/5B encoding on multimode fiber.

T-carrier TDM transmission method usually referring to a line or cable carrying a DS-1 signal.

TCP (Transmission Control Protocol) Connection oriented transport layer protocol that provides reliable full-duplex data transmission. TCP is part of the TCP/IP protocol stack.

TCP/IP (Transmission Control Protocol/Internet Protocol) Common name for the suite of protocols developed by the U.S. DoD in the 1970s to support the construction of worldwide internetworks. TCP and IP are the two best-known protocols in the suite.

TCU (trunk coupling unit) In Token Ring networks, a physical device that enables a station to connect to the trunk cable.

TDM (time-division multiplexing) Technique in which information from multiple channels can be allocated bandwidth on a single wire based on preassigned time slots. Bandwidth is allocated to each channel regardless of whether the station has data to transmit.

TDR (time domain reflectometer) Device capable of sending signals through a network medium to check cable continuity and other attributes. TDRs are used to find physical layer network problems.

telco Abbreviation for telephone company.

telecommunications Term referring to communications (usually involving computer systems) over the telephone network.

telephony Science of converting sound to electrical signals and transmitting it between widely removed points.

telex Teletypewriter service allowing subscribers to send messages over the PSTN.

Telnet Standard terminal emulation protocol in the TCP/IP protocol stack. Telnet is used for remote terminal connection, enabling users to log in to remote systems and use resources as if they were connected to a local system. Telnet is defined in RFC 854.

termid SNA cluster controller identification. Termid is meaningful only for switched lines. Also called *Xid*.

terminal Simple device at which data can be entered or retrieved from a network. Generally, terminals have a monitor and a keyboard, but no processor or local disk drive.

terminal adapter Device used to connect ISDN BRI connections to existing interfaces such as EIA/TIA-232. Essentially, an ISDN modem.

terminal emulation   Network application in which a computer runs software that makes it appear to a remote host as a directly attached terminal.

terminal server   Communications processor that connects asynchronous devices such as terminals, printers, hosts, and modems to any LAN or WAN that uses TCP/IP, X.25, or LAT protocols. Terminal servers provide the internetwork intelligence that is not available in the connected devices.

terminator   Device that provides electrical resistance at the end of a transmission line to absorb signals on the line, thereby keeping them from bouncing back and being received again by network stations.

TFTP   (Trivial File Transfer Protocol) Simplified version of FTP that allows files to be transferred from one computer to another over a network.

TH   (transmission header) SNA header that is appended to the SNA basic information unit (BIU). The TH uses one of a number of available SNA header formats.

Thinnet   Term used to define a thinner, less expensive version of the cable specified in the IEEE 802.3 10Base-2 standard.

throughput   Rate of information arriving at, and possibly passing through, a particular point in a network system.

TIA   (Telecommunications Industry Association) Organization that develops standards relating to telecommunications technologies. Together, the TIA and the EIA have formalized standards, such as EIA/TIA-232, for the electrical characteristics of data transmission.

TIC   (Token Ring interface coupler) Controller through which an FEP connects to a Token Ring.

time-out   Event that occurs when one network device expects to hear from another network device within a specified period of time but does not. The resulting time-out usually results in a retransmission of information or the dissolving of the session between the two devices.

TN3270   Terminal emulation software that allows a terminal to appear to an IBM host as a 3278 Model 2 terminal. The Cisco TN3270 implementation allows users to access an IBM host without using a special IBM server or a UNIX host acting as a server.

token   Frame that contains control information. Possession of the token allows a network device to transmit data onto the network.

token bus   LAN architecture using token passing access over a bus topology. This LAN architecture is the basis for the IEEE 802.4 LAN specification.

token passing   Access method by which network devices access the physical medium in an orderly fashion based on possession of a small frame called a token.

Token Ring   Token passing LAN developed and supported by IBM. Token Ring runs at 4 or 16 Mbps over a ring topology. Similar to IEEE 802.5.

TOP   (Technical Office Protocol) OSI-based architecture developed for office communications.

topology   Physical arrangement of network nodes and media within an enterprise networking structure.

TP0   (Transport Protocol Class 0) OSI connectionless transport protocol for use over reliable subnetworks. Defined by ISO 8073.

TP4   (Transport Protocol Class 4) OSI connection based transport protocol. Defined by ISO 8073.

traffic policing   Process used to measure the actual traffic flow across a given connection and compare it to the total admissible traffic flow for that connection. Traffic outside of the agreed upon flow can be tagged (where the CLP bit is set to 1) and can be discarded en route if congestion develops. Traffic policing is used in ATM, Frame Relay, and other types of networks. Also known as *admission control, permit processing, rate enforcement,* and *UPC (usage parameter control).*

traffic shaping   Use of queues to limit surges that can congest a network. Data is buffered and then sent into the network in regulated amounts to ensure that the traffic will fit within the promised traffic envelope for the particular connection. Traffic shaping is used in ATM, Frame Relay, and other types of networks. Also known as *metering, shaping,* and *smoothing.*

trailer   Control information appended to data when encapsulating the data for network transmission.

transaction   Result oriented unit of communication processing.

transaction services layer   Layer 7 in the SNA architectural model. Represents user application functions, such as spreadsheets, word-processing, or electronic mail, by which users interact with the network. Corresponds roughly with the application layer of the OSI reference model.

transit bridging   Bridging that uses encapsulation to send a frame between two similar networks over a dissimilar network.

translational bridging   Bridging between networks with dissimilar MAC sublayer protocols. MAC information is translated into the format of the destination network at the bridge.

transmission control layer   Layer 4 in the SNA architectural model. This layer is responsible for establishing, maintaining, and terminating SNA sessions, sequencing data messages, and controlling session level flow. Corresponds to the transport layer of the OSI model.

transmission group   In SNA routing, one or more parallel communications links treated as one communications facility.

transparent bridging   Bridging scheme often used in Ethernet and IEEE 802.3 networks in which bridges pass frames along one hop at a time based on tables associating end nodes with bridge ports. Transparent bridging is so named because the presence of bridges is transparent to network end nodes.

transport layer   Layer 4 of the OSI reference model. This layer is responsible for reliable network communication between end nodes. The transport layer provides mechanisms for the establishment, maintenance, and termination of virtual circuits, transport fault detection and recovery, and information flow control. Corresponds to the transmission control layer of the SNA model.

trap   Message sent by an SNMP agent to an NMS, console, or terminal to indicate the occurrence of a significant event, such as a specifically defined condition or a threshold that has been reached.

tree topology   LAN topology similar to a bus topology, except that tree networks can contain branches with multiple nodes. Transmissions from a station propagate the length of the medium and are received by all other stations.

trunk   Physical and logical connection between two ATM switches across which traffic in an ATM network travels. An ATM backbone is composed of a number of trunks.

TTL   (Time To Live) Field in an IP header that indicates how long a packet is considered valid.

TUD   (trunk up-down) Protocol used in ATM networks that monitors trunks and detects when one goes down or comes up. ATM switches send regular test messages from each trunk port to test trunk line quality. If a trunk misses a given number of these messages, TUD declares the trunk down. When a trunk comes back up, TUD recognizes that the trunk is up, declares the trunk up, and returns it to service.

tunneling   Architecture that is designed to provide the services necessary to implement any standard point-to-point encapsulation scheme.

twisted pair   Relatively low speed transmission medium consisting of two insulated wires arranged in a regular spiral pattern. The wires can be shielded or unshielded. Twisted-pair is common in telephony applications and is increasingly common in data networks.

TWS   (two-way simultaneous) Mode that allows a router configured as a primary SDLC station to achieve better utilization of a full-duplex serial line. When TWS is enabled in a multidrop environment, the router can poll a secondary station and receive data from that station while it sends data to or receives data from a different secondary station on the same serial line.

Type 1 operation   IEEE 802.2 (LLC) connectionless operation.

Type 2 operation   IEEE 802.2 (LLC) connection oriented operation.

U

UART   (Universal Asynchronous Receiver/Transmitter) Integrated circuit attached to the parallel bus of a computer that is used for serial communications. The UART translates between serial and parallel signals, provides transmission clocking, and buffers data sent to or from the computer.

UB Net/One   (Ungermann-Bass Net/One) Routing protocol developed by UB Networks that uses Hello packets and a path-delay metric, with end nodes communicating using the XNS protocol. There are a number of differences between the manner in which Net/One uses the XNS protocol and the usage common among other XNS nodes.

UBR   (unspecified bit rate) QoS class defined by the ATM Forum for ATM networks. UBR allows any amount of data up to a specified maximum to be sent across the network, but there are no guarantees in terms of cell loss rate and delay.

UDP   (User Datagram Protocol) Connectionless transport layer protocol in the TCP/IP protocol stack. UDP is a simple protocol that exchanges datagrams without acknowledgements or guaranteed delivery, requiring that error processing and retransmission be handled by other protocols. UDP is defined in RFC 768.

UL   (Underwriters Laboratories) Independent agency within the United States that tests product safety.

ULP   (upper layer protocol) Protocol that operates at a higher layer in the OSI reference model relative to other layers. ULP is sometimes used to refer to the next-highest protocol (relative to a particular protocol) in a protocol stack.

unbalanced configuration   HDLC configuration with one primary station and multiple secondary stations.

UNI (User-Network Interface) ATM Forum specification that defines an interoperability standard for the interface between ATM based products (a router or an ATM switch) located in a private network and the ATM switches located within the public carrier networks. Also used to describe similar connections in Frame Relay networks.

unicast Message sent to a single network destination.

unicast address Address specifying a single network device.

uninsured traffic Traffic within the excess rate (the difference between the insured rate and maximum rate) for a VCC. This traffic can be dropped by the network if congestion occurs.

unipolar Literally meaning one polarity, the fundamental electrical characteristic of internal signals in digital communications equipment.

unity gain In broadband networks, the balance between signal loss and signal gain through amplifiers.

UNIX Operating system developed in 1969 at Bell Laboratories. UNIX has gone through several iterations since its inception. These include UNIX 4.3 BSD (Berkeley Standard Distribution) developed at the University of California at Berkeley, and UNIX System V, Release 4.0 developed by AT&T.

unnumbered frames HDLC frames used for various control and management purposes, including link startup and shutdown and mode specification.

UPC usage parameter control.

URL (Universal Resource Locator) Standardized addressing scheme for accessing hypertext documents and other services using a WWW browser.

USENET Initiated in 1979, one of the oldest and largest cooperative networks with more than 10,000 hosts and a quarter of a million users. Its primary service is a distributed conferencing service called news.

UTP (unshielded twisted-pair) Four-pair wire medium used in a variety of networks. UTP does not require the fixed spacing between connections that is necessary with coaxial type connections. There are five types of UTP cabling commonly used: Category 1 cabling, Category 2 cabling, Category 3 cabling, Category 4 cabling, and Category 5 cabling.

V

V.24 ITU-T standard for a physical layer interface between DTE and DCE. V.24 is essentially the same as the EIA/TIA-232 standard.

V.25 bis ITU-T specification describing procedures for call setup and teardown over the DTE-DCE interface in a PSDN.

V.32 ITU-T standard serial line protocol for bidirectional data transmissions at speeds of 4.8 or 9.6 Kbps.

V.32 bis ITU-T standard that extends V.32 to speeds up to 14.4 Kbps.

V.34 ITU-T standard that specifies a serial line protocol. V.34 offers improvements to the V.32 standard, including higher transmission rates (28.8 Kbps) and enhanced data compression.

V.35 ITU-T standard describing a synchronous physical layer protocol used for communications between a network access device and a packet network. V.35 is most commonly used in the United States and in Europe, and is recommended for speeds up to 48 Kbps.

V.42 ITU-T standard protocol for error correction using LAPM.

VBR (variable bit rate) QoS class defined by the ATM Forum for ATM networks. VBR is subdivided into a real time (RT) class and non-real-time (NRT) class. VBR (RT) is used for connections in which there is a fixed timing relationship between samples. VBR (NRT) is used for connections in which there is no fixed timing relationship between samples, but that still need a guaranteed QoS.

VCC (virtual channel connection) Logical circuit made up of VCLs that carries data between two endpoints in an ATM network. Sometimes called a *virtual circuit connection*.

VCI (virtual channel identifier) A 16-bit field in the header of an ATM cell. The VCI, together with the VPI, is used to identify the next destination of a cell as it passes through a series of ATM switches on its way to its destination. ATM switches use the VPI/VCI fields to identify the next network VCL that a cell needs to transit on its way to its final destination. The function of the VCI is similar to that of the DLCI in Frame Relay.

VCL (virtual channel link) Connection between two ATM devices. A VCC is made up of one or more VCLs.

VCN (virtual circuit number) A 12-bit field in an X.25 PLP header that identifies an X.25 virtual circuit. Allows DCE to determine how to route a packet through the X.25 network. Sometimes called *LCI (logical channel identifier)* or *LCN (logical channel number)*.

vector Data segment of an SNA message. A vector consists of a length field, a key that describes the vector type, and vector-specific data.

VINES (Virtual Integrated Network Service) NOS developed and marketed by Banyan Systems.

virtual circuit Logical circuit created to ensure reliable communication between two network devices. A virtual circuit is defined by a VPI/VCI pair, and can be either permanent (a PVC) or switched (an SVC). Virtual circuits are used in Frame Relay and X.25. In ATM, a virtual circuit is called a *virtual channel*. Sometimes abbreviated *VC*.

virtualization Process of implementing a network based on virtual network segments. Devices are connected to virtual segments independent of their physical location and their physical connection to the network.

virtual path Logical grouping of virtual circuits that connect two sites.

virtual ring Entity in an SRB network that logically connects two or more physical rings together either locally or remotely. The concept of virtual rings can be expanded across router boundaries.

virtual route In SNA, a logical connection between subarea nodes that is physically realized as a particular explicit route. SNA terminology for virtual circuit.

VLAN (virtual LAN) Group of devices on a LAN that are configured (using management software) so that they can communicate as if they were attached to the same wire, when in fact they are located on a number of different LAN segments. Because VLANs are based on logical instead of physical connections they are extremely flexible.

VLI (virtual LAN internetwork) Internetwork composed of VLANs.

VLSM (variable length subnet mask) Ability to specify a different subnet mask for the same network number on different subnets. VLSM can help optimize available address space.

VPC (virtual path connection) Grouping of VCCs that share one or more contiguous VPLs.

VPI (virtual path identifier) An 8-bit field in the header of an ATM cell. The VPI, together with the VCI, is used to identify the next destination of a cell as it passes through a series of ATM switches on its way to its destination. ATM switches use the VPI/VCI fields to identify the next VCL that a cell needs to transit on its way to its final destination. The function of the VPI is similar to that of the DLCI in Frame Relay.

VPL (virtual path link) Within a virtual path, a group of unidirectional VCLs with the same endpoints. Grouping VCLs into VPLs reduces the number of connections to be managed, thereby decreasing network control overhead and cost. A VPC is made up of one or more VPLs.

VTAM (Virtual Telecommunications Access Method) Set of programs that control communication between LUs. VTAM controls data transmission between channel attached devices and performs routing functions.

VTP (Virtual Terminal Protocol) ISO application for establishing a virtual terminal connection across a network.

W

WAN (wide area network) Data communications network that serves users across a broad geographic area and often uses transmission devices provided by common carriers. Frame Relay, SMDS, and X.25 are examples of WANs.

watchdog packet Used to ensure that a client is still connected to a NetWare server. If the server has not received a packet from a client for a certain period of time, it sends that client a series of watchdog packets. If the station fails to respond to a predefined number of watchdog packets, the server concludes that the station is no longer connected and clears the connection for that station.

watchdog spoofing Subset of spoofing that refers specifically to a router acting for a NetWare client by sending watchdog packets to a NetWare server to keep the session between client and server active.

watchdog timer (1) Hardware or software mechanism that is used to trigger an event or an escape from a process unless the timer is periodically reset. (2) In NetWare, a timer that indicates the maximum period of time that a server will wait for a client to respond to a watchdog packet. If the timer expires, the server sends another watchdog packet (up to a set maximum).

waveform coding Electrical techniques used to convey binary signals.

wildcard mask A 32-bit quantity used in conjunction with an IP address to determine which bits in an IP address should be ignored when comparing that address with another IP address. A wildcard mask is specified when setting up access lists.

wiring closet Specially designed room used for wiring a data or voice network. Wiring closets serve as a central junction point for the wiring and wiring equipment that is used for interconnecting devices.

workgroup Collection of workstations and servers on a LAN that are designed to communicate and exchange data with one another.

workgroup switching   Method of switching that provides high speed (100-Mbps) transparent bridging between Ethernet networks and high speed translational bridging between Ethernet and CDDI or FDDI.

wrap   Action taken by an FDDI or CDDI network to recover in the event of a failure. The stations on each side of the failure reconfigure themselves, creating a single logical ring out of the primary and secondary rings.

WWW   (World Wide Web) Large network of Internet servers providing hypertext and other services to terminals running client applications such as a WWW browser.

WWW browser   GUI based hypertext client application, such as Mosaic, used to access hypertext documents and other services located on innumerable remote servers throughout the WWW and Internet.

X

X.121   ITU-T standard describing an addressing scheme used in X.25 networks. X.121 addresses are sometimes called *IDNs (International Data Numbers)*.

X.21   ITU-T standard for serial communications over synchronous digital lines. The X.21 protocol is used primarily in Europe and Japan.

X.21 bis   ITU-T standard that defines the physical layer protocol for communication between DCE and DTE in an X.25 network. Virtually equivalent to EIA/TIA-232.

X.25   ITU-T standard that defines how connections between DTE and DCE are maintained for remote terminal access and computer communications in PDNs. X.25 specifies LAPB, a data link layer protocol, and PLP, a network layer protocol. Frame Relay has to some degree superseded X.25.

X.28   ITU-T recommendation that defines the terminal-to-PAD interface in X.25 networks.

X.29   ITU-T recommendation that defines the form for control information in the terminal-to-PAD interface used in X.25 networks.

X.3   ITU-T recommendation that defines various PAD parameters used in X.25 networks.

X3T9.5   Number assigned to the ANSI Task Group of Accredited Standards Committee for their internal working document describing FDDI.

X.400   ITU-T recommendation specifying a standard for electronic mail transfer.

X.500   ITU-T recommendation specifying a standard for distributed maintenance of files and directories.

X.75   ITU-T specification that defines the signaling system between two PDNs. X.75 is essentially an NNI.

XID   (exchange identification) Request and response packets exchanged prior to a session between a router and a Token Ring host. If the parameters of the serial device contained in the XID packet do not match the configuration of the host, the session is dropped.

XDMCP   (X Display Manager Control Protocol) Protocol used to communicate between X terminals and workstations running UNIX.

XNS   (Xerox Network System) Protocol suite originally designed by PARC. Many PC networking companies such as 3Com, Banyan, Novell, and UB Networks used or currently use a variation of XNS as their primary transport protocol.

XRemote   Protocol developed specifically to optimize support for X Windows over a serial communications link.

XStream   Major public PSN in the United States operated by MCI. Formerly called *TYMNET.*

X terminal   Terminal that allows a user simultaneous access to several different applications and resources in a multivendor environment through implementation of X Windows.

XWindows   Distributed, network-transparent, device-independent, multitasking windowing and graphics system originally developed by MIT for communication between X terminals and UNIX workstations.

Z

zero code suppression   Line coding scheme used for transmission clocking. Zero line suppression substitutes a one in the seventh bit of a string of eight consecutive zeros.

ZIP   (Zone Information Protocol) AppleTalk session layer protocol that maps network numbers to zone names.

ZIP storm   Broadcast storm that occurs when a router running AppleTalk propagates a route for which it currently has no corresponding zone name. The route is then forwarded by downstream routers, and a ZIP storm ensues.

zone   In AppleTalk, a logical group of network devices.

INDEX